Studies in Surface Science and Catalysis 96

CATALYSIS AND AUTOMOTIVE POLLUTION CONTROL III

Studies in Surface Science and Catalysis

Advisory Editors: B. Delmon and J.T. Yates

Vol. 96

CATALYSIS AND AUTOMOTIVE POLLUTION CONTROL III

Proceedings of the Third International Symposium (CAPoC 3), Brussels, Belgium, April 20-22, 1994

Editors

A. Frennet and J.-M. Bastin
Catalyse Hétérogène, Université Libre de Bruxelles, Brussels, Belgium

1995
ELSEVIER
Amsterdam — Lausanne — New York — Oxford — Shannon — Tokyo

ELSEVIER SCIENCE B.V.
Sara Burgerhartstraat 25
P.O. Box 211, 1000 AE Amsterdam, The Netherlands

ISBN 0-444-82019-1

This book is printed on acid-free paper.

Printed in The Netherlands

CONTENTS

FOREWORD

This Congress is the third edition of a series of Symposia devoted to the use of catalysis for the depollution of the exhaust gases of motor vehicles.

As the first two of these meetings, it took place at the « Université Libre de Bruxelles », April 20-22, 1994.

In spite of the number of years that catalysts started to be developped for this purpose (from the early seventies) the important economical impact of the problem makes it still a topic of the day. This is namely revealed by the increasing number of submitted, accepted and published papers, as it appears in the following table:

Papers	Submitted	Accepted	Presented and published	General Lectures.
CAPoC 1	38	-	28	9
CAPoC 2	66	42	34	4
CAPoC 3	131	79	67	4

It has been a hard task for the paper selection committee to select the proposed presentations. The high quality of the submitted extended abstracts lead this committee to accept for presentation 79 papers from the 131 ones submitted. In practice, from the accepted papers, 12 were not delivered. This was due to the impossibility of some authors to attend the meeting. These papers are not published in the proceedings.

The very large number of accepted communications resulted in the organization of a poster session. Anyway, as announced in the circulars, no discrimination is made in these proceedings between the communications presented orally or as a poster.

In these proceedings, the papers are grouped into the following headings:

1.Model reactions and model catalysts.
2.Substrates and washcoat technologies
3.Gasoline catalysts technologies
4.Diesel catalyst technologies
5.Lean NO_x catalyst technologies
6.Catalyst ageing and poisonning
7.Alternative fuels and appropriate catalysts
8.Miscellaneous

It is interesting to notice that 25 presented and published papers are coming from industries, and 42 from Universities and Research Institutes.

On the other hand the number of registered and invited participants was of 279, within which the number of academic people (79 from Universities + 40 from Research Institutes) and industrial representatives (133) was about equivalent. Most of these participants came from western Europe: France (71), Belgium (51), Germany (33), Great Britain (27), The Netherlands (17), Sweden (14), Italy (10), Spain (4), Finland (4), Luxembourg (3), Denmark(1), Greece (1), Ireland (1).

Only 8 participants were from the former east block: Hungary (6), Russia(1), Roumania (1).

Eighteen participants came from the U.S.A. and finally let us mention 4 attendants from Australia, 2 from Japan and one from South Africa and from Argentina.

The large number of answers to the questionnaire distributed to the participants at the end of the Congress contained strong encouragement to continue the series of these « CAPoC » meetings.

The organizers.

J.-M. BASTIN

Secretary of the Organizing Committee

A. FRENNET

Chairman of the Organizing Committee

ACKNOWLEDGMENTS

The organizers are indebted to the authorities of the « Université Libre de Bruxelles » for their hospitality and for the welcome address made by the reprensentative of the Rector of the University.

They also greatly appreciate the cooperation of the members of the organizing committee and the paper selection committee, namely for their important contribution in selecting the proposed papers.

Special thanks are due to all lecturers of the introductory session: K.C.Taylor, K.P. de Jong, P. Z. Eyzat and D. Savey for the high quality of their contribution.

It is also a pleasant duty to thank all the members of the « Laboratoire de Catalyse Hétérogène » of the Université Libre de Bruxelles for their help in organizing this congress. We are in particular very much indebted to Béatrice Depuydt-Parmentier and Axel Löfberg for their continuous active involvement in solving various material problems.

The organizers want to recognize the important, but sometimes hidden, contribution of Mrs.Frennet to the succes of the symposium.

The scientific quality of the Congress is certainly for an important part due to the quality of the contributions of both the authors and the participants in the discussions. They all contributed to the success of the Congress. Thanks to all of them.

The organizers.

J.-M. BASTIN

Secretary of the Organizing Committee

A. FRENNET

Chairman of the Organizing Committee

FINANCIAL SUPPORT

The following companies have accepted to provide financial support to this Congress. The Organizers express their gratitude to these companies for their generosity.

AlliedSignal Inc.

Degussa AG

Ford Motor Co

General Motors Luxembourg

Johnson Matthey Ltd

ORGANIZING COMMITTEE

Chairman :
FRENNET A. Université Libre de Bruxelles

Secretary :
BASTIN J-M. Université Libre de Bruxelles

Members :
CAMPINNE M. Ecole Royale Militaire
CUCCHI C. ACEA
DEROUANE E. Facultés Universitaires Notre Dame de la Paix
ENGLER B. Degussa AG
ESPRIT M. Union Minière
GAGNERET P. Allied Signal Environmetal Catalysts
GERMAIN A. Université de Liège
GERRYN C.
HECQ W. Université Libre de Bruxelles
JANNES G. C.E.R.I.A. - Institut Meurice
KÖNIG A. Volkswagen AG
KUPE J. General Motors Luxemburg
LEDUC B. Université Libre de Bruxelles
NIEUWENHUYS B. Rijksuniversiteit Leiden
PENTENERO A. Université de Nancy
PONCELET G. Université Catholique de Louvain
PRIGENT M. Institut Français du Pétrole
SCHWEICH D. LSGC-ENSIC
WEBSTER D. Johnson Matthey LTD

PAPER SELECTION COMMITTEE

All members of the organizing committee, and
BELOT G. Peugeot S.A.
BRUNELLE J.P. Rhone Poulenc
GARIN F. Université de Strasbourg
LECLERCQ L. Université de Lille
PONEC V. Rijksuniversiteit Leiden

General Lectures

A. Frennet and J.-M. Bastin (Eds.)
Catalysis and Automotive Pollution Control III
Studies in Surface Science and Catalysis, Vol. 96
1995 Elsevier Science B.V.

Automotive and Environment: Towards a global approach

Dominique Savey

Deputy Managing Director
Automotive Division
PSA PEUGEOT CITROËN
75 avenue de la Grande Armée
75116 Paris

Thank you for inviting me to join you here at the third international congress on catalysis and automotive pollution control.

CATALYSEUR, KATALYSATOR, CATALYST, in the past few years, this equipment has reached a level of public awareness, exceptional for a purely technical device. The catalytic converter gained almost mythical status for a period. People in charge of marketing, people in charge of advertising and politicians seized on the concept to the point where cars were no longer judged on their real pollution emissions but on the one criterium of being fitted or not with a catalytic converter. And in some cases, vehicles bans have been implemented according to this criterium. Rarely has a technical problem of such major economic significance been handled so irrationally.

Today, passions have cooled and catalytic converters are no longer a controversial issue. So it is time to take a calm and collected look at environmental problems.

So I am pleased that you gave me the opportunity first of all to expose to you the studies that PSA PEUGEOT CITROËN is conducting in this area, secondly to present to you the global approach which we feel is necessary in order to integrate properly the passenger cars in the environment, and finally to tell you about the advances that we, carmarkers, are expecting from the catalyst experts that you are.

Since World War II, the usage of motor car has grown tremendously in democratic countries. However, that development raises concerns for the environment because of the nuisance caused by cars. Today, most of those nuisances from cars are on the decrease and this improvement should become more widespread in coming years.

I would like to speak of two major types of problems relating to the environment: firstly, air pollution and secondly the consequences on the climate and our natural resources, of the car fuel consumption.

I - On the first point, ***air pollution,*** major efforts have been made in the past 20 years to reduce exhaust emissions of new cars, and you have contributed much to this effort. Between 1970 and 1993, the level of legal emission standards in Europe has been devided by between 10 and 20, depending on the pollutant (Figure 1).

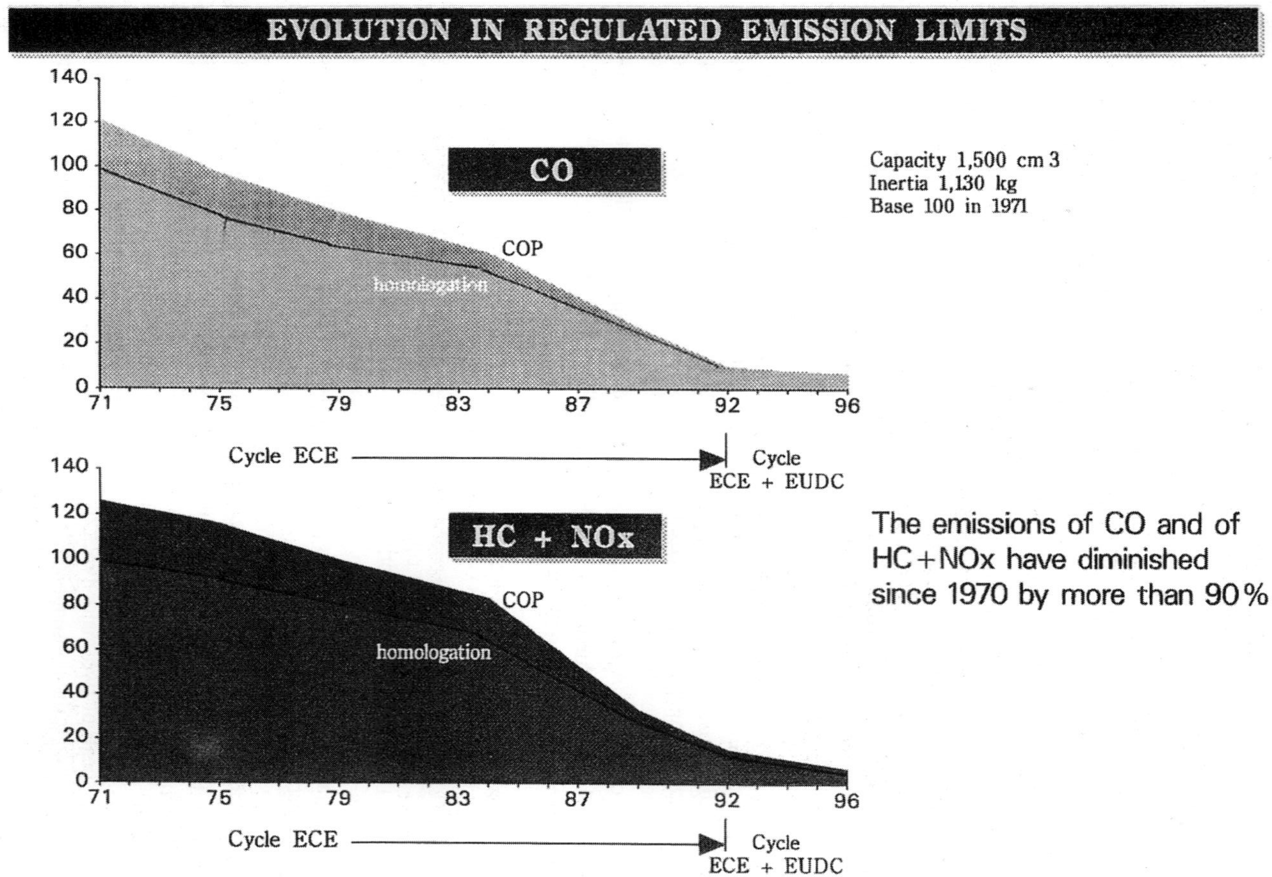
EVOLUTION IN REGULATED EMISSION LIMITS
CO
Capacity 1,500 cm 3
Inertia 1,130 kg
Base 100 in 1971
COP
homologation
140
120
100
80
60
40
20
0
71
75
79
83
87
92
96
Cycle ECE
Cycle
ECE + EUDC
HC + NOx
COP
homologation
Cycle ECE
Cycle
ECE + EUDC
The emissions of CO and of
HC+NOx have diminished
since 1970 by more than 90%,

Figure 1

However, to date, such efforts have only involved new vehicles, and it will take time before the full effects are felt. Since a car's average life is approximately 10 years, and given the time needed for a full-scale replacement of the in-use vehicles, the complete results of new measures are apparent only after about fifteen years, once the majority of old cars have been replaced. Thus, the catalytic converter, which became mandatory in 1993, will make its mark only gradually and will not be fully effective until year 2008 or 2010.

Several studies conducted by INRETS ([1]) for France, the European Union for the 12 European countries, and QUARG ([2]) for the United Kingdom have analysed this problem and have estimated the evolution of air pollution caused by cars in the coming years. All of them have reached similar conclusions but I will quote the INRETS Study for France ; it says that for most pollutants, the global volume of emissions of passenger vehicles on the roads increased regularly, before peaking a few years ago ; now it is declining and should fall even more quickly in the future as old cars will be replaced by new ones. For carbon monoxides (CO), hydrocarbons (HC), volatile organic compounds (VOC), oxides of nitrogen (NOx), lead, sulphur dioxides (SO^2) and aromatic hydrocarbons (HAP) the volume of car exhaust emissions will be cut by at least two thirds by the year 2010 (and some of them will be completely eliminated), despite the fact that road traffic will increase by an estimated 2% per year (see figures 2 and 2a), and without taking in account any further tightening of the European regulation.

As a result, the quality of the air in almost all our cities is already improving, due among other things, to the reduction in car exhaust emissions. The improvement is bound to become more widespread in the next 15 years.

So today, there is no point imposing ever-stricter standards on new cars which already have extremely low emissions values, and any new measures would be counterproductive in terms of cost-effectiveness because of the almost imperceptible gain in environment that would be achieved.

On the contrary, it may be desirable to speed up the pollution reduction process by working on existing cars and to solve any remaining, localised problems by means of better focused measures. I'd like to comment on four types of measures:

- incentives to speed up the rate of renewal of the existing vehicle fleet,
- better control of emissions of cars currently on the road,
- improve fuel quality,
- treatment of urban pollution on a local level, for example through the development of electric and natural gas-powered vehicles for city use or through improved traffic flow management.

---

1 *Institut National de Recherches sur le Transport et la Sécurité Routière* (National Transport and Road Safety Research Institute)

2 Quality of Urban Air Review Group

EVOLUTION IN GLOBAL POLLUTANT EMISSIONS OF PASSENGER CARS IN FRANCE BETWEEN 1970 AND 2010

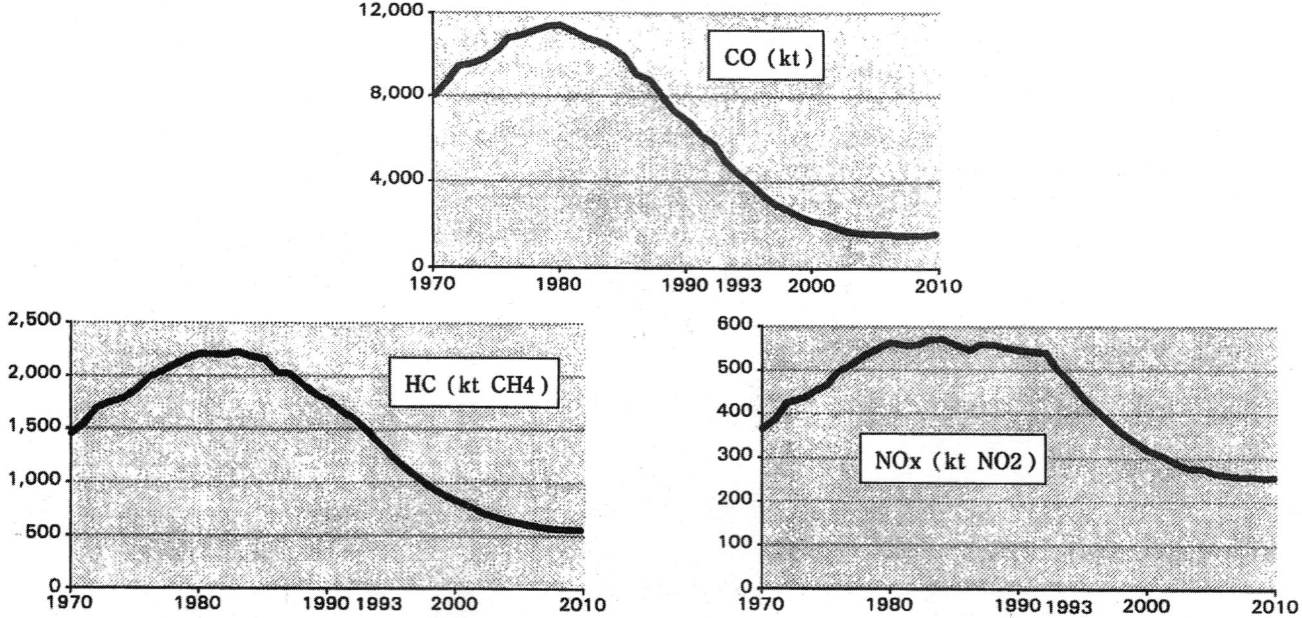

Reneval of the parc by low-emitting vehicles will bring a rapid drop in all pollutants, with an eventual stabilisation at very low levels.

Source : INRETS report 143 (July 1991)

Figure 2

EVOLUTION IN THE EMISSIONS OF CO2, SO2, LEAD AND SMOKE IN FRANCE - DGXI STUDY - EUR13854 - 1991 (scenario A)

PSA PEUGEOT CITROËN

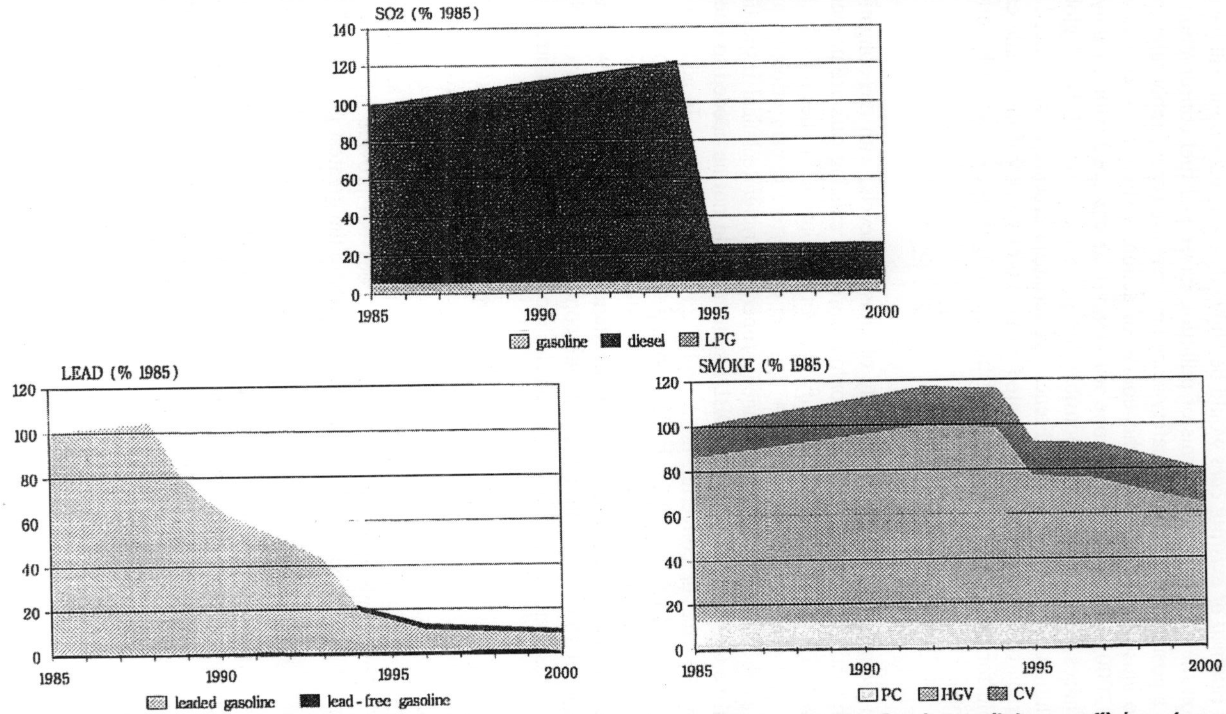

The emissions of CO2 will stabilise and the improvement in fuel qualities will lead to a strong reduction of SO2, lead and smoke emissions.

Figure 2bis

To ***speed up the renewal of existing cars*** would be highly desirable because it will have a positive impact on the environment in terms of pollution, noise and fuel consumption. It would also have a positive effect on road safety since new cars are considerably safer than older models. To illustrate the effects of this course of action, a study by Arco showed that by the year 2000, 30% of cars in Europe will not yet be equipped with a catalytic converter and thoses cars would be responsible for more than two-thirds of global automotive pollution (Figure 3). The French government recently announced a programme whereby consumers would be granted a FFr 5,000 subsidy if they trade cars over ten years old for new models. This type of measure ought to be adopted by other European countries.

Stricter periodic technical inspection, especially of emission levels, with the obligation for car owners to rectify any anomalies. This would ensure that existing cars are properly maintained and would make it possible to identify the worst polluters. Studies on vehicles in-use equipped with catalytic converters show that the converters remain effective and reliable as long as they are used and maintained properly. Stricter technical inspection would mean that the On-Board Diagnostic (OBD) system would not be needed to check the functioning of the catalytic converter, the λ probe, ignition or canister. In addition to the fact that OBD is not 100% reliable, it is also too expensive for what it is supposed to do given the reliability of today's pollution reduction systems. This has been proven by a number of studies, including one carried out in Switzerland between 1988 and 1990 which examined emissions of 13,000 vehicles equipped with controlled catalytic converters. The study showed that less than one percent of the vehicles monitored exceeded legal limits.

Another measure that involves existing vehicles is the ***improvement of fuel quality.*** This would immediately and drastically reduce emissions since all vehicles on the road would be affected. European carmakers and oil companies are working together in a European research programme called the European Programme on Emissions Fuels and Engine Technologies (EPEFE). Unfortunately, the programme does not provide for trials on vehicles that do not meet current standards, since this is where the greatest gains could be made.

The fourth line of action would be to handle ***urban pollution problems at the local level*** and to tailor solutions to the individual problems of each city. This would require analysing and dealing with pollution problems on a needs-specific basis in order to achieve better results. Several solutions will have to co-exist in order to attain maximum efficiency. I shall discuss three of them:

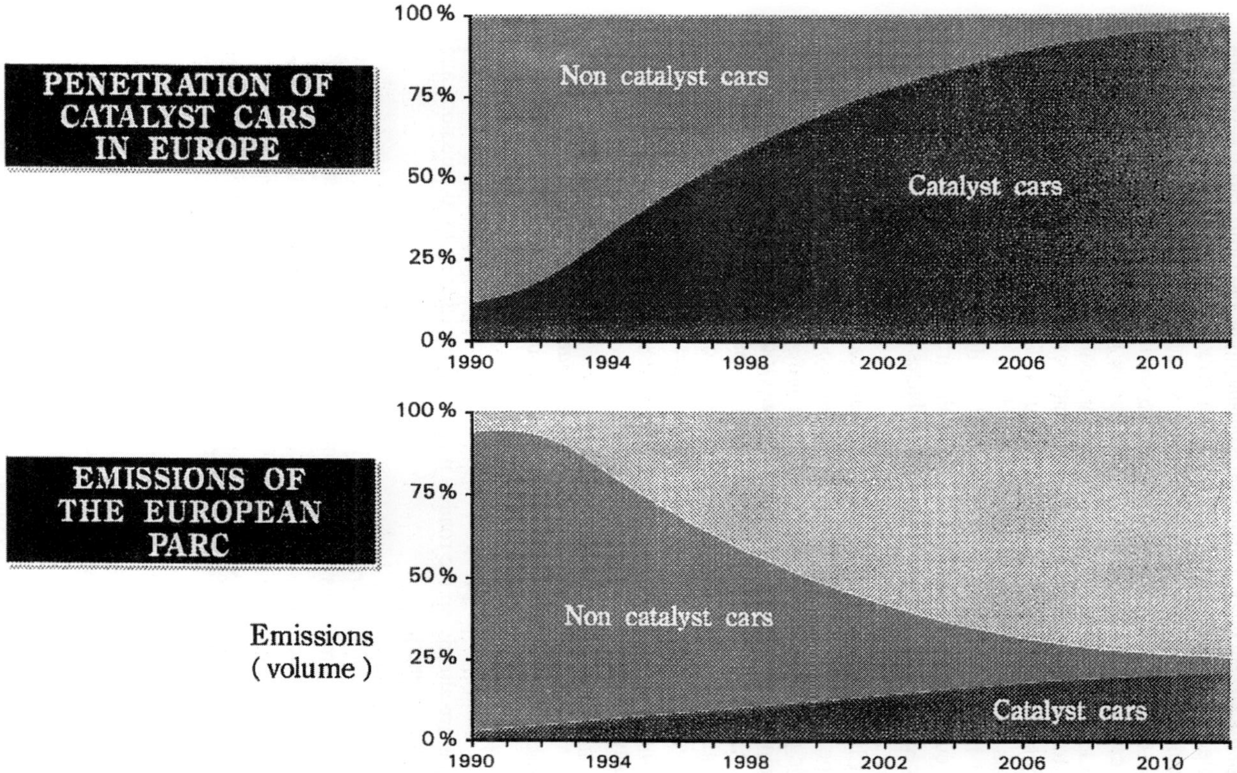

By the year 2000, the non catalyst cars will emit 70% of the total emissions while they will constitute only 30% of the parc.

Source ARCO

Figure 3

1) In cities and towns, the ***electric vehicle*** reduces both pollution and noise while still giving the urban dweller mobility. In addition to being quiet and clean, electric vehicles rely on available energy since batteries would generally be recharged at night, when there is low demand for electricity. Naturally, the vehicles' effect on the environment depends on the type of powerplant that generates the electricity: hydro-electric and nuclear plants cause no pollution, emit no greenhouse gases and protect our natural resources. This is not the case with fuel oil or coal-fired plants. If one million electric vehicles were to be used in Europe, the demand for electricity would equal the production of just one-half of a 900 MW nuclear unit. Therefore, current capacities in off-peak periods would be more than sufficient. Even if, on a European scale, the electricity used by these vehicles were generated by different types of powerplants according the current production parttern, overall carbon dioxide (CO_2) emissions would be reduced by 20% compared to gasoline vehicles.

   PSA Peugeot Citroën was the first among the major car manufacturers to study and develop this type of vehicles in cooperation with component suppliers. Already in 1989, we began selling electric commercial vehicles to city councils and fleet owners in France ; beginning early next year, we will bring electric versions of the Peugeot 106 and the Citoën AX to market for thoses clients and the private customers. And we plan to launch an entirely new, specifically designed, electric model before the end of the century. Moreover, in the city of La Rochelle, we have been working since last year with the city council and France's electricity utility EDF to test fifty 106s and AXs, which are actually driven by everyday users.

2) In urban areas, ***vehicles powered by compressed natural gas*** can also make an important contribution since they cause less pollution than gasoline cars ; because of their greater range, they are more flexible than electric vehicles, but they are not as clean or as quiet. Although the gas solution is less innovative - thousands of natural gas powered vehicles are already on the roads - it is useful in reducing local emissions and diversifying sources of energy.

3) The third solution is to ***improve traffic flow***. Traffic jams cause pollution, increase fuel consumption and waste time. This problem could be improved in two ways:

   - The launch of new road infrastructure programmes: this includes building new motor ways, city bypasses and car parks since insufficient parking space causes additional traffic as drivers are searching for a place to park.

   - The development of traffic management: new technologies offer promising prospects which would optimise the use of road infrastructures by managing traffic flows. This field is extremely vast. It includes :

     ⇨ automatic tolls that vary according to day and the time,
     ⇨ real-time information on traffic conditions for drivers,
     ⇨ information on available parking with remote reservation capability,
     ⇨ interactive traffic management to suggest optimised routes based on prevailing traffic conditions. PSA Peugeot Citroën has studied a traffic management system called Isis and is ready to set up pilot operations for validation.

II -The second major problem related to environment and caused by passenger cars is ***the emission of carbon dioxide (CO^2) and the usage of non-renewable energy***. In fact in Europe cars represent only about 12 % of the emissions of carbon dioxide, but it is essential that everyone make his best effort to reduce these emissions. Car manufacturers have been working for over 15 years to decrease fuel consumption in order to reduce CO_2 emissions and to preserve nonrenewable resources. Efforts have produced significant results since, at PSA Peugeot Citroën for example, average fuel consumption for cars sold in Europe fell by 17% between 1980 and 1990. However, in recent years, anti pollution legislation has opposed our efforts by making mandatory the use of controlled catalytic converters, which increase fuel consumption in gasoline cars. In future, providing that no new legislation gets in our way, further and substantial progress should be possible by expanding the use of diesel in passenger cars, (this is the most effective method of reducing fuel consumption and CO_2 emissions in the short term), by making cars lighter, improving the efficiency of both gasoline and diesel engines and redesigning tyres.

PSA Peugeot Citroën has set itself a goal for the year 2000 : average fuel consumption for its cars sold in Europe must be lowered to 6 litres/100 km (47 mpg), i.e. a 15% gain over 1993. In the longer term, we should be able to make even greater progress, and possibly within the next ten years cars consuming less than 3 litres per 100 km (94.2 mpg) will come on the market.

Finally, given the expected increase in road traffic, we estimate that fuel consumption for passenger cars will remain globally stable or will decrease over the next fifteen years. Naturally, most of the measures I have presented to speed up the improvement of air quality would also have positive effects on reductions in fuel consumption and CO_2 emissions.

❖❖❖

I'd like now to come back to catalytic converters and present the advances that car manufacturers and customers are expecting from you ; I can see four:

- substantial cost reduction
- development of the Denox catalytic converter
- better light-off from a cold start
- elimination of hydrogen sulfide (H_2S) release

In terms of ***cost reduction:*** you know that the use of three-way controlled catalytic converters caused a substantial increase in the price of gasoline cars. In a country such as France, where 80% of new gasoline cars were equipped with a carburettor, the average price increase was 7%. Even though car manufacturers did not pass on the entire cost through to the customer, these higher prices certainly contributed to the current crisis in the car market in Europe. Car is a mass consumer product and it must be affordable to as many consumers as possible. As the passenger car market has developed, so value for money has continuously improved. The catalytic converter costs ten times more per kilogram than the average car on a cost for weight basis. Therefore, it can hardly be considered as an item of equipment for passenger cars since its value for money is far too poor for the customer. It is also fair to question its current environmental cost-effectiveness: between the purchase price and the additional fuel consumption, the global annual cost of the catalytic converter and its associated equipments is around 7bn Ecu for carbuyers in the European Union. Therefore, the cost of the three-way catalytic converter must be reduced drastically through technical improvements, and eventually by replacing precious metals with less costly materials.

The second area where improvement must be made is in the use of the ***Denox catalytic converter***. You know that one of the drawbacks in the use of a catalytic converter is an increase in fuel consumption. You also know that reducing fuel consumption is one of the primary objectives of European car manufacturers both for environmental reasons and because of customer demand. This can be achieved at the engine level by using diesel technology (indirect injection and more probably direct injection) and by improving combustion in gasoline engines, notably by leaning the fuel/air mixture. In both cases, exhaust levels of oxides of nitrogen (NOx) are increased. The Denox catalytic converter is therefore a key to reducing fuel consumption. Europe must not fall behind in this area given the importance of the diesel engine in our countries. It is unfortunate that there is not a stronger movement on the European level to increase research in this area, as is now the case in Japan.

Finally, I would like to mention two problems of a more technical nature: light-off and smells.

Under certain conditions, the catalytic converter does not begin operation early enough. This is particularly true in short trips around town, when the engine is cold. Thus, the catalytic converter's effectiveness from a cold start must be improved by speeding-up the light-off process. We must do this without affecting its durability and avoiding solutions which are too complex or costly such as heating the catalytic converter.

Since we began equipping our cars with catalytic converters, our customers complain of unpleasant smells primarily during short trips when the car is new. We know that this is caused by hydrogen sulphide (H_2S) emissions when the sulphur that accumulates in the catalytic converter is released, and an appropriate solution must be found.

❖❖❖

To conclude this brief overview, I would like to make two comments:

- Changes over the past ten years and, even more so, prospects for the future show that the motor car can adapt to environmental requirements. However, it is now time that we tackle problems less emotionally and develop a more rigorous approach:

  - ⇨ a global approach where governments stop imposing legislation in one area without considering its impact in others,
  - ⇨ an approach based on cost-effectiveness aimed at organising priorities and selecting appropriate measures,
  - ⇨ a forward-looking approach so as not to exclude promising technical developments.

- As far as the catalytic converter is concerned, much progress can still be made and new avenues be explored. Specialists in catalytic converters must be committed to offering significant results as quickly as possible.

If we combine these efforts with a global approach, we will produce positive results both for our customers and for the environment.

A. Frennet and J.-M. Bastin (Eds.)
Catalysis and Automotive Pollution Control III
Studies in Surface Science and Catalysis, Vol. 96
1995 Elsevier Science B.V.

DEVELOPMENTS IN GASOLINE REFORMULATION AND THE ENHANCEMENT OF REFINERY MTBE PRODUCTION

K.P. de Jong[a], W. Bosch[b] and T.D.B. Morgan[c]

[a]*Koninklijke/Shell Laboratorium, Amsterdam (Shell Research B.V.), Postbus 38000, 1030 BN Amsterdam, The Netherlands.*
[b]*Shell Internationale Petroleum Maatschappij (B.V.), Postbus 162, 2501 AN The Hague, The Netherlands.*
[c]*Shell Research Ltd, Thornton Research Centre, P.O.Box 1,Chester, Cheshire CH1 3SH,UK*

ABSTRACT

Automotive pollution control calls for joint attention to the three main aspects involved, i.e the fuel, the engine and the exhaust catalyst. In order to establish the potential contribution that gasoline composition can make to reduced emissions in Europe, we present recent data on the relationship between gasoline composition (MTBE, aromatics, olefins) and bulk emissions (HC, CO, NO_x). More work is needed to differentiate between the composition and physical properties of the gasoline in these intricate relationships. The relationship between the hydrocarbon emissions ant the gasoline composition, however, is straightforward: for C_{5+} species the molecular composition (fingerprint) of the exhaust gas resembles that of the fuel, whereas lighter hydrocarbons in the exhaust gas are combustion products. This conclusion migth be important for further development of exhaust catalysts to cope with the compounds most difficult to convert.

The manufacture of MTBE,; one component of US reformulated gasoline is addressed. It is argued that the skeletal isomerisation of n-butenes to isobutene is an attractive route to produce enhanced amounts of MTBE in the oil refinery or chemical complex. Although the rate of MTBE production worldwide has grown rapidly (to 20 Mt/a in 1993) it is still modest in relation to the world demand for gasiline (800 Mt/a). This and the high energy intensity of producing « synthetic » gasoline components such as MTBE make a « wel-to-wheel » approach mandatory in optimizing environmental and economic factors.

1. INTRODUCTION

The growing concern about the environment has led to a drastic reduction of emissions from the transport sector. Taking 1968 as a base level for gasoline-fuelled cars, the European commission standards for nitrogen oxides (NO_x) and hydrocarbons (HC) have

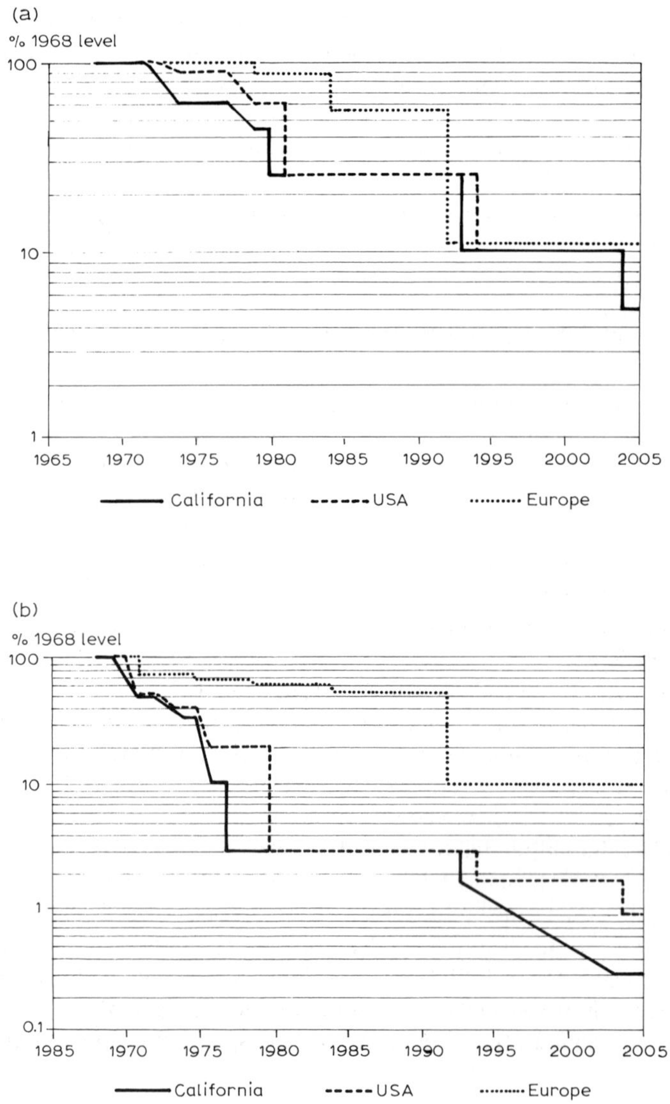

Figure 1. (a) NO_x and (b) hydrocarbons emission standards for cars.

both decreased by 90% (Fig. 1). In the USA, and particularly in California, even more severe standards are being introduced. Between 1970 and 1990 emissions were reduced by the introduction of exhaust catalysts which required unleaded gasoline. However, air quality is still not acceptable in many cities so further measures are needed. The new "Clean Air Act Amendments" of 1990 require both lower vehicle emission limits and the introduction of so-called "reformulated gasoline" in major cities. In Europe there is an ongoing "Tripartite" debate under the leadership of the European Commission and involving the Oil and Motor industries. Its intention is to establish new optimised vehicle emission limits and fuel specifications for the year 2000 in order to improve air quality.

In this paper we describe some recent work done by Shell Research (which is submitted to the Tripartite Group) on the effects of gasoline composition and properties on vehicle emissions. In addition the manufacture of MTBE, a relatively new gasoline component and a major feature of US reformulated gasoline is reviewed. Catalysts appear to play an important role in both gasoline manufacture (illustrated for MTBE production) and gasoline end use (exhaust catalysis).

2. DEVELOPMENTS IN GASOLINE REFORMULATION — USA AND EUROPE

In recent years the continuing refinement of engine management and emission control systems has made a significant contribution to lowering tail pipe emissions from gasoline fuelled cars. In the USA, the penetration of cars with catalyst converters is effectively complete. In Europe penetration is around 20% as only recently there has been a requirement for all new cars to be equipped with an exhaust gas catalyst system.

Despite these improvements, there is still pressure for further reductions in exhaust emissions. This will require further improvements to vehicles and emission control systems, but attention is also being focused on how changes in the composition and properties of fuels can contribute.

In the USA the new 1990 Clean Air Act Amendments require both reduced vehicle emission limits and the introduction of so-called "reformulated gasoline" in cities which do not meet atmospheric ozone targets. Following this legislation, a joint Auto/Oil industry test programme, the Air Quality Improvement Research Programme (AQIRP), has examined the effects of changes in key gasoline properties on emissions from both older and current US vehicles (all catalyst equipped). Using information from this and other programmes, the US Environmental Protection Agency has now published further legislation defining the requirements for reformulated gasoline. This is very complex (some 3000 pages!) but is essentially as follows. Reformulated gasoline must contain a minimum of 2% oxygen, a maximum of 1.0% benzene and no heavy metals but must have detergent additives, while levels of T90, olefins and sulfur must not be higher than each refiners' 1990 average figure (see Table I). In addition, it must meet specified emission targets for VOCs,

Table I
Reformulated gasoline exemplified — USA: simple model 1995–1997; annual averages (region south)

Parameter	Value
Benzene, %vol	< 0.95
RVP, psi	< 7.1
Oxygen, %w	> 2.1 *
Sulfur, ppmw	< 100 **
T90, vol% evaporated	< 100 **
Olefins, %vol	< 100 **

* 2.7 %wt maximum.
** Reference 1990 level (%).

air toxics (benzene, 1,3-butadiene, formaldehyde and acetaldehyde) and NO_x emissions. These targets are in the form of specified percentage reductions in emissions as compared to the 1990 baseline. The emissions performance of gasolines are calculated on the basis of mathematical models developed using data from the AQIRP and other programmes. The "Simple model", which must be used from 1995–97, specifies fixed RVP limits (assumed to give a 15% reduction in evaporative VOC emissions) and comprises relatively simple equations to calculate air toxic levels. The "Complex model", which must be used from 1998, is a complex set of equations including terms for all major fuel properties, and in many cases second-order terms as well. For further details we refer to [1].

In Europe there is also concern to improve air quality, both local as in the USA and global, i.e. reduced CO_2 emission. One sub-group of the Tripartite Group has reviewed the available information (including AQIRP) on the effects of fuel properties, both gasoline and diesel, on vehicle emissions. This has led to the development of a European test programme (EPEFE, the European Programme on Emissions, Fuels and Engine technology), which will cover both gasoline and diesel vehicles and look at effects of specific fuel properties on emissions from advanced technology vehicles. The results from the EPEFE programme are intended to provide a valuable technical input to the European legislative process, which will develop new standards for the year 2000.

Shell Research has carried out a major test programme to investigate fuel effects on non-US vehicles. The results from this work have been used as input to the Tripartite investigation and are summarised, together with the results from some other programmes, in the next section.

3. GASOLINE COMPOSITION — EMISSIONS

The work carried out has included two Shell studies. The first was a preliminary study [2] in which 4 cars (1 with a catalyst) were tested over the FTP-75 cycle for regulated and speciated HC emissions on 4 fuels ranging from a conventional to a severely reformulated (research) gasoline. Subsequently, a multi-laboratory study was carried out within the Shell Group to investigate the effect of better controlled step wise changes in gasoline on regulated and speciated HC emissions behaviour, for 24 cars of different engine/catalyst sophistication (and including noncatalyst-equipped cars). The main driving cycle used was the extended European cycle. The work on the regulated emissions has been reported in detail [3].

Figures 2 and 3 summarise the main effects on regulated emissions of varying the key compositional variables, MTBE, aromatics and olefins using paraffins to balance. The designed variations in fuel composition (Table II) inevitably led to changes in other fuel properties especially distillation and stoichiometric air/fuel ratio. The results are given in fleet-averaged form for subsets of noncatalyst- (Fig. 2) and catalyst-equipped cars (Fig. 3). The overall findings show that:

(i) Inclusion of MTBE (and removal of paraffins) yields HC and CO reduction benefits in both noncatalyst- and catalyst-equipped cars.

(ii) MTBE effects on NO_x are small and no clear trend is apparent.

(iii) The apparent effect of reducing the aromatics content is to reduce tail pipe HC and CO emissions further, but the accompanying volatility changes in the fuels make the

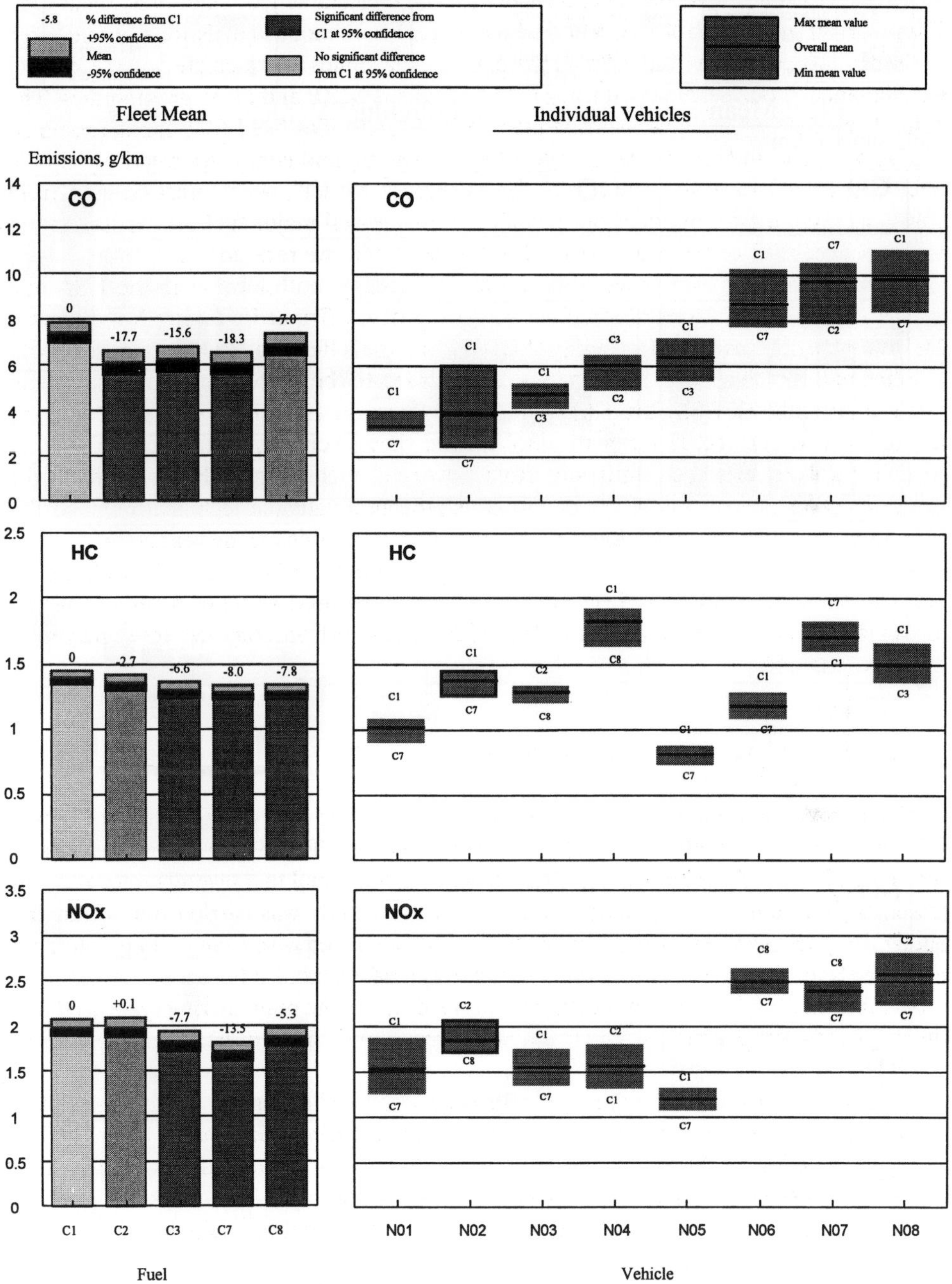

Figure 2. Fuel effects on regulated emissions; extended ECE cycle; noncatalyst-equipped cars.

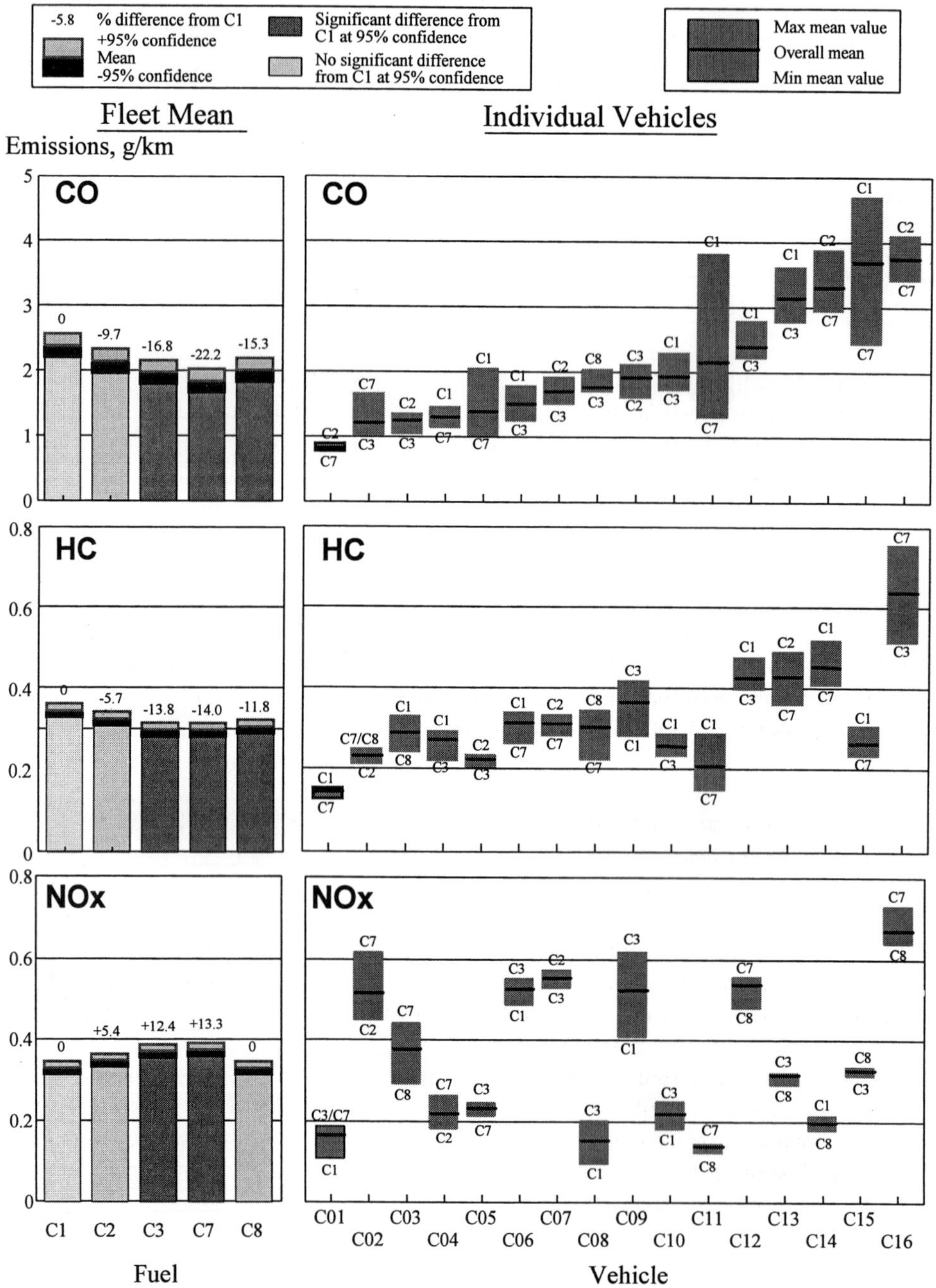

Figure 3. Fuel effects on regulated emissions; extended ECE cycle; catalyst-equipped cars.

Table II
Composition and properties of test fuels

	Fuel No.							
	1		2		3		7	8
Aromatics, % mass	45		45	→	25		25	45
MTBE	0	→	15		15		15	8
Olefins	15		15		15	→	5	0
T90, °C	150		151		140		133	146
E70, %vol	27		30		34		42	39

reasons for this effect uncertain. It should be noted also that separation of aromatics/mid-range volatility effects is an objective in the EPEFE gasoline study.

(iv) Reducing aromatics content reduces NO_x emissions from the noncatalyst-equipped cars, most probably due to an effect on peak combustion temperature, but increases NO_x emissions from catalyst-equipped cars, due to reduced catalyst efficiency.

Figures 2 and 3 show the car-to-car variability of tail pipe regulated emissions performance, a feature masked by fleet averaging. Variability arises not only in the absolute values of emission levels but also in the response to fuel property changes.

The effects of gasoline sulfur on the capacity of the three-way catalyst to control regulated emissions was one of the issues highlighted by the AQIRP programme in which fleet-averaged effects of fuel sulfur were found to be detrimental to the catalytic removal of all three pollutants (VOC, NO_x and air toxics). In Fig. 4 we have reproduced data from a recent European study [4] which shows that, as fuel sulfur is increased, HC emissions become higher, CO emissions remain virtually unchanged while NO_x emissions are reduced. It is also important to recognise that, even at the 90% confidence interval, statistically significant emission changes are rarely evident. Furthermore, the catalyst composition (its nature and probably also loading of the metals) has a considerable impact. From the foregoing it is clear that the question of sulfur sensitivity requires further clarification, particularly under European conditions, and this is a key technical objective of the EPEFE.

From [2] we present further information on the nature (so-called speciation) of the hydrocarbons emissions in the form of the relationship between fuel and exhaust (engine-out) composition (Fig. 5). It is evident that the fingerprints for compounds of C_5 and higher (C_{5+}) are comparable on a fuel-to-exhaust basis. Thus unburned fuel is a major contributor to the C_{5+} engine-out profile. Lighter hydrocarbons (C_{4-}) are produced by the breakdown of larger molecules and other partial oxidation products will also form (e.g. aldehydes and ketones, which are not considered here). Figure 6 provides a more quantitative example of exhaust/fuel comparisons for C_{5+} components.

A more fundamental approach was followed in a study by the "Ricardo Oil Industry Consortium" using model fuels and developing fuels/exhaust gas composition relationships [5]. An example of this is given in Fig. 7 for toluene, isooctane and diisobutylene as model gasolines. Both pre- and post-catalyst data on HC speciation profiles are given (by group though individual species can be treated) for engine tests at a variety of steady state engine conditions. Again, the extremely different fuels produce comparable amounts of total HC emissions but their nature is determined by the type of fuel. When toluene

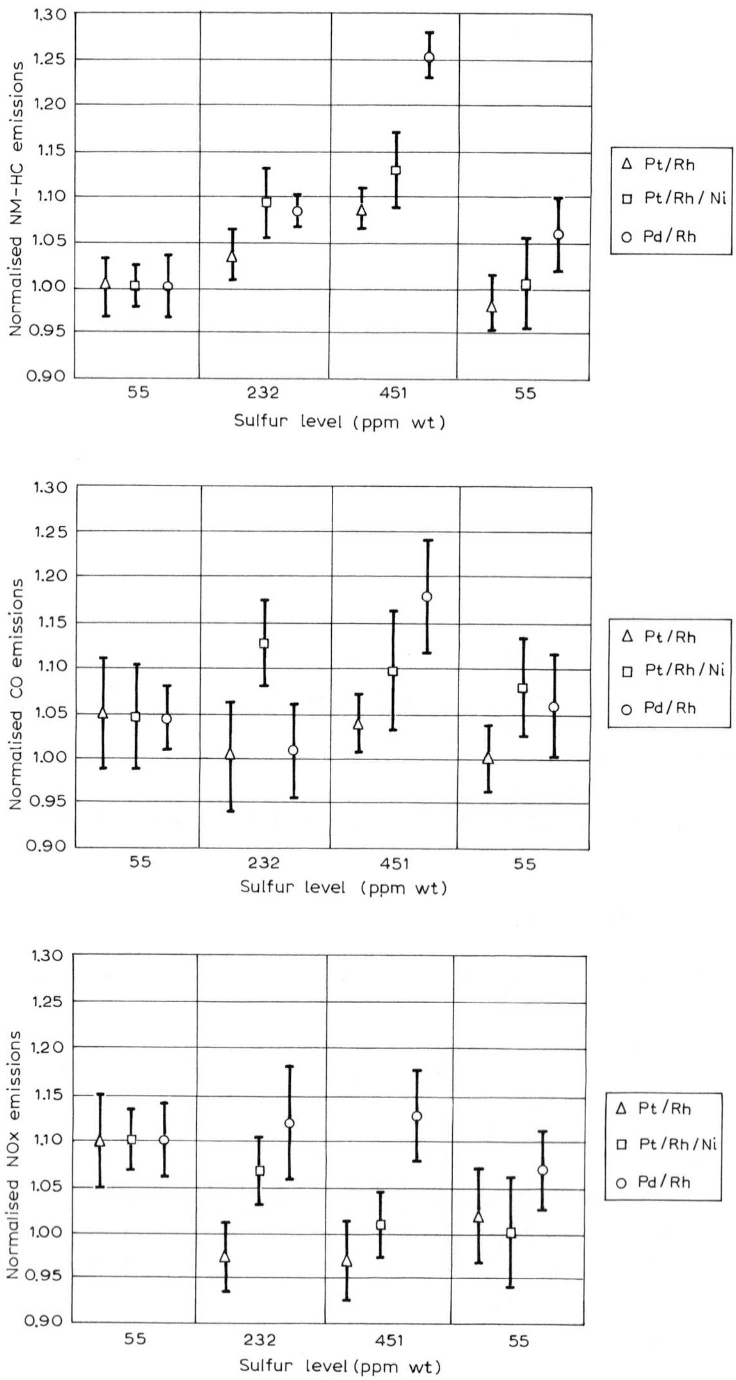

Figure 4. Effects of sulfur on regulated emissions for different exhaust catalyst compositions.

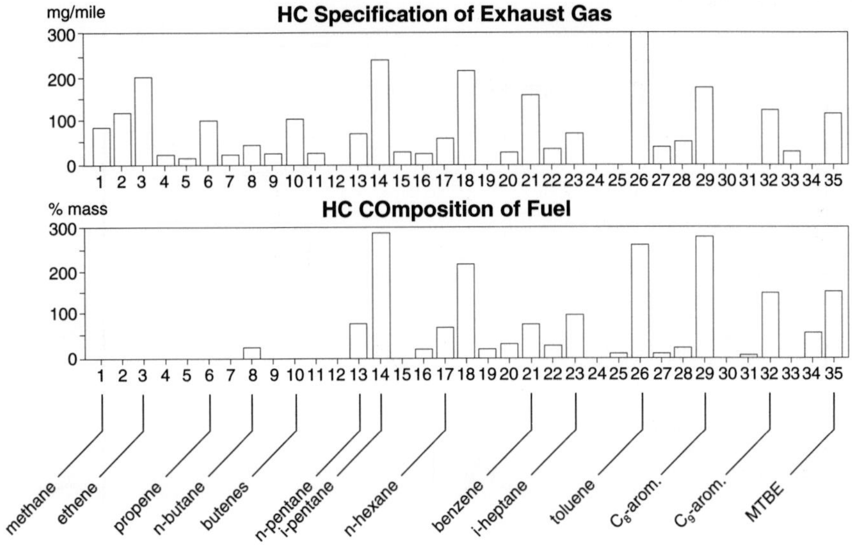

Figure 5. Molecular composition of exhaust gas versus fuel.

is used as a fuel, unburned hydrocarbons (i.e. toluene itself) predominate whereas the aliphatic fuels (both paraffinic and olefinic) produce significant amounts of lighter olefinic products as a result of cracking reactions. It may be concluded that, depending on the (major) fuel components, the exhaust catalyst is exposed to a different HC compound and may require different characteristics. In our opinion, this concept of tuning fuel and catalyst for maximum HC conversion is relevant to arrive at further emission abatement.

4. GASOLINE PRODUCTION — MTBE MANUFACTURE

Gasoline is a complex mixture of over 200 hydrocarbons and possibly oxygenates. The gasoline pool in a refinery is made up of a number of process streams with a variety of properties. Gasoline from the fluid catalytic cracking (FCC) unit and the catalytic reformer are the largest contributors followed by alkylate, butanes and (C_5, C_6) paraffin and isomerate. Changing the overall composition of the gasoline pool may require major adaptation of the refinery lay out.

Taking MTBE as an example (see section 2) we note that for the USA in the coming years a doubling of the MTBE content of the total gasoline pool is expected as a result of the oxygen mandate for reformulated gasoline. In considering the manufacture of MTBE a number of important issues will be dealt with, such as the requirement for new process options and the energy intensity of the processes applied.

First we consider briefly the manufacture of ethers from iso-olefins and alcohols (Fig. 8). MTBE is made by the commercially widely applied etherification of isobutene with methanol, using acidic ion exchange resin catalysts. In general, methanol is made from natural gas via synthesis gas and is readily available. Clearly, expansion of the MTBE manufacturing capacity calls for more isobutene feedstock.

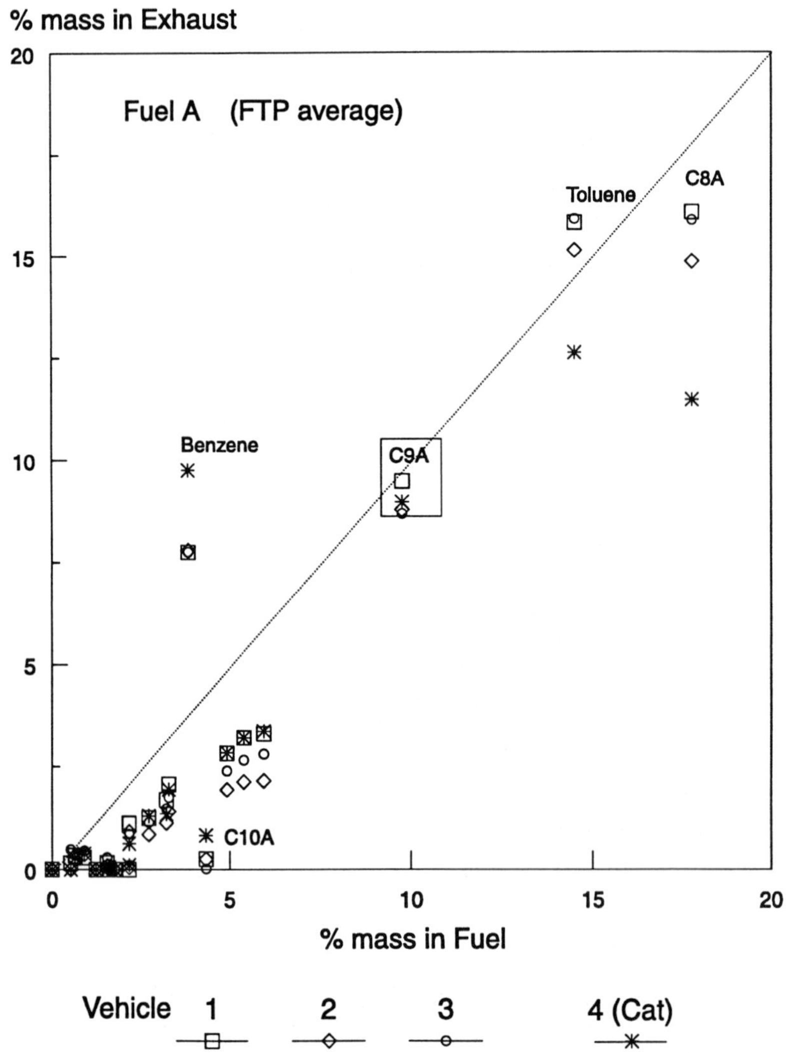

Figure 6. C_{5+} hydrocarbons; exhaust gas versus fuel.

In Fig. 9 we sketch two process routes to isobutene. Starting from *n*-butane, isomerisation to isobutane is followed by dehydrogenation to isobutene. Both process steps have been widely operated commercially. The second step involves a capital intensive dehydrogenation reaction. Because of the intensity large plants are called for (economy of scale). An alternative and in principle less capital-intensive route is the skeletal isomerisation of *n*-butenes to isobutene. Previously this was not economically attractive because of the limited lifetime of the catalyst and the limited yield of isobutene. The latter drawbacks were due to the high temperatures applied previously (>400°C) and the related fast coke formation.

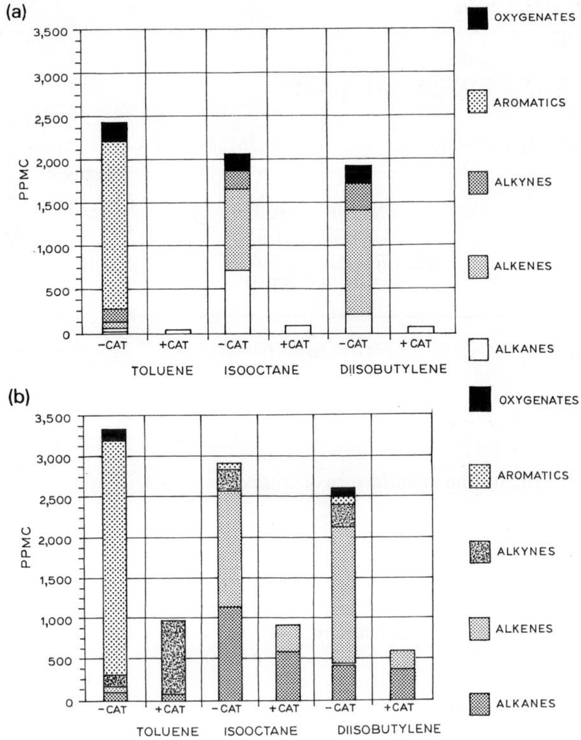

Figure 7. Pre- and post-catalyst hydrocarbon type emissions for model gasolines and different steady-state engine conditions. (a) 2400 rev/min, part. load; (b) 4000 rev/min, wide open throttle.

Isobutene + Methanol → MTBE
Isobutene + Ethanol → ETBE
Isopentene + Methanol → TAME

Figure 8. Manufacture of ethers from iso-olefins and alcohols.

Research at the Shell laboratories has focused on the use of shape-selective zeolites, notably ferrierite (FER), for the latter conversion. The data in Fig. 10 show that at temperatures as low as 350 C the maximum isobutene yield (thermodynamics) can be achieved.

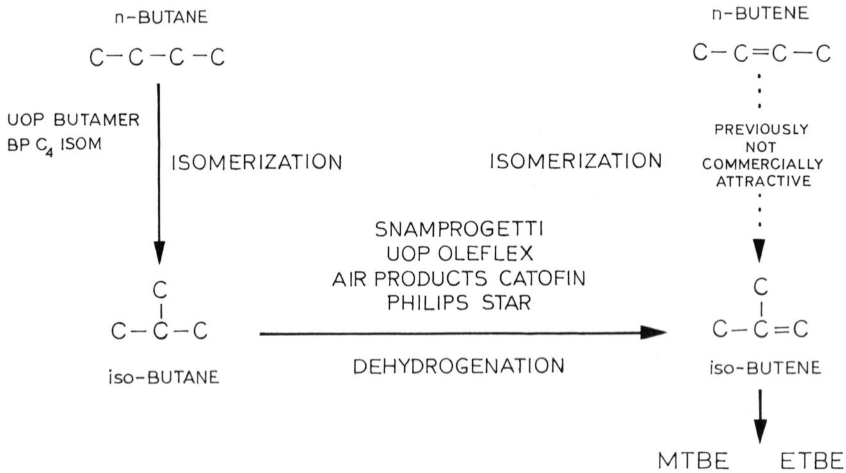

Figure 9. Routes to isobutene for the manufacture of ethers.

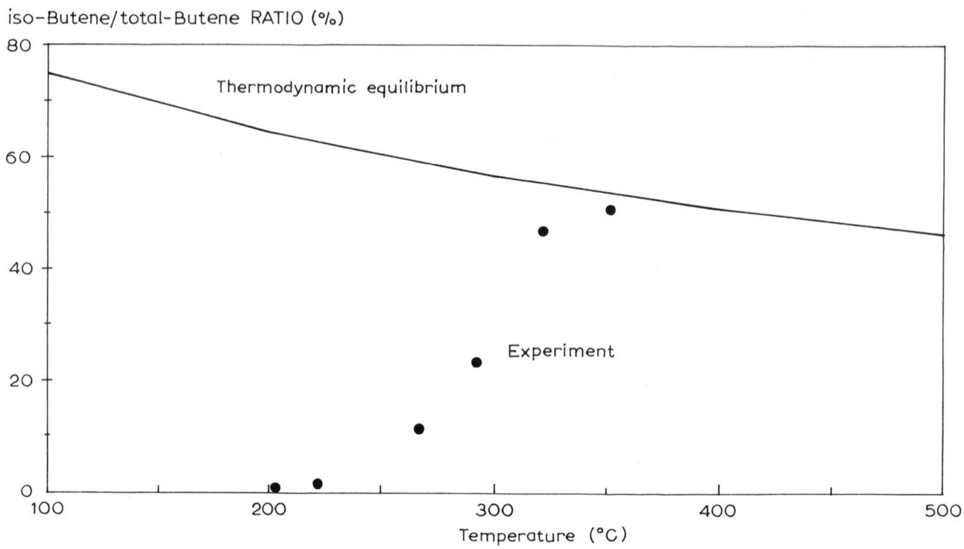

Figure 10. Skeletal isomerisation of *n*-butenes; effect of temperature (catalyst: ferrierite, feedstock: 1-butene).

Comparison with other molecular sieves (Fig. 11) shows that the yields obtained with FER are very high indeed. As elaborated upon elsewhere [6–8], we have proposed that the isomerisation involves a bi-molecular mechanism in which e.g. di-methylhexene isomers crack selectively to isobutene and *n*-butene (Fig. 12). The mono-molecular mechanism requires the energetically unfavourable primary carbenium ions. Molecular modelling [7] has provided support for this mechanism in that the branched octenes can be formed in the intra-crystalline voids of FER but their diffusion out of the pores is hindered.

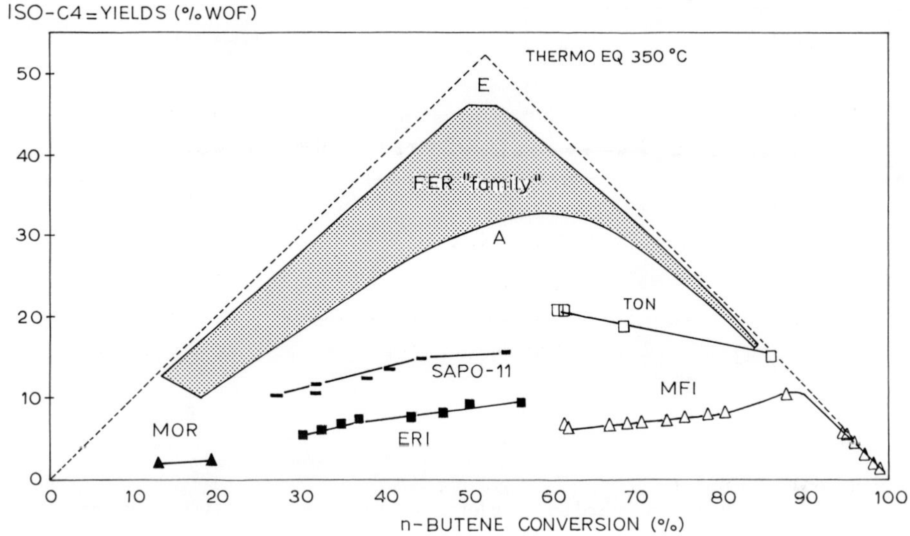

Figure 11. Skeletal isomerisation of *n*-butenes; effect of molecular sieves.

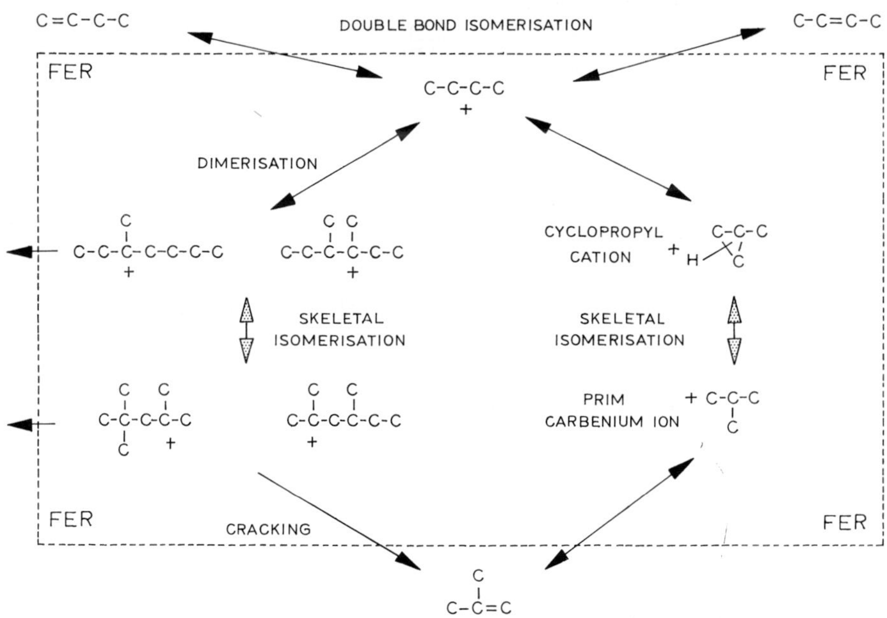

Figure 12. Proposed mechanism for the isomerisation of *n*-butenes.

For refinery applications the excellent stability of FER is important for the isomerisation of butenes (Fig. 13). The refinery or chemical complex process line-up to utilize the new isomerisation technology as shown in Fig. 14 involves a selective hydrogenation unit (to

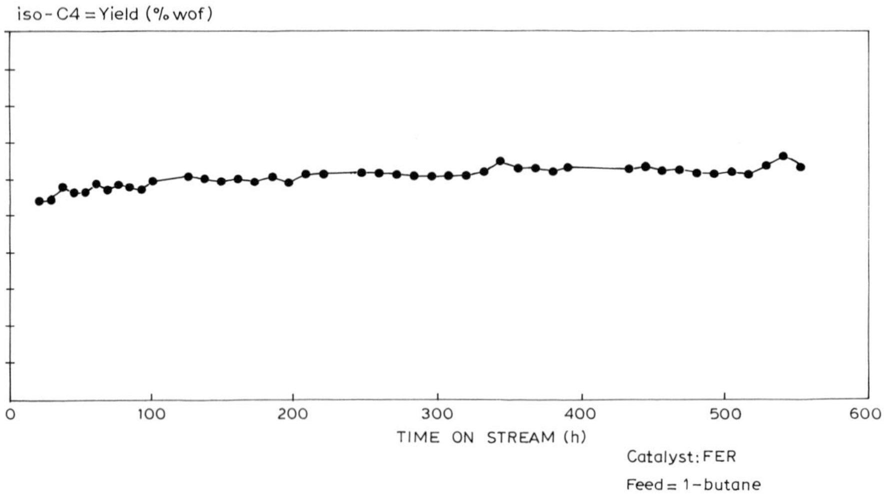

Figure 13. Yield and stability of optimised ferrierite for the isomerisation of 1-butene to isobutene.

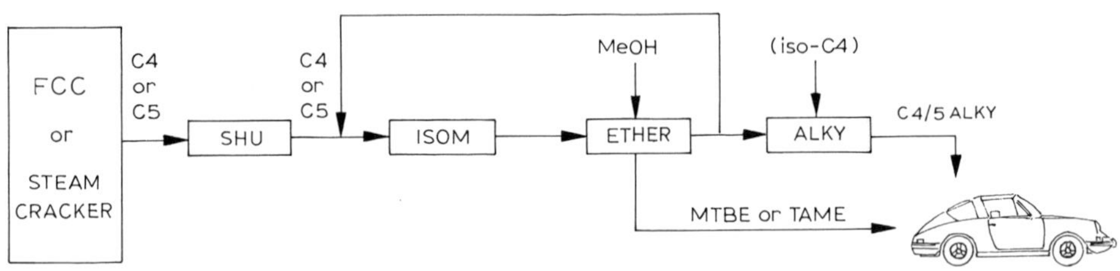

Figure 14. Possible process line up in a refinery or chemical complex including the isomerisation of C_4/C_5 olefins.

remove traces of dienes) and an etherification plant. The unconverted olefins from the latter are either recycled to the isomerisation reactor or are used in e.g. an alkylation process. The isomerisation process has been further developed by Lyondell/CD Tech and has been demonstrated at a scale of 3000 barrels per day.

A further enhancement of isobutene yields can be achieved in the refinery by adding the zeolite ZSM-5 to the FCC catalyst inventory. The ZSM-5 addition leads to secondary cracking of aliphatics in the gasoline range to lower olefins. From the extreme example [9] summarized in Table III it appears that the butenes yield can be doubled in this way.

All the routes to MTBE mentioned so far as well as others have contributed to the 1993 world capacity of 20 Mt/a (Fig. 15). Although this may seem impressive, the amount is modest in view of the world 1993 gasoline demand of 800 Mt/a. A full reformulation of the gasoline pool worldwide with for instance 15% MTBE would require a production of 120 Mt/a, which is 6 times the current capacity. Severe feedstock shortages (butanes, butenes) would occur should such huge amounts of MTBE be mandatory.

Finally, we comment from a wider perspective on the energy efficiency of producing MTBE. In the route of producing MTBE from field butanes and methanol, a typical figure

Table III
FCC product yields

	Base case	Maximum light olefins
Riser temperature, °C	523	549
ZSM-5 additive, % inventory	0	20
Conversion, %vol	79	89
Yields, %vol		
Propane	3.2	4.3
Propene	7.9	22.0
Butanes	7.8	9.8
Butenes	8.8	18.6
Gasoline	55.9	47.1

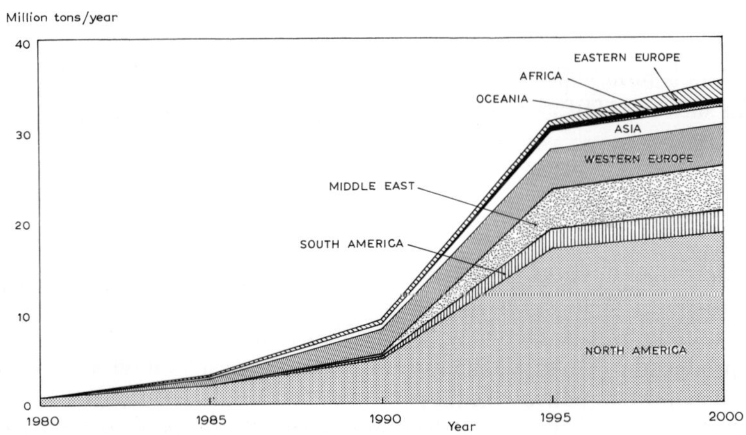

Figure 15. Projected growth of the MTBE capacity by region (source: SRI International).

for the energy consumption of the process steps involved (cf. Fig. 9) amounts to 5 GJ per tonne of MTBE produced. As the thermal efficiency of producing methanol from natural gas by modern technology is 67% (basis lower heating value) it can be calculated in a straightforward manner that the overall thermal efficiency for the production of MTBE from natural gas and butanes is 80% (see data in Table IV). Note that the route to MTBE from refinery butenes undoubtedly will be more efficient than that based on butanes since the processes involved are less energy intensive. Anyway, the direct use of butanes (as LPG) or methane (CNG) in the transport sector is much more efficient than their use after conversion to MTBE. Furthermore, refining oil to produce a variety of products (mainly LPG, gasoline, kerosine, gas oil and fuel oil) is a very efficient operation, with thermal efficiencies of typically 90% when an FCC unit is part of that refinery

Table IV
Thermal efficiencies of energy conversion processes

Operation/product	Efficiency (% LHV of feedstock)
Oil refining *	90
MTBE **	80
Methanol ***	67

LHV = lower heating value.
* Refinery including FCC unit.
** From butanes and natural gas (via methanol).
*** From natural gas.

(Table IV). Of course, the latter comparison is a gross simplification since oil refining gives rise to a product mix whereas MTBE is a high-octane quality gasoline component. A more comprehensive study (so-called "well-to-wheel") comparing gasoline components proper indicates that the use of MTBE is attractive from an overall CO_2 emission point of view (relative to high-octane oil-derived components). The latter observation is due to the hydrogen rich raw materials used (methane and butanes) rather than the energy efficiency of the processes involved. We therefore conclude that direct use or modest refining of energy carriers is more efficient than their extensive conversion to "synthetic" gasoline components such as MTBE.

5. ALTERNATIVE FUELS

Summarising what has been discussed in section 4, we present the global production of some alternatives in Table V. The scale of demand for gasoline calls for huge investment to replace gasoline by, e.g., methanol. Economic factors as well as the low efficiency of methanol production (Table IV) indicate that careful consideration is required before vigorously pursuing alternatives like these.

A global perspective of the use of alternatives in the transport sector is provided in Fig. 16. LPG and CNG are used to some extent. Special circumstances (Brazil — ethanol; South Africa — gasoline from coal) have led to larger contributions in some parts of the world.

Table V
Manufacture of (alternative) gasoline components — World, 1993

Component/product	Amount (Mt/a)
Motor gasoline, total	800
MTBE	20
Methanol	20
(Hydrogen	40)

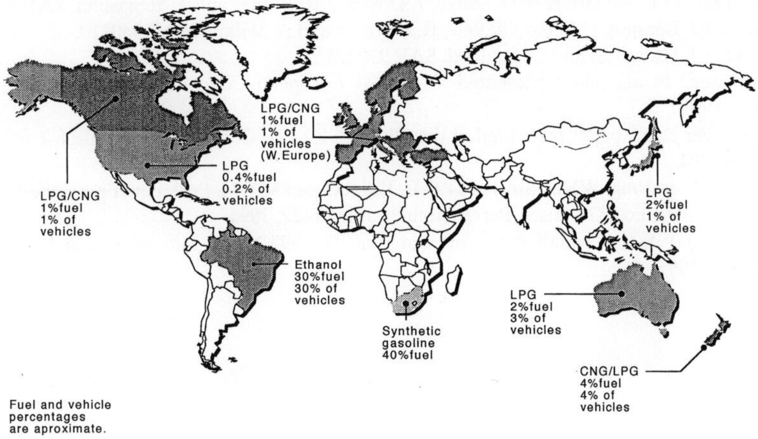

Figure 16. Main users of alternative fuels in the transport sector.

6. CONCLUSIONS

- Significant reductions of emissions from gasoline-fueled cars have been achieved by introduction of the exhaust catalyst in combination with unleaded gasoline.
- The effects of gasoline properties and composition on regulated emissions (HC, CO and NO_x) are complex and there are large variations in sensitivity between vehicles. Much work has been done, but there is a clear need for further work in Europe, especially on sulfur effects and aromatics/distillation. This will be covered in the new EPEFE programme.
- The molecular composition of the exhaust hydrocarbon species closely reflects the fuel composition for C_{5+} compounds, whereas lower hydrocarbons in the exhaust gas are mainly combustion products of fuel aliphatics.
- One of the most "refractory" hydrocarbons present in the exhaust gas is methane thus presenting a challenge to the exhaust catalyst developers.
- A more comprehensive study (so-called "well-to-wheel") is required to properly indicate the attractiveness of the use of MTBE from a CO_2 and energy point of view, addressing both local and global environmental aspects.
- MTBE manufacture from FCC-produced butenes can at least be doubled by skeletal isomerisation of normal butenes to isobutene. Addition of ZSM-5 to the FCC catalyst inventory may be applied to quadruple the MTBE output.

REFERENCES

1. A.K. Rhodes, Oil & Gas Journal, January 17th, 1994, p. 16–20.
2. G.J. den Otter, R.E. Malpas and T.D.B. Morgan, SAE 930372.

3. T.D.B. Morgan, G.J. den Otter, W.W. Lange, J. Doyon, J.R. Barnes and T. Yamashita, SAE 932678.
4. P. Beckwith, P.J. Bennett, C.L. Goodfellow, R.J. Brisley and A. Wilkins, SAE 940310.
5. P.R. Shore, D.T. Humpries and O. Hadded, SAE 930373.
6. H.H. Mooiweer et al., paper presented at the 1st European Congress on Catalysis, Montpellier, September 1993.
7. H.H. Mooiweer et al., paper presented at the 10th Int. Zeolite Conference, Garmisch-Partenkirchen, July 17–22, 1994.
8. J.E. Naber, K.P. de Jong, W.H.J. Stork, H.P.C.E. Kuipers and M.F.M. Post, paper presented at the 10th Int. Zeolite Conference, Garmisch-Partenkirchen, July 17–22, 1994.
9. T.E. Johnson and A.A. Davidan, paper presented at the 1993 NPRA Annual Meeting, San Antonio, March 21–23, 1993.

A. Frennet and J.-M. Bastin (Eds.)
Catalysis and Automotive Pollution Control III
Studies in Surface Science and Catalysis, Vol. 96

INTERNAL COMBUSTION ENGINES PROBABLE EVOLUTIONS AND TRENDS

P. Eyzat

ENSPM - Director for Combustion Engines and Hydrocarbon Utilisations

1. INTRODUCTION

The first time I participate with a group of specialists, to forecast the type of converter to be used in the coming years was in 1964. The S.A.E. entitled the session "Should we have a new engine ?". By itself, the question mark indicates how confident the organiser was for the possible quality of the reply. Naturally, we have been more often wrong than right. The mistakes classicaly are made of over optimistics evaluations of new contenders and over pessimistic ones for the adjustment potential of the standard engines. Having this lesson in mind, I forecast that the present "4 stroke-fuel" ticket will maintain its leadership for at least a decade. This is to say that the market for the electric car use will be limited to certain over populated area, where a consensus will be agreed on the extra-cost sharing between citizens and vehicle users for the sake of local pollution abatement. Another big consequence of this hypothesis is that the thermal engine will still have to be adjusted and judged in city conditions : low speed, low torque, short trip. Secondly substitute fuels, biomass or others will have to adjust to basic standard engines or to limit their use to dedicated local fleets.

As will be seen later on, the 4 stroke-fuel ticket might be very different from the present average : a lot of innovations are in sight. In sight, is by no means, equivalent to predictable : who would have, 15 years ago, forecast the huge discrepancies between developped countries on the diesel share market ?

2 STROKE SPARK-IGNITION ENGINES

2.1. Standard systems

The most common engines use a lot of fixed system which might well become variable and adjusted more and more often. Let us particularly look at

- Connection pipes : inlet and exhaust
- Valves lift, number and profile
- Compression and expension ratio
- Supercharging

All these characteristics, whenever constant, result from very accute balancing and compromise. Naturally low cost is the goal pursued by technicians. Sometimes the market allows some more degree of freedom in some niche : fixed items become variable ones. If users are satisfied, and if extracost becomes manageable for a bigger production line, then the solution is extended. Japanese firms are often in the front line for engine modifications, naturally not all of them are destined to spread and become a world standard.

The other big problem created by an item which is modified from stable and fixed to variable is the real time control of this system. You need a CPU, pick-up sensors, actuators. You must define a strategy, create hysteresis, when not natural, in order to avoid instability. Furthermore you must never forget that the engine behavior is mainly transcient. The time constant of interest ranges from tenths of milliseconds, (noise and vibrations) to milliseconds, (cycle to cycle variation) and seconds or even minutes. The average trip is so short (1 or 3 km) that when you stop your car neither the water nor the oil are at their equilibrium temperature. When you remember that oil viscosity plays a major role in engine friction and is very sensitive to temperature (you need a log-log scale to draw a straight line for viscosity versus temperature law). You realise how difficult it must be to define a strategy and to check it.

2.2. Inlet Pipe adjustment

Already performance car use inlet pipe of variable accoustic lengths. A well localised butterfly can be automatically actuated, offering two and sometimes three different design lengths. The origin of the variation of the maximum mass of air trapped inside a cylinder, a quantity directly connected to the torque, is easy to understand. Imagine a pipe closed at 1 by a valve and connected to the atmosphere at 0 (fig.1). If you create a brief suction at 1 you will observe your pipe oscillate in quarter wave, e.g. after a time equal to $+\frac{2L}{C}$, with C the sound speed, you will have a pressure increase at your valve. Naturally, if you close your valve around $t_0 + \frac{2L}{C}$ you will benefit from an increase of air filling. But later at $t_0 + \frac{4L}{C}$ it will be the reverse, you will empty your cylinder. So at high running conditions, you need a short pipe and at low running conditions a long

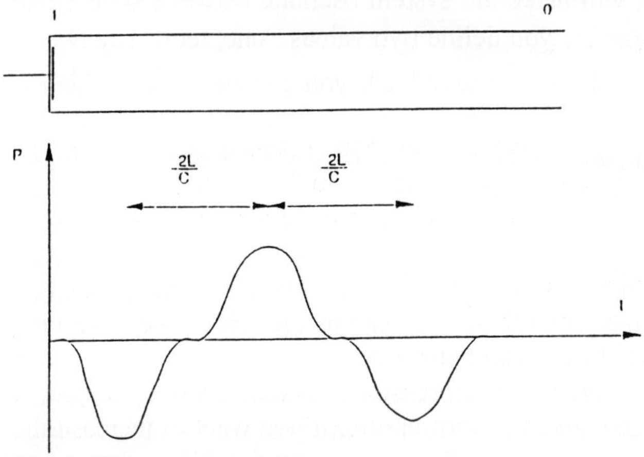

Fig.1 Wave propagation

pipe. Naturally the real phenomenon is for more comple. There is a volume effect, (Helmotz resonator) friction losses at the boundary layer, and losses due to curvature radius of pipes and to abrupt change of surface.

Admission can also be thermally heated : in accordance to a thermal sensor you can heat up inlet gas during cold starting by ensuring a heat transfer from the exhaust gases. Or create an electrically heated plate in order to compensate for the latent heat of vaporisation of the fuel. The cooling potential of this effect can be very significant : about 30 K for gasoline but more than 200 K with methanol.

Hysteresis effects are described in the (fig.2).

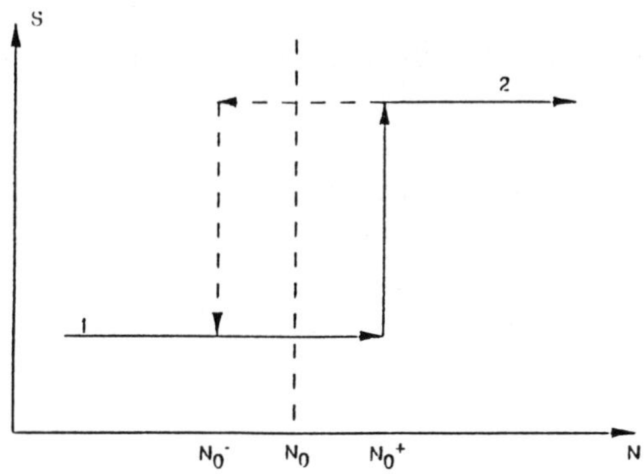

Fig.2 Hysteresis Tigger

If you decide that you will change your configuration for a given value of a parameter, e.g. running condition N_O, and you try to stabilize at this point, any perturbation, however small, will make the system oscillate between state 1 and state 2. To avoid this phenomenon, you define two values : one, let us say N_o^+ is selected when N is going up and on the way back you change at N_o^- which is lower than N_o^+.

In the same circumstances, you can have two different equilibrium conditions, you create a bias in test reproducibility.

2.3. Exhaust pipe adjustment

The waves behave in the exhaust as they do in the inlet. They are only travelling at higher speeds due to exhaust temperatures. So you can also imagine a variable exhaust pipe which creates a depression at exhaust valve closure, if you want to increase the torque, and an overpressure, if you want to increase the internal exhaust gas recirculation very efficient means for NO_X abatement. However the cost of an exhaust butterfly is much higher than its inlet equivalent and they will probably be reserved for E.G.R. and may be for prompt E.G.R., an interesting idea which is described in the (fig.3).

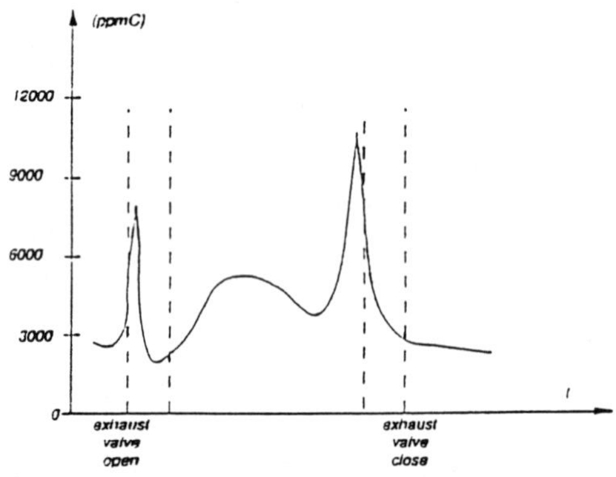

Fig3 Prompt E.G.R.:Recycling the last part of the exhaust

If you analyse the concentration of unburned HC in an exhaust burst, you will find that the instantaneous value can be far higher than the average ones. Particularly at the end of the exhaust, at a time when the piston has scraped all the boundery layers, you see a pulse of high HC concentration. If you imagine this part of the exhaust being recycled then you will be more efficient : you will obtain the normal effect on NO_X you are looking for and you will give HC a second chance to burn. In some cases, you can recycle 35 % of the HC with only 10 % E.G.R.

2.4. Cam profiles and number

It is obvious that at high running conditions the filling of the cylinder will be limited by the surface offered by the inlet valve when open. One of the biggest technological innovations for SI engines has been the 4 valve-per-cylinder concept. However at low running conditions, too high a surface, means lower gas speed and consequently smaller pressure wave amplitudes in the inlet duct. The suction phase of (fig.2) is smaller in accordance with $p+\frac{1}{2}\rho V2=cte$, and the gain of pressure at $\Delta t=\frac{2L}{C}$ later is smaller even for a well tuned inlet pipe.

It might be of interest to have a variable number of valves according to the running conditions. The possibility of modifying the cam lift profile can be fruitfully utilized for combustion speed adjustment. The combustion, expressed in percent burned per crank-angle rotation of the engine diminishes with an increase in both gas dilution (air excess or E.G.R.) or running conditions. To improve your combustion speed, when it is needed, you must increase the turbulence in your combustion chamber. One way of doing this is to vary the tumble (an air rotation whose axis is parallel to the shaft) versus the swirl (an air motion of perpendicular axis). The creation of turbulence near top dead center (TDC) where you need it, comes from the tumble which is destroyed when the piston comes close to the head. So you can specialize your valves for tumble and swirl, and by adjusting their lift with the conditions you will correct the otherwise detrimental evolution of combustion speed. Finally, if you add the gas dynamic effect which, in connection with pipe length, explains why you must open your inlet valve earlier in the cycle and close it later at high running conditions than at low running conditions. You can imagine easily how fruitfull an adjustable cam mechanism can be.

Some examples of cam mechanism adjustment and uses will be briefly described now.

<u>2.4.1 Mechanical systems</u>

The Mitsubishi system uses two cams, one for low the other for high speed range. Cams are driven, or not driven, according to the position of a hydraulic piston (fig.4). Here you can have three situations. At low torque and low running conditions (fig.5) two cylinders are disconnected (their valves remain closed).

In such circumstances a gain in efficiency mainly from the reduction of pumping losses, is obtained. Apart from this case, one or other of the cams is driven. The Honda system is of the same type.

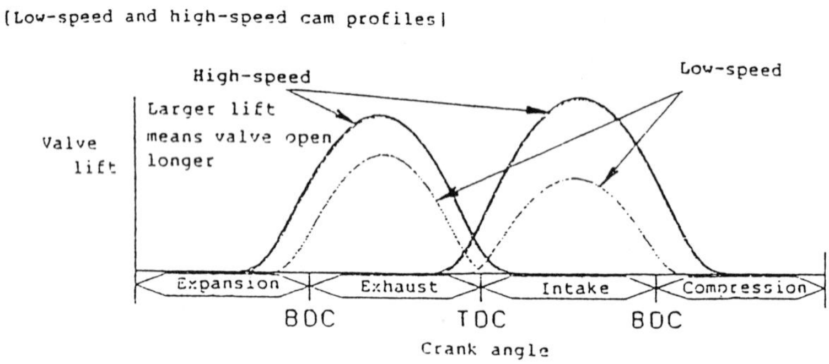

Fig.4 Variable valve Timing

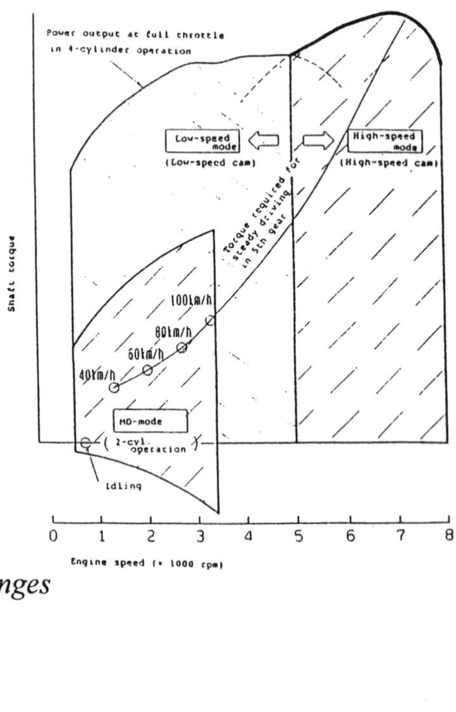

Fig.5 Operating ranges

<u>2.4.2 Continuously adjusting system</u>

The potential of hydraulic and electronic control can be pushed a little bit further. Siemens (fig.6) for example uses a solenoïd to control the oil pressure in an actuator located between cam and valve.

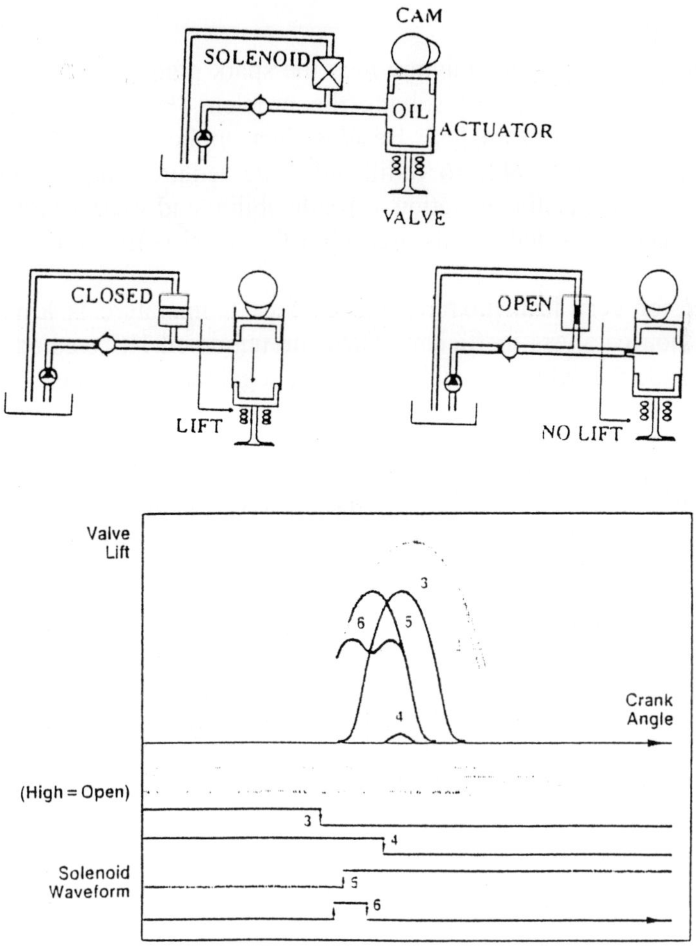

Fig.6 Siemens variable valve actuation

Considerable development work is being done to electronically govern very high injection systems for Diesel engines. Such a system might easily be used for valve motion. It offers a fully flexible and continuous lift profile adjustment, but its cost remains too high at present.

2.5. Lean burn concept

Lean burn is a very old topic in engine research. The basics are well understood. Gains come mainly from pumping losses, which are those of the Diesel engine, and from thermodynamic advantages linked to the more favorable Cp/Cv ratio of air compared to a stoechiometric mixture.

2.5.1 Stratification

A stoïchiometric or rich mixture near the spark plug ; air in excess close to the boundaries is one way of solving the problem (Baudry process, Texaco, Proco are well known examples). Besides fleet applications for Texaco, these solutions have not been able to fulfill industrial requirements. Problems arose from high pressure gasoline injection, reproducibility and wear, (a lot of progress has recently been reported in this area) and the quasi perfection of combustion needed.

In fact for very lean mixtures the exhaust temperature is low enough to make oxidation catalysts inefficient. Then the unburned HC created by gas-gas coinching inside the cylinder may be far above legal treshold limits.

2.5.2 Homogeneous lean burn engines

The goal is no longer to have an unthrotlled engine but to burn a nearly homogeneous mixture, at a point where NO_x emissions are compatible with legislation, at least until NO_x catalytic systems working in lean conditions and fulfilling automotive requirements are available.

A famous example is the Honda VTEC-E engine (fig.7) Another type is the Mitsubishi (fig.8) which is a compromise between purely stratified charge engines and homogeneous lean burn.

The Honda engines clearly demonstrate that the ambition of this manufacturer to create a"Diesel Killer" e.g. an engine emitting less CO_2 than their best Diesel competitor is not out of reach (fig.9).

However, with the breakthrough created for the sake of the two stroke engines in gasoline direct injection, notably by Siemens, or with air injection, (orbital or IAPAC) the 4 stroke lean burn engine is clearly a potential winner for tomorrow's green house engines. Already existing results evidence this forecast.

2.5.3 Heavy EGR engines

Instead of diluting by air you can think of diluting by burned gases. From a thermal stand point lowering burned gas temperatures can be equally attained by oxygen excess or CO_2 excess. Drawbacks are linked to CO_2 dissociation at high temperature, an obviously endothermic evolution, and kinetic effects. A comparison of both solutions is given in (fig.10). This clearly shows that an optimized piloted E.G.P. can offer a drastic reduction of NO_x (this solution is

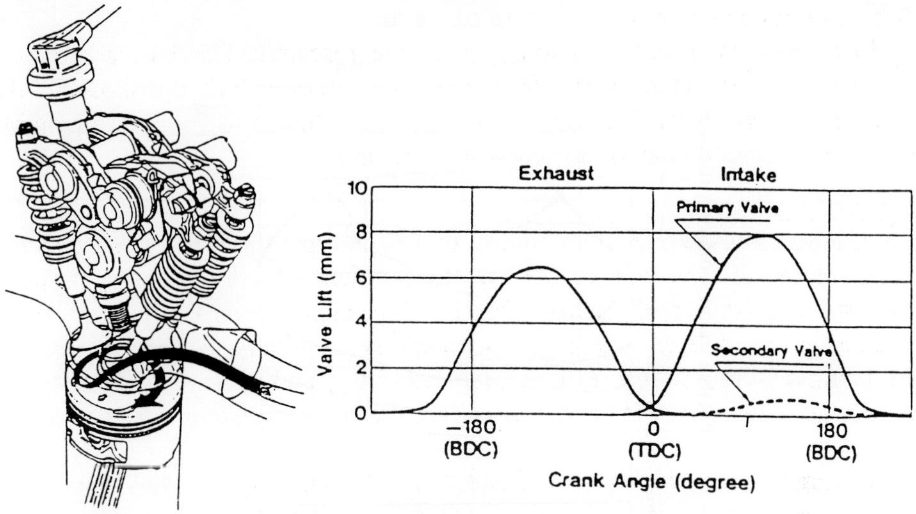

Fig.7 Honda lean burn VTEC-E

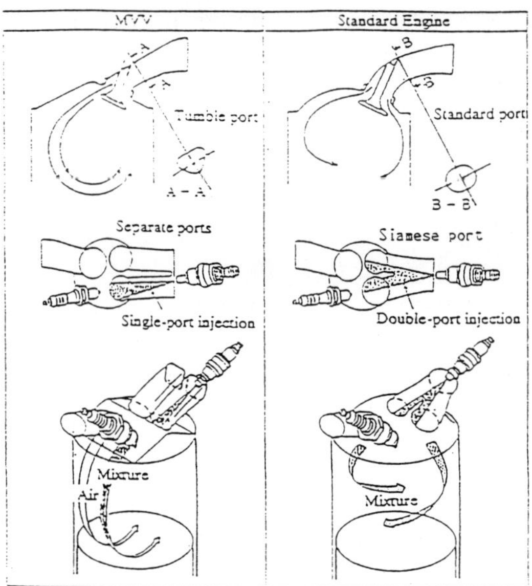

Fig.8 Mitsubishi Vertical Vortex Lean burn engine

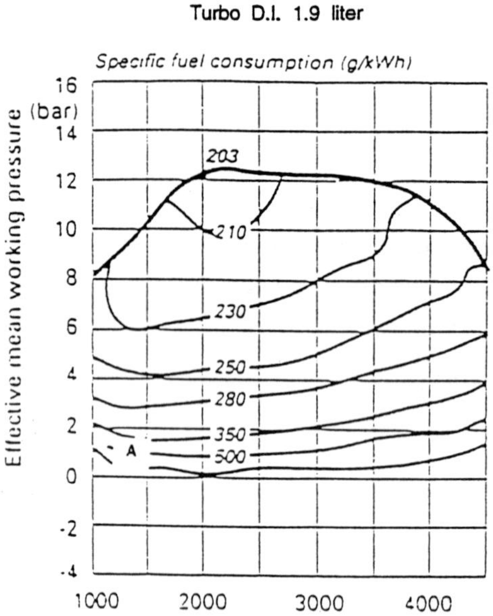

Point	Type	RPM	PME	CS (g/kwh)
A	Gasoline Lean Burn	1500	1	490

Fig.9 Honda MTEC-E in a Diesel map

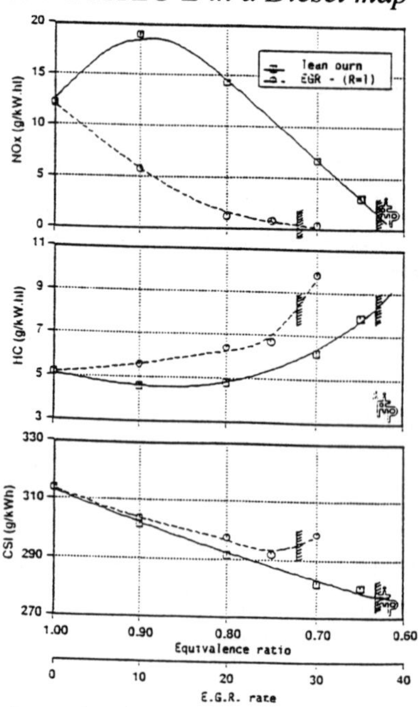

Fig.10 Lean burn versus E.G.R. 2000 rpm IMEP 4 bar

compatible with 3-way catalytic requirements) even if CO_2 optimization remains in favor of air dilution.

2.6. Compression ratio

2.6.1 Mechanical adjustment

Changing the shaft to head distance is not easy the system must sustain heavy forces and adjusting time must be low to follow the car transcient. In spite of these difficulties researchs is pursued : potential gains at part lead justify sustained effort in this direction (fig.11).

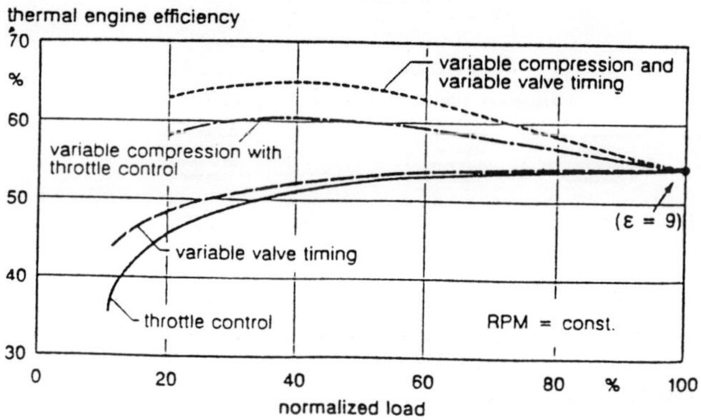

Fig.11 Variable compression ratio engine. Theoretical gains(VW)

2.6.2 Cam profile adjustment

The work done by gases occurs during the expansion stroke. Mechanical systems used today make compression equal to the expansion stroke. It is possible to avoid this combination by using a special cam profile adjustment along with the well known Miler principle. If, during the first part of the upward movement of the piston you maintain the admission valve open, you are in fact reducing the compression stroke. Mazda is said to be marketing such a solution soon (fig.12).

2.7. Engine control

Closed loop control used to maintain the fuel/air ratio within narrow limits is well known. A lambda sensor delivers a very sensitive signal to oxygen concentration around the stoechiometric ratio. This signal is used to pilot the injection in order to correct observed deviations.

Besides limitations coming from time constants, (those of the lambda sensor and the injection system) the liquid gasoline film deposited in front of the inlet

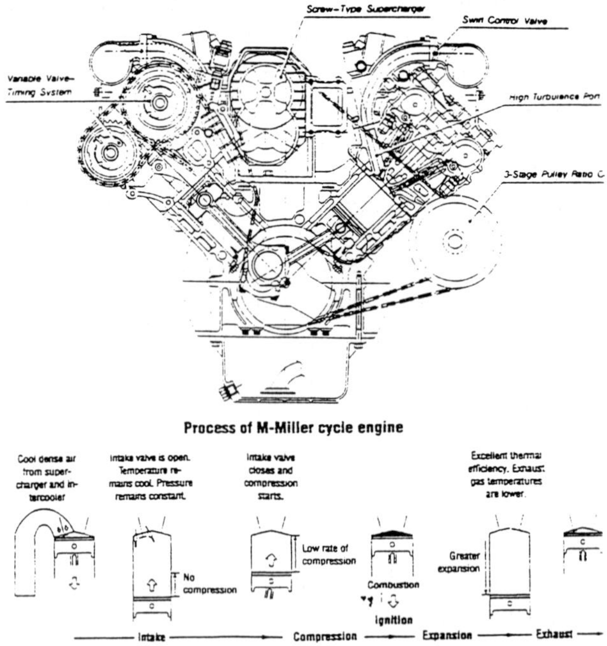

Fig 12 V6 Mazda

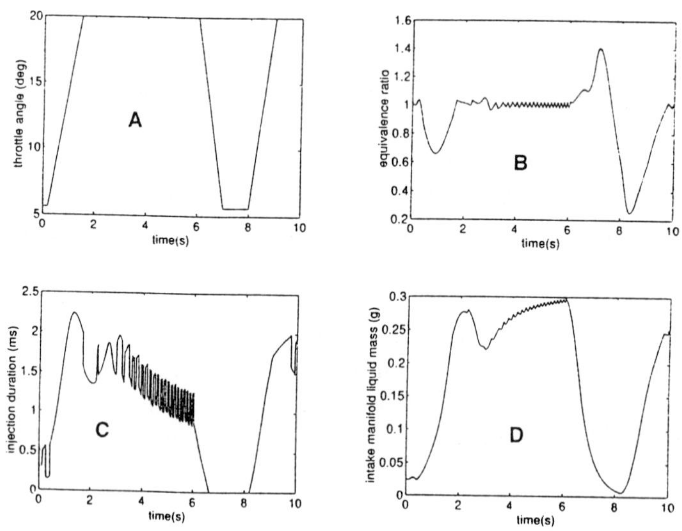

Fig.13 simulation of a simple closed lopp control(E.N.S.P.M.simulator)

valve can seriously alter the accuracy of the control in transcient conditions. Part of the fuel injected is condensed on the pipe wall. This film evaporates and due to the high air turbulence intensity, secondary droplets are also created. Then a given cycle is fed with part of the injected quantity and with part of the fuel film —a phenomena completely outside the simple closed loop control logic—. A simulator of the car and engine system has been developed and placed on a work station at ENSPM. It will help us to understand the strange behavior of an engine with a simple control system. Imagine you force (fig.13) the throttle to follow the law described in A.

The instantaneous equivalence ratio will follow the curve drawn in B. The drop of 40 % which occurs in the first second is well known (air delay is smaller than injection delay). The very wide oscillation occuring between 6 and 9 s deserves some comments. At the beginning of the throttle opening rich mixture is measured by the lambda sensor. Naturally (C) the injection duration is adjusted. But even with a smaller injection duration the film continues to fill the cylinder (D). Then the C.P.U. interprets such behavior by continually reducing the injection duration. Just before 7 seconds, the injection has been halted and the equivalence ratio remain rich ! This lasts until the the liquid film disappears (D).

As is shown in (B) the cycle by cycle equivalence ratio oscillates between .3 and 1.4. This is why the control system must be corrected for transcients. One way most often used today is to enter empirical laws in the C.P.U. Another way is to model the non linear fuel film system and to enter it in the C.P.U. of the engine : besides lower emissions, a more stable engine running is obtained (fig.14). Obviously the direct injection S.I. engine definitely solves the fuel film trouble. As has been said earlier, on the lean burn engines, technological limitations of high injection pressure gasoline systems have now been mostly overcome.

For those involved in the development of catalytic muflers the problem of engine control is of the upmost importance. Bearing in mind what happens with simple engines, one can easily imagine what problems can arise in the transcients of an engine in which duct length, cam profile and compression ratio are adjusted through non linear algorithms stored in the C.P.U. Lastly, optimisation will also have to include the exhaust pipe and its catalytic bed which add its own transcients : oxygen storage, thermal inertia, pressure drop...Comparing two catalytic systems at a ULEV level might not be possible without perfect knowledge of the C.P.U. program.

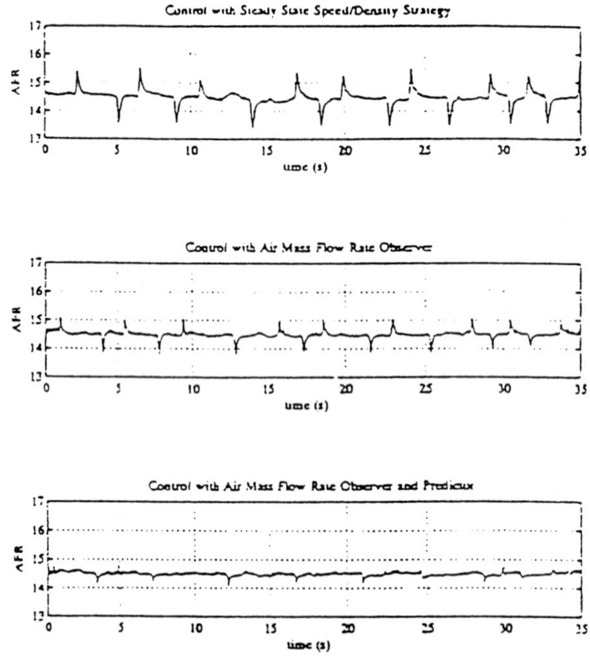

Fig.14 (From Ricardo ISATA 93EN032)

	% of total car			
	1990	1991	1992	1993
Germany	10,8	11,8	14,8	15
Italy	7,3	5,7	7,6	10
Spain	14,7	12,7	16,6	21
U.K.	6,4	8,7	12,5	17
France	33,0	38,4	39,0	44
Europe(17 countries)	14,1	14,6	17,0	19-21

Fig.15 European Diesel market share

3. DIESEL ENGINES

The Diesel engine share of the automotive market has strongly increased in Europe during the last few years (fig.15). Peculiar drawbacks of this type of engine are well known : noise, particulates and NO_X. Diesel engine specialists have made impressive progress in the last decade on the noise side. In front of catalyst specialists I will not comment further on particulate filter regeneration and NO_X reduction in lean conditions ; these are two topics of research in your field. I will just remind you that diesel specialists will have a sigh of relief if you can offer them an industrial catalyst solution !

Contrary to their S.I. competitor, progression Diesel engines is not easily identified from outside by non specialists. Their technological breakthroughs appear mainly in clearance, reproducibility, machining precision. A bad engine is not very different from a good engine. Basic solutions for tomorrow are already on the market : prechamber of the Ricardo or Mercedes type, and direct injection engines (fig.16).

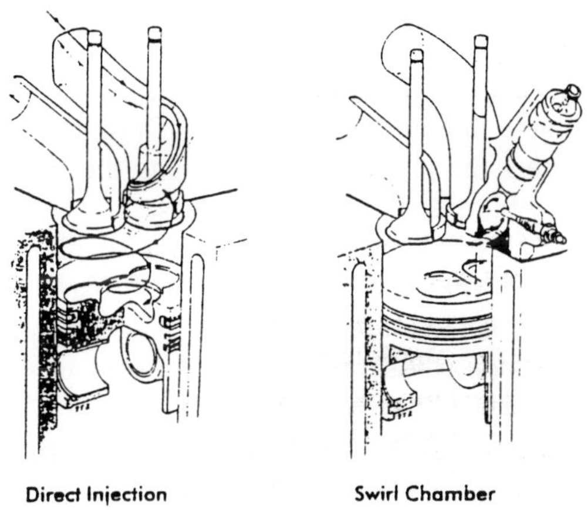

Fig.16 Diesel combustion chambers

Recent progress in prechamber engines (I.D.I.) has come from the prechamber detailed optimization (localisation of the glow plug, maintained in action after the engine starts during an adjusted time, flow guidance at the outlet of the prechamber into the piston). Electronic control of the injection system can, and probably will be developped in future engines.

Regarding the D.I. engines, noise reduction occurs through mechanical and combustion improvements. For mechanical noise the piston slap phenomenon (a vibration created by the piston to cylinder shock when the piston travels, near T.D.C., from left to right) decreases after better adjustment of clearance, shape and weight. Combustion noise is closely connected to fuel injected during the self-ignition delay. With a given fuel, one way of reducing the delay period is to use a small pilot injection. A technological break-through enables such an injection with a two-spring nozzle (fig.17).

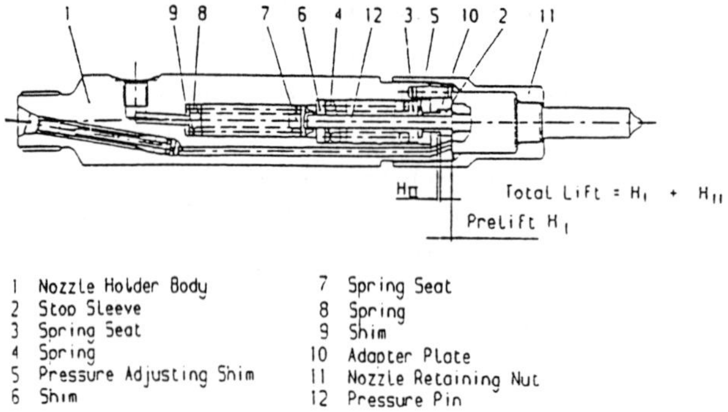

Fig. 17 - Bosch two spring nozzle

Fig.17 Bosch two spring nozzle

A very small prelift (some hundredths of mm) is limited by a second spring. The increase of force needed to complete the lift creates a delay. During this time the small amount of fuel injected during the prelift self-ignites and the main part of the fuel injected later in an already burning mixture burns less abruptly than with normal injection. If you add the fact that in order to obtain smaller droplets, injection pressure is increased (1200 bars today, 1500 to 1700 bars tomorrow) and at the same time nozzle hole diameters are reduced, you can understand where Diesel progress lies : in details, whose industrial management is the key point for competitivity.

For newcomers, in the Diesel field, the question is : will it be an NO_x regulation situated between what is possible for S.I. engines and what is not possible for Diesel engines ? Nothing of this kind is foreseeable in the coming years and on the contrary the Diesel market share is increasing even in countries where there is no fiscal advantage for its use. However such a threat in an area where technological knowhow needs time to be created, acts as a brake.

4. Stroke Engines

The more advanced contender of the standard 4-strokes for car powering deserves some comments even if, as it has been assumed in the introduction, the

4-strokes has sufficient potential to remain the winner. And, furthermore, the two strokes has its own market for two and three wheeler vehicules.

Reasons for two-stroke efficiency gains lie in reduction of friction losses and pumping losses at part load (fig.18). Friction losses are mainly related to cylinder capacity and running conditions : the two stroke which creates more power from the same capacity has lower friction losses. Improvement in pumping losses benefits from the fact that the S.I. four strokes functions like a vacuum pump during the suction stroke at part throttle. Gains in efficiency occur during city driving conditions : a 15 % reduction can be attained. Another potential big advantage of the two stroke is size reduction, a quality that designers appreciate even more than weight reduction for the freedom they gain in hood shaping (fig.19).

To obtain these gains, engineers must have solved the fuel short circuiting which makes the standard carburetor 2 stroke a high pollutant and low efficiency engine. In these engines, the burned gas is scavenged by the fuel air mixture which pushes it out. If we do this, part of the fresh mixture goes directly to the exhaust without having participated at all in the combustion phase. Injection offers the solution : scavenging will be done with pure air, and fuel injection occurs close to or after the exhaust port closure. Technological solutions are numerous and they can be examined through their injection system :

- Direct injection systems, mainly developed by Toyota, Chrysler, PSA and Renault to cite some of them. Key points are the high injection pressure system and the control of the fuel air mixing which must be optimized in a very short time.

- Air assisted injection systems. Two sub-classes must be defined :

  - medium pressure, small amount of air, most often with an air compressor. The most famous name is ORBITAL, and a lot of licences has been taken by manufacturers. This is probably the closest to industrial car use of all solutions ;

  - low pressure, amount of air compressed below the piston in the crankase, typical of the IAPAC system.

Air injection systems offer a finer atomization which make combustion and mixing phenomena a little easier to solve.

However the lubrication of a dry crankase engine is not easy to solve with lost oil : low enough for pollution, high enough to prevent wear of moving parts. Toyota has avoided this by using a standard 4 stroke technology : 4 valves per

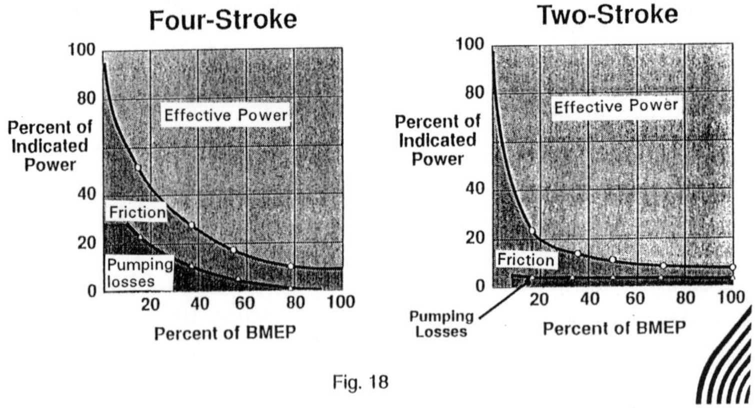

Fig. 18

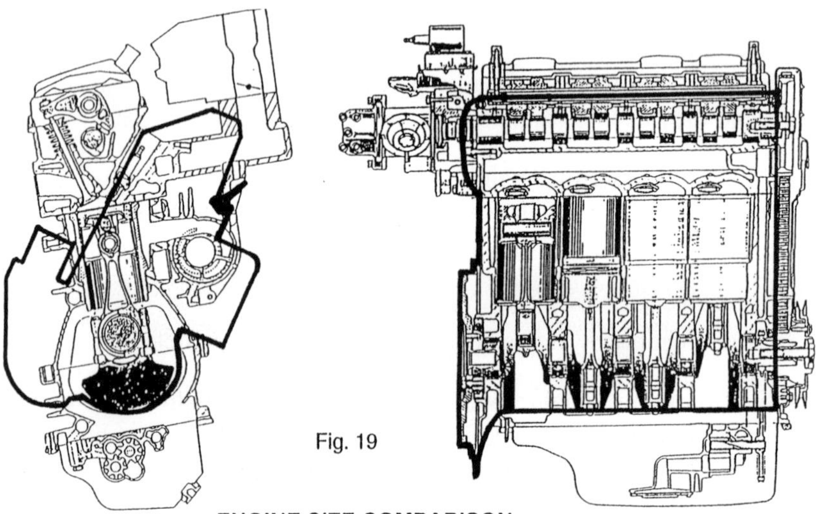

Fig. 19

ENGINE SIZE COMPARISON

4 - STROKE PEUGEOT TU3 1360 cc 50,5 kw/5600 RPM - 109 mN/3000 RPM

2 - STROKE IAPAC 1230 cc 55 kw/5000 RPM - 125 mN/3000 RPM

ifP

cylinder, normal crankase, external blower for exhaust gas scavenging. NO_x emissions are obviously the limiting factor.

5. CONCLUSION

Petroleum based fuels will remain the main sources of energy for transportation. In the car sector, the four stroke engine has a very good chance to maintain its predominance. Electronics offers it a great potential for real time optimisation of a lot of possible adjustable technologies : cams, ducts, compression ratio, firing cylinder with E.G.R. or lean combustion are some candidates for flexibility increases. The Diesel engine will maintain its impetus and the two stroke will enter into the game if efficiency and CO_2 reduction become a greater priority (fig.20).

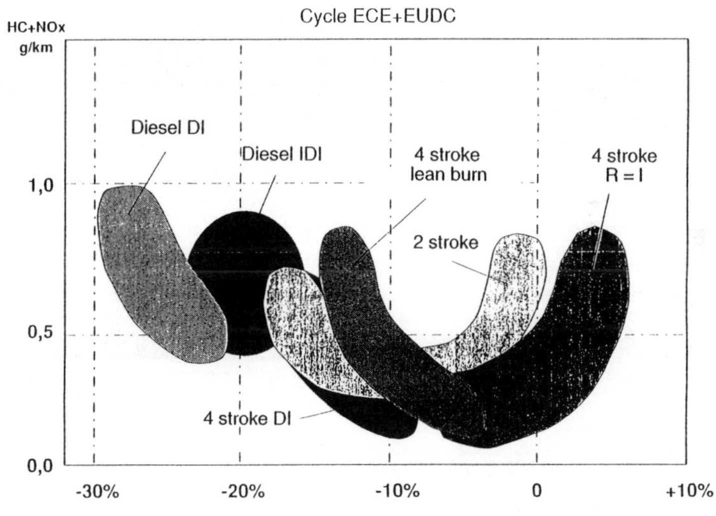

Fig.20 converters comparision

However unless an NO_x catalytic bed working with oxygenated exhaust gas is available, an NO_x threshold may be defined which condemns them. Even if very unlikely, such a hypothesis cannot be dismissed.

While these trends define the backbone of the technological evolutions, as we see it, they leave some place for niche markets . However, and we learnt this with unleaded gasoline, the idea of having a new dedicaded source of energy for a new fuel seems hopeless. A new fuel or an extension of its use will benefit from the naturally more flexible engine of tomorrow, it may be the case for L.P.G. for example. Although Natural gas suffers from the constraint of its onboard storage, its availability and CO_2 bonus explain the strong political pressure for its use elsewhere. Combustion characteristics of natural gas, low combustion speed and long self-ignition delay, create an incentive for a dedicated engine. It is forecasted that starting with fleet use, the number of filling stations will increase and this will help to spread the use of gas engine.

Model Reactions and Model Catalysts

A. Frennet and J.-M. Bastin (Eds.)
Catalysis and Automotive Pollution Control III
Studies in Surface Science and Catalysis, Vol. 96
1995 Elsevier Science B.V.

LABORATORY DATA FOR THREE-WAY CATALYTIC CONVERTER MODELLING

D. Schweich

Laboratoire de Génie des Procédés Catalytiques
CNRS-CPE, 43 Bd du 11 Novembre, BP 2077
F-69616 Villeurbanne Cedex

1. INTRODUCTION

Numerical simulations could be advantageously substituted for some expensive experiments with an engine bench. However, the reliability of the numerical results depends on the assumptions made and on the quality of the physical and chemical parameters involved in the model. Presently, converter models work well for specified experimental conditions, but there is no decisive proof of their predictive ability in a wide range of conditions (Pattas et al., 1994). This is because four important topics deserve further attention:

Kinetics,
External heat and mass transfer processes,
Internal diffusion in the wash-coat layer,
Flow distribution in the monolith.

The experimental and interpretation problems related to these topics are reviewed in the next sections.

2. Kinetic data

Converter models require reliable kinetic expressions that account for the composition and temperature dependence of the reaction rates. The temperature dependence is essential for predicting light-off performance, whereas the consequences of the concentration dependence are more difficult to assess.

Most models rely on the historical data provided by Voltz et al. (1973) which were obtained under oxidizing conditions on a platinum catalyst. The rate expressions are of the Langmuir-Hinshelwood (LH) type and account for the inhibition due to NO. Unfortunately, there is no mechanism supporting the expressions that should be considered as fitting rate laws. For instance, the heat of adsorption of NO has the wrong sign.

The first rate expressions concerning the simultaneous oxidation of CO and reduction of NO were provided by Subramanian and Varma (1985). Here again, the expressions are of the LH type, but they also involve fractional and positive orders which are of empirical nature. It seems that there are no available rate expressions for NO reduction by hydrocarbons.

To point out the main problems that arise in kinetic investigations we will focus on the simple example of CO oxidation for which reliable kinetic data are available. Using a detailed model we will generate sample "experimental" results corresponding to various experimental conditions. We will show how these results can be interpreted back with kinetic expressions, and what are the difficulties of this "inverse problem". However, let us mention that numerical calculations require that the rate remains bounded. This forbids negative orders in empirical rate expressions. Strong inhibition by reactants or products must be accounted for by a Langmuir-Hinshelwood type expression.

2.1. The example of CO oxidation

There are many kinetic expressions for CO oxidation which are derived from realistic mechanisms. Many of them, together with the expressions from Voltz et al. (1973) and from Subramanian and Varma (1985), reduce to:

$$r = \frac{k\, x_{CO}\, x_{O2}}{(1 + K x_{CO})^2} \quad \text{mole.m}^{-3} \text{ of wash-coat.s}^{-1} \tag{1}$$

where x_j is the mole fraction of species j. K and k obey van't Hoff and Arrhenius laws respectively. Based on Heck et al. (1976) the kinetic parameters are:

$$k_0 = 4.14\,10^{16} \text{ mole.m}^{-3}.\text{s}^{-1} \qquad K_0 = 65.5$$
$$E_a / R = 12600 \text{ K} \qquad \Delta H_{ads} / R = -961 \text{ K} \tag{2}$$

E_a is the apparent activation energy that is composed of the contributions of the intrinsic activation energy and of some heats of adsorption. Equations (1) and (2) were used to generate sample Light-Off (LO) curves corresponding to an inlet fluid containing 2% mole CO, 1% mole O_2 diluted in an inert gas, and a GHSV, based on total reactor volume, of 75000 h^{-1} STP. The reactor is either a monolith with square channels of width D_h = 1.1 mm, with a wash-coat thickness w = 10 mm, and a void fraction e_M = 0.76, or a packed bed of crushed catalyst particles diluted by an inert solid with an intergranular void fraction e_B = 0.36. The volume of catalyst particles per unit volume of intergranular gas is $f = 4we_M/(D_h e_B) = 0.077$. This ensures that the volume of catalytic material per unit volume of reactor is identical in the monolith and in the packed bed (See appendix).

2.2. The light-off curve

LO curves represent most laboratory data obtained with either a monolith or a fixed bed of crushed catalyst. Figure 1 shows simulated LO curves in an adiabatic monolith where the heat and mass transfer processes compete with the catalytic reaction, and where heat conduction through the solid is neglected.

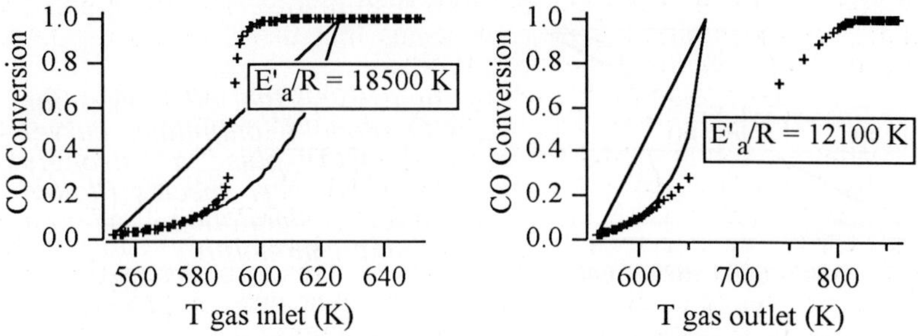

Figure 1: Light-off curve (crosses) versus inlet and outlet gas temperature. Fitting (dotted curve) the leading part of the curve yields an apparent activation energy. Simulation of an adiabatic monolith, no axial heat conduction.

Usually, one draws the curve versus the inlet gas temperature. At low temperature the reaction rate is proportional to the conversion, and one may obtain an apparent activation energy, E'_a, from an Arrhenius plot. Remark first that equations (1) and (2) imply that:

$$12600\ K < E'_a/R < 14500\ K \quad (3)$$

according to whether the reaction is strongly inhibited by CO or not. Figure 1 shows that the estimated E'_a are not in the required interval. More generally, the LO curve gives an overestimate of E'_a, when plotted versus the inlet gas temperature. Consequently, the leading part of the LO curve for a monolith has no simple kinetic interpretation.

The bias is essentially due to the progressive temperature increase along the catalyst bed that accelerates the reaction. The inlet temperature underestimates the average catalyst temperature, and using this temperature to determine E'_a yields an overestimate. Conversely, the outlet temperature overestimates the average temperature, and E'_a is then underestimated.

2.3. Presence of an adsorbed and non reacting species

The presence of an adsorbed species is also responsible for misleading results. Figure 2 illustrates the effect of a constant trace amount of an adsorbable inert species that increases the light-off temperature (T_{LO}) by 35 K. Mabillon et al. (1994) gave an even more convincing example. They showed that acetylene preadsorption can shift the LO curve by 100 K. After light-off, acetylene burns, and a second experience, without acetylene preadsorption, yields a lower T_{LO}.

In both cases, the LO curve shifts because catalytic sites are occupied by strongly adsorbed species that inhibit the reaction. According to LH theory, the inhibition is essentially governed by the adsorption characteristics at equilibrium without reaction. Consequently, adsorption isotherms under non reacting condition must be measured. They yield the adsorption constants involved in a LH rate expression.

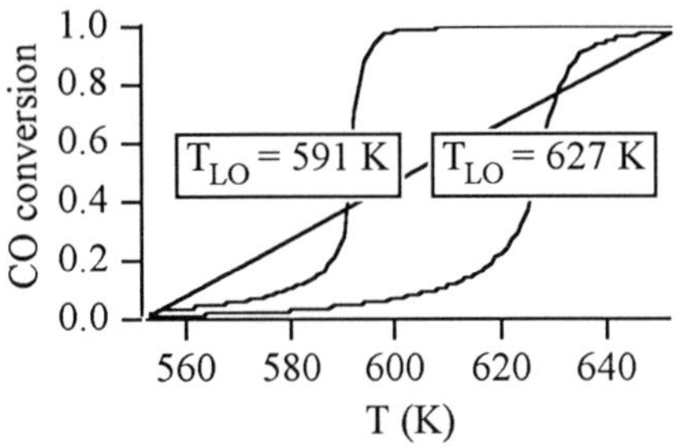

Figure 2: Light-off curves for CO alone (Continuous curve) and CO plus a strongly adsorbed inert species (dotted curve). Simulation of an adiabatic monolith.

2.4. Presence of other reacting species

One often performs LO experiments with a mixture of more than two reacting species, for example CO, O_2, and a hydrocarbon. On a Pt-Rh-CeO_2 catalyst, hydrocarbon LO takes place when CO is almost completely converted. For a given inlet temperature, the gas and solid phase temperatures are thus much higher in the presence than without CO in the feed. This implies that the hydrocarbon T_{LO} measured by the inlet temperature is higher in the absence than in the presence of CO. Consequently, the light-off temperatures for multiple reactions are technical data that have no fundamental interpretation. Kinetic studies must be performed with simple mixtures.

2.5. Heat and mass transfer

One of the drawbacks of the monolith is the heat and mass transfer resistance between the bulk gas phase and the wash-coat. This resistance can be lessened when using a fixed bed packed with crushed catalyst. Figure 3 illustrates the LO curves when the heat and mass transfer processes are not the rate determining steps. Two cases are considered depending on the efficiency of heat removal from the catalytic bed. Either the efficiency is low and the bed is close to adiabatic condition, or the efficiency is high and all the phases are at the temperature of the inlet gas.

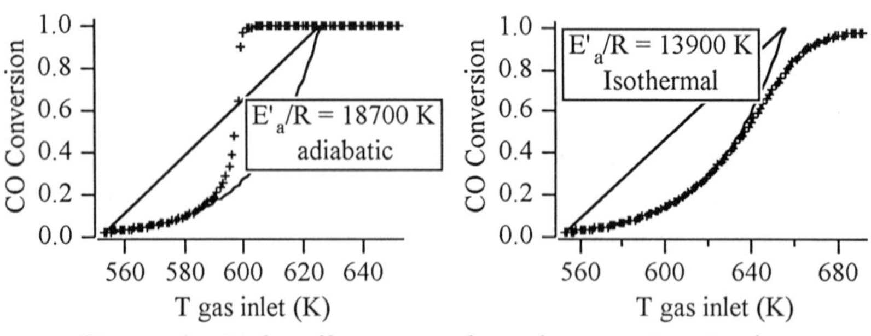

Figure 3: Light-off curves when the reaction is the rate determining step (no heat and mass transfer limitation). Left: adiabatic reactor, no axial heat conduction; right: isothermal reactor (gas and solid at the temperature of the inlet gas).

We observe first that the LO curve of the adiabatic case is closer to typical experimental results than the curve of the isothermal case. This suggests that even in laboratory fixed beds, adiabatic condition can prevail. Second, we observe that E'_a is a biased measure of the apparent activation energy in the adiabatic case, whereas it is not (condition (3) fulfilled) in the isothermal case.

The adiabatic plug flow behavior can be avoided using a recycle loop reactor at the same GHSV. Since the GHSV per pass is much higher, the heat and mass transfer resistances can be low. A recycle rate higher than 10 is recommended. Thermocouples located at both faces of the bed can indicate whether the gas phase is isothermal or not. If not, the recycle rate must be increased. Finally, the recycle loop reactor behaves as a continuous stirred tank reactor where the reaction rate is proportional to the conversion at steady state. Thus, it is the best reactor for kinetic investigations under steady state conditions. Finally, remark that a recycle loop reactor needs not to operate at the inlet temperature. Since the temperature and the composition inside the reactor are homogeneous, the LO curve versus the outlet temperature is meaningful. It remains to know how the type of flow affects the LO curve.

2.6. Effect of backmixing

Figure 4 compares the LO curves assuming either plug flow or complete mixing in an adiabatic reactor. Under complete mixing condition, the temperatures of the solid and gas phases and the composition of the gas are uniform throughout the reactor. Complete mixing prevails in a recycle loop reactor with a high recycle rate. Generally, light-off takes place at a lower temperature when complete mixing prevails. This is because the heat generated by the reaction warms the catalyst everywhere. Conversely, in a plug flow reactor, the catalyst located close to the inlet is not warmed enough by the reaction and ignition, if any, occurs downstream. Complete ignition, i.e., light-off, thus occurs at a higher inlet temperature.

Light-off represents a transition from a low conversion state to a high conversion state due to temperature. When complete mixing prevails, light-off occurs suddenly at a given feed temperature. The LO curve is thus discontinuous. In plug flow, a light-off front appears at the outlet and moves continuously upstream when the feed temperature increases. The LO curve is thus continuous, and smoother than in mixed flow.

Except the recycle loop reactor, real reactors, either packed beds or monoliths, are neither plug flow nor mixed flow reactors. However, in small laboratory reactors, heat conduction through the solid phase probably makes the temperature to be uniform as in mixed flow. Conversely, the concentration profiles are those of plug flow. Real LO curves are thus intermediate between those of plug and mixed flow.

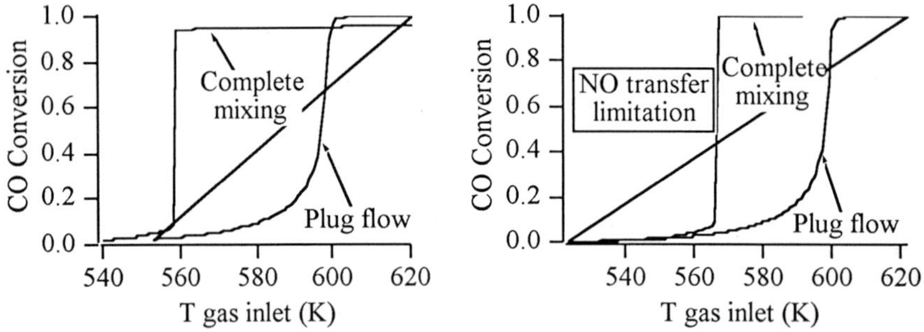

Figure 4: Light-off curves for a plug flow reactor and a recycle loop reactor. Left: reaction and transfer are competing. Right: no heat and mass transfer limitations.

Figure 5 illustrates the apparent activation energies extracted from the leading part of the LO curve obtained in an adiabatic recycle loop reactor when heat and mass transfer limitations take place or not. When the LO curve is plotted versus the feed temperature, E'_a is biased. Conversely, when the outlet temperature is used, E'_a is satisfactory. Remark that about T = 550 K and at low conversion, the inhibiting term in equation (1) is much larger than unity. Consequently, E'_a must be close to its upper bound.

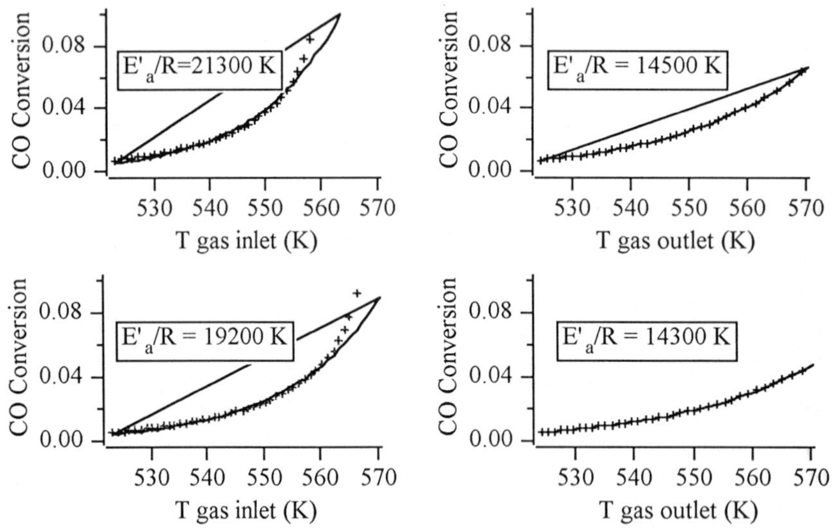

Figure 5: Best fit of the leading part of the LO curve in a recycle loop reactor. Upper curves: with transfer limitations. Lower curves: no transfer limitations.

Consequently, in a recycle loop reactor, the leading part of the LO curve plotted versus the outlet temperature gives a meaningful apparent activation energy. Mass and heat transfer limitations do not induce bias.

2.7. The zero-order model

Figure 6 illustrates the effect of the reaction order on the LO curve. The zero-order rate expression, r = k'(T), used in the figure is derived from (1) and (2) assuming that the mole fractions of CO and O_2 are those at the inlet of the catalyst bed and that the adsorption constant is calculated at the inlet gas temperature T_0:

$$k'(T) = \frac{k(T)x_{CO,0}x_{O2,0}}{(1+K(T_0)x_{CO,0})^2} = k'_0 \exp\left(-\frac{E_a}{RT}\right) \qquad k'_0 = \frac{k_0 x_{CO,0}\, x_{O2,0}}{(1+K(T_0)x_{CO,0})^2} \tag{4}$$

$$k'_0 = 1.16\ 10^{11}\ \text{mole.m}^{-3}.\text{s}^{-1} \text{ at } 553\text{ K} \qquad E_a/R = 12600\text{ K} \tag{5}$$

where $x_{j,0}$ is the inlet mole fraction, and k'_0 the zero-order frequency factor. At a given temperature and low conversion, this makes the zero-order and the LH kinetic rates to be identical.

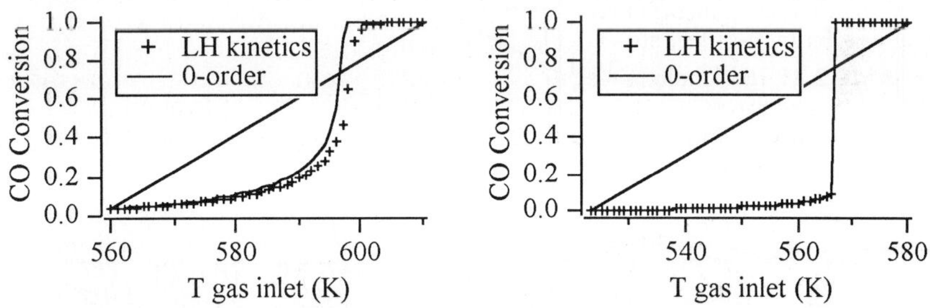

Figure 6: Comparison between the light-off curves assuming either a zero-order or the Langmuir-Hinshelwood kinetic expression. Adiabatic case, no mass and heat transfer limitation. Left: plug flow. Right: complete mixing.

The curves of Figure 6 are very close to each other. This means that light-off is poorly sensitive to the concentration dependence of the kinetic expression for a given inlet composition. The shape of the LO curve is essentially due to the feedback effect of temperature on conversion through Arrhenius law. Since the LO curve is poorly sensitive to the concentration dependence of the rate law, it can be fitted using a zero-order rate expression. Figure 7 shows the agreement between the sample LO curve and the calculated LO curve in plug flow and mixed flow. The adjusted E'_a is close to E_a, and the zero-order frequency factor is close to that given by (5). The model curve for adiabatic plug flow is given by the solution, X_{CO}, of (see appendix):

$$\frac{C_0}{t_0}\frac{dX}{dz} = \phi k'(T)H(1-X) \qquad T = T_0 + \Delta T_{ad}X \qquad k'(T) = k'_0 \exp\left(-\frac{E'_a}{RT}\right)$$

$$z = 0,\ X = 0 \qquad z = 1,\ X = X_{CO} \tag{6}$$

H(1-X) = 1 when X < 1 and H(X-1) = 0 when X = 1. C_0 and t_0 are respectively the inlet concentration and the residence time of the fluid at a reference temperature, and ΔT_{ad} is the adiabatic temperature rise. When complete mixing prevails, the model curve is given by the solution of:

$$\frac{C_0}{t_0} X_{CO} = \phi k'(T); \qquad k'(T) = k'_0 \exp\left(-\frac{E'_a}{RT}\right); \qquad T = T_0 + X_{CO} \Delta T_{ad} \tag{7}$$

$X_{CO} = 1$ when the solution of (7) is greater than unity. Equation (6) and (7) are easily solved on any microcomputer, and k'_0 and E'_a can be determined using any fitting method from standard libraries.

Consequently, when there are no mass and heat transfer limitations, a light-off curve can be interpreted assuming a zero-order reaction rate. Then, correlating the apparent zero-order kinetic constant to the inlet composition should give insight in the true kinetic expression. Since the exhaust gas coming from an engine has a composition that lies in a relatively narrow range, the zero-order approximation explains why empirical and early published rate expressions lead to good modelling results provided that the activation energies and frequency factors are suitably adapted.

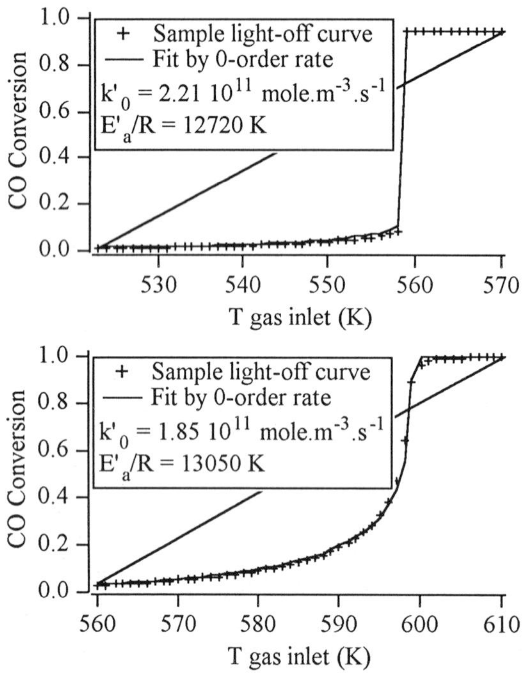

Figure 7: Best fit of the sample light-off curve by the zero-order model. Upper curves: complete mixing. Lower curves: plug flow. No mass and heat transfer limitations, adiabatic case.

One may thus wonder why further kinetic data are necessary since a zero-order rate expression fits well the LO curves. Let us remark first that, since heat and mass transfer limitations take place in the monolith, the temperatures and compositions are different in the wash-coat layer and in the bulk gas. Under mass transfer control, the concentration in the wash-coat layer is smaller than in the gas. Consequently, the zero-order rate constant based on the gas composition becomes meaningless. Second, the crucial problems are met after light-off when conversion and temperature are high. At high temperature one may argue that the monolith works under transfer control and that reaction kinetics become secondary. However, Leclerc and Schweich (1993) showed that the reaction becomes the rate determining step at high conversion in a stoichiometric mixture.

Finally, oxygen storage in and release from ceria affect conversion in the transient state. Thus, there is probably a need for kinetic data at high temperature and conversion, and certainly a need for kinetic data in the transient state. In the next subsections we deal briefly with the required data and the associated interpretation problems.

2.8. The quasi steady-state assumption

LH rate expressions often rely on the quasi steady state assumption that stipulates that one elementary step is rate determining whereas the other steps are at equilibrium. The assumption is debatable in the wide temperature range encountered in the monolith, because the rate determining step may be different according to the temperature level. It would thus be useful to know whether the rate expressions obtained at moderate temperature are still reliable at the high temperature and low reactant concentrations observed after light-off has occurred. However, care must be taken of the possible transfer limitations, and of the possible catalytically induced homogeneous reactions.

2.9. Characteristic reaction time constants

A reaction time constant, either of an overall reaction or of an elementary step is defined by:

$$t_{R0} = \frac{C_0}{r_0}$$

where C_0 (mol.m^{-3} of fluid) is a reference concentration (for example the inlet concentration of CO at T_0), and r_0 (mol.m^{-3} of fluid.s^{-1}) the reaction rate at the reference composition and temperature. Remark that the rate of reaction of a heterogeneous reaction must be referred to the volume of fluid and not to the volume (or mass) of catalyst. Let t_{R0i} be the time constant of the ith step of a LH reaction mechanism. If one time constant is much larger than the other, then it defines the rate determining step and the other steps are at equilibrium.

Without mass and heat transfer limitations, the conversion, X, decreases with the ratio t_0/t_{R0}. In an isothermal plug flow reactor X decreases linearly for a zero-order reaction, exponentially for a first-order reaction, etc. (See standard textbooks on chemical kinetics and chemical reaction engineering). Thus, the analysis of X versus t_0 (i.e., GHSV) allows estimating t_{R0}. When the reaction order is not known, a very rough approximate of t_{R0} is given by t_0 when $X = 0.5$. In other words, the GHSV and the T_{LO} yield an estimate of an overall reaction time. For example, Figure 2 gives a light-off temperature about 590 K that yields $t_0 = 7.3$ ms (from GHSV = 75000 h^{-1}) which is in agreement with $t_{R0} = 5.3$ ms (from equations (1) and (2)). For obtaining the time constants of elementary steps, Temporal Analysis of Products (See Ansell et al., 1994) is probably the best method.

Oxygen Storage Capacity measurements can give the amount of oxygen available in the catalyst but nothing is known concerning the rate at which this oxygen is available. Rhodium is known to be more or less oxidized according to the composition of the gas phase, and the oxidation state of rhodium affects the reduction rate of NO. Consequently, it would be useful to know the time constants of oxygen storage in and release from ceria, and of the oxidation/reduction reaction of rhodium.

If the above time constants are much smaller than the time constants of the main reactions, then the rates of the latter can be assumed to depend on the current composition of the gas phase. If not, the instantaneous activity of the catalyst depends on its history, and the rates of the main reactions depend on the fluid composition and on some composition variable of the solid phase (available stored oxygen, oxidation state of rhodium for example).

Few papers are devoted to the previously mentioned transient surface processes. Herz (1987) proposed the first simple and efficient model for oxygen storage in ceria and transient deactivation of the noble metal. Another oxygen storage model was recently proposed by Pattas et al. (1994). It seems that there is nothing about oxidation and reduction of rhodium. We will just outline what could be done for NO reduction over partially oxidized rhodium following Herz's approach. Let q_{Ox} be the fraction of oxidized rhodium, and r_{Red} be the rate of NO reduction on fully reduced rhodium. We assume that r_{Red} depends only on the gas phase composition. The rate of NO reduction on partially reduced rhodium can be:

$$r = (1-q_{Ox})r_{Red} \tag{8}$$

Let t_{Red} be the reduction time of rhodium. Assuming a first-order rate for rhodium reduction, we have:

$$\frac{d\theta_{Ox}}{dt} = -\frac{\theta_{Ox} - K_{Ox}f(P_{O2})}{t_{Red}} \tag{9}$$

where K_{Ox} is a pseudo-equilibrium constant of rhodium oxidation by O_2. This does not mean that Rh - Rh_2O_3 forms a solid mixed phase, but rather that the amount of reduced rhodium can depend on oxygen partial pressure, at least in a certain range defined by function $f(P_{O2})$. Equations (8) and (9) define the rate of NO reduction under unsteady conditions. Steady state experiments at various oxygen concentrations and temperatures would give $K_{Ox}(T)$. Temporal Analysis of Products (TAP) experiments at various temperatures would probably give $t_{Red}(T)$. TAP experiments at various oxygen concentrations would also tell whether the first-order law (9) is valid or not. The same approach could be used to model the effect of SO_2.

According to the time dependence of P_{O2}, the quasi steady state assumption holds or not. When P_{O2} is constant and $q_{Ox} = K_{Ox}f(P_{O2})$, (9) is identically fulfilled. When P_{O2} varies periodically with a period, P, much larger than t_{Red}, the solution to (9) is again close to $q_{Ox} = K_{Ox}f(P_{O2})$. In these two cases, r depends only on the gas phase composition and the quasi steady state assumption holds. Conversely, if P is smaller than or of the order of t_{Red}, then q_{Ox} is no longer proportional to P_{O2} because of (9), and r becomes implicitly time-dependent. This example shows that a transient surface process becomes important when its characteristic time is greater or equal to the characteristic time of some external phenomenon. Since composition or temperature variations with a frequency higher than 1 Hz are considerably damped by hydrodynamic and transfer processes, and by the engine itself, it is expected that quasi steady state can be assumed for surface processes which have characteristic times smaller than one second.

Simple transient state experiments in laboratory plug flow reactors can qualitatively tell whether the surface processes mentioned above are instantaneous or not. For example, a step composition change can lead to two types of results. Either the outlet composition follows what is expected from the steady state rate equations or not. In the former case, one may assume quasi steady state. In the latter case, one is facing a slow surface step (oxidation/reduction of rhodium, oxygen storage/release, transient deactivation by SO_2, etc.) that affects the main reaction mechanism.

Intrinsically fast reactions can serve as probes to measure the reaction time of a transient surface process. According to equations (1) and (2), CO oxidation has a characteristic time smaller than 0.1 s above 520 K. Switching from a lean CO/O_2 feed to a rich feed, and then studying the conversion of CO versus time gives the amount and the characteristic time of the oxygen released from the catalyst (See Herz (1987), Smedler et al. (1993)). In such an experiment care must be taken first with the water gas shift reaction when water is in the feed, and second with the possible heat and mass transfer limitations.

Measuring reaction time constants is a raw approach to the mechanism and rate expression. However, since t_{R0} is proportional to the reciprocal of the rate, it must obey Arrhenius law. This enables one to measure the reaction time at moderate temperature and then to extrapolate the result to a realistic temperature range to know whether the surface process is at equilibrium or not.

Experiments made with an oscillating feed composition mimic what can happen in a real converter. However, their mechanistic interpretation is often difficult. It is sometimes argued that if the result of such an experiment is different from the result of an experiment performed at the average composition, then there is some "storage" phenomenon that is overlooked at steady state. Let us show that this conclusion can be wrong with the example of CO oxidation in a plug flow reactor. Consider a square wave feed made of 4% CO, 0% O_2 in the firts half period and 0% CO, 2% O_2 in the second. The average feed is 2% CO, 1% O_2 that leads to the LO curves shown above. In the oscillating regime, when the adsorption and desorption steps are instantaneous CO is never in contact with O_2 and the conversion is zero whatever the temperature. This clearly means that the oscillating feed and the average feed do not give the same result when the quasi steady state assumption holds. Conversely, if and only if there is an adsorption or desorption time constant that is larger than the cycle period, then CO conversion is nonzero. In the latter case we may speak of an O_2 or CO storage effect either on the metal or in ceria.

A reaction time estimate is reliable only when the feed system, the reactor, and the detectors have an overall time constant smaller than the reaction time under study. If not, no conclusion can be drawn. This is illustrated by the oscillating feed experiment performed in a recycle loop reactor when the period is smaller than the residence time of the fluid. In the latter case, the two alternating feeds are mixed and CO conversion is nonzero even under quasi steady state conditions. The same problem arises when the feed signal is not a perfect square wave owing to mixing in the inlet tubes. A plug flow instead of a recycle loop reactor is then recommended, and any dead volume or by-pass must be eliminated from the flow system. Furthermore, it must be kept in mind that CO or NO detectors can have intrinsic time constants of a few seconds that often mask smaller reaction times. Finally, the heat storage time constant can be large (it is responsible for the delayed light-off at start-up of the engine!). Conversely, TAP allows the observation of reaction times shorter than one second in an isothermal catalytic bed.

2.10. Further remarks

Reversible poisoning by SO_2 is a typical transient storage/release phenomenon that depends on the oxidizing ability of the fluid. Here again, determining poisoning time constants versus temperature and SO_2 and O_2 contents of the feed would be helpful.

If the zero-order method of section 2.6 reveals to be efficient, then it should be used to compare fast and slow oxidizing hydrocarbons (alkenes and methane for example). It is expected that the activation energy is constant in a homologous series of hydrocarbons. Then, the ratio of the rates of reaction should be close to the ratio of the zero-order kinetic constants determined with the LO curves. This fortunate situation would enable one to scale the rates of oxidation of various hydrocarbons with respect to a standard hydrocarbon (propylene for example).

Finally, let us recall that kinetic investigations with a complex mixture that involves multiple reactions are an overwhelming task. It is better to work with simple mixtures for obtaining reliable rate expressions. Then, mechanistic considerations should indicate how the rate expressions generalize for the complex mixture. However, the simple mixture must at least contain water that is known to interact with the catalyst. This means that the water gas shift and the steam reforming reactions are unavoidable. Knowing the kinetic importance of the latter reactions remains a challenge.

3. External Mass and heat transfer data in the monolith

Heat and mass transfer processes have been extensively dealt with in many papers (See a brief review in Schweich and Leclerc, 1993). In most models, heat and mass transfer processes are accounted for by coefficients that are given by non-dimensional numbers:

$$Sh = \frac{k_D D_h}{D_m} \qquad Nu = \frac{h D_h}{\lambda} \tag{10}$$

Sh and Nu are the Sherwood and Nusselt numbers, k_D and h the mass and heat transfer coefficients, D_h the hydraulic diameter of a channel, D_m and l the molecular diffusivity and the heat conductivity of the fluid. Models for simultaneous laminar flow and transverse diffusion yield for a long monolith:

$$Nu \text{ Å } Sh \text{ Å } 3 \tag{11}$$

However, based on experimental results, Votruba et al. (1975) pointed out that Nu and Sh are probably smaller than 3. More recently, Ullah et al. (1992) and Bennett et al. (1992) gave other experimental evidence based on CO and C_3H_8 oxidation respectively. For a monolith 15 cm long operated at Re = 250 (close to the upper limit), Ullah and Bennett correlations give Sh Å 1 and Nu Å 0.2 respectively. These low values were obtained by fitting the experimental results with a first-order rate expression. Under reaction control, the apparent rate constant is the reaction rate constant, whereas under mass transfer control it gives the Sherwood number. The dependence of the apparent rate constant on temperature yields the rate determining step. At low temperature, chemical regime prevails and the rate constant obeys Arrhenius law. At high temperature, mass transfer controls and the apparent activation energy is close to zero. Ullah et al. reported that they worked in a temperature range where the apparent activation energy was close to zero with no further quantitative details. Conversely, Bennett et al. reported an apparent activation energy about 36 kJ.mole^{-1} in the mass transfer regime instead of 90 kJ.mole^{-1} in the chemical regime. The latter authors finally attributed the low Sherwood numbers to the possible presence of homogeneous reactions. Figure 8 illustrates another possible explanation using CO oxidation in an adiabatic plug flow reactor.

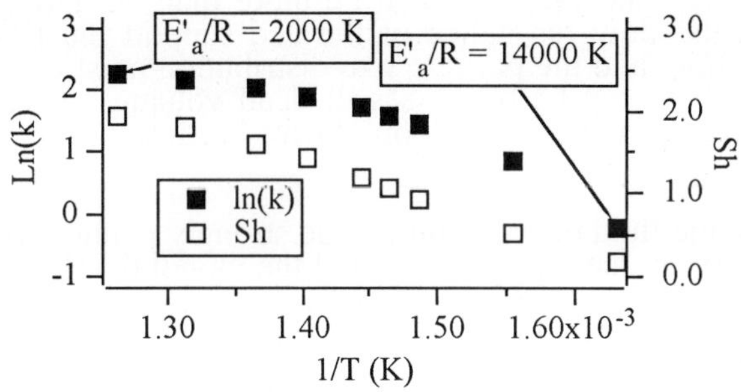

Figure 8: Arrhenius plot of the apparent first-order rate constant (full dots) and interpretation of the rate constant as a Sherwood number (open dots). Temperature range (340 - 520°C) above light-off temperature.

The catalyst is ignited on the full range of temperature. At low temperature (340°C), the rate constant obeys Arrhenius law with a consistent apparent activation energy. The corresponding Sherwood number is very low and it does not make sense. At high temperature (520_C), E'_a is still nonzero and the corresponding Sherwood number is still smaller than 3. This shows that obtaining full mass transfer control is difficult, and that unusually low Sherwood numbers can be due to a partial kinetic control.

4. INTERNAL MASS TRANSFER RESISTANCE

This problem was dealt with by Schweich and Leclerc (1991) and Leclerc and Schweich (1993). Most models assume that there is no internal diffusion limitation because the wash-coat layer is "thin". Presently, it is not known whether this assumption holds, and decisive experiments with different wash-coat thicknesses at a given metal loading should be undertaken.

The experiments should be carefully performed to avoid confusing internal and external mass transfer resistances. A dedicated reactor should probably be designed. It could be a continuous stirred tank reactor (mechanical stirring) containing a sheet of wash-coated substrate. If the rate of reaction depends on the wash-coat thickness, and not on stirring speed, then internal diffusion resistance prevails.

5. HYDRODYNAMIC CONDITIONS

The consequences of the flow pattern on the experimental results are strikingly illustrated by Figure 4. It is thus of prime importance to control the flow pattern or at least to be aware of the possible hydrodynamic problems.

Experiments with laboratory monoliths of small cross-section area can lead to biased results due to an uneven flow distribution in the channels, especially close to the reactor wall. The wash-coat of the outer broken channels should be scraped away, and the void between the reactor wall and the monolith should be carefully plugged. To minimize wall effects, the diameter of the monolith should be ten times the channel diameter at least. Plug flow must prevail in a packed bed of crushed catalyst. The bed length and radius should be more than 50 and 10 particle diameters respectively, the flow resistance of the bed support must be uniform throughout its cross-section, and the particle size distribution must be as narrow as possible. Otherwise, there can be by-passes or dead volumes. These hydrodynamic problems are overcome in a recycle loop reactor because the same physical and chemical conditions prevail everywhere.

In commercial converters, the fluid distribution can be strongly nonuniform owing to the short inlet cone between the exhaust pipe and the monolith (Howitt and Sekella, 1974; Wendland and Matthes, 1986). Unfortunately, few are known concerning the dependence of the flow maldistribution on monolith properties, working temperature, and flowrate. Li et al. (1991) provided some flow field characteristics using PHOENICS software. Bella et al. (1991) included the oxidation reactions in the description of the flow field. These papers describe case-by-case simulations, and no correlations between the operating conditions and the fluid distribution are given. Further experimental and simulation works on this subject would be welcome.

6. CONCLUSIONS

Kinetic expressions are probably the most crucial data for model reliability. These data are certainly different from one catalyst to the other. However, the reactions are the same, and the catalyst formulations are similar. This suggests that the structures of the rate expressions are probably independent of the catalyst, whereas the parameters are dependent. Elucidating the structure of the rate expressions at steady state, estimating the time constants of transient surface processes and reversible poisoning, knowing whether internal diffusion is a limiting process would be more significant contributions than accumulating light-off curves obtained with complex mixtures. Although these curves are technically meaningful, they cannot be interpreted from a scientific point of view, especially when the reactor behavior is not well controlled. This is demonstrated by Figures 1 to 7 which show that the light-off temperature is somewhere between 560 and 630 K depending on the reactor behavior (hydrodynamics and transfers).

The second challenge concerns hydrodynamic data in commercial converters. The "flow distribution index" of Wendland and Matthes (1986) and Bella et al. (1991) is a first approach to the problem. However, further work is necessary to correlate this index to the operating conditions. As long as flow maldistribution is not accounted for in the models, simulation of vehicle tests will not be reliable.

LIST OF SYMBOLS

C_0	inlet concentration
D_h	hydraulic diameter of a channel
D_m	molecular diffusivity
E_a	apparent activation energy
E'_a	fitted activation energy
GHSV	gas hourly space velocity
H(x)	Heaviside step function
h	heat transfer coefficient
K	adsorption constant
K_0	pre exponential factor of the adsorption constant
K_{Ox}	rhodium oxidation equilibrium constant
k	kinetic rate constant
k'	apparent kinetic constant
k_0	frequency factor
k'_0	apparent frequency factor
k_D	mass transfer coefficient
L	monolith or bed length
Nu	Nusselt number
P_{O2}	oxygen partial pressure
R	ideal gas constant
Re	Reynolds number
r	rate of reaction
r_{Red}	NO reduction rate on fully reduced rhodium
Sh	Sherwood number
T	temperature
T_{LO}	light-off temperature
T_0	inlet temperature
T_w	wash-coat temperature
t	time
t_0	residence time of the fluid
t_{R0}	reaction time at inlet conditions
t_{Red}	reduction time of rhodium
u_B, u_M	fluid velocity in the packed bed, in the monolith
w	wash-coat thickness
X	conversion
X_w	conversion in the wash-coat layer
x_j	mole fraction of species j
$x_{j,0}$	mole fraction at inlet
z	reduced distance
z'	distance

Greek symbols

DH_{ads}	adsorption enthalpy
DT_{ad}	adiabatic temperature rise
e_M	open frontal surface area of the monolith
e_B	intergranular void fraction in the packed bed
f	volume of catalyst per unit volume of intergranular gas
q_{Ox}	fraction of oxidized rhodium
l	gas heat conductivity

LITERATURE CITED

Ansell G.P., S.E. Golunski, J.WX. Hayes, and A.P. Walker: The mechanism of the lean NO_x reaction over Pt-based catalysts. CAPoC 3 Symposium, April 1994, Brussels, Belgium.

Bella G., V. Rocco, M. Maggiore, F. Stella, and F. Succi: Automotive catalytic converter performance evaluation: a computational approach. ATA Ingegneria Automotoristica, 44, 242, 1991.

Bennett C.J., R.E. Hayes, S.T. Kolaczkowski, and W.J. Thomas: An experimental and theoretical study of a catalytic monolith to control automobile exhaust emissions, Proc. R. Soc. Lond. A, 439, 465, 1992.

Froment G.F., and K.B. Bischoff: Chemical reactor analysis and design, J. Wiley, New York, second Ed.

Heck R.H., J. Wei, and J.R. Katzer: mathematical modelling of monolithic catalysts, AIChE J., 22, 3, 477, 1976.

Hegedus L.: Temperature excursions in catalytic monoliths, AIChE J., 21, 849-853.

Herz R.K.: Dynamic behavior of automotive three-way emission control systems. In "Catalysis and Automotive Polution control", A. Crucq and A. Frennet Ed., Elsevier, 1987, pp. 427-444.

Howitt J.S., and T.C. Sekella: Flow effects in monolithic honeycomb catalytic converters, SAE paper 740244, Automotive Engineering Congress, Detroit, 1974.

Leclerc J.P., and D. Schweich: Modelling catalytic monoliths for automobile emission control, in "Chemical reactor technology for environmentally safe reactors and products", H.I. de Lasa et al. Ed., Kluwer Academic Pub., the Netherlands, p. 547-576, 1993.

Li P., G. Chui, and J.D. Pakko: A numerical study of automotive catalytic converter internal flow. In "4th International PHOENICS User Conference", CHAM Ed., p 189-230, 1991.

Mabilon G., D. Durand, and Ph. Courty: Inhibition and poisoning of post-combustion catalysts by alkynes: a clue for understanding their behavior under real exhaust conditions. CAPoC 3 Symposium, April 1994, Brussels, Belgium.

Pattas K.N., A.M. Stamatelos, P.K., Pistikopoulos, P.C. Koitsakis, and P.A. Konstandinidis: Transient modeling of 3-way catalytic converters. SAE Paper 940934, 1994.

Schweich D., and Leclerc J.P.: Flow, heat and mass transfer in a monolithic catalytic converter. In "Catalysis and Automotive Pollution Control II", A. Crucq Ed., Studies in surface science and catalysis, Elsevier, p. 437, 1991.

Smedler G., S. Eriksson, M. Lindbald, H. Bernier, S. Lundgren, and E. Jobson: Deterioration of three-way automotive catalysts, Part II - Oxygen storage capacity at exhaust conditions. S1E Paper 930944, 1993.

Subramanian B., and A. Varma: Reaction kinetics on a commercial three-way catalyst: the CO-NO-O_2-H_2O system, Ind. Eng. Chem. Proc Des. Devel., 24, 512, 1985.

Ullah U., S.P. Waldram, C.J. Bennett, and T. Truex: Monolithic reactors: mass transfer measurements under reacting conditions. Chem. Eng. Sci., 47, 9, 1992.

Voltz S.E., C.R. Morgan, D. Liederman, and S.M. Jacob: Kinetic study of carbon monoxide and propylene oxidation on platinum catalysts, Ind. Eng. Chem. Proc. Res. Dev., 12, 295, 1973.

Votruba J., O. Mikus, K. Nguen, V. Hlavacek, and J. Skrivanek: Heat and mass transfer in honeycomb catalyst - II. Chem. Eng. Sci., 30, 201, 1975.

Wendland D.W., and W.R. Matthes: Visualization of automotive converter internal flow. SAE paper 861554, International Fuels and Lubricants Meeting and Exposition, Philadelphia, 1986.

APPENDIX

Using the film model and assuming plug flow, the mass balance equations for a reactant are (See Heck et al., 1976, Froment and Bischoff, 1990 for example):

$$C_0 u_B \frac{dX}{dz'} = r(X_w, T_w)\phi \quad \text{in a packed bed} \tag{A1}$$

$$C_0 u_M \frac{dX}{dz'} = r(X_w, T_w)\frac{4}{D_h} \quad \text{in a monolith} \tag{A2}$$

where C_0u_M or C_0u_B is the specific molar feed flowrate of reactant, C_0 the inlet concentration of reactant, $u_{M/B}$ the interstitial velocity, X the reactant conversion in the gas phase, X_W and T_W the conversion and the temperature in the wash-coat layer. When there are no mass and heat transfer resistances, $X_W = X$, and $T_W = T$. Since $C_0u_{M/B}$ is independent of the temperature, C_0 and $u_{M/B}$ can be calculated at any reference temperature. The GHSV obeys:

$$\text{GHSV} \approx \frac{\varepsilon_M u_M}{L} = \frac{\varepsilon_B u_B}{L} \tag{A3}$$

where $e_{M/B}$ is the void fraction in the monolith or catalyst bed. The proportionality constant depends on the time units. A fixed bed and a monolith operated at the same feed composition, temperature and GHSV will have similar behaviors when equations (A1) and (A2) become identical. This implies:

$$\frac{4w\varepsilon_M}{D_h} = \varepsilon_B \phi \tag{A4}$$

The behaviors of the monolith and of the packed bed are not identical in general because X_W and T_W are not still specified. If there are no heat and mass transfer limitations, and if the gas temperature is the same in the two reactors, then they behave identically.

Let L be the length of the catalytic system and z=z'/L be the reduced distance. (A1) and (A2) become respectively:

$$\frac{C_0}{t_{0M}}\frac{dX}{dz} = r(X_w, T_w)\frac{4w}{D_h} \qquad \frac{C_0}{t_{0B}}\frac{dX}{dz} = r(X_w, T_w)\phi \tag{A5, A6}$$

where $t_{0M/B} = L/u_{M/B}$ is the residence time of the fluid in the gas phase. Assuming no heat and mass transfer limitations gives $X_W = X$ and $T_W = T$. When the reactor is adiabatic, the gas temperature increases proportionally to the conversion (Hegedus, 1975):

$$T = T_0 + \Delta T_{ad} X \tag{A7}$$

Equations (A6) and (A7) give (6) when r(X,T) = k'(T) H(1-X). In the recycle loop reactor, the derivative dX/dz is replaced by the ratio of the variables, i.e., X_{CO}. This gives (7).

A. Frennet and J.-M. Bastin (Eds.)
Catalysis and Automotive Pollution Control III
Studies in Surface Science and Catalysis, Vol. 96
1995 Elsevier Science B.V.

REACTIVITY OF STEAM IN EXHAUST GAS CATALYSIS. PART II : SINTERING AND REGENERATION OF Rh AND PtRh CATALYSTS IN PROPANE OXIDATION

J. Barbier Jr. and D. Duprez

Laboratoire de Catalyse en Chimie Organique URA CNRS 350
40 av. du Recteur Pineau , 86022 POITIERS Cedex FRANCE
Tel : (33) 49 45 39 98.

ABSTRACT

PtRh catalysts were prepared on different supports composed of Al_2O_3, CeO_2, ZrO_2 and $NiAl_2O_4$. The variations of the activities in propane oxidation and steam reforming were used to obtain some indication concerning the surface state of these catalysts after thermal treatments at high temperature in an oxidizing and in a reducing medium. Cyclopentane hydrogenolysis was also carried out to observe the changes in the rhodium surface state. Platinum was the metal which catalysed the direct oxidation of propane while rhodium was the key-component in steam reforming catalysis.

The treatment at 800°C in an **oxidizing medium** induces a very important decrease of Rh area linked to the fact that rhodium in its oxidized form (Rh^{3+}) can diffuse into the support. This phenomenon is more marked on Al_2O_3 and, to a lesser extent on CeO_2-Al_2O_3, than on the other supports. On the other hand, oxidative treatments lead to an increase of the particle size of platinum, and temporarily to an enhancement of the oxidation activities. After a treatment in a **reducing medium** at high temperature (T>700°C), the steam reforming activities can be recovered by extraction of rhodium from the support, with two exceptions however : Rh/CeO_2-Al_2O_3 and PtRh/ZrO_2 catalysts. This treatment has pratically no impact on platinum activities in oxidation except for the catalysts supported on $NiAl_2O_4$ which are deactived. Cyclopentane hydrogenolysis confirm all the above results.

1. INTRODUCTION

PtRh catalysts are commonly used in catalytic converters for eliminating pollutants (CO, hydrocarbons, NOx) from exhaust gases [1]. In the first part of our work [2], we investigated, on Pt(1wt.-%), Rh(0.2wt.-%) and

Pt(1wt%)Rh(0.2wt%) catalysts supported on Al_2O_3 and CeO_2-Al_2O_3, carbon monoxide and propane oxidation by oxygen (direct oxidation), by steam (water gas shift and steam reforming) and by a mixture of oxygen and steam (oxy-WGS and oxy-steam reforming).

Steam can be considered a cor-eactant of oxidation during rich-phases (lean in O_2) [3-5]. In oxy-steam conversion of propane, we showed (fig.1) that propane oxidation was catalyzed by platinum (between 200 and 350°C) while rhodium was the key-component in the catalysis of steam reforming (between 350 and 600°C). Ceria was an excellent promotor of steam reactions [3, 6], particularly when this reaction was carried out in the presence of oxygen. Therefore, the steam reforming activity is an excellent indicator of the rhodium surface state since the activity systematically decreases when the metallic rhodium area decreases [7]. On the other hand, oxidation activity is a more complex indicator of platinum surface state because there exists an optimum dispersion [8, 9].

Metal sintering constitutes a very significant cause of loss in catalyst activity owing to the reduction in the metallic area [10,11].

The aim of this work is to study the sintering of Rh and PtRh catalysts by means of oxidation and steam reforming activities in propane conversion. Changes in the steam reforming activity have been compared with those obtained in cyclopentane hydrogenolysis, a reaction extremely sensitive to the state of Rh in the catalyst [12].

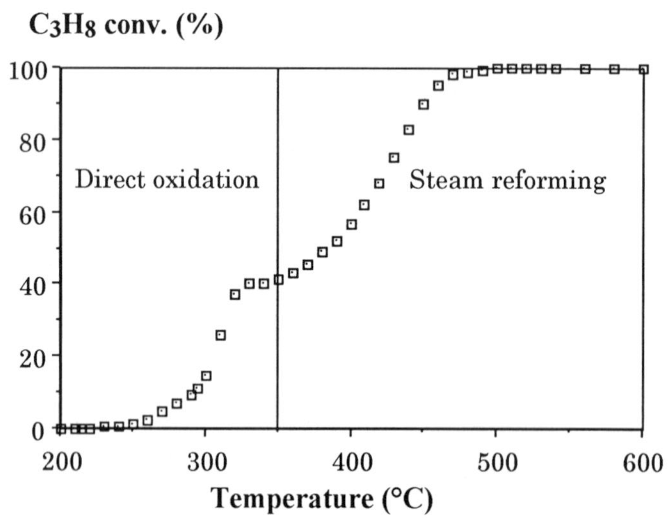

Fig 1 : Oxy-steam reforming of propane on PtRh/CeO_2-Al_2O_3

2. EXPERIMENTAL

2.1 Catalysts

Five different supports were used : (i) a gamma-alumina ($120m^2\ g^{-1}$) which was impregnated with an aqueous solution of cerium nitrate to obtain (ii) a 12wt.-%CeO_2-Al_2O_3 after calcination ($100m^2\ g^{-1}$); (iii) a zirconia ($40m^2\ g^{-1}$) supplied by Degussa, (iv) a support prepared by coimpregnation of a $Ni(NO_3)_2$ and an $Al(NO_3)_3$ aqueous solution on an alumina ($200m^2\ g^{-1}$) so as to obtain after calcination (1000°C, air, 24h) a 8.5wt.-%$NiAl_2O_4$-Al_2O_3 support and this support is impregnated with an aqueous solution of cerium nitrate to obtain (v) a 12wt.-%CeO_2-8.5wt.-%$NiAl_2O_4$-Al_2O_3 support after calcination. These supports (Al_2O_3, CeO_2-Al_2O_3, ZrO_2, $NiAl_2O_4$-Al_2O_3 and CeO_2-$NiAl_2O_4$-Al_2O_3) were crushed and sieved to 0.1-0.2 mm. They were used to prepare two series of catalysts by impregnation or coimpregnation with aqueous solutions of rhodium chloride and chloroplatinic acid. The catalysts were dried at 120°C then calcined at 500°C under an air flow and prereduced in H_2 at 450°C. The metal loadings were Rh(0.2wt.-%) and Pt(1wt.-%)Rh(0.2wt.-%). A Pt(1wt.-%) on gamma-alumina ($120m^2\ g^{-1}$, 0.1-0.2mm) supplied by I.F.P. (French Institute of Petroleum) was also used for certain experiments.

Table 1 : Catalyst compositions

Symbolic Name	Pt (wt.-%)	Rh (wt.-%)	Support
Pt/A	1	0	Al_2O_3
Rh/A	0	0.2	Al_2O_3
Rh/CeA	0	0.2	Al_2O_3 +CeO_2 (12wt.-%)
PtRh/A	1	0.2	Al_2O_3
PtRh/CeA	1	0.2	Al_2O_3 +CeO_2(12wt.-%)
PtRh/Z	1	0.2	ZrO_2
PtRh/ANi	1	0.2	Al_2O_3+Al_2O_4Ni(8.5%)
PtRh/CeANi	1	0.2	Al_2O_3+Al_2O_4Ni(8.5%)+CeO_2(12%)

2.2 Catalytic reaction

Before each run the catalyst samples were heated in air for 1h at 450°C. Propane oxidation was carried out in a flow reactor under the following conditions : (i) catalyst bed : 40mg diluted in 360 mg of cordierite (0.1-0.2 mm). (ii) Feed gas (in vol-%) : C_3H_8, 0.4 ; O_2, 0.8; N_2, 98.8. (iii) Gas flow-rate : 380cm^3 min^{-1} (volume space velocity : 250,000h^{-1}). Temperature-programmed reactions were carried out from 150 to 900°C using a 4°C min^{-1} temperature ramp (1 atm.). Analyses were carried out by gas chromatography : CO_2, CH_4 and C_3H_8 on Porapak Q (0.7m, 1/4 in.; 25°C or 100°C, carrier gas H_2), CO, O_2, N_2 and CH_4 on molecular sieve 5A (0.4m, 1/4 in.; 25°C, carrier gas H_2), H_2 on molecular sieve 5A (1m, 1/4 in.; 25°C, carrier gas N_2). C_3H_8 conversions were determined from the mass balance of carbon-containing products and verified by the disappearance of the C_3H_8 peaks in the chromatograms. The main reactions considered here were direct oxidation (1), C_3H_8 steam reforming (2) and W.G.S. (3).

$$C_3H_8 + 5O_2 \Longrightarrow 3CO_2 + 4H_2O \qquad (1)$$
$$C_3H_8 + 3H_2O \Longrightarrow 3CO + 7H_2 \qquad (2)$$
$$CO + H_2O \Longrightarrow CO_2 + H_2 \qquad (3)$$

Reactions (2) and (3) occurred via the water produced in reaction (1) : even in the absence of steam in the inlet gases, typical light-off curves like those in Fig.1 were obtained.

Specific activities and activation energies in oxidation and steam reforming were determined at low conversion. Specific activities were calculated at 200°C for direct oxidation and at 300°C for steam reforming.

2.3 Sintering and regenerating conditions

The oxy-steam reforming reaction was carried out on :

- fresh catalysts (450°C; O_2 (3vol.-%); 1h)) (FRESH)

- oxidized catalysts (650°C or 800°C; O_2 (3vol.-%); 1h) (OX650 or OX800),

- oxidized (800°C; O_2 (3vol.-%); 1h) and then reduced catalysts (700, 800 or 900°C; H_2 (3vol.-%); 1h)(RED700 or 800 or 900).

2.4 Cyclopentane hydrogenolysis

This model reaction was carried out "in situ" on fresh or sintered catalysts (OX800 and RED900).

The catalysts were prereduced in H_2 (1h, 300°C, $30cm^3\ min^{-1}$). Cyclopentane hydrogenolysis was performed in a pulse flow reactor under the following conditions : (i) catalyst bed : 40mg diluted in 360mg of cordierite (0.1-0.2 mm); (ii) cyclopentane injection : 11μmole per pulse; (iii) hydrogen flow rate: $30cm^3min^{-1}$; (iv) temperature range 170-330°C. Under these conditions, the only reaction product was n-pentane analyzed by gas chromatography on a reoplex 400 column (2m, 1/8 in., 50°C, carrier gas H_2). The specific activities in cyclopentane hydrogenolysis of catalysts were calculated at 200°C.

3. RESULTS AND DISCUSSIONS

3.1 Rh/A catalyst

Figure 3 shows the light-off curves of Rh/A after different treatments. On this catalyst, the activities in steam reforming cannot be determined because the two regions (oxidation and steam reforming) are not sufficiently discrete : Rh being a poor oxidation catalyst, the two reactions occur pratically at the same temperature. Increasing the severity the treatment in O_2 induces a very important deactivation of the catalyst (FRESH, OX650 and OX800 in Fig.2). At low temperatures (T<650°C), an oxidizing atmosphere leads only to the total oxidation of rhodium into Rh_2O_3 without creating a significant decrease of the surface active area [10, 13]. Above 650°C, the treatment leads to a decrease of the accessible surface of rhodium on Al_2O_3, mainly linked to the fact that rhodium in its oxidized form (Rh^{3+}) can diffuse easily inside the alumina matrix [10, 13-18]. However this diffusion of rhodium ions in alumina to form a "diffuse oxide phase" seems to be limited to a subsurface layer of about 20 Å [10].

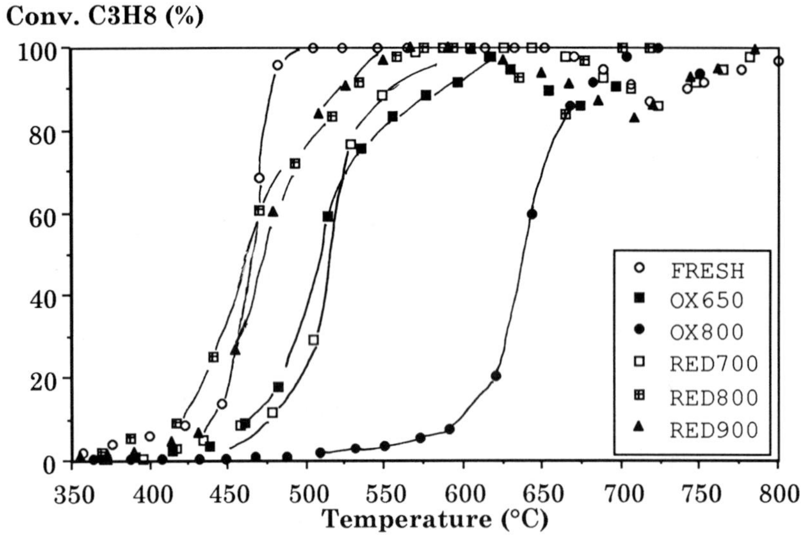

Fig 2. : Effects of thermal treatments on the activities of Rh/A Catalyst in propane conversion (0.4%C_3H_8 + 0.8%O_2)

After treatment in a reducing medium, the catalytic activity can be recovered progressively by extraction of the rhodium from the alumina [10, 13, 16]. But contrary to the surface rhodium oxide (Rh_2O_3), the Rh^{3+} ions contained in the "diffuse oxide phase" are difficult to reduce at low temperature (<500°C) [13,15]. Recently Beck et al. [19] have showed that rhodium deactivation, in oxidizing atmosphere, could be due to the formation, at the catalyst surface, of an unusual rhodium oxide which is very difficult to reduce. Whatever the cause of deactivation ("diffuse oxide phase" or refractory rhodium oxide), high temperature treatments in H_2 (900°C in Fig.3) are required to regenerate the catalyst.

3.2 Rh/CeA catalyst

Figure 3 compares the light-off temperatures determined on Rh/A and Rh/CeA catalysts. CeO_2 is known to stabilize noble metals [18] and probably limits here the rhodium migration into the support. Nevertheless, contrary to what was observed on Rh/A, a reducing treatment increases the deactivation of Rh/CeA. The most probable explanation is that a $RhCe_xO_y$ mixed oxide could be formed at high temperature in a reducing medium [20, 21]. This mixed oxide should be inactive both in oxidation and in steam reforming.

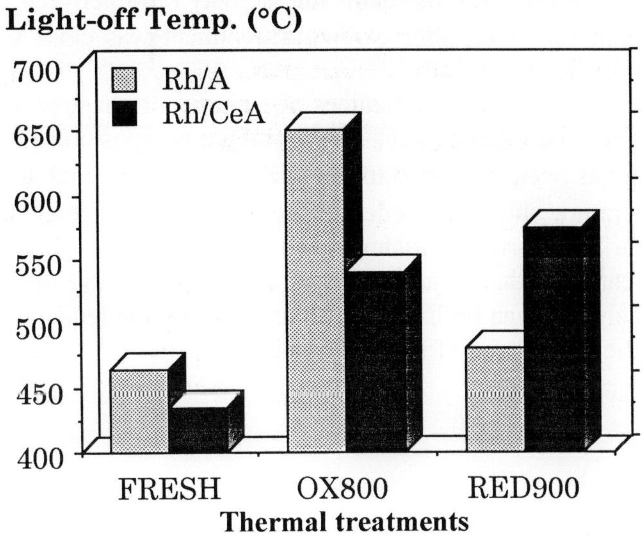

Fig 3. : Light-off temperatures of Rh/A et Rh/CeA after thermal treatment.

3.3 PtRh/A and PtRh/CeA catalysts

The presence of Pt increases the oxidation activity so that the two regions (oxidation and steam reforming) are now clearly separated as shown in Fig.1.

The activities in oxidation at 200°C (fig. 4) and steam reforming at 300°C (fig 5) are determined for the two bimetallic catalysts before and after thermal treatments. These figures show the following points :

(i) Fresh catalysts : in accordance with the literature, CeO_2 is a promotor of steam reforming [5] and an inhibitor of C_3H_8 oxidation [2, 8, 9]

(ii) OX800 catalysts : this treatment induces an enhancement of oxidation activities and a strong deactivation in steam reforming more particularly on Al_2O_3. An oxidizing treatment (O_2 or NO) at high temperature leads to a significant increase in the size of the platinum particles [22], which favors the oxidation activity [8, 9]. On the other hand, this treatment induces a surface segregation [22-24] depending essentially on the atomic ratio between Rh and Pt. Kacimi and Duprez [11] have shown, by isotopic exchange techniques, that in a $PtRh/Al_2O_3$ bimetallic series treated at 900°C in 1%O_2 there was a definite enrichment in rhodium for an atomic content of Rh/Pt+Rh greater than 45%. Below this content, the opposite phenomenon occurs because most of the rhodium ions migrate on (and probably in) the support to give the non-reducible form of rhodium that is inactive in catalysis [11].

(iii) CeO_2 stabilizes the rhodium : the activity ratio between PtRh/CeA and PtRh/A in steam reforming after oxidizing treatment was close to 100 while it was only 2 in the fresh catalysts.

(iv) the reducing treatment induces no significant changes in the oxidation activities. In fact, the measurement of the surface composition of a $PtRh/Al_2O_3$ catalyst which has been submitted to treatment in a reducing medium (H_2 or CO) at high temperature shows a moderate increase of the size of the crystallites without any striking surface enrichment up to 1000°C [22, 23]. But contrary to the Rh/CeA catalyst, this treatment leads to a regeneration of rhodium in the PtRh/CeA catalyst. When both Rh and Pt are present, the tendency to form PtRh bimetallic particules instead of $RhCe_xO_y$ is favored.

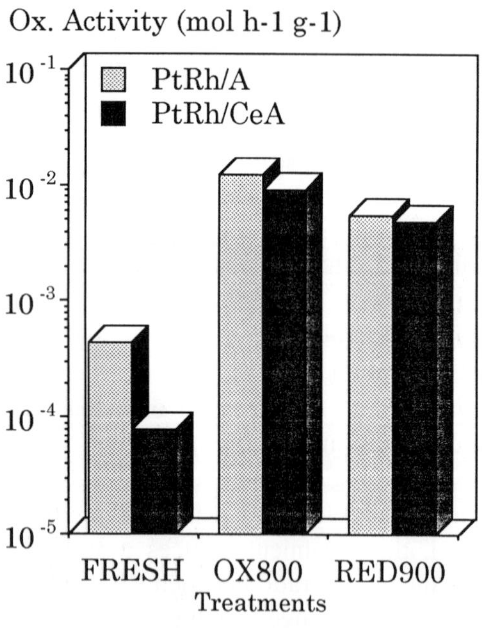

Fig 4 : Oxidation activities of PtRh catalysts: Effects of thermal treatments

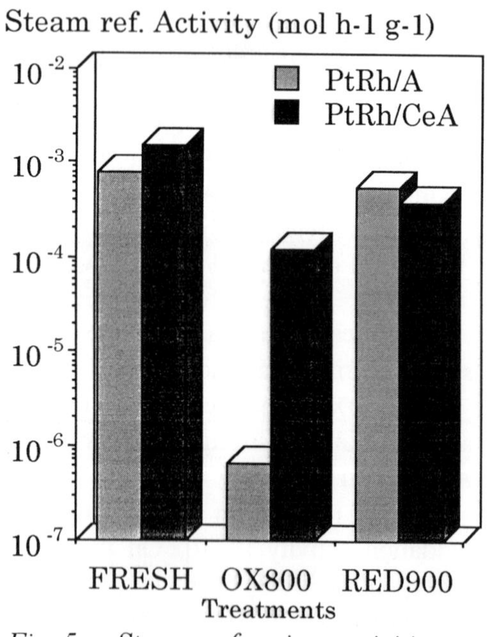

Fig 5 : Steam reforming activities PtRh catalysts:
Effects of thermal treatments

3.4 Cyclopentane hydrogenolysis

Figure 6 represents the curves “conversion” vs “temperature” for the fresh Pt/A, Rh/A, Rh/CeA, PtRh/A and PtRh/CeA catalysts. Table 2 gives the activities calculated at 200°C for the fresh (X_{f200}) and treated catalysts (X_{200}).

Rh being much more active than Pt, the activities X_{200} can be used to observe the rhodium deactivation. The results corroborate those obtained in steam reforming of propane :

(i) a very important deactivation in hydrogenolysis after an oxidizing treatment at 800°C, more particularly on alumina,

(ii) a regenerative effect of the reducing medium for all the catalysts except for the Rh/CeA catalyst.

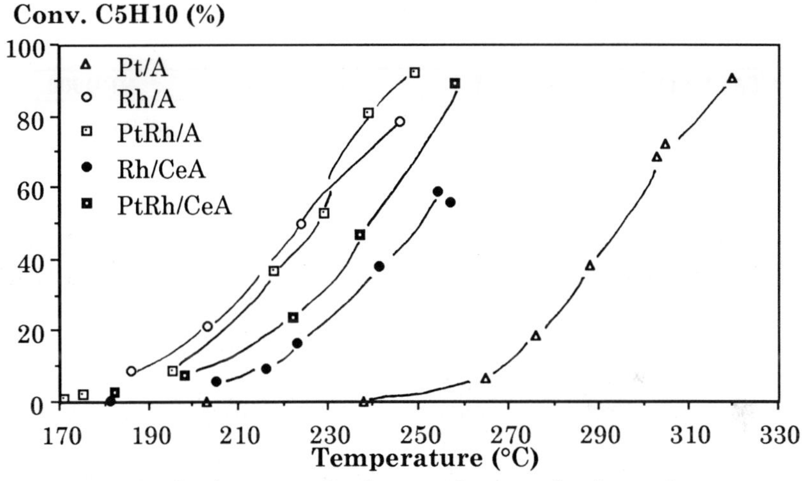

Fig 6 : Cyclopentan Hydrogenolysis on fresh catalysts.

CATALYSTS	$X_{f\,200}$	X_{200}	
		OX800	RED900
Pt/A	1.25	0.011	0.009
Rh/A	42.5	0.0034	4.3
PtRh/A	35.0	0.1	7.0
Rh/CeA	11.25	0.23	0.11
PtRh/CeA	22.5	1.1	3.4

Table 2 : Cyclopentan hydrogenolysis activities (μmol C_5H_{10} per pulse per gram at 200°C) on fresh and treated catalysts

3.5 Other supports

In order to stabilize noble metals during rich or lean excursions at high temperature, the role of three other supports have been studied. Tables 3 and 4 give the activities in oxidation and steam reforming of propane of PtRh bimetallic catalysts.

CATALYSTS	**FRESH**	**OX800**	**RED900**
	mmol g^{-1} h^{-1}	mmol g^{-1} h^{-1}	mmol g^{-1} h^{-1}
PtRh/CeA	0.076	9.0	4.2
PtRh/Z	1.2	1.8	0.54
PtRh/ANi	0.6	16.0	0.06
PtRh/CeANi	0.6	9.0	0.3

Table 3 : Oxidation activities (at 200°C) of fresh and treated catalysts

CATALYSTS	**FRESH**	**OX800**	**RED900**
	mmol g^{-1} h^{-1}	mmol g^{-1} h^{-1}	mmol g^{-1} h^{-1}
PtRh/CeA	1.56	0.12	0.39
PtRh/Z	1.78	0.48	0.024
PtRh/ANi	0.26	0.072	0.26
PtRh/CeANi	0.6	0.18	0.54

Table 4 : Steam reforming activities (at 300°C) of fresh and treated catalysts

In an oxidizing medium, zirconia seems to be the best support for stabilizing noble metals, there are low changes of the oxidation and steam reforming activities. But the steam reforming activity is strongly affected by a reducing treatment.

During the rich or lean excursions, the two catalysts containing $NiAl_2O_4$ seem to present a very good stability of Rh with a minimal loss of steam reforming activity after regeneration in H_2. But for those catalysts, platinum oxidative activity is very affected by a reducing treatment.

4. Conclusion

In Al_2O_3 supported catalysts, both ceria and platinum limit the deactivation of rhodium at high temperatures in an oxidizing medium. Nevertheless, while the rhodium in PtRh/CeA can be regenerated by a reducing treatment, Rh/CeA catalyst cannot be regenerated.

Zirconia is a very efficient support for stabilization of metals and particularly rhodium in an oxidizing medium but the latter is deactived by a reducing treatment. On the contrary, rhodium on $NiAl_2O_4$ presents a good stability both in an oxidizing and in an reducing atmosphere. But platinum is deactivated on this support after a reducing treatment.

ACKNOWLEDGEMENTS

This work was supported by the "Groupement de Recherche sur les Catalyseurs de Post-combustion" (IFP, ADEME, PIRSEM, CNRS). Thanks are due to IFP and ADEME for a grant (J.B.).

REFERENCES

1 K.C. Taylor, in A. Crucq et A. Frennet (editors), Catalysis and Automotive Pollution Control, Proc. 1st. Int. Symp., CAPOC 1, Brussels, Sept 8-11, 1986, Stud. Surf. Sci. Catal., Vol.30, Elsevier, Amsterdam (1987), 97.
2 J. Barbier Jr. and D. Duprez, Appl. Catal. B : Environmental, 3, (1993), 61.
3 J. Barbier Jr. and D. Duprez, Appl. Catal. A : General, 85, (1992), 89.
4 J.C. Schlatter, SAE Techn. Pap. Ser. n° 780199.
5 G. Kim, Ind. Eng. Chem. Prod. Res. Dev., 21, (1982), 267.
6 J.C. Schlatter and P.J. Mitchell, Ind. Eng. Chem. Prod. Res. Dev., 24, (1985), 43.
7 D. Duprez, Appl. Catal., 82, (1992), 111.
8 Y.-F. Yu Yao, Ind. Eng. Chem. Prod. Res. Dev., 19, (1980), 293.
9 Y.-F. Yu Yao., J. Catal., 87, (1984), 152.
10 D.D. Beck and C.J. Carr, J. Catal., 144, (1993), 296.
11 S. Kacimi et D. Duprez, in A. Crucq (editor), Catalysis and Automotive Pollution Control II, Proc. 2nd. Int. Symp., CAPOC 2, Brussels, Sept 10-13, 1990, Stud. Surf. Sci. Catal., Vol.71, Elsevier, Amsterdam, (1991), 581.
12 G. Leclercq and R. Maurel, Bull. Soc. Chim. Fr., (1971), 1234.
13 H.C. Yao, S. Japar and M. Shelef, J. Catal., 50, (1977), 407.
14 H.C. Yao, H.K. Stepien and H.S. Gandhi, J. Catal., 61, (1980), 547.
15 R.H. Hammerle and C.H. Wu, SAE Techn. Pap. Ser. n° 840549.
16 J.C. Summers and L.L. Hegedus, Ind. Eng. Chem. Prod. Res. Dev., 18, (1979), 318.
17 T.H. Ballinger and J.T. Yates Jr., J. Phys. Chem., 95, n°4, (1991), 1694.
18 B. Harrison, A.F. Diwell and C. Hallet, Plat. Met. Rev., 32, (1988), 73.
19 D.D. Beck, T.W. Caperhart, C. Wong, and D.N. Belton, J. Catal. 144, (1993), 311.

20 J.C. Summers and S.A. Ausen, J. Catal., 58, (1979), 131.
21 F.L. Williams and M. Boudart, J. Catal., 33, (1973), 438.
22 I. Onal, in C.H. Bartholomew and J.B. Butt (editors), Catalyst deactivation 1991, Stud. Surf. Sci. Catal., Vol 68, Elsevier, Amsterdam, (1991), 621.
23 W.B. Williamson, H.S. Gandhi, P. Wynblatt, T.J. Truex and R.C. Ku, AIChE Symp. Ser., n°201, 76, (1980), 212.
24 F.C.M.J.M. Van Delft, B.E. Nieuwenhuys, J. Siera and R.M. Wolf, Iron and Steel Institute of Japan International, n°7, 29, (1989), 550.

A. Frennet and J.-M. Bastin (Eds.)
Catalysis and Automotive Pollution Control III
Studies in Surface Science and Catalysis, Vol. 96

AN INFRARED STUDY OF CO AND NO ADSORPTION ON Pt, Rh, Pd 3-WAY CATALYSTS

R.L. Keiski[a], M. Härkönen, A. Lahti, T. Maunula, A. Savimäki and T. Slotte

Kemira Metalkat Oy, Catalyst Research, P.O. Box 171, FIN-90101 Oulu, Finland
[a]Department of Process Engineering, University of Oulu, Linnanmaa, FIN-90570 Oulu, Finland

ABSTRACT

Adsorption of CO, NO, CO-air, CO-NO, and CO-NO-air on Al_2O_3 and CeO_2-Al_2O_3 loaded with Pt, Rh, or Pd was studied using an in-situ infrared chamber installed under a DRIFT-measuring unit. The chamber allowed heating of a sample and introduction of various gases. The noble metal loading (Pt, Rh, Pd) on the washcoat varied in the range of 1.4 to 2.9 %. The samples were reduced at 300 °C by 5 % H_2/Ar prior to the FT-IR measurements. The reaction gases were introduced into the chamber in different sequences and at different temperatures.

According to the CO adsorption results the adsorbed gem-dicarbonyl species ($Rh^1(CO)_2$) on Rh and linearly bonded species on both Pt and Pd and bridged CO species on Pd were clearly identified. When CO pressure was increased stepwise no adsorbed CO was seen on the noble metal free washcoat, whereas linearly adsorbed CO was seen on Pt/Al_2O_3. Assignment of bands to different species of NO was difficult due to the numerous NO adsorption states. On Rh/Al_2O_3 and Rh/CeO_2-Al_2O_3 adsorbed NO showed a band caused by Rh-NO^+ at around 1910 cm^{-1}. Rh-NO and Rh-NO^- species could not be seen clearly. NO had a strong interaction with alumina giving adsorption bands in the range of 1600 to 1200 cm^{-1}. Furthermore, NO was the dominant species compared to CO in occupying the adsorption sites on the surface of the noble metals examined.

In kinetic studies the oxidation of CO in a CO-air mixture over a Rh/Al_2O_3 catalyst started at around 200 °C. Mixtures of CO and NO or CO, NO, and air led to the formation of an isocyanate complex. In the presence of CeO_2 the formation of an isocyanate complex required that the catalyst surface was first occupied by NO. It is concluded that the carbonyls on the noble metals probably reacted with adsorbed nitrogen and oxygen to isocyanate complexes and CO_2, respectively.

1. INTRODUCTION

One of the most important reactions in automobile exhaust catalysis is the reaction between NO and CO

$$NO(g) + CO(g) \rightarrow CO_2(g) + \frac{1}{2}N_2(g) \quad (1)$$

Rhodium, platinum and palladium are mostly used to catalyze this reaction. These noble metals are proposed to catalyze the NO dissociation, reaction of NO with adsorbed N atoms or combination of adsorbed N atoms to form N_2 and reaction of CO with adsorbed O atoms to form CO_2 [1]. Reaction (1) is proposed to proceed through these three steps. However, the interaction of NO and CO on Rh, Pt, Pd, and Rh-Pt catalysts is not yet clear. Many uncertaintities concerning the behavior of the surface species still remain to be solved.

The formation and reactivity of surface intermediates over three-way catalysts are important subjects in designing automobile exhaust catalysts. The high temperature NO-CO and NO-CO-O_2 reactions produce isocyanate surface intermediates (-NCO) and release N_2O intermediate products depending on, for example, how the catalyst surface is pretreated before the NO and CO adsorption [2]. Infrared spectroscopy is an excellent tool to investigate the formation of these kind of surface intermediates. With IR very low concentrations of surface compounds can be detected under reaction conditions.

Table 1 summarizes the most important surface complexes formed when NO and CO are adsorbed on noble metal catalysts. According to the literature NO and CO are adsorbed as nitrites, nitrates and carbonates on alumina [2]. The most important surface complexes for CO and NO adsorption on rhodium are a gem-dicarbonyl ($Rh(CO)_2$) and a linear Rh-NO^+ complex [1]. However, tricarbonyl and bridged Rh_x-CO complexes have been proposed to be formed and different kinds of linear Rh nitrosyl complexes are possible [1-4]. The adsorption of CO on Pt and Pd catalysts depends much on the oxidation stage of Pt and Pd [5]. CO adsorption on Pt forms mostly linear and bridged carbonyls [6-11]. NO is adsorbed linearly on Pt [12]. In the case of Pd the most common surface complexes are linear carbonyls [13], strong multilaterally-bonded carbonyls, bridged carbonyls [5,14,15] and triply-bonded CO [5]. Isocyanate, nitrous oxide or nitrogen dioxide are proposed to be connected to the reaction mechanism of the NO-CO reactions [2,16-19].

To obtain more information on the reaction mechanism over the three-way catalysts the CO-air, NO-CO, and NO-CO-air reactions over washcoats with and without noble metals were studied using in-situ infrared spectroscopy. The goal was to explain the reaction mechanisms and the effect of the preadsorption of the reaction gas components on the catalyst surface.

2. EXPERIMENTAL

The infrared instrument used in the investigation was a Perkin-Elmer 1760 FT-IR spectrometer. The chamber (Environmental Chamber, Specac 19930) allowed heating of a sample and introduction of various gases. The reaction gases were introduced into the chamber through a gas manifold which allowed the introduction of fixed amounts of different gas components separately. The different modes of measurements were used in this study. In some experiments, hereinafter called static experiments, the spectra were recorded at successive temperatures using a standard collection software whereas in others, called GC-IR experiments, the spectra were recorded at constant temperature using a GC-IR software. The spectra were recorded at 4000-600 cm^{-1} using a resolution of 4 cm^{-1}.

Table 1. NO and CO adsorption on noble metal catalysts.

Surface Species*)	Wavenumber, cm^{-1}/ref./ Rh	Pt	Pd
CO_3^{2-}, formate	1590,1395 /2,21/	1570 /21/	1580-1337 /15,21/
NO_2(g)	1618,1328,750 /27/		
N_2O(g)	2224,1286,589 /2/		
NO_3-, NO_2-, NH_3	1550,1295,1240 /2/		
M-NCO	2269,2265, 2240 /16,21/	2267 /21/	2264 /64/
$M\text{-}NO^+$	1912,1905,1890 /1,16,17/	-	-
M-NO	1830,1817 /1,16,19/	1812,1780 /12/	1751 /21/
$M\text{-}NO^-$	1740,1640 /16/	-	-
$M(NO)_2$	1830,1825(sym) /1,16,17/ 1743,1740,1716(asym)	-	-
M(NO)(CO)	2100,1755 /1/	-	-
M^{4+}-CO	-	2180 /9/	-
M^{2+}-CO	2125 /25/	2130 /3/	-
M^+-CO	-	2140,2124 /11/	2140,2120 /15/
$M(CO)_2$	2100,2098(sym) /20/ 2030,2025(asym)	2152,2133 /7/	2060,2050 /15/
$M(CO)_3$	2120,2078,2026 /24/	-	-
M-CO (linear)	2080,2060 /1,2,22/	2090,2080,2070 /11,21,26/	2030,2000 /5/
M_x-CO (bridged)	1870,1860 /1,11/	1850,1840 /11,26/	1930,1844 /15/ 1987,1965 /5/
M_3-CO (triply bonded)	2120,2078,2026 /3/	-	1880-1800/5/

*) M is a noble metal

Catalyst samples containing noble metals and different washcoats were studied. A flat metal foil was coated with the washcoat, thus allowing the analysis of a real exhaust catalyst. The washcoat layer on the metal substrate comprised Al_2O_3 or CeO_2-Al_2O_3. The noble metal loading of Pt, Rh, Pd and Pt-Rh on the washcoat was in the range of 1.4 to 2.9 %. The thickness of the washcoat was typically 15-40 μm.

The catalyst samples, all in fresh condition, were reduced at 300°C by 5% H_2/Ar prior to the FT-IR measurements. The reaction gases were introduced into the chamber in different sequences and at different temperatures. Static experiments were done by introducing NO or CO first into the chamber at room temperature, after which the second of the two components was added. In the case of a three component gas mixture the gases were added in the following sequences: NO-CO-air or CO-NO-air. Then the chamber was closed and the temperature was increased stepwise to 100, 150, 200, 300, and 400 °C at which temper-atures the spectra were measured.

GC-IR measurements were done at 200 and 300 °C. NO or CO was first introduced into the chamber at the reaction temperature and the gas was allowed to adsorb on the catalyst surface during the first 10 minutes. Then the second gas, CO or NO, was introduced into the chamber and this mixture was allowed to react for the next 10 minutes after which the chamber was evacuated. In the case of a three component mixture the second gas (NO or CO) and the third gas (air) were dosed into the chamber after 7 and 14 minutes from the introduction of the first gas. The spectra were recorded at 0.5 minutes intervals. The whole experiment lasted 25 minutes.

The partial pressures of NO, CO and air used in the experiments were 60, 90 and 180 mbar, respectively. NO was introduced into the chamber as a 10 % NO/N_2 gas (600 mbar).

3. RESULTS AND DISCUSSION

3.1. NO-CO co-adsorption on alumina washcoat

Nitric oxide was adsorbed on the washcoat already at room temperature giving rise to absorption bands at 1630, 1455 and 1230 cm^{-1}. These bands are proposed to be caused by Al-NO_2 species [2]. Carbon monoxide gave weak absorption bands at 1590 and 1395 cm^{-1} in the temperature range of 200 to 300 ° C. These bands were caused by the Al-CO_3 species [2]. According to Burkett et al. [15] and Haaland and Williams [26] absorption bands at 1580 - 1337 and at 1700-1200 cm^{-1} include also bicarbonates and formates due to the interaction of CO and hydroxyl groups on the alumina surface. Ce slightly weakened the interaction of NO and CO with the washcoat. Absorption bands were, however, found at the same wave-numbers when different washcoat compositions were used.

3.2. CO and NO adsorption and CO-air and NO-air co-adsorption on Rh/Al_2O_3

According to static experiments CO was adsorbed on reduced Rh catalysts mostly as gem-dicarbonyl species which were observed at around 2088 and 2022 cm^{-1} (Fig.1). At 300 to 400 °C traces of linearly bonded carbonyls at around 2060 cm^{-1} are proposed to arise. Increasing of the sample temperature did not have any effect on the band positions of the gem-dicarbonyl complex. The band intensities increased as the temperature was increased, which may be caused by the increasing adsorption ability of CO. At 400 °C the asymmetric stretching of the C-O bond (2022 cm^{-1}) increased while the symmetric stretching (2088 cm^{-1}) decreased.

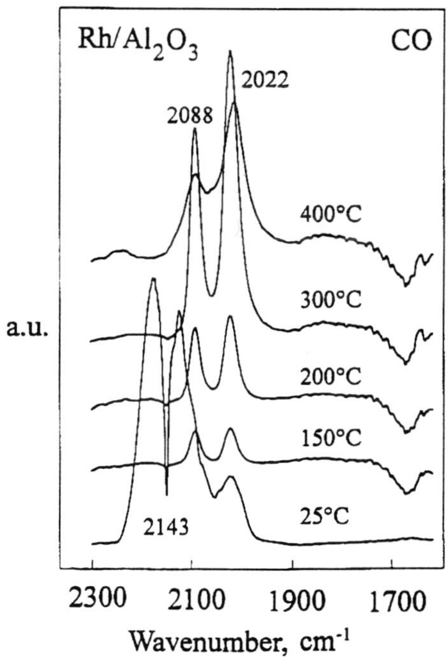

Figure 1. CO adsorption on Rh/Al_2O_3.

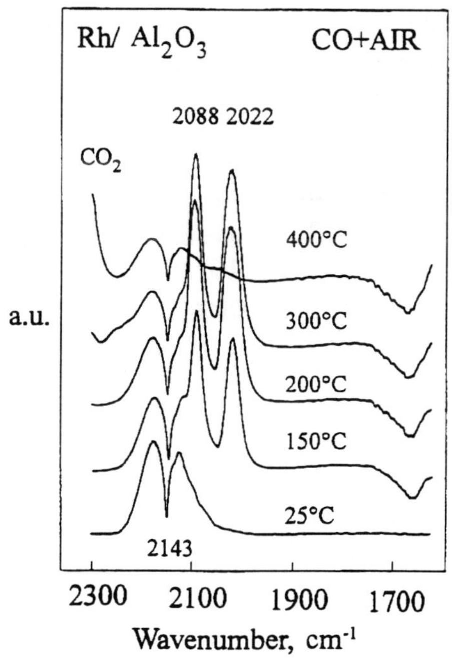

Figure 2. CO and air co-adsorption on Rh/Al_2O_3.

When air was introduced into the chamber with CO, Rh gem-dicarbonyl bands appeared at 150 C (Fig.2). The formation of CO_2 was observed at 200 C. At 400 °C gem-dicarbonyl bands were no more present indicating the reaction of CO to CO_2. The possible reaction path for the CO-O_2 reaction is proposed to be the following:

$$Rh(CO)_2 + O_2(g) \rightarrow Rh^+ + 2CO_2(g) \qquad (2)$$

In the literature [4] the following reaction path with the formation of a bridged carbonyl is also proposed

$$2Rh(CO)_2 + O(ads) \rightarrow Rh_2\text{-}CO + CO_2(g) + 2CO(g) \qquad (3)$$

At room temperature after NO exposure into the chamber, only gaseous NO was observed (1877 cm^{-1}) in most cases. However, in some cases a strong absorption band at around 1750 cm^{-1} was observed indicating the formation of Rh-NO^- species [1,16]. At 300 °C NO was adsorbed on Rh forming linear Rh-NO^+ absorption bands at 1903 cm^{-1}. These bands are typically found at 1900-1910 cm^{-1} [1,16,17]. The dissociation of NO could not be confirmed in these experiments.

3.3. NO-CO and NO-CO-air co-adsorption on Rh/CeO_2-Al_2O_3

According to the results the adsorption of the binary gas mixture (NO-CO) on Rh catalysts appeared as Rh gem-dicarbonyl bands at 300 °C (Fig.3). At this temperature a band at 1906 cm^{-1} was observed, indicating the formation of a Rh-NO^+ surface complex. At lower temperatures the adsorption of NO and CO on the washcoat was observed at 1600-1200 cm^{-1}. A band at around 1630 cm^{-1}, which showed as a shoulder in Figures 3 and 4, may be due to a Rh-NO^- complex, which is typically seen at 1648-1640 cm^{-1} [16,17]. At 300 and 400 °C the formation of an isocyanate complex became more and more obvious (2253-2245 cm^{-1}).

When gases were introduced into the chamber in the following order CO-NO the gem-dicarbonyl complexes could not be seen clearly (Fig.4). Gaseous NO was observed clearly at 1877 cm^{-1} whereas the bands for gaseous CO were very weak. The formation of a Rh-NO^+ complex became more and more obvious at 300 and 400 °C giving rise to the absorption band at 1896 cm^{-1}. The formation of the isocyanate surface complex was not as intense as in the case with NO-CO addition into the chamber.

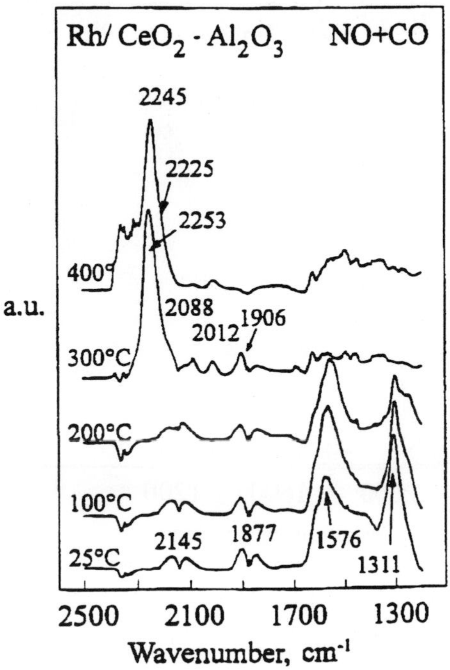

Figure 3. NO and CO co-adsorption on Rh/CeO$_2$-Al$_2$O$_3$.

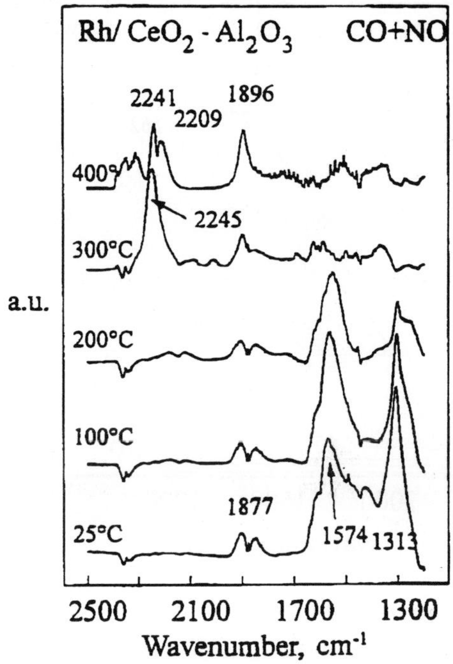

Figure 4. CO and NO co-adsorption on Rh/CeO$_2$-Al$_2$O$_3$.

The GC-IR spectra for the NO-CO-air and CO-NO-air adsorption on a Rh catalyst are shown in Fig.5 and 6. NO was adsorbed on Rh as a linear nitrosyl ($Rh\text{-}NO^+$ at 1911 cm^{-1} (Fig.5) and at 1894 cm^{-1} (Fig.6)) whereas CO formed gem-dicarbonyl surface complexes on Rh ($Rh(CO)_2$ at 2092 and 2020 cm^{-1}, Fig.6). The isocyanate surface complex formation was strong when air was added into the chamber after NO and CO (Fig.5). In the case of CO-NO-air, the absorption bands for the isocyanate surface complex were very weak. This proves the proposed reaction path for the Rh-NCO formation [2]. Isocyanate surface complex formation is possible when the catalyst surface is preadsorbed with NO whereas if CO is preadsorbed on the surface the reaction intermediate is nitrous oxide:

$$Rh \xrightarrow{NO} Rh\text{-}NO \xrightarrow{CO} Rh(NO)(CO) \xrightarrow{-CO_2} Rh\text{-}N \xrightarrow{CO} Rh\text{-}NCO \qquad (4)$$

$$Rh \xrightarrow{CO} Rh\text{-}CO \xrightarrow{NO} Rh(CO)(NO) \xrightarrow{-CO_2} Rh\text{-}N \xrightarrow{NO} Rh\text{-}N_2O \qquad (5)$$

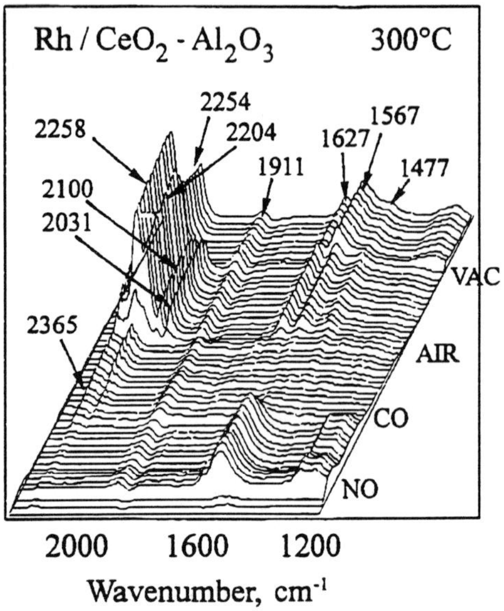

Figure 5. NO, CO, and air co-adsorption sorp-on Rh/CeO_2-Al_2O_3 at 300 ℃.

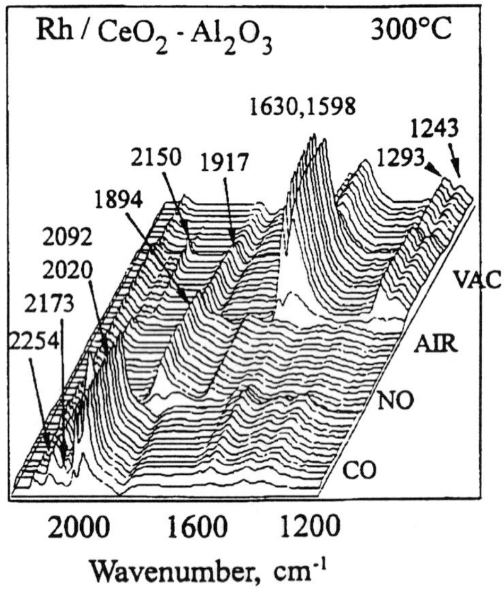

Figure 6. CO, NO, and air co-adsorption on Rh/CeO_2-Al_2O_3 at 300 ℃.

3.4. NO-CO co-adsorption on Pt/CeO_2-Al_2O_3

In GC-IR experiments CO was adsorbed on a reduced Pt/CeO_2-Al_2O_3 mostly as linear carbonyls which were observed at 2062 cm^{-1} (Fig. 7). A band at 2173 cm^{-1}, which was observed in some experiments, was probably due to the Pt^{n+}-CO complex formation [7]. Weak bands at 1588, 1481, and 1385 cm^{-1} were assigned to surface carbonate groups. When NO was added into the chamber linear Pt-CO bands disappeared and gaseous NO species at 1877 cm^{-1} and linear Pt-NO species at 1775 cm^{-1} were formed. Linear Pt-NO bands are typically observed at around 1812-1780 cm^{-1} [12]. The formation of Pt-NCO species was observed at 2258 cm^{-1} but the absorption bands were quite weak. However, when the cham-ber was evacuated strong absorption bands arose at 2258 and 2181 cm^{-1} indicating the isocyanate and Pt^{4+}-CO complex [9] formations, respectively.

3.5. NO-CO co-adsorption on Pd/Al_2O_3

The GC-IR spectra of NO-CO and CO-NO adsorption on Pd/Al_2O_3 at 300 °C are shown in Figures 8 and 9. NO formed linear Pd-NO bands at 1780 cm^{-1} and the adsorption of NO on the washcoat as nitrates and nitrites was observed at1650-1200 cm^{-1} (Fig.8). When CO was added into the chamber linear Pd^{+}-CO bands appeared at 2133 cm^{-1} and the formation of Pd-isocyanate complexes started. The Pd-NCO band at 2254 cm^{-1} was strong. When CO was first introduced into the chamber linear (Pd_x-CO at 2065 cm^{-1}) and bridged (Pd_x-CO at 1936 cm^{-1}) Pd-CO complexes were observed (Fig.9). The wavenumber of these bands is typically at 2080-2050 and 2000-1800 cm^{-1}, respectively [5,14,15]. Adsorption of CO on the washcoat as carbonate gave absorption bands at 1588 and 1469 cm^{-1}. When NO was added into the chamber only weak bands of the isocyanate complex were observed. The linear and bridged Pd-CO species disappeared and a linear bond between Pd and NO was formed (Pd-NO at 1750 cm^{-1}). When the chamber was evacuated the absorption band of the isocyanate complex was formed.

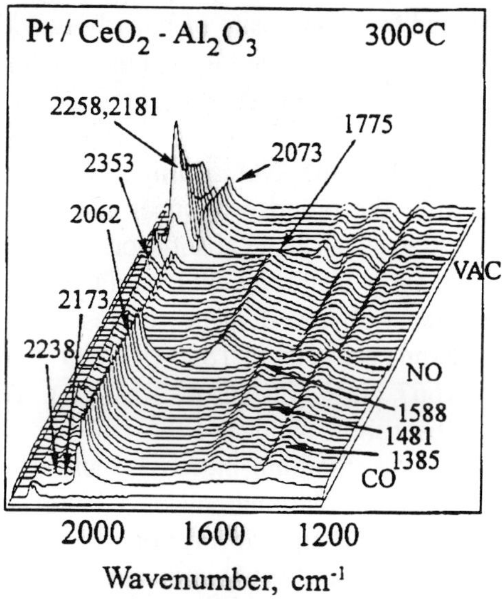

Figure 7. CO and NO co-adsorption on Pt/CeO_2-Al_2O_3 at 300 °C.

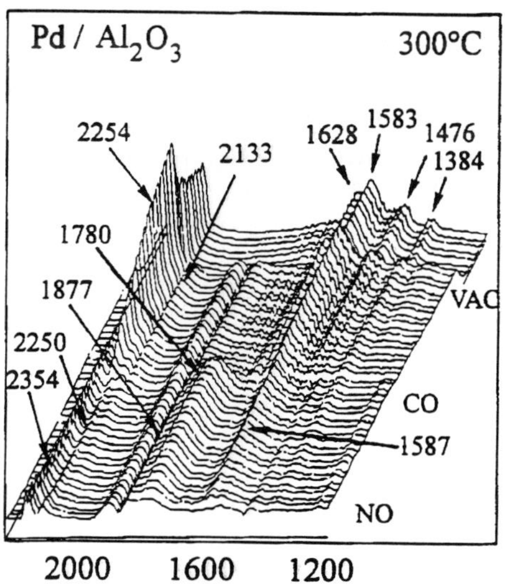

Figure 8. NO and CO co-adsorption on Pd Al_2O_3 at 300 °C.

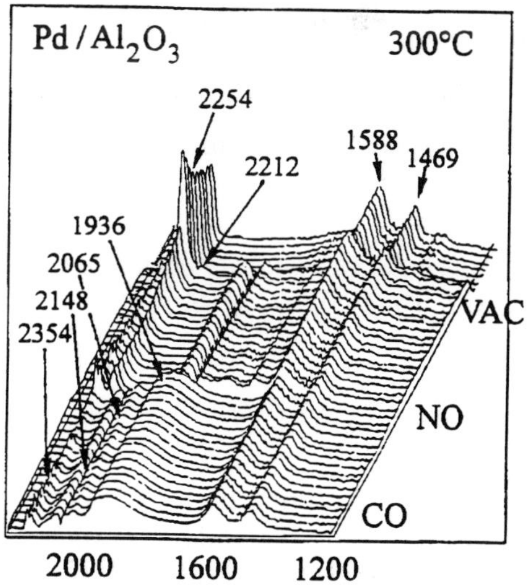

Figure 9. CO and NO co-ad sorption Pd/Al_2O_3 at 300 °C.

3.6. NO-CO co-adsorption on Pt-Rh/CeO_2-Al_2O_3

The GC-IR spectra of CO on Pt+Rh catalysts showed bands at 2092 and 2023 cm^{-1}. These bands are due to the rhodium gem-dicarbonyl species (Fig.10). The maximum at around 2065 cm^{-1} is assigned to a composite of linearly-bonded Rh and Pt carbonyls [20]. The band at 2173 cm^{-1} is assigned to carbonyl species on oxidized forms of platinum and rhodium (Rh^{2+}-CO and Pt^{+}-CO, Pt^{2+}-CO, and Pt^{4+}-CO, Table 1).

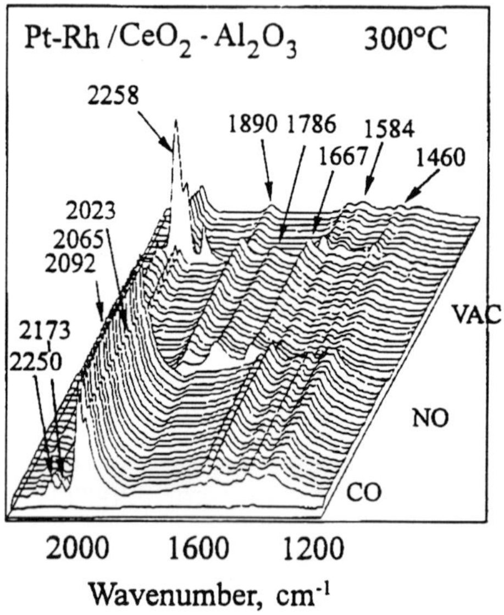

Figure 10. CO and NO co-adsorption on Pt-Rh/CeO_2-Al_2O_3 at 300 °C.

When the NO addition was done after the preadsorption of CO on the catalyst surface, absorption bands of linearly adsorbed NO on Pt and Rh were observed. The corresponding bands are seen at 1890 (Rh-NO^{+}) and 1667 cm^{-1} (Pt-NO or Rh-NO^{-}). The absorption band at 1786 cm^{-1} is due to the rhodium nitrosyl band (Rh-NO^{-}) or platinum nitrosyl band (Pt-NO). Strong absorption bands at around 2258 cm^{-1} indicate the presence of isocyanate intermediates (Pt-NCO and Rh-NCO) which appeared after the evacuation. If NO was introduced into the chamber before CO the isocyanate complex was formed earlier. NO and CO were also adsorbed strongly on the washcoat giving absorption bands below 1600 cm^{-1}.

4. CONCLUSIONS

CO adsorption on Rh/CeO_2-Al_2O_3 formed rhodium gem-dicarbonyl species ($Rh^{1}(CO)_2$) on the catalyst surface. Linearly-bonded species on both Pt and Pd and bridged CO on Pd were clearly identified. On Rh/Al_2O_3 and Rh/CeO_2-Al_2O_3 adsorbed NO showed a band caused by Rh-NO^{+} at around 1910 cm^{-1}. Rh-NO and Rh-NO$^-$ species could not be seen as clearly as Rh-NO$^+$ species. Absorption bands in the range of 1600 to 1200 cm^{-1} indicated a strong interaction between NO and the washcoat. Furthermore, NO was the dominant species compared to CO in occupying the adsorption sites on the surface of the noble metals examined.

Kinetic studies over Rh/Al_2O_3 showed that the amount of adsorbed CO increased on the Rh surface as a function of temperature and the oxidation of CO was observed at 200 °C burning the adsorbed CO off quickly. After the reaction only gaseous CO_2 remained in the chamber.

According to this study the most probable mechanism for the NO-CO reaction when NO/CO>1 is the formation of isocyanate surface complexes. Isocyanate formation was also seen when a three-component gas mixture (NO-CO-air) was introduced into the chamber. When CeO_2 was present in the catalyst the formation of isocyanate surface complexes required that the surface of the catalyst was first occupied by NO. If the addition of CO was done first the isocyanate formation could be seen only after the evacuation of the sample. The carbonyls on the noble metals probably reacted with adsorbed oxygen and nitrogen forming CO_2 and isocyanate, respectively.

REFERENCES

1 R. Dictor, J. Catal. 109 (1988) 89.
2 H. Arai and H. Tominaga, J. Catal., 43 (1976) 131.
3 D.A. Storm, F.P. Mertens, M.C. Cataldo and E.C. DeCanio, J. Catal., 141 (1993) 2, 478.
4 F. Solymosi, J. Raskó and J. Bontovics, Catal. Lett., 19 (1993) 257.
5 K.I. Choi and M.A. Vannice, J. Catal., 127 (1991) 465.
6 J. Mink, T. Szilagyi, S. Wachholz and D. Kunath, J. Mol. Struct., 141 (1986) 389.
7 L. Marchese, M.R. Boccuti, S. Coluccia, S. Lavagnino, A. Zecchina, L. Bonneviot and M.Che, Stud. Surf. Sci. Catal., 48 (1989) 653.
8 T. Hattori, E. Nagata, S. Komai and Y. Murakami, J. Chem. Soc., Chem. Comm., 15 (1986) 1217.
9 M.S. Brogan, J.A. Cairns and T.J. Dines., Royal Chem. Soc., 114 (1992) 282.
10 S.I. Abasov, V.Yu. Borovkov and V.B. Kazansky, Catal. Lett., 15 (1990) 269.
11 J.A. Anderson and C.H. Rochester, J. Chem. Soc. Faraday Trans., 87 (1991) 9, 1479.

12 F. Boccuzzi, G. Ghiotti, A. Chiorino and E. Guglielminotti, Surf. Sci., 269/270 (1992) 514.
13 V. Pitchon, M. Primet and H. Praliaud, Appl. Catal., 62 (1990) 317.
14 J.L. Duplan and H. Praliaud, Appl. Catal., 67 (1991) 325.
15 H.D. Burkett, S.D. Worley and C.H. Dai, Chem. Phys. Lett., 173 (1990) 5/6, 430.
16 E. Novák and F. Solymosi, J. Catal., 125 (1990) 112.
17 J. Liang, H.P. Wang and L.D. Spicer, J. Phys. Chem., 89 (1985)26, 5840.
18 F. Solymosi and J. Raskó, J. Catal., 65 (1980) 235.
19 W.C. Hecker and A.T. Bell, J. Catal., 84 (1983) 200.
20 J.A. Anderson, J. Catal., 142 (1993) 1, 153.
21 M.L. Unland, J. Catal., 31 (1973) 459.
22 P.B. Rasband and W.C. Hecker, J. Catal., 139 (1993) 2, 551.
23 P. Basu, D. Panayotov and J.T. Yates Jr., J. Phys. Chem., 91 (1987)12, 3133.
24 H.P. Wang and J.T. Yates Jr., J. Catal., 89 (1984) 79.
25 J.P. Wey, W.C. Neely and S.D. Worley, J. Catal., 134 (1992) 1, 378.
26 D.M. Haaland and F.L. Williams, J. Catal., 76 (1982) 450.
27 K. Nakamoto, Infrared Spectra in Inorganic and Coordination Compounds.John Wiley & Sons. New York 1970.

A. Frennet and J.-M. Bastin (Eds.)
Catalysis and Automotive Pollution Control III
Studies in Surface Science and Catalysis, Vol. 96

COMPARATIVE BEHAVIOUR OF Pd SUPPORTED CATALYSTS FOR THE REDUCTION OF NO BY CO IN THE PRESENCE OF GAS COMPLEX MIXTURE INCLUDING O_2, CO_2, H_2O AND HYDROCARBONS

A. Lemaire[a], J. Massardier[a], H. Praliaud[b], G. Mabilon and M. Prigent[c]

[a]*Institut de Recherches sur la Catalyse, C.N.R.S.*
2, Av. A. Einstein, 69626 Villeurbanne Cedex, France

[b]*Laboratoire d'Application de la Chimie à l'Environnement, Unité Mixte, C.N.R.S-U.C.B,*
43, Boulevard du 11 Novembre 1918, 69622 Villeurbanne Cedex, France

[c]*Institut Français de Pétrole,*
1-4 Av. du Bois Préau, 92506 Rueil Malmaison, France

ABSTRACT

On different Pd based catalysts (Pd/Al_2O_3, Pd/ZrO_2, Pd/Al_2O_3-BaO, Pd/Al_2O_3-La_2O_3), it has been shown that the introduction of hydrocarbons inhibits the catalytic reduction of NO by CO either in the presence or in the absence of O_2. This inhibition effect has been explained either by carbon poisoning of the active metal when the CO-NO reaction is performed without oxygen (or with moderate amounts of O_2) or by oxygen poisoning for higher oxygen concentrations.

Key words: Pd based catalysts, NO reduction, CO-NO-O_2-hydrocarbons reactions.

1. INTRODUCTION

The elimination of the main pollutants in automobile exhausts requires both the reduction of NO and oxidation of carbon monoxide and unburnt hydrocarbons [1]. Moreover, it is well known that rhodium is active and selective for the NO reduction and that the standard three-way catalysts (TWC) contain both platinum and rhodium with a Rh/Pt ratio higher than in the Pt-mine.

Therefore an important economic objective is to reduce the use of rhodium which is expensive and scarce. Among the precious metals, Pd is relatively abundant and less expensive than Pt or Rh. So, Pd-based catalysts have been considered as potential substitutes for Rh [2-8]. Indeed, Pd has been reported to carry out both reduction and oxidation reactions in the same CO-NO-O_2 feedstream than on the Rh based samples.

This paper deals on the influence of the addition of hydrocarbons (C_3H_6 or C_3H_8) to CO-NO and CO-NO-O_2 mixtures, simulating the composition of the exhaust gas, on the NO reduction by CO performed with different Pd based catalysts. The influence of the addition of CO_2 and H_2O is briefly mentioned.

2. EXPERIMENTAL

The samples differ by the nature of the supports. Four supports were used: γ (δ) Al_2O_3, ZrO_2 (mixture of tetragonal and monochimic structure), Al_2O_3-BaO and Al_2O_3-La_2O_3. These two last supports were prepared by impregnation of c Al_2O_3 with the adequate amounts of Ba or La nitrates followed by a calcination under an oxygen (or air) flow at 723 K. The BET areas and the respective contents of Ba and La of the mixed supports are presented on table 1.

The catalysts were prepared by impregnation of the supports by various Pd salts: $Pd(C_5H_7O_2)_2$ for Pd/Al_2O_3, H_2PdCl_4 for Pd/ZrO_2, $Pd(NO_3)_2$ for Pd/Al_2O_3-BaO and Pd/Al_2O_3-La_2O_3. Then, the Pd salts were decomposed under O_2 (or air) at temperatures ranging between 673 and 723 K. In some cases a reduction at 673 K was performed with a well defined heating rise (~ 2 $K.mn^{-1}$. The amounts of Pd and the dispersions of the metallic phase are shown on table 1. The behaviour of the Pd based samples was also compared to that of a Pt-Rh/Al_2O_3 considered as the reference catalyst.

Table 1: Physical characteristics of the samples: specific surface area S, metallic and additive contents, metallic dispersion measured by hydrogen chemisorption after reduction.

Samples	S(BET) $m^2.g^{-1}$	Ba or La (wt%)	Pd (wt%)	Dispersion
Pd/Al_2O_3	120		0.61	0.37
Pd/ZrO_2	62		0.79	0.26
Pd/Al_2O_3-BaO	112	12	0.91	0.17
Pd/Al_2O_3-La_2O_3	80	10.5	0.85	0.17
Pt-Rh/Al_2O_3			1% for Pt 0.2% for Rh	0.5

Catalytic experiments were carried out at atmospheric pressure in a gas flow microreactor with He as a diluent. The gas concentrations (NO, CO, O_2, hydrocarbons) have been chosen to be representative of the exhaust gas mixtures of spark ignition engines. Analysis were performed by gas chromatography with a dual column (porapak and molecular sieve) and a TCD detector for O_2, N_2, CO, CO_2, N_2O, and a flame ionization detector for hydrocarbons. NO and NO_2 were analyzed on-line by IR spectrometry (Rosemount analyzers).

The experimental procedure was as follows: a small amount of the catalyst (10 mg diluted with 40 mg of inactive α Al_2O_3) was used in order to prevent the mass and heat transfer limitations, at least for the low conversions. After heating in a flow of N_2 up to 423 K the catalyst was contacted with the reactant gases (between 12 and 22 lh^{-1}, hourly space velocity between 120 000 and 220 000). The analysis was performed at increasing and decreasing temperatures between 423 and 773 K with programmed rates of 2 K/mn. The stoichiometry of the feedstream was defined by the "s ratio" = $2(O_2) + (NO)/(CO) + (2x + y/2)(C_xH_y)$. No significant effect of a prereduction was observed on the catalytic activity. NO_2 was not detected and the amounts of N_2O remained small (O to 20% of the total NO converted).

3. RESULTS AND DISCUSSION

First of all, the catalytic behaviour of the different samples in the presence of the standard CO-NO-O_2 mixture (s~1.03) is illustrated by the light off temperatures for NO, CO and O_2 (temperature required to leave 50% conversion of the considered pollutant) which are presented on table 2.

Table 2: Light-off temperature (K) for the NO reduction, the CO total conversion and the CO oxidation by O_2 (CO-NO-O_2 mixture, s=1.03, flow rate 17.4 lh^{-1}).

Samples	NO-reduction	CO total conversion	CO-O_2
Pd/Al_2O_3	571	550	553
Pd/ZrO_2	591	605	607
Pd/Al_2O_3-BaO	562	524	524
Pd/Al_2O_3-La_2O_3	549	538	540
Pt-Rh/Al_2O_3	576	559	559

It is noteworthy that on Pd/ZrO_2 the NO reduction occurs at lower temperatures than the CO oxidation.

In a first step, in order to precise the influence of each additive, CO_2, H_2O, C_3H_6 or C_3H_8 were added separately to the CO-NO and CO-NO-O_2 mixtures.

In the presence or not of oxygen, the NO reduction by CO is not changed by the addition of 10 vol% CO_2, whatever the catalyst. In the same way, the influence of H_2O is weak and depends on the support, leading either to a small deactivation (Al_2O_3 support) or to a small enhancement of the activity (ZrO_2 support). In this case, this effect is explained by the contribution of reactions induced by the H_2O presence such as the water gas shift and the steam reforming reactions in the presence of hydrocarbons [9].

At the opposite, an important inhibition of the NO reduction is observed by the addition of propane or propene to CO-NO or CO-NO-O_2 mixtures. The extent of the inhibition depends both on the nature of the hydrocarbon and on the presence or not of oxygen.

3.1. Influence of the hydrocarbon addition on the NO reduction by CO in the absence of O_2

The figures 1a to 1c illustrate the influence of C_3H_6 or C_3H_8 addition to the initial CO-NO mixture on the NO reduction with Pd/Al_2O_3, Pd/ZrO_2 and Pt-Rh/Al_2O_3 catalysts. The C_3H_6 and C_3H_8 contents have been chosen in order to obtain strong reducing mixtures. The "s" ratio is thus equal to 0.1 considering that the hydrocarbons are totally oxidized according to the reaction:

$$C_xH_y + (2x + y/2)NO \rightarrow xCO_2 + y/2H_2O + (x + y/2)N_2$$

Whatever the samples, the NO reduction is strongly inhibited by C_3H_6 in spite of the reducing character of the reagent mixture. The inhibition is weaker with C_3H_8 than with C_3H_6 and is mainly observed on Pd/Al_2O_3. Such a hydrocarbon inhibition has already mentioned on Pd catalysts by Muraki et al [5]. This effect is clearly explained by carbon deposits whose formation is easier with olefins than with saturated hydrocarbons [10] and an additional treatment with O_2 restores the initial activity as shown on table 3.

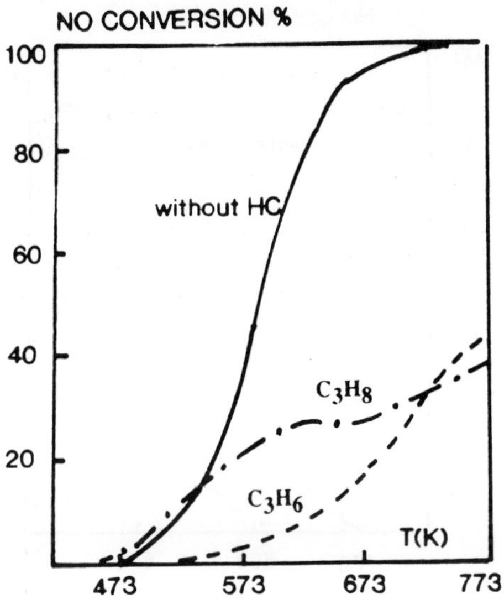

1a) Solid Pd/Al$_2$O$_3$

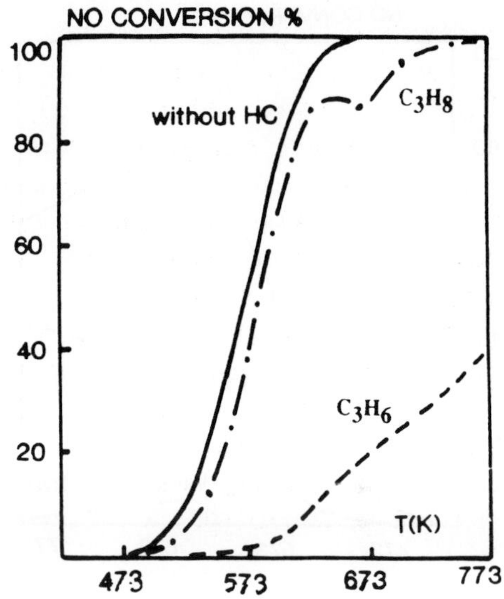

1b) Solid Pd/ZrO$_2$

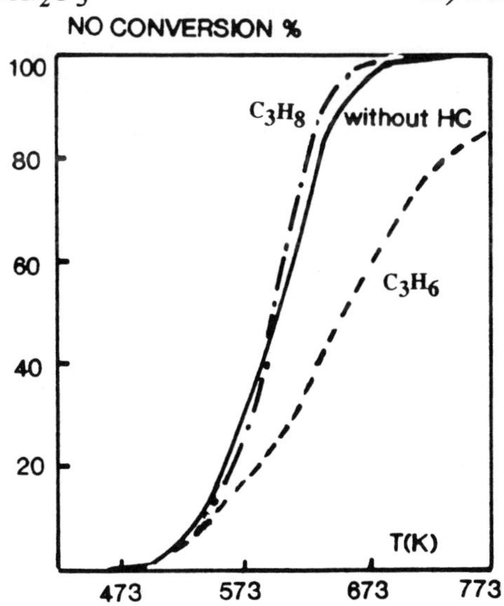

1c) Solid Pt-Rh/Al$_2$O$_3$

Figure 1: Influence of the hydrocarbon addition on the NO conversion in the absence of O_2 as a function of the reaction temperature. Flow rate 17.4 lh^{-1}, 1000 vpm NO 1000 vpm CO and 927 vpm C_3H_6 or 912 vpm C_3H_8.

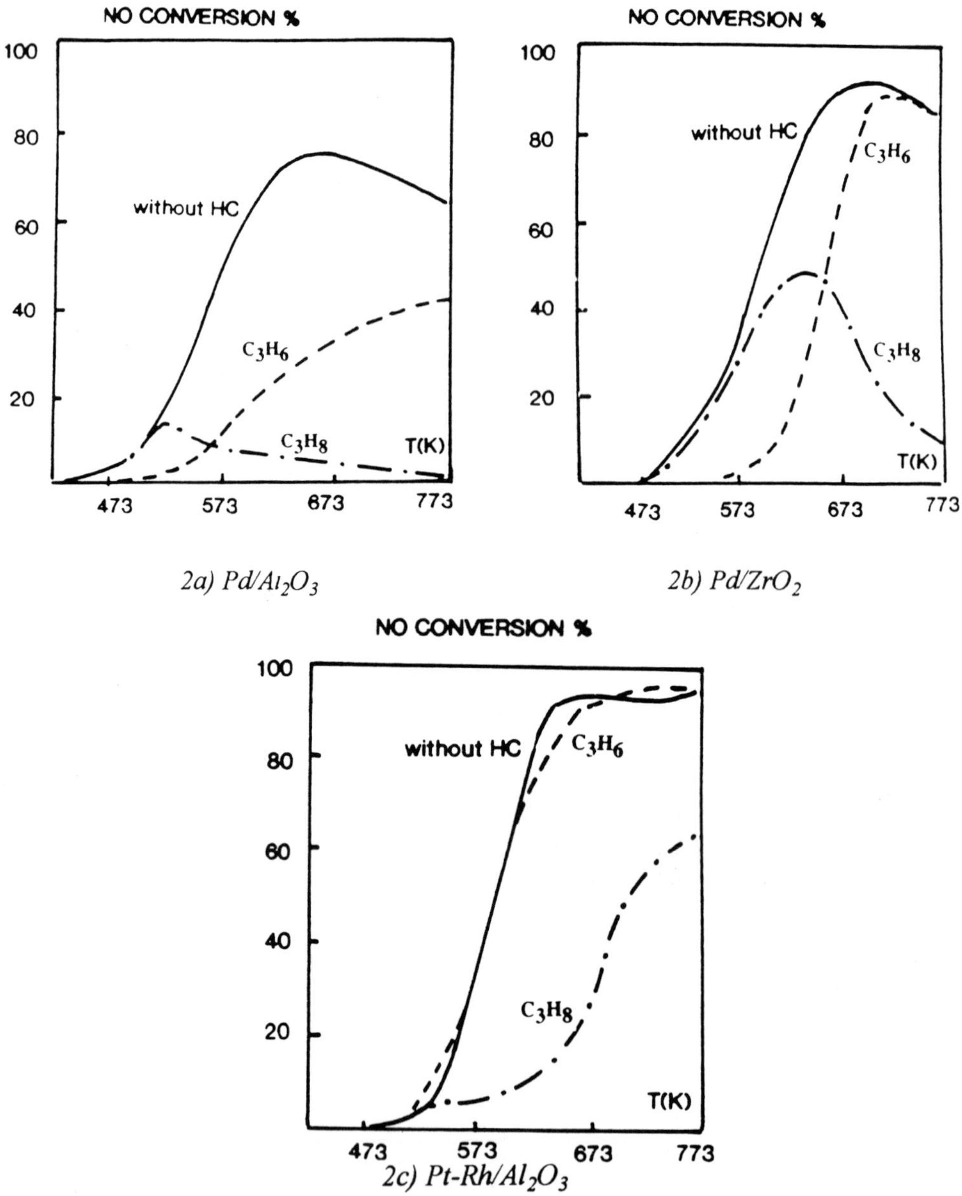

Figure 2. Influence of the hydrocarbon addition on the NO conversion in the presence of O_2 as a function of the reaction temperature. Flow rate 17.4 lh^{-1}, 5930 vpm CO, 788 vpm NO, 1000 vpm C_3H_6 or C_3H_8, O_2 to obtain theorically s = 1.03.

Table 3: Amounts of carbon deposits after 1h reaction at 773 K and relative activities A/A_O(A_O activity for the simple CO-NO mixture (1000 vpm NO, 1000 vpm CO) measured near the light-off temperature, A activity in the presence of the hydrocarbon, 927 vpm C_3H_6 or 912 ppm C_3H_8, at the same temperature).

Treatments		Carbon deposits - [A/Ao] (weight %)		
Gas composition and samples	s ratio	Pd/Al_2O_3	Pd/ZrO_2	Pd/Al_2O_3-BaO
Initial catalyst		0.1		
[CO + NO](1)	1	0.1-[1]	0.09-[1]	0.49-[1]
[CO + NO + C_3H_6](2)	0.1	0.4-[0.075]	0.17-[0.1]	1.03-[0.675]
[CO + NO + C_3H_8](3)	0.1	0.2-[0.5]	0.11-[0.875]	4.42[a]-[0.58]
(2) + O_2(573 K) then (1)	1	~ - [0.825]		

[a] after 6 h reaction at 773 K

This inhibition effect is lower on Pd/Al_2O_3-BaO and Pd/ZrO_2 than on Pd/Al_2O_3 showing the influence of the nature of the support, i.e., the influence of the acidity [9]. It is also lower on the Pt-Rh/Al_2O_3 solid, phenomenon which shows the influence of the nature of the metal.

Nevertheless, taking account of the carbon amounts measured by chemical analysis on Pd/Al_2O_3-BaO, a strong inhibition would be expected, which is not observed. However, on these samples, the C deposits are explained by the formation of carbonate species as evidenced by IR measurements [9].

3.2. Influence of the addition of hydrocarbons on the reduction of NO by CO in the presence of O_2

With O_2 in the reagent mixture, two main reactions occur almost simultaneously:

- the direct oxidation of CO by O_2: $CO + 1/2O_2 \rightarrow CO_2$
- the reduction of NO by CO: $CO + NO \rightarrow CO_2 + 1/2N_2$

The first reaction is followed by the O_2 consumption while the second is directly measured from the NO and N_2 analysis.

Figures 2a to 2c give the NO conversion in the absence and in the presence of the hydrocarbon (C_3H_6 or C_3H_8) for a quasi stoichiometric mixture on Pd/Al_2O_3, Pd/ZrO_2 and Pt-Rh/Al_2O_3.

In the presence of O_2 (s~1.03 calculated from the assumption that the hydrocarbon is completely oxidized) the reduction of NO into N_2 is decreased by the hydrocarbon addition, and the inhibition is more marked with propane than with propene. This inhibition is mainly attributed to O_2 which is present in large excess at temperatures for which the oxidation of hydrocarbons (C_3H_6 or C_3H_8) does not occur. According to this explanation, it may be expected that the excess of O_2 and therefore the importance of the inhibition would be greater with C_3H_8 than with C_3H_6. Moreover, the extent of the O_2 inhibition is expected to be more important if the gap between the starting temperatures for carbon monoxide and hydrocarbons oxidations is larger.

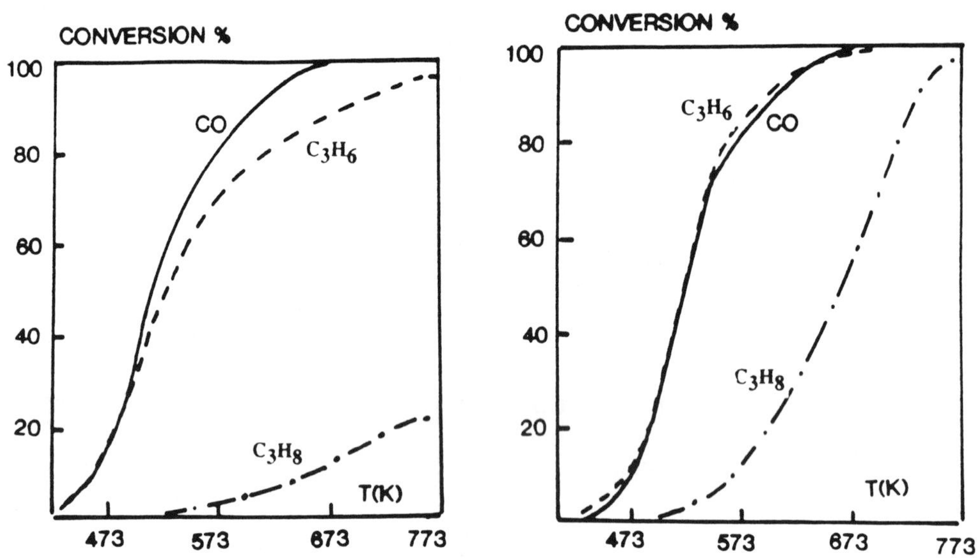

3a) Pd/Al_2O_3 *3b) $Pt\text{-}Rh/Al_2O_3$*

Figure 3. Oxidations of CO, C_3H_6 and C_3H_8 by O_2 in the absence of NO as a function of the reaction temperature (s ratio = 1).

Actually, with Pd/Al_2O_3, $Pd/Al_2O_3\text{-}La_2O_3$ and Pd/ZrO_2, the oxidation temperatures T of the reducing agents by O_2 follow the sequence: $TCO < TC_3H_6 << TC_3H_8$, as seen on the figures 3a and 3b. The difference between the light-off temperatures for CO and C_3H_6 varies between 20 and 40 K. Therefore, the more efficient reducting agent for NO, i.e.., CO, may be partly or completely oxidized whereas the hydrocarbon oxidation has not yet occurred, and the excess of O_2 is important. On the contrary, with Pd/Al_2O_3-BaO and Pt-Rh/Al_2O_3, the sequence becomes $TC_3H_6 \leq TCO << TC_3H_8$, the difference between the light-off temperatures for C_3H_6 and CO reaching 30 K with the last solid. The NO reduction must be therefore less inhibited for these solids, at least with C_3H_6, as observed (comparison of figures 2a and 2c).

The order of oxidation by O_2 (TCO, $TC_3H_6 << TC_3H_8$) qualitatively agrees with the stronger inhibiting effect observed with C_3H_8 than with C_3H_6. Nevertheless taking account of the small differences between the oxidations of CO and C_3H_6 by O_2, the inhibition by C_3H_6 would be very small. However, these oxidations by O_2 have been carried out in the absence of NO which is known to poison the oxidations [11]. This poisoning effect could be more or less strong according to the reagent. The oxidation of C_3H_6, which is a large molecule, could be more disturbed by NO than the oxidation of CO.

It may be added that a specific behaviour is observed with Pd/ZrO_2, the inhibition by C_3H_8 being very weak in the low temperature range (Fig. 2b). This peculiar evolution is explained by the occurrence of the CO oxidation by NO at lower temperatures than the COoxidation by O_2.

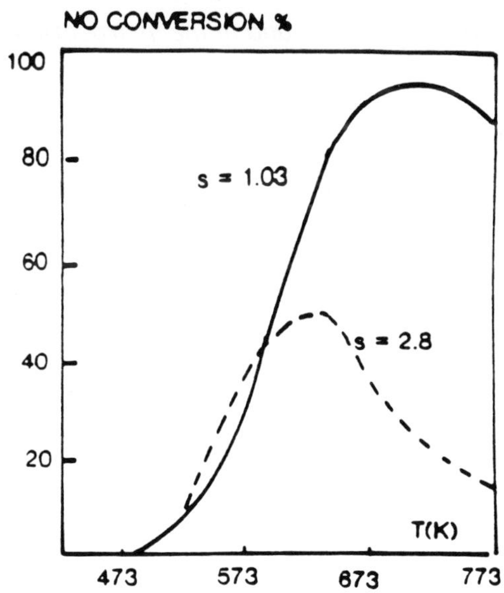

Figure 4. Influence of an O_2 excess in the absence of hydrocarbon in the NO conversion as function of the temperature, for the Pd/ZrO_2 solid. NO-CO-O_2 mixtures s = 1.03 and s = 2.8

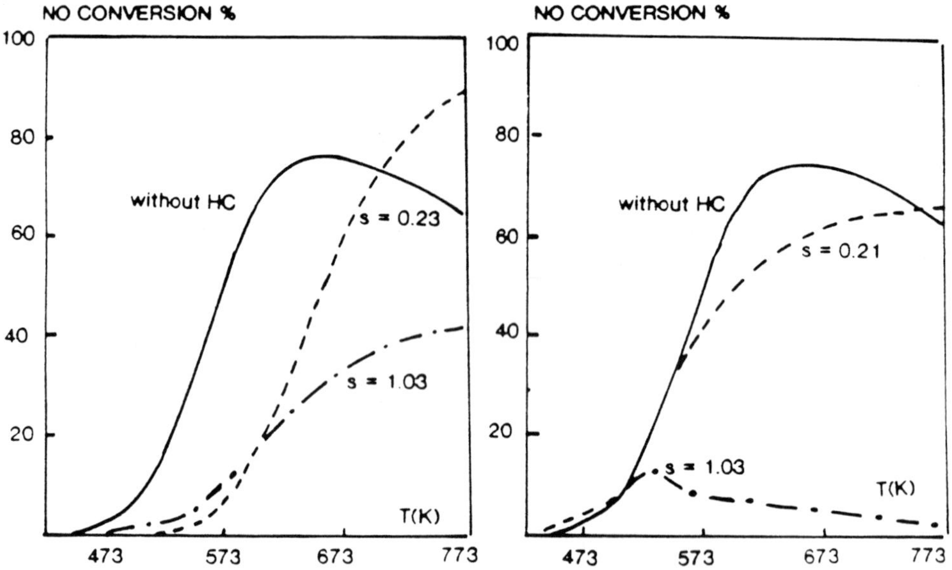

5a) influence of C_3H_6 *5b) influence of* C_3H_8

Figure 5. Influence of the O_2 *amount and of the hydrocarbon introduction on the NO conversion into* N_2 *as a function of the temperature for the* Pd/Al_2O_3 *solid. Flow rate 17.4* lh^{-1}*, 5930 vpm CO, 788 vpm NO, 2672 vpm* O_2 *without hydrocarbon,* O_2 *to obtain s = 0.21-0.23 or s = 1.03 in the presence of hydrocarbons.*

It is noteworthy that the inhibition by an excess of oxygen is clearly evidenced when the O_2 concentration is increased in the CO-NO-O_2 base mixture, as seen in the figure 4 in the case if Pd/ZrO_2. The NO conversion is thus similar (figure 2b) to that observed with the CO-NO-O_2-C_3H_8 mixtures, apparently "quasi-stoichiometric" (s ~ 1.03) and obtained adding O_2 in order to balance the hydrocarbon assumed to be completely oxidized with the mixture CO-NO-O_2-C_3H_8 (s = 1.03).

In the presence of moderate amounts of O_2 leading to strongly reducing mixtures (s~0.23), the effect of the hydrocarbons addition is expected to be weakened. Indeed, the inhibition effect attributed to the O_2 excess in the quasi stoichiometric mixture (s~1.03) is partly ruled out whereas the carbon deposits observed in the absence of O_2 which would be partly oxidized.

The influence of the O_2 amounts in relation with the nature of the hydrocarbon is illustrated on figure 5 in the case of Pd/Al_2O_3. Keeping in mind the importance of the inhibition by the two hydrocarbons in the absence of O_2 (Fig. 1a) and in the presence of an excess of O_2 (Fig. 2a) it can be concluded that a moderate amount of O_2 is beneficial for the NO reduction in the presence of hydrocarbons.

From all these results concerning the influence of each compound added separately to the "standard" CO-NO or CO-NO-O_2 mixtures, a sequence for the NO reduction activity of the different catalysts can be given:

$$Pd/Al_2O_3 < Pd/Al_2O_3\text{-}La_2O_3 < Pd/ZrO_2 \leq Pd/Al_2O_3\text{-}BaO \leq PtRh/Al_2O_3.$$

When the compounds are added simultaneously the NO reduction follows the order suggested above, as seen on figure 6.

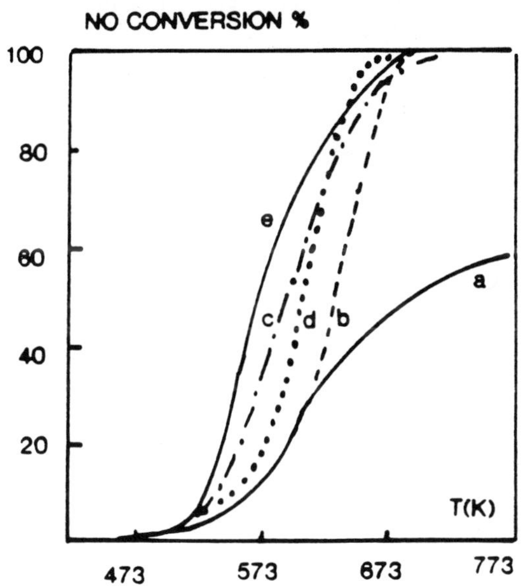

Figure 6. Conversion of NO into N_2 for the various solids in the presence of a complex mixture: NO, CO, O_2, CO_2, C_3H_6, H_2O.
Flow rate 22 lh^{-1}, 10 vol. % CO_2, 10% H_2O; s = 1.03

a) Pd/Al_2O_3
b) Pd/Al_2O_3-La_2O_3
c) Pd/BaO-Al_2O_3
d) Pd/ZrO_2
e) Pt-Rh/Al_2O_3

4. CONCLUSION

The influence of CO_2 and H_2O on the reaction rates is negligible or weak. The inhibition by the hydrocarbons depends both on the nature of the hydrocarbon and on the O_2 content. For low O_2 contents and especially with propene it is explained by the formation of carbon deposits. For high O_2 contents and especially with propane it is explained by the excess of unreacted O_2. The best catalysts would be those which are able to limit the formation of carbon deposits, to oxidize CO and the hydrocarbons in the same temperature range or to oxidize CO by NO more easily than by O_2. With the complex mixtures (CO, NO, O_2, CO_2, H_2O, hydrocarbons), the best catalysts, Pd on zirconia and Pd on Ba-modified alumina, which obey on these requirements, have an activity nearly comparable to that of the reference Pt-Rh/Al_2O_3 solid.

5. ACKNOWLEDGEMENTS

This work has been carried out with the financial support of the "Groupement de Recherches Catalyseurs d'épuration des gaz d'échappement automobile" funded by the "Centre National de la Recherche Scientifique", the "Institut Français du Pétrole" and the PIRSEM (Programme Interdisciplinaire de Recherches Scientifiques pour l'Energie et les matières premières).

REFERENCES

1 K.C. Taylor and J.C. Schlatter, J. Catal., 63 (1980) 53.
2 J.C. Summers, W.B. Williamson and M.G. Henk, SAE Technical Paper Series, 880281 (1988).
3 J.C. Summers, J.J. White and W.B. Williamson, SAE Technical Paper Series, 890794 (1989).
4 J.C. Summers, J.F. Skowron, W.B. Williamson and K.I. Mitchell, SAE Technical Paper Series, 920558 (1992).
5 J. Muraki, K. Yokota and Y. Fujitani, Appl. Catal., 48 (1989) 93.
6 J.L.Duplan and H. Praliaud in "Studies in Surface Science and Catalysis", A. Crucq et al., Eds., Elsevier, 71 (1991) 667.
7 R.G. Silver, J.C. Summers and W.B. Williamson, in "Studies in Surface Sciences and Catalysis" A. Crucq et al., Eds., Elsevier, 71 (1991) 167.
8 A. El Hamdaoui, J. Massardier, G. Bergeret and A. Renouprez in "Proceedings of the 10th Intern. Congress on Catalysis", L. Guczi et al., Eds., Elsevier, Part C (1993) p.2709
9 A. Lemaire, H. Praliaud and J. Massardier, unpublished results.
10 J. Barbier, Appl. Catal., 23 (1986) 225.
11 S.H. Oh and J.E. Carpenter, J. Catal., 101 (1986) 114.

A. Frennet and J.-M. Bastin (Eds.)
Catalysis and Automotive Pollution Control III
Studies in Surface Science and Catalysis, Vol. 96
1995 Elsevier Science B.V.

XPS/TPR STUDY OF THE REDUCIBILITY OF M/CeO_2 CATALYSTS (M=Pt, Rh): DOES JUNCTION EFFECT THEORY APPLY?

J.P. Holgado and G. Munuera.

Dpto. de Química Inorgánica & Instituto de Ciencia de Materiales.
Univ de Sevilla–CSIC
P.O Box 553 c/ Prof. García González s/n 41071–SEVILLA (Spain).

ABSTRACT

The study of the reduction of a set of high surface area M/CeO_2 catalysts (M= Rh, Pt and Au, S_{BET} ca. 90 m^2g^{-1}) prepared from several precursors, either containing or not chlorine, has been carried out using TPR/MS and XPS/Ar^+–etching techniques combined with volumetric adsorption of H_2. Reduction up to 773K leads in all the samples to formation of the stable $CeO_{1.82}$ phase of ceria. Incorporation of hydrogen occurs in the oxygen vacancies present in this phase competing with chlorine when it is present in the samples. The results are discussed in relation to the "junction effect" theory proposed by Frost [1] for metal–oxide catalysts concluding that it does not apply to this type of catalysts due to high ionic mobility of the oxygen in ceria and the possibility of a levelling of the Schottky metal–oxide barriers due to hydrogen incorporation at the metal–ceria interface.

1.INTRODUCTION

The nature of the interaction of noble metals (e.g. Rh, Pt, Pd etc) with CeO_2 has been examined in a great detail in most recent years [2-4] owing to the importance that this oxide has reached (as promoter) in the formulation of "Three Way Catalysts" (TWC). The role of CeO_2 as an "oxygen reservoir" is assumed to be directly related to its reducibility to form non–stoichiometric CeO_{2-x} phases under reach conditions which are readily reoxidized under lean (net oxidizing) conditions.

In principle, these redox processes should be determined by "junction effects" as recently proposed by Frost [1] for Cu/ZnO and other M/MO systems used in methanol synthesis from syngas. In fact, this author proposes that the mode of action of such catalysts involve metal promotion of the oxide productivity by a dramatic enhancement (10^3) of the equilibrium concentration of ionized oxygen vacancies (V_0^{++}) at the surface. The origin of this fact is the large contribution to the enthalpy of formation of a V_0^{++} in an oxide of the energy required to rise the two electrons to the oxide conduction band. So, a metal which forms a "Schottky" junction with a given oxide must produce more active catalysts because in such cases the conduction band edge lies higher than the metal Fermi level and therefore the enthalpy of formation of a doubly ionized vacancy should be reduced by allowing electrons to move to the metal Fermi level.

In principle, Frost suggests that the high activity observed by Lambert et al [5]with a Cu/CeO_2 catalyst in the synthesis of methanol is due to this metal–oxide "junction effect". However, these authors have tried to prove unsuccessfully this relationship using Ag/CeO_2 [6] and Au/CeO_2 [7]catalysts where the "junction theory" predicts that the reactivity should be even higher than for Cu/CeO_2.

The reducibility of metal unloaded CeO_2 has been studied recently in detail by several authors[8,9]. Thus, Mooi at al [10] have found that, in an initial stage (up to ca. 773K), only surface reduction seems to occur on the basis of the observed relationship between the TPR hydrogen consumption and the BET surface area of different CeO_2 samples. This agrees with recent Catlow et al [11] calculations for the energy of formation of oxygen vacancy (V_0^{++}) in CeO_2 using atomistic simulation techniques which indicate that they are more readily formed either at the 111 or 110 planes at the surface than at the bulk of CeO_2 a condition that is also implicit in the Frost's model to account for its dramatic increase at the oxide surface.

On the basis of the "junction effect" theory it should be expected that Rh and Pt, having work functions (4.8 and 5.7 eV) higher than Cu (4.65), would lead to a deeper reduction of the CeO_2 when they are supported on this oxide if similar reduction conditions are used, particularly in the case of Pt. So, with this goal in mind, we have examined both Rh/CeO_2 and Pt/CeO_2 systems which, on the other hand, are of major importance in the TWC formulation. Some of the results of such study are presented in this paper.

2. EXPERIMENTAL

2.1 Materials

Rh/CeO_2 and Pt/CeO_2 catalyst precursors were prepared by incipient wetness impregnation of a high surface area CeO_2 (Rhône–Poulenc, 120 m^2/g) using aqueous solutions of either $Rh(NO_3)_3$ and $[Pt(NH_3)_4](NO_3)_2$ or $RhCl_3$ and

H_2Cl_6Pt to get non–chlorinated and chlorinated materials (labelled hereafter as N and Cl–samples) with metal loading of 3% Rh and 6% Pt (e.g. 300 μmol/g or ca. 1.5 atoms/nm^2). The impregnated precursors were dried in air at 423K for 24 h. and then calcined at 673K in a 100 ml min^{-1} dry air flow for 4 h. and finally stored in a desiccator until their use. BET surface areas were 91±2 m^2g^{-1} for all four samples.

Gases (H_2, O_2, Ar, He etc) with a purity better than 99.999, were supplied by S.E.O. In all reduction experiments the gases were passed through deoxo units to remove traces of oxygen.

2.2 Methods

TPR (300K–1200K) of the four precursors were carried out, after outgassing the samples at 673K for 1 h., in a TCD–system connected to an HP–3054–DL unit for data storage and processing. The purified (deoxo) carrier gas stream (H_2 5% in Ar) was passed at 50 ml min^{-1} through ca. 0.15 g of sample at 300K until base line equilibrium was reached (ca. 30 min) and then heated using a constant heating rate of 10 $Kmin^{-1}$ until a temperature of 1200K was reached remaining at this temperature for 30 min. The evolved gases were pased through a trap with glass beds at 200K, to condense water, before reaching the TC–detector and then analysed either by GC/MS. Hydrogen consumption was calibrated using CuO (99.999) as previously described [12]. Several reoxidation cycles were carried out using the same sample by flowing 100 ml min^{-1} O_2/He (20%) on the sample heated at 473K for 2 h. after each TPR run up to 773K.

Hydrogen isotherms at 300K were measured up to 100 torr on all four samples after reduction at 573K for 1 h. under a flow of 300 ml g^{-1} min^{-1} of pure H_2 followed by outgassing up to 10^{-6} torr at least 8 h. either at 300K or 573K to remove the hydrogen incorporated to the CeO_2. [13] Equilibrium at each point was reached very slowly, so, the data were recorded after ca. 12 h in contact with H_2, once the pressure, recorded with a MKS–Baratron manometer, remained constant for ca. 30 min.

XPS spectra were recorded in an LHS–10 spectrometer (Leybold) working in the pass energy constant mode at 50 eV and using the Mg Kα line as excitation source. B.E. reference was taken at the lowest K.E. Ce(3d) peak at 917.0 eV. In this conditions spurious carbon C(1s) peak in the original sample appears at 284.6 eV. The samples, as pressed pellets, were placed on a Mo holder that could be heated resistively while controlling the temperature by a thermocouple spot–welded at the rear of the holder. All the treatments were carried out in the pretreatment chamber of the spectrometer (base pressure < 10^{-8} torr) while Ar^+ sputtering was carried out with a penning source working at 3.5 Kv. Calibration using a Ta_2O_5/Ta foil leads to a sputtering rate of 1.2 nm/min in these conditions.

Reduction treatments in the XPS preparation chamber consisted of heating at 773K for 1 h of the calcined precursors, first under vacuum, to remove most of the contaminants adsorbed on the samples (H_2O, CO_2, etc), and then under 10 torr of H_2 for another hour at the appropriate temperatures in the range 300K–773K. The spectra were directly recorded and stored in a HP– Vectra 386 computer on line with the spectrometer where they were submitted to background subtraction (Shirley base line) and Factor Analysis to calculate % of Ce^{3+} in the reduced samples as will be described elsewher[14].

3.RESULTS

3.1 TPR of Rh and Pt loaded CeO_2.

TPR profiles up to 1200K of the four samples studied in this work have been plotted in Fig. 1 together with the profile for the unloaded CeO_2 support. Several points are worthy of note

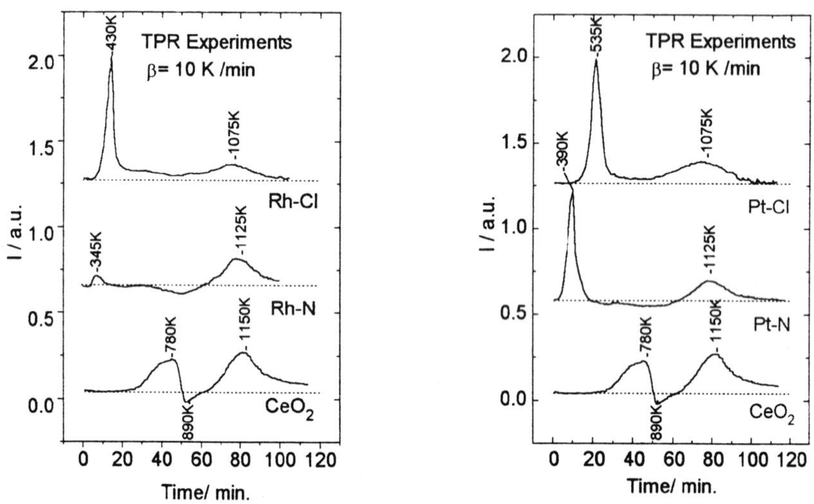

Fig.1. TPR of M/CeO_2 Catalysts (N and Cl):Right: M=Rh; Left: M=Pt. TPR of CeO_2 included for comparison

from this figure. First, the low temperature peak in the CeO_2, at ca. 780K, ascribed by Yao et al. and several other authors [8,9] to surface reduction, is greatly modified by the presence of the metals appearing now as a sharp peak except for Rh/CeO_2–N (see below) always at temperatures lower than ca. 573K. This shift to lower temperatures was particularly important in the N–samples.

Secondly, the negative peak, due to desorption of hydrogen incorporated into the reduced CeO_2 [15] was not observed in the Cl–samples though still is present in the N–samples. Meanwhile, the higher temperature TPR peak at ca. 1150K, normally ascribed to bulk reduction of CeO_2, remains in all four samples though it is now broader and displaced towards lower temperatures, particularly in the Cl–samples. After a first TPR run up to 773K followed by reoxidation at 473K, new TPR runs show completely flat profiles for the two N–samples. As shown for Rh/CeO_2–N in Fig. 2, this is due to a shift of the reduction peak to temperatures lower or near 300K. A similar shift was observed in the Cl–samples though in these case the peaks still appear in the TPR in the range 373–473K.

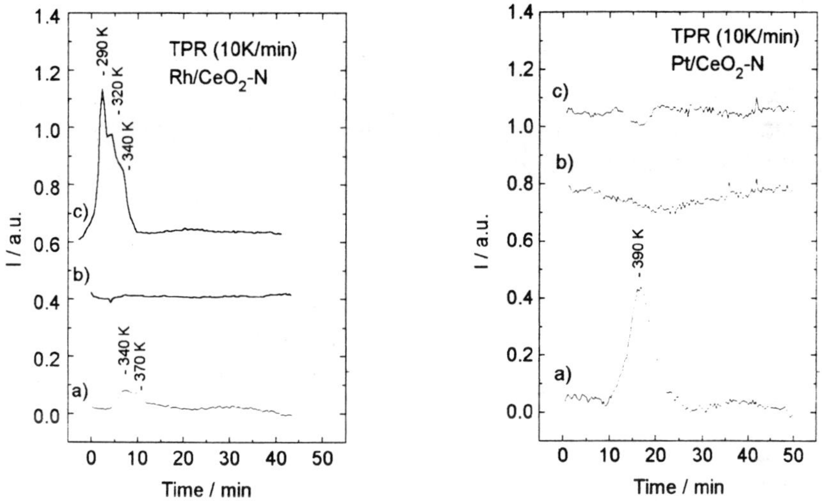

Fig.2. successive TPR runs up to 773K followed by reoxidations at 473K: a)first TPR; B) second TPR; c) third TPR. Left: Rh/CeO_2-N sample. (run c) started at ca. 200K). Right: Pt/ CeO_2-N sample.

In the N–samples Hydrogen consumptions, in μmol H_2/g of sample, calculated from the TPR profiles up to 950K in the N–samples and up to 773K in the Cl–samples (thus, corresponding to the first peak minus the small negative peak due to hydrogen desorption) have been included in Table 1 (together with those for a Au/CeO_2–Cl sample prepared on the same CeO_2 support). When they are calculated as H/M ratio (M=Rh, Pt or Au), values in the range 6–9 were obtained clearly indicating that part of this hydrogen should be consumed in the reduction of the CeO_2.

Data from XPS in Fig. 3 for the two N–samples show that, after outgassing the calcined precursors at 673K, the noble metals are mainly as Rh^+/Rh^0 and Pt^{2+}

but H_2 at 573K leads to a complete reduction in these two samples while the Ce(3d) spectra also show an important reduction of the CeO_2 support. Reoxidation at 473K after a reduction up to 773K restores the Rh^+ /Rh^0 and Pt^{2+} oxidation state of the metals while all Ce^{3+} completely disappears.

Similar results were obtained for the two Cl–samples though in this case the reoxidation leaves some Ce^{3+} in the samples, while the Cl(2p) peak after the reduction become sharper and even more intense than in the original Cl–precursors (See below). MS–analyses during the reduction did not show any loss of chlorine from the samples.

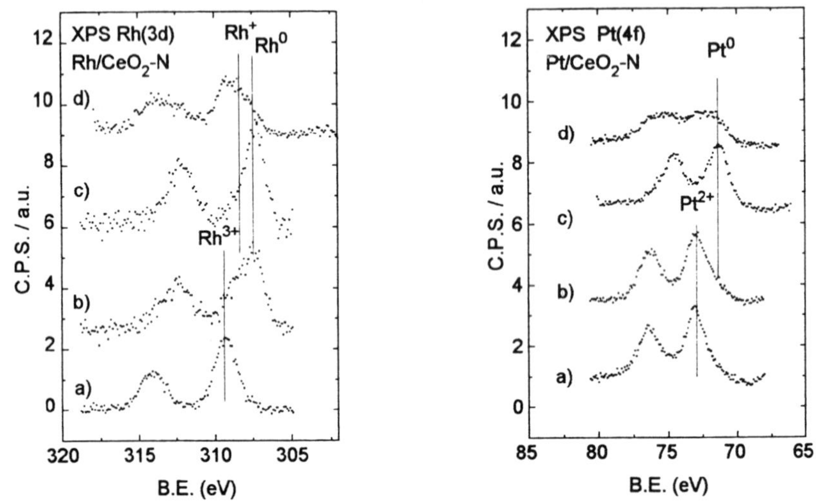

Fig. 3.X'S of the metallic phase in samples M/CeO$_2$-N: a) Original b) outgassed at 673K; c) H$_2$ reduced at 373K; d) reoxided at 473K. *<u>Left</u>:M=Rh. <u>Right</u>: M=Pt*

*Table 1 TPR H_2 consumptions for M/CeO_2 samples up to 950K**

Sample	μmol H_2/ g T ≤950K	Total H/M ratios	Corrected** H/M ratios
Rh/CeO_2–N	1054	7.0	6.5
Pt/CeO_2–N	1237	8.3	6.3
Rh/CeO_2–Cl	1092	7.3	6.8
Pt/CeO_2–Cl	1372	9.1	7.1
Au/CeO_2–Cl	980	6.5	6.5
CeO_2 (support)	450	–	(3.0)

* Up to 773K in Cl–samples

**Considering the metal oxidized up to Rh^+/Rh^0 (1:1), Pt^{2+} and Au^0.

It is noteworthy that, if we subtract from the H/M values in Table 1 the amounts of hydrogen required to reduce Rh^+/Rh^0 or Pt^{2+} (and Au^0) to the metallic state, the last column of this table can be calculated corresponding to hydrogen consumed in the reduction of the ceria support. Finally it is remarkable that, for all the samples, including the unloaded CeO_2 support, the total amount of hydrogen consumed up to 1200K in the TPR was also very similar (ca. 1400±200 μmol H_2/g) corresponding to a reduction of ca. 48±5% Ce^{3+} in all the samples (including the CeO_2 support).

3.2 XPS Study of the Reduction of CeO_2

Since the first reduction peak at ca. 780K in the TPR profiles of high surface area CeO_2 has been previously ascribed by several authors [8-10] to surface reduction, we have examined in further detail by XPS the reduction of all our samples step by step from 300K to 773K. Fig.4 shows the percentages of Ce^{3+}calculated from the Ce(3d) spectra using Factor Analysis for the two N–samples together with an example of the fitting obtained in the case of Rh/CeO_2–N. As shown in this figure for the two N–samples the percentage of reduction first increases up to ca. 38– 40% and then decreases to 35±3% at T>573K. Data obtained for the Cl–samples were very similar giving slightly higher values

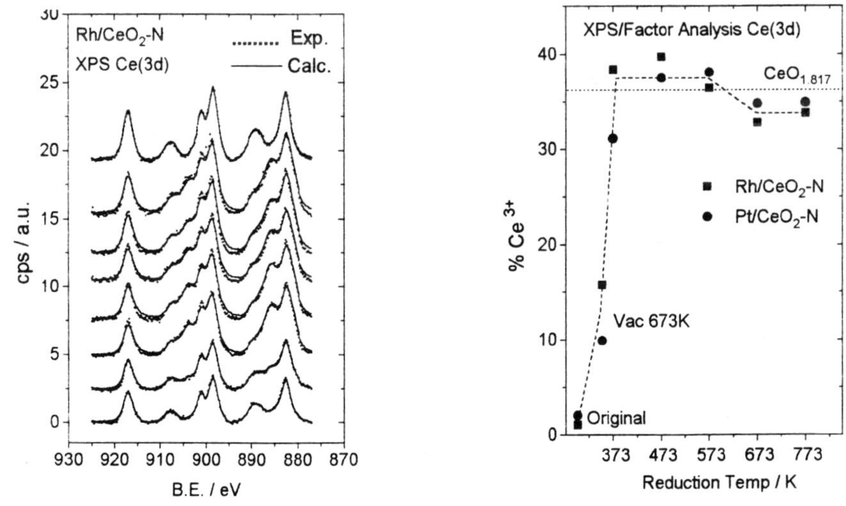

Fig.4.Study by XPS of H_2 reduction up to 773K of M/CeO_2-N (M=Rh,Pt). Left:Example of Factor Analysis fitting of Ce(3d) spectra.Right:Calculated % Ce^{3+} in both samples.

(ca.40±2%), after reduction up to 773K. All these data have been included in Table 2 together with those calculated (as %Ce^{3+}) from the last column in Table 1. As can be seen, both set of data show a remarkable coincidence.

*Table 2 Reduction of M/CeO_2 samples up to 950K**

Sample	% Ce^{3+} (XPS)	% Ce^{3+} (TPR)
Rh/CeO_2–N	33.8	33
Pt/CeO_2–N	34.9	32
Rh/CeO_2–Cl	40.0	34
Pt/CeO_2–Cl	41.4	36
Au/CeO_2–Cl	41.0	33
CeO_2 (support)	23.8	17

* Up to 773K in Cl–samples

3.3 Hydrogen incorporation to CeO_{2-x}.

The results above seem to suggest that reduction up to 773K, leads in all the M/CeO_2 samples to the same final state of the CeO_2 whatever the work function of the noble metals (including Au!, 5.10 eV). To get more information on the incorporation of hydrogen after the first reduction step of the CeO_2 we have

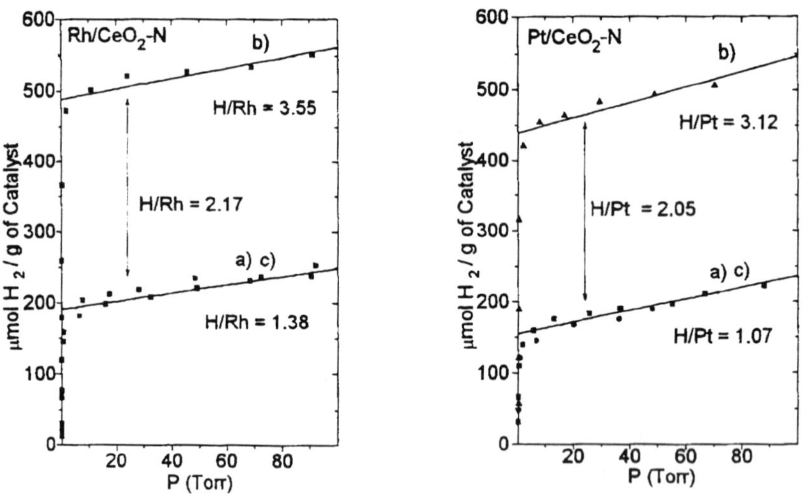

Fig.5. H_2 adsorption at 300K on M/CeO_2-N samples reduced at 573K: a) outgassed at 300K after reduction;b) outgassed at 573K; c) outgassed again at 300K.Left: M=Rh. Right: M=Pt

studied the adsorption of hydrogen at 300K on the two sets of samples (N and Cl) after reduction for 1 h. in flowing H_2 at 573K, just after the first TPR–peak. Isotherms for the two N–samples, shown in Fig. 5, are very similar indicating again that the metals (Rh or Pt) have only a very small influence in the final reduction state of the samples. When the samples were cooled down to 300K after the reduction, and then the excess of hydrogen was carefully pumped out at this temperature, the isotherms give extrapolated values close to H/M = 1.1–1.4 while outgassing at 573K leads to H/M of ca. 3.1–3.5 clearly indicating that at least part of the hydrogen removed at 573K should be reincorporated to the reduced CeO_2. Similar set of experiments using the Cl–samples (Rh and Pt) give isotherms that always extrapolate to H/M=0.9±0.1 while in the case of the Au/CeO_2–Cl sample H_2 adsorption was negligible in the same conditions. These results again suggest that most of the "excess" of hydrogen incorporated in samples N outgassed at 573K (ca. H/M =2.1–2.2) belongs to the reduced CeO_2 support and that chlorine seems to prevent the incorporation of this type of hydrogen to the ceria.

3.4 Incorporation of Chlorine to the CeO_2.

Since the TPR profiles in Fig. 2 suggest that a stable steady state is reached after the second redox cycle we have examined, using XPS combined with Ar^+ etching, the profiles for Chlorine in the CeO_2 support after the set of three

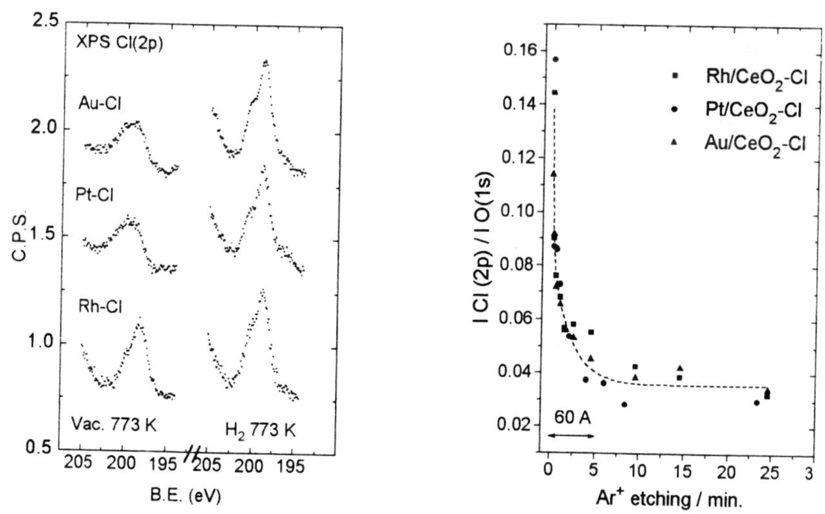

Fig.6. Evolution of XPS Cl(2p) signal in M/CeO_2-Cl samples. *Left: Cl(2p) before and after H_2 reduction. Right: XPS/Ar^+ -etching profile analysis for Cl(2p).*

TPR/reoxidation runs shown in that figure. Fig. 6 shows the Cl(2p) peaks for all three M/CeO_2–Cl samples before and after reduction at 773K. We can see that after reduction the intensity of the peak increases in all the samples and becomes much better defined. Meantime, the plot of the Cl(2p)/O(1s) ratios indicate a rather similar distribution of the Cl in all the samples.

4. DISCUSSION

Contrary to what expected from Frost's "junction effect" theory, the reduction reached up to 773K (first TPR–peak) in TPR and XPS experiments (Table 2) are clearly coincident and close to ca. 36±4% for all our M/CeO_2 samples leading to one stoichiometry close to $CeO_{1.82}$ for the reduced support in all of them. This level of reduction attained in all our M/CeO_2 samples agrees with that calculated from Mooi's model for samples with ca. 90–110 m^2g^{-1} BET surface assuming that 50% of the oxygen at the surface are removed, except just in the case of the unloaded CeO_2 support where it is about half that amount. This seems to indicate a more difficult reduction of the ceria surface in the absence of metals. However, the coincidence of the results obtained independently with TPR and XPS in our M/CeO_2 samples allows us to conclude that the V_0^{++} (and Ce^{3+}) initially created at the surface of the small ceria particles in the reduction process must be homogeneously distributed in the whole particle of ceria and not exclusively at its surface as previously has been suggested by different authors [8,9] and, in particular, is assumed in Mooi's reduction model [10]and also stated, as a premise, in the Frost's model [1]for "Junction effects". In fact, a simple calculation using the value for the mean free path, for the Ce(3d) electrons (λ= ca. 1.5 nm) indicates that if the 9% V_0^{++} (e.g. 36% Ce^{3+}) obtained for the first peak reduction in TPR were at the upmost surface layer, a percentage of Ce^{3+} >50% would be observed by XPS. So, we must assume that during the reduction with hydrogen up to 773K diffusion of V_0^{++} (and Ce^{3+}) from the surface towards the bulk of the CeO_2 readily occurs thus leading to such homogeneous distribution. In fact we can see in Fig. 4 that the %Ce^{3+} first sharply increases with the reduction temperature up to ca. 38±2% and then drop to a steady level at ca. 35% at T>573K probably due to an increase of the rate of V_0^{++} diffusion at these higher temperatures. In the case of the Cl– samples the final reduction level from XPS seems to be somewhat higher (ca. 40%) what can be in part ascribed to incorporation of the Cl into V_0^{++}, thus preventing diffusion. It is worthy of note that, in the case of the CeO_2 the reduction at 773K calculated from XPS data is somewhat higher than that measured from TPR, what may be due to a slower rate of diffusion of the generated V_0^{++} towards the bulk in this case. This difference agrees with our previous observations that hydrogen incorporated as hydride–like species into partially reduced TiO_2 by "spillover" from the metal during the reduction of Rh/TiO_2 and Pt/TiO_2 catalysts greatly enhances TiO_x mobility in the TiO_2 support [16,17].

In principle, our results cast some doubts on Mooi's (and others) assumption that the origin of the first TPR peak is the reduction of the upmost surface layer of the small particles of ceria, at least in these high surface area samples. In fact the Ce–O phase diagram in Fig.7 shows that the $CeO_{1.82}$ composition attained in all our samples corresponds to one of the well known[18] defective stable ceria phases generated by "clustering" of oxygen vacancies in this oxide (Ce_nO_{2n-2}, n=11) thus forming a superstructure in which the oxygen vacancies associated to Ce^{3+} (ca. 9% V_0^{++}) interact giving "pairs" in the [111] direction of the fluorite structure of ceria. As shown also in the same figure, reduction up to this defective phase could be expected to occur at 773K on the basis of free energy calculations for the CeO_2–H_2–H_2O system for ours P_{H2O}/P_{H2} << 1 conditions.

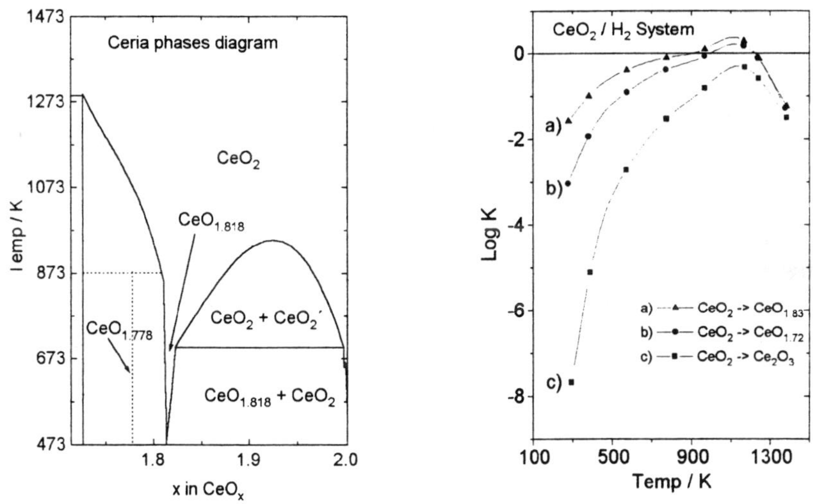

Fig.7. Some thermodynamic data for CeO_2. Left: Phase diagram for Ce-O. Right: Equilibrium conditions for CeO_2 reduction

It is noteworthy that this defective phase is formed in all our M/CeO_2 samples whatever the metallic phase used (Rh, Pt or even Au) and either if chlorine is present or not, thus suggesting a thermodynamic control of the whole reduction process. Moreover, reduction up to 1200K leads in all our samples, including now also the CeO_2 support, to a stoichiometry close to $CeO_{1.78}$ that, according to the data shown in the same figure, corresponds to a new phase of the same homologous series (with n=9 , V_0^{++} = 11%).

Some differences are observed of the behaviour of the generated $CeO_{1.82}$ phase at in the two sets of samples (N, Cl). In principle, the "Cl–profiles" in Fig. 6 for all the Cl–samples are very similar and indicate that the Cl is concentrated at the surface while the amount of hydrogen incorporated into the ceria (as hydride–like species[19]) is very different in both set of samples. Data in Fig.5 suggest that hydrogen is incorporated to the $CeO_{1.82}$, into the V_0^{++} existing throughout the whole structure of the defective phase. The amount calculated from this figure (using H/M ca. 2.1) corresponding to the filling of ca. 80% of these vacancies with hydrogen which is only removed at T>573K. However, incorporation of H_2 does not occur when Cl is present at the ceria support indicating that chlorine at the surface of these samples is able to prevent the access of this hydrogen into the defective $CeO_{1.82}$ structure. This is probably due to the incorporation of the Cl itself to the V_0^{++} thus stabilizing Ce^{3+} species as could be detected by XPS after reoxidation and is also suggested by the sharpening of the Cl(2p) signal. Moreover, the higher reduction observed by XPS in these Cl–samples should be ascribed to this fact, which might prevent V_O^{++} diffusion.

Let us now ask why "junction effects" do not seem to control the reduction of M/CeO_2 and, probably, methanol synthesis on these catalysts? In our view two points should be considered to answer this question. First, Frost in his model makes the assumption that "vacancies are concentrated at the surface of the thin oxide overlayers" (i.e layers of the 0.5–5 nm thickness). Clearly, this is not the case in our samples. Though V_0^{++} (and Ce^{3+}) should be initially created at the surface, where according to recent calculations by Catlow et al [11] they are more easily formed (particularly in the less stable 110 planes) they readily diffuse at T > 573K to the bulk (while lattice oxygen ($O^{=}$)moves toward the surface as could be expected for the well known behaviour of ceria as an ionic conductor) thus decreasing the surface concentration of V_0^{++} probably by one or two order of magnitude. At least for our high surface area ceria, formation of the non–stoichiometric stable Ce_nO_{2n-2} phases seems to act as the driving force again surface segregation of the oxygen vacancies.

Second, during hydrogen reduction as well as during the synthesis of methanol when, besides CO/CO_2, an excess of hydrogen is in contact with the catalysts, we cannot discard that the generated $M–CeO_2$ "Shottky" junctions were reduced to the same value, determined by the hydrogen work function at the $M–CeO_2$ interface and by the electron affinity of the ceria support, thus leading to a similar final reduction state of the ceria for all metals (as actually observed in our samples). This fact has been observed by Aspnes et al [20] in Rh/TiO_2, Pt/TiO_2 (and several others) metal–semiconductor pairs and by ourselves in Rh/TiO_2, and Pt/TiO_2 catalysts [21] where EPR shows that a complete change of contact character from rectifying (Schottky) to Ohmic occurs upon exposure to hydrogen at 300K as detected by generation of Ti^{3+} species that readily disappear when hydrogen is removed from the gas phase. Though hydrogen doping of the TiO_2 [22] (and CeO_2) can not be ruled out as the origin of this ambient gas–induced change, Aspnes et al [20] conclude from electrical measurements that the general mechanism, that bears on all contacts studied, is the change in the surface dipole component of the metal work function upon changing the ambient gas.

So, we must conclude, in agreement with Lambert et al [7], that "the most natural explanation for the high activity in methanol synthesis of their Cu/CeO_2 catalyst, and the lack of activity of Ag/CeO_2 and Au/CeO_2, is that active sites involving copper species are responsible for the reaction" while the crucial role for ZnO, ThO_2, or CeO_2 should be to raise the effective partial pressure of hydrogen on the catalysts under reaction conditions as suggested by Burch et al. [23]and as we had also previously reported for Rh/TiO_2 catalysts [24] where we have shown that hydrogen, incorporated as hydride–like species into V_0^{++}, acts as a "reservoir" for suppling hydrogen in the reaction to produce methanol.

5. ACKNOWLEDGMENT

We thank the CICYT (Project MAT91–1080–CO3–01) and CEE (Project SC1*–CT91– 0704(TSTS)) for financial support

REFERENCES

1 J.C. Frost, Nature,334 (1988) 557
2 B. Harrison, A.F. Diwell and C. Hallett, Platinum Metals Rev., 32 (1988) 73
3 A. Laachir, V. Perrichon, A. Badri, J. Lamotte, E. Catherine, J.C. Lavaley, J. El Fallah, L. Hillarie, F. Le Normand, E. Quemere, G.N. Sauvion and O. Touret, JCS Faraday Trans., 87 (1991) 1601.
4 A.Trivarelli, G. Dolcetti, C. Leitenburg, J. Kaspar, P. Dinetti and A. Santoni, JCS Faraday Trans. 88 (1992) 1311.
5 R.M. Nix, T. Rayment, R.M. Lambert, J.R. Jennings and G.Owen, J.Catal.,106 (1987) 216.
6 E.A. Shaw, T. Rayment, A.Walker, R.M. Lambert and J.R. Jennings, J.Catal., 126 (1990) 219.

7 E.A. Shaw, A.P. Walker, T. Rayment and R.M. Lambert, J. Catal., 134 (1992) 747
8 H.C. Yao and Y.F. Yao, J. Catal. 86 (1984),254.
9 J. Soria, A. Martinez–Arias, J.C. Conesa, G. Munuera and A.R.Gonzalez–Elipe, Surf. Sci. 251/252 (1991) 990.
10 M.F.L. Johnson and J. Mooi, J. Catal. 103,(1987),502; ibid 140 (1993) 612
11 T.X.T. Sayle, S.C. Parker, and C.R.A. Catlow, JCS Chem. Comm. (1992) 977.
12 P. Malet, A. Caballero, J. Chem. Soc., Faraday Trans. 1, 84 (1988) 2369.
13 J.M. Rojo, J. Sanz, J. Soria and J.L.G. Fierro, Z. Phys. Chem (NF) 152 (1987) 149.
14 J.P. Holgado , R. Alvarez and G. Munuera (to be published)
15 L. Tournayan, N.R. Marcilio and R. Frety Appl. Catal. 78 (1991) 31.
16 G.Munuera, A.R.González–Elipe, J.P.Espinós, J.C.Conesa, J.Soria and J.Sanz, J.Phys.Chem.,_91 (1987) 6625.
17 A.R.González–Elipe, P.Malet, J.P.Espinós, A.Caballero and G.Munuera, Stud.Surf.Sci. and Catal. 30 (1989) 427.
18 Chapman & Hall (eds.) Dictionary of Inorganic Compounds , London 1984 (and references therein)
19 S. Bernal, J.J. Calvino, G.A. Cifredo, J.M. Gatica, J.A. Perez–Omil and J.M. Pintado, JCS Faraday Trans. 89 (1993) 3499.
20 D.E. Aspnes and A. Heller, J. Phys. Chem. 87 (1883),4919
21 G.Munuera, A.R.González–Elipe, A.Muñoz, A.Fernández, J.Soria, J.C.Conesa and J.Sanz Sensors and Actuators 18 (1989) 337.
22 J.Conesa, J.Soria, J.M.Rojo, J.Sanz and G.Munuera Zeit Phys.Chem. 152 (1987) 83.
23 R. Burch, S.E.Golunski and M.S. Spencer, Catalysis Letters 6 (1990), 152.
24 A.Muñoz, A.R.González–Elipe, G.Munuera, J.P.Espinós and V.Rives–Arnau. Spectrochimica Acta, Part A, (Molecular Espectroscopy) 43 (1987) 1599 .

A. Frennet and J.-M. Bastin (Eds.)
Catalysis and Automotive Pollution Control III
Studies in Surface Science and Catalysis, Vol. 96

ENHANCEMENT OF THE REACTION OF NITRIC OXIDE AND CARBON MONOXIDE BY HYDROGEN AND WATER OVER PLATINUM AND RHODIUM-CONTAINING CATALYSTS

R. Dümpelmann[a], N.W. Cant[a] and D.L. Trimm[b]

[a]*School of Chemistry, Macquarie University, Sydney NSW 2109, Australia*

[b]*School of Chemical Engineering and Industrial Chemistry, University of New South Wales, Kensington NSW 2033, Australia*

ABSTRACT

The addition of hydrogen or water to a stoichiometric mixture of CO and NO over supported platinum and rhodium catalysts can significantly increase the conversion of both NO and CO. Theselectivity of NO reduction is also affected in a remarkable manner.

With Pt/Al_2O_3 at 210°C, for example, addition of ≈900 ppm H_2 increases the NO conversion from 6% to 48% and that of CO from 5% to 33%. The nitrogen atom of NO is converted quite selectively to NH_3 while the oxygen atom forms CO_2 rather than water. Addition of ≈1400 ppm water also enhances both CO and NO conversion over Pt/Al_2O_3 but N_2 rather than NH_3 is the favoured product. Hydrogen and water also produce significant rate enhancements with Rh/Al_2O_3 and $Rh/CeO_2/Al_2O_3$ but the product distributions are somewhat different from those with $Pt/\ Al_2O_3$. Possible reasons for the enhanced activities and the observed selectivities are discussed.

1. INTRODUCTION

There is considerable literature concerning catalysis of the NO + CO and NO + H_2 reactions over Rh and Pt in various forms. A general conclusion is that the latter reaction is substantially faster than the former with Rh [1,2] and especially Pt [1-4] under equivalent conditions. With respect to NO removal, the presence of CO inhibits the NO + H_2 reaction [1-6], see also [7]. Rather surprisingly there appears to be no definitive studies of the mixed NO, CO, H_2 reaction system [7] even though all three gases are simultaneously present in automobile exhaust gases.

There is also little detailed work on the effect of water on the catalysed reaction of CO with NO. One might expect any effects to arise via hydrogen formation through the water gas shift reaction (WGSR) between CO and H_2O. However, appreciable formation of NH_3 has been found with CO, NO and water mixtures under conditions of little or no WGS activity [8] which indicates that water can act in other ways. In this context it may be noted out that current promoters such as ceria are associated with a high WGS activity [9,10] and that water may also act as an reoxidant of reduced ceria [11] with hydrogen evolution [12].

Promotional effects by both hydrogen [13-15] and water [15] have been reported for the oxidation of CO by oxygen. The exact mechanism is unclear.

The present work describes the effect of hydrogen and of water on the NO + CO reaction over Rh and Pt at temperatures typical for the warm-up phase of catalytic converters. The data obtained demonstrate pronounced effects on reaction rates and product distributions which have apparently not been reported both in previous laboratory studies of the binary systems (NO + CO and NO + H_2) and when using simulated exhaust gas mixtures.

2. EXPERIMENTAL

Catalyst: The catalysts were prepared by incipient wetness of powdered Al_2O_3 (Condea alumina washcoat grade, surface area ≈140m^2/g) with aqueous H_2PtCl_6 and $RhCl_3$ solutions to yield nominal contents of 1 wt% Pt and 0.53% Rh (same molar content). The slurries were dried at 50°C under mild vacuum in a rotary evaporator, dried further at 100°C overnight and subsequently calcined for 4 h at 500°C. The ceria containing catalyst was prepared by first impregnating with a solution of $Ce(NO_3)_3$, followed by drying and calcination as above, and then impregnated with the rhodium salt. The resulting powders were pressed, crushed and a sieve fraction of 106-180 mm was used in subsequent experiments. The pretreatment consisted of a temperature programmed reduction (1% H_2/He, 9°C/min up to 500°C) followed by 30 min equilibration with 2000 ppm of CO and NO at 500°C and cooling in CO/NO to reaction temperature.

Reactor: The experiments were performed in a continuous flow tubular reactor (Pyrex, 8 mm OD, 5 mm ID). The catalyst sample was held by a plug of quartz wool on top of a thermocouple. All gases and gas mixtures were of high purity grade and not further purified. Calibrated mass flow controllers assured the desired concentrations.

Analysis: The analyses were performed on-line by infrared spectroscopy (IR) and mass spectrometry (MS). The former method used a dispersive infrared spectrometer (Perkin-Elmer 580B) with a multiple path cell (2.4m total path length) and a control computer. A routine was developed by which the absorbances of CO (2117 cm^{-1}), NO (1877 cm^{-1}), N_2O (1300 cm^{-1}), NH_3 (965 cm^{-1}) and CO_2 (679 cm^{-1}) were repeatedly acquired and stored on a cycle time of ≈ 7 minutes. The mass spectrometer (VG300SX) was used for the detection of H_2 (m/e =2), CO_2 (m/e =22) and NO (m/e =30). Signals at m/e =44 can stem from either N_2O or CO_2. Therefore, carbon dioxide was reliably measured by use of its doubly charged ion at m/e =22, which has a low background signal and is not interfered with N_2O (no signal at m/e = 22 detected). In one instance (Figure 1) CO was also calculated from the MS signal at m/e =12 knowing the fragmentation patterns of CO and CO_2. Formation of N_2 and H_2O were calculated by nitrogen and hydrogen balances. The gases CO, NO and N_2O were calibrated through use of diluted standard mixtures. Known concentrations of CO_2 and NH_3 were obtained by passing CO and excess O_2, or NO and excess H_2, over the catalyst at 300°C. In the latter case, concurrent nitrogen formation was measured by GC analysis and the absence of N_2O demonstrated by IR.

Conditions: The catalyst weight was 75 mg and the flow rate 100 ml/min (STP) giving a GHSV of 50,000 h^{-1} (STP). The effect of addition of hydrogen and/or water on the reaction of CO and NO was determined under isothermal conditions. Each experiment was performed in the sequence a,b,c,d,a,c,b,d,a with a = only CO and NO, 2000 ppm each; b = a + ≈ 890 ppm H_2; c = a + ≈ 1400 ppm (0.14%) H_2O ; d = a + 900 ppm H_2 and 1400 ppm H_2O. Steady-state reaction rates were obtained by running the corresponding experimental conditions for at least 21 minutes (3 IR analysis of the 5 gases) and taking the average of the last two measurements. Periods as long as one hour were required under some conditions to achieve a steady-state. The sequences were executed over several days at random temperatures in the range 190 - 290°C.

3.RESULTS

3.1 Example run

Figure 1 provides an example of the changes in concentration observed during the first half of the standard sequence - in this case for reaction over Pt/Al_2O_3 at 290°C. It should be noted that the data are not internally normalized, e.g. MS versus IR data, which would be impossible without some assumptions. The apparently negative N_2 concentrations in some cases simply reflect the limits of analysis under conditions of small formation of N_2. The conclusions which follow are based not solely on Figure 1 but on the examination of many data sets for a range of conditions.

As shown in the top panel, introduction of H_2 (a → b) induces a drop in NO signal (i.e. enhanced NO conversion) and an increase in CO_2 formation as determined by both MS and IR. Thus CO conversion has increased as well. Substitution of H_2 by

H_2O (b → c) gives ≈ 2000 ppm CO_2 (i.e. complete CO conversion) and the NO signal drops to zero. Addition of both H_2 and H_2O (c → d) gives a similar result while deletion of both (d → a) returns the concentrations close to the original values, after a delay. The subsequent a → c step results in similar steady state concentrations to those seen after the successive a → b → c steps. The distribution between the various nitrogen containing products (lower panel) is also interesting.

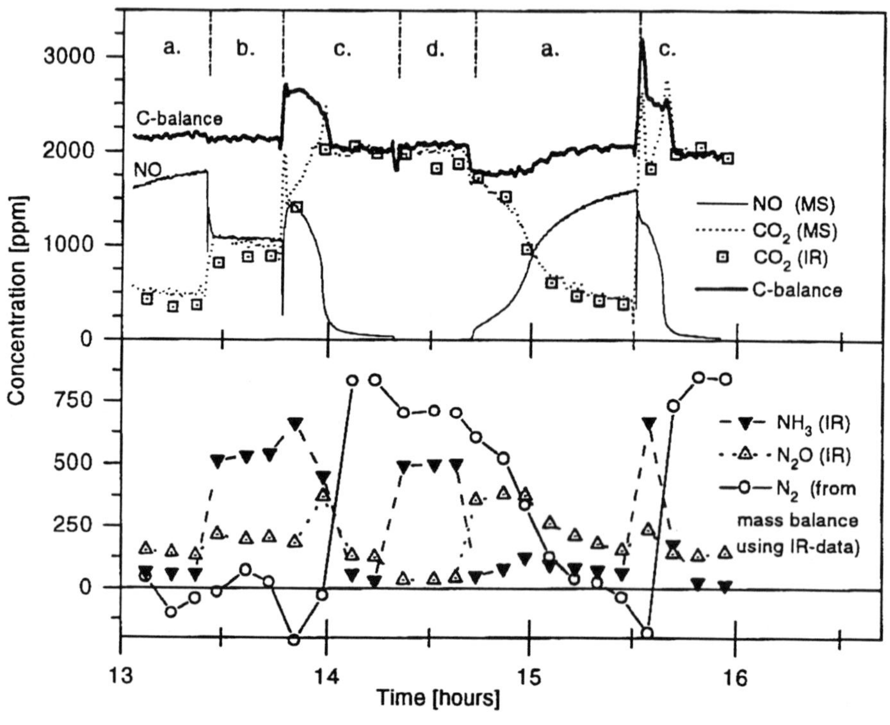

Fig.1 Concentration changes upon addition of H_2 and/or H_2O using Pt/Al_2O_3 at 290°C. a=2000 ppm CO+NO, b.=a+890 ppm H_2, c=a+1400 ppm H_2O, d=a+H_2+H_2O

Ammonia is dominant when hydrogen alone is included (b) whereas nitrogen dominates on addition of water alone (c). Both NH_3 and N_2 are formed when H_2 and H_2O are included together (d). It should be noted here that the quantities of NH_3

detected (≈ 540 ppm) are sufficient to account for most of the 890 ppm of H_2 added. Conversely, very little hydrogen can be converted to water (< 60 ppm by H balance).

The total carbon balance traces in the top panel reveal interesting transient effects. Addition of water (b → c or a → c) results in periods during which the carbon balance is in excess of the input quantity (2000 ppm). This indicates that a surface species has been reacted to produce carbon dioxide. Peaks in ammonia evolution occur over the same periods and the calculated N_2 concentrations are then strongly negative as expected if the quantities of NH_3 (plus N_2O) being produced exceed that expected from the NO conversion. Deletion of water (d → a) is followed by a period of deficient carbon balance (i.e. uptake to form a surface species). The significance of these transient effects is commented upon later. It should be stressed here that both the steady state concentrations and the transient effects were quite reproducible in response to other step-changes in the input concentration in the remainder of the standard sequence (i.e. c → b → d → a).

3.2 Effects of temperature

Similar sequences of experiments to those illustrated in Figure 1 were carried out at other temperatures and with the Rh/Al_2O_3 and $Rh/ CeO_2 /Al_2O_3$ catalyst. Figure 2 shows concentration versus temperature plots extracted from the data set. The method of data collection results in some scatter (since determinations at different temperatures may be long separated in time) but minimises bias (since measurements with different compositions at the same temperature are close in time). The following conclusions may be drawn

1. With Pt/Al_2O_3 inclusion of H_2 and/or H_2O greatly enhances CO_2 production (i.e. CO conversions) at all temperature (top panel, left). The enhancement by H_2 exceeds that by H_2O below 250°C but the reverse is true at higher temperatures. A similar pattern of enhanced CO conversion is evident with Rh/Al_2O_3 (top panel, right) but its extent is less pronounced because of the higher intrinsic activity of rhodium for the reaction of CO with NO.
2. The principal nitrogen-containing product over Pt/Al_2O_3 when H_2 is added is ammonia (second panel, left). However enhancement by water occurs largely with nitrogen formation (third panel, left). There is moderate N_2O formation under all conditions (bottom panel left) and it is the most important nitrogen containing product for the reaction of CO and NO alone.
3. With Rh/Al_2O_3 enhancement by H_2 (or by H_2 and H_2O together) gives more N_2O than NH_3 or N_2 at the lowest temperatures but the relative amounts of each tend to equalise in terms of nitrogen content above 240°C. As with Pt/Al_2O_3 nitrogen is the favoured product when water alone is added (third panel, right).

Experiments with a catalyst containing 6% ceria ($Rh/CeO_2/Al_2O_3$), not shown, exhibit similar responses to the addition of H_2 and H_2O as the undoped Rh/Al_2O_3.

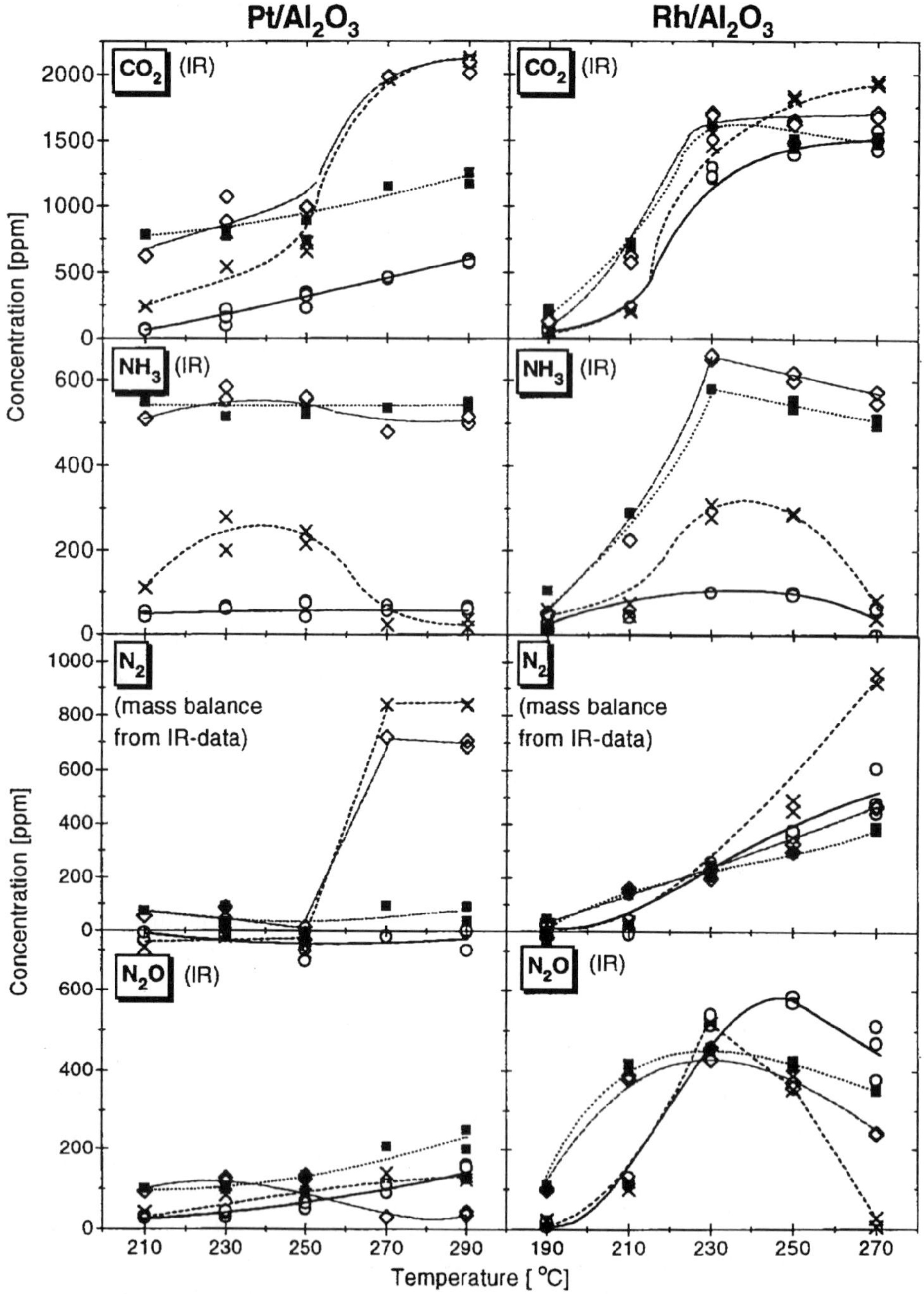

Figure 2. Product concentrations as a function of temperature for reactions of:
(—o—) CO + NO (2000 ppm each) (----x----) CO + NO + H_2O (1400 ppm)
(·····■·····) CO + NO + H_2 (890 ppm) (—◇—) CO + NO + H_2 + H_2O

Differences in activity and selectivity between the two catalysts were also minor indicating that ceria has little effect on reaction characteristics under the conditions used here.

4. DISCUSSION

Tables 1 and 2 summarise the most striking findings of the present work for the effect of H_2 and H_2O on the reaction of CO and NO. With Pt/Al_2O_3 at 210°C addition of hydrogen (Table 1) increases the conversion of NO by a factor of eight and that of CO similarly. Hydrogen conversion is close to complete. Over 90% of this hydrogen reacts with the nitrogen atom of NO to form ammonia, the dominant nitrogen-containing product, rather than with the oxygen atom to give water. Conversely the oxygen from NO is converted to CO_2 with very high selectivity. In stoichiometric terms the reaction can be represented as

$$CO + NO + 1.5H_2 \rightarrow CO_2 + NH_3 \qquad (1)$$

At the same temperature, hydrogen enhances NO conversion over Rh/Al_2O_3 to a similar extent. However the conversion of H_2 is incomplete and the reaction tends to produce less ammonia and more N_2O and N_2. As a result the NO conversion is increased to a greater extent than the CO conversion.

The increase in CO conversion on hydrogen admission appears very peculiar and to our knowledge has not been reported previously. Assuming that the reaction of NO proceeds by dissociation of adsorbed NO on vacant sites (*) i.e.

$$*NO + * \rightarrow *N + *O \qquad (2)$$

then the adsorbed oxygen is seen to prefer to react with adsorbed carbon monoxide to form carbon dioxide

$$O* + *CO \rightarrow CO_2 + 2* \qquad (3)$$

rather than with hydrogen to form water

$$O* + H* \rightarrow *OH + * \qquad (4a)$$
$$*OH + H* \rightarrow H_2O + 2* \qquad (4b)$$

The most probable explanation is that steps (3), (4a) and (4b) are all fast but that the quantity of adsorbed CO is much greater than that of hydrogen. As a result, the probability of an individual O* reacting according to (3) is very high. This is similar to the situation prevailing during the reaction of $CO/H_2/O_2$ mixtures on platinum group metals [13,14] where it is possible selectively to oxidise CO in a large excess of H_2 because the former predominates on the metal surface.

There is no simple explanation for the enhanced NO conversion observed when hydrogen is present. Step (2) above is generally thought to be the initial step [16,17]. It seems unlikely that hydrogen could act by removing adsorbed O since the concentration of this species is believed small due to its efficient removal by reaction with adsorbed CO. In any case this reaction should lead to water which is not the favoured hydrogen containing product. Removal of adsorbed N to form ammonia (as observed) with creation of vacant sites could increase the rate of reaction (2) but only in situations where the nitrogen coverage was relatively high. This is just conceivable with rhodium [7] but is most unlikely with platinum where a very large fraction of the surface is occupied by carbon monoxide [17,18].
However, removal of either adsorbed O or N by hydrogen may lead only temporarily to a large increase in vacant sites, because thus sites would be rapidly occupied by CO. A different situation arises if the removal of N or O itself is rate determining, which seems rather unlikely.

Another possibility is that NO dissociation is `hydrogen assisted' as proposed by Hecker and Bell [19] for the reaction between NO and H_2 i.e.

$$*NO + H* \rightarrow *N + *OH \tag{5}$$

Table 1 Enhancement by hydrogen

Catalyst (temperature)	Inlet composition	Conversion (%) NO	Conversion (%) CO	Selectivities (%) of underlined atoms to $N\underline{H_3}$ [a]	$C\underline{O_2}$ [b]	$\underline{N_2}$ [c]	$\underline{N}H_3$ [d]
Pt/Al_2O_3 210°C	2000 ppm CO + NO	6	5	-	100	e	-
	2000 ppm CO + NO + 890 ppm H_2	47	33	97	96	0	75
Rh/Al_2O_3 210°C	2000 ppm CO + NO	16	12	-	100	44	-
	2000 ppm CO + NO + 890 ppm H_2	69	35	82	89	45	13

[a] selectivity of the hydrogen atom, $H \rightarrow NH_3 = 3NH_3/(H_2$ consumed)
[b] selectivity of the oxygen atom, $N(O) \rightarrow CO_2 = CO_2/(NO$ consumed - N_2O formed)
[c] selectivity of the nitrogen atom, $(N)O \rightarrow N_2 = 1-(NH_3+2N_2O)/(NO$ consumed)
[d] selectivity of the nitrogen atom, $(N)O \rightarrow NH_3 = NH_3/(NO$ consumed)
[e] not measured accurately at low conversions but N_2O dominant

This path is favoured by recent theoretical calculations [20] which show that the activation energy is less than that for unassisted dissociation (reaction (2)) on both Pt (25 versus 54 kJ/mol) and Rh (13 versus 29 kJ/mol). An apparent problem with step (5) in $CO/NO/H_2$ mixtures is that it produces OH which one might expect to react to form water which is a minor product especially with Pt. A rapid reaction between CO and OH, or between water and support-bound isocyanate species as discussed later, would be required to produce CO_2.

`Hydrogen assisted' NO dissociation could also proceed via

$$*NO + H* \rightarrow *NH + *O \qquad (6)$$

Although high activation energies were calculated for this reaction 20 (Pt: 96 kJ/mol, Rh: 67 kJ/mol) it would produce directly the adsorbed oxygen required for reaction with CO (3). In recent discussions on the NO + H_2 reaction Hirano *et al.* [7] found no experimental support for hydrogen assisted dissociation of NO. They argued, as an alternative, that NO dissociation requires empty sites adjacent to adsorbed NO molecules and that hydrogen could replace some weakly bound NO. They thought that this might provide additional sites for dissociation because of the small size of hydrogen. In summary there are a variety of possible explanations for the rate enhancement caused by hydrogen but none are very convincing with the current state of knowledge.

Table 2 Enhancement by water

Catalyst (temperature)	Inlet composition	Conversion (%)		Selectivities[a] (%) of N-atom of NO to		
		NO	CO	N_2	N_2O[b]	NH_3
Pt/Al_2O_3 290°C	2000 ppm CO + NO	24	21	c	83[c]	-
	+ 1400 ppm H_2O	98	100	77	10	12
Rh/Al_2O_3 270°C	2000 ppm CO + NO	100	75	69	30	-
	+ 1400 ppm H_2O	100	100	98	1	2

[a] for N_2 and NH_3 defined as in Table 1
[b] $(N)O \rightarrow N_2O = 2N_2O/(NO\ consumed)$
[c] not measured accurately, but selectivity to N_2O is at least 83%

Water can enhance the reaction of CO with NO to a similar extent as shown in Table 2. The effects are apparent at somewhat higher temperatures and are again more pronounced with Pt/Al_2O_3 than with Rh/Al_2O_3. It is most unlikely that the promotion arises through the water gas shift reaction since this would produce hydrogen and large amounts of ammonia would then be expected on the basis of the results in Table 1. Experimentally very little ammonia is observed using Pt/Al_2O_3 at temperatures above 270°C. Nitrogen is the dominant nitrogen-containing species and the main reaction during water promotion can be represented simply as

$$2CO + 2NO \rightarrow N_2 + 2CO_2 \quad (7)$$

With Rh/Al_2O_3 the principal effect of water is to favour nitrogen production at the expense of nitrous oxide

$$CO + 2NO \rightarrow CO_2 + N_2O \quad (8)$$

CO conversion is increased as a result of either reaction (7) or (8).

The formation of nitrogen over Pt and Rh catalysts at temperatures below 327 °C is now thought to proceed via a surface N_2O intermediate under conditions where there are vacant sites [21]. It is difficult to see how this could be accelerated by water. One faint possibility was suggested by Muraki *et al.* [15] to explain water promotion of the reaction between CO and O_2 on platinum catalysts. They suggested that water adsorbed in competition with carbon monoxide in such a way as to reduce the strong inhibitory effects present when carbon monoxide coverages are very high. It is difficult to see how this is possible given that the heat of adsorption of CO on Pt(111) for example, (134 kJ/mol) is very much greater than that of H_2O (46 kJ/mol) [20]. The heat of adsorption of CO falls very steeply at high coverages (which might allow competition) but even then it is hard to see how this would increase the number of vacant sites as required to increase the rate of NO dissociation, reaction (2).

Alternative explanations assuming that reaction is confined to the metal would require the supposition of `catalytic cycles' such as

$$H_2O + 2{*} \rightarrow {*}OH + H{*} \quad (9a)$$
$${*}NO + {*}H \rightarrow N{*} + {*}OH \quad (9b)$$
$${*}N + {*}NO \rightarrow {*}N_2O \quad (9c)$$
$${*}N_2O + {*}H \rightarrow N_2 + {*}OH + {*} \quad (9d)$$
$${*}CO + {*}OH \rightarrow CO_2 + H{*} + {*} \quad (9e)$$

in which a small quantity of adsorbing water results in the dissociation of many NO molecules. Reaction (9e) would need to be fast to prevent buildup of *OH. Recent calculations [22] indicate that the activation energy for reaction (9e) is low on platinum group metals (e.g. 4kJ/mol for Pd). While this cycle is very speculative, it can account for some of the differences between H_2 and H_2O promotion. Dissociation of water according to reaction (9a) would be slower than hydrogen dissociation necessitating a higher temperature, as observed experimentally. As a consequence the concentrations of adsorbed H and OH would be lower because of faster removal by reactions (9b), (9d) and (9e), leading to low ammonia formation - again as observed.

A final possibility which warrants mention is the potential involvement of isocyanates (NCO) as intermediates. Such species are very unstable on platinum group metals [23] but their spillover to supports is well documented [24]. Support bound isocyanates react readily with water even at room temperature to give ammonia and carbon dioxide [24]. Some evidence for the presence of NCO and their diffusion to the support is provided by the carbon balance and NH_3 traces in Figure 1. The periods of `excess' carbon and ammonia evolution after water introduction are equivalent to about 30 mmol. For comparison the total quantity of Pt is 3.9 mmol and the number of support sites is 150 mmol (assuming 10^{19} sites/m^2). Thus the data is consistent with the view that roughly 20% of the alumina surface is covered by isocyanate groups which are hydrolysed to NH_3 and CO_2 on introduction of water. The corresponding periods of carbon deficit observed on deletion of water are almost step changes, indicating that the spillover and transport processes required to refill empty sites are quite fast at 290°C. It is unlikely that hydrolysis of isocyanates is a major product route under steady state conditions with water present since little ammonia is formed. The process could act as a way of converting water to ammonia during enhancement by hydrogen but more direct experimental methods, such as infrared spectroscopy, are needed to evaluate this possibility.

5. CONCLUSIONS

1. Hydrogen enhances the rate of the reaction of CO and NO on alumina supported platinum and rhodium. The major pathway involves combination of hydrogen with the nitrogen of NO to form ammonia while the oxygen released is taken up by CO.
2. Water promotes the reaction of CO with NO. It does not seem to participate directly in the reaction but affects the characteristics of the reaction in a way which favours nitrogen production.
3. A variety of explanations can be advanced to explain the effects of hydrogen and water but none are very convincing given current knowledge.
4. Support bound isocyanate species may be involved in transient effects seen when water is introduced and subsequently deleted.

ACKNOWLEDGMENT

This work was supported by a grant from the Australian Research Council. The financial support of the Swiss National Foundation for one of the authors (R.D.) is gratefully acknowledged.

REFERENCES

1 T.P. Kobylinski and B.W. Taylor, J. Catal., 33 (1974) 376.
2 H. Shinjoh, H. Muraki and Y. Fujitani, Stud.Surf.Sci.Catal., 30 (1987) 187.
3 K.C. Taylor and R.L. Klimisch, J. Catal., 30 (1973) 478.
4 M. Shelef and H.S. Gandhi, Ind. Eng. Chem. Prod. Res. Develop., 11 (1972) 393.
5 J.H. Jones, J.T. Kummer, K. Otto, M. Shelef and E.E. Weaver, Env.Sci.Technol., 5 (1971) 790.
6 L. Heezen, V.N. Kilian, R.F. van Slooten, R.M. Wolf and B.E. Niewenhuys, Stud.Surf.Sci.Catal. 30, (1987) 381.
7 H. Hirano, T. Yamada, K.I. Tanaka, J. Siera and B.E. Nieuwenhuys, in: *New Frontiers in Catalysis*, Guczi et al. (eds.), Proc. 10th Int. Congress on Catalysis, Elsevier, (1993) 345, see also the included Question/Answer section.
8 M.L. Unland, J. Phys. Chem., 77 (1973) 1952.
9 G. Kim, Ind. Eng. Chem. Prod. Res. Dev., 21 (1982) 267.
10 B. Harrison, A.F. Diwell and C. Hallett, Platinum Metals Rev., 32 (1988) 73.
11 R.K. Herz, Ind. Eng. Chem. Prod. Res. Dev., 20 (1981) 451.
12 C. Padeste, N.W. Cant and D.L. Trimm, Catal. Lett., 18 (1993) 305.
13 S.E. Oh and R.M. Sinkevitch, J. Catal., 142 (1993) 254.
14 J.R. Stetter and K.F. Blurton, Ind. Eng. Chem. Prod. Res. Dev. 19 (1980) 214.
15 H. Muraki, S. Matunaga, H. Shinjoh, M.S. Wainwright and D.L. Trimm, J.Chem. Tech. Biotechnol., 52 (1991) 415.
16 B.E. Nieuwenhuys, Surf. Sci., 126 (1983) 307.
17 R.M. Lambert and C.M. Comrie, Surf. Sci., 46 (1974) 61.
18 R.F. van Slooten and B.E. Nieuwenhuys, J. Catal., 122 (1990) 429
19 W.C. Hecker and A.T. Bell, J. Catal., 92 (1985) 247.
20 E. Shustorovich and A.T. Bell, Surf. Sci., 289 (1993) 127.
21 H. Hirano, T. Yamada, K.I. Tanaka, J. Siera, P. Cobden and B.E Nieuwenhuys, Surf. Sci., 262 (1992) 97.
22 E. Shustorovich and A.T. Bell, Surf. Sci., 253 (1993) 386.
23 J. Raskó and F. Solymosi, J. Catal., 71 (1981) 219.
F. Solymosi, L. Völgyesi and J. Raskó,Z. Phys. Chemie (Neue Folge), 120 (1980) 79.

24 M.L. Unland, J.Catal., 31 (1973) 459.
25 R.J.H. Voorhoeve and L.E. Trimble, J.Catal., 54 (1978) 269.
26 D.A. Lorimer and A.T. Bell, J.Catal., 59 (1979) 223.
27 M. Shelef and H.S. Gandhi, Ind.Eng.Chem.Prod.Res.Develop., 13 (1974) 80.
28 W.F. Egelhoff, in: *The chemical physics of solid surfaces and heterogeneous catalysis*, D.A. King and D.P. Woodruff (Eds.), Elsevier, Vol. 4 (1982) 397.
29 J.J. Jones, J.T. Kummer, K. Otto, M. Shelef and E.E. Weaver, Env.Science & Tech. 5 (1971) 790. .
30 H. Shinjoh, H. Muraki and Y. Fujitani, in: *Catalysis and Automotive Pollution Control*, A. Crucq and A. Frennet (Eds.), Elsevier, 187 (1987).
31 L. Heezen, V.N. Kilian, R.F. van Slooten, R.M. Wolf and B.E. Niewenhuys, in: *Catalysis and Automotive Pollution Control*, A. Crucq and A. Frennet (Eds.), Elsevier, 381 (1987).
32 Y. Amenomiya and T. Tagawa, 8th Proc.Int.Cong.Catal.,Berlin, Vol. 2 (1984) 557.

A. Frennet and J.-M. Bastin (Eds.)
Catalysis and Automotive Pollution Control III
Studies in Surface Science and Catalysis, Vol. 96

SIMULTANEOUS NO_X REDUCTION AND SOOT ELIMINATION FROM DIESEL EXHAUST ON PEROVSKITE-TYPE OXIDE CATALYSTS

V. Duriez, L. Monceaux and P. Courtine

Division Physico-chimie et Analyse des Procédés Industriels, Université de Technologie de Compiègne, B.P. 649, 60206 COMPIEGNE Cédex, France

ABSTRACT

Series of partially substituted manganates belonging to perovskite structure family is prepared and characterized with a view to use them as catalyst for reduction of nitric oxides with carbon black. Catalytic tests show that the series $La_{0.8}Sr_{0.2}Mn_{1-x}Cu_xO_{3\pm 1}$ have the best activity among the perovskite samples. La_2CuO_4, which belongs to K_2NiF_4 structure family, is more active than the others, but unfortunately under test conditions it is partially reduced into La_2O_3 and CuO. Some interpretation of the high activity of these phases are given, based on intrinsic properties of the studied phases.

1 INTRODUCTION

In gasoline engine exhaust gas, the nitrogen oxides are eliminated with the help of carbon monoxide which is present in sufficient proportion. The composition of diesel exhaust gas is, on the contrary, low in CO and high in oxygen, and contains soot particulates. In the present work, an attempt is made to study the feasibility of using soot particulates as a reducing agent of NO. For this purpose, to simplify the system, soot is simulated by carbon black and the gas phase is exclusively composed of nitrogen and NO.

Various substituted lanthanum manganates having the perovskite ABO_3-type structure, as well as La_2CuO_4 and simple copper oxides are studied. In the present study, are reported the synthesis, characterization of the various phases and catalytic test results.

2. EXPERIMENTAL

2.1. Catalyst preparation

A series of lanthanum, partially substituted by strontium or cerium, manganese perovskites are synthetized. Some of them are also substituted by noble metals (Pt and/or Pd) or by copper in B-site. To improve homogeneity and to lower the phase formation temperature, a sol-gel process is used for the synthesis.

Citric acid, rare earth nitrates, other metal nitrates (or acetates for copper compounds), and, when necessary, hexachloroplatinic acid in the required proportions are dissolved in ethylene glycol. After complete dissolution under energetic and constant stirring, the gel is formed by heating slowly. Then, it is transferred into a porcelain dish and again heated to obtain a black solid which is slowly fired in a muffle furnace under air from room temperature to 600°C (5°C/min) and afterwards maintained at 600°C during 6 hrs. Cerium containing samples are heated at 600°C during 6 hrs, then 800°C during 4 hrs and 900°C during 4 hrs, and samples of the series $La_{0.8}Sr_{0.2}Mn_{1-x}Cu_xO_{3\pm1}$ are calcined at 600°C (6 hrs), then 700°C (6 hrs) and 900°C (4 hrs).

La_2CuO_4 sample has been prepared differently. Lanthanum nitrate and copper acetate in stoichiometric quantity are dissolved in water. Then the solution is evaporated under vacuum by heating at 60-70°C, until a gel is obtained. After desiccation the precursor is heated slowly under air (5°C/min) to 300°C and maintained at this temperature during 3 hours and then calcined at 800°C during 5 hours.

Before testing the powder is carefully ground with carbon black (10 wt %) and then compacted into pellets (3 mm diameter and height).

2.2. Characterization

X-Ray diffraction

X-Ray diffraction experiments (XRD) are performed on a curve position sensitive detector INEL CPS 120 which allows a simultaneous collection of diffracted beams in the range $5° < 2\theta < 125°$, with the monochromatized $CuK\alpha_1$ radiation. Powder samples are manually compacted on windowed aluminum holders. After typically one hour of accumulation, patterns are analyzed with the DIFFAC-AT Siemens software program. α-alumina is used as an internal standard.

Specific area

Specific area measurements are performed with a "Quantasorb Jr." apparatus by N_2 adsorption in liquid nitrogen temperature (standard multipoint BET method).

Manganese valency measurement

To determine manganese valency, the method previously described by Bloom et al. [1] is employed. As it uses the ability of Fe^{2+} ions to reduce Mn^{3+} and Mn^{4+} in Mn^{2+} ions, it may not be employed for cerium and copper containing compounds as long as the valencies of these ions in the sample are not determined. Owing to the determination of Mn^{4+} content it is then possible, using electrical neutrality equations, to find the value of the non-stoichiometry in oxygen l.

2.3.Catalytic Test

Experimental set-up

The catalytic activity is tested in a vertical fixed bed reactor made of stainless steel and placed in a tubular furnace. 2 cm^3 of pellets composed with a mixture of carbon black (10 wt %) and catalyst are placed in the reactor. To clean the catalyst and carbon black surfaces the pellets are pre-treated by heating at 300°C during 1.5 hr under nitrogen. Then during the cooling of the solid to ambient temperature, a gas mixture containing 1020 ppm of NO in nitrogen is prepared in the by-pass circuit. As soon as the gas stoichiometry is well stabilized it is allowed to flow through the reactor which is heated from ambient temperature to 500°C at 4°C/min, then the temperature is kept constant during 12 hours. The total gas flow rate is 50 l/hr (VVH = 25,000 hr^{-1}), each flow rate being adjusted by means of massic flow rate controllers. Gas composition is continuously measured on-line by various analyzers : CO_2 and CO contents are determined by IR absorption, NO or NO_x by chemiluminescence. N_2O is not measured but oxygen balances allow us to think that N_2O may not be formed in quantity above 100 ppm.

3.RESULTS

3.1.Characterization

Results of characterization are given in Table 1. X-ray analysis shows that in all cases (except La_2CuO_4, CuO and Cu_2O), a perovskite type phase is formed with CeO_2 for cerium containing phase and $SrCO_3$ (traces) for B-site substituted compounds.

For the series $La_{0.8}Sr_{0.2}Mn_{1-x}Cu_xO_{3\pm\lambda}$, it is important to note that La_2CuO_4 appears only for $0.4 \leq x \leq 0.6$. The highest content in La_2CuO_4 is found for $x = 0.6$. Nevertheless, measurement of relative intensities of the most important XRD peaks in La_2CuO_4 and the perovskite shows that the concentration in La_2CuO_4 must not be more than 9 %. On an other hand, for x^{3} 0.7 the perovskite (without any trace of La_2CuO_4) crystallises in the orthorhombic system whereas all the others may be indiced in the hexagonal system.

La_2CuO_4 is synthetized as a single phase. Its structure is derived from the perovskite one and known as K_2NiF_4 structure [2]. It may be considered to be built up of alternating layers of perovskite (ABO_3) and rock salt (AO) structures [3]. These compounds, as for perovskite oxides, may accept ions in substitution and have a certain degree of non-stoichiometry. A tolerance factor, comparable to Goldschmitt tolerance factor for perovskite, has been defined for K_2NiF_4 structure and it must be noted that the one of La_2CuO_4 is very close to the upper limit [2].

Concerning the determination of Mn^{4+} content and the evaluation of the non-stoichiometry in oxygen our results for the two first compounds of Table 1 are in perfect accordance with those of Takeda et al. [4] who have also found that the introduction of a deficiency in A cation leads to a decrease in oxygen stoichiometry, whereas the Mn^{4+} content remains constant. This may be explained by the increase in cationic vacancies. In the same way, the lack of B cation leading to an enhancement in Mn^{4+} content and at the same time to a very important decrease in oxygen stoichiometry, may only be understandable by the formation of cationic vacancies.

Lanthanum substitution by strontium has been widely used to improve catalytic performance [5-7] arguing the increase in Mn^{4+} content for charge compensation. In fact, the determination by titration of Mn^{4+} content shows, on the opposite, a decrease in Mn^{4+} content accompanied by a decrease in oxygen stoichiometry. This is in accordance with the results of Kuo et al. [8] who determined oxygen stoichoimetry by TGA measurements as well as other works based on Mn^{4+} titration [9, 10].

Table 1 Catalysts characterization

	Structure	Specific area (m^2/g)	Mn^{4+} (%)	l
$LaMnO_{3\pm\lambda}$	P	20.6	46	0.23
$La_{0.8}Sr_{0.2}MnO_{3\pm\lambda}$	P	15.9	36	0.08
$La_{0.9}Ø_{0.1}MnO_{3\pm\lambda}$	P	19.2	46	0.08
$La_{0.9}Ce_{00..1}MnO_{3\pm\lambda}$ *	P + εCeO_2	8.9	-	-
$La_{0.8}Sr_{0.2}Mn_{0.9}Ø_{0.1}O_{3\pm\lambda}$	P + $\varepsilon SrCO_3$	26.65	55	- 0.03
$La_{0.8}Sr_{0.2}Mn_{0.999}Pt_{0.001}O_{3\pm\lambda}$	P + $\varepsilon SrCO_3$	22.4	42	0.11
$La_{0.8}Sr_{0.2}Mn_{0.999}Pd_{0.001}O_{3\pm\lambda}$	P + $\varepsilon SrCO_3$	23.0	50	0.15
$La_{0.8}Sr_{0.2}Mn_{0.998}Pt_{0.001}Pd_{0.001}O_{3\pm\lambda}$	P + $\varepsilon SrCO_3$	22.3	50	0.15
$La_{0.8}Sr_{0.2}Mn_{0.9}Cu_{0.1}O_{3\pm\lambda}$**	P	6.2	-	-
$La_{0.8}Sr_{0.2}Mn_{0.8}Cu_{0.2}O_{3\pm\lambda}$ **	P	5.2	-	-
$La_{0.8}Sr_{0.2}Mn_{0.7}Cu_{0.3}O_{3\pm\lambda}$ **	P	5.0	-	-
$La_{0.8}Sr_{0.2}Mn_{0.6}Cu_{0.4}O_{3\pm\lambda}$ **	P + εLa_2CuO_4	4.2	-	-
$La_{0.8}Sr_{0.2}Mn_{0.5}Cu_{0.5}O_{3\pm\lambda}$ **	P + εLa_2CuO_4	4.7	-	-
$La_{0.8}Sr_2Mn_{0.4}Cu_{0.6}O_{3\pm\lambda}$ **	P + εLa_2CuO_4	3.02	-	-
$La_{0.8}Sr_{0.2}Mn_{0.3}Cu_{0.7}O_{3\pm\lambda}$ **	P'	2.2	-	-
$La_{0.8}Sr_{0.2}Mn_{0.2}Cu_{0.8}O_{3\pm\lambda}$ **	P'	1.5	-	-
$La_{0.8}Sr_{0.2}Mn_{0.1}Cu_{0.9}O_{3\pm\lambda}$ **	P'	1.8	-	-
$La_{0.8}Sr_{0.2}CuO_{3\pm\lambda}$ **	P'	0.6	-	-
La_2CuO_4**	K_2NiF_4	0.83	-	-
CuO	CuO	4.3	-	-
Cu_2O	Cu_2O	0.8	-	-

* T° calcination : 600°C (6 h), 800°C (4 h), 900°C (4 h)
** T° calcination : 600°C (6 h), 700°C (6 h), 900°C (4 h)

P : perovskite structure, hexgonal system
P' : perovskite structure, orthorombic system

The introduction of noble metal in B-site substitution does not have any influence on oxygen stoichiometry.

3.2.Catalytic Activity

First of all, in order to evaluate the specific contribution to the reaction of the catalyst on the one hand, and the carbon black on the other hand, experiments are performed either with catalyst and without carbon black or with carbon black in silica as an inert material. In both cases there is no conversion of NO. On the opposite, if catalyst and carbon black are both present, the evolution of the composition of the exhaust gas shows that there is effectively an oxidation of carbon black into CO_2 accompanied by NO reduction. On Figure 1, are represented typical curves showing the variation of NO and CO_2 concentration as a function of time and temperature. These curves are divided into three main parts. In the first zone (zone A) there is an important drop of NO concentration due to the entering of the gas flow through the reactor and to NO adsorption. The second part (zone B) corresponds to NO desorption due to the temperature increase, with two main peaks : one around 170°C and the other around 350°C, the exact value depending on the type of catalyst. These desorption peaks are not apparent on La_2CuO_4 curve because of its low specific area. It must be noticed that CO_2 production begins generally during the first part of this second peak and this is the real beginning of the reaction between NO and C. In all experiments, no CO has been detected. In the third zone (zone C), NO concentration drops more or less drastically and, jointly, CO_2 concentration increases. Then, as the carbon black is consumed and as there are more and more diffusion phenomena inside the pellets, $deNO_x$ reaction, though being still effective, is less important.

For all the experiments, oxygen balance has been verified and the reproducibility of the results is about 5 %.

In order to compare the various catalysts it is necessary to take into account the catalyst specific area and the amount of carbon black introduced as the pellets are not always compacted exactly alike. Therefore, the activity is given as the number of moles of NO converted per unit of surface of catalyst and unit of mass of the solid mixture (catalyst + carbon black).
Results are given in Table 2.

4.DISCUSSION

Results of the catalytic tests show that it is possible to reduce NO with the help of carbon black. NO is probably adsorbed on the surface of the catalyst and then either just after desorption, or as it is still adsorbed, it reacts with carbon black to give CO_2 and N_2. The exact mechanism is still not completely understood.

The activity of $LaMnO_3$ is very low compared to all the other studied catalysts, and it does not seem to be directly connected with the Mn^{4+} content contrarily to what was established for the reduction of NO by CO [11]. On the opposite, the results confirm the benefic effect of substitution, particularly in A-site (Sr or Ø) which may be explained by the modification of the binding energies A-O. As Voorhoeve has shown for the reduction of NO by CO, the progression of the activity follows the order : $LaMnO_3$ < $LaØMnO_3$ < $LaSrMnO_3$.

Table 2 Catalytic activity of the tested catalysts

Catalyst	Activity* x 10^4	Catalyst	Activity* x 10^4
$LaMnO_3$	0.29	$La_{0.8}Sr_{0.2}Mn_{0.9}Cu_{0.1}O_3$	2.89
$La_{0.8}Sr_{0.2}MnO_3$	1.30	$La_{0.8}Sr_{0.2}Mn_{0.8}Cu_{0.2}O_3$	3.66
$La_{0.9}Ø_{0.1}MnO_3$	0.65	$La_{0.8}Sr_{0.2}Mn_{0.7}Cu_{0.3}O_3$	2.88
$La_{0.9}Ce_{0.1}MnO_3$	0.40	$La_{0.8}Sr_{0.2}Mn_{0.6}Cu_{0.4}O_3$	1.99
$La_{0.8}Sr_{0.2}Mn_{0.9}Ø_{0.1}O_3$	0.74	$La_{0.8}Sr_{0.2}Mn_{0.5}Cu_{0.5}O_3$	6.84
$La_{0.8}Sr_{0.2}Mn_{0.999}Pt_{0.001}O_3$	1.70	$La_{0.8}Sr_{0.2}Mn_{0.4}Cu_{0.6}O_3$	2.91
$La_{0.8}Sr_{0.2}Mn_{0.999}Pd_{0.001}O_3$	2.36	$La_{0.8}Sr_{0.2}Mn_{0.3}Cu_{0.7}O_3$	2.38
$La_{0.8}Sr_{0.2}Mn_{0.998}Pt_{0.001}Pd_{0.001}O_3$	2.54	$La_{0.8}Sr_{0.2}Mn_{0.2}Cu_{0.8}O_3$	4.39
		$La_{0.8}Sr_{0.2}Mn_{0.1}Cu_{0.9}O_3$	1.82
		$La_{0.8}Sr_{0.2}CuO_3$	2.82
		La_2CuO_4	30.4
		CuO	18.43
		Cu_2O	0.17

* (moles NO converted/m^2.g)

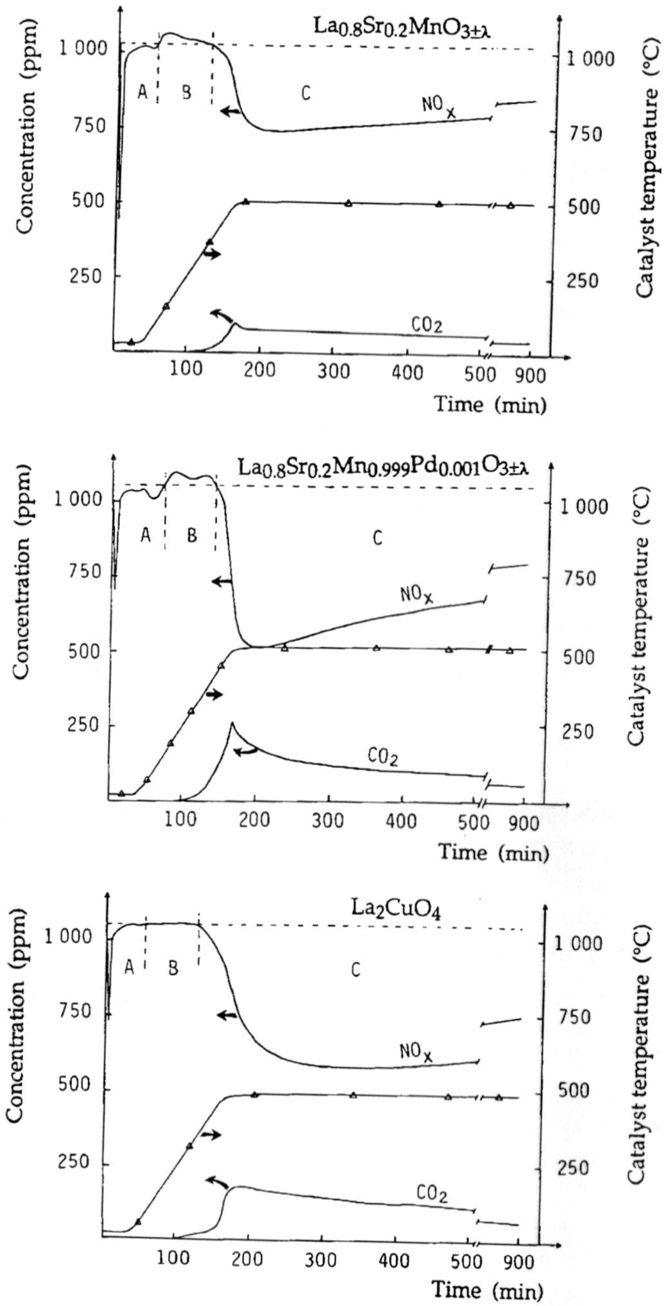

Figure 1. Evolution of NO_x and CO_2 concentrations
(- - - : inlet NO_x)

When very little quantities of noble metals (Pt, Pd) are added, the activity is again increased. This may be related to the work of Johnson et al [12] who observed an enhancement of the activity of platinum when inserted into the perovskite matrix for the reaction of oxidation of CO. However, as noble metals are introduced in a very low proportion, no modification of the XRD patterns showing the insertion of Pd or Pt in the perovskite lattice can be evidenced. On the opposite, using SEM observations, no cluster of palladium or platinum is detected ; therefore, to be really sure that the noble metals are in the lattice, more investigations are needed.

The second series of tested compounds are the copper containing phases : $La_{0.8}Sr_{0.2}Mn_{1-x}Cu_xO_{3\pm\lambda}$ ($0 \leq x \leq 1$), La_2CuO_4 and simple copper oxides : CuO and Cu_2O. Theses phases are tested under the same conditions in order to try to evidence the influence of the host matrix on the activity and stability of copper. Concerning the stability, it is important to note that perovskite compounds are the only ones the structure of which remains unchanged under operating conditions. On the contrary XRD analysis of the samples after test shows that La_2CuO_4 decomposes into a mixture of La_2CuO_4, La_2O_3 and CuO, that CuO is partially reduced in Cu_2O and at last Cu_2O is also partially reduced in metallic copper. It seems, therefore, that the insertion of copper in the perovskite lattice leads to a certain stabilization as it was already reported in literature for $LaMn_{1-x}Cu_xO_{3\pm\lambda}$ compounds [13].

$La_{0.8}Sr_{0.2}Mn_{1-x}Cu_xO_{3\pm\lambda}$ and La_2CuO_4 phases are more active for the studied reaction than copper free perovskite and therefore copper seems to play a very important role. This may be explained by the fact that in the perovskite-type phase, the CuO_6 octahedra are distorted due to Jahn-Teller effect, and therefore there are two longer Cu-O bonds, inducing a lower binding energy, and making possible the creation of oxygen vacancies which are often considered as being preferential adsorption sites of NO [14].

The evolution of the activity as a function of copper substitution is given on Figure 2 : the shape of the curve may be compared to the one obtained by Chan et al. [9] for the oxidation of CO over copper substituted lanthanum manganates although these authors tested neither $x = 0.5$ nor $x > 0.8$. For $x = 0.5$ there is a maximum in activity, which may be related to the maximum in NO adsorption as reported by Mizuno et al. [15].

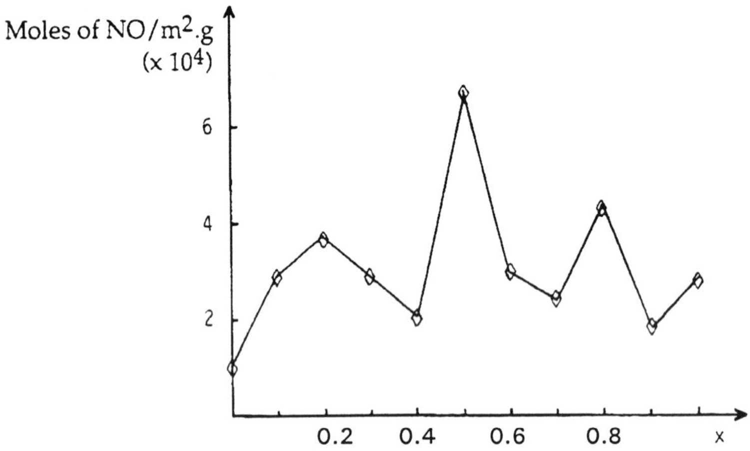

Figure 2. Activity of $La_{0.8}Sr_{0.2}Mn_{1-x}Cu_xO_{3\pm\lambda}$ as a function of copper substitution

The highest activity is observed for La_2CuO_4. In this type of structure, copper ions are at the centre of distorted CuO_6 octahedra as in perovskite phase, but the coordination number of lanthanum is 9 instead of 12. Kudo et al. [16] have also noted a high activity of this phase in the case of the reduction of NO with NH_3. An explanation could be that the value of the tolerance factor t of La_2CuO_4 is very close to the upper limit of stability of K_2NiF_4 structure and therefore, to stabilize the compound in this structure copper ions are able to modify their ionic radii toward a more favourable value of t by disproportionation of Cu^{2+} in Cu^+ and Cu^{3+} or more probably by forming charge density waves [2].

CONCLUSION

This study has shown the feasibility of catalytic NO reduction with carbon black, on various substituted rare earth transition metal oxides having perovskite-type structure or related structure (K_2NiF_4). The most active catalysts are those containing copper, particularly $La_{0.8}Sr_{0.2}Mn_{0.5}Cu_{0.5}O_3$ and La_2CuO_4. The origin of the very important activity of these phases may be found in the intrinsic properties of copper ion which is in the centre of a distorted octahedra with two weak Cu-O bonds. Therefore, oxygen vacancies, which are supposed to be preferential NO adsorption sites, may be easily created. Concerning La_2CuO_4, the hypothesis of formation of charge density waves which has been reported by Ganguly may also enhance the activity. However, more investigations concerning NO adsorption, copper ion valency measurements and stability of the various phases are needed. Jointly, tests with partial pressure of oxygen are in progress.

REFERENCES

1 E. Bloom Jr. and T.Y. Komenati, J. Inorg. Nucl. Chem., 40 (1978) 403-405.
2 P.Ganguly and C.N.R. Rao, J. Sol. State Chem., 53 (1984) 193-216.
3 D.Balz and K.Pleith, Z. Electrochem., 59 (1955) 545.
4 Y. Takeda and S. Nakai, Mat. Res. Bull., 26 (1991) 153-162.
5 T. Nitadori and S. Kurihara, J. Catal., 98 (1986) 221-228.
6 R.J.H. Voorhoeve, Adv. Mat. Catal., (1977) 129-181.
7 J.L.G. Fierro, Catal. Today, 8 (1990) 153-174.
8 J.H. Kuo and H.U. Anderson, J. Sol. State Chem., 83 (1989) 52-60.
9 K.S. Chan, J. Ma, S. Janicke and G.K. Chuah, Applied Cata. A, 107 (1994) 201-227.
10 G.H. Jonker and J.H. van Stanten, Physica XVI, 3 (1950) 337-349.
11 R.J.H. Voorhoeve, J.P. Remeika, L.E. Trimble, A.S. Cooper, F.J. Disalvo and P.K. Gallagher, J. Sol. State Chem., 14 (1975) 395.
12 D.W. Johnson, P.K. Gallagher, G.K. Werther and E.M. Vogel, J. Catal., 48 (1977) 87-97.
13 M.L. Rojas, J.L.G. Fierro and L.G. Tejuca, A.T. Bell, J. Catal., 124 (1990) 41-51.
14 Y. Teraoka and H. Fukuda, Chem. Lett., (1990) 1-4.
15 N. Mizuno, Y. Fujiwara and M. Misono, J. Chem. Soc., Chem. Commun., (1989) 316-318.
16 T. Kudo, T. Gejo and K. Yoshida, Am. Chem. Soc., 12 (1978) n° 2, 185-189.

A. Frennet and J.-M. Bastin (Eds.)
Catalysis and Automotive Pollution Control III
Studies in Surface Science and Catalysis, Vol. 96
1995 Elsevier Science B.V.

HOW A THREE-WAY CATALYST IS AFFECTED UNDER TRANSIENT CONDITIONS : A STUDY OF Pt-Rh/Al_2O_3 CATALYST

C.Howitt, V.Pitchon, F.Garin and G.Maire

LERCSI, Laboratoire d'Etudes de la Réactivité Catalytiques, des Surfaces et Interfaces, URA 1498 du CNRS - Institut le Bel, Université Louis Pasteur, 4, rue Blaise Pascal, 67070 Strasbourg, FRANCE.

ABSTRACT

This publication describes results obtained for CO, NO and C_3H_8 conversion over alumina supported Pt/Rh, Pt and Rh catalysts operating under transient conditions. We show that, for the bimetallic catalyst, oxygen plays an important role in changing the surface composition. By performing activity testing experiments under cycling conditions and different gas mixtures, along with 'step-change' experiments, we suggest that, under fuel-rich and stoichiometric conditions, both Pt and Rh exist on the surface in a reduced state. However, under fuel-lean conditions, we show that Rh segregates from the Pt forming an inactive surface oxide, thought to be Rh_2O_3. It is shown that this redox process occurs within 10 seconds.

1. Introduction

Since their introduction to production vehicles in 1975, a great deal of research has been undertaken to try to understand the reactions occuring over, and the mode of operation of, three-way automotive exhaust catalysts [1-3]. Since, under 'real' conditions, these catalysts never operate under steady-state conditions [4] one of the best ways to study the operation of these catalysts has been to use a system similar to that designed by Schlatter *et al.* [5,6], which facilitates the study of three-way responses to transient changes in gas composition.

In recent years, numerous workers have studied the binary reactions CO + O_2 and NO + CO over precious metal catalysts. However, whilst it is widely accepted that the reaction of CO and O_2 proceeds via a Langmuir-Hinshelwood mechanism involving the reaction between adsorbed CO and O [7-9], and the

NO/CO reaction involves the adsorption/dissociation of NO [8,10-13], the literature is still unclear as to the state of the active surface of the catalyst [14-16]. These binary reactions have been studied over Pt-Rh/SiO_2 catalysts [17,18], different crystallographic faces [19] and using different techniques [20,21]. In all cases they show that some degree of surface reorganisation occurs during the reactions. When NO and O_2 are in excess it has been suggested that Rh segregates from the Pt due to the high Rh-O bond strength compared to Pt-O, it is also shown that the catalyst surface behaves differently towards the NO/CO reaction than to the CO/O_2 reaction [17,18].

This study uses low loaded Al_2O_3 supported Pt, Rh and Pt-Rh catalysts to study the reaction between NO, CO, O_2 and C_3H_8 (HC) as a full gas mixture of varying composition, or any combination of oxidising and reducing gases (e.g. the reactions between CO + O_2, NO + CO and NO + HC). The flexibility of the apparatus also allows the study of the response to so-called 'step-changes' in gas composition, described by H. and M. Kobayashi [22] and also in this laboratory by Weibel [23]. Thus, we present results which contribute significantly to our understanding of the mode of operation of three-way automotive catalysts under transient conditions.

2. EXPERIMENTAL

2.1. Catalysts

All of the catalysts used in this study were prepared by the Institut Français du Petrole (IFP) by impregnation of Al_2O_3 spheres (3 mm diameter), prior to testing the catalysts were crushed to obtain grains of diameter between 80 and 250 μm. The metal loadings of the catalysts used are 1.1% Pt, 1.0% Rh for the monometallic and 1.1% Pt, 0.2% Rh for the bimetallic.

2.2. Catalytic measurements

The experimental setup used for the catalytic tests was similar to that described by Schlatter *et al.* [5,6]. The system was designed such that the reaction gas streams could be as simple as binary mixtures, or as complex as the reaction between NO, CO, O_2 and C_3H_8 (HC). By choosing the correct flow rate of each of the reactant gases, the gas composition could be varied thus allowing the catalytic activity to be studied under oxidising or reducing media with varying strengths, S_N, defined by the following ratio :

$$S_N = (2\,[O_2] + [NO]) / ([CO] + 10\,[C_3H_8])$$

The flexibility of the apparatus, and the accompanying 'home-made' software allowed two distinct types of experiment to be conducted, namely '*light-off temperature*' measurements under *oscillating* and *steady-state* conditions, and so called '*step-change*' experiments [22,23], the details of each type of test shall now be described.

i) Light-off temperature measurements under oscillating conditions

The aim of this test was to measure the light-off temperatures for NO, CO and HC conversions, while exposing the catalyst to a flow of gases oscillating between two compositions under a constant heating rate. The gas composition could be chosen so as to oscillate about stoichiometric (S_N=1), reducing (S_N<1), or oxidising (S_N>1) media, with varying amplitudes.

The frequency of switching between the two gas compositions, was also variable and was controlled by two fast acting solenoid selector valves located at the top of the reactor. The selector valves allowed the frequency to be varied between 0.05 and 1 Hz.

Testing Procedure

A sample of the catalyst (100-400 mg) was placed in a straight silica reactor (10 mm i.d.), and mounted in the centre of a vertically held tube furnace. The point about which the reaction should oscillate was selected, along with the amplitude and frequency. The flow of gases was controlled by a series of eight mass flow controllers (Tylan FC260), and the switching was controlled by two solenoid selector valves (Burkert Type 330). The flow controllers, and the selector valves were driven by a 'home-made' computer program. The catalyst was heated to typically 450°C at a ramp rate of 4°C min^{-1} under a total gas flow of 200 cm^3 min^{-1}, after which time the catalyst was cooled back to room temperature, this first run was to 'activate' the catalyst. After activation the experiment was repeated, it is from this second run that the results are taken.

For analysis, CO, HC and CO_2 were measured using infrared detectors, NO by chemiluminescence, and O_2 by paramagnetism.

ii) Step-change measurements for the CO/O_2 reaction

The catalyst was initially stabilised under a reducing or oxidising stream under *isothermal* conditions. For the reduction step measurement, the catalyst was stabilised under oxidising conditions (i.e. O_2) for a chosen length of time, after which the gas stream was switched to reducing conditions for a length of time, before being switched back to O_2. During the oxidation step measurement the catalyst is stabilised under CO before being switched to O_2, and then back to CO. Each transition was separated by a 15 minute pulse of N_2 in order to remove the gas phase, and to remove weakly held species from the surface of the catalyst. The transient response to the step-changes was measured by monitoring the instantaneous evolution of CO_2 from the surface of the catalyst.

For the step-change experiments, 400 mg of the Pt-Rh catalyst were used. The temperature was increased up to 300°C at a rate of 4°C min^{-1} under the appropriate pretreatment gas (CO or O_2) at a flow rate of 200 cm^3 min^{-1}. The times for which the catalysts were exposed to each gas are given as footnotes to the results tables; where, t1 is the time for the pretreatment step, t2 is the oxidation/reduction step length, and tN is the time for which the system was purged in N_2 prior to changing gases.

3. RESULTS AND DISCUSSION

3.1. Activity studies of the Pt-Rh/Al_2O_3 catalyst

i) Variations in activity with catalyst mass for the full mixture under oscillatory operation

The activity of the system for CO, NO and HC conversion was measured using 100, 200 and 300 mg of the catalyst. Figure 1 shows typical variations in conversion with reaction temperature for CO, NO and HC. It is to be noted from this figure that in the NO profile a peak centred around 250°C is observed. This peak has been shown to be due to the formation of N_2O arising from the partial reduction of NO.

Typical results obtained from this series of experiments, along with the reaction conditions are given in Table 1. Normally, in the absence of diffusion limitations, we would expect the activity of a catalytic system to increase with increased catalyst mass, however, it can be seen from this Table that up to 50% conversion, no such correlation is observed. In order to investigate this phenomenon, it is necessary to study, in isolation, the simple binary reactions occurring in the full gas mixture.

Table 1. Temperatures (°C) for 50% conversion of CO, NO and HC over Pt-Rh/Al_2O_3[a]

	100 mg	200 mg	300 mg
CO	233	245	239
NO	299	330	342
HC	283	323	334

[a] Oscillating about S_N=1, +/- 0.1 unit; f=0.05 Hz; Mean concentrations (ppm): CO, 5600; NO, 2050; O_2, 6000; HC,900.

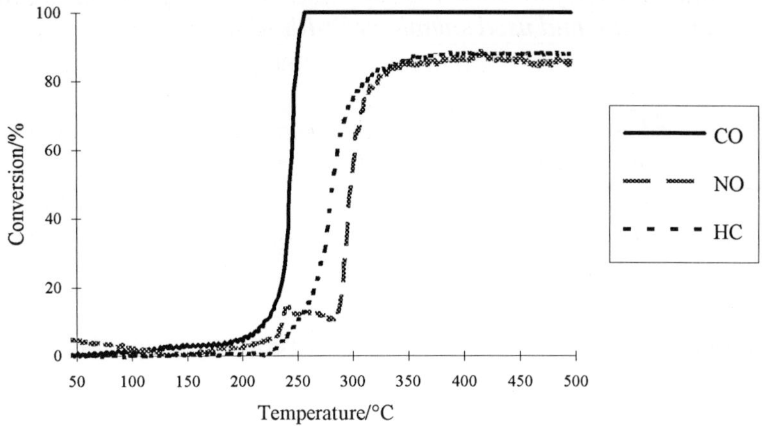

Figure 1 : Conversion vs. temperature for CO, NO, HC on Pt-Rh/Al_2O_3 under oscillatory operation

(ii) Variations in activity as a function of mass for the binary reactions

The binary reactions occurring in the full reaction scheme are : $CO + 1/2\ O_2$, NO + CO and C_3H_8 + 10 NO.

The results, and conditions, corresponding to this series of experiments are given in Tables 2-4. It can be seen that for the CO/O_2 and NO/HC reactions, the 50% conversion temperature does indeed decrease with catalyst mass. An interesting phenomenon is noted for the NO/CO reaction (Table 3); when the catalyst is fresh, the 50% conversion temperature decreases with catalyst mass, in the manner expected. Once activated, however, we observe that the 50% conversion temperature is constant between 100 and 300 mg. Thus, as noted by several workers [17-21], we suggest that there is some degree of surface reconstruction occurring during the reaction. This very important concept could explain the lack of correlation between activity and mass when operating under the full gas mixture. However, further investigations are necessary, and are discussed in subsequent sections.

Table 2. Temperatures (°C) for 50% CO conversion during the CO/O_2 reaction over Pt-Rh/Al_2O_3[a]

	100 mg	200 mg	300 mg
CO	221	213	200

[a] Oscillating about S_N=1, +/- 0.25 unit; frequency 0.05 Hz; Mean concentrations (ppm) - CO, 11000; O_2, 5300.

Table 3. Temperatures (°C) for 50% NO and CO[a] conversions during the NO/CO reaction for a fresh, and used sample of Pt-Rh/Al_2O_3[b]

	100 mg	200 mg	300 mg
Fresh Catalyst	297	284	264
Used Catalyst	289	291	293

a Conversion of NO and CO was simultaneous.
b Oscillating about S_N=1, +/- 0.25 units; frequency 0.05 Hz; Mean concentrations (ppm) - NO, 3600; CO, 3600.

Table 4. Temperatures (°C) for 50% NO and HC conversions during the NO/HC reaction over Pt-Rh/Al_2O_3[a]

	100 mg	200 mg	300 mg
NO	355	348	339
HC	380	352	356

a Oscillating S_N=0.33, +/- 0.06 units; frequency 0.05 Hz; Mean concentrations (ppm) - NO, 2600; HC, 900

(iii) The effect of frequency and oxidising/reducing strength on the catalytic activity

Varying the frequency of switching between reaction media of different oxidising/reducing strengths provides important information on how the activity of the system responds to excursions into rich and lean reaction environments [5,6].

Table 5 shows the results obtained by operating the reaction under the full gas mixture under deeply reducing (S_N=0.45), stoichiometric (S_N=1.00) and deeply oxidising (S_N=1.45) media using different cycling frequencies, and also under steady-state conditions. From the results presented in this Table it can be clearly seen that, under cycling conditions, CO and HC oxidation reactions under deeply oxidising conditions are *less* favoured than under stoichiometric conditions, indicating that excess O_2 inhibits these reactions.

In order to investigate this phenomenon we compare the 50% conversion temperatures at 0.05 Hz, 1 Hz and under steady-state conditions (Table 5). It can be seen that the oxygen inhibition effect is *only* observed when we operate under cycling conditions (even at frequencies as high as 1 Hz), implying that the effect associated with oxygen inhibition occurs very rapidly. We suggest that this effect could have two origins:

a) A transient form of adsorbed oxygen which desorbs slowly from the surface of the catalyst during rich excursions,

b) Excess oxygen is consumed by forming inactive surface oxides of the noble metal(s).

This latter explanation would support the earlier suggestion that there is some degree of surface reconstruction occurring during the reaction.
It has to be noted that the increase of the steady-state light-off temperature of the hydrocarbon from 289 to 311°C when going from stoichiometric to lean conditions does not implie the same phenomenom, indeed we think that the hydrocarbon reaction involves two different pathways, i.e. oxydation by both O_2 and NO.

Table 5. Temperatures (°C) for 50% conversion of CO, NO and HC over Pt-Rh operating under different frequencies and activities[a]

Frequency (Hz)	Conversion of	S_N=0.45[b]	S_N=1.00[c]	S_N=1.45[d]
0.05	CO	250	233	245
0.05	NO	270	299	-
0.05	HC	550	283	328
1	CO	268	228	239
1	NO	277	310	-
1	HC	568	300	345
Steady-state	CO	251	229	217
Steady-state	NO	260	297	-
Steady-state	HC	29%[e]	289	311

[a] Oscillating about chosen point +/- 0.1 unit;
[b] Concentrations (ppm) - CO, 12400; NO, 2600; O_2, 3700; HC, 980.
[c] Concentrations (ppm) - CO, 5000; NO, 2150; O_2, 6500; HC, 980.
[d] Concentrations (ppm) - CO, 3600; NO, 1900; O_2, 9400; HC, 1000.
[e] Maximum conversion obtained.

3.2. The activity of single metal catalysts as a function of mass

(i) 1% Pt/Al_2O_3

The variation in activity with catalyst mass, for the full gas mixture operating under cycling conditions, over a single metal Pt catalyst was measured to try to assess the roles of the different metals in the bimetallic catalyst. The results obtained from these experiments are presented in Table 6. It is clear that the temperatures for 50% conversion of CO, NO and HC decrease with increased catalyst mass. In the absence of diffusion limitations this trend is expected, nevertheless, it was not observed for the parallel experiments with the bimetallic catalyst (Table 1). This result again provides evidence that surface rearrangement occurs on the Pt-Rh catalyst.

An interesting observation can be made by comparison of Tables 1 and 6 (for similar catalyst masses). It can be seen that the monometallic Pt catalyst is more active than the bimetallic system when operating about stoichiometry, this result suggests that the addition of Rh appears to deactivate the catalyst. The reason for the addition of Rh becomes apparent when studying the reaction under reducing conditions, and is discussed more fully in the following section.

Table 6. Temperatures (°C) for 50% conversion of CO, NO and HC over 1% Pt/Al_2O_3[a]

	100 mg	200 mg	300 mg
CO	216	208	198
NO	283	279	276
HC	277	271	266

[a] Oscillating about $S_N=1$, +/- 0.1 unit; frequency 0.05 Hz; Mean concentrations (ppm): CO, 5600; NO, 2050; O_2, 6000; HC, 900.

(ii) 1% Rh/Al_2O_3

From Table 7, we observe that the 50% conversion temperatures for CO, NO and HC decrease with increased catalyst mass as observed in the case of the monometallic Pt catalyst. This Table shows that, under stoichiometric conditions, Rh is considerably less active than the Pt and Pt-Rh catalysts. In the case of NO, the difference is somewhat exagerated since our Tables show temperatures corresponding to the complete reduction of NO to N_2. However, over the Rh catalyst, large amounts of N_2O formed from the partial reduction of NO are observed, an observation which has been noted previously over the Pt-Rh catalyst. Figure 2 shows the NO conversion profiles for Pt-Rh, Pt and Rh catalysts operating under stoichiometric conditions.

It is to be noted that, when operating under stoichiometric conditions, the Rh catalyst is considerably less active than both the Pt and Pt-Rh catalysts. The reason for the inclusion of Rh in the bimetallic system becomes clear when we study the activity under deeply reducing, and oxidising conditions [24]. Our results show that under deeply oxidising conditions Rh forms an inactive surface oxide (thought to be Rh_2O_3), whereas under deeply reducing media Rh greatly enhances the activity for CO, NO and HC conversion.

Table 7. Temperatures (°C) for 50% conversion of CO, NO and HC over 1% Rh/Al_2O_3[a]

	100 mg	200 mg	300 mg
CO	280	266	192
NO	445	399	386
HC	428	396	384

[a] Oscillating about S_N=1, +/- 0.1 unit; frequency 0.05 Hz; Mean concentrations (ppm): CO, 5600; NO, 2050; O_2, 6000; HC, 900.

Thus, we suggest that when the bimetallic catalyst operates under cycling conditions, lean excursions lead to the formation of an inactive surface oxide, which we suggest as being Rh_2O_3. During rich, and stoichiometric, excursions the metals exist in a fully, or partially, reduced state. this oxide does not form and the Pt-Rh alloy remains. It is this change in surface structure which is thought to be responsible for the phenomena noted in this publication., Results in Table 5 seem to show that the surface changes occur within at most 10 seconds because of the range of frequencies studied, but the time taken for these processes to occur is currently under investigation

In order to further investigate these surface rearrangement processes, the reaction of CO and O_2 has been studied using the 'step-change' procedure.

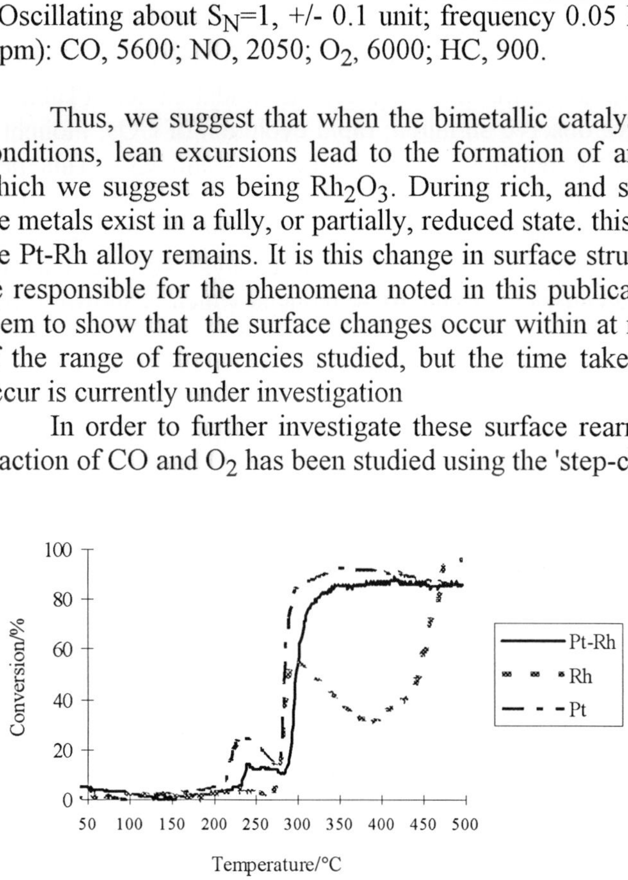

Figure 2 : Conversion of NO vs. temperature for the Pt-Rh, Rh and Pt catalysts

3.3. Step change experiments on Pt-Rh/Al_2O_3

i) Fresh catalyst, oxidation step (CO-O_2-CO)

During the first transition (CO-O_2), a rapid desorption of CO_2 is observed as soon as the gas is switched to O_2. The CO_2 is formed by the surface reaction of preadsorbed CO and O_2 via the well-known Langmuir-Hinshelwood mechanism. The exact amount of CO_2 liberated, and the peak baseline width, are given in Table 8. After t2, the catalyst is subjected to the second transition (O_2-CO), and a second CO_2 peak is observed. We suggest that during t2, all the preabsorbed CO reacts, and is replaced on the surface by O_2. After t2, CO_2 is formed by the surface reaction of preadsorbed O_2 with CO.

ii) Fresh catalyst, reduction step (O_2-CO-O_2)

Once again, two CO_2 peaks are observed, however, their characteristics are somewhat different to those observed during the oxidation test (Table 8). During the O_2-CO step we observe an initial, rapid evolution of CO_2, thought to be due to the rapid surface reaction of coadsorbed O_2 and CO. This peak however, takes nearly 48 minutes to return to the baseline (compared to 6.9 minutes for the O_2-CO step in the oxidation test). We suggest that the reason for this is that, during the oxygen pretreatment, the catalyst surface becomes oxidised; switching to CO causes the oxide to be reduced which is shown to be a long process.

iii) CO/O_2 reaction over a reacted catalyst

In order to further study the processes occurring on the catalyst surface after exposure to the full gas mixture, a sample of the catalyst was heated to 450°C, under the conditions given in the footnote to Table 1. After reaching 450°C, the catalyst was cooled to 300°C under N_2 and the step-change experiments were carried out as previously described. The results are reported in Table 8.

For the oxidation test, both the amounts of CO_2 formed, and the baseline widths, are different to those obtained for the fresh catalyst. This result suggests that exposure of the catalyst to the full reaction mixture causes a rearrangement of the surface structure. For the reduction test, we see that the CO_2 peak areas, and the corresponding baseline widths, for both the fresh and reacted catalysts are similar, showing that, as opposed to a reducing pretreatment, pretreating the catalyst in oxygen leads to a surface composition which is well defined, regardless of the state of the surface prior to the pretreatment.

Table 8. Peak areas, and baseline widths for the CO/O_2 reaction observed by the step-change procedure for the 'fresh' and 'reacted'catalysts

	CO-O_2-CO[a] Fresh	**O_2-CO-O_2[b] Fresh**	**CO-O_2-CO[a] Reacted**	**O_2-CO-O_2[b] Reacted**
First Peak Area (ppm.s)	1.1×10^4	1.0×10^5	2.4×10^4	1.0×10^5
First Peak Width (mins)	2.3	47.9	14.8	41.3
Second Peak Area (ppm.s)	3.3×10^4	1.8×10^4	4.9×10^4	1.9×10^4
Second Peak Width (mins)	6.9	2.5	11.9	4.5

[a] t1 = 10 mins; tN = 15 mins; t2 = 15 mins.
[b] t1 = 10 mins; tN = 15 mins; t2 = 60 mins.

4. CONCLUSIONS

It has been shown in this publication that interesting surface phenomena occur on the surface of a Pt-Rh/Al_2O_3 catalyst when exposed to mixtures of CO, NO, O_2 and C_3H_8. We have observed that, for the bimetallic catalyst, the 50% conversion temperatures do not increase with mass in the expected manner, whereas the expected correlation is observed for the Pt/Al_2O_3 and Rh/Al_2O_3 single metal catalysts, implying effects associated with the bimetallic system. By varying the oxidising/reducing strengths of the reaction media, and also the frequency of cycling between the two media we have provided evidence for an inhibiting effect of oxygen. By comparison with results obtained under steady-state conditions we have shown that this effect is a transient phenomenon associated with switching between different reaction compositions.

Using step-change experiments on fresh and aged catalysts we have shown that exposure to reaction conditions does indeed cause a surface rearrangement to occur. When the catalyst is exposed to a reducing or stoichiometric reaction environment, we observe an activity attributed to reduced surface metals, whereas under oxidising conditions, the catalytic activity is due entirely to Pt, with Rh forming the inactive surface oxide Rh_2O_3; this phenomenon occurs regardless of the state of the surface prior to oxidation.

Acknowledgements

This work was carried out within the 'Groupement de Recherche sur les Catalyseurs de Postcombustion'. Thanks are due to the Institut Français du Petrole for the grant for C.H.

References

1 H.S. Gandhi, M. Shelef, *Catalysis and Automotive Pollution Control*, A. Crucq and A. Frennet (Editors), Elsevier Science Publishers, Amsterdam, 1987, 199.
2 Proceedings from CAPoC1 and CAPoC2, *Catalysis and Automotive Pollution Control*, A. Crucq and A. Frennet (Editors), Elsevier Science Publishers, Amsterdam, 1987 and 1991.
3 B. Harrison, A.F. Diwell, C. Hallett, Platinum Metals Rev., 32(2), 1988, 73.
4 R.K. Herz, *Catalysis and Automotive Pollution Control*, A. Crucq and A. Frennet (Editors), Elsevier Science Publishers, Amsterdam, 1987, 427.
5 J.C. Schlatter, R.M. Sinkevitch, P.J. Mitchell, Ind. Eng. Chem. Prod. Res. Dev., 22, 1983, 51.
6 J.C. Schlatter, P.J. Mitchell, Ind. Eng. Chem. Prod. Res. Des. Dev., 19, 1980, 288.
7 T. Engel, G. Ertl, J. Chem. Phys., 69, 1978, 1267.
8 B.E. Niewenhuys, Surf. Sci., 126, 1983, 307.
9 T. Engel, G. Ertl, Adv. Catal., 28, 1979, 1.
10 W.C. Hecker, A.T. Bell, J. Catal., 59, 1979, 223.
11 S.H. Oh, G.B. Fisher, J.E. Carpenter, D.W. Goodman, J. Catal., 100, 1986, 360.
12 S.B. Schwartz, G.B. Fisher, L.D. Schmidt, J. Phys. Chem., 92, 1988, 389.
13 H.A.C.M. Hendrickx, A.M.E. Winkelman, B.E. Niewenhuys, Appl. Surf. Sci., 27, 1987, 458.
14 H. Abderrahim, D. Duprez, *Catalysis and Automotive Pollution Control*, A. Crucq and A. Frennet (Editors), Elsevier Science Publishers, Amsterdam, 1987, 359.
15 J.A. Anderson, J. Catal., 142, 1993, 153.
16 Se.H. Oh, C.C. Eickel, J. Catal., 128, 1991, 526.
17 A.G. van den Bosch-Dreibergen, M.N.H. Kieboom, A. van Dreumel, R.M. Wolf, B.E. Niewenhuys, Catal. Lett., 2, 1989, 73.

18 A.G. van den Bosch-Dreibergen, M.N.H. Kieboom, A. van Dreumel, R.M. Wolf, F.C.M.J.M. van Delft, B.E. Niewenhuys, Catal. Lett., 2, 1989, 235.
19 J.Siera, R. van Silfhout, F. Rutten, B.E. Niewenhuys, *Catalysis and Automotive Pollution Control*, A. Crucq and A. Frennet (Editors), Elsevier Science Publishers, Amsterdam, 1991, 395.
20 F.C.M.J.M. van Delft, G.H. Vurens, M.C. Angevarre-Gruter, B.E. Niewenhuys, *Catalysis and Automotive Pollution Control*, A. Crucq and A. Frennet (Editors), Elsevier Science Publishers, Amsterdam, 1987, 229.
21 W.B. Williamson, H.S. Gandhi, P. Wynblatt, T.J. Truex, R.C. Ku, A. I. Chem. Eng. Symposium Series, 76(201), 1980, 212.
22 H. Kobayashi, M. Kobayashi, Catal. Rev. Sci. Eng., 10, 1974, 138.
23 M. Weibel, Ph.D. Thesis, Université Louis Pasteur, Strasbourg, 1991.
24 C. Howitt, V. Pitchon, Results to be published.

A. Frennet and J.-M. Bastin (Eds.)
Catalysis and Automotive Pollution Control III
Studies in Surface Science and Catalysis, Vol. 96

COMPARISON OF Pt/MnO$_X$/SiO$_2$ AND Pt/CoO$_X$/SiO$_2$ CATALYSTS FOR THE CO OXIDATION WITH O$_2$ AND THE NO REDUCTION WITH CO

Y. J. Mergler, A. van Aalst, J. van Delft and B. E. Nieuwenhuys

Gorlaeus Laboratories, P.O. Box 9502, 2300 RA Leiden, The Netherlands.

ABSTRACT

Pt/MnO$_x$/SiO$_2$ and Pt/CoO$_x$/SiO$_2$ catalysts are efficient catalysts for the reduction of NO with CO and the oxidation of CO with O$_2$ at low temperature, with a much lower on-set temperature for the reactions compared with Pt/SiO$_2$. An oxidative pretreatment lead to a shift to higher light-off temperatures for O$_2$ and NO conversion. For a pure CoO$_x$/SiO$_2$ catalyst primarily N$_2$O was formed following an oxidation step.

Over Pt/CoO$_x$/SiO$_2$ CO is already oxidised at room temperature. Three possible models to account for the high activity of Pt/CoO$_x$/SiO$_2$ in the CO/O$_2$ reaction are presented and discussed.

The highest NO conversion was reached by using the Pt/MnO$_x$/SiO$_2$ catalyst. It was found that N$_2$O is an important intermediate compound in CO/NO reactions over the Pt/MnO$_x$/SiO$_2$ catalyst. The performance of Pt/CoO$_x$/SiO$_2$ in CO/NO reactions is only slightly better than Pt/SiO$_2$.

1. INTRODUCTION

The present three-way catalyst (TWC) consists of the precious metals Pt (or Pd) and Rh. [1, 2] Platinum is very active in converting CO and hydrocarbons (hc) into the harmless compounds CO$_2$ and H$_2$O. It is , however, not an efficient catalyst for the reduction of NO to N$_2$ since it produces NH$_3$ under net reducing conditions. Whether N$_2$, N$_2$O or NH$_3$ is formed as a product of NO reduction over Pt depends on the reaction conditions [3-5]. Rhodium is highly selective towards N$_2$ formation [4-6].

At present, more than 80% of the total worldwide production of rhodium is used in TWCs. According to stricter legislations to reduce automotive pollution, this percentage is expected to increase even further. As a result, rhodium may become scarce and more expensive. Another important point is that most of the hc and CO is emitted following the cold start of the engine [7]. This has led to an urgent need to develop catalysts with improved activity at lower temperatures.

On early research connected with automobile exhaust treatment many research groups studied the CO oxidation by O_2 and by NO over several supported metal oxides [8]. Shelef et al. [8] reported that the oxides of Co, Cu, Mn and Ni are the most active of the oxides examined in CO oxidation [9]. Supported metal oxides showed promising results in the treatment of automotive exhaust. A few years later, however, it was realised that supported noble metals performed better with respect to the conversion of CO, hc and NO_x, and are less susceptible to poisoning by S or P [1, 10].

Several studies were performed to explore the use of metal oxides in combination with supported Pt or Pd as a possible replacement for Rh. Regalbuto et al. and Gandhi et al. used WO_3 and MoO_3 respectively as an additive to Pt to improve their catalysts for CO/NO reactions [11, 12]. Muraki et al. and Halasz et al. added La_2O_3 and MoO_3 respectively to their Pd supported catalysts [13, 14].

Continuing attempts are also being made to develop a TWC without noble metals. For example, Stegenga et al. [15] studied the combination of Cu-Cr supported on alumina catalyst. At the moment, a fair number of studies involve the use of zeolites, especially for NO_x reduction under lean conditions. This concept is not new, because Fe-Y-Zeolites were already investigated by Fu et al. [16]. Cu-Zeolites also exhibit reasonable characteristics towards NO_x reduction [17-19]. However, it is doubtful if a catalyst with no noble metals in it can replace the noble metal containing TWC for the present generation of gasoline-fuelled engines, in view of the strict legislations concerning automotive exhaust gases and catalyst lifetime.

Recently, Pd-only catalysts have been introduced as commercial TWC [20]. These catalysts contain Pd as the only precious metal and, in addition, a number of non-classified oxides. This success demonstrates that the development of Rh-free TWC is no longer a utopia. However, it is unlikely that the Pd-only TWC can replace the Pt-Rh TWC on a large scale due primarily to the limited availability of Pd for TWC compared to Pt. In our studies, we have concentrated on the performance of Pt-only catalysts (Pt in combination with oxides) as a possible substitute for the Pt-Rh TWC.

In our laboratory it has already been established that Pt/MnO_x and Pt/CoO_x have promising catalytic properties for CO oxidation and NO reduction. In this paper we will compare the activity of $Pt/MnO_x/SiO_2$ and $Pt/CoO_x/SiO_2$ catalysts in CO oxidation by O_2 and NO reduction by CO. Special attention will be given to the low temperature oxidation of CO.

2. EXPERIMENTAL

$H_2Pt(OH)_6$, dissolved in hot nitric acid, was used as a precursor and added to a silica suspension. A 5 w% Pt/SiO_2 catalysts was made by urea decomposition [21]. After drying, this catalyst was reduced for 3 hours in flowing H_2 at 400°C. Part of this reduced catalyst was impregnated with a cobalt nitrate solution. Another part with a manganese nitrate solution. The catalysts contained, after calcination in air at 400°C, respectively 3 w% of Co_3O_4 (atomic ratio Pt:Co = 1:1.5) and 25 w% MnO_2 (atomic ratio Pt:Mn = 1:15). A 3 w% Co_3O_4/SiO_2 catalyst was made for comparison. In addition, a 50 w% CoO_x/SiO_2 was prepared for experiments using FTIR. These catalysts will be referred to as $Pt/CoO_x/SiO_2$, $Pt/MnO_x/SiO_2$ and CoO_x/SiO_2 respectively.

The catalysts were reduced in flowing H_2, or oxidised in air, at 400°C for 3 hours, prior to the activity measurements. The measurements were performed in an atmospheric flow apparatus. The gases used were 4 vol% NO/He, 4 vol% CO/He and 4 vol% O_2/He (Hoekloos). The flow rate could be adjusted with mass flow controllers to a maximum of 40 ml/min. The CO/O_2 and CO/NO ratios were varied from oxidising to reducing. Product and reactant analysis occurred by means of a mass spectrometer. The conversion of O_2 or NO was plotted against the temperature at reducing or stoichiometric gas mixtures, while the conversion of CO was used at oxidising gas mixtures. The temperature was raised with 3°C/min to 400°C.

X-Ray diffraction, CO chemisorption and Fourier transform infrared spectroscopy were used for characterisation.

Self-supporting disks were used for the IR measurements. Before pressing the catalyst into disks, the catalyst was first reduced in flowing H_2 for 3 hours at 400°C. The disks were placed in an IR cell, that could be evacuated to 10^{-5} mbar. The IR cell was placed in a sample compartment of a Fourier infrared spectrometer. (Galaxy 2020 from Mattson) This compartment, however, was not purged with an inert gas. As a result, no information about the possible formation of products like CO_2 could be extracted. Thirty-two scans were taken with a resolution of 4 cm^{-1} for one spectrum. The catalysts were all reduced or oxidised in-situ, with 100 mbar H_2 or O_2, followed by an evacuation at 10^{-5} mbar for 30 min. Background spectra were taken at several temperatures at 10^{-5} mbar. Sample spectra were taken at several temperatures after the addition of 5 mbar CO or NO or a mixture of 10 mbar CO and NO.

3. RESULTS AND DISCUSSION

3.1. Characterisation

Table 1 shows the characteristics of the catalysts as measured by CO chemisorption and X-Ray Diffraction.

Table 1. Characteristics of the catalysts

Catalyst	mole CO/g cat.	Disp. (%)	XRD (red)	XRD (ox)
Pt/SiO_2	$4.7\ 10^{-5}$	18	Pt, 160 Å	-
$Pt/MnO_x/SiO_2$	$1.9\ 10^{-5}$	10	Pt, 170 Å MnO, 150 Å	Pt, 200 Å MnO_2, 190 Å
$Pt/CoO_x/SiO_2$	$2.6\ 10^{-5}$	-	Pt, 55 Å	Pt, 70 Å

It was assumed that one CO molecule adsorbs per surface Pt atom. The dispersion of $Pt/CoO_x/SiO_2$ was not calculated because CoO_x also adsorbed CO. This can also be seen in the following FTIR spectra. By using X-ray diffraction, only the characteristic peaks of Pt^o were found, regardless the pretreatment.

Figure 1 shows the absorption bands when 5 mbar CO and 5 mbar NO were co-adsorbed on the Pt/SiO_2 catalyst. Only one band appeared, at 2070 cm^{-1}, which can be ascribed to linearly bonded CO on Pt^o [22]. There were no bands indicating NO adsorption. At higher temperatures no isocyanate complex bound to Pt was detected. (not shown). Figure 1 also shows the adsorption of 5 mbar CO on 50 w% CoO_x/SiO_2 after a reductive pretreatment. (FTIR experiments of 3 w% CoO_x/SiO_2 revealed hardly any bands of CO and NO adsorption). A small band at 2178 cm^{-1} was found, which disappeared after heating above 100°C. This indicates that CO is adsorbed to an ionic form of cobalt. Busca et al. [23] identified a band at 2180 cm^{-1} when they adsorbed 100 mbar CO on pure, outgassed cobalt oxide. They ascribed this band to CO bonded on unsaturated Co^{3+} ions. We assign the band at 2178 cm^{-1} to CO bonded to co-ordinative Co^{3+}.

The adsorption of 5 mbar NO on CoO_x/SiO_2 is also shown in figure 1. Two characteristic bands for NO bound to cobalt oxide were visible at 1858 and 1784 cm^{-1}. They were assigned to linear and bent NO, respectively [24, 25]. Co-adsorption of CO and NO led only to bands of linear and bent NO on cobalt oxide. No indications of the formation of isocyanate groups were found.

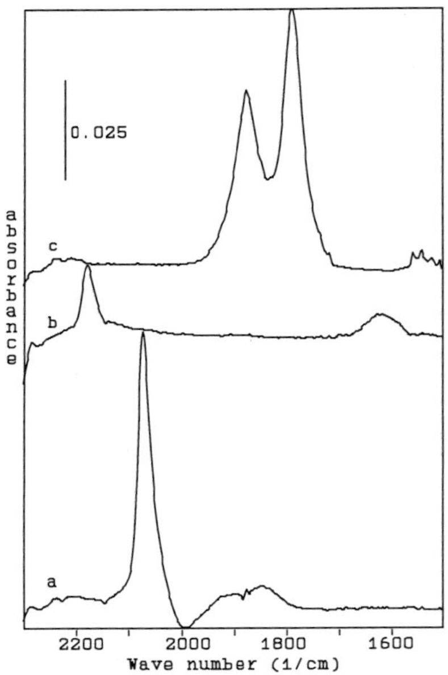

Figure 1.

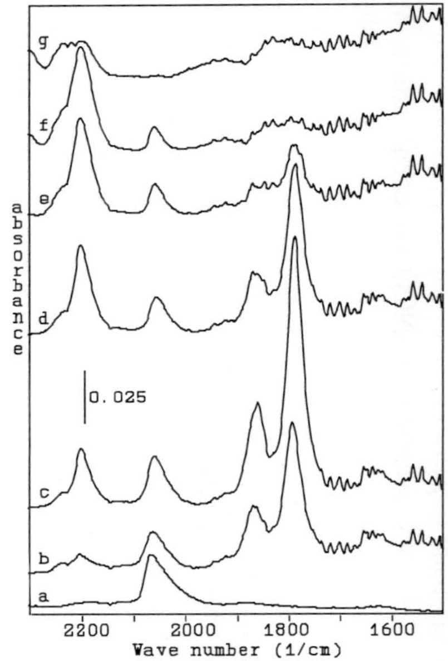

Figure 2.

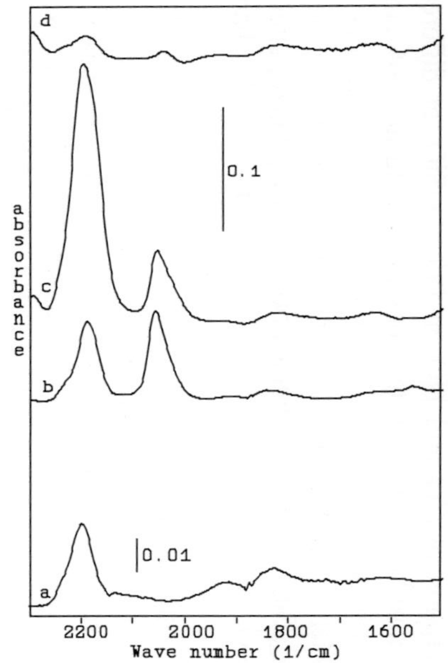

Figure 3.

Figure 1. FTIR spectra: a) Co-adsorption of 5 mbar CO with 5 mbar NO on Pt/SiO_2 at 100°C, b) Adsorption of 5 mbar CO on 50 w% CoO_x/SiO_2 at 25°C, c) Adsorption of 5 mbar NO on CoO_x/SiO_2 at 100°C.

Figure 2. FTIR spectra: a) 5 mbar CO introduced to $Pt/CoO_x/SiO_2$ at room temperature, b) Introduction of 5 mbar NO at 100°C, c) 150°C, d) 200°C, e) 250°C, f) 300°C, g) 350°C.

Figure 3. FTIR spectra: a) Co-adsorption of 5 mbar CO and 5 mbar NO on MnO_x/SiO_2 at 250°C, b) Co-adsorption of 5 mbar CO and 5 mbar NO on $Pt/MnO_x/SiO_2$ at 250°C, c) 350°C, t = 14 min, d) 350°C, t = 46 min.

Figure 2 shows the adsorption of CO and NO on $Pt/CoO_x/SiO_2$ at different temperatures. When CO was introduced at room temperature a band at 2066 cm^{-1} appeared, ascribed to linear CO bound to Pt^o. This band at 2066 cm^{-1} is broader at the base, but no distinct shoulder or band around 2020 cm^{-1} could be distinguished. This was different for a $Pt/CoO_x/SiO_2$ catalyst containing 50 w% CoO_x where, after a reductive pretreatment at 400°C, also a band at 2023 cm^{-1} was found. This band arose from linearly bonded CO on Co^o. (These results are not shown here). After the introduction of NO at 100 °C, bands of linear and bent NO were found at 1867 and 1792 cm^{-1}, respectively. The band at 2060 cm^{-1} is ascribed to linearly bonded CO on Pt^o. A new band at 2201 cm^{-1} showed up at this temperature, which increased in intensity when the temperature was raised to 250°C. This band at 2200 cm^{-1} was neither found upon the co-adsorption of CO and O_2 on the reduced $Pt/CoO_x/SiO_2$ catalyst, nor after the addition of CO to an oxidised $Pt/CoO_x/SiO_2$ catalyst. In conclusion, the band at 2200 cm^{-1} could not be ascribed to CO adsorbed on Co^{3+}. Yao and Shelef [26] found a small band at 2190 cm^{-1} on pure cobalt oxide when CO and NO were co-adsorbed, which they ascribed to -NCO on cobalt oxide. We therefore believe that the band at 2200 cm^{-1} originates from isocyanate on cobalt oxide.

When the temperature was increased from 200 to 250°C, the bands of linear and bent NO on cobalt oxide almost disappeared, while the -NCO band increased. More NO is dissociated at higher temperatures, leading to more N_a and more -NCO.

The formation of -NCO can proceed on Pt [22]. Migration of the -NCO group to the carrier is possible. This process is rapid or slow, depending on the carrier. Solymosi et al. and Anderson et al. found that the migration of -NCO to SiO_2 is a very slow process. On alumina, however, this process is so rapid that no isocyanate on Pt could be distinguished [27, 28]. No -NCO bands were found on Pt/SiO_2 upon the co-adsorption of CO and NO, up to 300°C. This can mean either that 300°C is too low to start the dissociation of NO or that the -NCO reacts further with NO to N_2 and CO_2 [29]. The mechanism by which the -NCO formation takes place is not yet clear. Is -NCO formed on Pt with a concomitant migration to cobalt oxide? Alternatively, does CO_a on Pt metal spill-over to N_a on cobalt oxide? Or does the reaction between N_a and CO_a take place only on cobalt oxide? The latter model may explain why the -NCO formation starts at temperatures as low as 100°C, while at 100°C the NO dissociation on Pt^o is unlikely due to the inhibition of Pt^o sites by CO. However, in our opinion, the third model is also rather unlikely since we did not find any -NCO upon the co-adsorption of CO and NO over our CoO_x/SiO_2 catalyst. It must be noted, however, that Yao and Shelef did detect -NCO on their pure cobalt oxide [26]. Model two is the most likely of all the models presented above. This means that an increase in the intensity of the 2200 cm^{-1} band will be accompanied by a decrease in the intensity of the Pt-CO band. Figure 2 shows that this is the case between 200 and 250 °C.

Isocyanate groups were formed above 250°C when CO and NO were co-adsorbed on MnO_x/SiO_2 (figure 3). When NO and CO were added to $Pt/MnO_x/SiO_2$, the formation of -NCO started at around 150°C. The isocyanate band grew in intensity when the temperature was raised to 350°C. The bands of Pt-CO and -NCO were not stable at this temperature, because they disappeared after 45 minutes. We detected no bands of linear or bent NO, or CO-bands on manganese oxide.

3.2. CO oxidation by O_2

Table 2 shows the results obtained for the Pt/SiO_2, $Pt/CoO_x/SiO_2$ and $Pt/MnO_x/SiO_2$ catalysts after a reductive pretreatment and after an oxidative pretreatment in CO oxidation by O_2.

Table 2. Temperature (°C) of 50 % O_2 (CO) conversion at several CO/O_2 ratios after a reductive or an oxidative pretreatment.*

		Ratio	$CO : O_2$	
Catalyst	Pretreatment	3 : 1	2 : 1	1 : 1
Pt/SiO_2	Reduction	204	210	175
	Oxidation	210	210	187
$Pt/CoO_x/SiO_2$	Reduction	29	26	15
	Oxidation	103	103	101
$Pt/MnO_x/SiO_2$	Reduction	125	114	94
	Oxidation	162	173	150

* Conversion of O_2 in case of reducing or stoichiometric gas mixtures and conversion of CO in case of oxidising gas mixtures.

The performance of $Pt/CoO_x/SiO_2$ is superior to $Pt/MnO_x/SiO_2$ and Pt/SiO_2. Over Pt/ CoO_x /SiO_2 CO is even oxidised at room temperature after a reductive pretreatment. Below 100°C all the CO (or O_2) was converted. The temperature of 50% CO (or O_2) conversion increased by 75°C to a higher temperature after an oxidative pretreatment. Hence, the performance of $Pt/CoO_x/SiO_2$ is still much better than those of $Pt/MnO_x/SiO_2$ and Pt/SiO_2. $Pt/MnO_x/SiO_2$ showed an intermediate behaviour compared to the performance of Pt/SiO_2 and $Pt/CoO_x/SiO_2$. An oxidative pretreatment increased the temperature of 50% conversion by only 25 - 50°C.

Oxidation or reduction as pretreatment had no significant effect on the CO oxidation by O_2 over Pt/SiO_2. Upon comparing the activity of the Pt/SiO_2 catalyst for the various CO/O_2 ratios, it was found that CO inhibition occurred. This effect was less marked for the $Pt/MnO_x/SiO_2$ and $Pt/CoO_x/SiO_2$ catalysts.

The question arises why $Pt/CoO_x/SiO_2$ shows such a good CO oxidation behaviour, especially after a reductive pretreatment. The $Pt/CoO_x/SiO_2$ catalyst is more active after a reduction than after an oxidation. After an oxidative pretreatment Pt^o was detected using X-ray diffraction with an average particle size of 70 Å (table 1). With higher weight percentages cobalt oxide, CoPt, $CoPt_3$ and Co^0 peaks could be detected following reduction at 400°C. (Not shown here). It is possible to form alloy particles of Pt and cobalt under our reductive pretreatment conditions, but in the case of the $Pt/CoO_x/SiO_2$ with 3 w% Co_3O_4 no alloy particles could be identified with XRD.

When they studied the CO oxidation reaction, Meunier et al. found that Co^{2+} ions in $Pt/CoO_x/Al_2O_3$ catalysts dissociate oxygen more easily than Co^{3+} [30]. Partially reduced CoO_x is apparently necessary for O_2 dissociation. Pande and Bell [31] found a high activity of Rh/TiO_2 for the NO reduction by H_2. The authors attributed the promoting effect of TiO_x to the presence of catalytically active TiO_x centres on the support as well as on the surface of the supported Rh crystallites.

Low-temperature CO oxidation by O_2 proceeds between 25 and 100°C over a Au/MnO_x catalyst, as was shown by Gardner et al. [32]. The Au/MnO_x catalyst performed even better than the commercially available Pt/SnO_x in CO oxidation reactions [32]. No mechanism was presented to account for the observed activities for these catalysts. It has been suggested by Nieuwenhuys [33] that MnO_x serves as dissociation centres for O_2. Oxygen atoms can spill-over from MnO_x to Au, in the presence of the reactant gases. Gold itself is not active in O-O bond scission. However, O adatoms are stable on Au at room temperature. In the presence of CO the O adatoms can easily react with CO due to the low Au-O bond strength [33].

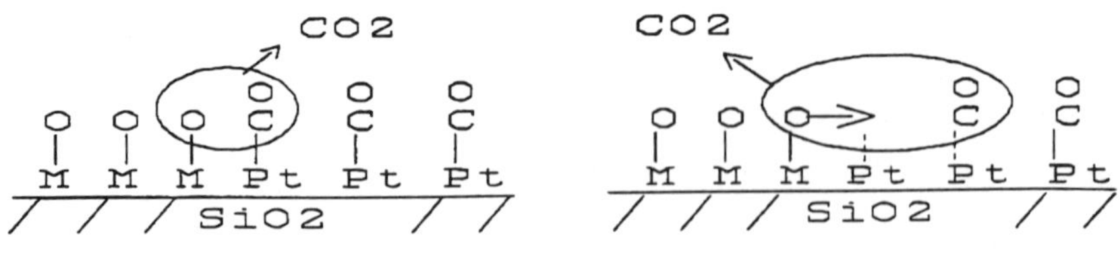

Figure 4. Second and third model for the reaction of CO with O_2 over $Pt/CoO_x/SiO_2$. (See text).

CoO_x may affect the adsorption of CO or O_2 on Pt. Since at low temperatures the reaction rate on Pt is determined by the slow adsorption of oxygen due to CO inhibition, it is most likely that CoO_x serves as O-supplier for the reaction. No influence of cobalt oxide on the CO adsorption on Pt was detected by IR measurements. If we assume that Pt-Co alloy formation does not play an important role in the CO/O_2 reaction over Pt/CoO_x/SiO_2, several models may account for the observed effects. According to our first model, cobalt cations enhance the adsorption of O_2 on Pt by an increased electron back-donation into the anti-bonding orbitals of O_2, which facilitates O_2 dissociation. The increased back donation may be induced by the electrical field of the cobalt cations. The second model is shown schematically in figure 4. CO is adsorbed on Pt. O_2 dissociates on CoO_x and the dissociation may be assisted by the presence of O-vacancies present on cobalt oxide. CO_a on Pt will react with O_a on cobalt. This reaction will then take place at the interface between Pt and CoO_x. It is also possible that O_a migrates from the CoO_x to Pt, in which case the reaction proceeds on the Pt surface (third model). The authors are in favour of the last two models since Pt itself is already able to dissociate O_2 around 100 K if free Pt sites are available (no CO inhibition) [33].

3.3. NO reduction by CO

The two possible reaction pathways are:

$$2CO + 2NO \dashrightarrow N_2 + 2CO_2 \qquad (1)$$
$$CO + 2NO \dashrightarrow N_2O + CO_2 \qquad (2)$$

N_2O is believed to be an important intermediate product in the CO/NO reaction [9, 34]. At higher temperatures N_2O may react further according to (3) or may decompose [34]. The following reaction can proceed:

$$CO + N_2O \dashrightarrow N_2 + CO_2 \qquad (3)$$

Since the masses of CO/N_2 and CO_2/N_2O are equal and the fact that our mass spectrometer is not sensitive enough to monitor masses 14 of N, 16 of O accurately, only the differences in activity of the Pt/SiO_2, 3 w% Co_3O_4/SiO_2, Pt/CoO_x/SiO_2 and Pt/MnO_x/SiO_2 catalysts could be measured.

Figures 5a to 5c and 6a to 6c show the results of the catalysts in the NO reduction by CO after a reductive or an oxidative pretreatment, respectively.

The sequence of the activity of the catalysts in the CO/NO reaction is different from that of the CO/O_2 reaction. Pt/MnO_x/SiO_2 is more active than Pt/CoO_x/SiO_2. Figure 6a also shows the NO conversion over CoO_x/SiO_2 after an oxidative pretreatment. It is noteworthy that the maximum in NO conversion was around 300°C for Pt/MnO_x/SiO_2 and for CoO_x/SiO_2 (fig 6a).

Reaction 2 is favoured at lower temperatures, because then the NO coverage is high and the N coverage low, resulting in a high rate of N_2O formation and a low rate of N_2 formation [35-37]. Figure 5a shows the presence of a maximum in the NO conversion around 300°C for the CoO_x catalyst. A similar maximum in NO conversion was also found in NO/H_2 reactions over Pt/CoO_x/SiO_2 catalysts [38]. This maximum corresponds to the maximum in N_2O formation. Thus it seems that the maximum in conversion observed for the CO/NO reaction around 300°C is also caused by N_2O formation. N_2O decomposes around 350°C [9, 38], leading to a maximum in the NO conversion curve. This suggests that the main reaction over the oxidised cobalt oxide catalyst is reaction 2 at lower temperatures.

The assumption that reaction 2 is important at low temperatures over oxide catalysts may also explain the excellent activity of Pt/MnO_x/SiO_2 at lower temperatures. At higher temperatures reaction 3 proceeds, but primarily over the Pt species of the Pt/MnO_x/SiO_2 catalyst

The NO conversion at lower temperatures of Pt/MnO_x/SiO_2 can be explained by the preferential formation of N_2O, as stated above. The activity of Pt/CoO_x/SiO_2 in CO/NO reactions was only slightly better than Pt/SiO_2. Since cobalt oxide is able to dissociate NO (fig 6a) at around 200°C, it is difficult to understand why no N_2O formation starts at around 200°C on Pt/CoO_x/SiO_2. In our opinion this is due to the formation of -NCO on cobalt oxide, which blocks the active sites for NO adsorption and dissociation. NO starts to dissociate on cobalt oxide but a rapid formation and build-up of -NCO retards the dissociation on cobalt oxide. Platinum is then the only species available for NO dissociation, leading to temperatures of 50 % NO conversion that are almost equal to the temperatures of 50 % NO conversion over Pt/ SiO_2.

Excess CO results in the formation of more -NCO [29]. With excess NO the reaction of isocyanate with NO is expected, leading to N_2 and CO_2 production [39]. Isocyanate is regarded as a spectator species [29]. The higher activity of the Pt/CoO_x/SiO_2 catalyst with excess NO, can be explained by a decrease in the blockade by -NCO of the active CoO_x species. Preadsorbed oxygen stabilises the -NCO band [40]. This may explain the small differences in the temperatures of 50 % NO conversion after a reductive or an oxidative pretreatment; after an oxidation the -NCO is stabilised and affects the activity by blocking cobalt oxide, resulting in higher temperatures of 50% NO conversion.

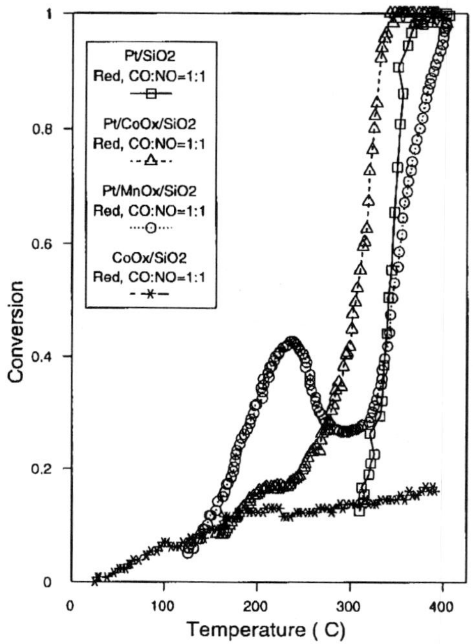

Figure 5a.

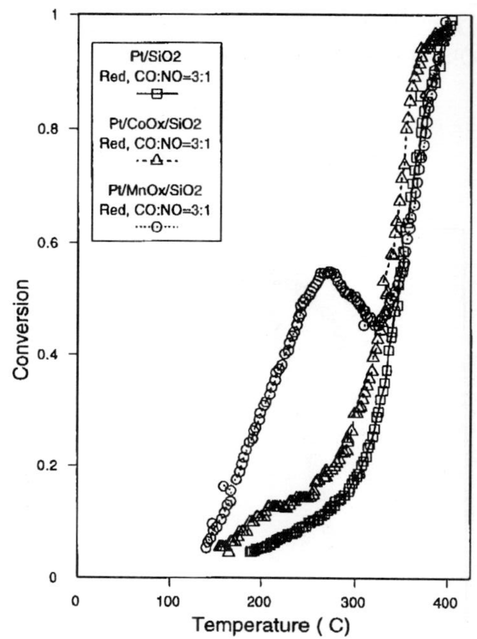

Figure 5b.

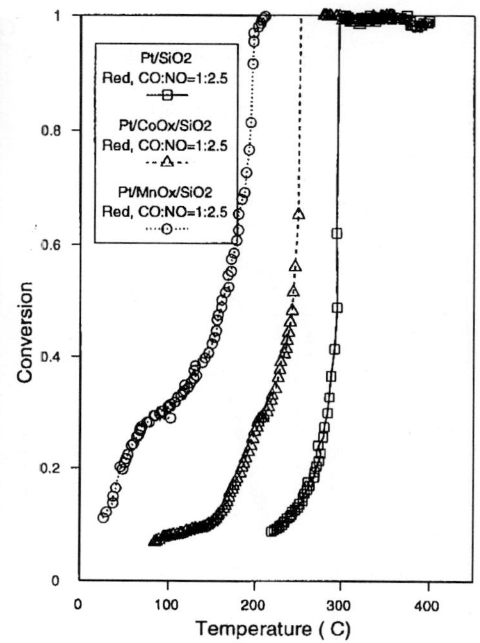

Figure 5c.

Figure 5. NO conversion versus temperature in the NO reduction by CO over Pt/SiO_2, $Pt/CoO_x/SiO_2$ and $Pt/MnO_x/SiO_2$ after a reductive pretreatment with a CO:NO ratio of a) 1:1, b) 3:1, and c) 1:2.5 (CO conversion).

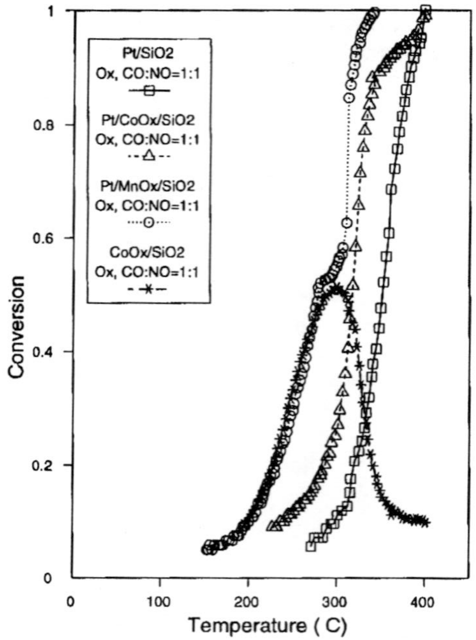

Figure 6a.

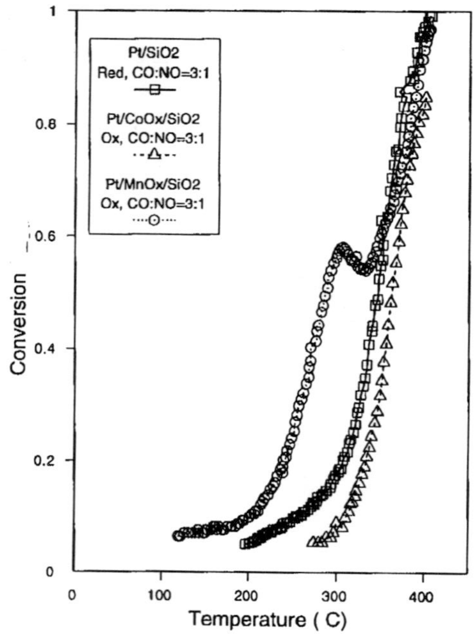

Figure 6b.

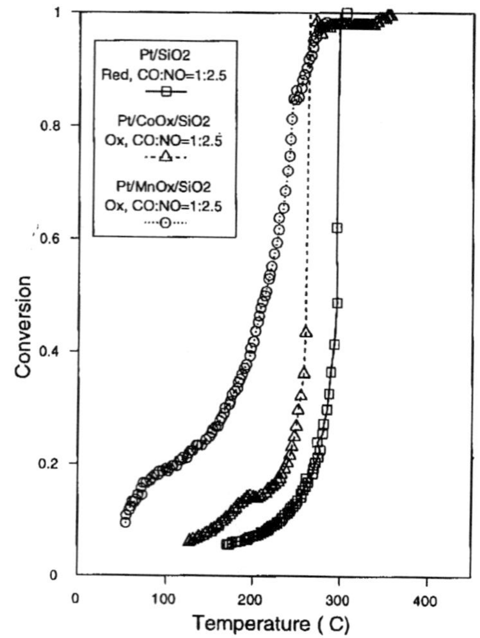

Figure 6c.

Figure 6. NO conversion versus temperature in the NO reduction by CO over Pt/SiO_2, $Pt/CoO_x/SiO_2$ and $Pt/MnO_x/SiO_2$ after an oxidative pretreatment with a CO:NO ratio of a) 1:1, b) 3:1, and c) 1:2.5 (CO conversion).

3.4. Comparison of $Pt/MnO_x/SiO_2$ and $Pt/CoO_x/SiO_2$ in CO/O_2 and CO/NO reactions

The results presented above show that a partially reduced catalyst performed better in CO oxidation and NO reduction. Partially reduced centres are apparently necessary to enhance low temperature O_2 and NO dissociation. In the case of Rh single crystal surfaces, Wolf et al. [35] found direct evidence that vacancies are required for the dissociation of NO molecules. A high coverage of molecularly adsorbed NO inhibits the reaction rate at low temperatures because of the limited availability of vacancies on the surface of the noble metal. Partially reduced metal oxides may provide for the dissociation centres for NO. This was also shown by Tomishige et al., who studied the promoting effect of Sn on NO dissociation and NO reduction with H_2 over Rh-Sn/SiO_2 catalysts [41]. They found that NO even dissociated at room temperature on a reduced Rh-Sn/SiO_2 catalyst and that the rapid dissociation stopped when one-third of surface Sn atoms was oxidised by NO.

An oxidative pretreatment increases the temperatures of 50 % conversion, but overall the activity is still better than that of Pt/SiO_2. Cobalt and manganese oxides can be reduced very easily, especially when Pt^o is present [41]. CoO_x is itself active in CO/NO, especially after an oxidative pretreatment.

$Pt/CoO_x/SiO_2$ is very good in low-temperature CO oxidation, while $Pt/MnO_x/SiO_2$ shows an intermediate behaviour between $Pt/CoO_x/SiO_2$ and Pt/SiO_2. In CO/NO reactions, however, $Pt/MnO_x/SiO_2$ is better than $Pt/CoO_x/SiO_2$ when the NO conversion activity is compared, but the selectivity is to N_2O at lower temperatures.

The isocyanate bands on $Pt/MnO_x/SiO_2$ and $Pt/CoO_x/SiO_2$ are not stable above 350°C. No bands of adsorbed CO or NO were found on manganese oxide, while linear and bent NO were clearly visible on cobalt oxide. The decoration with -NCO, resulting in an inhibition of the CO/NO reaction at lower temperatures, took place over $Pt/CoO_x/SiO_2$.

4. Conclusions

$Pt/MnO_x/SiO_2$ and $Pt/CoO_x/SiO_2$ catalysts have a lower onset temperature for the reaction of CO with O_2 or NO. An oxidative pretreatment led to a shift to higher temperatures for O_2 and NO conversion. In the case of a pure CoO_x/SiO_2 catalyst, primarily N_2O was formed following an oxidation step.

When a $Pt/CoO_x/SiO_2$ catalyst was used, CO was even oxidised at room temperature. A number of possible models have been discussed to explain the high activity of $Pt/CoO_x/SiO_2$ in the CO/O_2 reaction. In our opinion the models for CO oxidation by O_2 in which O-vacancies on the CoO_x play an important role as dissociation centres for O_2 are the most acceptable models.

$Pt/CoO_x/SiO_2$ is the most active catalyst in CO oxidation at low temperature, while $Pt/MnO_x/SiO_2$ had the highest NO conversion in CO/NO reactions. However, N_2O is an important primary reaction product, using the $Pt/MnO_x/SiO_2$ catalyst. The performance of $Pt/CoO_x/SiO_2$ in CO/NO reactions was rather disappointing. Its activity was only slightly better than Pt/SiO_2. This may be due to the formation of isocyanate groups on the cobalt surface, leading to blocking of the active sites needed for dissociation of NO.

ACKNOWLEDGEMENT

We would like to thank the Johnson Matthey Technology Centre (Reading, U.K.) for the loan of the Pt salt. This research was financially supported by SON/STW, project number 349 1878.

REFERENCES

1 K. C. Taylor, in "Automotive Catalytic Converters", Springer, Berlin (1984)
2 J. T. Kummer, J. Phys. Chem., 90 (1986) 4747
3 K. Otto, and H. C. Yao, J. Catal., 66 (1980) 229
4 W. F. Egelhof, Jr., in "Chemical Physics of Solid Surfaces and Heterogeneous Catalysis" 4, ch. 9. King, D. A. and Woodruff, D. P. (eds), Elsevier, Amsterdam, (1982)
5 B. E. Nieuwenhuys, J. Siera, K. I. Tanaka and H. Hirano, in "Proc. NO reduction Symposium, ACS", 1993, Denver, in press
6 L. Heezen, V. N. Kilian, R. F. van Slooten, R. M. Wolf and B. E. Nieuwenhuys, in "Stud. Surf. Sci. and Catal.", A. Crucq and A. Frennet (eds), (1991) 381
7 R. C. Rijkeboer, Catalysis Today, 11 (1991) 141
8 M. Shelef, K. Otto, and H. Gandhi, J. Catal., 12 (1968) 361
9 M. Shelef, and K. Otto, J. Catal., 10 (1968) 408
10 J. C. Summers, J. Catal., 57 (1979) 380
11 a) J. R. Regalbuto, T. H. Fleisch, and E. E. Wolf, J. Catal., 107(1987) 115;
b) J. R. Regalbuto, C. W. Allen, and E. E. Wolf, J. Catal., 108 (1987) 305;
c) J. R. Regalbuto, and E. E. Wolf, J. Catal., 109 (1988) 13
12 a) H. S. Gandhi, H. C. Yao, and H. K. Stepien, ASC Symp. Series, 178 (1982) 143;
b) H. S. Gandhi, and M. Shelef, Proc. CAPOC 1986
13 a) H. Muraki, K. Yokota, and Y. Fujitani, Appl. Catal., 48 (1989) 93;
b) H. Muraki, H. Shinjoh, and Y. Fujitani, Appl. Catal., 22 (1986) 325

14 I. Halasz, A. Brenner, and M. Shelef, Catal. Lett., 16 (1992) 311
15 S. Stegenga, Thesis, University of Amsterdam, Sept. 1991
16 C. M. Fu, M. Deeba, W. K. Hall, Ind. Eng. Chem. Prod. Res. Develop., 19 (1980) 299
17. M. Shelef, Catal. Lett., 15 (1992) 305
18 Y. Li, and W. K. Hall, J. Catal., 129 (1991) 202
19 M. Iwamoto, H. Yahiro, K. Tanda, N. Mizuno, Y. Mine, and S. Kagawa, J. Phys. Chem., 95 (1991) 3727
20 J. C. Summers, and Williamson, W. B., in: *Environmental Catal. ACS Symposium Series 552*, Armor, J. N., (eds.), 1994, 94
21 J. W. Geus, Dutsch Patent Application 6, 705 (1967) 259
22 G. M. Alikina, A. A. Davydov, and I. S. Sazonova, Kin. i Kataliz., 27 (1986) 875
23 G. Busca, R. Guidetti, and V. Lorenzelli, J. Chem. Soc. Faraday Trans., 86 (1990) 989
24 H. Niiyama, and E. Echigoya, J. Catal., 38 (1975) 238
25 T. Ishihara, K. Eguchi, and H. Arai, Chem. Lett., (1986) 1695
26 H. C. Yao, and M. Shelef, J. Phys. Chem., 78 (1974) 2490
27 F. Solymosi, L. Völgyesi, and J. Sárkány, J. Catal., 54 (1978) 336
28 J. A. Anderson, and C. H. Rochester, J. Chem. Soc. Faraday Trans., 86 (1990) 3809
29 W. C. Hecker, and A. T. Bell, J. Catal., 85 (1984) 389
30 G. Meunier, F. Garin, J. Schmitt, G. Maire, and R. Roche, Stud. Surf. Sci. and Catal., 30 (1987) 243
31 a) N. K. Pande, and A. T. Bell, J. Catal. 98 (1986) 7; b) Appl. Catal. 20 (1986) 109
32 S. D. Gardner, G. B. Hoflund, B. T. Upchurch, D. R. Schryer, E. J. Kielin, and J. Schryer, J. Catal. 129 (1991) 114
33 B. E. Nieuwenhuys, in "Elementary Reaction Steps in Heter. Catal.", R. W. Joyner and R. A. van Santen (eds), Kluwer, (1993) 155
34 B. K. Cho, B. H. Shanks, and J. E. Bailey, J. Catal. 115 (1989) 486
35 R. M. Wolf, J. W. Bakker, and B. E. Nieuwenhuys, Surf. Sci. 246 (1991) 135
36 H. Hirano, T. Yamada, K. I. Tanaka, J. Siera, and B. E. Nieuwenhuys, in "New Frontiers in Catalysis", Guczi, L. et al. eds., 345 (1993) (Proc. 10th Int. Congr. Catal., 1992, Budapest)
37 H. Hirano, T. Yamada, K. I. Tanaka, J. Siera, and B. E. Nieuwenhuys, Surf. Sci. 97 (1992) 262
38 Y. J. Mergler, A. van Aalst, and B. E. Nieuwenhuys, accepted for publication in: Proceedings of the ACS Symposium on NO_x reduction, San Diego, 1994
39 H. Niiyama, M. Tanaka, H. Iida, and E. Echiogoya, Bull. Chem. Soc. Jpn., 49 (1976) 2047
40 F. Solymosi, and J. Rasko, Appl. Catal., 10 (1984) 19
41 K. Tomishige, K. Asakura, and Y. Iwasawa, J. Chem. Soc., Chem. Commun., (1993) 184

A. Frennet and J.-M. Bastin (Eds.)
Catalysis and Automotive Pollution Control III
Studies in Surface Science and Catalysis, Vol. 96

REDUCTION BEHAVIOR OF Rh-Sn/SiO_2 BIMETALLIC CATALYSTS AND ITS CO OXIDATION ACTIVITY

Satoru Nishiyama, Ikuo Yamamoto, Masahiro Akemoto, Shigeru Tsuruya, and Mitsuo Masai

Department of Chemical Science and Engineering, Faculty of Engineering, Kobe University, Rokkodai, Nada, Kobe 657, Japan

ABSTRACT

Reduction behavior of Rh-Sn on SiO_2 was studied by temperature programmed reduction methods. During the reduction of Rh-Sn/SiO_2 preoxidized, Rh oxide was reducible in the range of 370-440 K first, then the reduction of Sn oxide proceeded at a higher temperature up to 470 K. X-ray diffraction powder patterns indicates that the higher temperature than 673 K was required in order to obtain well-mixed Rh-Sn bimetallic system. The catalyst reduced at 673 and 773 K indicated the high activity for CO oxidation reaction at low reaction temperatures. These results indicate the reduction behavior in the following; Rh oxide was reduced first and metallic Rh was formed, then Sn oxide was reduced by the spillover hydrogen which was formed on the metallic Rh particles. Finally, bimetallic system of Rh and Sn, e.g. Rh-Sn alloy, was formed at the higher temperature than 673 K. The obtained bimetallic surface was markedly active for CO oxidation reaction.

1. INTRODUCTION

Supported Rh- and Ru-Sn catalysts have shown high activity for reduction of NO by CO [1]. Especially, the effective removal of NO was observed in O_2-rich atmosphere. The high removal of NO in the oxidative atmosphere was attributable to a cleaning effect of Sn in the bimetallic system by spillover of oxygen atoms adsorbed on Rh or Ru sites to Sn sites [2]. Recently, we have also reported that oxidation of CO over the silica supported Rh catalysts was enhanced by incorporation of Sn [3]. The temperature which CO oxidation started was lowered by the addition of Sn [3]. These studies indicate that Sn is an effective additive in CO oxidation reaction itself as well as NO reduction. The XPS study [4] have indicated that the Sn in Rh-Sn/SiO_2 was reduced in a lower oxidation state, may be in Sn^0 - Sn^{2+}. The reduced Sn plays a significant role in activation of O_2 molecules [4]. Mixing state between Rh and Sn may be important for CO oxidation, because an oxidation state of Sn is strongly influenced by the adjacent Rh atoms. Preparation method and activation condition (temperature and atmosphere) seem to be very important to control the mixing state of the bimetallic system. Dautzenberg et al. [5] have reported the interaction between Pt and Sn over Al_2O_3. The TPR (temperature programmed reduction) of the oxidized Pt-Sn/Al_2O_3 indicated that the reduction of Pt and Sn was greatly affected in the presence of Sn. They have also reported that Sn was not completely reduced to metallic state over Al_2O_3. Burch has reported the details of TPR spectra of Pt-Sn/Al_2O_3 [6]. Srinivasan et al. [7, 8] have reported that the microelectron diffraction method in order to study micro crystals of Pt-Sn system on Al_2O_3 and SiO_2. They have found the alloy formation on Al_2O_3 support by the micro diffraction method, whereas no peak was observed in X-ray diffraction powder patterns. These results suggest that the alloy formation between Pt and Sn is not so difficult even over Al_2O_3, which strongly interacts with Sn species and may stabilize the state of Sn^{2+}. Even in those case, 70 % of Sn was not reduced to metallic Sn on Al_2O_3. The state of Sn was strongly affected by loading of Sn and Pt and condition of pretreatment, that is, atmosphere and temperature. Bacaud et al. have reported that the percent of reduction of Sn was strongly affected by the content of Pt and Sn [9]. The starting material of Sn and the preparation method were important for the reduction state of Sn as reported by Sexton et al. [10]. The silica is a suitable support for reduction of Sn species and alloy formation [10]. In our case, SiO_2 support, the high content of Rh and Sn (5 wt% each), and 773 K of reduction temperature will bring about the reduction to metallic Sn and the alloy formation might be expected.

In this paper, we report the detail of reduction behavior of Rh-Sn/SiO_2 by temperature programmed reduction method. The reduction temperature was one of the important factors to control the oxidation state of Sn in Rh-Sn bimetallic system. The relationship between the reduction behavior of Rh-Sn system and the activity for CO oxidation is studied.

2. EXPERIMENTAL

2.1. Catalyst preparation

The details of the preparation method of the catalysts was described in the previous paper [4]. Rh-Sn/SiO_2 was prepared by a conventional impregnation method. The silica gel was co-impregnated with a mixed aqueous solution of $Rh(NO_3)_3$ (Nakarai Tesque Inc., Kyoto Japan) and $SnCl_2$ (Nakarai Tesque Inc., Kyoto Japan). Loading of Rh was at 5 wt% with respect to the support and Sn was introduced in unity of Sn/Rh atomic ratio. It was reported that the efficient Sn/Rh ratio was unity for activation of oxygen molecules [4, 11]. The catalyst precursor prepared above was activated by calcination in flowing air followed by reduction in flowing H_2 at prescribed temperatures for 5 h, respectively for adsorption experiments and for CO oxidation reaction. For temperature programmed reduction, the precursor was used directly without any thermal treatment, that means as impregnated samples were used, in order to investigate effects of treatment temperature, given below.

2.2. Temperature programmed reduction (TPR)

Temperature programmed reduction was carried out in a flowing H_2 (5.14 vol%) diluted by N_2 with a flow system equipped with a thermo balance (Model DGC-40, Shimadzu Co., Kyoto). A furnace in the thermo balance was controlled by the thermal analyzer DT-40 (Shimadzu Co., Kyoto). The catalyst precursor was calcined in a flowing air at a prescribed temperature for 0.5 h in the system, then the sample was cooled in the flowing air to room temperature. The carrier gas, N_2 or H_2/N_2 mixture, was dehydrated by molecular sieve just before the TPR cell. The calcined precursor was heated in flowing H_2/N_2 mixture in a constant rate of 10 K/min. A typical heating pattern, the calcination was carried out at 773 K, is shown in Fig. 1. The obtained responses weight loss or gain (TG) and difference of temperatures between the sample cell in which the catalyst was placed and the reference cell in which the SiO_2 support was placed (DTA) were accumulated in a personal computer, and were processed with a soft ware supplied by Shimadzu Co. (Kyoto, Japan). It should be noted that the weight of the sample was kept in ca. 25 mg in order to obtain precise and reproducible results.

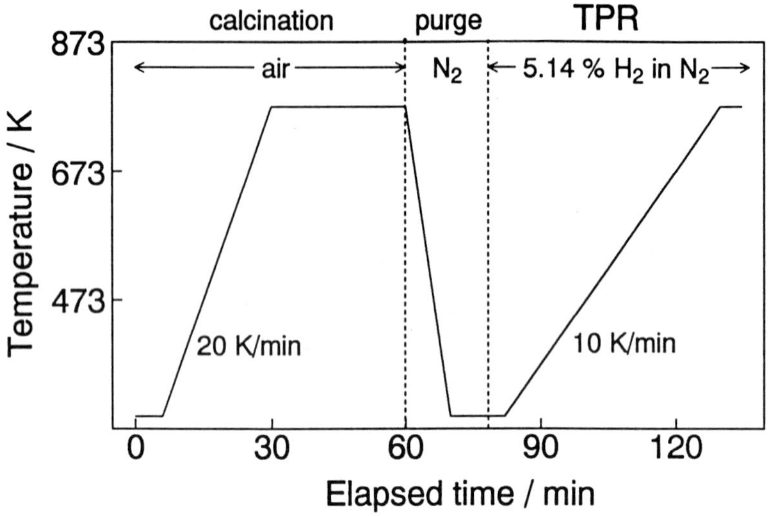

Figure 1. Typical heating pattern in the TPR experiment

2.3. Adsorption

Adsorption capacity of the catalysts for H_2 and CO was measured in a static vacuum system. The adsorption was carried out at room temperature in 10 kPa of each gas. The capacity was evaluated by the difference between total amount of adsorption and reversible adsorption. The *absorption* of H_2 at 373 K over the calcined precursor at was also measured in the system.

2.4. X-ray diffraction powder pattern

The mixing-state of supported Rh and Sn was investigated by X-ray diffraction powder patters. The X-ray powder pattern of the catalysts was obtained by using Ni filtered Cu-Ka.

2.5. CO oxidation

Oxidation of CO was carried out in a closed circulation system. The catalyst was reduced again in a reactor in 13 kPa of H_2 *prior to* the reaction. The reactant gas was 15.9 kPa of $CO\text{-}O_2$ mixture, which molar ratio was two in CO/O_2. The conversion of CO was calculated from decrease of total pressure in the system The reaction rate was evaluated by an initial rate method.

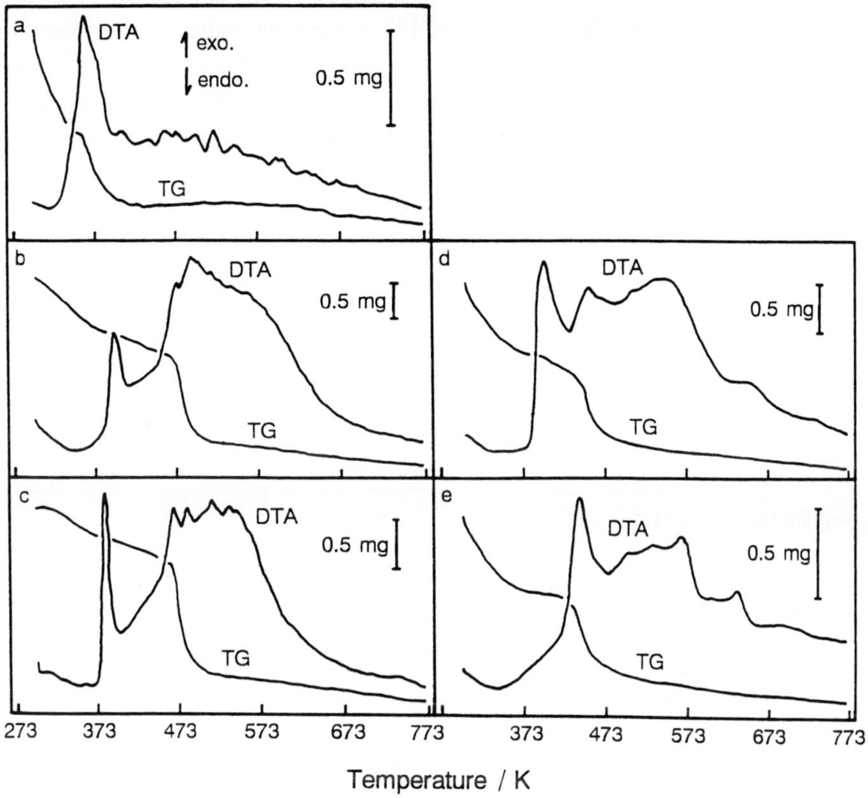

Figure 2. TPR spectra of Rh/SiO2 and Rh-Sn/SiO2 catalysts
a: Rh/SiO2 calcined at 773 K, b: Rh-SiO2 calcined at 573 K
c: at 673 K, d: at 773 K, e: at 873 K

3. RESULTS AND DISCUSSION

3.1. Reduction behavior of Rh-Sn/SiO_2

Temperature programmed reduction spectra of Rh/SiO_2 and Rh-Sn/SiO_2 calcined at different temperatures were shown in Fig. 2. The spectrum **a** indicates a strong exothermic peak at 360 K over Rh/SiO_2 (in DTA line) accompanying with the weight loss which began at 356 K (in TG line). This result indicates that

Rh oxide supported on SiO_2 was reduced at c.a. 360 K. The spectra **b** to **e** show DTA-TG lines over Rh-Sn/SiO_2. The DTA spectra indicate that two major peaks, one was sharp near 370 K, the other was a broad beyond 420 K. The first sharp peak was similar to reduction of Rh oxide over Rh-Sn/SiO_2, which accompanied no weight loss (TG). The formed H_2O was not desorbed from the catalyst below 400 K. The position of peaks in the DTA lines obtained over Rh-Sn/SiO_2 catalyst are summarized in Table 1.

Table 1 Temperature of first DTA peak and weight loss stated

Catalyst		Temperature first peak	Temperature weight loss started
Rh/Si O_2	773K*	360K	356K
Rh-Sn/SiO_2	573K	395K	469K
	673K	383K	465K
	773K	400K	442K
	873K	442K	428K

* calcination temperature

The temperature of the first peak except the spectrum **c**, which was calcined at 673 K, tended to shift toward higher temperature with increasing calcination temperature as shown in Table 2 and Fig. 2. The reducibility of Rh oxide was lowered with addition of Sn. The *H_2-absorption* capacity of the calcined catalysts at 773 K was measured at room temperature and 373 K in order to study the reducibility of the catalysts as shown in Table 2.

Table 2 H_2-absorptionof Rh and Rh-Sn/SiO_2

Absorption temperature		Amount of absorbed H_2 µmole/g-cat Rh/SiO_2	Rh-Sn/SiO_2
room temp.	1st dose	189	5.2
	2nd dose	136	2.8
	total	325	8.0
	1st dose	600	484
	2nd dose	111	158
	total	711	642

At room temperature, a significant amount of absorption was observed over Rh/SiO_2, whereas no hydrogen was absorbed over $Rh-Sn/SiO_2$ as shown in Table 2. At 373 K, 711 μmole/g of H_2 was absorbed over Rh/SiO_2, which was almost same as the theoretical value of H_2 consumption to reduce Rh^{3+} to Rh^0, 694 μmole/g-H_2 for 5 wt% Rh/SiO_2. Over $Rh-Sn/SiO_2$, the H_2 consumption was 642 μmole/g, which was identical to the theoretical value to reduce only Rh^{3+} to Rh^0, 658 μmole/g. These results indicate that Rh oxide was more easily reduced to metallic Rh in Rh/SiO_2 than in $Rh-Sn/SiO_2$ and that the reduction peak in the range of 370-440 K in TPR spectra was ascribed to the reduction of Rh oxide in $Rh-Sn/SiO_2$. No weight loss at 400 K in TPR spectra means that desorption of H_2O from the catalyst required much higher temperature than the reduction of Rh oxide. The absorption experiment also indicated that H_2 absorption was observed without H_2O evolution at room temperature and 373 K over Rh/SiO_2 and $Rh-Sn/SiO_2$. The calcination at a high temperature will result in a Sn-enriched surface of $Rh-Sn/SiO_2$ because the affinity of Sn for oxygen is greater than that of Rh. The enrichment of Sn on the surface would suppress the reduction of Rh oxide. Burch [6] has also reported that the peak corresponding the reduction of Pt oxide in the calcined $Pt-Sn/Al_2O_3$ was shifted toward higher temperature with increasing Sn content. The second broad peak over 420 K was ascribed to the reduction of Sn oxide, which was accompanied by the weight loss. The weight loss corresponded not only to reduction of Sn oxide but also to the decrease of weight by the desorption of H_2O which had been formed during the reduction of Rh oxide described above. Enough reduction of the calcined precursors were required reduction at 573 K as shown in Fig. 2. The weight loss was summarized in Table 3.

Table 3 Weight loss and corresponding oxygen eliminated

catalyst		weight loss (mg/g-cat)	corresponding O_2 molecules μmole/g-cat	O/Rh (atomic ratio)
Rh/SiO_2	773K*	15.8	0.49	2.1
$Rh-Sn/SiO_2$	573K	39.9	1.25	5.7
	673K	38.9	1.22	5.2
	773K	23.2	0.72	3.3
	873K	14.8	0.46	2.1

* calcination temperature

The corresponding eliminated-oxygen calculated from the weight loss was also shown in the table. The theoretical value of the *O/Rh* was estimated at **3.5** in order to reduce both Rh (III) and Sn (IV) to the corresponding metals. The larger values than 3.5 in Table 3 would be ascribed to the residual water which came from the carrier gas, although the carrier gas was dehydrated by molecular sieve just before the DTA-TG cell. The original adsorbed water on the catalysts would not affect the weight loss because the sample was calcined at 573 K before TPR experiment, which was higher than the desorption temperature (350-440 K) of H_2O as shown in Fig. 2. The larger values observed over Rh-Sn/SiO_2 than over Rh/SiO_2 suggested that the significant amount of Sn in Rh-Sn/SiO_2 was readily reduced to lower oxidation state, Sn (0). The previous paper [11] has also indicated that the reduced Sn/SiO_2 at 773 K did not absorb oxygen during oxidation experiment at 773 K. Tin in Sn/SiO_2 was hardly reduced at 773 K in H_2 [11]. The TPR spectra of Sn/SiO_2 (5 wt% of Sn) calcined at 773 K showed no reduction peak up to 773 K. These results indicate that the reduction of Sn oxide was catalyzed by the presence of an active component, Rh metal, that is, the spillover hydrogen which was formed on metallic Rh particles reduced the Sn oxide in Rh-Sn/SiO_2 catalyst. Recently, Aranda et al. [12] have reported that the reduction of Sn was catalyzed by the presence of Pt over Al_2O_3. They have also found that ca. 28 % of Sn (IV) was reduced to Sn (0). The similar behavior was observed over Nb_2O_5 support [12]. In the case of Al_2O_3, the interaction between Sn and Al_2O_3 may be too strong to reduce Sn oxide to metallic Sn at 773 K completely. Our results indicate that a part of Sn in Rh-Sn/SiO_2 is readily reduced to metallic Sn and some part of reduced Sn might be alloyed with Rh. Srinivasan and Davis have also reported the formation of Pt-Sn alloy on silica support by X-ray diffraction and electron micro diffraction [8].

3.2. Adsorption capacity

The adsorption capacity of the catalysts were shown in Tables 4 and 5 for H_2 and CO. Table 4 indicates the influence of reduction temperature on the capacity.

The adsorption capacity for H_2 and CO was drastically decreased by addition of Sn. Especially, the amount of H_2 adsorbed decreased by 10 to 20, whereas CO decreased by 4. This result was ascribed that hydrogen molecules required at least two adjacent atoms of Rh to be adsorbed dissociatively. This is well known as an *ensemble effect* for bimetallic catalysts. The adsorption capacity of the Rh-Sn/SiO_2 catalyst which was reduced at 873 K was larger than that reduced at 473 K twice for H_2 and CO adsorption. Table 5 indicates the

Table 4 Effect of reduction temperature on adsorption capacity

catalyst		H_2 (μmole/g-cat)	CO (μmole/g-cat)
Rh/SiO_2	773K*	52.5	103.2
$Rh-Sn/SiO_2$	473K	2.02	26.2
	573K	2.44	26.6
	673K	1.32	30.1
	773K	2.00	31.3
	873K	4.94	41.0

*reduction temperature

Table 5 Effect of calcination temperature on adsorption capacity

catalyst		H_2 (μmole/g-cat)	CO (μmole/g-cat)
Rh/SiO_2	773K*	52.5	103.2
$Rh-Sn/SiO_2$	573K	6.80	56.8
	673K	5.48	42.6
	773K	2.43	26.6
	873K	1.80	18.3

*calcination temperature

influence of calcination temperature on the capacity. Both the capacity for H_2 and CO was decreased with increasing calcination temperature. These results suggest that the higher reduction temperature brings about a Rh rich surface and that the higher calcination temperature brings about a Sn-rich surface. The Sn-enriched surface would suppress the reduction of Sn oxide itself as well as the reduction of Rh oxide as discussed above. The low value of oxygen consumption of Rh-Sn/SiO_2, calcined at 873 K, calculated from TPR spectra is ascribable to the Sn-enriched surface as shown in Table 3.

It should be checked whether Sn affects the dispersion of Rh particles or not. We have already studied the effect of Sn addition on the dispersion of Rh by comparing the XPS intensity and the adsorption capacity of H_2 and CO [11]. The relative XPS intensity of $Rh3d_{5/2}$ to a monolayer catalyst was larger than the theoretical value calculated from the adsorption capacity of H_2 and CO by the method of Kerkhof and Mouljin [13]. These results indicated that the decrease of the adsorption capacity of $Rh-Sn/SiO_2$ catalysts was ascribed not to the increase of isolated Rh particle size, but to surface composition of Rh-Sn bimetallic particles so called the *ensemble effect* and/or to chemical modification of Rh metals by adjacent Sn atoms so called a *ligand effect*.

3.3. X-ray diffraction powder pattern

Figure 3 shows the X-ray diffraction powder patterns of Rh-Sn/SiO_2 reduced at different temperatures. The catalyst which were reduced below 573 K showed no shift of Rh (111) diffraction peak as shown in Fig. 3-a. The reduction over 673 K brought about a shift toward lower diffraction angle, which indicates well-mixing between Rh and Sn [3, 4, 11]. The Sn oxide in the calcined precursor

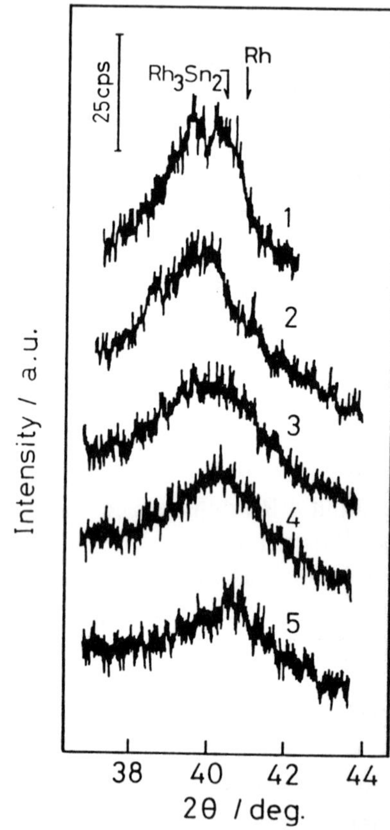

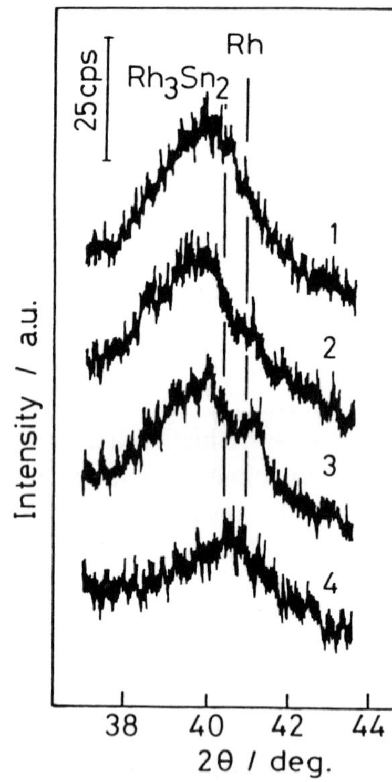

Figure 3. (left) XRD patterns of Rh-Sn/SiO_2 reduced at various temperatures 1: 873 K, 2: 773 K, 3: 673 K, 4: 573 K, 5: Rh/SiO_2 at 773 K

Figure 4. (right) XRD patterns of Rh-Sn/SiO_2 calcined at various temperatures 1: 873 K, 2: 773 K, 3: 673 K, 4: Rh/SiO_2 at 773 K

was enough reduced around 573 K as shown in Fig. 2. The mixing between the reduced Rh and Sn is required much higher temperature. Figure 4 shows the XRD powder patterns of the Rh-Sn/SiO_2 which was calcined at different temperatures followed by reduction at 773 K. The catalyst calcined at 673 K indicates both Rh and Rh-Sn structure. The calcination at a low temperature brought about the significant segregation of Rh and Rh-Sn. Although neither metallic Sn or Sn oxide was observed in the XRD patterns, a segregated Sn could not be excluded.

3.4. CO oxidation

Figure 5 shows the CO oxidation activity over Rh-Sn/SiO_2 catalysts which were reduced at different temperatures. The activity was evaluated with the apparent first order rate constant. The initial reaction rate for CO oxidation depended on partial pressure of O_2 in first order over Rh and Rh-Sn/SiO_2 described previously [3]. The dashed line indicates the activity over Rh/SiO_2. The activity over the catalyst reduced at 573 K was identical to that over Rh/SiO_2 as shown in Fig. 5.

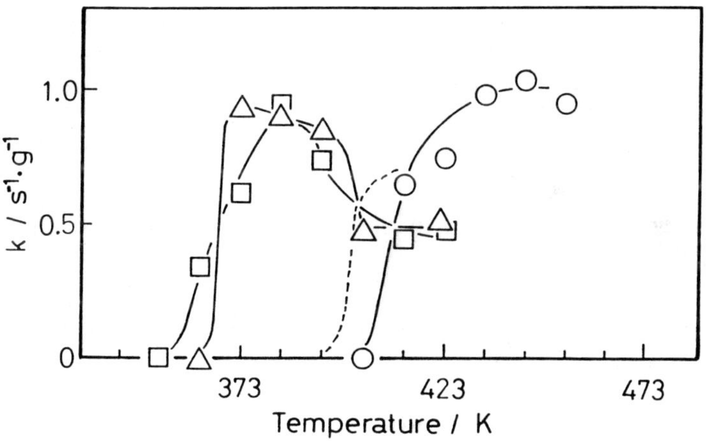

Figure 5. CO oxidation activity of Rh-Sn/SiO_2 reduced at various temperatures
O: reduced at 573 K, Δ: at 673 K, □: at 773 K
The dashed line means the activity of Rh/SiO_2 reduced at 773 K.

The catalysts reduced at 673 and 773 K indicated much higher activity at low reaction temperatures. The temperature which the activity appeared was lowered by 40-50 K. These results are well correlated to the peak shift in XRD powder patterns as shown in Fig. 3. The well-mixing of Rh and Sn brings about the high activity at low temperatures of Rh-Sn/SiO_2 for CO oxidation reaction. The surface composition and oxidation state would be changed from the state just after pretreatment because the catalysts were exposed to the oxygen containing atmosphere during the reaction, whereas only the metallic Rh phase (or Rh-Sn phase) was observed in the XRD patterns after the reaction. The unusual temperature dependence of the CO oxidation as shown in Fig. 5 has been reported [14] and would be ascribed to the modified surface composition of ad-layer, which was consisted of CO and O atoms, during the reaction discussed previously [3]. Although the change of the surface composition during the reaction seemed to be considered, the initiation temperature of the reaction would be affected directly by the original surface composition and electronic state.

3.5. State of Rh-Sn on SiO_2 and CO oxidation activity

The reduction behavior of Rh-Sn/SiO_2 can be summarized below.

1)Rhodium oxide of the calcined catalyst was reduced first in the range of 360 to 420 K.

2)Tin oxide was reduced over 440 K with catalyzing by metallic Rh.

3)The alloying between Rh and Sn proceeded at a higher temperature than 673 K during the reduction.

For CO oxidation reaction, the most important factor was the alloying between metallic Rh and reduced Sn, because the high activity of the catalyst required the reduction at higher temperature than 673 K as shown Fig. 5. The reduction up to 573 K may gave metallic Rh and the reduced low valent Sn (some part of Sn was reduced to Sn^0), which was separated each other. The separated Rh and Sn indicated the same activity for CO oxidation as Rh/SiO_2. The reduction at higher than 673 K would induce the mixing between Rh and Sn. The well-mixed Rh-Sn on SiO_2 seems to be markedly active for CO oxidation.

REFERENCES

1 M. Masai, K. Murata and M. Yabashi, Chem. Lett., (1979) 989.
2 M. Masai, K. Nakahara, K. Murata, S. Nishiyama, and S. Tsuruya, Stud. Surf. Sci. & Catal., 17 (1983) 89.
3 S. Nishiyama, M. Akemoto, I, Yamamoto, S. Tsuruya, and M. Masai, J. Chem. Soc. Faraday Trans., 88 (1992) 3483.
4 S. Nishiyama, H. Yanagi, H. Nakayama, S. Tsuruya, and M. Masai, Appl. Catal., 47 (1989) 25.
5 F. M. Dautzenberg, J. N. Helle, P. Biloen, and W. M. H. Sachtler, J. Catal., 63 (1980) 119.
6 R. Burch and L. C. Garla, J. Catal., 71 (1981) 360.
7 R. Srinivasan, L.A. Rice, and B.H. Davis, J. Catal., 129 (1991) 257.
8 R. Srinivasan and B.H. Davis, Appl. Catal. A, 87 (1992) 45.
9 R. Bacaud, P. Bussière, and F. Figueras, J. Catal., 69 (1981) 399.
10 B.A. Sexton, A.E. Hughes, and K. Foger, J. Catal., 88 (1984) 466.
11 S. Nishiyama, H. Yanagi, S. Tsuruya, and M. Masai, Proc. 2nd Intern. Conf. Spillover (Leipzig, 1989), (1989) pp 116-126.
12 D.A.G. Aranda, F.B. Noronha, M. Schmal, and F.B. Passos, Appl. Catal. A, 100 (1993) 77.
13 F.P.J.M. Kerkhof and J.M. Mouljin, J. Phys. Chem., 83 (1979) 1612.
14 G.K. Boreskov, in Catalysis, ed. J.R. Anderson and M. Bourdart, Springer-Verlag, Berlin, 1982, vol. 3, p.39.

A. Frennet and J.-M. Bastin (Eds.)
Catalysis and Automotive Pollution Control III
Studies in Surface Science and Catalysis, Vol. 96
1995 Elsevier Science B.V.

PREPARATION OF Pt-Rh / Al_2O_3-CeO_2 CATALYSTS BY SURFACE REDOX REACTIONS

L.Pirault [a], D. El Azami El Idrissi [a], P. Marécot [a], J.M. Dominguez [b], G. Mabilon [c], M. Prigent [c] and J. Barbier [a] .

[a] *Université de Poitiers . URA CNRS 350, Laboratoire de Catalyse en Chimie Organique, 40 Avenue du Recteur Pineau, 86022 Poitiers Cedex, France .*

[b] *Instituto Mexicano del Petroleo, I.A. Apdo Postal 14-805, Eje Central L. Cardenas 152, 07730 DF Mexico .*

[c] *Institut Français du Pétrole, 4-6 Rue du Bois Préau, 92506 Rueil Malmaison Cedex, France .*

ABSTRACT :

Three-way automotive Pt-Rh catalysts were prepared either by coimpregnation of the two noble metals (C.I. catalysts) or by an original method of successive impregnations with a reduction step after platinum impregnation (S.I. catalysts) . It was shown that addition of rhodium can lead to opposite effects on the activity of platinum deposited on alumina-ceria for the reaction of propane oxidation, according to the preparation procedure : the coimpregnation of the two metals induces an inhibition of platinum activity when the addition of rhodium by successive impregnations leads to the reverse effect . The differences between the two preparations are greater on oxidized samples after aging at high temperature (900°C). These results are explained by the formation of alloy particles on coimpregnated catalysts while rhodium added by successive impregnations would be selectively deposited on surface cerium, avoiding the formation of Pt-Rh alloy . Energy dispersive spectroscopy fitted to a STEM unit allowed to bear out such hypothesis.

1. INTRODUCTION

Typical commercial three-way catalysts contain both platinum for CO and hydrocarbon oxidation, and rhodium for NO reduction . However, an intimate interaction between platinum and rhodium with alloy formation results in decreased activity for hydrocarbon or CO oxidation [1-4] . This detrimental effect is generally observed over platinum-rhodium bimetallic catalysts prepared by coimpregnation of the two noble metals . Indeed, this preparation procedure leads to alloy formation and to the enrichment of metallic particle surface with relatively inactive rhodium oxides after high-temperature aging under oxidizing conditions [1-8], resulting in poor oxidation activity . Recently, we developped a preparation procedure of bimetallic Pt-Rh / Al_2O_3-CeO_2 catalysts by an original technique of successive impregnations (SI) in the course of which rhodium deposition would occur by a redox reaction between surface cerium atoms and rhodium ions [9] .

In this paper we investigated the properties of three-way automotive Pt-Rh / Al_2O_3-CeO_2 catalysts prepared either by simultaneous impregnation of the two noble metals (coimpregnation = CI) or by the technique of successive impregnations (S.I.) . Their catalytic performances were measured after reduction or calcination at 500°C, for the oxidation of a propane-propene mixture under lean conditions and for the reduction of NO by CO under stoichiometric mixture . Monometallic platinum catalysts and an alumina supported bimetallic platinum-rhodium catalyst were also prepared for a comparative study .

2. EXPERIMENTAL

2.1. Catalysts preparation

The supports used were a γ-Al_2O_3 with a BET area of 100 m^2 /g and the same alumina modified by addition of cerium chloride in order to obtain an alumina ceria with 12 wt % ceria after calcination at 450°C . Monometallic catalysts were prepared by impregnation of chloroplatinic acid . After drying, the different samples were calcined at 450°C and reduced under hydrogen at 500°C for 4 h and 8 h, respectively. Bimetallic catalysts were prepared by impregnation of chloroplatinic acid and rhodium chloride following two procedures : the first one was the classical coimpregnation of the support with aqueous solutions of platinum and rhodium salts (C.I. catalysts) . After drying at 120°C, the catalyst samples were calcined in air flow at 450°C for 4h, then reduced under hydrogen at the same temperature for 8 h . The latter method employs a two-step impregnation procedure (impregnation of platinum followed by impregnation of rhodium) with air-drying (120°C overnight), calcination (450°C, 4 h) and reduction (500°C, 8 h) steps between successive noble metal impregnations (S.I. catalysts). Following final impregnation, the S.I. catalyst was dried in air overnight at 120°C and reduced at 500°C for 8 h. The different catalysts were either dechlorinated at 500°C for 10 h in a stream of N_2 + 10% H_2O or thermally aged at 900°C for 16 h in a stream of 1% O_2, 10 % H_2O, 10 % CO_2, N_2 and then reduced at 500°C. The dechlorination treatment was carried out in order to avoid the inhibiting effect of chlorine on the activity of catalysts for hydrocarbon oxidation [10]. The chlorine content of dechlorinated samples was below 0.2wt%. Before activity measurements, catalysts were either reduced under hydrogen at 500°C or oxidized at the same temperature. Metal and ceria loadings of the different catalysts are reported in Table 1 .

Table 1 Metal and ceria loadings of the catalysts (CI : coimpregnated catalyst ; SI : catalyst prepared by successive impregnations) .

Catalyst	Ceria loading wt %	Platinum loading wt %	Rhodium loading wt %
Pt / Al_2O_3	/	1.0	/
Pt / Al_2O_3-CeO_2	12.0	1.0	/
PtRh / Al_2O_3 (CI)	/	1.0	0.1
PtRh / Al_2O_3-CeO_2 (CI)	12.0	1.0	0.1
PtRh / Al_2O_3-CeO_2 (SI)	12.0	1.0	0.1

2.2. Hydrocarbon oxidation activity measurements

Hydrocarbon oxidation was performed in a flow reactor system equipped with a flame ionization detector (FID) . The reactant mixture was composed of 0.2% propene and 0.2 % propane in N_2 with 2 % oxygen (5 % excess oxygen) . The gas flow rate was set at ca. 15000 ml h^{-1} . The catalyst weight was typically 50 mg diluted in 250 mg α-Al_2O_3 .

Catalysts were evaluated by studying their light-off behaviour with a constant flow of the gas mixture from 100° C to 500°C at a heating rate of 3°C min^{-1} .

2.3. Reaction of CO +NO

Reaction of CO + NO was conducted in a flow reactor at a space velocity of 20000 h^{-1} with a stoichiometric mixture containing 2.0 mole % NO and CO in helium. Catalysts were evaluated by studying their light-off behaviour in temperature programmed experiments from 20°C to 500°C at a heating rate of 2°C min^{-1} . Before the catalytic run, catalyst samples were either calcined or reduced under H_2 at 500°C for 2 hours . The analysis of the gas mixture was performed by gas chromatography using a CTR1 column provided by Alltech which allows to separate N_2, NO, CO, N_2O and CO_2 .

2.4. Elemental analysis (energy dispersive spectrometry EDS)

All the microanalysis work was carried out in a JEOL-100 CX electron microscope fitted with a STEM unit, an X-ray detector and a Tracor-Northern-5500 console . This system has a series of image processing routines such as filtering and particle recognition programs, as well as the standard programs and files for elemental analysis : SMTF and MICRO Q .

The compositional microanalysis was done by bombarding a sample region with an electron beam at 40 kV, thus provoking X-ray emission which carries out both the characteristic radiation and the continuous background from the region of interest . The X-ray signals which are generated in the microscope chamber are detected and transmitted by a Si / Li detector . The information is processed in the Tracor-Northern system . A typical spectrum is obtained in a couple of minutes from 0 to 40 keV ; in this energy region all the primary (K lines) or secondary (L,M,...lines) emission lines appear sharply . As each element of the periodic table has its own emission line family, a straightforward qualitative analysis may be done . The quantitative analysis requires a further treatment of the data by means of the computer programs . This is usually done by use of the standard metallurgical thin film or the MICRO Q programs, which provide the atomic and weight percents .

The EDS analysis may be done either from a relatively large region (average analysis) or from a 200 Å-wide region (point analysis), therefore one can obtain the compositional fluctuation around an average value . All the analysis are based on the direct relationship between the intensity of the X-ray signal and the concentration of the element under analysis : $I(A) / I(B) = K(AB)C(A) / C(B)$, but in most of the cases a further correction is applied for absorption and fluorescence effects [11 - 13] . The following X-ray lines were used for analysis : Al - Kα at 1.487 keV, Ce - Lα at 4.840 keV, Pt - Lα at 9.441 keV and Rh - La at 2.696 keV) .

3. Results

3.1. Propane - propene oxidation

Catalysts were evaluated by their light-off temperature in temperature programmed experiments under lean conditions (5% excess oxygen).

On dechlorinated samples, calcination at 500°C before hydrocarbon oxidation induces an inhibiting effect on ceria containing catalysts for propene oxidation (Fig 1a) . On the other hand, the inhibiting effect of ceria appears already after reduction for propane oxidation (Fig 1b)

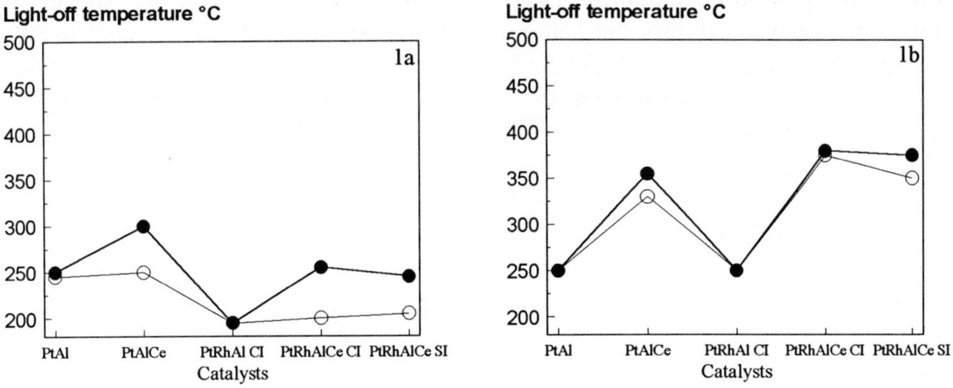

Figure 1. Effects of the pretreatment at 500°C (calcination (●) or reduction (○)) on the light-off temperatures of dechlorinated catalysts for propene (1a) and propane (1b) oxidation.

. Thus, dechlorinated catalysts exhibit the same activity for propane oxidation after calcination or reduction at 500°C . We must note also that the catalytic properties of bimetallic PtRh / Al_2O_3 - CeO_2 catalysts are not affected by their preparation procedure .

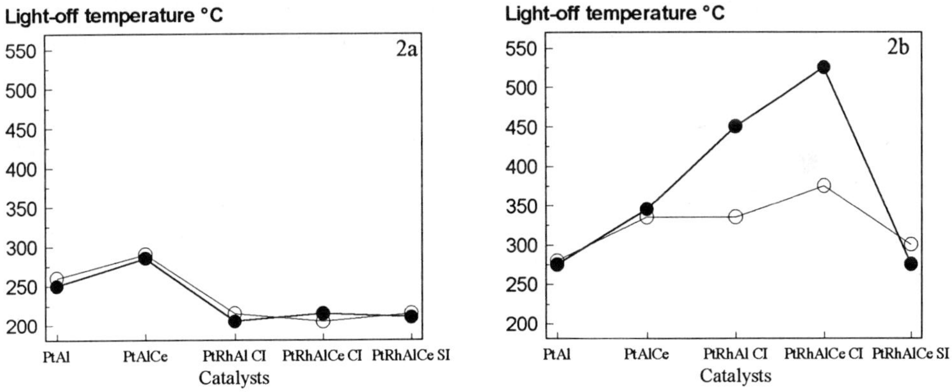

Figure 2. Effects of the pretreatment at 500°C (calcination (●) or reduction (○)) on the light-off temperatures of thermally aged (900°C) catalysts for propene (2a) and propane (2b) oxidation.

On thermally aged catalysts, the pretreatment (calcination or reduction at 500°C) does not affect catalyst activity for propene oxidation (Fig 2a) . On the other hand, the results reported in Fig 2b show that coimpregnated bimetallic platinum-rhodium catalysts endure an increase in their light-off temperature with respect to propane oxidation after an oxidizing treatment at 500°C . The same effect is observed when prereduced samples are submitted to a second oxidation cycle after a first cycle up to 850°C under lean conditions (Fig 3) . As for the bimetallic platinum-rhodium / alumina-ceria catalyst prepared by successive impregnations, it exhibits the same activity for propane oxidation whatever the nature of the pretreatment (Figs 2 - 3) .

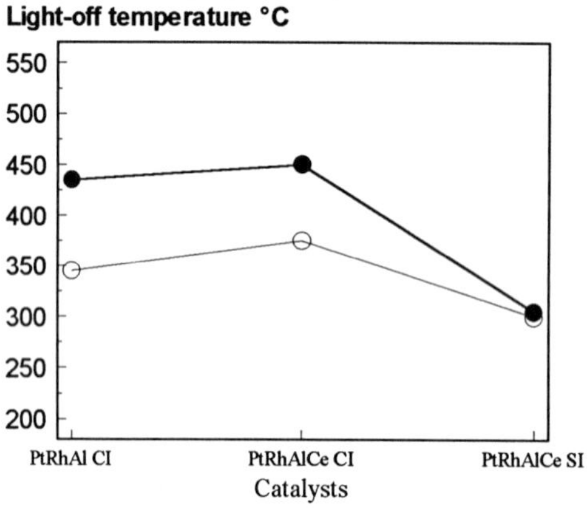

Figure 3. Effects of successive oxidation cycles under lean conditions on the light-off temperatures of prereduced thermally aged catalysts for propane oxidation (○ : first cycle; ● : second cycle ; final temperature of the first cycle : 850°C)

3.2. CO + NO reaction

The activity of the different catalysts was evaluated by the temperatures at 50% conversion for NO and CO. The selectivity of NO reactions was characterized by the amount of NO transformed into N_2O in the course of the temperature programmed experiment and by the temperature at 50% conversion of NO into N_2.

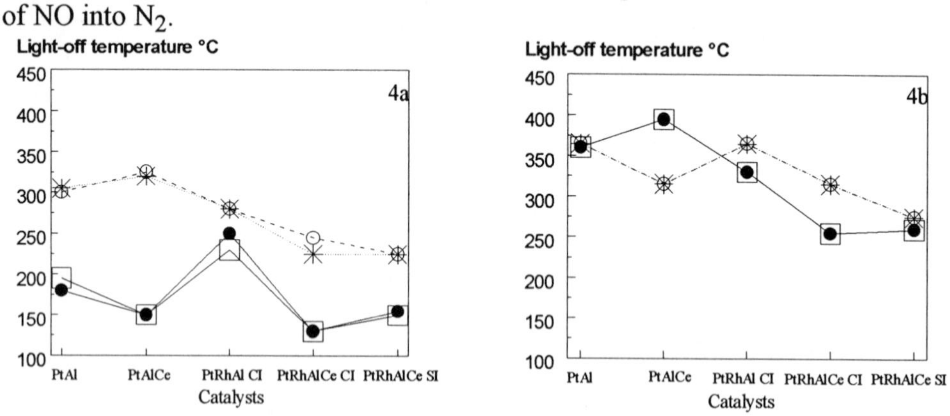

Figure 4. Effects of the pretreatment at 500°C (calcination (○ ,✳) or reduction (●, □)) on the light-off temperatures of dechlorinated (4a) and thermally aged (4b) catalysts for NO reduction (✳, □) and CO oxidation (○ , ●).

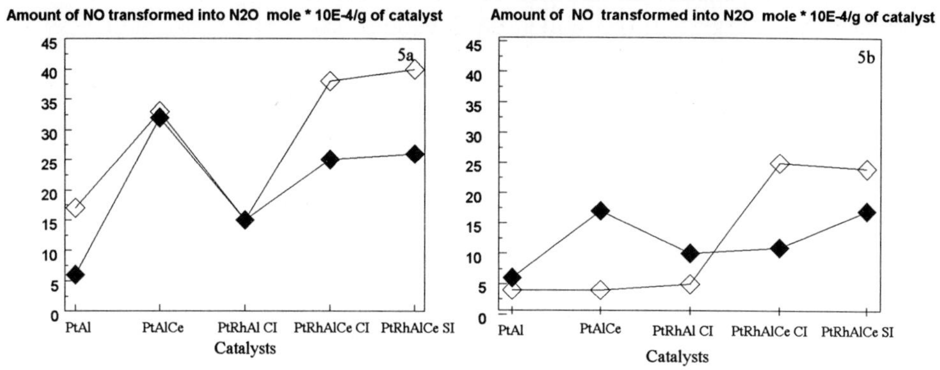

Figure 5. Effects of the pretreatment at 500°C (calcination (◆) or reduction (◇)) on N_2O formation by dechlorinated (5a) and thermally aged (900°C) (5b) catalysts.

The results reported in Figs 4-5 show that the preparation procedure does not affect the catalytic properties of dechlorinated and thermally aged PtRh / Al_2O_3 - CeO_2 catalysts after initial calcination or reduction at 500° C . Thus, these catalysts are generally more active than the monometallic catalysts or the PtRh / alumina catalyst for the CO + NO reaction . However, PtRh / Al_2O_3-CeO_2 catalysts produce more N_2O, particularly after a reducing pretreatment .

3.3. EDS microanalysis

EDS microanalysis were conducted on thermally aged PtRh / Al_2O_3-CeO_2 catalysts prepared either by coimpregnation or successive impregnations . Table 2 gives the mean ratio of microanalysis of platinum and rhodium carried out on large particles (near 200 Å) of the CI and SI catalysts . The results indicate that the particles examined on the SI catalyst contain only platinum while mainly bimetallic particles were observed on the CI catalyst . Moreover the analysis of the particle contour showed rhodium rich regions on the SI catalyst while platinum and rhodium were not detected in the vicinity of the large bimetallic particles on the CI catalyst (Fig 6) .

Table 2 EDS microanalysis of metallic particles on thermally aged CI and SI PtRh / Al_2O_3-CeO_2 catalysts (CI : coimpregnation ; SI : sucessive impregnations).

Catalyst	$\frac{Rh}{Pt}$ (mean atomic ratio)
PtRh / Al_2O_3-CeO_2 CI	0.17
PtRh / Al_2O_3-CeO_2 SI	0.02

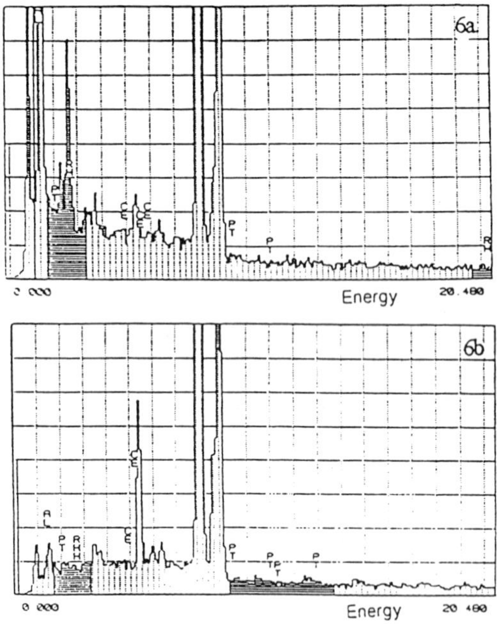

Figure 6. EDS microanalysis of the particle contour on SI (6a) and CI (6b) thermally aged Pt-Rh/Al_2O_3-CeO_2 catalysts

4. Discussion - Conclusion

The results presented in this paper indicate that ceria inhibits catalyst activity for hydrocarbon oxidation, particularly after an initial calcination step at 500°C . This inhibiting effect of ceria can be explain by a charge transfer from precious metals to surface cerium resulting in higher oxidation state in the metals and therefore in lower activity [14, 15] . However, we must note that the inhibition of catalyst activity by ceria for propene oxidation disappears when the sample is reduced at 500°C before the catalytic run . For the reduction of NO by CO, dechlorinated catalysts are far less active after an oxidizing pretreatment at 500°C. Thus, catalyst activity for the CO + NO reaction is also lowered when the oxidation state of the metals is higher . However the initial treatment (calcination or reduction) modifies the oxidation state of surface cerium (Ce^{4+} or Ce^{3+}) which is involved in the CO + NO reaction [16 - 17] . Finally, the preparation procedure does not affect the catalytic properties of dechlorinated PtRh / Al_2O_3-CeO_2 catalysts for the reactions studied in this work .

The main differences between the two preparation procedures appear for propane reaction on oxidized thermally aged samples (1 % O_2, 10 % H_2O, 10 %

CO_2, N_2 ; 900°C, 16h) . Thus, the activity of the PtRh / Al_2O_3-CeO_2 catalyst prepared by successive impregnations is not sensitive to the pretreatment before the oxidation reaction while coimpregnated PtRh catalysts are far less active when they are preoxidized at 500°C (Fig 2b). This result is in accordance with the conclusions of previous work [18] which state that, after oxidizing thermal aging, the coimpregnation would lead to the formation of Pt-Rh alloys with surface enrichment in relatively inactive rhodium oxides . On the other hand, rhodium added by successive impregnations, with a reduction step after platinum impregnation, would be selectively deposited on surface cerium in the vicinity of platinum particles, avoiding the formation of PtRh alloy . EDS microanalysis of thermally aged PtRh / Al_2O_3-CeO_2 catalysts bear out this conclusion since the large particles examined on the SI catalyst contain only platinum while mainly bimetallic PtRh particles were observed on the CI catalyst .

ACKNOWLEDGEMENTS

This work was carried out within the "Groupement de recherche sur les Pots Catalytiques" funded by the "Centre National de la Recherche Scientifique", the "Institut Français du Pétrole" and the "ADEME" (Agency for Environment and Energy Savings) .

REFERENCES

1 W.B. Williamson, H.S. Gandhi, P. Wynblatt, T.J. Truex and R.C. Ku, Aiche Symp. Ser. n 201, 76 (1980) 212 .

2 S.H. Oh and J.E. Carpenter, J. Catal, 98 (1986) 178 .

3 F.C. Van Delft, B.E. Nieuwenhuys, J. Siera and R.M. Wolf, Iron and Steel Institute of Japan International, n°7, 29 (1989) 550 .

4 I. Önal in "Catalyst Deactivation" (C.H. Bartholomew and J.B. Butt, Eds.), Amsterdam, (1991), 621 .

5 B.R. Powell, Appl. Catal., 53 (1989) 233 .

6 S. Kim, M.J. d' Aniello, Appl. Catal. , 56 (1989) 23 .

7 S. Kim, M.J. d' Aniello, Appl. Catal. , 56 (1989) 45 .

8 S. Kacimi and D. Duprez in "Catalysis and Automotive Pollution Control II" (A. Crucq, Ed.), Amsterdam (1991) 581 .

9 P. Marécot, L. Pirault, G. Mabilon, M. Prigent and J. Barbier, Appl. Catal. B, 5, (1994), 57

10 P. Marécot, A. Fakche, B. Kellali, G. Mabilon, M. Prigent and J. Barbier Appl. Catal. B, 3, (1994), 283
11 J.M. Dominguez, G.W. Simmons and K. Klier, J. Mol. Catal., 20 (1983) 369 .
12 J.M. Dominguez and D.R. Acosta, 8th Int. Congress on Catalysis, Berlin, (1984) DECHEMA, Vol 5, pp 287 - 298 .
13 G. Del Angel, S. Alerarool, J.M. Dominguez, R.D Gonzales and R. Gomez, Surf. Sci. 224 (1989) 407 .
14 Y.F. Yu Yao, J. Catal. 87 (1984) 152 .
15 J.G. Nunan, H.J. Robota, M.J. Cohn and S.A. Bradley, J. Catal. 133 (1992) 309 .
16 B. Harrison, A.F. Diwell and C. Hallett, Platinum Metal review 32 (1988) 73 .
17.G. Leclercq, C. Dathy, G Mabilon and L. Leclercq, in "Catalysis and Automotive Pollution Control II" (A. Crucq, Ed.), Amsterdam, (1991), 181 .
18 P Marécot, A Fakche, L Pirault, C. Géron, G Mabilon, M. Prigent and J. Barbier . Appl. Catal. B, 5, (1994), 43

A. Frennet and J.-M. Bastin (Eds.)
Catalysis and Automotive Pollution Control III
Studies in Surface Science and Catalysis, Vol. 96
1995 Elsevier Science B.V.

REACTIVITY OF PEROVSKITES AS AUTOMOTIVE CONVERTERS

Laure SIMONOT, François GARIN and Gilbert MAIRE

Laboratoire d'Etudes de la Réactivité Catalytique, des Surfaces et Interfaces, URA 1498 CNRS - ULP - EHICS, Institut Le Bel, 4, rue Blaise Pascal, 67070 Strasbourg cedex, France

ABSTRACT

The reactivity of three perovskites containing cobalt in three different oxidation states, Co(II), Co(III) and Co (IV) has been studied under two types of gas flow composition : i) binary mixtures containing either $CO + O_2$, $CO + NO$ or $C_3H_8 + O_2$. ii) a complete gas mixture with $CO + C_3H_8 + NO + O_2$. In both cases nitrogen was the balance gas. The reactivity order is found always to be the same, whatever the reaction : $Co_3O_4 > LaCoO_3/Co_3O_4 \approx LaCoO_3 > BaCoO_{2.x} > Ba_2CoWO_6$, and is correlated to the reducibility of the samples followed by TPR. The first four catalysts exhibit the same reactivity for the $CO + O_2$ reaction as classical Pt-Al_2O_3 and Pt-CeO_2/Al_2O_3 catalysts, their light-off temperatures being between 430 and 500 K. These perovskites give reproducible results after the surface has been changed by the reactant gases, this phenomenon has been observed by the presence of a hysteresis, for Co_3O_4 , under the $CO + O_2$ gas mixture.

1. INTRODUCTION

Since 1970 perovskite-type oxides (ABO_3) have been suggested as substitutes for noble metals in automotive exhaust catalysts [1]. These oxides are efficient for oxidation reactions when for reduction the results obtained from the literature are dissimilar [2], mainly due to huge differences in the experimental conditions. The properties of perovskite-based catalysts are a function of the spin and the valence state of the metal in the B site cation, which is surrounded octahedrally by oxygen. The A site cation is located in the cavity made by these octahedra. For some perovskite-type oxides, their electronic structures have been pointed out to be similar to those of transition metals on the basis of theoretical

calculations and valence band structures [3-6]. Thus one of the catalytic features of these systems is to be active for hydrogen-involving reactions such as, for example, the hydrogenolysis of hydrocarbons [7].
As correlations have been observed, for catalytic oxidation reactions, between the activity and the electronic state of the transition metal [8], or the non stoichiometry of the perovskite [9], we have chosen, in our study, three perovskites containing cobalt in three different oxidation states, i) Co (II) in Ba_2CoWO_6 , ii) Co (III) in $LaCoO_3$, iii) Co (IV) in $BaCoO_3$, and we have compared their reactivity with the spinel Co_3O_4.
The aim of this study is to assess the catalytic behaviour of these perovskites as automotive converters.

2. EXPERIMENTAL

2. 1. Preparation of the catalysts

All the catalysts have been prepared by a sol-gel method, starting from the oxide, acetylacetonate and carbonates : La_2O_3, $Co(CH_3COCHCOCH_3)_3$, $BaCO_3$, $Co(CO_3)$. Each precursor was dissolved in propionic acid, in such a way to get a concentration of 0.1 mol.dm^{-3}, and mixed together in the desired proportions.
For Ba_2CoWO_6, the ammonium metatungstate could not be dissolved in the propionic acid, and was thus suspended in this acid before mixing.
After evaporation of the acid, the gel obtained was frozen, ground and calcined. All the calcinations were performed under air, with two temperature rates : one of 2 Kmin^{-1} up to 525 K, and the other of 3 Kmin^{-1} up to the final temperature, except for Ba_2CoWO_6 which was put directly in the furnace at 1225 K. Temperature and time of calcination are listed in Table 1.

2. 2. Characterization

BET : the measurements were performed by the use of krypton at a temperature of 77 K with a P_0 = 3.276 mBar, the cross sectional area of krypton was taken equal to 19.5 Å^2. All the results are listed Table 1.

SEM : The average of the measured diameters of the particles are from 0.1 to 0.2 mm.

TPR/TPO measurements were obtained under 1.5% CO/He and 1% O_2/He respectively. Oxygen was analysed by thermal conductivity detectors (TCD), CO and CO_2 by IR.
About 0.015 g of the catalyst was placed in a U-shaped reactor, and was heated at a constant rate of 5 $Kmin^{-1}$ under a gas flow of 40 cm^3min^{-1} from room temperature to 1075 K. The results are discussed in the next section.

X-RAY analysis : the identification of the catalysts by X-Ray diffraction (K_a Cr, D5000 Siemens) did not show phases other than those desired, i.e. perovskite or spinel phases. For $BaCoO_3$, the phase obtained is a non-stoichiometric perovskite, a mix between $BaCoO_{2.8}$ and $BaCoO_{2.x}$.

MICROANALYSIS was performed at Vernaison in the CNRS laboratory. The results obtained (table 1) showed a good stoichiometry in the perovskite phase between the ions in A and B sites, and gave us exactly the ratio between $LaCoO_3$ and Co_3O_4 in the sample containing the two oxides : this ratio is Co/La = 2.7.

Table 1 Different characteristics of the catalysts

Catalysts	Calcination		BET surface	Microanalysis
$LaCoO_3$	1075 K	4 hours	3.2 m^2g^{-1}	Co/La = 1.02
$LaCoO_3/Co_3O_4$	1075 K	4 hours	3.3 m^2g^{-1}	Co/La = 2.70
$BaCoO_{2,x}$	1125 K	3 hours	1.5 m^2g^{-1}	Co/Ba = 1.03
Ba_2CoWO_6	1225 K	5 hours	1 m^2g^{-1}	Co/W = 0.92
Co_3O_4	1075K	4 hours	2.6 m^2g^{-1}	Co/O = 0.95 *

2. 3. Reactivity Testing

The catalyst powder (0.8 g) was placed in a tubular quartz reactor. The composition of the gas flow was obtained by mixing the different components (CO, O_2, NO, C_3H_8) diluted in N_2 with the use of gas flowmeters. For all experiments the ratio (oxydant species/reductive species) was taken equal to 1. Before each reaction, the reactor is bypassed in such a way as to know precisely the composition of the gas flow before reaction.

* This ratio is surprising, by the fact its value is higher than for the pure stoechiometric oxide. But the X-Ray data doesn't show any other pick as for Co_3O_4. If there is an other phase (cobalt suboxide), this phase is not crystallised.

The analysing section is composed of the following detectors: IR for CO, C_3H_8, and CO_2 , chemiluminescence for NO, paramagnetism for O_2 .
The experiments consist of increasing gradually the temperature with a rate of $4Kmin^{-1}$ and following the concentration of the gas flow. In all experiments, *except when hysteresis reactions were performed,* the catalyst was cooled under N_2. When more information about the NO conversion was needed, we used a similar apparatus in which the analysing section is composed of three IR detectors for NO, N_2O, NO_2. The composition of the other gases was determined by gas chromatography. Two sorts of gas flow composition were used: i) a *binary mixture* containing either CO + O_2, CO + NO or C_3H_8 + O_2 . ii) a *complete gas mixture* contaning CO + C_3H_8 + NO + O_2 . All experiments were performed four times. The first one is considered as the activation process. The results given in the tables are an average of the three last experiments.

3. Results and Discussion

3.1 Temperature programmed reduction

Figure 1 shows the TPR profiles for the three following compounds : Co_3O_4, $LaCoO_3$ and $LaCoO_3$ / Co_3O_4. We can observe the presence of one broad reduction peak at 640 K for Co_3O_4, in contrast two reduction peaks appear for the two other compounds; their reduction temperatures are 640 K and 750 K for $LaCoO_3$ / Co_3O_4 and 720 K and 855 K for $LaCoO_3$. The presence of metallic cobalt at the end of the experiment was verified by X-ray diffraction. In $LaCoO_3$ we can observe that one cobalt species is more difficult to reduce. The last TPR peak appears at 855 K and can be attributed to an hindered reduction process of the cobalt due to the close presence of lanthanum oxide. This peak is shifted to lower temperature for $LaCoO_3$ / Co_3O_4 because less lanthanum oxide is present in this sample, and finally for Co_3O_4 only one peak appears shifted by 80 K towards lower temperature.
For Co_3O_4 similar results were obtained by Arnoldy and Moulijn [10]. They observed one sharp reduction peak for Co_3O_4 at 595 K, the TPR experiment was performed under hydrogen at a temperature rate of 10 $Kmin^{-1}$. In our case we observed one reduction peak in the same range of temperature. For the perovskite, two reduction peaks were formed corresponding to Co (III) —> Co (II) and Co (II) —> Co (0), as already observed by Crespin and Hall [11].

For $BaCoO_{2.x}$ TPR cannot be interpreted due to carbonation of this perovskite under CO.
For Ba_2CoWO_6, the TPR showed a little CO consumption at 655 K and can be related to adsorption-desorption phenomena.

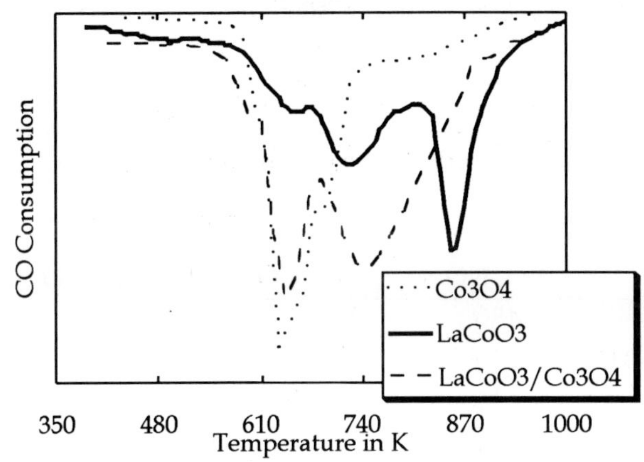

Figure 1 : CO consumption in TPR experiments as a function of the temperature in K. The scale for the consumption is related to the ppm of CO consumed per mole of cobalt in the catalyst.

3. 2 Temperature programmed oxidation

The X-ray diffractions performed in a Guinier chamber (K_a Fe) have shown that, in all cases, *TPO experiments performed after a TPR reproduced the initial compound.* Only one oxidation peak was observed at 625 K for $LaCoO_3$ and $LaCoO_3$ / Co_3O_4 and at 725 K for Co_3O_4 ; no significant peak was observed for Ba_2CoWO_6.

3. 3 Catalytic activity under binary mixtures

The results are listed in Tables 2a, 2b and 2c, and illustrate the temperatures at which the conversion, calculated on the basis of the disapearance of the polluant, is equal to 10%, 50% and 90%.

Table 2a Temperature of 10% conversion in binary mixture (K)

	CO + NO		$CO + O_2$	$HC + O_2$
	T_{10} NO	T_{10} CO	T_{10} CO	T_{10} HC
Co_3O_4	385	415	385	490
$LaCoO_3/Co_3O_4$	445	485	405	520
$LaCoO_3$	445	520	410	550
$BaCoO_{2.x}$	495	525	425	635
Ba_2CoWO_6	645	655	475	680

Table 2b Temperature of half conversion in binary mixture (K)

	CO + NO		$CO + O_2$	$HC + O_2$
	T_{50} NO	T_{50} CO	T_{50} CO	T_{50} HC
Co_3O_4	480	545	430	550
$LaCoO_3/Co_3O_4$	570	595	453	595
$LaCoO_3$	615	600	440	640
$BaCoO_{2.x}$	660	640	475	775
Ba_2CoWO_6	760	770	525	855

Table 2c Temperature of 90% in binary mixture (K)

	CO + NO		$CO + O_2$	$HC + O_2$ *
	T_{90} NO	T_{90} CO	T_{90} CO	T_{80} HC
Co_3O_4	545	590	450	590
$LaCoO_3/Co_3O_4$	745	640	470	670
$LaCoO_3$	710	640	450	730
$BaCoO_{2.x}$	765	700	490	< 80%
Ba_2CoWO_6	865	885	575	< 80%

* for HC we do not reach more than 85% conversion, the results listed are for 80% conversion, when it was reached.

From these tables we can observe the following points :

- The reactivity order is always the same, whatever the reaction and conversion, Co_3O_4 > $LaCoO_3/Co_3O_4$ ≈ $LaCoO_3$ > $BaCoO_{2.x}$ > Ba_2CoWO_6. The two latter compounds have very small TPR peaks, implying low oxygen mobility, which may explain in part their low reactivity.

- Concerning the two oxidation reactions i.e CO + O_2 and HC + O_2, the former occurs at the lower temperature than the latter. The CO + NO reaction takes place in between these two ranges of temperature.

- In the CO + NO reaction, the NO transformation is faster than the CO oxidation. To understand this point we have performed the experiments on the apparatus where IR detectors for NO, N_2O and NO_2 are present. We have noticed that the NO reduction does not give directly N_2 but rather N_2O, as can be seen in figure 2.

- When we compare the oxidation temperatures for CO in the CO + O_2 and CO + NO reactions, we notice that the temperature is always lower in the former reaction. NO might be a poison of these catalysts, but on the other hand the apparent activation energy for CO + NO is higher than for CO + O_2 . From the work by Oh et al. [12] on rhodium catalysts they noticed that the apparent activation energy for CO + O_2 was equal to 14 kcal/mol and was lower by 17 kcal/mol than for CO + NO reaction. Halasz et al. [13] observed an apparent activation energy of 13 kcal/mol for the catalytic oxidation of CO over Co_3O_4. These results may suggest that the same slow step occurs on rhodium and Co_3O_4, and it was observed that the surface oxides can be formed either on Rh(111) or on supported rhodium, when they are treated under an oxidising environment [14,15].

From these results we can observe that Co_3O_4 is the more reactive system, but we notice that this oxide is the only one to be modified after the activation step. This modification was observed by XRD, and we showed the presence of two oxides : CoO and Co_3O_4. This result led us to study this oxide more carefully.

3 .4 Co_3O_4 reactivity

One question which may arise in these experiments concerns the reversibility* of these reactions. We studied the CO + O_2 reaction by increasing and then decreasing the temperature under the same gas mixture. In figure 3 we

* The term reversible has to be taken with care. In our case it does not imply the thermodynamic reversibility.

can see the presence of a hysteresis loop. After returning to room temperature, the next experiment follows the first curve obtained when the temperature was increased.

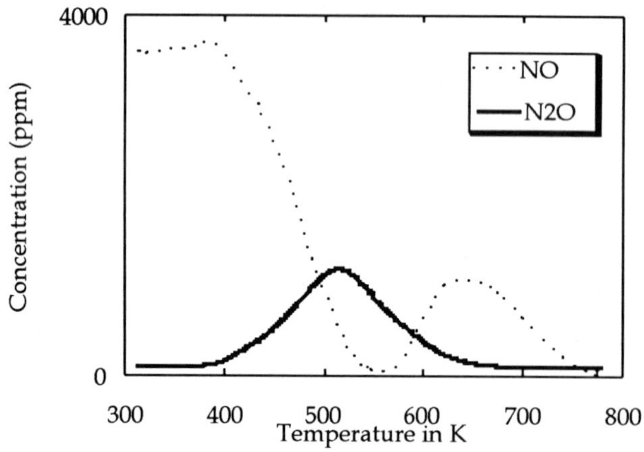

Figure 2 : CO + NO reaction over $LaCoO_3/Co_3O_4$

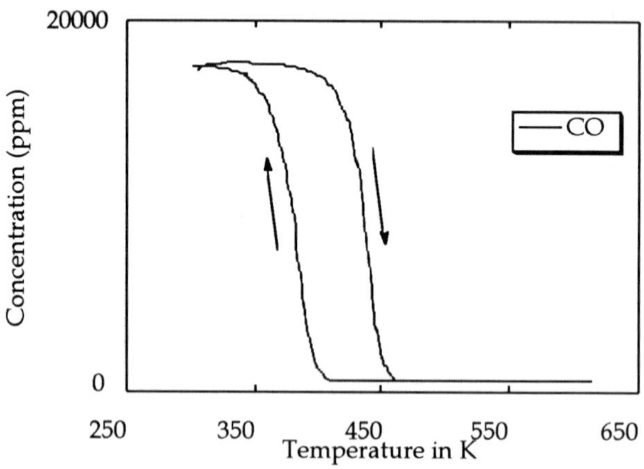

Figure 3 : Hysteresis experiment for Co_3O_4 under the binary mixture $CO + O_2$

Similar behaviour is also observed when the return to room temperature occurs under nitrogen instead of being performed under the reactive gas..
This result points out that either two different sites may exist at the surface, their presence being a function of the reaction temperature, or the nature and the amount of the adsorbed molecules are different versus the reaction temperature
This phenomenon has also been observed with alumina supported catalysts, and rules out the hypothesis that hysteresis is related to a change in the oxidation state of the cobalt

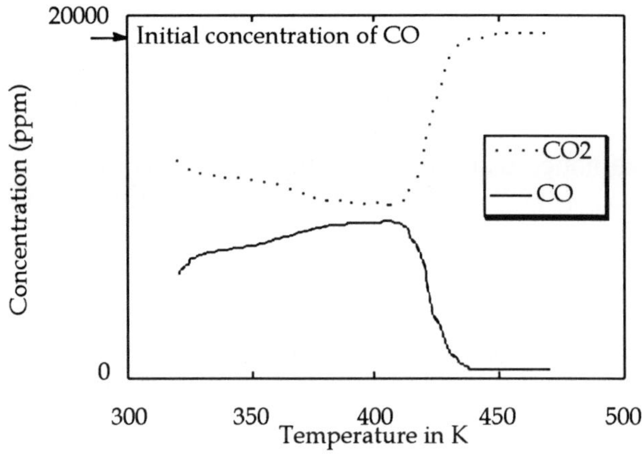

Figure 4 : The $CO + O_2$ reaction over Co_3O_4

To confirm this idea we increased the cooling rate of the catalyst. At the end of a $CO + O_2$ reaction, when the catalyst was at 573 K, it was rapidily cooled to room temperature in ten minutes under N_2 and put again in contact with $CO + O_2$, the CO concentration being 18 000 ppm. We have observed that this oxide was able to transform $CO \longrightarrow CO_2$ at room temperature with a conversion of 50% as shown in figure 4.

This result may be explained by the quenching of the high temperature surface sites.

3. 5 Catalytic activity under the complete gas composition

The results are listed in table 3, and show the temperature at which the conversion is equal to 50%.

Table 3 Temperature of half conversion in complete mixture (K)

	T_{50} NO	T_{50} CO	T_{50} HC
Co_3O_4	<50%	435	595
$LaCoO_3/Co_3O_4$	<50%	440	635
$LaCoO_3$	<50%	465	665
$BaCoO_{2,x}$	<50%	500	>520
Ba_2CoWO_6	<50%	575	910

First of all we can notice that the temperatures of half conversion for NO, CO, and C_3H_8 are always higher than in the experiments performed with binary mixture.We never reach more than 15% NO conversion under a complete gas flow : conversion starts between 575 - 675 K, but in excess of this temperature the NO conversion activity is seen to decrease to zero. The small conversion of NO can be explained by the fact that the more rapid reaction CO + O_2 occurs first, then all the CO is transformed in CO_2 and NO has to react with hydrocarbon which needs very higher temperature.

CONCLUSION

These catalysts give reproducible results after the activation process which effects a change in the catalyst surface. The catalysts have been found to be very active for oxidation reactions, however NO transformation occurs only to a small extent. The cobalt oxide Co_3O_4 and the perovskite $LaCoO_3/Co_3O_4$ are the more reactive catalysts. They show a reversible modification of the surface when the temperature is increased, when compared to Pt-Al_2O_3 and Pt-CeO_2/Al_2O_3 catalysts [16,17] where the platinum loading is around 2%, the light-off temperatures for all the catalysts were found to be in the same range (around 500 K) for the CO + O_2 reaction, under similar experimental conditions.

ACKNOWLEDGEMENTS

The authors would like to thank Dr C.Petit and Prof. M.H.Simonot-Grange for their contributions.

REFERENCES

1 W.F.Libby, Science. 171 (1971) 499
2 R.J.H.Voorhoeve, J.P.Remeika and L.E.Trimble, Mat.Res.Bull. 9 (1974) 1393
3 J.B.Goodenough and P.M.Raccah, J.Appl.Phys. 36 (1965) 1031
4 L.F.Mattheiss, Phys.Rev.B. 6 (1972) 4718
5 V.G.Bhide, D.S.Rajania, G.Rama Rao and C.N.R.Rao, Phys.Rev.B. 6 (1972) 1023
6 E.A.Kraut, T.Wolfram and W.Hall, Phys.Rev.B. 6 (1972) 1499
7 K.Ichimura, Y.Inoul and I.Yasumori, Bull.Chem.Soc.Japan. 53 (1980) 3044, ibid 55 (1982) 2313
8 R.J.H.Voorhoeve, J.P.Remeika and L.E.Trimble, Ann.New York Acad.Sci. 272 (1976) 3
9 B.Viswanathan, Properties and applications of perovskite type oxides, 271 (1993)
10 P.Arnoldy and J.A.Moulijn, J.Catal. 93 (1985) 38
11 M.Crespin and W.K.Hall, J.Catal. 69 (1981) 359
12 S. H.Oh, C.B.Fisher, J.E.Carpenter and D.W.Goodman, J.Catal., 100 (1986) 360
13 I.Halasz, A.Brenner, M.Shelef and K.Y.Simon Ng, J.Catal., 134 (1992) 737
14 T.Matsushima, J.Catal., 85 (1984) 98
15 S.H.Oh and J.E.Carpenter, J.Catal., 80 (1983) 472
16 C.Serre, F.Garin, G.Belot and G.Maire, J.Catal., 141 (1993) 1
17 C.Serre, F.Garin, G.Belot and G.Maire, J.Catal., 141 (1993) 9

A. Frennet and J.-M. Bastin (Eds.)
Catalysis and Automotive Pollution Control III
Studies in Surface Science and Catalysis, Vol. 96

EFFECT OF THE CeO2 DISPERSION ON ALUMINA ON ITS REACTIVITY FOR CO AND NO CONVERSION

R. Cataluña, A. Arcoya, X.L. Seoane, A. Martínez-Arias, J.M. Coronado, J.C. Conesa, J. Soria and L.A. Petrov*

Instituto de Catálisis y Petroleoquímica, CSIC, Campus Universitario de Cantoblanco, 28049 Madrid, Spain

ABSTRACT

The reactivity of CeO_2, pure or supported on alumina, in NO adsorption and catalytic activity for the CO + NO + O_2 reaction (in the absence of metal), has been tested using IR, ESR and temperature-programmed desorption/reactivity techniques. Significant differences are found in alumina-supported ceria in comparison with pure CeO_2; its lower activity can be connected to the difficulty of generating on well-dispersed CeO_2/Al_2O_3 the associated oxygen vacancy centers observed on CeO_2, which are able to activate reductively NO to give N_2O. The same centers might play also a role in enhancing the reduction of surface nitrite species by CO simultaneously with the combustion of the latter molecule.

1. INTRODUCTION

Cerium oxide has been used as promoter in many car exhaust three-way catalyst formulations. Its presence is known to affect several aspects of the catalyst state and performance: active noble metal dispersion, oxygen storage capacity, alumina support stability and generation of new active centers (or modification of some of those existing in its absence) for catalyzing some of the reactions occurring during the normal catalyst operation [1]. In many cases, theseeffects have been studied examining the behaviour of metal-containing

*On leave from the Institute of Kinetics and Catalysis, Bulg. Acad. Sci., Sofia 1113, Bulgaria

catalysts where the support is either CeO_2, Al_2O_3 or mixtures thereof, in order to deduce by comparison the role of the ceria component; less frequent is the examination of the supports alone (i.e. without noble metal loaded), which can have interest in order to better isolate the specific behaviour of the oxide support material in the different environmental conditions. This is especially the case for the reactions of adsorption/desorption of O_2 and other (potentially) oxidizing components of the reacting gas feed, since one of the recognised roles of ceria in the system is the uptake and release of oxygen molecules; these processes are controlled to a large extent by the formation and presence of anion vacancies in this reducible oxide, and the reactivity of these centers by themselves must be well understood in order to master adequately the behaviour of the catalyst. We have tried to obtain some additional information about several of the redox processes originated by adsorption on those vacancies, comparing in this respect the properties of a typical ceria/alumina support with those of the pure oxide.

2. EXPERIMENTAL DETAILS

CeO_2 from Rhône-Poulenc and with $S_{BET} \approx 110$ m^2/g (sample A) was used as received. Ceria/alumina samples with nominal CeO_2 loads of 30% (sample B) and 10% w/w (sample C) were prepared by incipient wetness method using alumina (Condea, high purity Puralox grade, S_{BET}=210 m^2/g, mixture of γ and δ forms; this material is called here sample D) and an aqueous solution of $Ce(NO_3)_3$ (Fluka); the resulting solid was first dried 24 h in air at 395 K, then calcined under flow of dry air at 773 K during 3 h. In the preparation of the catalyst for experiments in the gas flow reactor, the alumina used was in form of small spheres, ca. 1 mm dia.; for the specimens used in spectroscopic measurements it was in powder form. Samples B and C presented after calcination S_{BET}= 165 and 195 m^2/g respectively.

O_2 and N_2 gases used were from SEO, NO and CO from Carburos Metálicos (all of them in "high purity" grade, although the NO used in temperature-programmed experiments contained a small NO_2 impurity, around 1%). Gases used in spectroscopic measurements were previously purified with liquid N_2 condensation techniques.

A high vacuum manifold (base pressure <10^{-4} Torr) was used for outgassing and adsorption treatments on the samples prior to spectroscopic experiments. Special quartz probe-cells provided with double greaseless stopcocks were used for handling the samples in the ESR measurements, which were obtained at 77 K using a Bruker X-band 200 D spectrometer with ESP-1600 data system, double rectangular cavity and DPPH standard for g-value calibration. IR spectra were obtained on self-supporting sample wafers (ca. 13 mg in weight), prepared by pressing the powders at 2500 N/cm^2 and handled in standard greaseless cells; spectra were taken at room temperature and 4 cm^{-1} resolution using a Nicolet 5ZDX FT-IR spectrometer. Powder X-ray diffractograms were obtained with a Seifert 3000 P diffractometer.

Temperature programmed experiments were carried out using a glass and stainless steel gas flow reactor system; the analysis of the feed and outlet gas streams was performed using a Perkin Elmer FT-IR spectrometer mod. 1725X, coupled to a multiple-reflection transmission cell (Infrared Analysis Inc. "long path gas minicell", 2.4 m path length, ca. 130 cm^3 internal volume); O_2 was determined with a paramagnetic analyzer (Servomex 540 A). N_2 could not be detected in this setup, and was used as carrier gas. Before the NO adsorption/desorption and catalytic tests, the samples were subjected in this system to a standard pretreatment, consisting of heating under a 3% O_2:N_2 flow at 673 K during 1 h, cooling in the same gas to room temperature and then purging briefly (5 min) with N_2.

3. Results and Discussion

3.1. X-ray diffraction

The XRD patterns of samples B and C (not shown) present some broad lines due to CeO_2; this indicates that, even for sample C, in which the CeO_2 load (equivalent to 2 Ce atoms per nm^2 of alumina) is below that corresponding to a monolayer coverage on the alumina support, some CeO_2 crystals have been formed; it can be concluded that a large fraction of total sample surface corresponds to alumina not covered by ceria. These XRD lines appear slightly narrower for sample B, which suggests that this latter contains somewhat larger ceria crystals.

3.2. ESR data

The ESR spectra of the samples outgassed at different temperatures T_v in the 373-773 K range show that sample A only presents, for T_v ³ 673 K, a narrow symmetric signal I (g = 2.003, linewidth ΔHpp = 5 G), its intensity growing with T_v; samples B and C show additionally a sharp axial signal II, with $g_{\perp}$ = 1.967, $g_{\circ}$ = 1.937 and almost constant intensity for all T_v, and another signal III, nearly isotropic, centered at g = 2.005 and showing a six line hyperfine structure with A = 83 G. Signal III is due to Mn^{2+} impurities, as indicated by its g value and hyperfine splitting (for Mn, I = 5/2). Signals I and II have been assigned to electrons trapped in oxygen vacancies, stabilized by structural defects in the case of signal I or by aliovalent cation impurities (which in the present materials could be Mn^{3+} or Al^{3+} ions) in the case of signal II [2]. Signals I and II are reversibly broadened by O_2 adsorption, which indicates that the corresponding centers are placed at the sample surface, but no implication of these in reactions with the gases used in adsorption has been detected.

As reported in detail elsewhere [3], vacancy centers associated to reduced ceria, generated by outgassing on its surface, can be revealed through ESR study of the O_2^- radicals produced upon O_2 adsorption on them. An example of the spectra obtained is given in 0b,d. Such results show that two basic types of adsorption centers can be distinguished: in one of them, the signals obtained (called O_I-type) are nearly axial and have the lowest g feature located at g_1=2.011 (the other parameters are g_2=2.011-2.013, g_3=2.031-2.037), while the other type (giving signals called O_{II}-type) have their lowest g around g_1=2.008-2.006 (other parameters are g_2=2.010-2.015, g_3=2.038-2.052). Use of ^{17}O-labelled O_2 confirmed that O_2^- species (with nonequivalent oxygens in the second

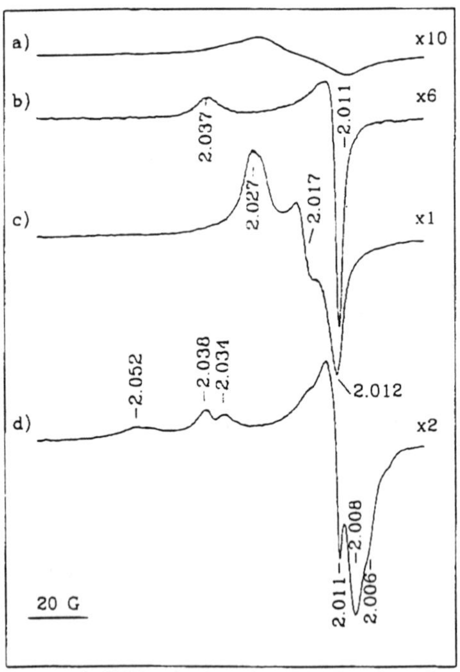

Figure 1 ESR spectra (at 77 K) after O_2 adsorption at 295 K on samples C (a, c) and A(b,d) previously outgassed at temperatures T_v=473 K (a, b) and 773 K (c, d)

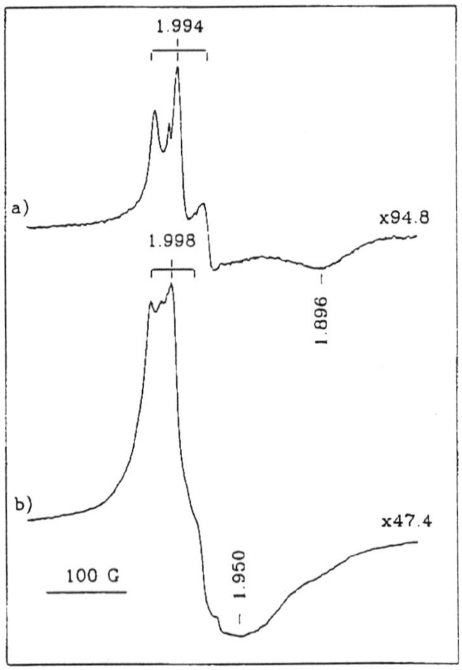

Figure 2: ESR spectra (at 77 K) after NO adsorption at 77 K on samples outgassed at 673 K: a) CeO_2; b) CeO_2/Al_2O_3.

type of species) originate both kinds of signal [3]. The first signal type, which appears already for outgassing temperatures T_V= 373 K, was ascribed associated surface vacancy sites where a larger density of excess electrons is accumulated [4]. In both cases, the fact that the two lowest g values observed deviate clearly from those predicted [5] by the ionic model for O_2^- species (g_1=2.003, g_2=2.009) can be ascribed to covalency effects, in which the π orbitals of the radical donate electron density to 4f orbitals of the Ce ions [6]. The smaller extent of deviation found for O_{II} type signals suggest then that a higher electron density (due to reduction) on the Ce ions at those sites decreases this covalency effect.

ESR study of NO adsorption on CeO_2 was also carried out. As reported in detail elsewhere [7,3b], in agreement with previous work by other authors [8] adsorption of NO at 77 K on pure ceria which had been outgassed at high temperature produced a well-resolved ESR signal (0a) showing hyperfine splitting (due to coupling to the N nucleus with spin I = 1) and g_1=1.996, g_2=1.9925, g_3=1.910, A_2=33 G (A_1 and A_3 being unresolved); this signal can be assigned to adsorbed NO molecules (which are paramagnetic if in the neutral state) located on coordination vacant sites on Ce ions. Subsequent O_2 adsorption at 77 K showed that the associated vacancy centers existing prior to NO adsorption disappeared by reaction with NO, while the isolated ones were not eliminated, but only modified, so that the g values of the resulting species are significantly changed [3b].

When similar experiments are performed on the CeO_2/Al_2O_3 samples, it is worthwhile to note that the O_{II}-type signals are no longer observed after adsorption of O_2 on samples which have been pre-outgassed at T≤773 K (0a,c). For low T_v a species (called here O_A) is detected which is characterized by parameters $g_\perp$= 2.025, $g_\parallel$= 2.011; for T_v ≥573 K another additional species $O_{A'}$ appears having g_1= 2.012, g_2= 2.017 and g_3= 2.027. These signals are also ascribable to Ce ion-bonded O_2^- species, but the different g parameters measured for them indicate that they are placed in a different bonding or crystal field environment; these results suggest that the dominant adsorption sites (anion vacancies also) are now located at or near to the ceria-alumina interface. The even higher deviation of the g parameters of these signals (in comparison with those found on pure CeO_2) from the values predicted by the ionic model suggests that the less basic character of the $O^=$ ions in a mixed ceria-alumina environment might be leading to a higher degree of electron density displacement from the O_2^- radical to the surface Ce ions, i.e. to larger covalency in the O_2^--Ce ion bonding. It is worth mentioning that these species, in comparison to those obtained on CeO_2, are less strongly adsorbed, as they disappear upon room temperature outgassing, contrarily to the behaviour observed for pure ceria. Only a small residual signal of O_I type remained, which probably was not detected in the previous spectra due to its smaller intensity; it corresponds probably to the small amount of CeO_2 crystals present (as detected by XRD).

As to the effect of NO adsorption on outgassed ceria-alumina, the ESR spectra obtained after contact of these samples with NO at 77 K present again signals due to adsorbed NO radicals (0b), now less resolved and with superposition of two types of species bonded respectively to Ce and Al ions (the latter being characterized by a high-field g value g_3=1.95); the intensity of the signals (in particular, thet of Ce-bonded NO) decrease substantially, however, after warming the sample at 295 K, indicating that in these samples, as was found for CeO_2, NO on Ce ions undergoes electron transfer reactions at room temperature to give diamagnetic products. Subsequent adsorption of O_2 gives now O_A but not $O_{A'}$ species, indicating again that the reaction of NO occurs mainly with the centers formed by outgassing at higher temperatures, i.e. those with higher amount of coordination vacancies on the Ce ions, even though these would have, in the case of CeO_2/Al_2O_3, a level of electron density lower than that existing in the corresponding centers on CeO_2, according to the above given discussion about the O_2^- ESR signals.

3.3. IR data

The FT-IR spectra of sample A contacted with NO (P= 40 Torr) at room temperature after having been outgassed at temperatures T_v= 373, 473 and 773 K are presented in 2. These curves were obtained by subtraction of the spectra taken for the sample outgassed at the corresponding T_v from those measured after subsequent NO adsorption; this procedure allows to minimize the interference of small bands, appearing around 1070 cm^{-1} and due to surface carbonates, with those originated by NO adsorption. Other carbonate bands with substantially higher intensity, appearing in the range 1350-1600 cm^{-1}, were more difficult to cancel, and therefore that spectral range is not used in this work. As observed in this figure, for sample A the spectrum obtained for T_v = 373 K shows only a band at ≈1175 cm^{-1}, which for T_v = 473 K appears with larger intensity while another band is observed at 1105 cm^{-1} as well as two smaller ones at 1270 and 825 cm^{-1}. These two latter small bands, together with that at 1175 cm^{-1}, can be assigned to chelating nitrite (NO_2^-) groups [9], while that at 1105 cm^{-1} can be ascribed to *trans*-hyponitrite ($N_2O_2^{=}$) species [10, 11]. For T_v = 773 K the spectrum shows in addition bands at 1015 and 955 cm^{-1}, ascribable to *cis*-hyponitrites [11], at 1305 and 1060 cm^{-1}, due to nitrate groups [12], and at 1255 and 2240 cm^{-1} (the latter outside the spectrum range included in 2), assignable to adsorbed (neutral) N_2O [13] (this species was already discernable, with small intensity, in the spectrum for T_v= 473 K); gas phase N_2O molecules also appear, and can be detected through their characteristic bands at 2225 and 1285 cm^{-1} if the sample is removed from the IR beam pathway.

The simultaneous formation of nitrites and hyponitrites can be understood on the basis of the redox properties of Ce ions. Thus nitrites may form when NO is adsorbed on Ce^{4+} ions with anion vacancies in their coordination spheres, leading to reduction of the cerium cation:

Erreur! Signet non défini.□ O-----N

| | |

Ce^{4+}-$O^{=}$ + NO → Ce^{3+}-O (1)

where □ represents an oxygen vacancy. Hyponitrites can be formed by adsorption of two NO molecules at closely located Ce^{+3}-oxygen vacancy sites [10, 3b], through a process of the type

$$2\ Ce^{3+}\text{-}\square + 2\ NO \rightarrow Ce^{4+}\text{-}(ONNO)^{=}\text{-}Ce^{4+} \quad (2)$$

Part of these $N_2O_2^{=}$ groups may be unstable at 295 K and decompose to produce N_2O (that can remain adsorbed or desorbe into the gas phase) and O^{2-} anions close to the reoxidized cerium cations. Since N_2O is observed in coincidence with the *cis* type of $N_2O_2^{=}$ species, it seems probable that these latter are mor unstable and reactive than those of the *trans* kind. The fact that such hyponitrites (and subsequently N_2O) can be detected for T_v lower than that necessary to observe O_{II} signals upon O_2 adsorption, even though the formation of both species should require in a similar way the existence of closely located oxygen vacancies, may be related to the fact that, in the latter case, it is necessary to generate also via outgassing the required excess electrons, while in the NO adsorption experiments these will be generated already during nitrite formation as sketched above. These steps can therefore be described as:

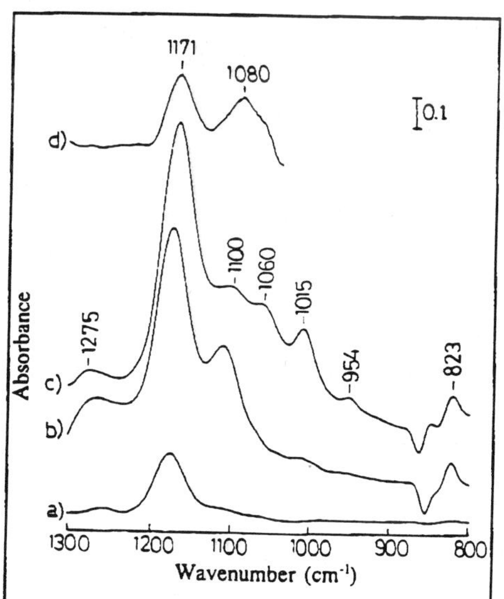

Figure 3 IR spectra after NO adsorption on samples outgassed at different temperatures T_v. CeO_2: a) 373, b) 473, c) 773 K; d) 10% CeO_2/Al_2O_3, T_v= 773 K.

$$2\ Ce^{3+}\text{-}\square + 2\ NO \rightarrow Ce^{4+}\text{-}(ONNO)^{-}\text{-}Ce^{4+} \dashrightarrow Ce^{4+}\text{-}O^{=} + Ce^{4+}\text{-}\square + N_2O \quad (3)$$

The resulting reoxidized Ce ions can act as new NO adsorption sites, leading e.g. to the formation of new chelating nitrites. Nitrates can conceivably be formed, on the other hand, by many different paths, most of them involving transformation of nitrite species and, in some cases, with electron transfer to the underlying oxide:

$$2\ NO_2^{-} \rightarrow NO_3^{-} + NO^{-} \quad (4)$$
$$NO_2^{-} + O^{=} + 2\ NO \rightarrow NO_3^{-} + N_2O_2^{=} \quad (5)$$
$$NO_2^{-} + N_2O \rightarrow NO_3^{-} + N_2 \quad (6)$$
$$Ce^{4+}\text{-}(NO_2)^{-} + Ce^{4+}\text{-}O^{=} \rightarrow Ce^{3+}\text{-}(NO_3)^{-} + Ce^{3+}\text{-}\square \quad (7)$$
$$Ce^{4+}\text{-}(NO_2)^{-} + Ce^{4+}\text{-}(ONNO)^{-}\text{-}Ce^{4+} \rightarrow Ce^{4+}\text{-}(NO_3)^{-} + Ce^{3+}\text{-}\square + (N_2O..Ce^{3+} \quad (8)$$

In the case of CeO_2/Al_2O_3, nitrites and hyponitrites (at least, of the *trans* kind) are also formed by NO adsorption (2d); the formation of nitrates and *cis*-hyponitrites could not be ascertained, because of the low IR transmission by the alumina support in the range below 1050 cm^{-1} (not depicted in 2 for the spectrum of this sample). It was however possible to observe that N_2O (both adsorbed and in gas phase) was formed in rather smaller amounts than for CeO_2, suggesting that *cis*-$N_2O_2^{=}$ are formed in this case in lower amounts; the reason for this could be a need of higher proximity of associated vacancies, i.e. a higher concentration of them than available on dispersed CeO_2, for generating the *cis* form.

3.4. Temperature-programmed desorption after NO adsorption

NO adsorption/thermodesorption tests were made by contacting the samples (previously subjected to the above mentioned standard treatment) with a $NO:N_2$ stream (0.7% NO v/v, 750 cm^3/min flow rate) at room temperature during 1 h, followed by briefly purging with N_2 (5 min) at the same flow rate and heating under the same N_2 flow at a 12.5 K/min rate up to 800 K. The amounts of evolved gases detected during such TPD run for samples A (Ce O_2), C (10% Ce O_2/ Al_2O_3) and D (pure Al_2O_3), using a 10 ml volume of catalyst (corresponding to sample quantities of 15.0, 6.2 and 5.5 g respectively), are plotted against temperature in 4a-c. These plots show that, for Ce O_2 (Figure 4a), significant NO evolution is observed in three distinct peaks, centered respectively around 350, 550 and 770 K. N_2O is detected in much smaller amount; an initial desorption tail is just hinted at the beginning of the experiment, followed some 100 K later by a small, broad peak. On the contrary, O_2 and NO_2 are clearly evolved nearly in coincidence with the high-temperature NO desorption peak. The behaviour of sample C (Figure 4b) is similar, except that the first NO desorption peak is just barely discernable as a small shoulder, and the proportion of NO_2 evolved in the high-temperature region is larger. Sample B (data not shown) behaves rather similarly to sample C, although a more noticeable NO desorption peak appears around 350 K, and the proportions between the products appearing in the high temperature range are intermediate between those of the A and C samples. In the case of the pure alumina (sample D, Figure 4c) the intermediate NO desorption peak is again observed, plus the set of desorption processes above 600 K in which the NO_2/NO proportion is larger than in any of the preceding cases.

The various features observed in these data must correspond to the stability or reactivity of the different surface species generated upon NO adsorption on the samples. Easiest to interpret are the desorption peaks observed at $T \geq 673$ K; because of the evolution of both NO_2 and O_2 (together with NO), this can only be interpreted as due to the decomposition of surface nitrate species, which can form both on CeO_2 and on Al_2O_3. The higher NO/NO_2 proportion observed in presence of cerium corresponds probably to the existence, in the stages previous to the beginning of this desorption peak, of a certain amount of Ce^{3+} which can be reoxidized in the process, decreasing the amount of NO_2 liberated. On the other hand, the intermediate peak includes most probably the decomposition of nitrites, which appear also easily on both types of oxides. On a support with scarce redox capability such as alumina, where this process is clearly observed, the disproportionation of nitrites into NO and nitrates is the most likely reaction to explain this feature, since, to our knowledge, the formation of surface species containing N in low (formal) oxidation degree, such as hyponitrites, upon NO adsorption has not been reported to occur in substantial amounts. Where ceria is present, other nitrite reactions can occur, grouped around the same temperature (corresponding to that of N-O bond breaking in NO_2^-) but involving perhaps Ce^{3+} and/or hyponitrite groups (which have been clearly detected on this oxide by IR); indeed, on sample A this desorption peak appears broader and with some incipient shoulder in the high-temperature side.

As to the NO desorption peak observed below 373 K, which appears only for sample A, it is reasonable to relate it with other specific observations on this sample. In particular, this sample was the only one able (if pre-outgassed at $T_v \geq 473$ K) to produce after NO adsorption at room temperature N_2O species (and, in fact, a tail of the desorption of this molecule can be detected at the beginning of the plot in 3a), the generation of which on ceria has been observed to coincide with the observation of the less stable *cis*-hyponitrite form. It seems therefore that the difference may lie in

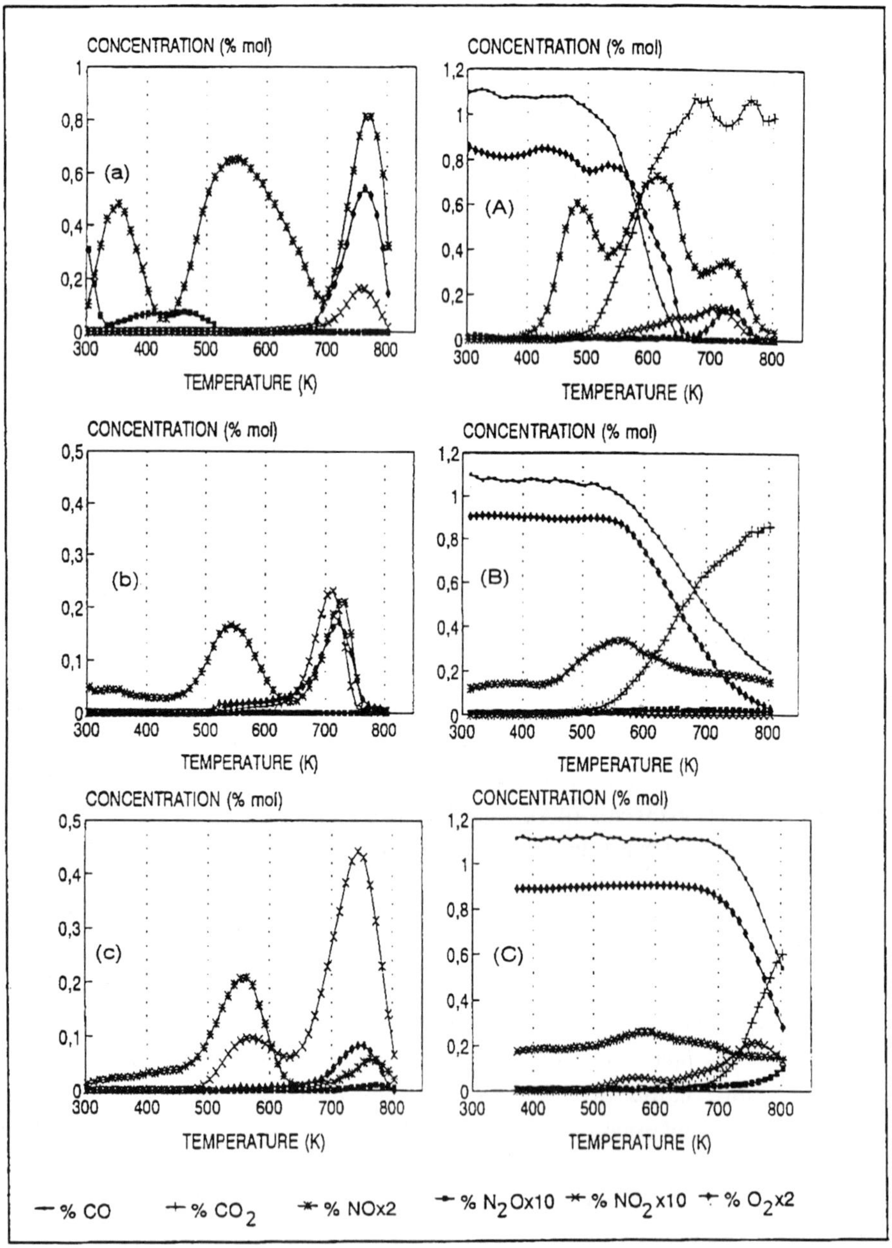

Figure 4 Data of (a-c): TPD after NO adsorption at 295 K; (A-C): TPReaction with CO + NO + O_2, obtained for (A,a): CeO_2; (B,b): 10% CeO_2/Al_2O_3; (C,c): Al_2O_3.

the higher capacity of pure ceria (in comparison with the other samples), when thermally pretreated at 573 K or above, to activate more deeply at low temperature the NO molecule so as to generate unstable $N_2O_2^{=}$ species which decompose easily into N_2O. This in turn can be related to the ESR data, showing that associated anion vacancy sites, of the kind giving the O_2^- signal O_{II} upon O_2 adsorption, are detected only for sample A. An interesting conclusion thus appears: the higher ability to activate reductively NO (and perhaps other molecules also) even at low temperatures depends on the capability to form these associated vacancy sites, which may exist on pure CeO_2 but be more difficult to generate on CeO_2/Al_2O_3 supports when ceria is well dispersed on the alumina substrate.

3.5. Temperature-programmed reaction experiments

The catalytic activity of the samples was probed also in a temperature-programmed way. In this case, after the standard pretreatment, a gas mixture containing ca. 1% CO, 0.45% O_2 and 0.1% NO in the N_2 carrier (i.e. of stoichiometric redox composition) was fed on the sample at a rate equivalent to 10^4 h^{-1} space velocity; after a short time (3 min) on stream (in order to ensure a steady gas flow on the catalyst), heating of the oven at a 10 K/min rate up to 800 K was started. The concentrations of the different gases measured at the reactor outlet for the A, C and D samples are plotted against the sample temperature in Figure 4A-C. These results show that, while for pure alumina CO and O_2 begin to be consumed in a quite parallel way, and CO_2 is produced (as corresponds to a process dominated by CO combustion), at temperatures around 670 K, for the Ce-containing samples the same process begins at distinctly lower temperature, which is nearly the same for the different samples (540 K), indicating that the onset of the CO combustion process is catalyzed by ceria in a relatively similar way in the different samples (although it is enhanced at higher temperatures in the pure CeO_2, probably due to the higher total amount of exposed ceria surface present in this case).

The NO profile is more complex. It is noteworthy that, for sample A, NO is not detected at the beginning of the temperature rise (even though it is present in the feed gas), which means that even after the initial interval of contact between the sample and the reacting mixture the NO molecules introduced are still being completely adsorbed by the sample, no other gas composition change being observable. NO does appear a $T \geq 380$ K, its concentration rising for a significant time above that of the incoming gas, which evidences clearly a substantial desorption of the initially adsorbed NO. For sample C, although such features of initial consumption and subsequent desorption of NO are less marked, they can be still clearly ascertained; actually, they appear to a small extent even in sample D. In the cases of samples B-D such NO adsorption/desorption phenomena (observed in the range $T \leq 700$ K) seem however to occur without significant involvement of CO in the process; the absence for sample C of the high-temperature peaks associated to decomposition of nitrate species suggests therefore that CO (which is not completely consumed in this experiment) may be acting at these higher temperatures to block the evolution of NO_2 and O_2 (a process in which the ceria component would be an active catalyst, since for pure alumina the NO_2 and O_2 evolution peaks, revealing the presence of nitrate groups, are still observed).

The mentioned evolution of NO to give concentrations in the gas flow above those in the incoming gas is observed to occur for sample C in a broad range betweeen 473 and 673 K; this coincides with the range for NO evolution due to nitrite decomposition observed in the TPD runs. For sample A, however, the extra NO evolution in this range appears split in two peaks. This splitting might be related to the small additional decrease in the CO concentration which can be observed around 500 K for this sample but not for sample C. On the other hand, the marked fall in CO concentration and simultaneous NO evolution observed around 600 K for sample A suggests that a reduction of NO_2^- by CO may be operating at this stage. In any case, the elucidation of the nature of the transient steps in this temperature range is not straightforward, and may need further experiments in order to be adequately clarified. Finally, at temperatures above ca. 750 K the NO concentration falls again below that in the feed gas, coincidently with the high level of consumption of CO. The new burst of NO evolution which still appears for sample A around 723 K together with production of some O_2 and NO_2 (clearly visible also for sample D) will be due to nitrate decomposition, and as said above indicates that, both on pure ceria and on alumina, surface nitrates can withstand the reaction conditions of this experiment up to temperatures of 700 K; in the first case, because previous reaction has completely depleted the gas stream from the reductant CO, and in the second one probably because of the low catalytic activity of Al_2O_3 against NO reduction. Reactivity seems thus to be in general rather smaller for the supported ceria samples than for the pure CeO_2 (the TPD and TP reaction data of sample B, not shown in the figure, although intermediate between those of samples A and C are much closer to those of the latter specimen).

ACKNOWLEDGMENTS

The authors thank the CICYT (project nr. MAT91-1080-C03-02) and the EEC SCIENCE program (contract nr. SCI-CT91-0704 TSTS) for financial support given to this work, and to Prof. G. Munuera for providing a CeO_2

sample. A. M.-A., J.M. C. and R. C. give thanks respectively to Comunidad de Madrid, the PFPI program and CSIC for PhD fellowships, under which their contributions to this work were carried out. Prof. L.A. P. gives thanks to the Spanish Ministerio de Educación y Ciencia for support to a sabbatical year leave.

References

1. a) G. Leclercq, C. Dathy, C. Mabilon and L. Leclercq, in A. Crucq (ed.), Catalysis and Automotive Pollution Control II (Stud. Surf. Sci. Catal. vol 71), Elsevier, Amsterdam, 1991, p. 181; b) B. Harrison, A.F. Diwell and C. Hallett, Plat. Met. Rev., 32 (1988) 73.
2. J. Soria, A. Martínez-Arias and J.C. Conesa, Vacuum, 43 (1992) 437.
3. a) J. Soria, A. Martínez-Arias and J.C. Conesa, submitted for publication. b) A. Martínez-Arias, PhD Thesis, Universidad Autónoma de Madrid, 1994.
4. J. Soria, A. Martínez-Arias, J.C. Conesa, A. Munuera and A.R. González-Elipe, Surf. Sci., 251/252 (1991) 990.
5. N. Känzig and M.H. Cohen, Phys. Rev. Letters, 3 (1959) 509.
6. M. Che, J.F.J. Kibblewhite, A.J. Tench, M. Dufaux and C. Naccache, J. Chem. Soc. Faraday Trans. 1, 69 (1973) 857.
7. A. Martínez-Arias, J. Soria , J.C. Conesa , R. Cataluña, A. Arcoya and X.L. Seoane, in preparation
8. J.H. Lunsford, J. Chem. Phys., 46 (1967) 4347.
9. E.L. Kugler, A.B. Kadet and J.F. Cryder, J. Catalysis, 41 (1976) 72.
10. M. Niwa, Y. Furukawa and Y. Murakami, J. Coll. Interf. Sci., 86 (1982) 260.
11. L. Cerruti, E. Modone, E. Guglielminetti and E. Borello, J. Chem. Soc. Faraday Trans. 1, 70 (1974) 729.
12. G. Busca and V. Lorenzelli, J. Catalysis, 72 (1981) 303.
13. C. Morterra, F. Boccuzzi, S. Coluccia and G. Ghiotti, J. Catalysis, 65 (1980) 231.

A. Frennet and J.-M. Bastin (Eds.)
Catalysis and Automotive Pollution Control III
Studies in Surface Science and Catalysis, Vol. 96

MECHANISM OF CHEMICAL ACTIVATION OF Pt-Rh ALLOY AND Pt/Rh BIMETALLIC SINGLE CRYSTAL SURFACES

Hiroyuki Tamura*, Akira Sasahara and Ken-ichi Tanaka

The Institute for Solid State Physics, The University of Tokyo 7-22-1 Roppongi, Minatoku, Tokyo 106
**KAO Corp. Wakayama Research laboratory, Wakayama 640*

ABSTRACT

Pt0.25Rh0.75 (100) alloy, and Pt/Rh(100) and Rh/Pt(100) bimetal surfaces underwent chemical restructuring at T > 400 K by heating in ca.10^{-7} Torr of NO or O_2, and the surfaces were covered with a common hybrid overlayer of Rh-O/Pt-layer. The Rh-O/Pt-layer gives a characteristic p(3x1) LEED pattern and is active for the reaction of NO + H_2. As a result, Pt-Rh(100) alloy and Pt/Rh(100) and Rh/Pt(100) bimetal surfaces have almost equal catalytic activity. Pt(110) surface was poorly active compared to Pt(100) surface for NO + H_2 reaction, but it was shown that Rh/Pt(110) and Rh/Pt(100) have almost equal catalytic activity. Therefore, it is concluded that a common Rh-O/Pt hybrid overlayer prepared by chemical restructuring is responsible to the prominent catalytic activity of the Pt-Rh catalysts for NO_x reduction.

1. INTRODUCTION

Catalytic reactions on metal surfaces are often remarkabily influenced by foreign metals. The Pt-Rh catalyst used for automotive exhaust gas control is a typical example. The Pt-based catalyst is markedly improved by adding Rh, but the roles of Rh are still controversial. In order to throw light on such fundamental problems, we studied the adsorption of NO and the reaction of NO with H_2 on Pt-Rh(100) alloy surface [1,2,3].
The catalytic reaction of NO with H_2 may be discussed by dividing into the following two steps.

i) $NO + H_2$ --- $N(a) + H_2O$

ii) $N(a) + H_2$ --- NH_3, N_2

When the first step i) is rapid but the step ii) is slow, N(a) intermediates will be accumulated on the catalyst surface.

In fact, when $NO + H_2$ reaction was performed on Pd(100) and Rh(100) surfaces, accumulation of N(a) intermediates occurs on theses surfaces although Pd(100) and Rh(100) surfaces are inactive for the chemisorption of N_2 [1,4]. However, no accumulation of N(a) intermediates was attained on Pt(100) and on Pd(110) surfaces although they catalyze the $NO + H_2$ reaction. It is noteworthy that the accumulation of N(a) does not occur on Pd(110) surface but it does on the Pd(100) surface [4]. This fact may suggest that $NO + H_2$ reaction will be a structure sensitive reaction.

Taking these results into account, the question is really interesting whether N(a) intermediates are accumulated on Pt-Rh(100) surface or not [1,2,3]. A layer-by-layer analysis of a Pt-Rh alloy tip surface by atom probe FIM showed that the alloy surface annealed at 700 C is composed of a Pt enriched topmost layer and Pt depleted 2nd layer but Rh enriched 2nd layer [5]. Therefore, Rh atoms are mainly extracted from the 2nd layer by reacting with oxygen and make an ordered arrangement on Pt enriched layer, which is the origin of the characteristic p(3x1) LEED pattern [3]. That is, such a hybrid surface as Rh-O/Pt-layer may be responsible for the p(3x1) pattern on the Pt-Rh(100) surface[6,8]. It should be pointed out that the Pt enriched layer may not be disturbed by diffusion at T < 900 K [7], so that the Pt-layer will be fixed during the catalysis at around 500 K. If this hybrid surface model is acceptable, the same p(3x1) surface should be formed on Rh deposited Pt(100) surface. In fact, we got a very sharp p(3x1) LEED pattern on Rh/Pt (100) surface by heating in NO or O_2 at 400 K [8]. In this paper, it wil be shown that a common hybrid surface composed of Rh-O/Pt-layer given by chemical restructuring is responsible to the prominent activity of the Pt-Rh catalyst for NO reduction.

2. EXPERIMENTAL

Electrochemical deposition of Pt ion on Rh(100) and Rh ion on Pt(110) was performed in a small volume cubic cell branched out the UHV chamber. Rh(100) and Pt(100) surfaces were subjected to a cleaning procedure consisting of Ar ion sputtering, O_2 treatment ($5x10^{-8}$ Torr) at 800 - 1000 K for 10 min, and annealing at 1000 - 1200 K for 20 min. After the cleaning, the crystal was transfered into a cubic cell for electrochemical deposition. The cell was filled with one atmospheric pressure of Ar gas, and then an electrochemical glass cell with electrolite solution of 0.05 M H_2SO_4 + 5x10 $^{-5}$ M $PtCl_4$ (or $RhCl_3$) was lifted up to make contact with the one side of the crystal disk. After the deposition of Pt or Rh ions, the crystal was washed with highly pure water and was transfered back to the UHV chamber for characterization by LEED and XPS as described before [8]. After the characterization of the surface, the sample was transfered again into the cubic cell for the reaction and/or! adsorption.

3. RESULTS AND DISCUSSION

Pt and Rh atoms may be randomly distributed on Pt-Rh alloy surfaces, but the depth distribution of Pt and Rh atoms depends strongly on the annealing temperature. So far, it has been accepted that the higher the annealing temperature, the higher the surface enrichment of Pt atoms. In fact, layer-by-layer analysis of a Pt-Rh alloy tip annealed at 700 C (ca. 1000 K) showed Pt-enrichment on the topmost layer and Pt-depleted (Rh-enriched) 2nd layer [5].

However, a careful experiment performed on a $Pt_{0.25}Rh_{0.75}(100)$ surface showed opposed result between 950 K and 1300 K. That is, the Pt fraction of the surface is lowered by raising annealing temperature, and it takes almost bulk composition at 1300 K [7]. It was also ponted out that the Pt-Rh(100) surface hardly attains to equilibrium distribution at $T < 900$ K. Therefore, annealing at higher T than 950 K is required to get equilibrium distribution of Pt and Rh atoms on Pt-Rh alloy surface.

In contrast to this, chemical restructuring proceeds easily even at ca. 400 K by NO or O_2 [2,3]. This fact suggests that the chemical restucturing is an unambiguously important process which should be generally considered on metals as well as on alloy surfaces during catalysis.

Rh deposited Pt(100) surface gives very sharp p(3x1) LEED pattern when it is heated in 10^{-7} Torr of O_2 or NO at $T > 400$ K [8]. From this fact, it was deduced that the Pt-Rh(100) alloy surface giving p(3x1) structure may be covered with the same hybrid stucture as Rh-O/Pt-layer on Rh/Pt(100) surface. In order to confirm the chemical restructuring, Pt deposited Rh(100) surface (θ_{Pt} = ca. 4 monolayer) was annealed in O_2. Segregation of Rh atoms took place by heating in $1x10^{-7}$ Torr of O_2 at 780 K as shown in Fig. 1. As increasing the fraction of Rh on the surface, LEED pattern changed from c(2x2) to p(3x1) pattern.

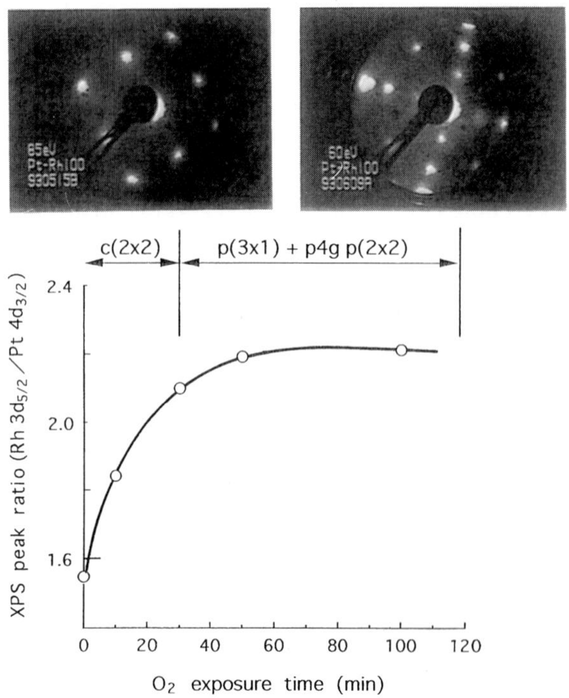

Figure 1 Segregation of Rh atoms on Pt/Rh(100) bimetallic surface by heating in O_2 of $1x10^{-7}$ Torr at 780 K. LEED pattern at 60 eV shows p(3x1)+p4g spots .

It is interesting that formation of the p(3x1) on the Rh/Pt(100) surface (θ_{Rh} = 0.4 - 1.0 monolayer) took place far more rapidly compared to that on the Pt/Rh(100) surface. That is, a streaky LEED pattern appeared at 340 K and it changed to a very sharp p(3x1) pattern at about 400 K by heating in 10^{-7} Torr of NO or O_2 [8]. However, the p(3x1) surface on these two surfaces showed almost equal reactivity to H_2 at room temperature. Taking these facts into account, thick Pt layer (ca. 4 monolayers) deposited on Pt/Rh(100) surface may be responsible to the slow formation of p(3x1) structure. Therefore, the p(3x1) pattern indicates the formation of a common hybrid Rh-O/P-layer on Pt/Rh(100), Rh/Pt(100) and Pt-Rh(100) alloy surfaces. Fig. 2 shows the characteristic behaviour of Pt/Rh(100) surface.

The hybrid structure of Rh-O/Pt-layer undergoes decomposition at 1000 K and the LEED pattern changes from the p(3x1) to p(1x1) patern on Pt-Rh(100) alloy as well as on Rh/Pt(100) and Pt/Rh(100) bimetal surfaces.

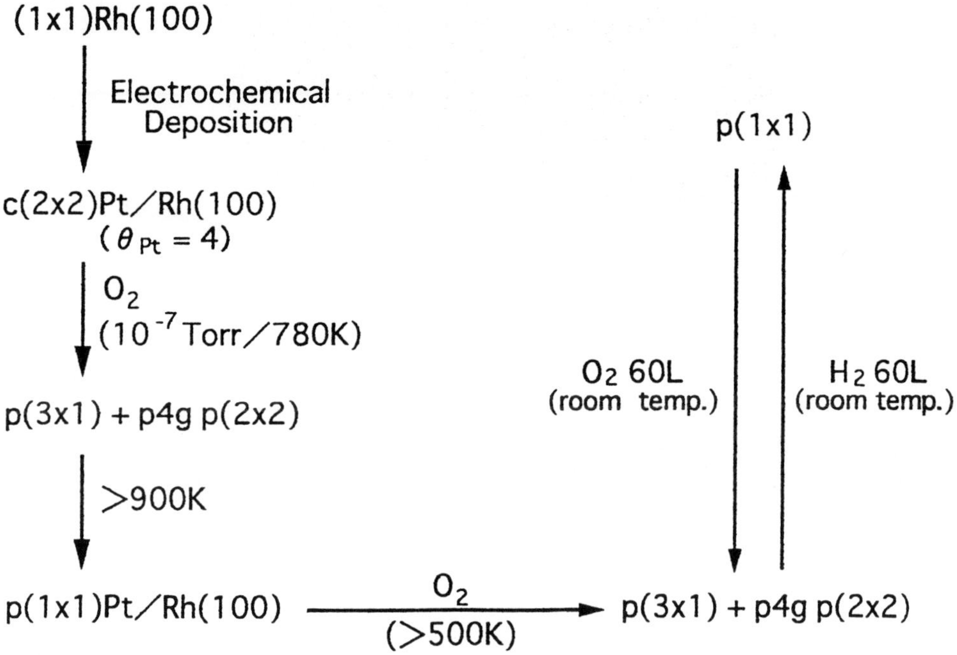

Figure 2 Schematic discription of the reversible change between p(3x1) and p(1x1) on Pt/Rh(100) at room temperature by H_2 or O_2 .

We are now capable to control Pt/Rh bimetal surfaces as well as Pt-Rh alloy surface by chemical restructuring. Based on these facts, our interest is focussed on the catalysis over the controlled surfaces. Catalytic reaction of NO + H2 (NO 1x10-6 Torr, H2 2x10-6 Torr) was performed on p(1x1) Pt/Rh(100) and p(3x1) Rh-O/Pt/Rh(100) surfaces.
the Pt/Rh(100) surface is self-activated during the catalysis at temperatures higher than ca. 600 K. On the other hand, the p(3x1) hybrid surface having Rh-O/Pt/Rh(100) construction promotes the reaction even at ca. 400 K as shown in Fig. 3 curve (ii), where the p(3x1)Rh-O/Pt/Rh(100) surface was prepared by heating the p(1x1)Pt/Rh(100) surface in O_2 (10^{-7} Torr) at 780 K for 10 min. Curve (iii) is the repeated run after run (ii). It is noteworthy that the ratio of NH_3/N_2 of run i) was 0.233 at 780 K, which was very close to the value of 0.277 on the activated p(3x1) surface given by run (ii). Therefore, we can conclude that the Pt/Rh(100) surface undergoes self-activation during catalysis by chemical
Fig. 3 (i) shows the reaction performed on a p(1x1) surface Pt/Rh(100) having Pt enriched topmost layer by annealing at 1000 K for 5 min. It is clear that the surface is poorly active for the formation of N_2 at T < ca. 600 K, but the surface

becomes active at temperaure higher than ca. 600 K, which seems to suggest that restructuring, and the catalysis takes place on this hybrid surface of Rh-O/Pt-layer.

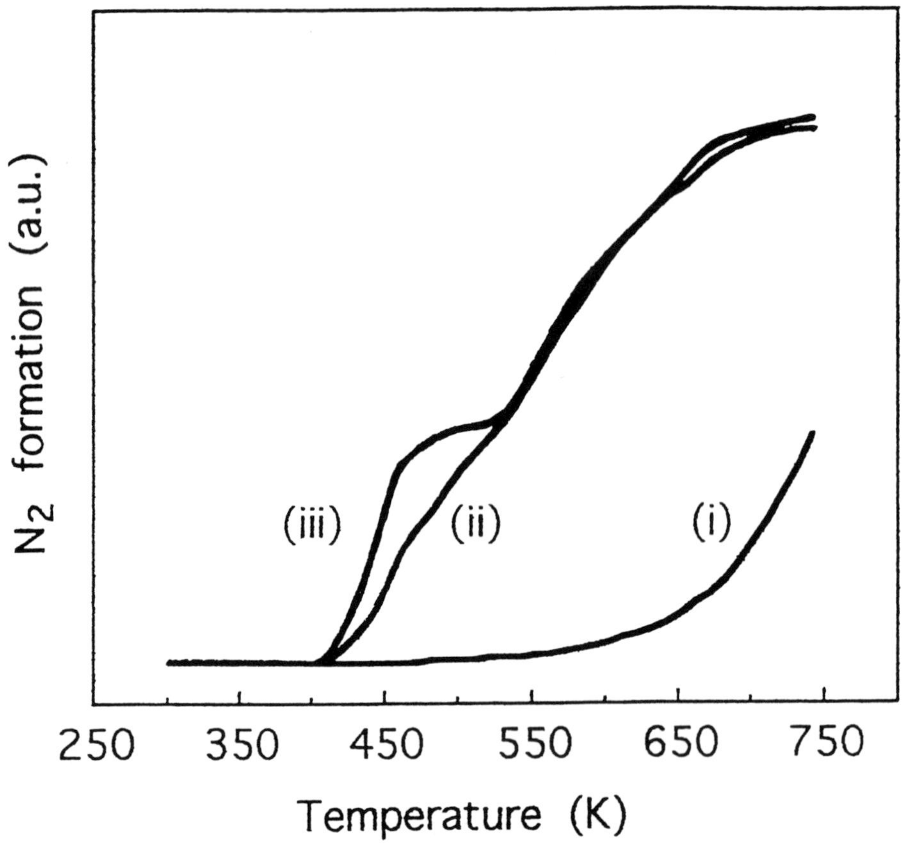

Figure 3 Catalytic reaction of NO ($1x10^{-6}$ Torr) + H_2 ($2x10^{-6}$ Torr) on Pt/Rh(100) surfaces with different constructions. (i); p(1x1)Pt/Rh(100), (ii); p(3x1)Rh-O/Pt/Rh(100), (iii); repeated run after the run (ii).

From the view point of the hybrid surface formation, NO + H_2 reaction on Pt(110) and Rh/Pt(110) surfaces are very interesting, because Pt(110) surface is poorly active compared to Pt(100) surface as shown in Fig. 4, that is, this reaction is a typical structure sensitive reaction. To compare the Rh/Pt(110) surface to Rh/Pt(100), we prepared Rh/Pt(110) surface by depositing Rh ions on a missing row p(1x2) Pt(110) surface. Rh deposited Pt(110) surface did not give any LEED spots because too thick Rh layer was deposited. Rh atoms were removed from the surface by Ar discharge sputtering and heating in O_2 at 770 K, and the first LEED pattern appeared was c(2x2) it changed to p(1x2) pattern when the thickness of theRh atoms were lower than one monolayer. The p(1x2) Rh/Pt(110) surface gives very high catalytic activity for the reaction of NO + H_2 as shown in Fig. 4. Although the reaction of NO + H_2 on Pt(110) and Pt(100) surfaces is

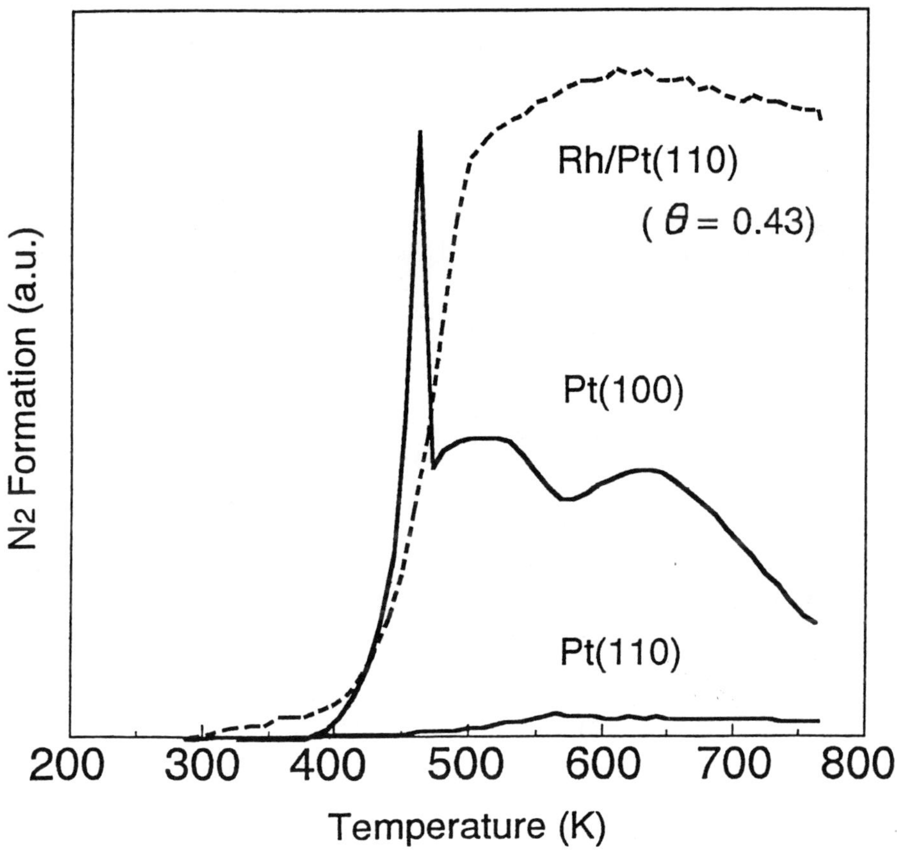

Figure 4 TPR of NO + H_2 on Pt(110), Pt(100), and Rh/Pt(110) surfaces.

highly structure sensitive as being Pt(100) >> Pt(110), Rh/Pt(110) and Rh/Pt(100) surfaces give almost equal catalytic activity. From these results, we can conclude that the Pt-Rh catalyst is self-actived, and a common hybride surface is built up on Pt-Rh(100) alloy as well as on Pt/Rh(100), Rh/Pt(110) bimetallic surfaces during the catalytic reduction of NO_x.

ACKNOWLEDGEMENT

This work was supported by the Grant-in-Aid for Scientific Research (05403011) of the Ministry of Education, Science and Culture of Japan, and one of the authors (Tamura) acknowledges KAO Corporation for giving a chance to study this work.

REFERENCES

1 T. Yamada and K. Tanaka, J. Am. Chem. Soc., 113 (1991) 1173 ; H. Hirano, T. Yamada, K. Tanaka, J. Siera, P. Cobden and B.E. Nieuwenhuys, Surf. Sci.,262 (1989) 97
2 H. Hirano, T. Yamada, K. Tanaka, J. Siera and B.E. Nieuwenhuys, Vacuum 41 (1990) 134
3 H. Hirano, T. Yamada, K. Tanaka, J. Siera, and B.E. Nieuwenhuys, Surf.Sci., 222 (1989) L804; H. Hirano, T. Yamada, K. Tanaka, J. Siera, and B.E. Nieuwenhuys, Surf. Sci.,226 (1990) 1
4 I. Matsuo, J. Nakamura, H. Hirano, T. Yamada, K. Tanaka and K. Tamaeu, J. Phys. Chem., 93 (1989)7747; K. Tanaka, Progress of Theoretical Physics, 106 (1991)419; K. Tanaka and T. Yamada, Research on Chemical Intermediates, 15 (1991) 213
5 T.T. Tsong, D.M. Ren, and M. Ahmad, Phys. Rev. B., 38 (1988) 7428 ; D.M.Ren, and T.T. Tsong, Surf. Sci., 184 (1985) L439.
6 H. Tamura, A. Sasahara, and K. Tanaka. to be published in Surf. Sci.Letters.
7 J. Siera F.C.M.J.M. van Delft, and B.E. Nieuwenhuys Surf. Sci. 264 (1992) 435
8 M. Taniguchi, E. Kuzembaev and K. Tanaka, Surf. Sci., 290 (1993) L711

A. Frennet and J.-M. Bastin (Eds.)
Catalysis and Automotive Pollution Control III
Studies in Surface Science and Catalysis, Vol. 96

CHANGES IN MICROSTRUCTURE AND CATALYTIC ACTIVITY EFFECTED BY REDOX CYCLING OF RHODIUM UPON CeO_2 AND Al_2O_3

J. Cunningham, D. Cullinane, F. Farrell, M. A. Morris, A. Datye* and D.Lalakkad*

Chemistry Department, University College, Cork, Ireland
**Farris Engineering Center, University of New Mexico, Albequerque, NM 87131-1341, USA*

ABSTRACT

Prior calcination at $T_{ox} > 673$ resulted in Rh_{ox}/CeO_2(A) materials having activities for R_o-type o.i.eq. and for R_2-type o.i.x. much enhanced relative to that over CeO_2 but in which, unlike Rh/Al_2O_3, the oxidised Rhodium component could not be detected by HRTEM or XRD. Observations in reducing conditions were likewise consistent with differences in the nature of Rhodium upon/within CeO_2 relative to that dispersed on Al_2O_3.

INTRODUCTION

Despite the acknowledged importance of rhodium as a component of three way catalytic converters (TWC's), some uncertainties remain concerning changes in its microstructural characteristics and the occurence of various valence states during various stages of TWC operation [1,2]. Controversy also continues concerning the question as to whether or not catalysts featuring rhodium dispersed on ceria are susceptible to inhibition of their activity by "decoration-encapsulation"-type Strong Metal Support Interaction (SMSI), or alternatively by rhodium loss through burial as ions or atoms within the oxide support [3,4]. Various experimental approaches have been adopted in this study in efforts to resolve some of these uncertainties: namely (i) the use of XRD and HRTEM directed at clarification of microstructural changes induced by redox treatments; (ii) studies by TPR of how *ease of rereduction* of oxidised rhodium species depended upon severity of prior oxidation; (iii) utilization of structure-sensitive hydrogenolyses as probe reactions to gain information concerning the SMSI susceptibility of rhodium in metallic form when supported on Al_2O_3 or CeO_2; and (iv) comparisons of the oxygen-handling capabilities of the oxidised catalysts using oxygen isotope equilibration and exchange.

EXPERIMENTAL

Materials

In order to minimise possibilities for spurious metal-support effects associated in the literature with use of $RhCl_3$ as precursor of ceria-dispersed rhodium [5], the materials used in the present study were prepared using rhodium(III) acetylacetonate (Aldrich) as precursor. Wet impregnation at 0.5, 2 or 4 wt% onto the surfaces of the oxide powders was achieved from solution in high purity methanol or tetrahydrofuran, after which samples were dried and calcined in O_2 for 2 hr at 823 K. Ceria available from Aldrich (CeO_2(A), 16 m^2m^{-1}) or Rhone-Poulenc (CeO_2(r.p.),110 m^2 g^{-1}), and Al_2O_3 (Conte, 200 m^2 g^{-1}) were used as supports.

Material Characterizations

These were made by temperature programmed reduction in 3% H_2+97% Argon (TPR), by high resolution electron microscopy (HRTEM) and by powder X-ray diffraction (XRD) on aliquots of the materials both in their "as prepared" states and after preoxidations (HTO_x) and/or prereductions (HTR). Powder XRD patterns were collected at RT using a Philips MCD diffractometer featuring optoelectronic control to ensure precise measurement of Q values. Peak width, positions and areas were evaluated by computer fitting of profiles. Particle sizes, where quoted, were calculated using the Scherrer formula. Detailed comparisons were made between patterns for samples after ageings in air at temperatures up to 1423 K and/or after reductions in 3% CO/He up to 573 K. HRTEM was done at the High Temperature Materials Laboratory, Oak Ridge National Laboratory, using a JEOL 4000 EX microscope. Transmission electron microscopy was also done at the University of New Mexico using a JEOL JEM 2000-FX microscope. The samples for TEM were made by simply dipping the holey carbon grid in the powder and shaking the excess off. No solvent was used in any stage of the sample preparation. Elemental analysis was done by energy dispersive spectroscopy (EDS) using a Tracor Northern System.

Reactivity Studies

Since structure sensitivity in the hydrogenolysis of n-butane over oxide-supported group VIII metals had been reported [6], this was utilised as a suitable catalytic probe reaction with which to compare activities of the catalysts in net reducing conditions. Particular interest attached to the question as to whether prior high temperature reductions (e.g. $HTR_{773\ K}$) would bring about similar strong inhibition of hydrogenolysis activity of Rh/Al_2O_3 or Rh/CeO_2 after LTR_{473}. Such reactivity studies were carried out in quartz-tube differential microreactors and samples of the exit gases were analysed by gas chromatographs fitted with appropriate columns and flame ionisation detectors. A reactant-flow composition ratio of $He:H_2:C_4H_{10}$ equal to 20:20:1 sccm was used. Between runs the removal of any carbonaceous deposit was effected by flowing (H_2+He) over catalysts to regenerate them. Experiments aimed at comparing the oxygen-handling activity of preoxidised aliquots of the materials, whilst ensuring that surfaces remained in an oxidised condition throughout (e.g. $Rh^{n+}{}_{ox}/CeO_2$), were carried out in a recirculatory reactor system under low pressures of an ($^{16}O_2+^{18}O_2$) equimolar mixture.

RESULTS AND DISCUSSION

TEM

Photo (a) in fig. 1 shows the low magnification transmission electron micrograph of Rh/Al_2O_3 in its precalcined state. The rhodium particles (shown by arrows) have a rounded morphology and are seen uniformly all over the catalyst and the average particle size is »300 Å. EDS confirmed a uniform distribution of rhodium in this catalyst. Photo (b) in fig. 1 shows a high magnification picture of the same catalyst. The lattice spacings on the large particle (shown by the arrow) indicate that the rhodium in this catalyst is present in the form of Rh_2O_3. Treatment of this catalyst in H_2 reduced the Rh_2O_3 to metallic Rh accompanied by a small decrease in particle size, but otherwise the distribution of particle sizes or the morphology of the catalyst was unchanged. Photo (c) of fig. 1 shows a high magnification view of Rh/CeO_2 in the oxidised state. In this case too, EDS confirmed the presence of rhodium in amounts similar to those seen on Rh/Al_2O_3. However, there is a distinct absence of a particulate rhodium phase on this catalyst, both in the oxidized catalyst which is shown, and the same catalyst reduced in H_2 at 473 K. This agrees with our previous study [7], where it was shown that on a CeO_2 support, Pt was difficult to observe in the TEM, at least in the fresh catalyst, but in that case, the Pt particles could be readily seen after several oxidation-reduction cycles. Murrell et al have previously demonstrated the existence of a dispersed oxidised phase of Pt on Pt/CeO_2 [8]. It is possible that such a dispersed oxidised phase of Rhodium here formed on or within CeO_2 thereby making the oxidised Rhodium component difficult to detect. [9].

XRD 4% Rh/CeO_2: No peaks due to rhodium or its compounds were visible after ageing acac-derived samples at 823, 923 and 1023 K. After a 1123 K treatment however, peaks readily ascribable to Rh_2O_3 could be easily detected (plot a, fig. 2). Between 1123 K and 1423 K the peaks sharpened consistent with average particle size having increased from ca. 300 to 650 Å. Other noteworthy points to emerge from particle size measurements included: **Firstly,** that below 1123 K no rhodia-related peaks were detectable, thereby indicating sizes below the XRD detection limit even

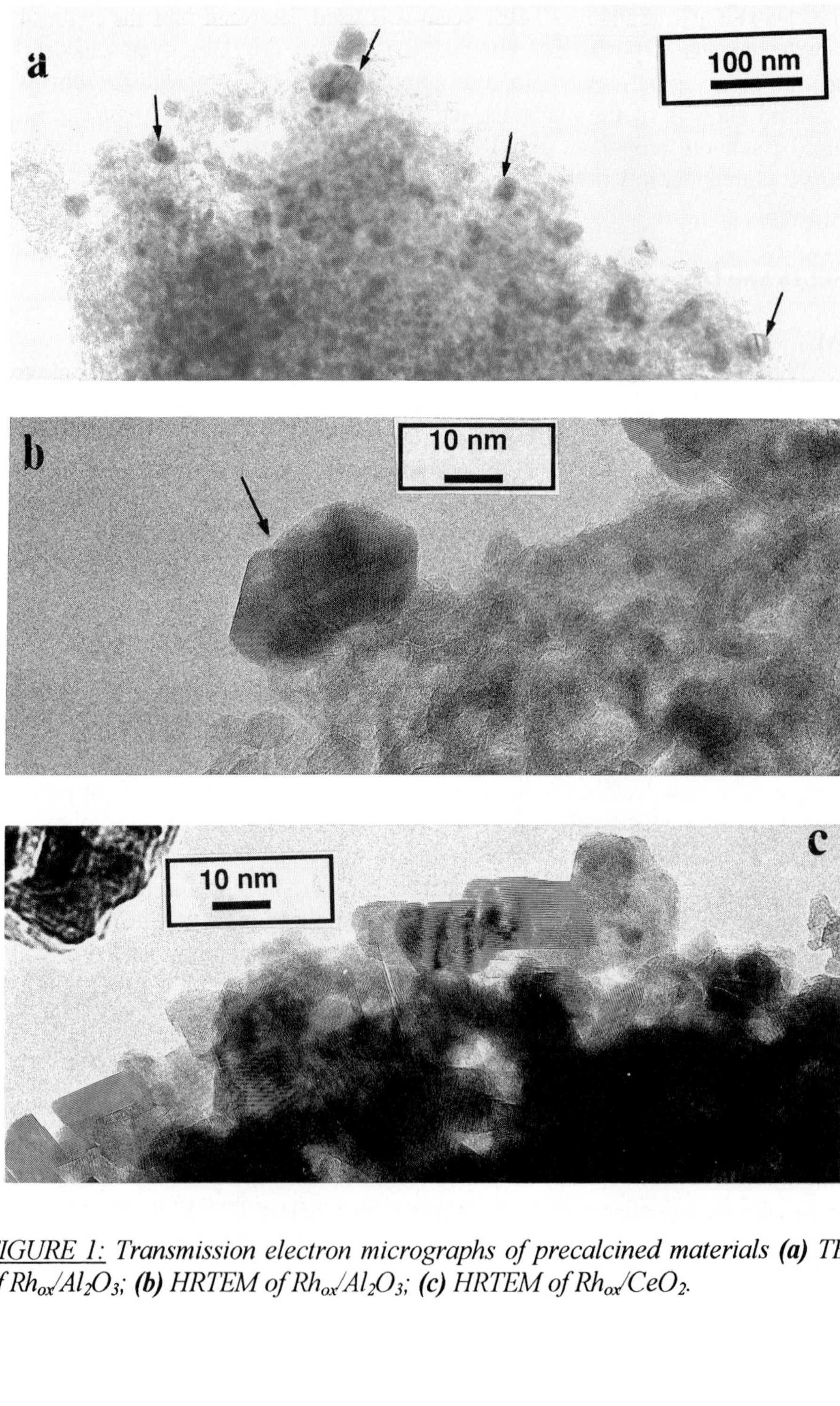

<u>FIGURE 1:</u> *Transmission electron micrographs of precalcined materials* ***(a)*** *TEM of* Rh_{ox}/Al_2O_3; ***(b)*** *HRTEM of* Rh_{ox}/Al_2O_3; ***(c)*** *HRTEM of* Rh_{ox}/CeO_2.

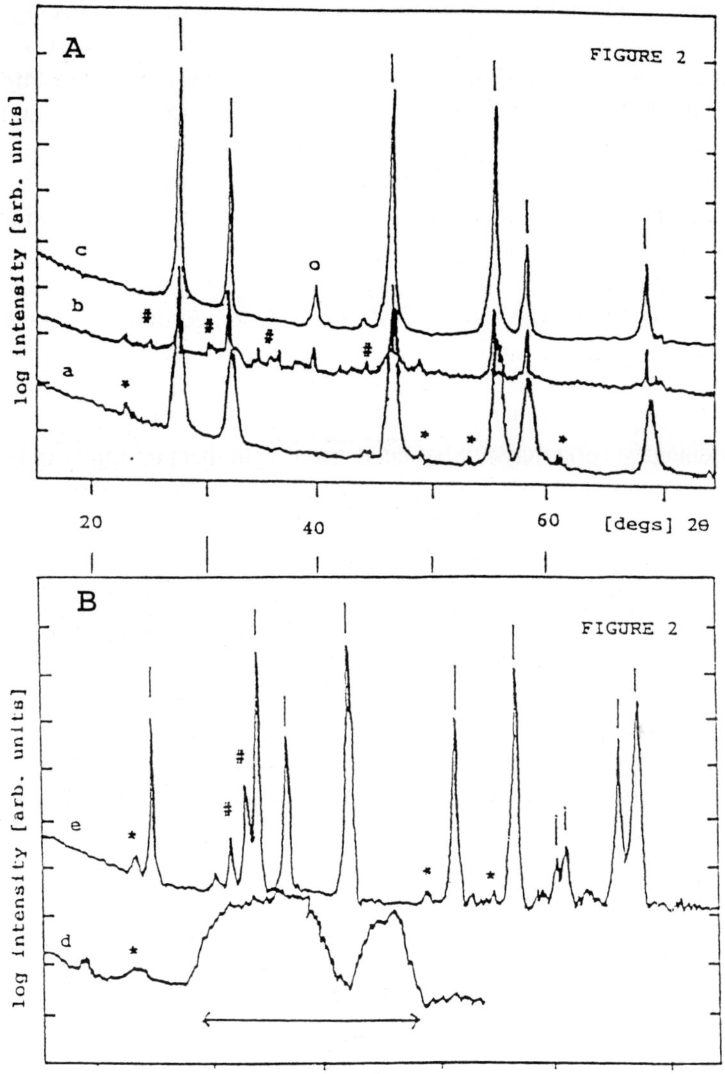

<u>FIGURE 2:</u> Powder X-ray diffractograms of 4% Rh/CeO_2 (Part A) and 4% Rh/Al_2O_3 (Part B) after treatments as follows: ***<u>Part A:</u>*** *Rh/CeO_2 with CeO_2 features indicated by vertical lines, metallic Rh by O, Rhodia by * and mixed Rhodium-Ceria phase by #: Plot* ***(a)*** *calcined @ ll23 K; plot* ***(b)*** *aged @ l423 K; plot* ***(c)*** *reduced in CO @ 523 K.* ***<u>Part B:</u>*** *Rh/Al_2O_3 with α-Al_2O_3 indicated by vertical lines, Rhodia by * and mixed Rhodium-alumina phase by #: plot* ***(d)*** *calcined @ 823 K; only 2Q angles 15°+0.55° are displayed since above this the complex γ-Al_2O_3 (and related structures) is too complex to reveal additional features;*
(e) *full 2Q scan after calcination @ 423 K. Note the change of γ-Al_2O_3 to* α-Al_2O_3.

for this substantial rhodium loading. As may be seen later, whenever metallic Rh was formed from this same loading it could be detected as particles around 50 Å. On that basis it appeared that whatever rhodia particles were present after calcinations below 1123 K must be smaller than 50 Å. That conclusion is strongly supported by the above-mentioned HRTEM observations upon Rh/CeO_2 (cf. photo c of Fig. 1) which gave no evidence for rhodia particles, although ceria (111) lattice planes with spacings of 3.1 Å were readily resolved. **Secondly,** in many experiments it was found that sizes calculated from different reflections spanned a wide range e.g. after calcination at the highest temperature (1423 K) results ranged 650±50 Å, whereas after 1123 K calcination the calculated sizes ranged 200-700 Å (average 300 Å). These results may be rationalised if rhodia particles were polycrystalline assemblies of individual crystallites having morphologies not corresponding to hemispherical. It has been suggested for Rh/Al_2O_3 that raft-like particles may be formed [10] and present XRD results for rhodia upon ceria may not be inconsistent with this possibility in respect of individual microcrystallites. After ageing at 1423 K new peaks are seen in the diffractogram not attributable to either ceria or rhodia (plot b, fig. 2). Peaks due to Rh metal presumably formed via the high temperature loss of oxygen - are indicated on the plot 2b. This agrees with recent work on rhodia by Carol and Mann [11]. However, other new peaks (also indicated) point to the formation of a completely new phase which has not yet been identified, but might be the Rhodium analogue of a surface-related, mixed phase such as reported previously for Pt upon CeO_2 [12]. When the 1423 K aged sample was reduced in CO at 523 K and re-oxidized at 573 K complete removal of peaks due to the mixed phase was achieved (but not of peaks due to metal, which rather increased in size during the reduction process cf. plot c fig. 2). However, the metallic particles produced by this 523 K reduction with CO appeared resistant to reoxidation at 573 K (cf. TPR data below).

XRD 4% Rh/γ-Al_2O_3: After ageing at 823 K peaks due to γ-Al_2O_3 were observed together with others attributable to the presence of rhodia (plot d, fig. 2). As a result of ageing at temperatures between 823 K and 1423 K rhodia particle size increased from ca. 50 (±30) to ca. 400 (±50) Å. Lack of consistency between particle-size calculated from different reflections was evident, consistent with observed rhodia particles being polycrystalline assemblies within which individual crystallites may not be equiaxed. The contrast between this facile detection of rhodia upon Al_2O_3 and its non-observance upon the ceria-supported sample indicates that one effect of the ceria was to produce/maintain oxidised Rhodium in hyper-dispersed form at temperatures up to 1023 K. However, after calcination at 1123 K the size of the rhodia particles on alumina was about 120±50 Å which was smaller than observed for Rh/CeO_2 similarly-pretreated. After ageing at 1423 K dramatic changes in the pattern occurred with the appearance of new peaks due to metallic Rh, as well as others due to an unidentified phase, and yet others attributable to a-alumina and rhodia (plot e, fig. 2). Yates et al [13] have suggested that during the high temperature processing of Rh/Al_2O_3 there is not only some solution of the rhodium oxide to give a surface mixed phase but also an encapsulation of rhodia particles as the $\gamma \rightarrow \alpha$ alumina phase transformation occurs. In our work the collapse at the g-structure was observed after ageing at 1400 K, thereby lending support to this hypothesis. However, it should be pointed out that the formation of the mixed phase is not dependent upon support structure collapse, since heating a physical mixture of Rh_2O_3 and α-Al_2O_3 at 1223 K in air for 2h rapidly produces the mixed-component phase.

Redox Cycles (TPR-T_{rox})

Comparisons between the TPR profiles at 10 K min^{-1} from aliquots of rhodium-loaded and rhodium-free CeO_2 and Al_2O_3 supports in their 'as prepared' condition (i.e. after calcination at 823 K) demonstrated that a reduction feature onsetting at ca. 330 K and exhibiting a maximum at 390 K originated from the rhodia component. Following that initial (and subsequent) TPR runs up to T_R-873 K, aliquots were retained in-situ whilst being cooled to 295 K in 3% H_2/Ar, after which they were exposed to a flow of pure, pre-dried O_2 for 2h at a reoxidation temperature,
T_{rox}, before being cooled to 295 K in O_2, flushed with 3% H_2/Ar and another TPR run made in standard conditions. Repetition of that sequence at progressively higher T_{rox} yielded, for each of the rhodium-loaded materials, a set of TPR profiles from which the following features emerged concerning the extent to which 2h at various T_{rox} restored the H_2-reduced rhodium component to oxidised form(s) reducible in the range 350-430 K: **(i)** no such reoxidation was detectable after T_{rox} @ 373 K, whereas 18%, 46% and 100% of that ultimately attainable was achieved by T_{max} @ 473, 573 and 673 K respectively; **(ii)** whilst position of T_{max} of the main TPR feature shifted progressively from 350 K after T_{rox} @ 473 to 430 K after T_{rox} @ 1073 K, continued existence of a lower-temperature shoulder at ca. 373 K (which became clearly resolved after T_{rox} @ 1073 K) suggested the continued co-existence of a minor, more easily reduced form of rhodia (possibly surface or near surface) together with the major and more difficult to reduce form (possibly within bulk) achieved by $T_{rox} \geq 773$ K. A set of such profiles similarly compiled for 4% Rh/Al_2O_3 again appeared consistent with two distinguishable forms of rhodium after $T_{rox} \geq 973$ K. However in that material the more easily reducible form ($T_{max} \sim 373$ K) was the major component after $T_{rox} \leq 873$ K - an observation which may be correlated with the TEM and XRD results above demonstrating the existance of sizeable rhodia particles upon that material after ageing at such temperatures. Another point made clear by the sets of TPR profiles for Rh/CeO_2 and Rh/Al_2O_3 is that, irrespective of T_{rox} values within the

range 673 → 1023 K, subsequent exposure to H_2 at $T_R \geq 450$ K sufficed for rapid conversion of the rhodium content to its reduced form:

Oxygen-handling Properties of Preoxidised Materials

Advantages in the use of an equimolar, isotopically non-equilibrated (*i.n.eq*) mixture of $^{16}O_2$ + $^{18}O_2$ to probe the oxygen-handling properties of oxidic materials include capabilities to detect: **(a)** activity of the surfaces at 295 K for R_o-type homophase isotopic equilibration (*o.i.eq*) as per eqn (0), via the intermediacy of very weakly held four-oxygen-atom surface species [14]; and **(b)** activity at higher temperatures for thermally activated heterophase oxygen isotope exchange (*o.i.x*) of types R_1 and R_2 as per eqns. 1 and 2, possibly via the intermediacy of strongly bonded O^- and O^- species respectively [15].

$$R_o\text{-type} \quad P(^{18}O_2)_{0.5} + P(^{16}O_2)_{0.5} \leftrightarrow P(^{16}O^{18}O)_{0.5} + P(^{16}O_2)_{0.25} + P(^{18}O_2)_{0.25} \qquad \text{Eq. (0)}$$

$$R_1\text{-type} \quad ^{18}O_2(g) + {}^{16}O^{2-}(s) \rightarrow {}^{16}O^{18}O(g) + {}^{18}O^{2-}(s) \qquad \text{Eq. (1)}$$

$$R_2\text{-type} \quad ^{18}O_2(g) + 2^{16}O^{2-}(s) \rightarrow {}^{16}O_2(g) + 2^{18}O^{2-}(s) \qquad \text{Eq. (2)}$$

The R_o-process was promoted at 295 K by preoxidised samples of ex Rh.ac.ac. $Rh_{ox}/CeO_2(A)$ samples, but not by the rhodium-free supports nor by ex-$RhCl_3$ materials. Thermal activation was necessary to bring about *o.i.x.*, as evidenced by onset of variations in partial pressures of $^{18}O_2$, $^{16}O_2$ and $^{16}O^{18}O$ during upward ramping of reactor temperature. Over rhodium-free CeO_2, no R_o-type *o.i.eq.* occurred, with the result that when *o.i.x.* did onset at ca. 773 K it operated on the *i.n.eq.* mixture and resulted in increases m/z = 32, equivalent decreases at m/z = 36, but with very little increase at m/z = 34. Those changes are consistent with operation of an R_2 or place-exchange mechanisms involving vacancy-pair creation and removal.

Figure 3 illustrates the thermally activated changes in isotopic composition of the gas phase over Rh_{ox}/CeO_2 and demonstrates onset of *o.i.x.* at ca. 523 K, i.e. ca. 250° below that over CeO_2 alone.

Since the $Rh/CeO_2(A)$ material already exhibited R_0-type activity at 295 K, operation of *o.i.eq.* had effected a change to the isotopically equilibrated composition as per eqn. O during the period of increase from 295 to 523 K. Slopes of the relative rates of change away from that composition, evident in fig. 3 as a result of the onset of o.i.x. at 523 K, are consistent with operation of a mainly R_2-type o.i.x. process. This observation that oxidised species within or upon Rh_{ox}/CeO_2 made such o.i.x. possible at much lower temperature than for Rhodia-free Ceria is qualitatively very similar to reports of o.i.x. promotion over Sr^{2+}/La_2O_3 relative to La_2O_3 [16], and a similar interpretation seems appropriate, viz. in terms of a promoting effect of an enhanced concentration of oxygen vacancies caused by incorporation of lower valent cationic dopants.

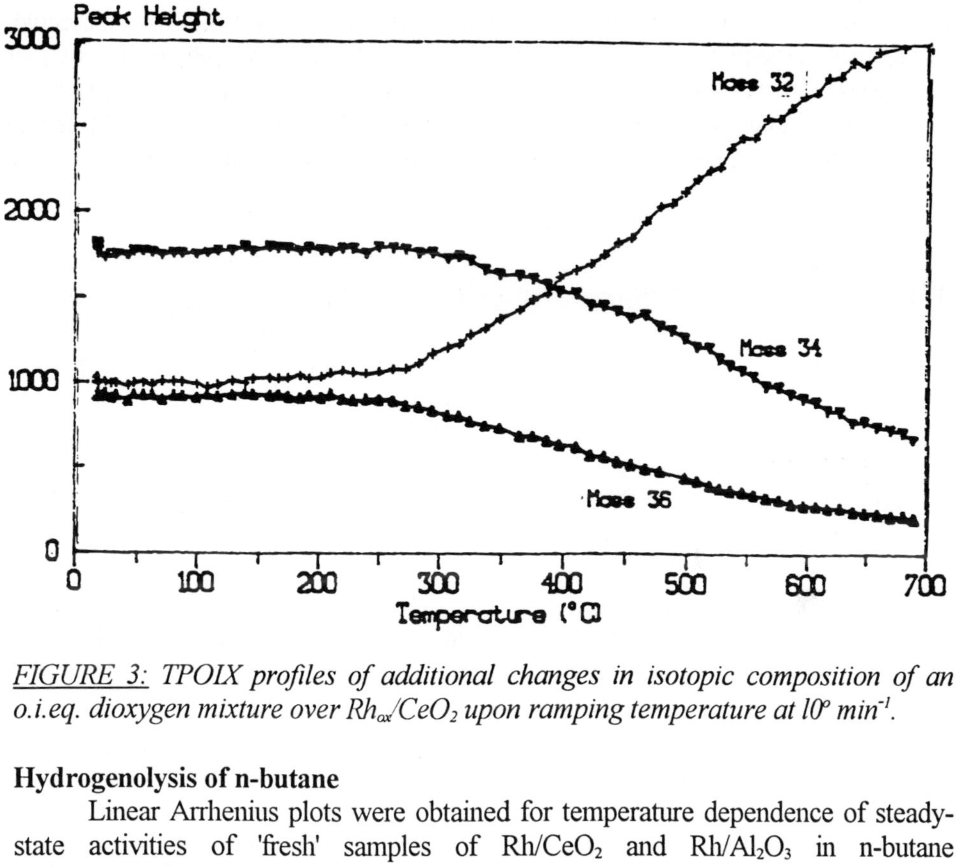

FIGURE 3: TPOIX profiles of additional changes in isotopic composition of an o.i.eq. dioxygen mixture over Rh_{ox}/CeO_2 upon ramping temperature at $10°$ min^{-1}.

Hydrogenolysis of n-butane

Linear Arrhenius plots were obtained for temperature dependence of steady-state activities of 'fresh' samples of Rh/CeO_2 and Rh/Al_2O_3 in n-butane hydrogenolysis at T_{rxn} 423-473 K. These plots indicated closely similar activation energies of 38 and 39 kcal mol^{-1} respectively, but showed hydrogenolysis rates over Rh/Al_2O_3 to be uniformly nine times larger than over Rh/CeO_2 at equal T_{rxn}. Results showing the extent to which Rh/CeO_2 hydrogenolysis activity, and the selectivity towards ethane,

could be modified by subsequent oxidation reduction cycles at higher temperatures are summarised in Table 1. Changes in activity are normalised relative to that observed for the fresh aliquot at the same T_{rxn}, and data in the Table make clear that whereas oxidation-reduction cycles caused significant reversible changes in the

ethane selectivity of Rh/Al_2O_3, the ethane selectivity of Rh/CeO_2 was little affected by such cycles.
The reported values for ethane selectivity were recorded under temperatures where the methane/propane product ratio was approximately 1.3, and did not change with catalyst pretreatment. The C_1-C_3 ratio near unity indicates that hydrogenolysis of the n-C_4 involves cleavage of a single C-C bond during the turnover of a n-C_4 molecule. Differences in the product distribution can then be related to cleavage of the n-C_4, at the central bond versus the terminal bond scission. On Rh/Al_2O_3, it is seen that the mole fraction of ethane in the product drops from ~ 0.5 in the fresh state to » 0.18 due to oxidation and this drop is reversed by high temperature reduction whereas no such changes occurred for Rh/CeO_2.

Table 1: *Effects of HTO_x and HTR upon n-butane Hydrogeneolysis Activity and Selectivity*

Pretreatment Sequence	Normalized Activities		Ethane Selectivity	
	Rh/Al_2O_3	Rh/CeO_2	Rh/Al_2O_3	Rh/CeO_2
1. Fresh (LTR_{673K})	1.00	1.00	0.484	0.401
2. 1st $HTO_{x\ 773\ K}$	0.69	1.82	0.218	0.488
3. 1st $HTR_{773\ K}$	0.075	0.058	0.398	0.488
4. 2nd $HTO_{x\ 773\ K}$	0.78	1.75	0.184	0.441

Those contrasting observations may be understood within the context of observations and interpretations developed in previous studies of relationships between ethane selectivity and rhodium particle size upon Al_2O_3 and SiO_2 supports [6]. On low weight loading Rh catalysts, where the particles were so highly dispersed that they were invisible by the TEM, oxidation-reduction cycles did not cause any change in either the activity or the ethane selectivity. On higher weight loading catalysts where the rhodium was present as well defined particles, the oxidation-reduction cycles caused significant changes in the activity and ethane selectivity that resembled those on macroscopic single crystals [8]. In that work it was argued that by virtue of their size, the surface structure of the small particles (< 10 Å) could not be altered because of insufficient room to pack the atoms in different ways. In the present study, the Rh/Al_2O_3 with particle size of » 30 nm showed reversible changes in ethane selectivity during n-butane hydrogenolysis, while the Rh/CeO_2 catalysts with no detectable particulate Rh phase exhibited no such cycling. This would be consistent with the rhodium particles on the CeO_2 support being very highly dispersed and/or having formed a complex that 'locks' them in a state where they did not undergo significant morphological changes under the pretreatments used in the hydrogenolysis studies.

ACKNOWLEDGEMENTS

Financial support for the research at UCC came in part from Contract SC1-CT91-0704 with DGXII of The European Commission: that for work performed at The University of New Mexico was provided by The Petroleum Research Fund of The American Chemical Society. High resolution TEM was performed at the High Temperature Materials Laboratory, a user facility supported by the U.S. Department of Energy at Oak Ridge National Laboratory, with additional work being performed at the Electron Microscope Laboratory within the Department of Geology, University of New Mexico.

REFERENCES

1 L.D. Schmidt and K.R. Krause, Catalysis Today, 12 (1992),264.
2 C.Z. Wan and J.C. Dettling, CAPoCI (eds. A. Crucq and A. Frennet,P359)
3 (a) J. Cunningham, S. O'Brien, J. Sanz, J.M. Rojo, and J.L.G. Fierro, J.Mol.Catal.,57, (1990), 379.
(b) J. Cunningham, D. Cullinane, J. Sanz, J.M. Rojo, J.A. Soria, and J.L.G. Fierro, J.Chem.Soc.Farad.Trans., 88, (1992), 3233.
4 (a) S. Bernal, F.J. Botana, J.J. Calvino, M.A.Cauqui,G.A.Cifredo, A. Jobacho, J.M. Pintaro and J.M. Rodriquez-Izquierdo,J.Phys.Chem.,97 (1993) 4118.
(b) J.El.Fallah, S. Boujana, H. Dexpert, A. Kiennemann, J. Majerus, O. Touret, F. Villain and F. Le Normand, J.Phys.Chem., 98 (1994), 5522.
5 H. Abderrahim and D. Duprez, Proc.9th Intern.Cong.Catal. (eds. M.J. Phillips & M. Ternan) Vol. 3, p.1296.
6 D.S. Kalakkad, S.L. Anderson, A.D. Logan, E.J. Braunschweig, J. Pena, C.H.F. Peden and A.K. Datye, J.Phys.Chem. 97,(1993), 1437.
7 D.S. Kaladdad, A.K. Datye and H.J. Robota, Appl.Catal.B., 1, (1992), 191.
8 L.L. Murrell, S.J. Tauster and D.R. Anderson, 10th North American Catalysis Lexington, KY (1991).
9 H.C. Yao, H.K. Stephen and H.J. Gandhi,J.Catal.,61(1980),547.
10 D.J.C. Yates, and E.B. Prostridge, J.Catal., 106 (1987), 549.
11 L.A. Carol and G.S. Mann, Oxid.Metals, 34 (1990) 1.
12 T.M. Bollinger and J.T. Yates, J.Phys.Chem., 95 (1991), 1694.
13 J.G. Chen, M.L. Collainni, P. Chen, J.T. Yates and G. Fisher, J.Phys.Chem., 94 (1990), 5059.
14 J. Cunningham in "Surface and Near-Surface Chemistry of Oxide Materials" (eds. J. Nowotny and L-C. Dufour) Materials Science Monograph 47, Elsevier, Amsterdam, 1988, P.345.
15 (a) J. Novakova, Catal.Rev., 4, (1970), 77;
(b) E.R.S. Winter, J.Chem.Soc., (1968), 2889.
16 Z. Kalenik and E.E. Wolf
(a) Catal.Lett. 11 (1991), 309, and
(b) Catalysis Today, 13 (1991), 255.

A. Frennet and J.-M. Bastin (Eds.)
Catalysis and Automotive Pollution Control III
Studies in Surface Science and Catalysis, Vol. 96

CATALYTIC OXIDATION OF PROPANE OVER PALLADIUM SUPPORTED ON ALUMINA AEROGEL. EFFECTS OF THE PRETREATMENT ON THE ACTIVITY AND INVESTIGATION OF THE STATE OF PALLADIUM BY GRAZING-INCIDENCE-X-RAY DIFFRACTION

C. Hoang-Van[a], R. Harivololona[a] and S. Fayeulle[b]

[a]*URA au CNRS "Photocatalyse, Catalyse et Environnement", Ecole Centrale de Lyon, B.P. 163, 69131 Ecully Cédex, France.*
[b]*URA au CNRS "Laboratoire Matériaux-Mécanique Physique", Ecole Centrale de Lyon, B.P. 163, 69131 Ecully Cédex, France.*

ABSTRACT

The effects of the pretreatment conditions on the activity in propane oxidation of Pd and Pt supported on Al_2O_3 aerogel catalysts and on their structure have been investigated by temperature-programmed oxidation and GIXD analysis. The results suggested that both PdO and Pd° were active sites for Pd catalysts. It was found that PdO generated under an oxidizing reaction mixture was particularly active for initially reduced samples, whereas for aged catalysts the dense faces on large crystallites were likely the active phase. In contrast, the active sites on Pt catalysts very likely consisted of Pt°. Furthermore, a large decrease in activity for aged catalysts tended to show that Pt was considerably less resistant to sintering in oxidizing environments than Pd. Under an oxidizing reaction mixture, the half conversion temperature for a Pd/Al_2O_3 -aerogel was slightly lower than those observed for both a Pt/Al_2O_3-aerogel and a Pt-Rh/Al_2O_3 catalyst.

1. INTRODUCTION

Palladium catalysts have long been recognized as having desirable performance properties in automotive emission control. In addition to its ready availability and low cost relative to platinum and rhodium, palladium is superior to platinum for CO oxidation and oxidation of unsaturated hydrocarbons [1].

However, improvements need to be made, particularly in the Pd-catalyzed conversion of both NO_x and saturated hydrocarbons, if Pd-only three-way catalysts are to be developed.

In a previous study, we found that Pd (1wt%) catalysts supported on finely divided alumina aerogels were more active, in the oxidation of CO by NO and O_2, than both Pd and Pt-Rh supported on commercial aluminas [2]. More recently, a study of propane oxidation over Pd supported on alumina-based aerogel catalysts led to the conclusion that favorable ensembles of active sites contituted by Pd^o atoms and Pd^{n+} ions might exist for this reaction [3].

The present work is concerned with the influence of the pretreatment of an alumina-aerogel supported Pd catalyst on its activity in propane oxidation. The state of Pd and the crystalline state of alumina aerogel were investigated by grazing-incidence X-ray diffraction (GIXD). To our knowledge, this method is applied for the first time to supported metal catalysts. In addition, a comparison of the structural and catalytic oxidation properties of Pd, Pt and Pt-Rh supported on alumina carriers was made.

2. EXPERIMENTAL

2.1. Catalysts

The alumina aerogel (490 $m^2.g^{-1}$, 1.3 $cm^3.g^{-1}$ micro-mesopore volume) was prepared from aluminum sec - butoxide dissolved in sec - butanol by hydrolysis with a stoichiometric amount of water. The alcogel was then dried in an autoclave under supercritical conditions with respect to sec - butanol [4]. Two conventional alumina xerogels (Degussa : 110 $m^2.g^{-1}$ and R.P. : 100 $m^2.g^{-1}$) were also used as supports of Pd and Pt-Rh catalysts.

Pd and Pt catalysts were obtained by impregnation of the supports with methanolic solutions of palladium or platinum acetylacetonates. The solids were then dried at 383 K for about 15 h. The Pt-Rh/Al_2O_3 three-way catalyst was prepared by impregnation of the support with a solution of H_2 $PtCl_6$ and $RhCl_3$ followed by a drying in air at 673 K [5].

2.2. Textural characterization

Surface area measurements and precious metal dispersion determinations by CO chemisorption were reported previously [2]. The chemical composition of the catalysts and their textural characteristics are given in Table 1. Prior to the BET or CO chemisorption measurements, the samples were either reduced by H_2 at

773 K for 3h (fresh catalysts) or first treated at 1273 K in flowing O_2 containing 10% of water vapour for 4h and then reduced in H_2 at 773 K (aged catalysts).

Table 1 Chemical composition and textural characteristics of catalysts

Catalyst	Surface area ($m^2.g^{-1}$)		Metal dispersion (%)	
	Fresh catalyst	Aged catalyst	Fresh catalyst	Aged catalyst
Pd/Al_2O_3-A a)	280	134	22 d)	17 e)
Pd/Al_2O_3-D a)	110	97	25 d)	7.5
Pt/Al_2O_3 -A a)	325	159	52	11
$Pt-Rh/Al_2O_3$ b)	100	- c)	55	- c)

a) Metal loading : ~1 wt% - A : aerogel - D : Degussa (gamma structure)
b) Conventional dechlorinated three - way catalyst : 1% Pt - 0.2% Rh/Al_2O_3[5].
c) - : not measured
d) Pd particles diameters in the range 3-9 nm as observed by TEM
e) Pd particles in the range 16 - 80 nm as observed by TEM. A few big alumina particles (0.8-7 mm) contained up to 5 wt % Pd as revealed by STEM analysis.

2.3. Pretreatment conditions

Before GIXD examinations or activity measurements, the catalyst samples were submitted to a treatment under different conditions that are summarized in Table 2.

Table 2 Conditions of catalyst pretreatments

Pretreatment	Atmosphere	Temperature (K)	Duration (h)
reduction	hydrogen	773	3
calcination	air	773	3
activation under reactants	3000 ppm C_3H_8 +1.5% O_2 in N_2 (s = 1)	773	3
aging	O_2+10%H_2O	1273	4

2.4. GIXD analysis

The grazing incidence X-ray diffraction (GIXD) measurements were performed using a 2-circle vertical goniometer which was specially designed in the laboratory. Scheme of the apparatus is shown on figure 1. Copper radiation was used without monochromator. A lithium-doped silicon cristal was used as the detector (operating at liquid nitrogen temperature) and the CuKa radiation (l = 0.154 nm) was selected owing to a multichannel analyser. Soller slits were installed on the incident beam to get a parallel beam and on the diffracted beam to limit the instrumental width of Bragg peaks to about 0.15° (theta) measured on silicon reference sample. The zero incident angle between the surface of the sample and the incident X-ray beam was carefully determined by measuring the intensity of fluorescence on a pure iron sample (for which the critical angle of total reflection is equal to 0.38°). The exact position was then defined by using a laser beam reflection.

Powder samples were deposited on glass slides and carefully flattened to obtain as smooth as possible surfaces. The incident angle was usually ranged between 0.5 and 2°. The penetration depth calculated owing to Fresnel formula [6] was equal to 1.5 mm in pure alumina for an incident angle of 1°. X-ray spectra were recorded for 2q varying from 20 to 80° at a constant speed of 0.02°/s.

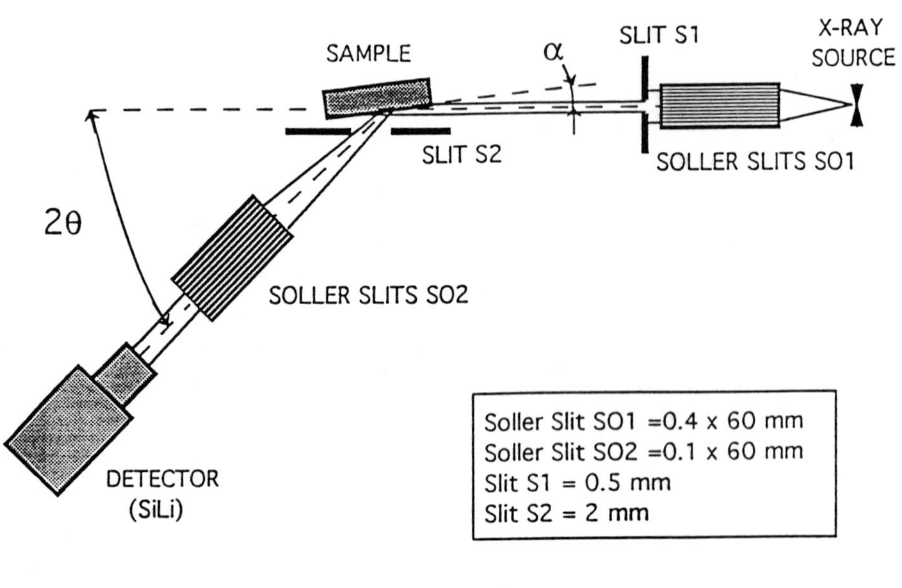

Figure 1. Scheme of the GIXD experimental set-up

2.5. Activity measurements

Propane oxidation was studied by using a flow reactor system equipped with a flame ionization detector (FID). The reactant mixture was composed of 3000 ppm C_3H_8 in N_2 with varying amounts of oxygen and the total flow rate was kept constant at 20 $L.h^{-1}$. The catalyst weight was 0.05 g diluted with 0.15 g of inactive α-Al_2O_3. The pretreatment was carried out under various conditions (see Table 2). The conversion of propane was measured at increasing temperatures (1.4 $K.min^{-1}$) in the range 423 - 773 K.

The stoichiometry of the feedstream was characterized by the ratio $s = (O_2)/5\ (C_3H_8)$ (stoichiometric mixture for s = 1 and oxidizing mixture for s > 1).

3. RESULTS AND DISCUSSION

3.1.GIXD results

Figure 2 shows the GIXD spectra of Pd/Al_2O_3-A catalyst previously reduced (sp. A), calcined (sp. B), aged (sp. C) or reduced after aging (sp. D). The conditions of these pretreatments are indicated in Table 2. Alumina aerogel support was poorly crystallized in the kappa form after reduction or calcination at 773 K (sp. A and B) whereas Pd° or PdO were distinctly detected after reduction (sp. A) or calcination (sp. B), respectively. After aging, PdO and δ-Al_2O_3 were observed (sp. C) and a subsequent reduction revealed Pd° and δ-Al_2O_3 (sp. D).

GIXD spectra (not shown here) were also recorded for initially reduced or calcined samples after the catalytic test under a stoichiometric or an oxidizing reaction mixtures (s = 1 or s = 4). In all cases, κ-Al_2O_3 was transformed into α-Al_2O_3. For the reduced sample, Pd° remained unchanged for the s = 1 mixture but was completely oxidized into PdO under the s = 4 mixture. The initial PdO phase in the calcined sample was partly reduced into Pd° (Table 3).

GIXD spectra concerning the Pt/Al_2O_3-A catalyst are shown in Figure 3. Distinct differences were observed between Pt/Al_2O_3-A and Pd/Al_2O_3- A catalysts : i) γ-Al_2O_3 was detected for the calcined catalyst (sp. A) instead of the kappa form for Pd/Al_2O_3 - A ; ii) PtO_2 and Pt° were present in the calcined sample (sp. A) but this oxide was completely reduced after a test under the stoichiometric mixture (sp. B) and iii) after aging, only Pt° was detected on the δ-Al_2O_3 support (sp. C). These differences may stem from the differences in the

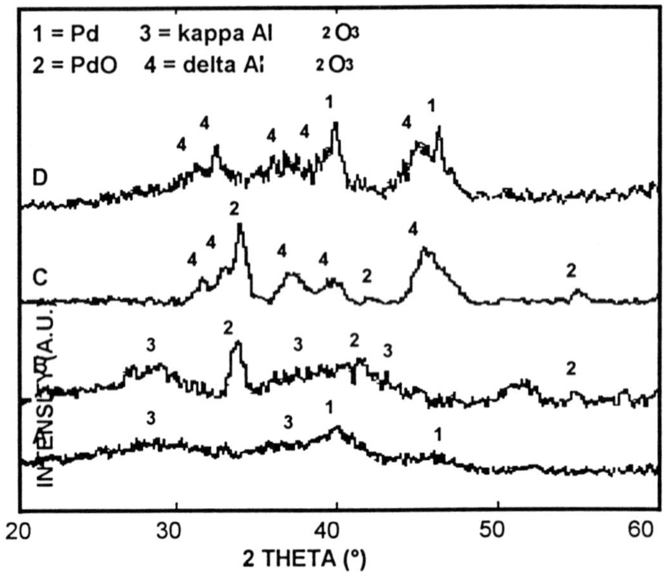

Figure 2. GIXD spectra of Pd/Al$_2$O$_3$ - A after : A : reduction - B : calcination - C: aging - D : aging + reduction (see Table 2 for conditions)

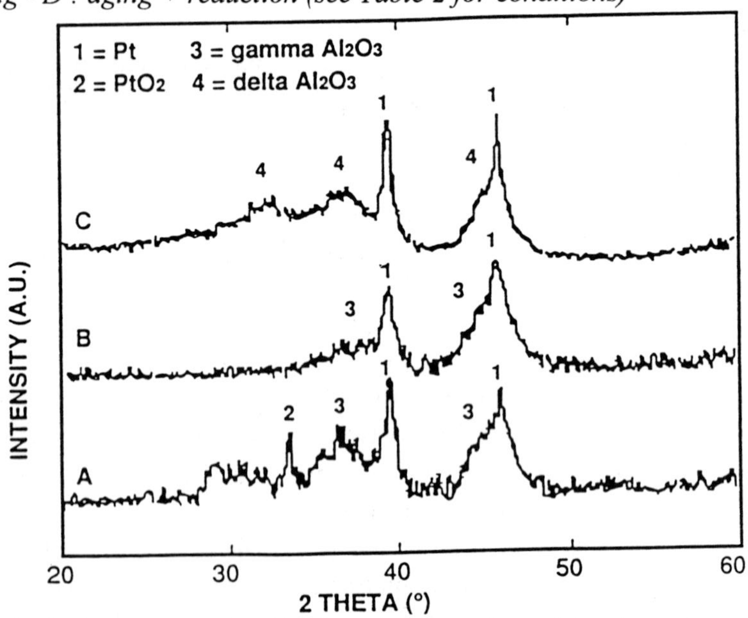

Figure 3. GIXD spectra of Pt/Al$_2$O$_3$ - A after : A : calcination - B : calcination + reaction (s=1) - C : aging (see Table 2 for conditions)

dispersions of Pt and Pd on the alumina support (Table 1) and in the intrinsic properties of these two noble metals. A higher dispersion of Pt (52% vs 22% for Pd) may enhance the metal-support interaction and leads to a deeper structural transformation of the Al_2O_3 aerogel carrier after the calcination treatment. On the other hand, Pt has a higher ionization potential (8.9 vs 8.3 eV for Pd) and its oxide has a lower stability. PtO_2 is known to be unstable at 973 K even under an oxidizing atmosphere [7]. Therefore, only Pt° was detected after a test under the stoichiometric mixture or after the aging treatment.

GIXD results are summarized in Table 3 which also indicates the phases detected for an aged Pd/Al_2O_3 - D (PdO and d-Al_2O_3) and an aged three-way Pt-Rh/Al_2O_3 catalyst (Pt° and d-Al_2O_3).

Table 3 Structure of Al_2O_3 support and state of deposited metal by GIXD analysis

catalyst a)	pretreatment b)	Al_2O_3	metal phase
Pd/Al_2O_3 - A	reduction	kappa	Pd°
	calcination	kappa	PdO
	aging	delta	PdO
	aging + reduction	delta	Pd°
	reduction + reaction (s=1)	gamma	Pd°
	reduction + reaction (s=4)	gamma	PdO
	calcination + reaction (s=1)	gamma	PdO + Pd°
Pd/Al_2O_3 - D	aging	delta	PdO
Pt/Al_2O_3 - A	calcination	gamma	Pt° + PtO_2
	aging	delta	Pt°
	calcination + reaction (s=1)	gamma	Pt°
Pt-Rh/Al_2O_3	aging	delta	Pt°

a) see Table 1 b) see Table 2

3.2. Propane oxidation

3.2.1. Pd/Al_2O_3 - A

The light-off curves of Pd/Al_2O_3 - A in the oxidation of propane under a stoichiometric reaction mixture are shown in Figure 4 A. Prior to the catalytic test, the catalyst was treated *in situ* at 773 K under different conditions (see Table 2). At conversions between 30 and 70%, the calcined sample showed the highest activity and the reduced one, the lowest.

The curves of Figure 4 B concern an aged Pd/Al_2O_3 - A sample whose activity was almost the same as that of a reduced sample (lower curves) whereas an aging-reducing combined pretreatment led to a distinctly higher activity (upper curve).

Figure 4 C illustrates the effect of oxygen concentration on the activity of pre-reduced Pd/Al_2O_3 - A samples. For reaction mixtures s = 1, s = 1.5, s = 2, s = 4 and s = 6, the concentration of propane was maintained at 3000 ppm, while that of oxygen was 1.5, 2.25, 3.0, 6.0 and 9.0%, respectively (N_2 diluent). The highest activity was observed for s = 4, which showed that a large excess of O_2 was beneficial to the activity of the reduced catalyst, but that a too large excess (s = 6) was detrimental.

3.2.2. Pt/Al_2O_3 - A

The effects of pretreatment and oxygen concentration in the reaction mixture on the activity of a Pt/Al_2O_3 - A catalyst are illustrated by the curves of Figure 5. In contrast with Pd/Al_2O_3 - A, the activity of Pt/Al_2O_3 - A was much lower under an oxidizing mixture (s = 4) than under a stoichiometric one (s = 1) for both samples, either reduced or aged and then reduced prior to the catalytic test.

Table 4 summarizes the results obtained in propane oxidation. The light-off temperature, T_{50}, (at 50% of C_3H_8 conversion), considered as representative of the catalyst activities, are given for a stoichiometric (s = 1) and an oxidizing (s = 4) mixtures (last columns). For comparison purposes, results obtained for a Pd/Al_2O_3 - D and a $Pt\text{-}Rh/Al_2O_3$ catalysts are also reported in this Table. The results showed that Al_2O_3 aerogel was a better support for palladium than the conventional Al_2O_3 - Degussa and, for the stoichiometric reaction mixture, Pt/Al_2O_3 - A was more active in propane oxidation than $Pt\text{-}Rh/Al_2O_3$ and Pd/Al_2O_3 - A. However, under an oxidizing mixture (s = 4), Pd/Al_2O_3 - A was slithtly more active than the other catalysts tested (Table 4, last column).

3.3. Correlation structure-activity

A tentative correlation between the structure and the activity of the studied catalysts can be made by comparing the results in Table 3 with those in Table 4.

For the Pd/Al_2O_3 - A catalyst, the light-off temperature (T_{50}) was decreased by 53 K when the originally reduced sample was tested under the oxidizing mixture (s = 4) instead of the stoichiometric one (s = 1) (Table 4). Therefore, PdO that was generated under the oxidizing mixture constituted very active sites, in comparison with Pd° still detected under the stoichiometric mixture (Table 3). This conclusion is in agreement with the work reported by Hicks et al. [8] which showed that the oxidation of an originally reduced Pd/Al_2O_3 catalyst formed new active sites associated with an increase in the rate of methane oxidation. On the other hand, after the aging treatment followed by a reduction prior to the catalytic

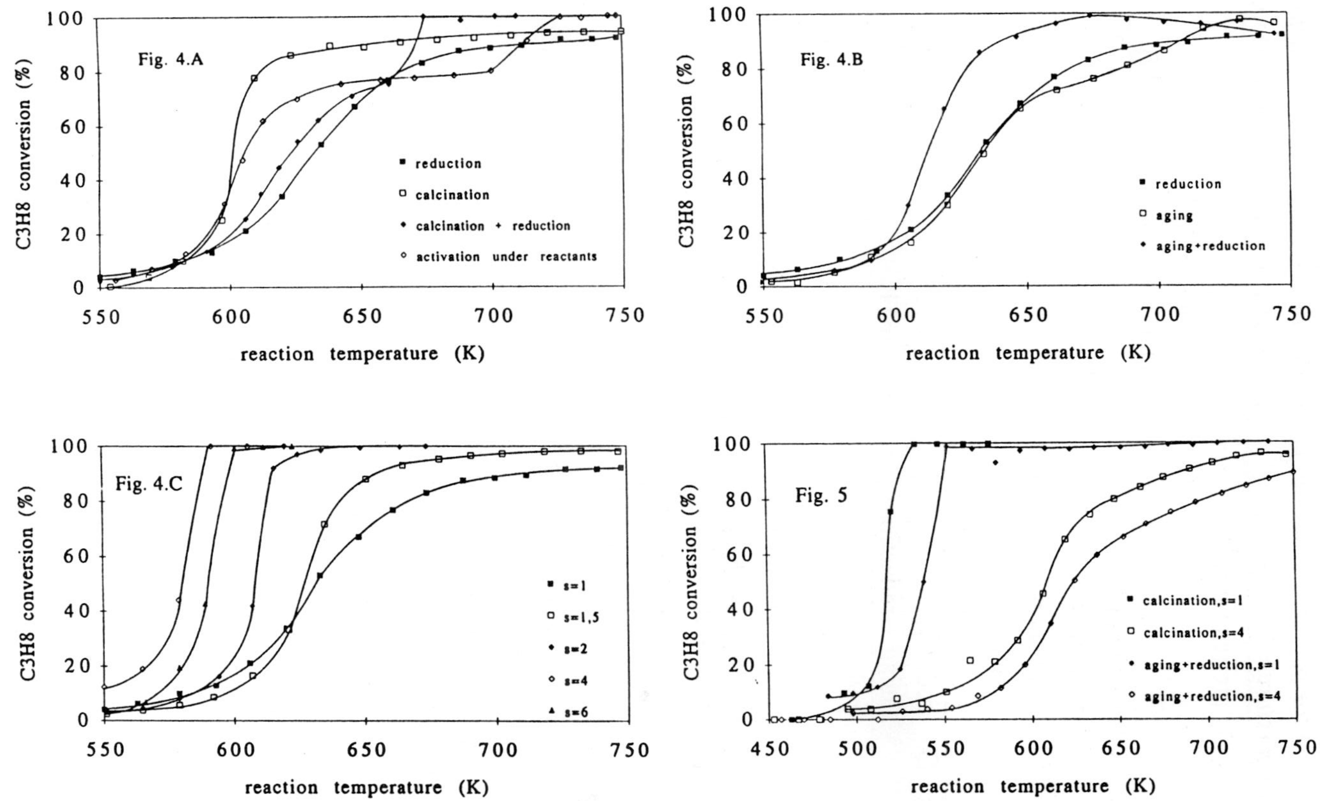

Figure 4. Propane oxidation over Pd/Al_2O_3 - A - (A). Effect of pretreatments (stoichiometric reaction mixture) - (B). Effect of aging (stoichiometrric reaction mixture) - (C). Effects of O_2 concentration in the reaction mixture reduced samples.

Figure 5. Propane oxidation over Pt/Al_2O_3 - A. Effects of pretreatments and O_2 concentration in the reaction mixture.

Table 4 Effect of catalyst pretreatment on the light-off temperature (at 50% conversion : T_{50}) for propane oxidation under stoichiometric (s=1) or oxidizing (s=4) reaction mixtures.

catalyst a)	pretreatment b)	T_{50}(K) stoichiometric mixture (s=1)	T_{50}(K) oxidizing mixture (s=4)
Pd/Al_2O_3 - A	reduction	633	580
	calcination	603	610
	activation under reactants	607	- c)
	aging	635	- c)
	aging + reduction	611	609
Pd/Al_2O_3 - D	reduction	676	589
	aging + reduction	611	616
Pt/Al_2O_3 - A	reduction	513	587
	calcination	513	608
	activation under reactants	513	- c)
	aging + reduction	539	624
$Pt\text{-}Rh/Al_2O_3$	reduction	571	589
	aging + reduction	565	- c)

a) see Table 1 b) see Table 2 c) Not measured

test under a stoichiometric mixture, the temperature of half conversion was lower by 22 K than that observed for the only reduced sample (611 vs 633 K, Table 4), although sintering of a number of Pd particles occurred during the aging treatment (Pd particles of 16-80 nm were observed by TEM). This is in favour of the hypothesis that the reaction occurs on highly active dense faces present on large Pd crystallites [9]. After the catalytic test, γ-Al_2O_3 was always detected for samples pretreated at 773 K. The aging treatment led to the formation of δ-Al_2O_3 . Baldwin et al. [10] found that Pd supported on δ-Al_2O_3 was much more active than its homologue supported on γ-Al_2O_3 in methane oxidation. The authors attributed that difference in activity to a difference in the morphology of palladium on various supports [10]. Similar support effect can also be advanced for Pd/Al_2O_3-A catalysts pretreated under different conditions.

For the $PtAl_2O_3$ - A catalyst, lower activities were always observed for the test under an oxidizing atmosphere (Table 4, compare the last two columns). Therefore, the active sites of this catalyst consisted of Pt° and even a partial (or superficial) oxidation of platinum led to a decrease in its activity. Furthermore, an

A. Frennet and J.-M. Bastin (Eds.)
Catalysis and Automotive Pollution Control III
Studies in Surface Science and Catalysis, Vol. 96

CHARACTERIZATION OF SURFACE AND BULK OXYGEN SPECIES OF THREE WAY CATALYSTS BY O_2 TPD AND H_2 TPR.

C. Bouly[a], K. Chandes[b], D. Maret[a] and D. Bianchi[b]

[a]*E.C.I.A, R&D Center, 25550 Bavans, France.*
[b]*L.A.C.E, Université Claude Bernard Lyon I, ISM,*
43 bd du 11 Novembre 1918, 69622 Villeurbanne, France

ABSTRACT

The deactivation of three way catalysts (TWC) can be studied by the determination of the change of some of their properties such as: specific surface area, catalytic activity or precious metal dispersion. Two qualitative and quantitative methods have been investigated in order to characterize TWC oxygen species: Temperature Programmed Desorption (O_2 TPD) and Temperature Programmed Reduction in hydrogen (H_2 TPR).

TPD experiments performed after oxygen adsorption at 873 K, showed four surface oxygen species. Three of them are related to the washcoat (CeO_2, Al_2O_3) and the last one to the precious metals (PM): Pt, Rh.

Two mains reduction processes are detected by H_2 TPR. The first one, at low temperature, shows two overlapped peaks of hydrogen consumption, respectively assigned to the reduction of metal oxides (T= 251°C), and to active dispersed CeO_2 in interaction with precious metals (T= 314°C). The second reduction process is related to bulk CeO_2 reduction and leads to a consumption of H_2 at temperatures higher than 400°C.

The evolution of the amounts of oxygen desorbed and of hydrogen consumed during the two temperature programmed experiments, are examined as a function of various thermal treatments of the catalysts.

1. INTRODUCTION

Increasingly stringent vehicle emissions standards make exhaust system (especially catalytic converter) operating conditions more and more severe. Most of the solutions studied to achieve such standards, at lowest cost, are focused on a faster light-off. This can be partially obtained by reducing the distance between the engine and the catalytic converter (close coupled catalyst) and by decreasing heat losses between these two parts. However, these solutions lead to higher operating temperatures which must be taken into account to find out the best trade-off regarding with more stringent durability requirements. This means that parameters causing catalyst deactivation (high temperatures, chemical composition of the gaseous mixtures,...) even if low, must be known very accurately. This knowledge will allow the use of fast development tools leading to catalysts aged exactly in the same way as on the vehicle. Therefore, it will be possible to judge the effect of the whole systems' design to meet the durability requirements and to achieve the best technical and economical compromise.

Information on the main parameters controlling the deactivation can be provided by the study of the change of various chemical and physical properties of the TWC such as: BET surface area, chemisorption of probe molecules and catalytic activity. In this paper, another way of catalyst characterization is presented: the analysis of catalyst oxygen species, which are likely to be involved in the catalytic process. It is indeed well known that the addition of promoters and stabilizers (as CeO_2) leads to an increase of the resistance to sintering of precious metals [1,2] and to the appearance of new chemical properties (in particular on the oxygen storage) [3-6]. Furthermore, some solutions to meet the requirements of OBDII are based on these oxygen storage properties. A synergy between the two PM is also considered on the CeO_2/Al_2O_3 solids [7].

Two methods of characterization of the various oxygen species have been investigated: Temperature Programmed Desorption (O_2 TPD) and Temperature Programmed Reduction (H_2 TPR) Finally, a comparison between fresh and various aged TWC is outlined.

2. EXPERIMENTAL

2.1. Catalysts.

TWC used in the present study are commercially available monoliths, from the same manufacturer, with around 20 weight % of washcoat (3.6 weight % of cerium) loaded with 40g/ft^3 of precious metals (PM), with a Pt/Pd/Rh ratio of 5/0/1. Three TWC were studied:by the various analytic methods (sections 2.2 and 2.3): a fresh catalyst (42 m^2/g), a 130 000 km vehicle aged, and a catalyst treated 24 h in air at 1100°C (3 m^2/g). Two samples were picked from the 130 000 km catalyst: at the inlet (17 m^2/g) and at the outlet section (15 m^2/g). Four solids were studied only by O_2 TPD and H_2 TPR: a washcoated cordierite without any PM, a TWC treated 4h under N_2 at 1050°C, a commercially available catalyst 0.5% Pt/Al_2O_3 and a sample of pure CeO_2 (4.4 m^2/g).

2.2. Light-off temperatures and vehicle emissions.

The light-off temperatures (25% of conversion) were determined on a laboratory catalytic evaluation bench, using a F.T.I.R. spectrometer as detector [8]. The composition of the reactive mixture was: 14%CO_2, 10%H_2O, 0.8%O_2, 0.4%CO, 950 ppm NO, 375 ppm C_3H_6, 125 ppm C_3H_8.

The emissions (expressed in g/km) of CO, NO_x and hydrocarbons (HC) during an ECE-EUDC cycle were measured on a 1.9L MPFI engine. The fresh TWC and the air aged catalyst have been previously treated on an engine test bench (4 h at 550°C, R=1).

2.3. O_2 TPD and H_2 TPR.

The gas flow regulator and control system, the quartz microreactor and the analytic system (mass spectrometer) are identical to those used previously described [9]. During the H_2 TPR experiments, a second liquid nitrogen trap is inserted after the reactor in order to remove the water.

The following procedures are used. The solids are dehydrated at 600°C under helium flow (30 cm^3/min). The temperature is increased, at 25°C/min, from room temperature to 600°C. The solids are treated 1h under O_2 at 600°C and cooled down in O_2 to 25°C. After a purge with pure helium, the TPD (25- 900°C, 25°C/min) is started and the desorption of O_2 is continuously recorded by mass spectroscopy (intensity of the peak m/e=32). The same procedure is used for the TPR experiments but after the helium purge at 25°C, a reactive mixture, either 3% or 10% H_2/He, is passed over the solids and the temperature is increased up to 900°C (25°C/min). The consumption of hydrogen is recorded by mass spectroscopy (intensity of the peak m/e= 2). The experimental data give the amounts of oxygen desorbed and hydrogen consumed expressed in micromole per gram of catalyst (respectively µmolO/g and µmolH/g).

3. RESULTS.

3.1. O_2 TPD.

3.1.1. Fresh catalyst.

The TPD spectrum carried out on a fresh catalyst, is shown in Fig. 1. Four oxygen desorption peaks are observed: a first small peak centred at T_m= 112°C, two distinct peaks at T_m= 478°C and T_m= 725°C, and a shoulder at T_m= 765°C. The temperatures T_m and the total amount of oxygen desorbed are indicated in Table 1 (column 2). The amounts of oxygen evolved in each peak (assumed symmetric) are respectively: 0.07, 5, 7 and 12 µmoleO/g. The cerium content of this solid is 3.6 weight % (around 250 µmoles of CeO_2 oxide/g of catalyst).

Therefore, it seems that the total amount of oxygen desorbed (24µmol/g) during the TPD (20% considering the formation of Ce_2O_3), is more representative of a desorption than a bulk decomposition of the support. This total amount of O_2 desorbed can be compared with the chemisorption data on TWC without nickel oxide [4].

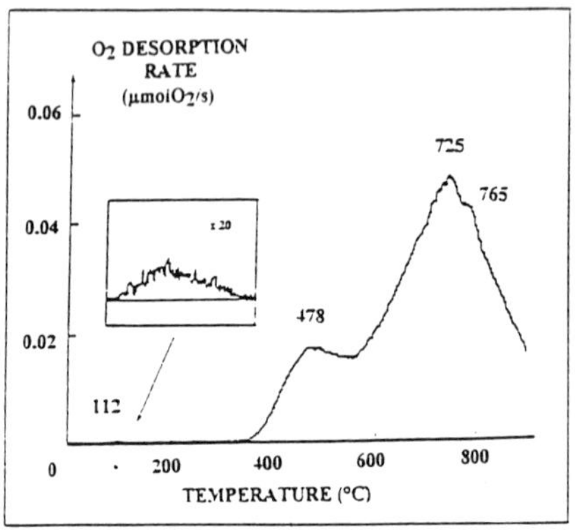

Fig. 1 : O_2 TPD spectrum of the fresh TWC

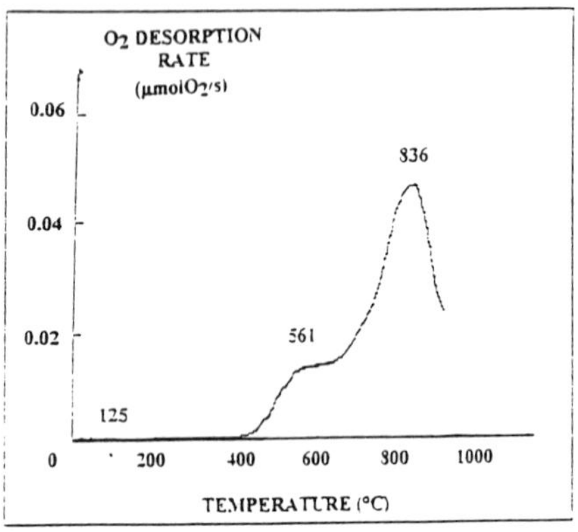

Fig. 2: O_2 TPD spectrum of the washcoat without PM.

In order to distinguish the oxygen species linked to the washcoat from those that may be assigned to the presence of metals, a TPD have been performed on the washcoated cordierite without any PM. The TPD spectrum (Fig. 2) shows three desorption peaks at T_m= 125°C, T_m= 561°C and T_m= 836°C. The amounts of oxygen evolved are respectively: 0.07, 2.1 and 9 µmolO/g. The main difference between the TPD spectra of the two solids (TWC and washcoat) is due to the disappearance of one of the two peaks at high temperatures.

Likewise, an oxygen TPD have been carried out on the 0.5% Pt/Al_2O_3 solid. The spectrum (Fig. 3) indicates two distinct desorption peaks centred respectively at T_m= 489°C (2.3µmolO/g.) and T_m= 687°C (11.6µmolO/g.). The peak at T_m= 489°C is similar to the peaks observed on the washcoat at T_m=561°C and on the fresh TWC at T_m=478°C. It can be observed that the TWC gives two peaks at high temperatures (Fig. 1) whereas the washcoat without PM (Fig. 2) and the

0.5% Pt/Al_2O_3 catalyst (Fig. 3) give only one main peak with different values of T_m (respectively 856°C and 687°C).

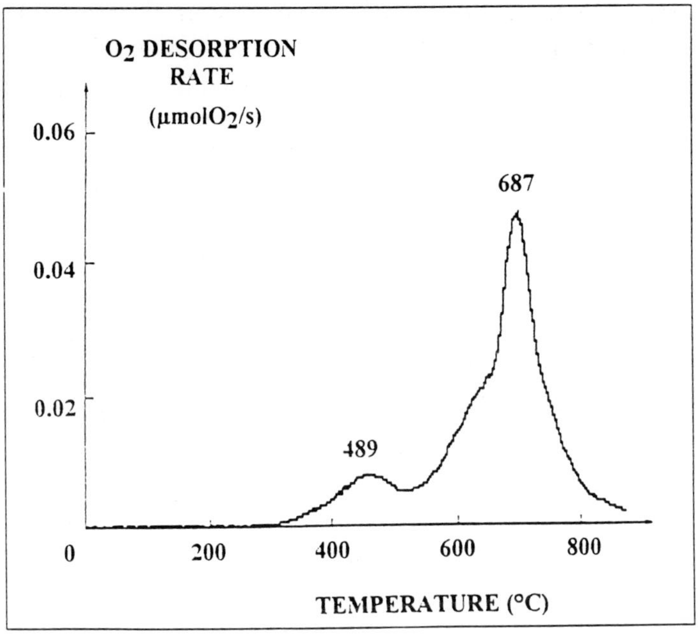

Fig. 3: O_2 TPD spectrum of the 0.5% Pt/Al_2O_3 solid

3.1.2. Aged catalyst.

The amounts of desorbed oxygen, and the temperatures T_m of the peaks for the aged solids are presented in Table 1. The two samples of the vehicle aged TWC give similar results. The ageing of the TWC leads to a) a decrease of the total amount of desorbed oxygen, especially on the air aged solid and b) a disappearance of the first peak. The amounts of oxygen in the second peak are: 2.9, 0.4 and 1.4 µmolO/g respectively for the N_2, air and vehicle aged TWC. On the air aged solid, the third peak appears as a small shoulde and the amount of O_2 is included in the last peak. The amounts of oxygen detected in this third peak are 2 and 2.3 µmolO/g respectively for the N_2 and the vehicle aged TWC. The amounts of oxygen evolved in the fourth peak are respectively: 5, 1.1 and 6.1 µmolO/g for the N_2, air and vehicle aged TWC.

3.2. H_2 TPR.

O_2 TPD experiments carried out on the various TWC, indicate the presence of four oxygen species for a total amount of 24µmol/g for a fresh TWC. This study is completed by H_2 TPR experiments which is an other method of characterization of the oxygen species .

3.2.1. Fresh Catalyst.

Figure 4 gives the H_2 TPR spectrum obtained on the washcoat without any PM with a 3% H_2/He mixture. Two major reduction processes are observed. The first one appears as a weak H_2 consumption peak centred at T_m= 115°C. The second process gives poorly resolved peaks and shoulders at T_m= 470°C, 695°C and 870°C which are due to various species exhibiting gradually increasing activation energies of hydrogenation. The reduction of the solid is incomplete at the final temperature (900°C).

Table 1 : Characterization of O_2 TPD spectra of fresh and aged TWC.

	Fresh TWC	4h in N_2 at 1050°C.	24h in air at 1100°C	130 000 km vehicle aged Inlet	130 000 km vehicle aged Outlet
Amounts of desorbed O_2 (µmolO/g)	24	9.9	1.5	9.8	9.8
Temperature (°C)					
1st peak	112	-	-	-	
2nd peak	478	572	500	480	
3rd peak	725	741	750	740	
4th peak	765	819	830	813	

In order to identify the cerium oxide contribution on the H_2 TPR spectrum of the washcoat (Fig. 4), a TPR have been carried out on pure CeO_2 oxide. The spectrum obtained (Fig. 5), indicates two reduction processes. The first one appears, as a well defined H_2 consumption peak at T_m= 604°C (169 µmolH/g). The second process is noticed over a wide range of temperatures (T> 650°C) and is incomplete at 900°C.

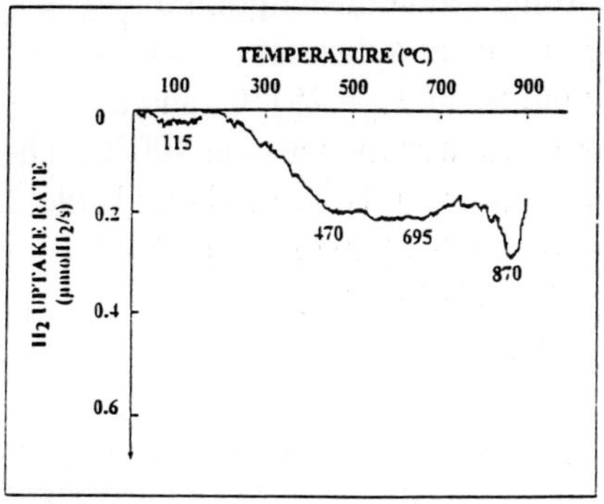

Fig. 4: H_2 TPR spectrum of the washcoat without any PM

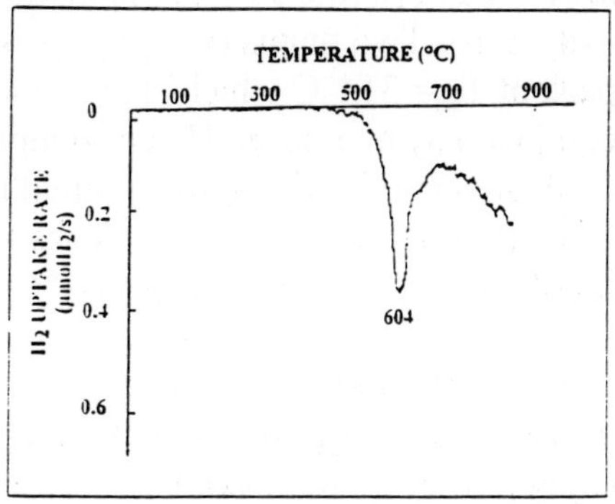

Fig. 5: H_2 TPR spectrum of the CeO_2 solid.

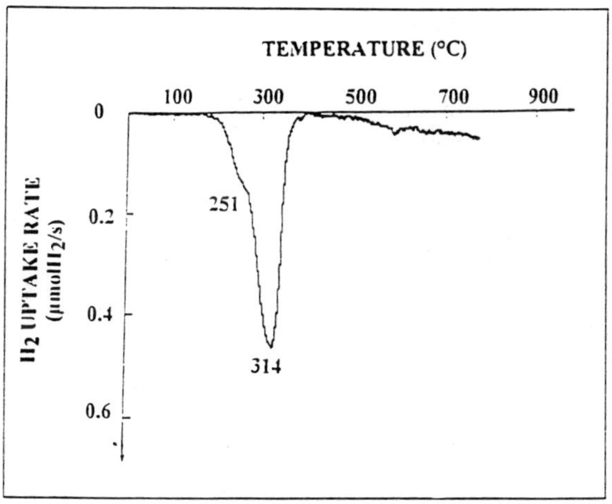

Fig. 6: H_2 TPR spectrum of the fresh TWC

Figure 6 gives the TPR spectrum of the fresh three-way catalyst using a 10%

H_2/He mixture (the 3% H_2/He mixture leads to the total consumption of hydrogen). Two mains reduction processes are noticed: a distinct consumption peak at T_m= 314°C which is preceded by a shoulder at T_m= 251°C, and a continuously increasing H_2 consumption at temperatures higher than 400°C. The total amount of hydrogen consumed in the first process is 160 µmolH/g (Table 2) and a deconvolution into two peaks (assumed symmetric) gives: 42 and 118 µmolH/g respectively at T_m= 251°C and T_m= 314°C (Table 2).

3.2.2 Aged Catalyst

Table 2 gives the amounts of consumed hydrogen (µmolH/g) and the temperatures at the maximum of the reduction peaks, for the various aged solids. It can be observed that the ageing treatments lead to a decrease of the total amount of H_2 consumed.

The first peak of the TWC either air or vehicle aged is located at a temperature T_m (around 100°C) lower than that observed on a fresh catalyst (251°C). Note that the two samples of the 130 000 km aged catalyst give different amounts of hydrogen consumed in the two peaks.

Table 2 : Characterization of H_2 TPR spectra of fresh and aged TWC.

	Fresh. TWC		4 h in N2 at 1050°C.		24h in air at 1100°C.		130 000 km vehicle aged Inlet		Outlet	
Peak n°	1st	2nd	1st	2nd	1st	2nd	1st	2nd	1st	2nd
T_m (°C)	251	314	267	362	95	330	85	360	117	355
H_2 consumption	42	118	29	21	1.5	13	4.5	41	10	71
(µmol H/g)	160		50		14.5		45.5		81	

3.3. LIGHT-OFF TEMPERATURES AND VEHICLE EMISSIONS.

Table 3 gives the light-off temperatures (25% of conversion) measured on a laboratory bench (8) with the reactive mixture given in section 2.2. The values of the emitted gases are an average of three ECE-EUDC cycles.

It can be observed that fresh and vehicle aged TWC (inlet section) have similar light-off temperatures (excepted for the conversion of C_3H_8). This is in agreement with the fact that the amounts of emitted gases are slightly different. The comparison of the fresh and air aged TWC reveals: a) the shift of the light-off temperatures to higher values (excepted for the conversion of C_3H_8) and b) the clear increase of the emissions measured during ECE-EUDC cycles.

Table 3: Light-off temperatures and vehicle emissions measured on fresh and aged TWC.

	Fresh TWC	Air aged TWC.	130 000 km vehicle aged TWC.
Emission (g/km)			
CO	1.15	3.06	1.23
HC	0.10	0.28	0.16
NO_x	0.04	0.29	0.07
T 25% (°C)			
CO	268	360	275
C_3H_6	313	370	305
C_3H_8	414	410	460
NO	318	355	290

4. DISCUSSION

4.1. O_2 TPD.

The TPD experiments carried out on the three-way catalyst shows the presence of 4 oxygen species detected by the peaks at T_m= 112°C, 478°C, 725°C and 765°C. Figures 1, 2 and 3 show that the three solids: TWC, washcoat without any PM and 0.5% Pt/Al_2O_3 give a similar peak respectively at T_m= 478°C, 561°C and 489°C. This suggests that such peak might be related to Al_2O_3, the common part of the three solids. It can be also observed that the washcoat without any PM and the 0.5% Pt/Al_2O_3 solid give only one main peak at high temperatures respectively T_m= 836°C and T_m= 687°C. Considering the order of appearance of the peaks on the TWC, it might be suggested that the peak at T_m=765°C is related to the washcoat part of the solid (cerium oxide) and that the peak at T_m= 725°C is due to the presence of the PM. This last peak can be due either to the PM oxides or to some oxygen species along the interface formed by the PM particles and the washcoat support.

The small TPD peak observed on the TWC at T_m= 112°C is also detected on the washcoat at T_m= 125°C and corresponds to the desorption of the same oxygen species adsorbed on the washcoat.

In section 3.1.1, we have considered that the total amount of O_2 detected on the fresh catalyst (Table 1) is rather representative of a desorption than a bulk oxide decomposition. The O_2 TPD of the aged solids seems to fit in with this interpretation because the decrease of the total amount of desorbed oxygen is closely related to the decrease of the specific surface area of the solids. For example, the ratios of the BET surface area of the aged TWC on the fresh TWC are 0.07 and 0.42 respectively for air and vehicle aged TWC. These values can be compared to the ratios of the total amounts of O_2 evolved, respectively 0.06 and 0.4 for the air and vehicle aged TWC.

4.2. H_2 TPR.

4.2.1. Washcoat without PM and CeO_2.

The TPR experiment carried out on the washcoat without any PM (Fig. 4) indicates an initial first small peak at T_m= 115°C, followed by three peaks and shoulders at T_m= 470°C, 695°C and 870°C. These observations are in agreement with TPR spectra obtained by Yao et al [3] on various CeO_2/Al_2O_3 solids (11 and 21 weight % of CeO_2). These authors have observed three overlapped peaks at around T_m= 520°C, 670°C and 870°C. The similarity of the data confirms that the spectrum obtained on the washcoat without PM corresponds to the reduction of the CeO_2/Al_2O_3 system. The small peak at T_m= 115°C can be assigned to the reduction of oxygen anions of Al_2O_3 [3].

The TPR spectrum of pure CeO_2 oxide (Fig. 5) is similar to those noticed on cerium oxides of various specific surface areas [3,10]. The first peak (T_m= 604°C) may correspond to the reduction of a surface oxygen species related to Ce^{4+} [3]. Johnson et al [10] have shown that the position of this peak shifts to lower values of T_m when the specific surface area of the solids increases. The hydrogen consumption detected at high temperatures (Fig. 5) was also noticed by the various authors [3, 10] and was attributed to the reduction of the bulk CeO_2 oxide. The differences in the TPR spectra of the washcoat without PM and of the CeO_2 oxide can be interpreted by the existence of interactions between CeO_2 and Al_2O_3 [3].

4.2.2. Fresh catalyst.

The TPR carried out on the fresh TWC (Fig. 6) reveals two reduction processes. The first one gives an intense H_2 consumption peak at T_m= 314°C with a shoulder at T_m=251°C. The second process gives a consumption of H_2 at temperatures higher than 400°C. The two first peaks are not recorded on the spectrum of the washcoat and therefore appear to be related to reactions which involve the noble metals.

The deconvolution (assuming symmetric peaks) gives 42 µmolH/g for the peak at T_m= 251°C and 118 µmolH/g for the peak at T_m= 314°C (Table 2). The value of the first peak is of the order of magnitude of the amount expected for the reduction of PtO_2 and Rh_2O_3 oxides (approximately 50 µmolH/g). This suggests that the peak at T_m=251°C is due to the reduction of the PM oxides[1]. The difference between the two values can be explained by the presence of bonds between PM and CeO_2/Al_2O_3 which can modify the oxide structures [11]. These interactions are mentioned in the literature in order to explain the stabilizing effect of CeO_2 on the metals. Tournayan et al [12] on a 0.6% Ir/CeO_2 catalyst observed during a H_2 TPR experiment, the presence of two reduction peaks (which are similar to the peaks in Fig. 6). A deconvolution gives an amount of H_2 consumed in the first peak leading, as in the present results, to a ratio O/metal lower than expected for the reduction of IrO_2 oxide. The two PM oxides seem to be reduced in the same peak at T_m=251°C. This is in agreement with the results of Kacimi et al [13] and J. Nunan et al [14], obtained on bimetallic catalysts: $Pt/Rh/Al_2O_3$ and $Pt/Rh/Al_2O_3/CeO_2$. Therefore, it seems that the TPR method, does not allow to differentiate the nature of the metal oxides..

The peak at 314°C (118 μmolH/g) must be related to the reduction of oxygen species of the CeO_2/Al_2O_3 part of the TWC. On the washcaot without any PM as well as on CeO_2 oxide, the first peaks are detected respectively at 470°C and 604°C. This decrease of the reduction temperature in presence of the precious metals, have been previously observed on various PM supported on CeO_2 containing solids [3, 14-16]. This peak is usually attributed to the reduction of oxygen species at the interface washcoat/PM particles. These species seem related to the numerous interactions mentioned between the CeO_2 and PM to explain the improvement of the stability of the PM as well as the promoter effect during the catalytic reactions [1-6, 17].

The reduction process noticed between 400°C and 900°C on the fresh catalyst leads to a progressive H_2 consumption which seems to correspond to the bulk reduction of CeO_2, in a possible interaction with Al_2O_3 [3]. However at temperatures higher than 700°C Al_2O_3 seems also consumes hydrogen [14].

4.2.3. Aged catalysts

The H_2 TPR spectra of the aged catalysts show that the temperatures T_m of the reduction peaks of the oxygen species and the amounts of hydrogen consumed (Table 2) are different according to the ageing process.

The comparison between the fresh and the nitrogen aged TWC shows that the amounts of hydrogen consumed in the two first peaks decrease, without a major change of the temperatures T_m. The deconvolution gives 29μmolH/g for the peak at T_m=267°C (instead of 42μmolH/g for a fresh solid). This decrease indicates a lower accessibility of the precious metal oxides to hydrogen reduction. This can be due to the decomposition of the PM oxides but it might also be considered a diffusion of the PM into the bulk of the washcoat as observed on Rh/Al_2O_3 catalysts after an oxygen treatment at high temperatures [18]. The second peak at T_m= 362°C is more strongly decreased 21μmolH/g instead of 118μmolH/g for a fresh catalyst. Taking into account the assignment of this peak (species at the interface PM/washcoat), this decrease can be explained by additional factors: the sintering of the PM particles and of the washcoat (decrease of the BET surface area). The increase of the size of CeO_2 particules have been considered to explain the decrease of the intensity of a similar H_2 TPR peak [14].

TPR spectra of air and vehicle aged TWC are different than those of the fresh and the nitrogen aged TWC. The first reduction peak observed previously at T_m= 251°C is no more detected and a new peak appears at T_m around 100°C (Table 2). A H_2 TPR peak in this range of temperature have been already mentioned on $Pt/Rh/CeO_2/Al_2O_3$ system, either reduced at high temperatures [10] or treated with exhaust gas mixture (rich) [14]. A similar peak was also observed on 10%Pt/Al_2O_3 solid [11]. According to these remarks, it seems that the detection of this peak is related to the reduction of PM oxides particles which are not in interaction with cerium oxide [11]. This peak can be considered as representative of a strong modification of the interactions between the PM and CeO_2 oxide. The decrease of the amounts of hydrogen consumed (compared to the fresh and the nitrogen aged TWC) can be interpreted by an increase of the factors considered above. However, the change in the morphology of the support (decrease of the BET surface area to 3 m^2/g for the air aged TWC) may be also considered to explain the inaccessibility of the PM oxides.

The amount of hydrogen consumed in the second reduction peak at around T_m= 350°C, previously assigned to oxygen species at the interface washcoat/PM is lower on the air aged solid (13μmolH/g) and on the 130 000 km catalyst (41μmolH/g and 71μmolH/g respectively for the inlet and outlet side) than on the fresh TWC (118 μmoH/g). This can be attributed to the same factors than those considered above.

4.3 Catalytic activity and temperature programmed characterizations.

O_2 TPD and H_2 TPR appear as two methods which allow to follow the changes of the surface and bulk oxygen species between a fresh and an aged TWC. However, the correlation with the catalytic activity as measured in Table 3 can be only roughly proposed. The differences of the light-off temperatures (as well as the emissions during cold ECE+EUDC cycles) between fresh and vehicle aged TWC are minor whereas O_2 TPD and H_2 TPD spectra present clearly different profiles. Only the strongly air aged TWC (3m^2/g) gives higher light-off temperatures and higher amounts of emitted pollutants. This might be interpreted as the fact that the modifications revealed by the temperature programmed analysis are not closely related to the catalytic activity. However, TWC used for the various ageing processes as well as the fresh catalyst are different monoliths. This fact must be also considered with a) the repeatability of the catalytic activity measurements and b) the range of values between a fresh and a strongly deactivated solid such as the air aged TWC (around 100°C for the light-off temperatures and a factor 3-4 for the emissions). This means that to establish a correlation between surface characterization and catalytic activities in exhaust gas conversion, the various properties must be studied on the same TWC monolith at different level of the ageing process.

5. CONCLUSION

The study carried out by O_2 TPD shows four oxygen species on a three way catalyst (24 μmolO/g for a fresh TWC). Three were attributed to the adsorbed species on the washcoat (CeO_2 and Al_2O_3), and the last one to the presence of precious metals.

H_2 TPR experiments indicate two major steps in the reduction of the solid. The first one, at T<400°C is related to the reduction of the PM oxides and of the washcoat at the interface with the precious metals (160μmolH/g). The second step, at high temperatures is assigned to the reduction of bulk CeO_2. The first reduction step is strongly modified by the various ageing treatments and a new TPR peak at around 100°C may appear.

6. References

1 A.F. Diwel, R.R. Rajaram, H.A. Shaw and T.J. Truex, in "Catalysis and Automotive Pollution control II", A. Crucq (ed), Stud. Surf. Sci. and Catal., Vol. 71, Elsevier, Amsterdam, (1991), 139.

2 M.G. Sanchez and J.L. Gazquez, J. Catal., 104 (1987) 120.

3 H.C. Yao and Y.F. Yu Yao, J. Catal., 86 (1984) 254.

4 P. Lööf, B. Kasemo and K.E. Keck, J. Catal., 118 (1989) 339.

5 C. Li, K. Domen, K. Maruya and T. Onishi, J. Catal., 123 (1990) 436.

6 Y.F. Yu Yao and J.T. Kummer, J. Catal., 106 (1987) 307.

7 S.H. Oh and J.E. Carpenter, J. Catal., 98 (1986) 178.

8 D. Bianchi, J.L Gass, C. Bouly and D. Maret, SAE, paper 910839

9 C. Bouly, J.L Gass, D. Maret and D. Bianchi, SAE paper 930778

10 M.F.L. Johnson and J. Mooi, J. Catal., 103 (1987) 502.

11 J.Z. Shyu and K. Otto, J. Catal., 115 (1989) 16.

12 L. Tournayan, N.R. Marcilio and R. Frety, Appl. Catal., 78 (1991) 31

13 S. Kacimi and D. Duprez, in "Catalysis and Automotive Pollution control II",A. Crucq (ed), Stud. Surf. Sci. and Catal., Vol. 71, Elsevier, Amsterdam, (1991), 581

14 J.G. Nunan, H.J. Robota, M.J. Cohn and S. A. Bradley, in "Catalysis and Automotive Pollution Control II", A. Crucq (ed)., Stud. Surf. Sci. and Catal., Vol. 71, Elsevier Amsterdam, (1991), 221.

15 R.K. Usmen, R.W. McCabe, G.W. Graham, W.H. Weber, C.R. Peters and H.S.Gandhi, SAE, paper 922336

16 B. Harrison, A.F. Diwel, and C. Hallet, Plat. Met. Rev., 32 (2) (1988) 73.

17 S.H. Oh and C.C. Eickel, J. Catal., 112 (1988) 543.

18 C. Wong and R.W., McCabe, J. Catal., 119 (1989) 47.

A. Frennet and J.-M. Bastin (Eds.)
Catalysis and Automotive Pollution Control III
Studies in Surface Science and Catalysis, Vol. 96

THE OXIDATION OF CARBON MONOXIDE BY OXYGEN OVER POLYCRYSTALLINE PLATINUM, PALLADIUM AND RODIUM FROM UHV TO NORMAL PRESSURE

S. Fuchs and T. Hahn

Institut für Chemische Verfahrenstechnik der Universität Karlsruhe, D-76128 Karlsruhe, FRG

ABSTRACT

The oxidation of CO by O_2 was studied over polycrystalline foils, stripes and wires of Pt, Pd and Rh as well as over Pt on different supports. The partial pressures were in the range 10-6 ≤ pi/mbar ≤ 40, the partil pressure ratio was varied in the range 0.01 ≤ pCO/pO_2 ≤ 50 and the temperature was varied between 180° and 600°C. For total pressure lower than 1 mbar the results agree perfectly with observatios of White et al [1]-[3] in the UHV range. For total pressure higher than 1 mbar the highest activities report so far could be observed using foils and supported catalysts. The measured reaction rates were controlled by mass transport with the exception of the CO-inhibition region even at low temperatures and high gas velocities of 10 m/s. No influence of Pt-catalyst structure or support could be observed.
In the region, where the reaction is first order with respect to CO and zeroth order with respect to O_2 one can approximately calculate the intrinsic reaction rates. They are a linear extrapolation from the UHV-results: impingement rates and sticking coefficients of the reactants are rate determining even at normal pressure.

1 INTRODUCTION

Since the classical work of Langmuir [4] the oxidation of carbon monoxide by oxygen catalysed by noble metals has been studied in great detail. The experimental conditions reported in the literature cover a broad range. With regard to the total pressure it is possible to divide the experimental work into two groups:

- studies at high or ultra-high vacuum conditions ($p<10^{-4}$ mbar) over model catalysts,

• studies at (or near) atmospheric pressure mostly over supported catalysts

The difference between the two groups is the called 'technological gap' [5], not only with respect to the total pressure but also with respect to surface structure and chemical composition.

The results reported in the literature are contradictory difficult to compare. In spite of the multitude of experimental studies there are only few cases where representative rate coefficients can be derived from the published data. Even then, the catalysts used in those investigations are quite different in nature and the domains of temperature and/or reactant concentrations explored remain small. It is therefore the scope of the present study to reexamine the oxidation of the carbon monoxide by oxygen over platinum, palladium and rhodium in a wide range of temperature and pressure and to compare the results with available literature data in order to solve the existing discrepancies.

2 Experimental

Two experimental set-ups were used

- a 'vacuum apparatus' in the range $10^{-8} \leq P_{reactor}/mbar \leq 1$,
- a jet-stirred loop reactor at atmospheric pressure and $0{,}3 \leq p_i/mbar \leq 40$ [6].

The reactants (CO,O_2) and the balance nitrogen (CO: 99,997 mol%,O_2:99,995 mol%, N_2: 99,999 mol%; Messer Griesheim) were fed to the systems from cylinders via leak valves or mass flow controllers. In the vacuum apparatus partial pressures were measured by use of a differentially pumped quadrupole mass spectrometer (QMG311, Balzers).

In the recirculation system concentrations were monitored by means of non-dispersive IR spectrometers (CO: Ultramat, Siemens; CO_2: Uras 2T, Hartmann & Braun) and a magnetic device (O_2: Magnos, Hartmann & Braun).

Two types of catalysts were used:

- polycrystalline foils, ribbons, stripes and wires of platinum, palladium and rhodium(Heraeus),
- different platinum catalysts supported on nonporous platelets of pyrex or quartz glass(Schott), zirconia (Friedrichsfed) and porous α-alumina (Condea Chemie, calcined at 1250°C/4h).

The foils, ribbons, stripes and wires were recrystallized in vacuum or in air at temperatures higher tan 800°C. The supported platinum catalysts were prepared from aqueous tetrammin nitrate to avoid chlorine contamination, dried (1h/70°C)and calcined (2h/500°C) in air. Dispersion D was determined by standard CO chemisorption at room temperature. Table 1 shows the data of the supported catalysts.

Reaction rates were obtained by a mass balance for the open systems in a steady state. They are normally related to the meta area F_{metal} of the catalyst:

$$r' = \frac{1}{F_{métal}} * \frac{d\xi}{dt} \quad ; \quad [r'] = \frac{mol}{cm^2 * s} \tag{1}$$

In the case of supported catalysts they are sometimes related to the outer geometrical area F_{geo} of the catalyst platelet

$$r_{geo} = \frac{1}{F_{geo}} * \frac{d\xi}{dt} \tag{2}$$

In both cases ξ represents the molar extent of the reaction

$$CO + 1/2\ O_2 \rightarrow CO_2 \tag{3}$$

Table 1 Data of supported catalysts

Catalyst	m_{Pt} / mg	F_{geo} / mm^2	F_{metal} / cm^2	D / %	L / mm
Pt-0,01-ZrO_2	0,0089	210	__*)	__*)	10,0
Pt-1,5-ZrO_2	1,5	392	30	0,72	14,0
Pt-10-ZrO_2	10,4	392	50	0,17	14,0
Pt-0,01-Al_2O_3	1,0	330	50	13	12,5
Pt-0,2-Pyrex	0,19	381	__*)	__*)	14,4
Pt-1,5-Pyrex	0,195	457	30		14,4
Pt-0,2-Quartz	0,135	310	__*)	__*)	13,0

__*) not mesurable

3 Results and Discussion

No measurable difference insteady state catalytic activity for the catalysts heated in vacuo or in air could be observed.

Figure 1 shows the reaction rate r' over palladium foil as a function of carbon monoxide partial pressure for different temperatures. The different (constant) values of the oxygen partial pressure are given in the upper part of the figure. The line represents the impingement rate of carbon monoxide calculated

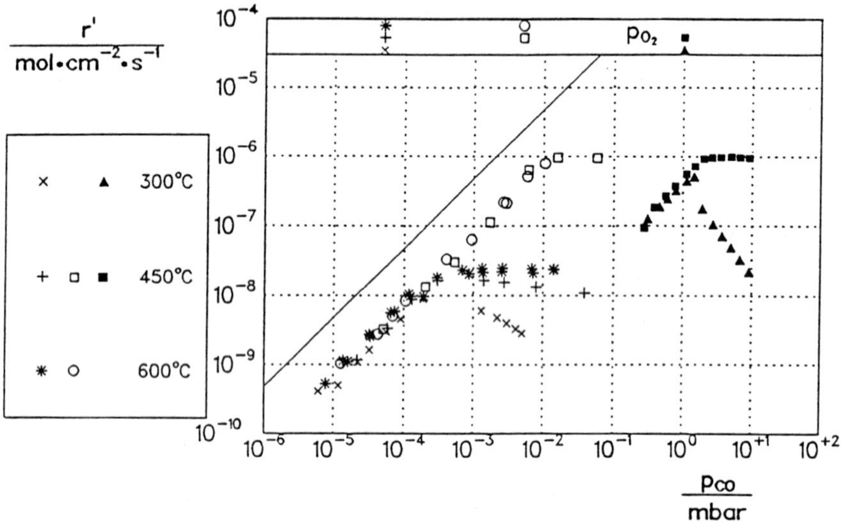

Figure 1 r'as a function of p_{CO} over palladium foil

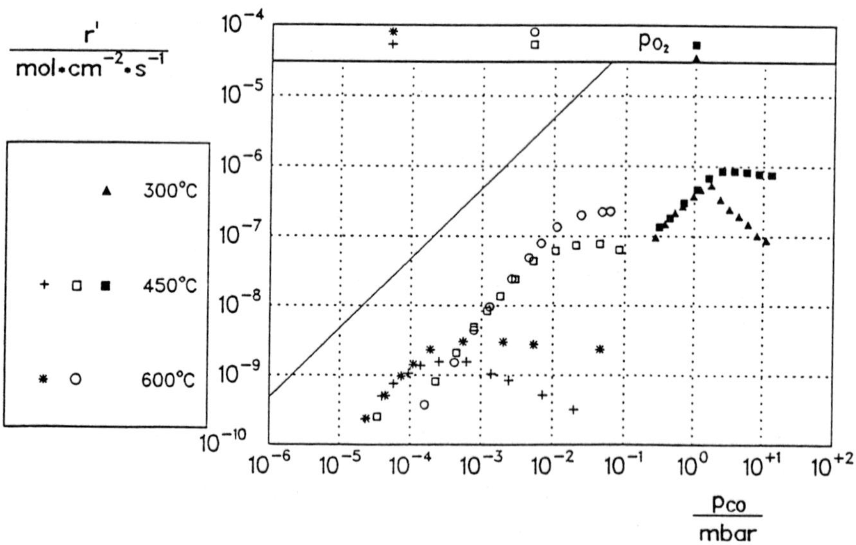

Figure 2 r'as a function of p_{CO} over rhodium foil

for a gas temperature of 25°C as an upper limit for r'. The results show the well known formal kinetics for CO oxidation on noble metals over the whole pressure range investigated.

For oxidising gas mixtures catalitic activity is first order with respect to p_{CO} and independent from pO_2. The temperature dependence of reaction rate is very small in the range $300 \leq \vartheta \,/\, °C \leq 600$. For reducing gas mixtures and 'low' temperatures the reaction is inhibited by excess CO,for 'high' temperatures rate is not dependent on p_{CO}. The order of reaction in the range of compositions is +1 with respect to oxygen. For oxidising conditions, that is in the region where the reaction is first order with respect to CO, one can describe reaction rate with

$$r' = k' * p_{CO}. \qquad (4)$$

A comparison of the k' values shows, that there is a difference of almost two orders of magnitude between the values of the rate constant k' calculated from low and high pressure conditions. Expressing the rate coefficient by the reaction probability b, the ratio of reacting and impinging molecules, one obtains in the low pressure region (p_{CO})< 0.1mbar) $b_{CO} \approx 0.15$,>0.1mbar the much lower value of $b_{CO} \approx 1.6\ 10^{-3}$.The results are quite similar for rhodium. In the latter

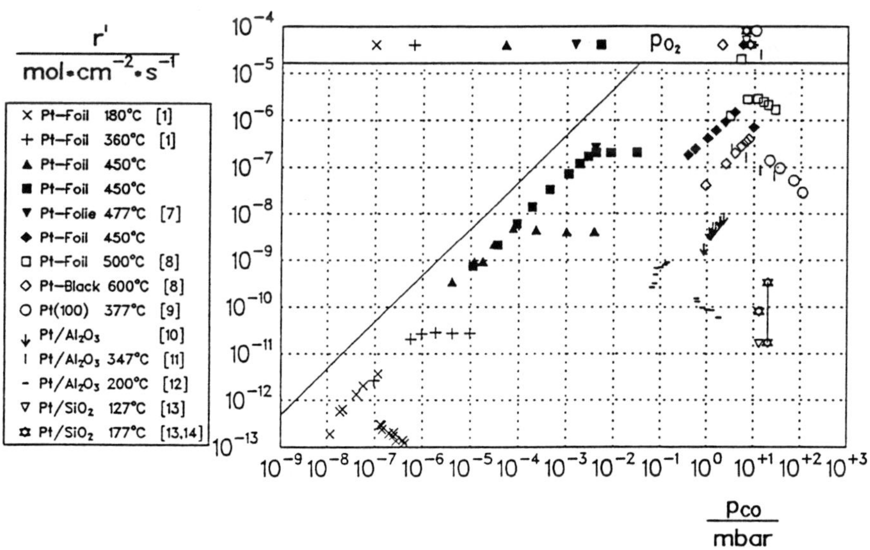

Figure 3 r'as a function of p_{CO} over different platinum catalysts

case there is an additional inhibition by excess oxygen at temperatures higher then 300°C, indicated by the parallel displacement of the 1st order line (figure 2). The differences observed in k'or b are of the same order of magnitude as in the case of palladium.

The rate pattern shown is well known. Figure 3 shows a comparison of literature data for platinum catalysts and values of CO partial pressure covering a range of 10 orders of magnitude. Once more the well known rate pattern is shown.

Our own results (foils) obtained at $p_{CO} < 5*10^{-2}$ mbar perfectly agree with the UHV results of Golchet and White [1]($p_{CO} < 10^{-5}$ mbar) and results obtained by Coulston and Haller [7] for $p_{CO} = 5 \, . \, 10^{-3}$ mbar. For oxidising gas composition, approximately 10 percent of the impinging CO molecules react to form CO_2

In the 'high pressure' region ($p_{CO} > 0.1$mbar) the rate data reported vary about 5 oders of magnitude. The highest rate values have been observed on platinum foils but even then the reaction probabilities are two orders of magnitude lower than in the 'low pressure' region.The findings are quite similar in the case of palladium or rhodium but the scatter of the literature data for palladium is even much higher then in the case of platinum or rhodium.

What may be the reason for this so called 'pressure - gap' problem - the differences between 'low' and 'high' pressure region, a problem also known for H/O over noble metals [15] - [17]? Several suggestions have been made in the literature:

- different mechanisms for 'low' and 'high' pressure,
- different phase compositions (esp. oxides and surface oxides),
- influence of surface structure (particle size),
- influence of support and
- influence of possible contaminants.

In the present study contamination effects could be avoided by using a jet reactor without moving parts [6]: The observed activities were constant for more than one week in every case. The influence of particle size and support were ruled out by the use of polycristalline foils.

On the other hand, the well defined geometry of the catalysts enabled us to take mass transport limitation into account, a fact often neglected in studies reported in the literature. The results of these considerations are presented underneath.

Figure 4 shows measured reaction rates over different noble metal foils as well as the calculated (lines) maximum molar flow to the catalyst surface, i.e.

$$\frac{1}{F_{geo}} * n_{i,max} = n'_{i,max} = \beta * c_{i,bul} \tag{5}$$

where $c_{i,bulk}$ is the concentration of component i ($CO;O_2$) in the bulk of the gas phase and β is an appropriate mass transfer coefficient obtained via the Sherwood number [6]. Therefore the horizontal line corresponds to the mximum flow of oxygen and the inclined line correspond to the maximum flow of carbon monoxide depedent on CO partial pressure.

It is obvious that the reaction rates measured are limited by mass transfer! This coclusion can be verified by variation of the characteristic length L of the catalysts and therefore the mass transfer coefficient β (figure 5).

Even if the value of L is reduced by factor of 80- i.e. β is enhanced by a factor of seven- the measured rates are still mass transfer limited. This results is not altered by using different supported catalysts (figure 6) even if the surface of the support platelet is not totally covered by noble metal (catalyst Pt-0,01-Zr O_2) It remains equally valid if the temperature is decreased to 180°C (for lower temperatures the conversion in the recirculation system is too low to be accurately determined): the measured rates are clearly transport limited in the region where the reaction rate is first order with respect ti the minor component.

However, we can exclude to some extent the existence of different elementary mechanisms in the 'high' or 'low' pressure region. In fact, for two first order steps in series - mass transfer and chemical reaction - theintrinsic rate constant k' can be easely calculated from the experimental one k'_{exp} and the appropriate transport coefficient β via

$$r' = \left(\frac{1}{k'} + \frac{1}{\beta}\right)^{-1} * c_{CO} = k'_{Exp} * c_{CO} \tag{6}$$

The calculated values of k' or b_{CO} are only limit values, but for the catalysts with the smallest characteristic length L (highest β values) the calculated values are comparable to the sticking probabilities for CO-adsorption! (Pt-wire, L=0,16 mm, b_{CO} =0,37,Pd-wire, L=0,16 mm, b_{CO} =0,24; Rh-foil, L=10 mm, $b_{CO} \approx 10^{-2}$).

This result indicates, that over Pt and Pd for oxidising compositions and temperatures higher than 180°C catalytic activity is limited by the rate of adsorption even at high pressures where mass transfer influences occur. For reducing gas phase compositions and 'low' temperatures (in the region of CO inhibition) the measured activities are not influenced by outer mass transfer. In that case the reaction rates have to be related to the metal surface area $F_{\text{the vlues}}$ and in fact, of r' are equal for all catalysts of the same metal used within a factor two. This clearly indicates that no significant influence of catalyst struture or support could be observed in the present study.

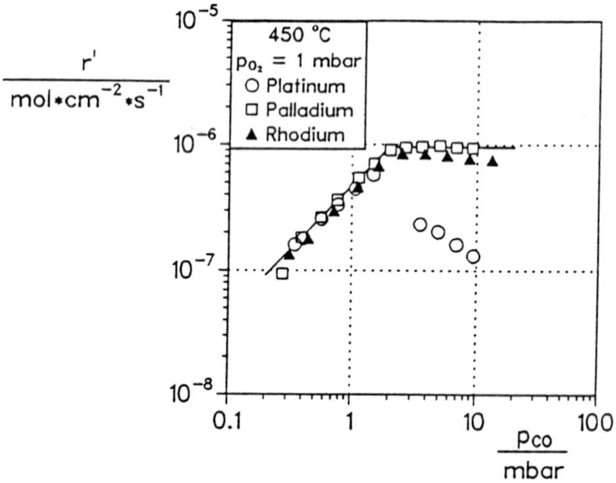

Figure 4. r' as a function of p_{CO} over Pt, Pd and Rh foils

4. CONCLUSIONS

In the region of gas compositions and temperatures, where the rate of reaction is zerth order with respect to the excess component and the first order with respect to the minor component, the values for reaction rates can be directly extrapolated from UHV results. For the region of CO inhibition no influence of catalyst structure and support can be observed for the platinum and the palladium catalysts used. For rhodium the results are not so clear, because a diminuation of the characteristic length by use of thin wires is not possible, but the rates observed are undoubtly higher than reported by Goodman et al [9]-[18]-[20] for single crystal surfaces. The implications of these results for modeling the CO/O2 system over Pt, Pd and Rh will be discussed later [21].

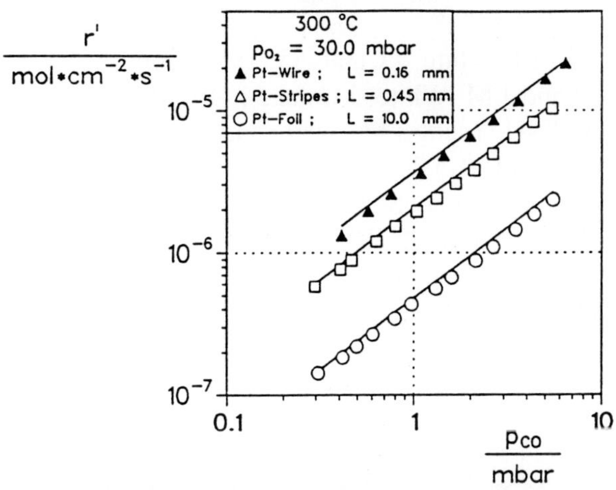

Figure 5. r' as a function of p_{CO} :variation of characteristic length

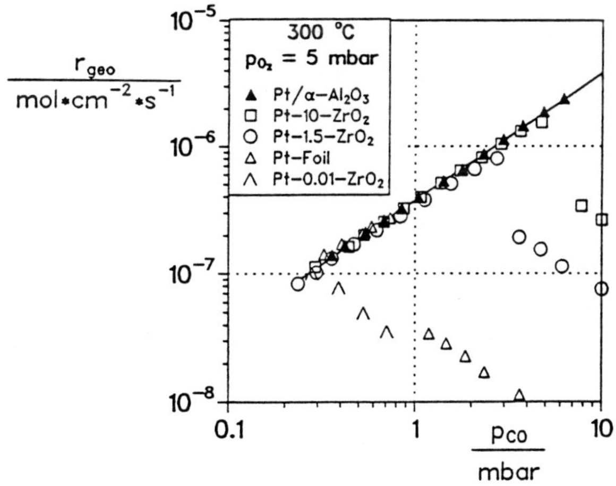

Figure 6. r' as a function of p_{CO} :influence of support and struture

REFERENCES

1 A. Golchet and J.M. White, J.Catal. 53 (1978) 266.
2 C.T. Campbell and J.M. White, J.Catal.54 (1978) 289.
3 J.R. Creighton, F.-H. Tseng,J.M. White and J.S. Turner J. Phys. Chem. 85. (1981) 703
4 I. Langmuir, Trans. Faraday SOC. 17 (1922) 621.
5 H.P. Bonzel, Phys. Blätter 32 (1976) 392.
6 S.Fuchs and T. Hahn, chem. eng. Proc. 32 (1993) 225.
7 G.W. Coulstonand G.L. Haller in D.W. Dweyer and F.M. Hoffmann (eds.), surface Science and Catalysis, ACS Symp. Ser. 482 (1992) Chapiter 4.
8 E. Häfele, Thesis, University of Karlsruhe 1986
9 P.J.Berlowitz, C.H.F. Peden and D.W. Goodman, J. Phys. Chem. 92 (1988) 5213.
10 E. McCarthy, J. Zahradnik, G.C. Kuczinsky and J.J. Carberry, J. Catal. 39 (1975) 29.
11 G.S. Zafiris and R.J. Gorte, J. Catal. 140 (1993) 418.
12 R.K. Herz and S.P. Marin, J.Catal. 65 (1980) 281.
13 N.W. Cant,P.C. Hicks and S.B. Lennon, J.Catal.54 (1978) 372.
14 N.W. Cant, J.Catal.62 (1980) 173.
15 M. Boudart,D.M. Collins,F.V. Hanson and W.E. Spicer, J. Vac. Sci. Technol. 14 (1977) 441.
16 V.P. Zhdanov, V.I. Sobolev and V.A. Sobyanin, Surf. Sci. 175(1986) L747.
17 L.-G.petersson and U. Ackelid, Surf. Sci.269/270 (1992) 500.
18 S.H. Oh, G.B. Fisher, J.E. Carpenter and D.W. Goodman J.Catal.100 (1986) 360.
19 C.H.F. Peden,D.W. Goodman,D.S. Blair,P.J. Berlowitz G.B. Fisher and S.H. Oh, J. Phys. Chem. 92 (1988) 1563.
20 C.H.F. Peden, P.J. Berlowitz and D.W. Goodman in M.J. Phillips and M. Ternan (eds.), Proc.9th Int. Congr. Catal., Chem. Inst. Canada (1988) 1214
21 S. Fuchs, T. Hahn and H.-G. Lintz, Chem. Eng. Proc. 33 (1994) 363.

A. Frennet and J.-M. Bastin (Eds.)
Catalysis and Automotive Pollution Control III
Studies in Surface Science and Catalysis, Vol. 96

OXIDATION AND DISPROPORTIONATION OF CARBON MONOXIDE OVER Pd/ZrO_2 CATALYSTS PREPARED FROM GLASSY Pd-Zr ALLOY AND BY COPRECIPITATION

S. Gredig, S. Tagliaferri, M. Maciejewski and A. Baiker

Department of Chemical Engineering and Industrial Chemistry, Swiss Federal Institute of Technology, ETH-Zentrum, CH-8092 Zürich

ABSTRACT

Palladium/zirconia catalysts prepared by oxidation of an amorphous $PdZr_3$ alloy were tested for the oxidation and disproportionation of CO. Their catalytic and structural properties were compared with those of a Pd/ZrO_2 catalyst of similar composition, prepared by conventional coprecipitation. For the preparation of Pd/ZrO_2, the glassy $PdZr_3$ alloy was either oxidized *in situ*, i.e. under CO oxidation conditions, or oxidation in air. The catalysts were characterized by thermal analysis (TG, DTA), XRD, electron microscopy, CO chemisorption and nitrogen physisorption measurements. *In situ* oxidation of the Pd-Zr alloy led to significantly higher BET and palladium surface areas than oxidation in air. The catalysts derived from the glassy Pd-Zr alloy exhibited considerably higher activities for both oxidation and disproportionation of CO at low temperature than the coprecipitated Pd/ZrO_2 catalyst and palladium powder. The higher activities are attributed to the extremely large interfacial area of palladium and zirconia phases in the alloy-derived Pd/ZrO_2 compared to the coprecipitated catalyst. For both alloy-derived catalysts the apparent activation energy for CO oxidation was 58 ± 3 $kJmol^{-1}$. In the absence of oxygen, disproportionation of CO with subsequent incorporation of carbon into the palladium lattice occurred with the alloy-derived catalysts readily at ca. 150°C, whereas no similar phenomenon was observed with the palladium powder up to 400°C. The interstitial carbon was found to be very reactive towards oxygen forming CO_2. The storage of carbon by formation of a solid solution with Pd was found to influence significantly the behaviour of the catalyst under forced cycling between CO and O_2 feed.

1. INTRODUCTION

The use of glassy metals as catalyst precursors has opened new routes for the preparation of supported metal catalysts with unusual chemical and structural properties [1,2]. Active metal/metal oxide catalysts have been prepared by controlled oxidation of suitable glassy alloys. Amorphous Pd-Zr alloys were found to be suitable precursors for the preparation of Pd/ZrO_2 catalysts, highly active in the hydrogenation of CO_2 [3] and trans-2-hexene-al [4], and the oxidation of CO [5]. The structural and chemical transformations occurring during exposure of the Pd-Zr alloy to an oxygen containing atmosphere, such as a CO oxidation reaction mixture, have been followed using *in situ* X-ray diffraction [6], thermoanalytical methods (DTA, TG) combined with mass spectroscopy [6], electron microscopy and photoelectron spectroscopy [7]. More recently, we found that in the absence of oxygen, as-prepared catalysts exhibit high activity for the disproportionation of CO (Boudouard reaction) [8]. Carbon formed by disproportionation is incorporated into the palladium lattice, as was observed by XRD.

The aim of the present work was to gain further knowledge about the catalytic behaviour of the catalysts derived from glassy Pd-Zr alloys. Various aspects, including: (i) CO oxidation under different oxygen partial pressures; (ii) the disproportionation of CO with subsequent carbon incorporation, and (iii) the role of carbon storage during alternative cycling of CO and O_2, have been studied. The behaviour of the alloy-derived Pd/ZrO_2 catalysts is compared with that of a similar catalyst prepared by coprecipitation and with an unsupported palladium powder.

2. Experimental

The amorphous $PdZr_3$ alloy was prepared from the pre-mixed melts of the pure constituents by rapid quenching using the technique of melt spinning. The 5 mm wide and 20-30 mm thick ribbon was ground into flakes of 0.1 - 1 mm under liquid nitrogen before use. The *in situ* oxidation of the glassy alloy and the catalytic tests were carried out in a fixed-bed continuous tubular reactor at atmospheric pressure. Reactant gas and product gas mixtures were analysed by means of infrared and mass spectroscopy. The standard reactant feed gas mixture contained 1730 ppm CO and 865 ppm O_2 in argon or nitrogen. The total gas flow rate was adjusted to 178.5 ml min^{-1} resulting in a gas hourly space velocity of 70000 h^{-1}.

Pd/ZrO_2 catalysts were prepared from the glassy $PdZr_3$ alloy by oxidation either *in situ*, (i.e. in the reactant gas) or in air. For the *in situ* oxidation (catalyst PdZr-i), the alloy was exposed for ca. 90 h to the reactant gas mixture ($CO : O_2$ = 1:1) at 280°C. The oxidation in air (catalyst PdZr-a) was carried out at the same temperature for 20 h. Before use, the alloy-derived catalysts were reduced in pure hydrogen at 300°C for 2 h and then kept in an argon stream at 150°C for 30 min. The coprecipitated catalyst (PdZr-c) with the same Pd content (22.4 wt%) was prepared by calcination of the water-insoluble hydroxides of the constituents. Pure palladium powder (Pd-p) was used as reference.

The CO disproportionation ($2\ CO \rightarrow CO_2 + C$) activity of the catalysts was studied in the absence of O_2, i.e. with 1730 ppm CO in argon as reactant feed. After characterizing the bulk structure by XRD, the incorporated carbon was removed by oxidation in an O_2/Ar (800 ppm O_2) stream.

3. RESULTS AND DISCUSSION

3.1 Textural and structural properties of catalysts

Palladium/zirconia catalysts were derived from the glassy $PdZr_3$ precursor by either *in situ* oxidation in the reactant CO/O_2 mixture (PdZr-i) or by oxidation in air (PdZr-a). The conditions of these oxidation pretreatments are described in the experimental part.

Figure 1 depicts the oxidation of the Pd-Zr alloy in air, as monitored by DTA and TG. The maximum weight was reached at 600°C, corresponding to a weight increase upon oxidation of 29.6%. This weight increase agrees well with that expected for complete oxidation of $PdZr_3$ to $(PdO)(ZrO_2)_3$ (29.5%).

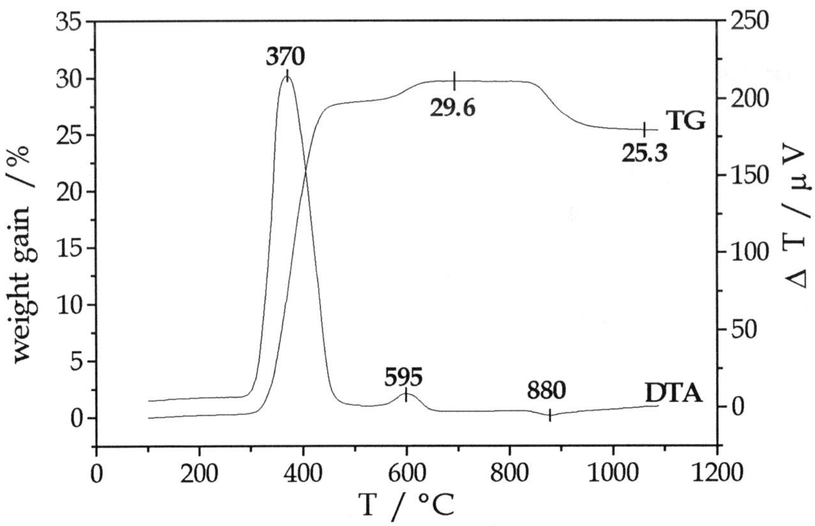

Figure 1. Oxidation of amorphous $PdZr_3$ alloy in air investigated by DTA and TG. Conditions: heating rate, 5°C min^{-1}; air flow rate, 30 ml min^{-1}; sample weight, 52 mg.

The loss of weight occurring at ca. 900°C is due to the decomposition of PdO.

The bulk structural changes of the glassy alloy were followed by XRD. Figure 2 depicts the XRD patterns of the precursor alloy before and after oxidation, and after use of the samples in long term CO oxidation tests.

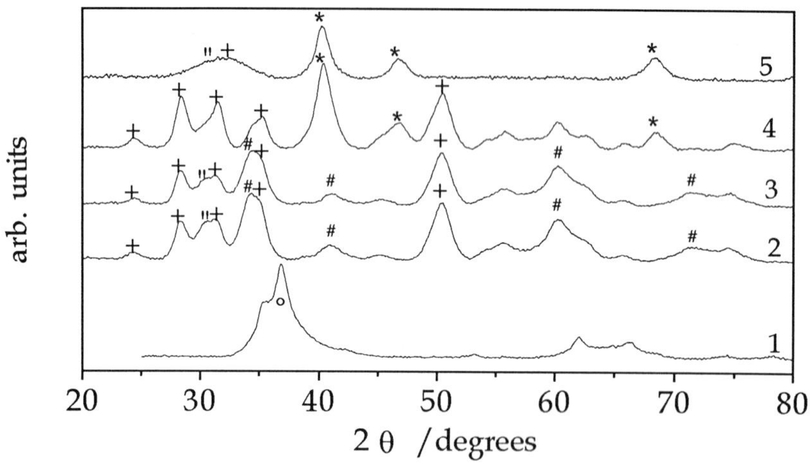

*Figure 2. Bulk structural properties of glassy $PdZr_3$ metal alloy and Pd/ZrO_2 catalysts. XRD patterns (CuKa) of glassy $PdZr_3$ alloy (trace 1); PdZr-i (trace 2); PdZr-a (trace 3); PdZr-i after reduction in hydrogen (trace 4) and PdZr-c after reduction (trace 5). Denotation of reflections: * Pd, # PdO, + ZrO_2 monoclinic, " ZrO_2 tetragonal, ° $PdZr_2$ and Zr.*

The XRD patterns of the *in situ* (PdZr-i, trace 2) and air-oxidized (PdZr-a, trace 3) alloys do not indicate significant bulk structural differences between these catalysts. After oxidation of the alloy under reaction conditions (*in situ*) as well as in air, a solid made up of PdO and monoclinic and tetragonal ZrO_2 was formed. After reduction in hydrogen only reflections due to metallic palladium and zirconia phases were detected by XRD (trace 4).

The XRD pattern of the coprecipitated catalyst (PdZr-c, trace 5) show palladium reflections indicating the presence of palladium crystallites, while the reflections due to zirconia phases are greatly broadened, which suggests amorphous phases. XRD line broadening and electron microscopy indicated that the catalysts prepared by oxidation of the glassy alloy were made up of small poorly crystalline palladium domains of about 5-7 nm lateral dimension. These domains were fully integrated in predominantly amorphous zirconia. Although the coprecipitated catalyst contained palladium particles of about similar size (8 nm), its structure differed significantly from the catalysts prepared from the glassy metal alloy. The palladium and zirconia domains were not intergrown as was observed with the alloy derived catalysts.

The bulk structural changes of the glassy Pd-Zr alloy induced by the oxidation were accompanied by large morphological and textural changes of the samples. Textural properties of the alloy-derived Pd/ZrO_2 catalysts (PdZr-i and PdZr-a) and the coprecipitated Pd/ZrO_2 (PdZr-c) are listed in Table 1 together with the corresponding properties of the palladium powder (Pd-p).

Table 1
Textural properties of the Pd/ZrO_2 catalysts prepared by oxidation of the glassy alloy (PdZr-i, PdZr-a) and by coprecipitation (PdZr-c). Pd-p denotes the pure palladium powder reference.

	BET-surface area m^2g^{-1}	Pd-surface area[a] m^2g^{-1}	dispersion[a] %	Pd content %wt
PdZr-i	61.3	5.7	5.7	22.3
PdZr-a	48.4	5.1	5.2	22.3
PdZr-c	96.2	9.5	7.9	30.4
Pd-p	4.0	1.7	0.4	—

[a] derived from CO chemisorption measurements as described in ref. [5].

Note that the coprecipitated Pd/ZrO_2 catalyst had the largest BET and palladium surface areas. The procedure used to oxidize the Pd-Zr alloy had a significant influence on the textural properties of the final Pd/ZrO_2 catalyst. Larger surface areas were obtained with *in situ* oxidation. Interestingly, with all three Pd/ZrO_2 catalysts the BET/Palladium surface area ratio was approximately 10:1. The relatively low palladium dispersion is not surprising considering the high palladium content of the samples.

3.2 Catalytic properties

CO oxidation: The temperature dependence of the CO oxidation activities of the catalysts are compared in Figs. 3A and B, and Fig. 4. The overall activities of the catalysts decreased in the order: PdZr-i ≅ PdZr-a >PdZr-c > Pd powder, as indicated by the Arrhenius plots depicted in Fig. 3A and the conversion versus temperature plots shown in Fig. 3B. Significant differences were also observed for the activation energies. For both alloy-derived catalysts the activation energy, determined from experiments with less than 10% conversion (Fig. 3A), was 58 ± 3 kJ mol^{-1}, whereas for the coprecipitated catalyst (PdZr-c) and the palladium powder it amounted to 36 ± 0.5 kJ mol^{-1} and 108 ± 1 kJ mol^{-1}, respectively.

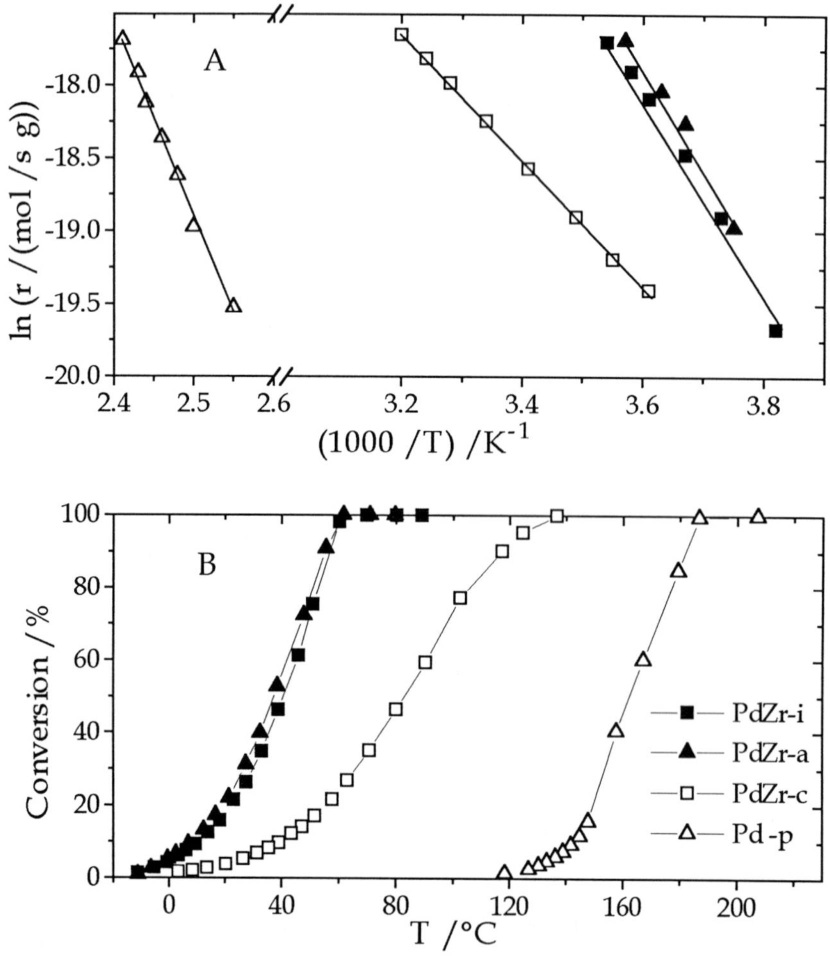

Figure 3. Temperature dependence of CO oxidation rate observed over Pd/ZrO_2 catalysts derived from glassy Pd-Zr alloy and reference catalysts. A) Arrhenius plots; B) CO conversion versus temperature. Conditions: standard feed gas mixture, 178.5 ml min^{-1}. Note different scale of the Arrhenius plots shown in Fig. A.

The activation energy determined for CO oxidation over the palladium powder is comparable to that reported for palladium wires, which is in the range of 101 kJ mol^{-1} [9] to 125 kJ mol^{-1} [10]. The oxidation of CO over palladium was shown to be structure-insensitive [11,12]. The comparative catalytic studies presented here indicate that the interfacial contact of the palladium particles with the zirconia exhibits a strong influence on the catalytic behaviour of the palladium phase. Note that pure zirconia showed no activity for CO oxidation at the conditions used in this work.

CO turnover frequencies (TOF) based on the determination of the accessible palladium atoms by CO chemisorption indicated roughly similar activities for the Pd in the alloy derived catalysts, while the TOF of the coprecipitated catalyst was much lower. This is illustrated in Fig. 4. This behaviour is attributed to the marked structural differences of these catalysts. In our previous work [5,7] we have shown that catalysts derived by *in situ* oxidation from a $PdZr_2$ alloy consisted of intergrown poorly crystalline domains of palladium and zirconia phases, resulting in an extremely large interfacial contact area between the active Pd and the zirconia phases. Such structural characteristics are not obtained with impregnation [5] or coprecipitation, as revealed by high resolution electron microscopy. The large interfacial area has been suggested to facilitate the oxygen transfer through the zirconia - palladium interface [5,7].

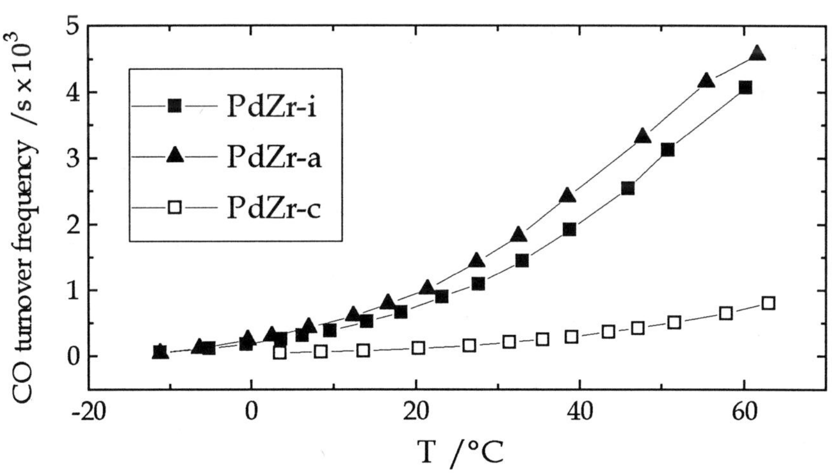

Figure 4. Temperature dependence of turnover frequencies (TOF's) for CO oxidation determined for Pd/ZrO_2 catalysts derived from Pd-Zr alloy and corresponding coprecipitated catalyst. For conditions, see fig. 3.

CO disproportionation: In the absence of oxygen, CO underwent disproportionation according to the equation 2 CO → CO_2 + C. As for CO oxidation, the alloy-derived Pd/ZrO_2 catalysts were the most active catalysts, with CO conversions of 57% (PdZr-i) and 47% (PdZr-a) at 170°C. Note that the CO conversion at this temperature was only 20% with PdZr-c, and the palladium powder (Pd-p) was inactive for CO disproportionation under these conditions. The carbon formed during disproportionation of CO was partly incorporated into the palladium lattice, as the XRD analysis of the catalysts used for disproportionation tests revealed. The interstitial carbon was very reactive and could easily be removed by subsequent exposure of the carbon loaded catalysts to an oxygen containing atmosphere (O_2/argon). Carbon dioxide is formed upon oxygen exposure at temperatures as low as 80°C [8]. However, the interstitial solid solution was found to be stable under a CO atmosphere up to about 400°C, at higher temperatures it decomposes to elemental palladium and carbon. XRD and DTG investigations indicated that at temperatures above 150°C, carbon was incorporated into the Pd lattice up to a maximum solubility corresponding to a $PdC_{0.16}$ phase [8]. Below 100°C there was no significant carbon incorporation into the catalyst PdZr-i. Figure 5 illustrates the carbon incorporation by comparing the XRD patterns of the alloy-derived catalyst PdZr-i before and after exposure to CO disproportionation conditions for 18 hours at 170°C.

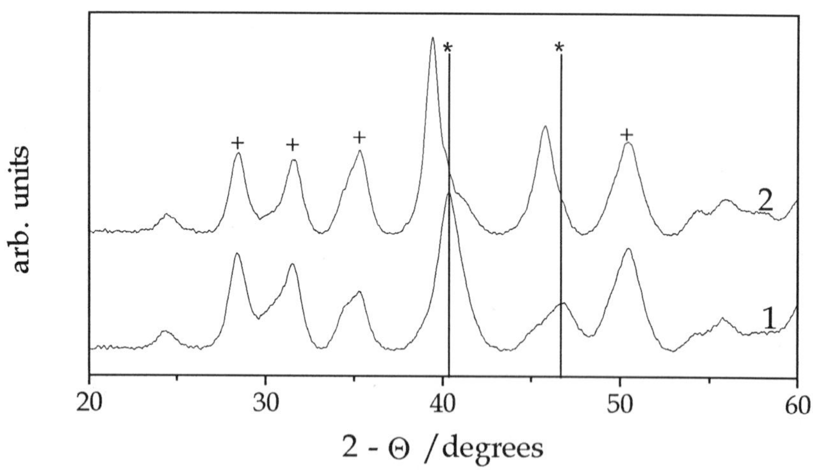

*Figure 5. XRD patterns (CuKα) of reduced catalyst PdZr-i before (trace 1) and after (trace 2) exposure to CO disproportionation conditions: 1730 ppm CO in argon, 170°C for 18 hrs; * Pd, + ZrO_2 monoclinic.*

PdZr-c showed significantly lower carbon incorporation under the same experimental conditions, whereas no CO disproportionation and carbon incorporation could be observed with the palladium powder below 400°C. Thus the intimate interaction of palladium and zirconia seems to be crucial for the high CO disproportionation activity observed with the catalysts derived from the glassy alloy.

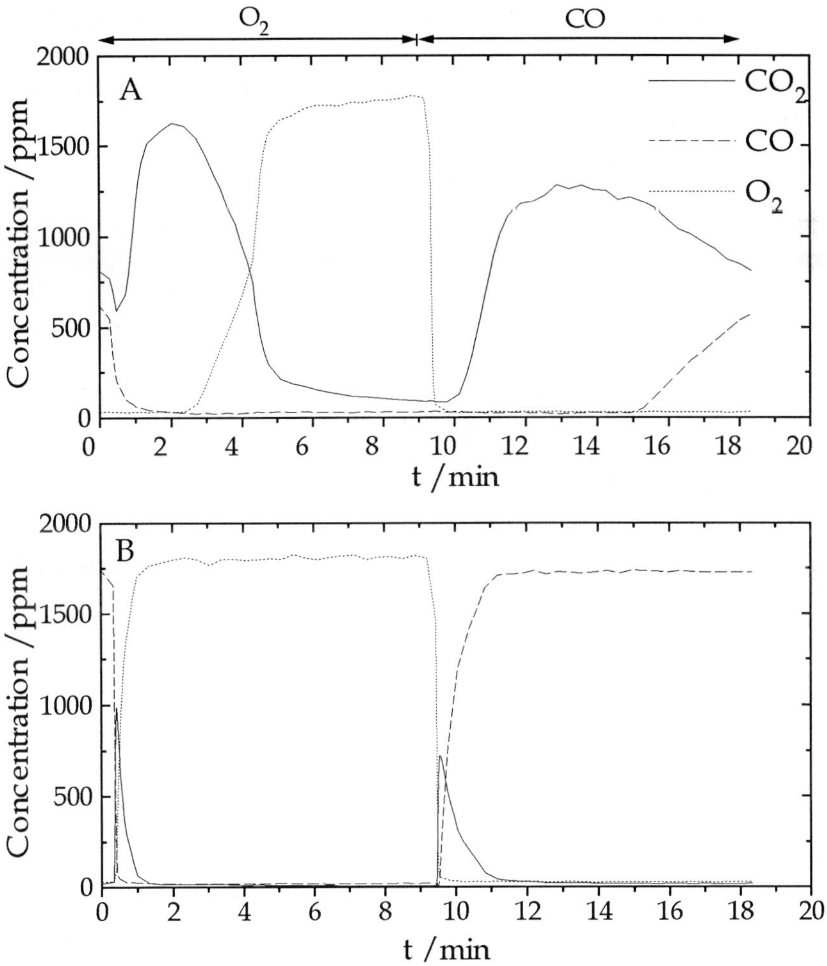

Figure 6. Behaviour of catalyst PdZr-i (A) and reference Pd powder catalyst (B) during alternate cycling between O_2/Ar and CO/Ar feed. Conditions: 150°C, 9 min exposure to 1730 ppm O_2 in argon, then 9 min to 1730 ppm CO in argon; flow rate, 178.5 ml min^{-1}.

CO-O_2 cycling: Figures 6 A and B illustrate the influence of the carbon storage on the behaviour of catalysts PdZr-i and the palladium powder (Pd-p) during forced cycling between CO and O_2 containing feed at 150°C. As with the disproportionation experiments performed under stationary conditions, highest storage capacity for carbon was also observed with the alloy-derived catalysts (PdZr-i and PdZr-a). Figures 6 A and B depict one period of the cycling. The reactant gas was changed in 9 minutes intervals from O_2/Ar to CO/Ar. The high CO_2 production over PdZr-i during the CO/Ar feed period is attributed to the presence of adsorbed oxygen on the surface and to the disproportionation of CO. The carbon produced by this reaction and stored either as surface or interstitial carbon was oxidized in the subsequent O_2/Ar period.

4. CONCLUSIONS

Palladium/zirconia catalysts prepared by oxidation of a glassy metal $PdZr_3$ alloy possess superior activity for both oxidation and disproportionation of CO compared to corresponding catalysts prepared by coprecipitation and palladium powder. With the alloy-derived Pd/ZrO_2 catalysts the disproportionation becomes significant at about 150°C, whereas with a coprecipitated palladium/zirconia catalyst and palladium powder much higher temperatures are necessary. CO-disproportionation is accompanied by carbon incorporation into the palladium lattice reaching a maximum solubility which corresponds to a $PdC_{0.16}$ phase [8]. The interstitial carbon is very reactive and easily removed by exposure of the sample to an oxygen containing atmosphere. These phenomena play an important role in explaining the dynamic behaviour of such catalysts under conditions, where the reactant feed changes from CO to O_2 rich periods, as realized during cycling between CO and O_2. The high activity of the Pd/ZrO_2 prepared from the glassy metal alloy is attributed to its unique structural properties which provide an extremely intimate contact between the active palladium metal and the zirconia phase.

REFERENCES

1 A. Baiker, in Topics of Applied Physics, Vol. 72 (Eds. H. Beck and H. J. Güntherodt), Springer Verlag, Berlin, 1994, 121-162
2 A. Molnar, G. V. Smith and M. Bartok, Adv. Catal. 36, 329 (1989)
3 A. Baiker and D. Gasser, J. Chem. Soc. Faraday Trans. I, 85, 999 (1989)
4 A. Baiker, J. De Pietro, M. Maciejewski and B. Walz, Stud. Surf. Sci. Catal. 67, 147 (1991)
5 A. Baiker, D. Gasser, J. Lenzner, and R. Schlögl, J. Catal. 126, 555 (1990)
6 A. Baiker, M. Maciejewski, S. Tagliaferri, Ber. Bunsenges. Phys. Chem., 97, 286 (1993)
7 R. Schlögl, G. Loose, G. Wesemann, A. Baiker, J. Catal. 137, 139 (1992)
8 M. Maciejewski and A. Baiker, J. Phys. Chem. 98, 285 (1994)
9 G. M. Schwab and K. Gossner, Z. Phys. Chem. Neue Folge, 16, 39 (1958)
10 Y. F. Yu Yao, J. Catal. 28, 1 (1979)
11 T. Engel and G. Ertl, Adv. Catal. 28, 1 (1979)
12 S. Ladas, H. Pappa and M. Boudart, Surf. Sci., 102, 151 (1981).

A. Frennet and J.-M. Bastin (Eds.)
Catalysis and Automotive Pollution Control III
Studies in Surface Science and Catalysis, Vol. 96

COMBUSTION OF M-XYLENE OVER Pd CATALYSTS DERIVED FROM AMORPHOUS $Pd_2Ni_{50}Nb_{48}$ ALLOY

László Borkó, Hua Zhu[a], Zoltán Schay, István Nagy, Antal Lovas[b] and László Guczi

Department of Surface Chemistry and Catalysis, Institute of Isotopes of the Hungarian Academy of Sciences, P. O. Box 77, Budapest Hungary, H-1525;

[a] *On leave from Institute of Coal Chemistry, Chinese Academy of Sciences, Taiyuan, Shanxi 030001, P. R. China;*

[b] *Research Institute for Solid State Physics of the Hungarian Academy of Sciences, P. O. Box 49, Budapest, Hungary, H-1525*

ABSTRACT

Oxidation of 0.05-0.1% m-xylene in 2.5-4.0% oxygen/nitrogen mixtures has been studied as a model reaction for the oxidation of aromatics in automotive exhaust gases over Pd catalysts derived from amorphous $Pd_2Ni_{50}Nb_{48}$ alloy. The alloy in the as quenched form was inactive because its surface was covered by an NbO_x layer. On etching in 40% HF the top oxide layer could be removed as well as Nb was leached from the subsurface layers resulting in a Raney-type surface layer of Pd and Ni. This layer had high initial activity but sintered during reaction. The final activity developed at 673 K in the reaction mixture during the initial 10-20 hours of operation as Pd segregated to the outermost layer of the ribbon. This resulted in a catalyst where Pd oxide was embedded in partly crystallized niobia-nickel amorphous matrix. In this high activity state light-off temperatures were 523-573 K at 50,000 h^{-1} space velocity depending on the xylene concentration and the temperature ramp. In some cases hysteresis loops were also observed.

1. INTRODUCTION

Metallic monoliths seem to be a good candidate for automobile exhaust catalysts because of their excellent thermal properties. However, it is difficult to fix a stable washcoat containing the active metals onto a metallic monolith. A solution could be the application of noble metals incorporated into the monolith and being activated before use. Amorphous alloys offer a possibility to prepare new catalysts of this type as they form a single phase solid solution in which the noble metal content can be varied continuously even outside the solubility limit in the equilibrium state. Catalytic properties of different types of amorphous alloys have recently been reviewed [1,2]. The catalysts derived from ribbons of rapidly quenched FeB and FeNiB amorphous alloys showed superior activity in the $CO+H_2$ reaction [3], nickel-base alloy catalysts were used for oxidation of CO by O_2 and NO [4].

In all of the cited works the as quenched amorphous ribbons needed some treatment as acid etching or in situ oxidation to obtain high catalytic activity. On the PdZr alloys a surface layer of highly dispersed Pd supported on nanocrystalline ZrO_2 was developed [5,6,7]. Similar transformation was observed on the NiZr alloys [8], showing that both Ni and Pd can be transformed into highly dispersed metals supported on porous ZrO_2.

Ni based amorphous alloy containing Pd in low concentration seems to be a good candidate for hydrocarbon oxidation as both metals are active in oxidation reactions [8]. In the present work from the possible Ni-based metal alloys the $Ni_{50}Nb_{50}$ composition has been selected as we have good deal of experience in its preparation. We replaced part of Nb by Pd and used $Pd_2Ni_{50}Nb_{48}$ amorphous alloy ribbons as starting material for the oxidation of xylene. The latter reaction has been selected as a model reaction for the oxidation of unburnt hydrocarbons in automobile exhaust gases.

2. EXPERIMENTAL

$Pd_2Ni_{50}Nb_{48}$ amorphous alloy ribbons of about 2 mm width and 0.02-0.03 mm thickness were prepared by melt spinning using a single copper roller. The ribbons were cut into pieces of 15-20 mm length, corrugated by tweezers, etched in a 40% HF solution for different times and rinsed in high purity water before the catalytic test.

About 0.17 g of the ribbons were placed between quartz wool plugs into a quartz tubular reactor of 8 mm diameter having a preheating and mixing zone of over 100 ml. Inconel shielded NiCr-Ni thermocouples were introduced directly into the gas phase before and after the catalyst. The temperature was controlled by a temperature programmer. In addition to the isothermal mode, temperature ramps of 2-20 K/min were applied both in heating and cooling modes. m-Xylene typicaly in 0.065% concentration was added by a syringe pump into a preheated

N_2 make up gas flow of 20 ml/min before the reactor. N_2 and O_2 flows were controlled by mass flow controllers. The contact time was varied between 0.3-0.03 s corresponding to gas flows of 100-1000 ml/min at a catalyst volume of 0.5 ml. A nondispersive IR analyser was used for CO and CO_2 and an FID for hydrocarbon analysis. All relevant signals were collected by on line computer. The amount of CO_2 formed was taken as a measure of the catalytic activity. XRD using a Philips machine equipped with Guinier camera and KRATOS XSAM XPS machine were applied for characterizing the catalyst before and after the catalytic test.

3. Results

The phases found by XRD are listed in Table 1. In the ribbon without any treatment only amorphous NiNb phase was detected in the form of a very broad and diffuse peak, typical for amorphous alloys without any trace of crystalline Pd in it. On the shortest etching in HF solution trace amount of Nb_2O_5 phase could already be detected. The amount of Nb_2O_5 showed no systematic change with the etching time. Etching for 15 min resulted in the appearance of broad Pd lines indicating the presence of small Pd particles. No attempt was made to estimate any crystallite size. In the catalyst which showed the highest catalytic activity diffraction lines of Nb_2O_5, crystalline NiNb compounds, traces of crystalline Pd as well as amorphous NiNb phase were found.

Table 1 XRD measurements of the $Pd_2Ni_{50}Nb_{48}$ alloy submitted to different treatments

Treatment	Phases determined by XRD
no	NiNb amorphous phase
5 min in HF before reaction	Nb_2O_5 (cryst) NiNb (cryst) & (am)
10 min in HF before reaction	the same as at 5 m
15 min in HF before reaction	Pd (cryst), Nb_2O_5 (cryst) NbNi (cryst) & (am)
15 min in HF after reaction	Nb_2O_5 (cryst) NbNi (cryst) & (am)

This indicates that the bulk of the ribbon remained in an amorphous state even after prolonged treatment in oxidizing atmosphere at 673-773 K.

XPS showed that the surface of the untreated ribbon was covered with an oxide layer in which only Nb in the +5 oxidation state could be detected. Nickel and palladium in the surface layer were present in trace amounts, although the main body of the bulk consisted of NiNb amorphous phase.

On the other hand, on etching for 5 and 10 min the major surface components were Ni and Nb in oxide state. Pd became the prevailing component only after 15 min etching in agreement with XRD studies. After reaction Pd remained the major component on the surface mainly in PdO form and the other components disappeared from the surface. However this was not in contradiction with the XRD measurements because the bulk could be a partially crystallized NiNb alloy embedded in amorphous NiNb alloy the surface of which was fully covered with highly dispersed Pd oxide.

The results of the catalytic tests are shown in Figures 1-5. In Figures 1-3 the effects of etching time, contact time and oxygen concentration are given during the first temperature ramp when the catalysts were in a low activity state. The effects of contact time and oxygen concentration are similar to those we observed on the palladium catalysts prepared by electroless deposition of Pd from $PdCl_2$ solution onto stainless steel sheets [10]. The only difference is the formation of CO even at high oxygen concentration at the low conversion side of the temperature ramp as given in Figure 4. The amount of CO decreases on increasing oxygen concentration as shown in Figure 3b. Note the small CO_2

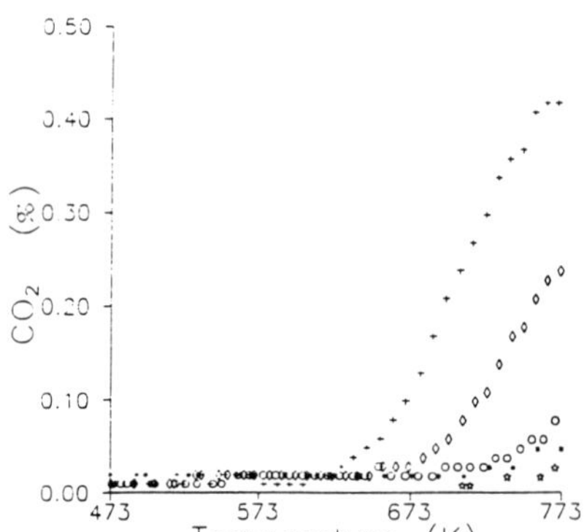

Figure 1. Effect of the etching time in the xylene oxidation to CO_2
+ 15 min; ◊ 10 min; o 5 min; ★ as received; ★ thermal oxidation. Contact time 0.06 s; temperature ramp 5 deg/min; xylene concentration 20 %

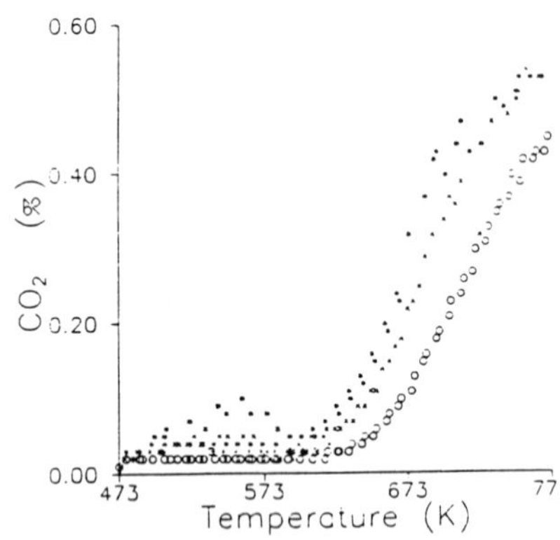

Figure 2. Xylene oxidation to CO_2 with varying contact time:
★ 0.3 s; x 0.15 s; o 0.03 s.
Etching time 15 mins. Temperature ramp 5 deg/min; xylene concentration 0.065 %, oxygen concentration 20 %

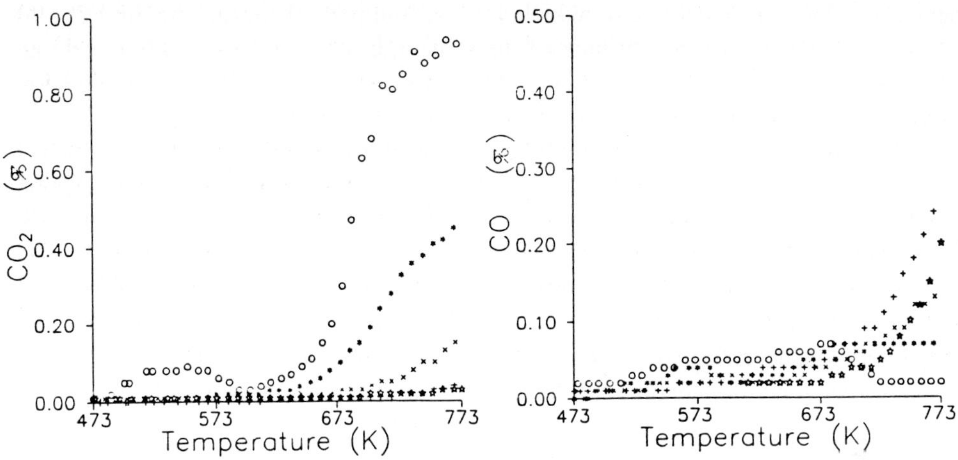

a b

Figure 3. Concentration of CO_2 (a) and CO (b) as a function of oxygen concentration:
o 40 %; ★ 20 %; x 10 %; + 5 %; ★ 2.5 %.
Etching time 15 mins. Temperature ramp 5 deg/min; xylene concentration 0.117 %, contact time 0.06 s.

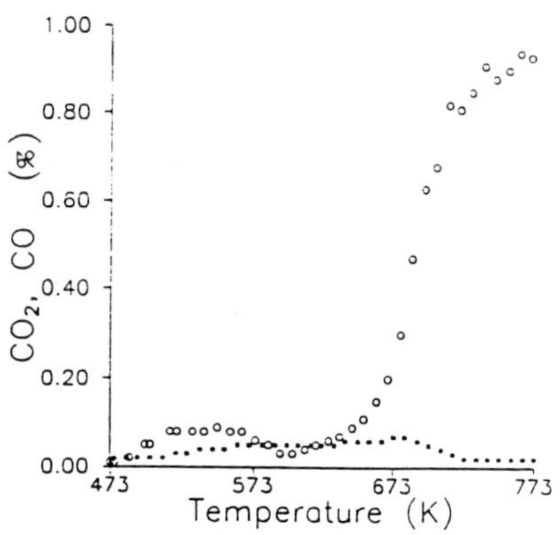

Figure 4. Typical curves of xylene oxydation on $Pd_2Ni_{50}Nb_{48}$ type catalyst.
o CO_2; ★ CO. Etching time 15 mins, temperature ramp 5 deg/min; xylene concentration 0.117 %, oxygen concentration 20 %

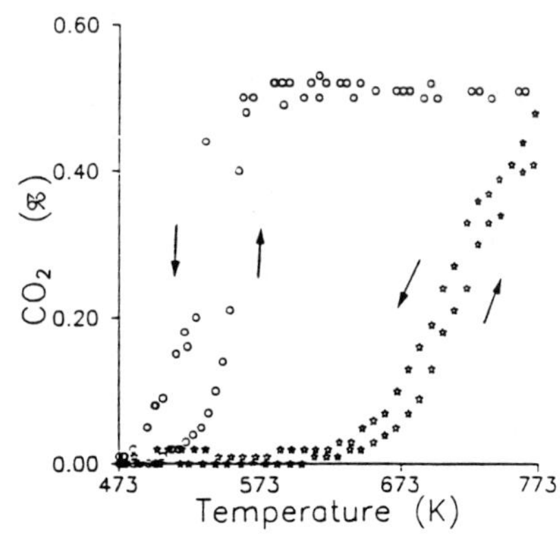

Figure 5. High and low activity state of etched $Pd_2Ni_{50}Nb_{48}$ catalyst.
o high activity; ★ low activity. Temperature ramp 5 deg/min; contact time 0.06 s; xylene concentration 0.065 %; oxygen concentration 20 %

peaks at 523 K in Figures 2-4, which typically appeared during the first cycles only, indicating the disappearance of sites of high activity present after etching. These sites are different from the ones that developed on extended use as given in Figure 5. As expected, on increasing etching time (Figure 1) the initial activity increases as well. Catalysts in the as quenched form or oxidized in air only are practically inactive. An increase of the etching time over 15 min resulted in no further increase in the catalytic activity. Extremely long etching time resulted in the desintegration of the ribbon leaving some Pd powder as residue. On repeated cycling or keeping the catalyst under the reaction mixture for long time at 673-773 K a high activity state develops. In Figure 5 the two states are compared as they were observed on the same catalyst. In both states hysteresis loops are observed between the heating and the cooling ramps, as indicated in the figure.

4. Discussion

In the as quenched form the $Pd_2Ni_{50}Nb_{48}$ amorphous alloy ribbons are inactive in the oxidation of xylene. In contrast to the PdZr and NiZr alloys, whose catalytic activity increased on oxidative treatment [5,8] the surface of the $Pd_2Ni_{50}Nb_{48}$ alloy is covered with a compact NbO_x layer, where the niobium is in the +5 oxidation state. This layer prohibits any further oxidation of the ribbon. It can be removed by etching in 40% HF, which dissolves oxides but does not attack metals during short etching. On increasing the etching time the initial catalytic activity increases but it still remains low. Some unstable active sites are also formed, giving rise to oxidation at 523 K but these sites soon disappear. The origin of the activity is probably the formation of a Raney-type surface layer consisting of high surface area Ni with some Pd acting as impurity. During the initial few hours on stream, the Raney-type Ni structure is sintered and at the same time NiO is formed as the active component of the catalyst, resulting in a relatively low catalytic activity and the disappearance of the activity at 523 K. With extended time on stream, a high activity state develops, which can be interpreted by enrichment of Pd in the outer layer as well as by the formation of a high surface area niobium oxide acting as a support for Pd and NiO. The bulk of the ribbon still remains in the amorphous state ensuring mechanical strength and a good heat conductivity.

The hysteresis loops observed are due to thermal effects as the width of loops in the high activity state is about 30 K in good agreement with the temperature increase along the catalyst bed measured by the two thermocouples placed before and after the catalyst. On cooling the heat of the reaction compensates for the decreasing inlet temperature and keeps the xylene conversion high resulting in a narrow hysteresis loop in contrast to the Pd/stainless steel catalyst, where the loop was of 100 K wide [10] and surface initiated gas phase reactions also contributed to the hysteresis.

ACKNOWLEDGEMENTS

The authors are indebted to Dr.E. Zsoldos for the XRD measurements and to the National Committee for Technological Development for financial support.

REFERENCES

1. J. C. Yoon, D. L. Cocke, J. Non-Cryst. Solids 79 (1986), 217.
2. A. Molnár, G.V. Smith, M. Bartók, Adv. Catal. 36, (1989) 329
3. L. Guczi, G. Kisfaludi, Z. Schay, A. Lovas, Appl. Surf. Sci. 35, (1989) 469.
4. K. Teruuchi, H. Habazaki, A. Kawashima, K. Asami, K. Hashimoto, Appl. Catal. 76, (1991) 79.
5. A. Baiker, D. Gasser, J.C.S. Faraday I 85, (1989) 999.
6. A. Baiker, D. Gasser, J. Lenzner, A. Reller, R. Schlögel, J. Catal.,126, (1990) 555.
7. H. Kimura, A. Inoue, T. Masumoto, Mater. Lett. 14, (1992) 232.
8. B. Qixun, Z. Baoying, L. Zhen, M. Enze, Appl. Catal. A. 85, (1992) 101.
9. G. I. Golodets, "Heterogeneous Catalytic Reactions Involving Molecular Oxygen", Elsevier, New York, 1983
10. L. Borkó, Z. Schay, L. Guczi, submitted to Appl. Catal.

Substrates and Washcoat Technologies

A. Frennet and J.-M. Bastin (Eds.)
Catalysis and Automotive Pollution Control III
Studies in Surface Science and Catalysis, Vol. 96

DESIGN AND PERFORMANCE OF A CERAMIC PRECONVERTER SYSTEM

S.T. Gulati, L.S. Socha and P.M. Then

Corning Incorporated, Corning, NY 14831 USA

ABSTRACT

The preconverter is an essential element of exhaust gas treatment to help meet the tighter emission standards of TLEV and LEV levels. Its design must be so chosen as to meet the simultaneous requirements of compactness, faster light-off, low back pressure, high temperature durability and low cost. This paper presents design options and performance data for a ceramic preconverter system which meets the above requirements. In addition, the high temperature physical durability data for selected preconverter systems are presented. These durability, back pressure, light-off and aging data presented here clearly demonstrate that a properly designed ceramic preconverter system is a viable and cost-effective approach to meeting TLVE and LEV emission standards.

1 INTRODUCTION

The present paper examines design considerations and presents performance data for a ceramic preconverter system which is being sought by the automotive industry, worldwide, to help meet tighter emission standards - notably those corresponding to TLEV and LEV standards. While Europe, Japan and North America have been experimenting with metallic converter systems for light-off application and have met with some success, they have uncovered certain limitations, namely lack of design flexibility, limited high temperature capability, rapid heat dissipation, and high cost [1-4]. Clearly, a ceramic alternative is needed to overcome these limitations.

The successful performance of cordierite ceramic converter system over the past 19 years in a myriad of applications - automotive, light duty vehicles and trucks - has amply demonstrated its suitability as emissions treatment device [5]. Furthermore, new developments in the area of ceramic converter systems have helped extend their applicability to small engines, both two-stroke, used in motorcycles [6-8]. In addition to design flexibility, the ceramic converter offers higher offers higher use temperature, improved heath retention and lower system cost.

The preconverter system must meet the following requirements:

i) it must be compact to fit the available space in the engine compartment;
ii) it must provide sufficient surface area for effective exhaust gas treatment;
iii) it must be light and well-insulated to reach the light-off temperature quickly;
iv) it must retain and prolong the effectiveness of precious metal catalyst to meet TLEV and LEV emissions standards;
v) it must endure high operating temperature (>1000°C) while minimizing temperature gradients;
vi) it must have high strength to withstand mechanical and thermal stresses under operating temperatures and vibrations;
vii) it must be efficient in transferring exothermic heat to the main converter for good conversion activity; and
viii) it must be easy to assemble in a low-cost durable package.

From automakers' point of view, a preconverter package designed and assembled in the above manner is readily implemented in the exhaust system. It also shortens the developmental cycle and reduces engineering costs—a most desirable combination for meeting legal and commercial requirements.

This paper focuses on design and performance data of a ceramic preconverter system as function of substrate composition, size, cell geometry, cell density and catalyst formulation. The high strength, low porosity cordierite ceramic composition permits a thin wall catalyst support with the advantages of larger hydraulic diameter and lower back pressure compared with those attainable with the standard automotive substrate. Both cylindrical and elliptic preconverters with volumes ranging from 0.29 liter to 0.74 liter have been catalyzed with two different catalyst formulations. Their light-off temperature, FTP NMHC emissions with bag 1 and steady state HC, CO and NO_x emissions, following 60 and 120 hours of aging in 900°C rich-lean cycle, are presented [9,10]. Finally, the measured pressure drop across the preconverter is presented as function of cell density, flow rate, gas temperature and converter size. Trade-offs in performance vs. durability vs. systems cost are reviewed.

2. DESIGN OF CERAMIC PRECONVERTER

The design flexibility offered by ceramic substrates stems from the availability of several cordierite compositions [11-13] and well-established extrusion technology [14] which permit the choice of microstructure for optimum interaction with alumina washcoat as well as the choice of cell geometry with the desired geometric and physical properties. Table 1 compares the pertinent properties of two different thin-wall substrates of low porosity and high strength composition, EX-22, with those of standard substrates of EX-20 and EX-32 compositions [6, 15]. The mechanical integrity factor MIF, thermal integrity factor TIF, open frontal area OFA, geometric surface area GSA and Hydraulic diameter D_h are performance parameters which depend on cell shape and size [16].

They are readily computed for a given cell geometry and are very useful for preliminary design. They have a direct impact on physical properties and converter performance. For example, the substrate strength σ_s is proportional to the product of MIF and wall strength σ_w, i.e.

$$\sigma_s = k\sigma_w(\mathrm{MIF}) \tag{1}$$

Similarly, the thermal shock resistance, TSR, of ceramic substrates is given by

$$TSR = \frac{\sigma_s \times TIF}{2E_w\alpha_w}, \qquad \text{square cell} \tag{2}$$

$$= \frac{\sigma_s \times TIF}{3E_w\alpha_w}, \qquad \text{triangular cell} \tag{3}$$

In EQs 1, 2, and 3 k is a constant, E_w denotes wall modulus and α_w its coefficient of thermal expansion. The pressure drop Δp across the preconverter depends on flow rate Q, hydraulic diameter D_h, and open frontal area OFA, as follows:

$$\Delta p = \frac{Q\ell}{\frac{\pi}{4}d^2 \times OFA \times D_h^2} \tag{4}$$

where d denotes substrate diameter and ℓ its length.

For a given space availability in the engine compartment, we assume a circular substrate of diameter d and length ℓ with three different compositions and four different cell geome-

Table 1
Nominal Properties of Standard and Thin-Wall Cordierite Substrates

	Composition			
	EX-20 (std)	EX-32 (std)	EX-22 (Thin-Wall)	EX-22 (Thin-Wall)
Cell structure	400/6.5	236/11.5	350/5.5	340/6
Cell shape	❑	Δ	❑	Δ
Wall thickness (mm)	0.188	0.292	0.140	0.152
Open porosity	35%	42%	20%	20%
Mean pore size (μm)	3.0	7.0	2.0	2.0
Wall strength (MPa)	20.3	14.3	43.1	43.1
Wall modulus (GPa)	26.1	20.9	42.9	42.9
MIF	0.025	0.034	0.012	0.013
Substrate strength*	73 k	70 k	75 k	81 k
TIF	6.9	7.1	9.7	11.3
TSR	110 k	110 k	290 k	254 k
OFA (%)	73	64	80	76
GSA (cm^2/cm^3)	27.4	22.0	26.4	28.9
Avg. CTE @ 800°C (10^{-7}/°C)	6.0	5.0	2.0	2.0
D_h (mm)	1.106	1.159	1.218	1.056
Wall density (g/cm^3)	1.61	1.42	2.0	2.0
Substrate density (g/cm^3)	0.43	0.51	0.39	0.49
Heat capacity of Substrate (cal/cm^3°C)	0.11	0.12	0.10	0.10
Thermal diffusivity (10^{-3} cm^2/s)	4.65	4.08	5.13	4.08

CTE: Coefficient of Thermal Expansion

* The constant k is a normalization constant to help compare strength of cordierite substrates with different porosity and cell structures.

tries as shown in Table 1. For optimum light-off performance it must offer the greatest geometric surface area, highest strength and lowest density, i.e. the quantity $\left(\frac{GSA \times \sigma_s}{\rho_s}\right)$ must be maximized. Similarly, for optimum thermal shock resistance, its TSR must be maximized and finally, for low pressure drop the quantity $(OFA \times D_h^2)$ must be maximized. Table 2 compares the calculated values of these parameters for the four substrates shown in Table 1. It shows clearly the superior performance of EX-22 substrates over that of standard substrates. Both the light-off performance and thermal shock resistance are 10 to 60% better than those of standard substrates. The back pressure reduction is significant for 350/5.5 ❑ while that for 340/6 Δ compares favorably with EX-32, 236/11 Δ structure.

4. PHYSICAL PROPERTIES OF LIGHT-OFF SUBSTRATES

The pertinent physical properties which impact the mechanical and thermal durabilities of substrate include biaxial compressive strength [17-19], high temperature modulus of rupture [20], elastic modulus, and thermal expansion curve as function of temperature [5, 6]. The biaxial compressive strength is a direct measure of substrate's ability to withstand radial mounting pressure during canning. To ensure long-term durability, the preconverter should be packaged in a tourniquet can with high and uniform biaxial compression facilitated by its circular contour. The biaxial compressive strength, which is critical for mechanical durability, was measured in a dual-band tourniquet fixture [19]. Five EX-22 substrates, 85.5 mm diameter x 120 mm long, having a volume of 790 cm^3 (0.79 liter), were wrapped in 3660 g/m^2, Series IV, intumescent mat and subjected to a uniform radial compressive stress until they failed. The choice of substrate size was dictated by both the generic space envelope in the engine compartment and the total surface area required for light-off conversion. The 0.79 liter substrate is deemed adequate for light-off conversion for engine displacements ranging from 2 to 3.5 liters. Table 3 summarizes the biaxial strength data for uncoated substrates. The strength values of 5 to 7 MPa for square cell substrate and 7 to 12 MPa for triangular cell substrate are nearly an order of magnitude higher than the mounting stresses during high speed canning. Furthermore, the biaxial strength of coated substrates which represent the final product is 30% higher than that of uncoated substrate thereby making the safety factor even higher [21, 22]. Finally, the circular contour of preconverter permits a uniform radial mounting pressure which, in turn, enhances its long-term durability.

Table 2
Comparison of Performance Parameters of Standard and Thin-Wall Cordierite Substrates

	Composition			
	EX-20 400/6.5❑	EX-32 236/11.5Δ	EX-22 350/5.5❑	EX-22 340/6.3Δ
$\frac{GSA \times \sigma_s}{\rho_s}$	4480	2900	5075	4790
TSR	110 k	105 k	290 k	255 k
$OFA \times D_h^2$	0.893	0.856	1.187	0.853

Table 3
Biaxial Compressive Strength of EX-22 Substrates
(85.5 mm dia. x 120 mm long)

	350/5.7 ❑ (MPa)	340/6.3 Δ (MPa)
Min.	4.9	7.4
Mean	5.7	10.1
Max.	6.8	12.6

The modulus of rupture (MOR) of above substrates was measured in four-point bending using rectangular bars, 2.5 cm wide x 1.25 cm thick x 10 cm long, cut from the substrates in both axial and tangential directions [23]. These measurements were carried out at 25°, 700° and 900°C with latter temperatures representing the maximum operating temperature of peripheral region of preconverter which experiences the highest thermal stress in service. The MOR data for 350/5.7 square cell substrate and 340/6.3 triangular cell substrate are summarized in Tables 4 and 5. Figure 1 which is a Weibull plot of axial MOR data at 25°C demonstrates the unimodal nature of strength distribution with Weibull slope of 13 equivalent to a standard deviation of 8% [24]. It should be noted in Tables 4 and 5 that MOR values increase with temperature due to both the absence of stress corrosion and healing of microcracks in the porous cordierite body at high temperature [25]. Such an effect allows higher operating stresses and is beneficial to preconverter's long-term durability. A comparison of Tables 4 and 5 shows the triangular cell structure to be 20 to 40% stronger than the square cell structure which affords it a superior mechanical reliability to combat canning, vibrational and impact loads in service. Finally, the MOR values will show further improvement of about 30% after the substrates are washcoated with high surface area γ-alumina washcoat [21,22].

Table 4
High Temperature MOR Data for
EX-22, 350/5.7 ▢ Substrates (Mean Values)

Temp. (°C)	No. of Specimens	Axial MOR (MPa)	Tang. MOR (MPa)
25	10	2.98	1.23
700	10	3.13	1.31
900	10	3.30	1.36

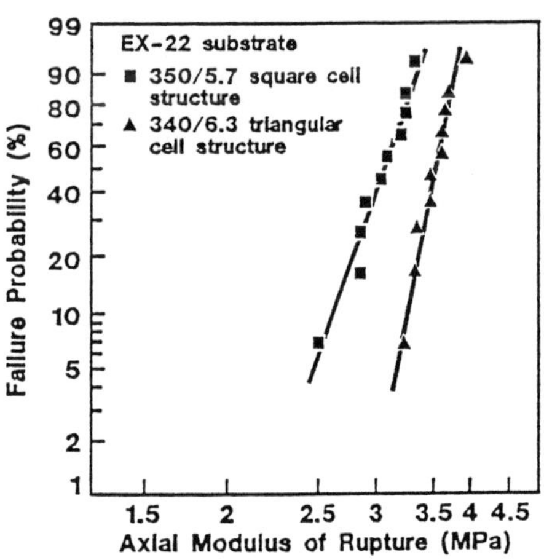

Figure 1. Weibull plots of axial MOR data at room temperature for EX-22 substrates with 350/5.7 ▢ cell and 340/6.3 Δ cell structure

The elastic moduli and thermal expansion curves in the axial direction of above substrates are shown in Figures 2 and 3, respectively, over a temperature range of 25°-1000°C. In view of its more rigid structure, the triangular cell substrate is approximately 10% stiffer than the square cell substrate. This is consistent with its higher MOR and biaxial compressive strength compared with those of square cell structure. It should be noted in Figure 2 that the moduli remain relatively constant up to 600°C and then increase with temperature due to stiffening associated with the healing of microcracks. The increase in moduli implies higher thermal stresses at operating temperature which are self-com-

Table 5
High Temperature MOR Data for
EX-22, 340/6.3 Δ Substrates (Mean Values)

Temp. (°C)	No. of Specimens	Axial MOR (MPa)	Tang. MOR (MPa)
25	10	3.50	1.32
700	10	3.70	1.49
900	10	4.55	1.53

pensated by the corresponding increase in MOR values noted earlier.

The expansion data in Figure 3 clearly show that both 350/5.7 square cell and 340/6.3 triangular cell structures have identical thermal expansions over the entire temperature range of interest (25°-1000°C). Furthermore, the expansion curves of EX-22 composition go through a minima at 300°C implying that the thermal stresses associated with temperature gradient from center to peripheral region will shift from tension at the center to tension at the periphery as the center temperature exceeds 300°C during light-off. At temperatures approaching 1000°C, the thermal stresses in the peripheral region will be dictated by radial temperature gradient, absolute thermal expansion, and elastic moduli. It should be noted, however, that the EX-22 composition has one-third the expansion coefficient of standard cordierite composition EX-20 [6]; see also Table 1. The low expansion behavior of EX-22 substrates makes them thrice as durable as EX-20 substrates and hence ideal for light-off application.

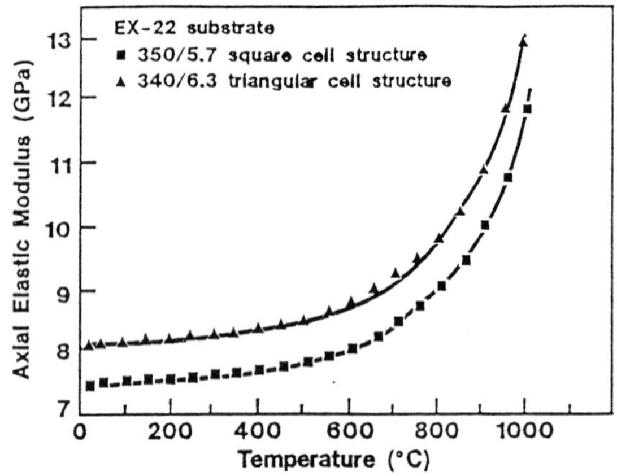

Figure 2. Temperature dependence of axial elastic moduli of EX-22 substrates with 350/5.7▢ cell and 340/6.3 Δ cell structure

5. PACKAGING DESIGN

Due to severe operating conditions in close-coupled light-off application, a robust mounting system is required for the preconverter to survive 160,000 vehicle kilometers. Such a system, shown in Figure 4 includes a new intumescent mat, Interam™ 100, which is protected on the front and back edges by high temperature seals. The seals are flexible during assembly and are attached to the mat to facilitate mounting and to ensure thermal integrity under operating conditions. The intumescent temperature range of this mat is broader than that of standard Interam™ mats used in underbody applications [7, 8, 25, 26]. This mat begins to expand at 40°C lower temperature than

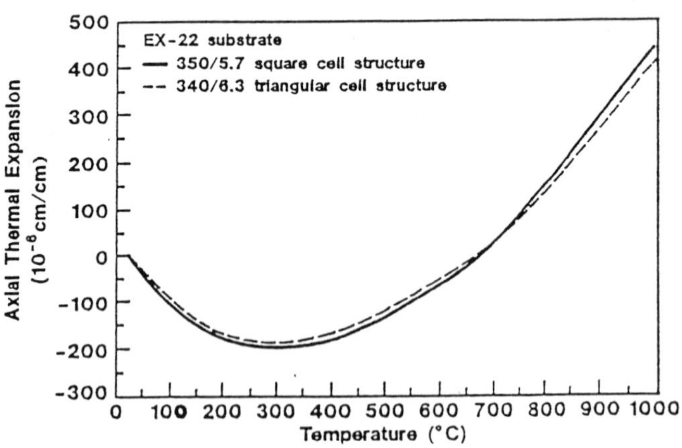

Figure 3. Axial thermal expansion curves for EX-22 substrates with 350/5.7 ▢ cell and 340/6.3 Δ cell structure

that for standard mats. Similarly, its upper temperature capability is 50°C higher than that of standard mats.

The high temperature seal rigidizes when it is heated during the first operating cycle of the converter and forms a strong barrier thereby protecting the mat from potential erosion caused by impingement of hot exhaust gases. The seal consists of glass frit, with low softening temperature, an organic binder and other inorganic fillers. The inorganic materials fuse together at a moderate temperature to form a solid body which is relatively impervious to exhaust gas pressure pulsations and which reduces the temperature of the mat edge.

A preconverter package of the design shown in Figure 4 was assembled with Interam™ 100 mat, compressed to a mount density of 1.2 g/cm^3, and subjected to durability testing discussed in next section.

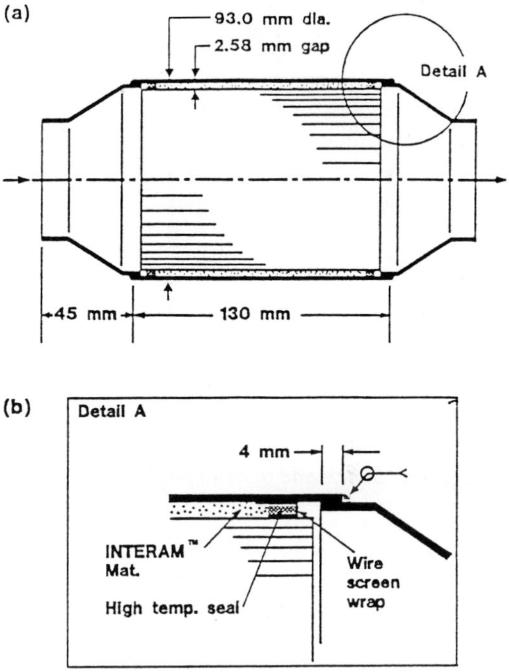

Figure 4. a) Schematic of packaging design for preconverter; b) detail of mat/seal attachment with wire screen wrap

6. PERFORMANCE DATA

In this section we will summarize the performance data consisting of

i) physical durability;
ii) back pressure across preconverter;
iii) emissions reduction with preconverter;
iv) effect of aging on FTP emissions.

The performance tests were carried out at three different independent laboratories using both the ceramic and metallic preconverter systems. The latter comprised of a commercially available Fe-Cr-Al alloy support with 400/2 ⌂⌂ cell structure and measured 93 mm diameter x 90 mm long. Both preconverters had identical catalyst loading (40 g/ft^3, TWC) with a Pt:Rh ratio of 5:1.

6.1. Physical Durability

The physical durability of EX-22, 350/5.7 ❑ ceramic preconverter was assessed in two different tests. The first of these was a high temperature vibration test, using an exhaust gas generator and electromagnetic vibration table [27], under the following conditions:

Exhaust Gas Temperature	1030°C
Acceleration	45 g's
Frequency	100 Hz

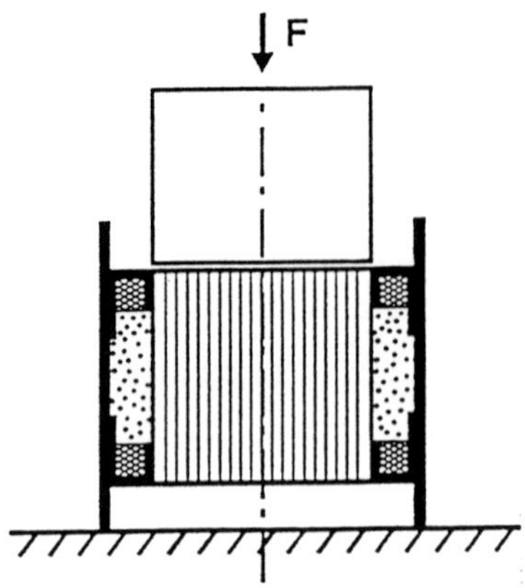

Figure 5. Schematic of axial push-out test

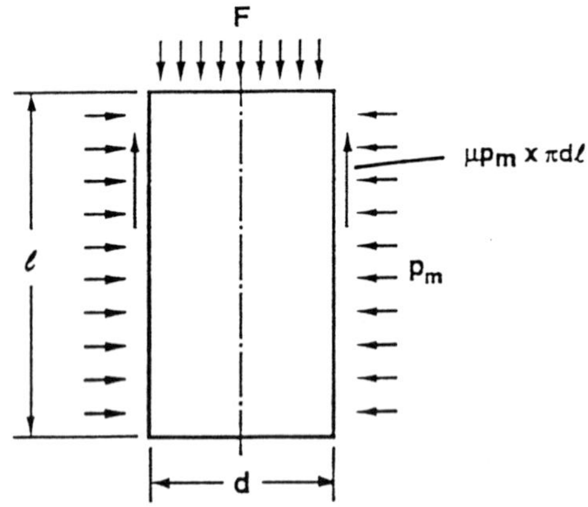

Figure 6. Force balance in axial direction during push-out test

The test was run for 100 hours and cycled back to room temperature 19 times with no failure detected in the mounting system. The ceramic substrate maintained its original position, the seals rigidized and remained attached to the mat, and the mat was not eroded.

The robustness of the mounting system was also verified via an axial push-out test shown schematically in Figure 5. The push-out force F required for relative motion between the substrate and mat is related to mounting pressure p_m, friction coefficient μ (~ 0.25) and substrate dimensions as shown in Figure 6. The push-out force F was measured as function of mat density ρ_m and the mounting pressure p_m was calculated from EQ 5. The push-out speed was held constant at 25 mm/min. The data, plotted in Figure 7, are consistent with the empirical equation relating mount density to mounting pressure [26]

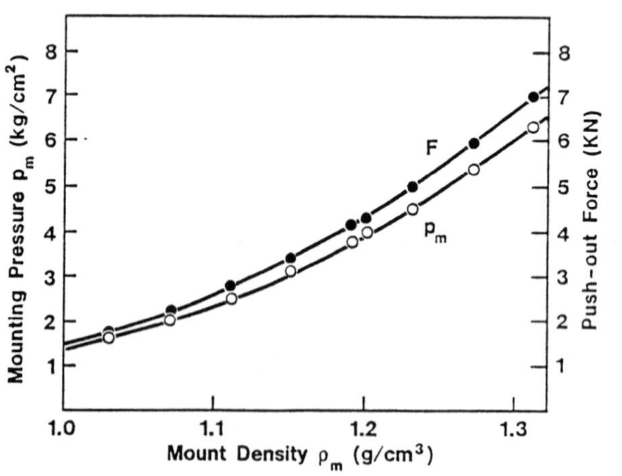

Figure 7. Axial force and mounting pressure during push-out test as function of mount density

$$p_m = \lambda e^{-6.7/\rho_m} \tag{5}$$

in which the constant λ depends on push-out speed. For the push-out speed of 25 mm/min., λ = 1060 kg/cm^2. At higher speeds λ can approach a value of 4500 kg/cm^2 and result in 4x higher pressure which, however, lasts for a very short time (few seconds) since the mat relaxes due to its viscoelastic behavior. The coated substrate can readily withstand a mounting pressure of 30 kg/cm^2 or 3MPa for short duration since

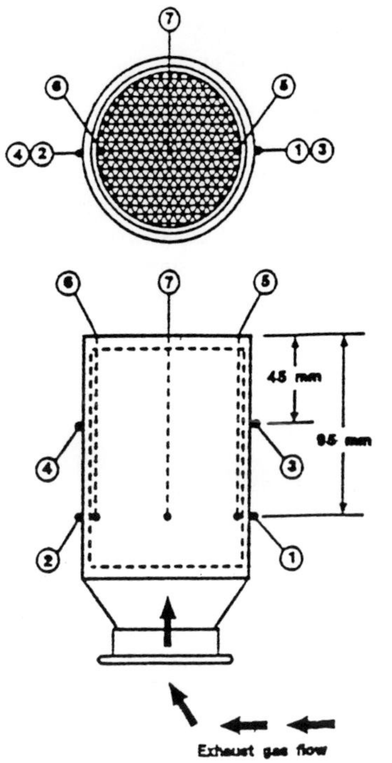

Figure 8. Thermocouple locations for measuring temperature distribution during exhaust gas simulation test

its minimum biaxial compressive strength is > 5MPa; see Table 3. The mat density of 1.2 g/cm^3 was selected to ensure a mounting pressure of 4-5 Kg/cm^2 which is needed for mechanical integrity.

The insulation property of Interam™ 100 mat is critical to the thermal integrity of ceramic preconverter evaluated by measuring the radial temperature gradient in the preconverter during the exhaust gas simulation test using compressed air and a natural gas burner; see Figure 8. The EX-22, 340/6.3 triangular cell substrate (85.5 mm diameter x 120 mm long) was packaged in a stainless steel can using a Interam™ 100 mat and the stuffing technique. The inlet gas temperature was 1030°C and the flow rate was 1.13 m^3/min. corresponding to a space velocity of 86,000/hr through the 0.79 liter substrate. The temperature vs. time output at the center and peripheral regions of preconverter and the stainless steel shell is plotted in Figure 9. According to Figure 9, a temperature gradient of 130°C was measured between the center and peripheral regions during exhaust gas simulation test. This is relatively low indicating excellent insulation property of Interam™ 100 mat. Furthermore, the heat retention afforded by the mat keeps the preconverter hot and efficient during cold-start conditions. The improved insulation property of the mat also maintains the shell temperature at a reasonably low value thereby minimizing shell distortion due to mounting pressure. The typical temperature drop across the mat during exhaust gas simulation test is 360°C as shown in Figure 9. This is a significant temperature drop which protects preconverter components, including the mat itself, from high temperature degradation.

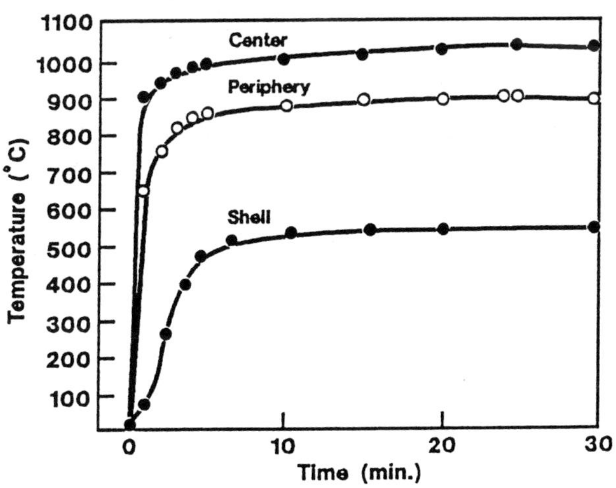

Figure 9. Time variation of temperature of shell and substrate periphery during exhaust gas simulation test

6.2. Back Pressure

The back pressure was measured for three different preconverter assemblies (all catalyzed) in a chassis dynamometer test (4 liter, 6 cyl. engine with PFI). The three preconverters had identical outside dimensions, namely 93 mm diameter x 90 mm long, but different cell structures and compositions:

i) 350/5.7 square cell structure of EX-22 cordierite ceramic composition;
ii) 340/6.3 triangular cell structure of EX-22 cordierite ceramic composition;
iii) 400/2 sinusoidal cell structure of Fe-Cr-Al metallic composition.

The back pressure was measured with the aid of H_2O monometers. In addition to pressure drop across preconverter, the total pressure drop from exhaust manifold to tailpipe was also recorded. The measured values are summarized in Table 6. It is interesting to note that under the specified test conditions the 350/5.7 ❑ ceramic preconverter yielded 3% lower back pressure than the 400/2 ⏢⏢ metallic preconverter. Similarly, the 340/6.3 Δ ceramic preconverter yielded 7% higher back pressure than the 400/2 ⏢⏢ metal preconverter. These differences are attributed to both the back pressure parameter, BPP, and frictional drag of catalyzed walls to gas flow which vary with cell structure. The former is defined by

Table 6
Back Pressure Data for Three Different Preconverters during Chassis Dynamometer Test @ 80 km/hr (93 mm diameter x 90 mm long preconverter with inlet gas temp. = 700°C)

Cell Structure	Back Pressure Across Preconverter ($cm\ H_2O$)	Total Back Pressure ($cm\ H_2O$)
350/5.7 ❑ ceramic	32.5	68.0
340/6.3 Δ ceramic	36.0	72.1
400/2 ⏢⏢ metallic	33.5	71.1

$$BPP = OFA \times D_h^2 \tag{6}$$

whose values for uncoated substrates appear in Table 2 and are inversely proportional to measured back pressure data. It should also be noted that the back pressure across all three preconverters is approximately 47 to 50% of the total back pressure from exhaust manifold to tailpipe.

6.3. Emissions Reduction with Preconverter

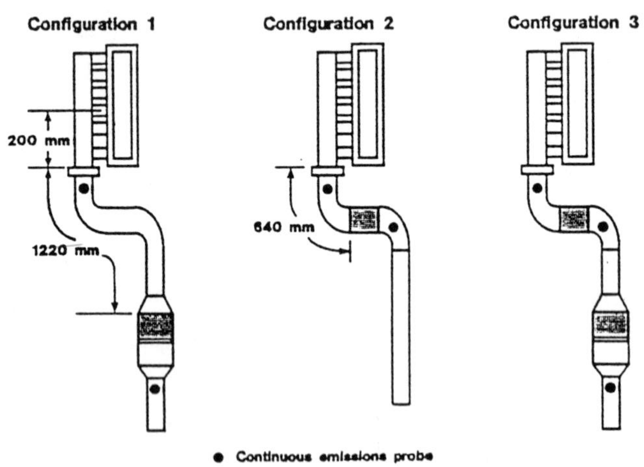

Figure 10. Exhaust configurations during FTP test: 1) main converter only, 2) preconverter only, 3) main converter and preconverter

The catalytic efficiency of three different preconverters was measured on a 1994 vehicle powered by a 4 liter, 6 cylinder engine with port fuel injection. This vehicle's engine-out emissions were 2.3 g/mile NMHC, 16.9 g/ mile CO, and 5.9 g/mile NO_x. Three different exhaust configurations were employed during emissions testing using the FTP cycle [28-30]; see Figure 10. The main converter consisted of two ceramic substrates with a total volume of 3.2 liters. One of these was catalyzed with a Pt/Rh catalyst and the other with Pd only catalyst. The characteristic dimensions and catalyst loading for the three preconverters are summarized in Table 7. Figure 11 compares the tailpipe HC emissions with and without the ceramic preconverter (350 □), i.e. for configurations 1 and 3. It shows that the preconverter contributes to emissions reduction during the 40-120 sec. interval. The FTP emissions are reduced significantly due to oxidation of HC and generation of exothermic heat for faster light-off of main converter. This benefit is attributed to higher surface area, lower thermal mass, and close proximity of the ceramic preconverter to exhaust manifold which reduce the time for light-off temperature to 40 sec. of cold-start. It is also clear in Figure 11 that after 120 secs, the HC emissions for both configurations are identical implying that the main converter is fully effective.

Table 7
Description of Ceramic and Metallic Preconverters

	350 □ ceramic preconverter	340 Δ ceramic preconverter	400 metallic preconverter
Substrate Diameter	85.5 mm	85.5 mm	90 mm
Substrate Length	80 mm	80 mm	80 mm
Outside Can Diameter	93 mm	93 mm	93 mm
Wall Thickness	0.145 mm	0.160 mm	0.05 mm
Catalyst Loading	40 g/ft^3	40 g/ft^3	40 g/ft^3
PM Ratio	5:1 Pt/Rh	5:1 Pt/Rh	5:1 Pt/Rh

Further reductions in the light-off time of preconverter, and hence in bag 1A emissions, are also possible by either moving it closer to the engine or using a different catalyst loading or formulation. For the tests described here the preconverter and main converter were located at a distance of 0.64 m and 1.22 m, respectively, from the exhaust manifold.

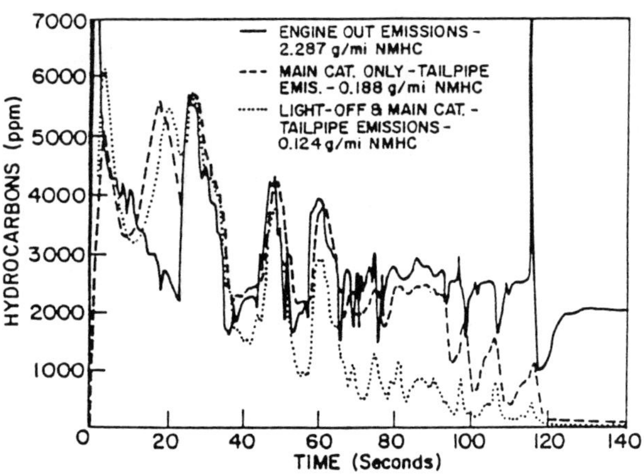

Figure 11. Continuous HC emissions during cold-start from engine and tailpipe with main catalyst only vs. main catalyst plus preconverter

To verify the onset of light-off conditions, the ceramic preconverter was instrumented with 10 chromel-alumel, Type K, thermocouples and the temperature-time history was recorded during the FTP cycle; see Figure 12. The temperature vs. time plot for the preconverter section, 2.5 cm from the entry face, is shown in Figure 13. It is clear from this figure that all of the five thermocouples are recording identical temperatures indicative of uniform, laminar, flow distribution. This is true for both the metallic and ceramic preconverter systems. Figure 14 shows the temperature-time curves for both ceramic and metallic preconverters prior to light-off. These data show that

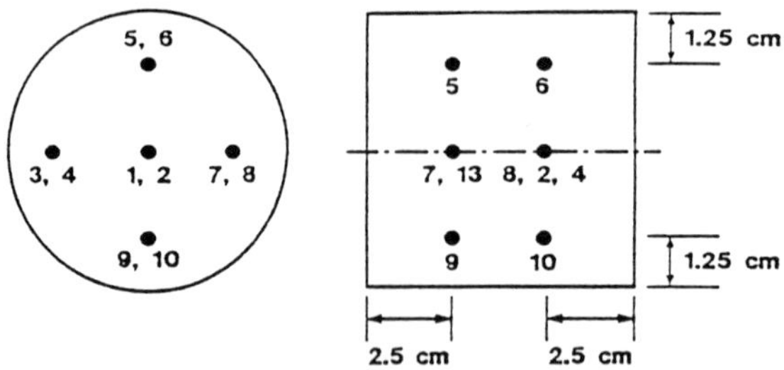

Figure 12. Thermocouple locations for monitoring preconverter temperature during cold-start

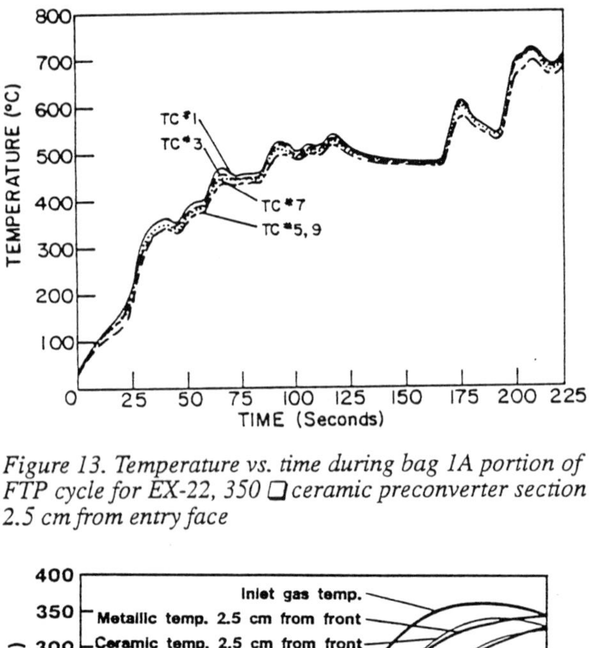

Figure 13. Temperature vs. time during bag 1A portion of FTP cycle for EX-22, 350 ❑ ceramic preconverter section 2.5 cm from entry face

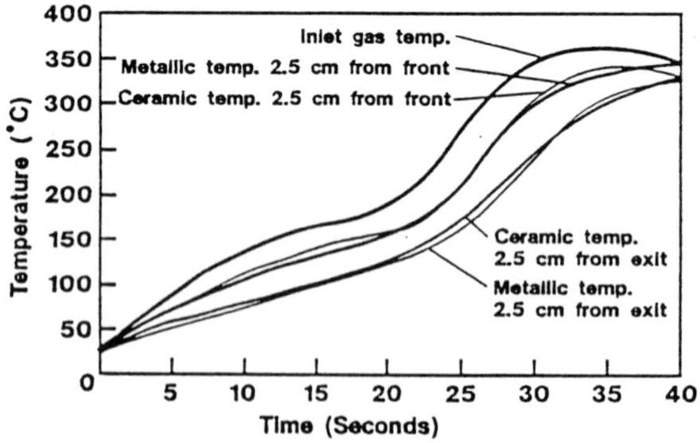

Figure 14. Temperature vs. time curves for inlet gas, ceramic preconverter and metallic preconverter prior to light-off

both preconverters heat up at a similar rate. Figure 15 shows the average internal temperature of ceramic and metallic preconverters. After the initial catalytic activity, the ceramic preconverter maintains an internal temperature only 20°C higher than that of metallic preconverter throughout the FTP cycle. However, Figure 16 shows that the average surface temperature of 409 S.S. can of ceramic preconverter is 200°C lower than that of metallic preconverter. This shows that the ceramic preconverter retains more heat in the exhaust stream thereby promoting early light-off of the main converter. The metallic preconverter, on the ohter hand, radiates heat to the surrounding environment with lower impact on light-off of main converter.

6.4. Effect of Aging on FTP Emissions

Four preconverters were aged simultaneously on a 7.5 liter, 8 cylinder engine with throttle body fuel injection. During aging the main converter was mounted in series following the preconverter. Uniform mixing of exhaust gas was achieved by using a 15.2 cm transition from the exhaust pipe to entry face of the preconverter. The same transition was also used at the exit end of main converter. Laminar flow elements, downstream of each converter, were used to balance the flow through each converter. The FTP emissions test used both the preconverter and main converter assemblies exactly as they were aged. The converters were aged for 60 and 120 hours using the aging cycle described in Table 8. This cycle was chosen because it exposes the catalyst to both lean and rich conditions for extended periods of time. It maintains a low inlet temperature to the catalyst minimizing thermal combustion in the exhaust system. Thermal reaction in the exhaust pipe is very difficult to control. With this cycle the temperature oscillations are created by the catalytic activity, which is a more appropriate and demanding test of the catalyst system.

The catalyst performance was evaluated by using the U.S. Federal Test Procedure. Since the key purpose of preconverter is to reduce the cold-start emissions, bag 1 of FTP cycle was split into two bags: bag 1A which represents 0 to 220 sec. and bag 1B which represents 220-505 sec. However, at t = 220 sec. the temperature of the main converter is relatively stable and sufficiently high to ensure rapid reaction kinetics.

The effect of aging on FTP emissions was measured using exhaust configuration 2 (see Figure 10) with the 350 ❑ ceramic preconverter and 400 ◬◬ metallic preconverter. The NMHC, CO and NO_x emissions output from each of these preconverters are compared as function of their

Table 8
Aging Cycle for Preconverters and Main Converter

Rich-Lean Aging	Inlet to Catalyst Lean	Inlet to Catalyst Rich	Out of Catalyst Lean	Out of Catalyst Rich
A/F (Spindt method)	15.66	12.38		
A/F (A/F analyzer)	13.65	12.50		
THC (ppm)	725	1500	0	925
CO (%)	3.75	6.00	0.027	5.64
CO_2 (%)	9.67	10.9		
NO_x (ppm)	1125	1440	1000	0
Oxygen (%)	4.2	0.7	1.5	0
Catalyst Bed Temperature (°C)	900	650	900	650

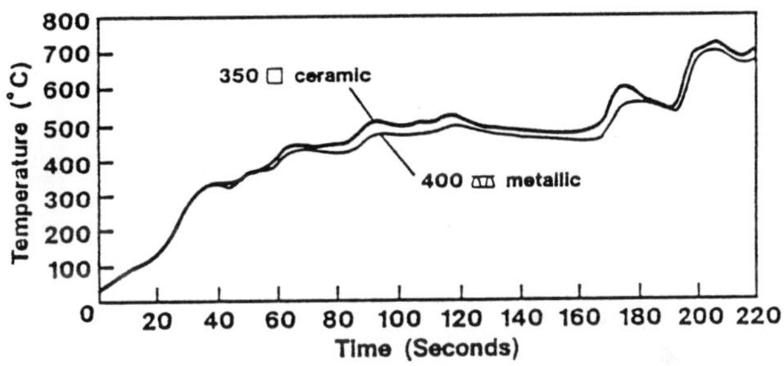

Figure 15. Average internal temperature for ceramic and metallic preconverters during FTP cycle (config. 2)

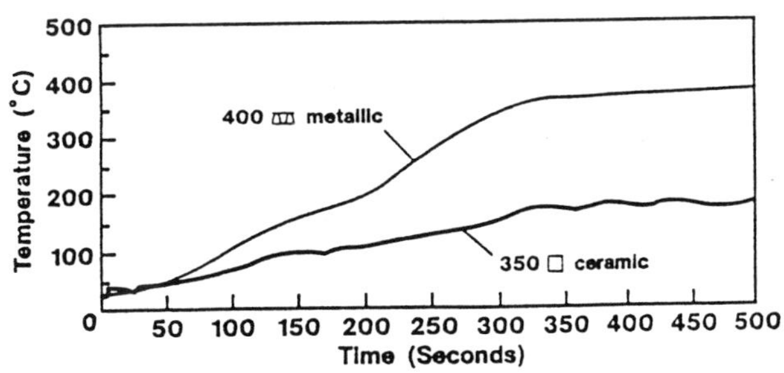

Figure 16. Average surface temperature of can for ceramic and metallic preconverters during FTP cycle (config. 2)

aging time in Figure 17. The ceramic and metallic preconverters demonstrate equivalent performance with only six hours of aging. At this level, the ceramic preconverter reduced NMHC emissions by additional 5% compared with metallic preconverter. However, after 60 hours of aging the metallic preconverter was 55% less effective than the ceramic preconverter in reducing NMHC. At 120 hours of aging the ceramic preconverter continued to outperform the metallic preconverter by 49% in NMHC reduction. This indicates that the metallic preconverter suffered a partial loss of conversion activity during the 6 to 60 hours of aging process. A possible explanation for this is that the continued expansion and contraction of metallic substrate could lead to partial loss of washcoat. Detailed analyses are underway to determine the precise reason for loss of conversion. The corresponding tests for 340 Δ ceramic preconverter are in progress and their results will be reported in the near future.

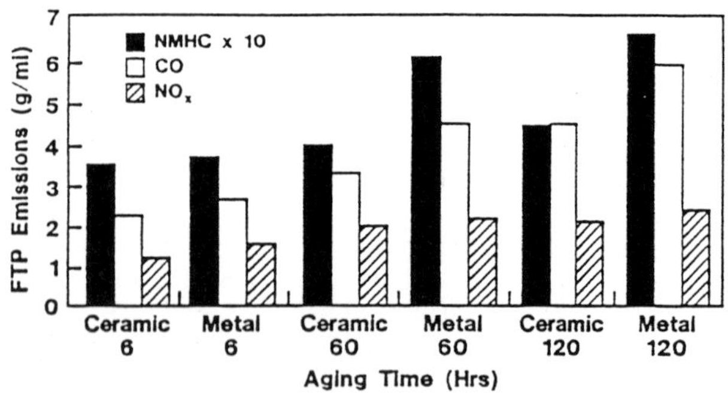

Figure 17. Comparison of FTP emissions from 350 □ ceramic vs. 400 ΔΔ metallic preconverter alone as function of aging time (config. 2)

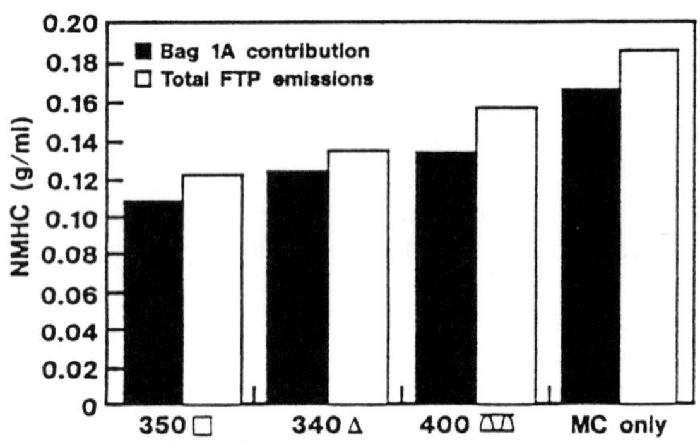

Figure 18. NMHC emissions for bag 1A and total FTP cycle with main converter alone (config. 1) vs. main converter plus three different preconverters (config. 3): all converters aged for 60 hours

The effect of aging on FTP emissions was also measured using exhaust configuration 3. Each of the three preconverters together with the main converter was aged for 60 hours prior to the FTP test. Both bag 1A and the total FTP emissions were monitored. The NMHC, CO and NO_x emissions from bag 1A as well as the total FTP emissions for each of the three preconverters in series with the main converter are compared with those from main converter only (configuration 1) in Figures 18, 19 and 20. Again, it is clear from these data that the 350 □ ceramic preconverter outperformed both 340 Δ ceramic and 400 ΔΔ metallic preconverters due to its lower thermal mass, high surface area and uniform flow distribution.

7. Summary

The primary objective of designing a ceramic preconverter system for light-off application has been met by proper selection of cordierite composition, cell geometry, substrate contour, size and volume, high temperature durable mat, and a robust packaging design.

The low porosity cordierite composition, EX-22, with two unique cell geometries, is recommended as the substrate of choice with inherent advantages of compactness, low mass, high surface area, high strength, low thermal expansion, and high temperature capability. These characteristics allow the ceramic preconverter to meet light-off requirements under variable operating conditions over the specified 160,000 vehicle kilometers. The triangular cell structure provides the added benefit of in-plane isotropy with 50 to 100% higher biaxial compressive strength than that of square cell structure, which permits a more robust packaging design.

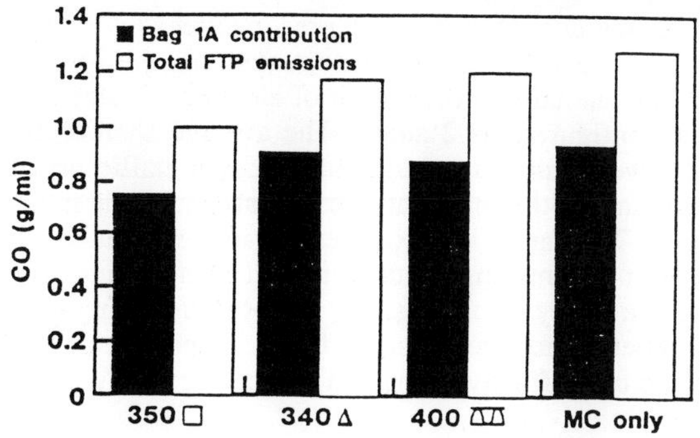

Figure 19. CO emissions for bag 1A and total FTP cycle with main converter alone (config. 1) vs. main converter plus three different preconverters (config. 3): all converters aged for 60 hours

The circular substrate contour is ideal for close-coupled light-off application in that it occupies small volume, facilitates optimum flow distribution and catalyst utilization, and is amenable to tourniquet packaging with uniformly high mounting pressure. Other contours, e.g. biradial oval, can also be tailored for preconverter application.

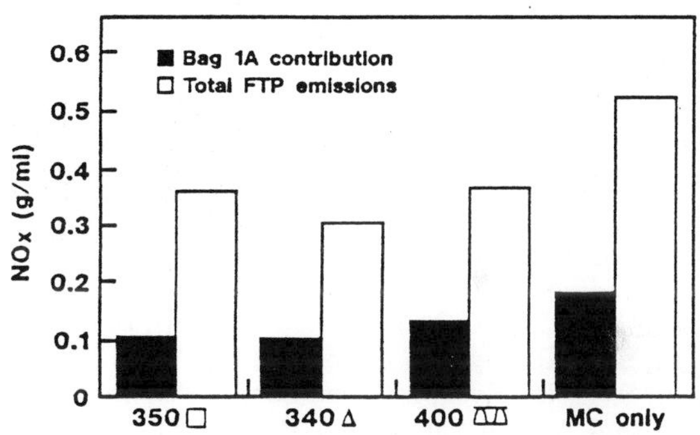

Figure 20. NO_x emissions for bag 1A and total FTP cycle with main converter alone (config. 1) vs. main converter plus three different preconverters (config. 3): all converters aged for 60 hours

The choice of special mat, Interam™ 100, with built-in high temperature edge seals ensures high mounting pressure and excellent erosion resistance required for durable packaging. In addition, its wide operating temperature range and superior insulation property relative to standard mats used in underbody application permit higher operating temperature while retaining bulk of the heat within the converter for efficient light-off operation and simultaneously limiting the shell temperature to acceptable values. The packaging design employing such a mat showed no degradation after 100 hours of high temperature vibration test and yielded uniformly high mounting pressure during axial push-out tests.

The back pressure measurements in the chassis dynamometer test have also shown that the EX-22 ceramic preconverter is a viable candidate for light-off application.

The light-off time and the heat-up rate as measured on a 4 ℓ, 6 cylinder engine, were identical for both ceramic and metallic preconverters. The ceramic preconverter was most effective in reducing cold-start emissions during the 40-120 sec. time interval of FTP cycle. The NMHC emissions, in particular, were 55% lower after 60 hours of aging and 49% lower after

120 hours of aging than those from metallic preconverter indicative of greater potential of washcoat degradation in metallic preconverter.

The FTP emissions test demonstrated that the 350 ❑ ceramic preconverter outperformed the 400 ⌂⌂ metallic preconverter of identical volume after both 60 and 120 hours of aging in exhaust configurations 2 and 3. The average shell temperature for ceramic preconverter is 200°C lower than that of the mantle for metallic preconverter indicating that the former retains more exothermic heat due to mat insulation and promotes early light-off of the main converter. The metallic preconverter radiates heat to surrounding environment and is less effective in transmitting exothermic heat to the main converter.

The durability, back pressure, light-off and aging data as measured in the tests described in the paper clearly demonstrate that a properly designed ceramic preconverter system is a viable and cost-effective approach to meeting TLEV and LEV emission standards.

ACKNOWLEDGEMENTS

The authors acknowledge the valuable assistance of Virginia Doud and Nancy Foster of Corning's Technology Division in the preparation of this paper. They are also grateful to Mark Sickels of Arvin Automotive, Phil Weber of Southwest Research Institute, and Paul Stroom of 3M Company for providing durability, back pressure and catalyst activity data.

REFERENCES

1. Howard, G., Automotive Visions, England; Oct. '93.
2. Maattanen, M. and Lylykangas, R.; SAE Paper No. 900505; February, 1990.
3. Gulati, S. T.; Reddy, K. P. and Thompson, D. F.; SAE Paper No. 902170; October, 1990.
4. Gulati, S. T.; Geisinger, K. L.; Reddy, K. P. and Thompson, D. F.; SAE Paper No. 910374; February 1991.
5. Gulati, S. T.; SAE Paper No. 850130; February, 1985.
6. Gulati, S. T.; CAPoC-2 Conference, Brussels, Belgium; September, 1990.
7. Gulati, S. T.; Ten Eyck, J. D. and Lebold, A. R.; SAE Paper No. 930161; March, 1993.
8. Gulati, S. T.; Ten Eyck, J. D. and Lebold, A. R.; SAE Paper No. 922252; October, 1992.
9. Lachman, I. M. and etal.; Ceram. Bull.; 1981.
10. Lachman, I. M. and Lewis, R. M.; "U.S. Patent 3885977;" May 27, 1975.
11. Day, J. P.; SAE Paper No. 902167; October, 1990.
12. Bagley, R. D.; "U.S. Patent 3790654;" February 5, 1974.
13. Gulati, S. T.; SAE Paper No. 881685; October, 1988.
14. Gulati, S. T.; Socha, L. S.; Then, P. M. and Stroom, P. D.; SAE Paper No. 940744; February 1994.
15. Gulati, S. T. and Reddy, K. P.; SAE Paper No. 930165; March, 1993.
16. Gulati, S. T. and Reddy, K. P.; SAE Paper No. 922333; October, 1992.
17. Reddy, K. P. and Gulati, S. T.; SAE Paper No. 932663; October, 1993.
18. Gulati, S. T. and Sweet, R. D.; SAE Paper No. 900268; February, 1990.
19. Gulati, S. T.; Summers, J. C.; Linden, D. G. and White, J. J.; SAE Paper No. 890796; February, 1989.
20. Gulati, S. T.; Cooper, B. J.; Hawker, P. N.; Douglas, J. M. and Winterborn, D.; SAE Paper No. 910372; February, 1991.
21. Gulati, S. T. and Schneider, G.; ENVICERAM '88, Cologne, W. Germany; December, 1988.
22. Weibull, W.; J. App. Mech., Vol. 18; 1951.
23. Gulati, S. T. and Helfinstine, J. D.; SAE Paper No. 852100; October, 1985.

24. Gulati, S. T.; SAE Paper No. 750171; February, 1975.
25. Gulati, S. T. and Merry, R. P.; SAE Paper No. 840074; February, 1984.
26. Stroom, R. D.; Merry, R. D. and Gulati, S. T.; SAE Paper No. 900500; February, 1990.
27. "Series 500 Shaker;" Umholtz-Dickie Corp., Wallingford, CT.
28. Socha, L. S. and Thompson, D. F.; SAE Paper No. 920093; February, 1992.
29. Socha, L. S.; Thompson, D. F. and Weber, P. A.; SAE Paper No. 930383; March, 1993.
30. Socha, L.S.; Thompson, D.F. and Weber, P.A.; SAE Paper No. 940468; February, 1994.

A. Frennet and J.-M. Bastin (Eds.)
Catalysis and Automotive Pollution Control III
Studies in Surface Science and Catalysis, Vol. 96

IMPROVEMENT OF THE THERMAL STABILITY OF CERIA SUPPORT.

M. Pijolat[a], M. Prin[a], M. Soustelle[a], O Touret[b], P. Nortier[b],

[a]*Département de Chimie-Physique des Processus Industriels, Ecole Nationale Supérieure des Mines, 158 cours Fauriel, 42023 Saint-Etienne Cedex (France)*

[b]*Synthèse Minérale, Centre de Recherche de Rhône-Poulenc, 52 rue de la Haie Coq, 93308 Aubervilliers Cedex (France)*

ABSTRACT

The experimental variations of the rate of crystallite growth in ceria at 943 K are reported as a function of the content (≤ 10 at %) in foreign cations or additives such as Mg^{2+}, Al^{3+}, Y^{3+}, and Si^{4+} ions. The process of surface area decrease due to crystallite growth is described by a kinetic model involving oxygen and cerium ions diffusion. From the calculation of the concentration in oxygen vacancies, electrons and cerium vacancies, the diffusion current can be obtained versus the oxygen partial pressure and the foreign cation concentration. The rate obtained from the model is compared to the experimental one.

1. INTRODUCTION

It is now well known that ceria promotes the performance of three-way catalytic converters used to remove carbon monoxide, nitrogen oxides and unburned hydrocarbons from automobile exhaust [1,2]. The high oxygen storage capacity of ceria improves the catalytic performances by acting on the composition of the gaseous mixture, storing oxygen under lean operating conditions and releasing it under rich conditions. Moreover, a great advantage of ceria as a catalytic support is that it stabilizes the dispersion of Pt particles [3]. In addition, the enhancement of the catalytic activity of a Rh supported on doped ceria has been recently demonstrated [4]. For these reasons the thermal stability of ceria itself has been investigated and several patents claim that the addition of cations like aluminum, zirconium, or silicium cations improves the stability of its surface area at high temperature [5]. However a comprehensive study of the behaviour of ceria at high temperature has not yet been achieved and the role of additives on its textural stability is not clearly understood.

We have done an experimental study and a modeling of the surface area decrease of ceria doped with several cations (Mg^{2+}, Y^{3+}, Al^{3+} and Si^{4+}) at 943 K. This temperature has been chosen in order to allow a kinetic study, but it is rather low for a catalytic application. To ascertain the results in practical conditions, we have verified the stabilizing effect of the most promising cations at higher temperatures such as 1073 and 1173 K.

2. Experimental Procedure

HSA ceria powder (135 m^2g^{-1}) was supplied by Rhône-Poulenc. The major impurities were La^{3+} ions (0.15 at %). In previous studies [6,7] we have shown that the surface area results from the contribution of a microporous surface and an external surface area. During thermal treatment at 943 K the surface area decreased, and after only two hours there was no longer any microporous surface. For this reason, we have chosen to follow the decrease of the external surface area by looking at the mean diameter of ceria crystallites. The amplitudes of the variations of the mean crystallite diameter of ceria are therefore a good indication of its thermal stability.

The mean crystallite diameter was obtained from the line broadening of the X-ray diffraction (XRD) pattern and using Warren's formula [7]. In the initial sample it was equal to 13.7 nm, then it increased up to 20 nm after five hours at 943K. This variation was found to be in agreement with observed decrease of the external surface area from 63 to 30 m^2g^{-1}.

Foreign cations were added to the powder according to the incipient wetness method from nitrate salts for Mg^{2+}, Y^{3+}, and Al^{3+}, and from tetraethyl ammonium silicate for Si^{4+}. A subsequent washing with water was achieved in order to eliminate nitrates from the powder surface. The residual concentration in nitrates was lower than 0.015 at %. The samples were then calcined at 723 K in the ambiant air for four hours. Final concentrations of the foreign cations added were fixed equal to 0.5, 1.0, 2.0 and 10.0 at %.

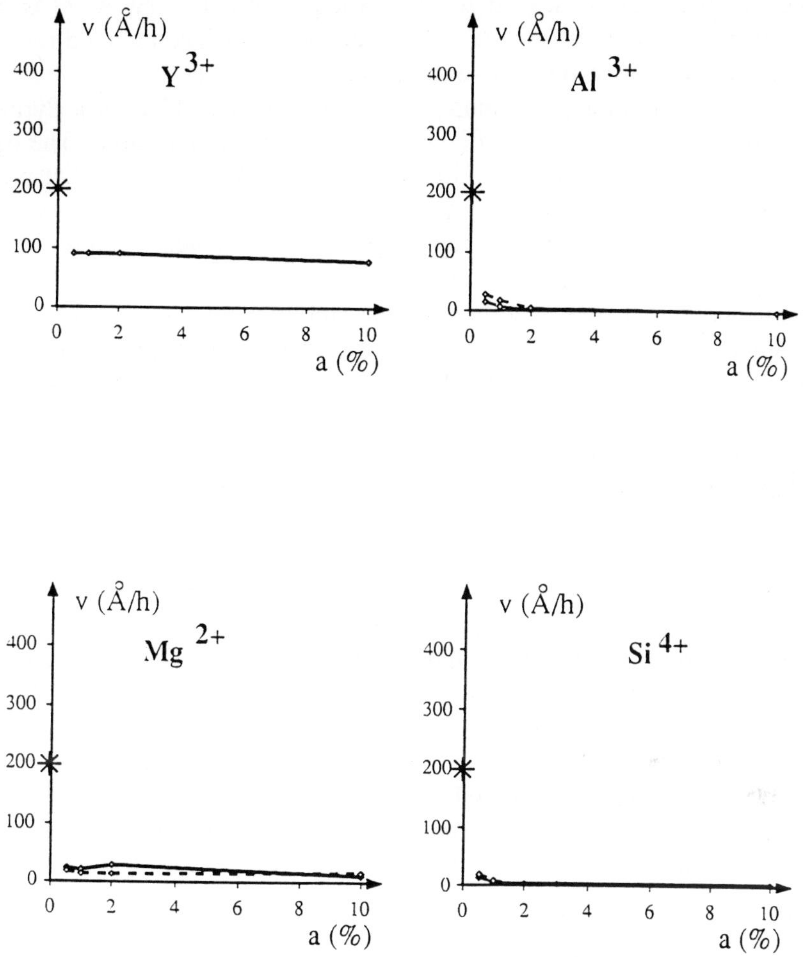

Figure 1: Experimental variations of the rate of crystallite size increase in doped ceria at 943 K, with P_{O_2} = 13.33 kPa (---) and 0.13 kPa (—), versus the additive concentration a (at %). The star corresponds to undoped ceria.

The samples were characterized by various methods (ESCA, ^{27}Al NMR, STEM, XRD) to verify the formation of a solid solution. The detailed results are not reported in this paper, but can be found in reference [7].

The doped (and undoped) samples were calcined at 943 K in a Pyrox B80 furnace under a flowing mixture of helium, oxygen and water vapor. The oxygen partial pressure varied from 0.2 to 30 kPa, and the water vapor partial pressure varied from 0.1 to 3.3 kPa. In order to avoid reproducibility errors, all the samples (undoped and doped with the same content) were calcined simultaneously.

The experimental curves of the mean crystallite diameter (D) versus time (t) of calcination were fitted by a mathematical function in order to calculate the experimental rate dD/dt. As previously explained [6,7], the variations of dD/dt, at a fixed value of D (14 nm in this study), were then obtained versus the oxygen partial pressure. No influence of water vapour was observed in the investigated range of pressure.

3. Experimental Results

Figure 1 shows the experimental variations of the rate of crystallite size increase in doped ceria versus the foreign cation concentration. The figure displays two sets of results: continuous lines corresponding to an oxygen partial pressure of 0.13 kPa and dotted lines corresponding to an oxygen partial pressure of 13.33 kPa. For comparison the rate measured without additive addition (about 200 $Å.h^{-1}$) has been reported in the figure. Figure 1 shows that the rate is strongly influenced by the addition of the foreign cation.

4. Modeling

The crystallite growth process occurs through an atomic transport at high temperature whenever two particles are in contact with each other, due to a concentration gradient which itself results from a curvature radius gradient: in the following "R>0" refers to surface areas of CeO_2 particles having a positive curvature radius, and "R<0" to those having a negative one. From an experimental study done on the influence of oxygen partial pressure on the rate of crystallite size increase [6,7], we could propose the following model using six elementary steps to describe the atomic migration of oxygen and cerium atoms :

1. Oxygen adsorption at "R > 0" surfaces :

$$\frac{1}{2}O_2 + V_{O_{R>0}}^{\cdot\cdot} + 2\,Ce_{Ce_{R>0}}^{'} = O_{O_{R>0}}^{x} + 2\,Ce_{Ce_{R>0}}^{x} \qquad (1)$$

2. Oxygen vacancy diffusion (and electron diffusion) :

$$V_{O_{R<0}}^{\cdot\cdot} \rightarrow V_{O_{R>0}}^{\cdot\cdot} \quad (O_{O_{R>0}}^{x} \rightarrow O_{O_{R<0}}^{x}) \text{ and } Ce_{Ce_{R<0}}^{'} \rightarrow Ce_{Ce_{R>0}}^{'} \qquad (2)$$

3. Vacancy creation at "R < 0" surfaces :

$$O_{O_{R<0}}^{x} + O_{O_{R<0}}^{x} = 2\,O_{O_{R<0}}^{x} + V_{Ce_{R<0}}^{''''} + 2\,V_{O_{R<0}}^{\cdot\cdot} \qquad (3)$$

4. Oxygen desorption at "R < 0" surfaces :

$$O_{O_{R<0}}^{x} + 2\,Ce_{Ce_{R<0}}^{x} = \frac{1}{2}O_2 + V_{O_{R<0}}^{\cdot\cdot} + 2\,Ce_{Ce_{R<0}}^{'} \qquad (4)$$

5. Cerium vacancy diffusion :

$$V_{Ce_{R<0}}^{''''} \rightarrow V_{Ce_{R>0}}^{''''} \quad (Ce_{Ce_{R>0}}^{x} \rightarrow Ce_{Ce_{R<0}}^{x}) \qquad (5)$$

6. Vacancy anihilation at "R > 0" surfaces :

$$V_{Ce_{R>0}}^{''''} + 2\,V_{O_{R>0}}^{\cdot\cdot} = 0 \qquad (6)$$

A complete explanation of this model is given in references [6,7]. It can be noticed that the intrinsic point defects involved in this model are cation vacancies, anion vacancies, and electrons which are trapped on cerium ions. These last two defects are known to be the predominating point defects in ceria [8].

To obtain the theoretical rate of the atomic transport, it is possible to assume that one of the six elementary steps is rate-determining. Then, the calculation needs the determination of the concentration of the intermediates, that can easily be achieved by using the Brouwer's approximation.

The insertion of foreign metallic cations in ceria gives rise to new point defects:

- substitution on cerium site: Mg'_{Ce}, Y'_{Ce}, Al'_{Ce}, and Si^{x}_{Ce},
- association between Mg'_{Ce} or Y'_{Ce} or Al'_{Ce} and oxygen vacancies. Such associations have clearly been shown by previous studies, done on ceria doped with trivalent cations [9,10]. Due to the neutral effective charge of the defects created by the insertion of tetravalent cations such as Si^{4+}, we will consider that the association with oxygen vacancies will not occur in this case,
- association between Mg'_{Ce} or Al'_{Ce}, or Si^{x}_{Ce} and electrons trapped on cerium ions. The association of a foreign cation with the electron bearing species Ce'_{Ce} has not yet been reported. The tendency to form such association may be related to the difference in electronegativity between the cation and cerium. In the case of yttrium there is no noticeable difference, and that kind of defect will not be considered.

All the extrinsic defects modify the concentration of the intrinsic ones compared to the undoped ceria and therefore they modify the rate of the process. In order to get a quantitative model, the concentrations of point defects in ceria must be theoretically expressed as function of the oxygen partial pressure, the amount of foreign cation and physical constants such as equilibrium constants and diffusion coefficients[7,11]. In the following, only two equilibrium constants will be considered:

- K_{AV}: equilibrium of formation of associated defects with oxygen vacancies,
- K_{Ae}: equilibrium of formation of associated defects with trapped on cerium ion electrons.

The rate calculation is based on the approximation of the rate-limiting step. Two kinds of elementary steps are involved in the model: diffusion steps ((2) and (5)) and reaction steps ((1), (3), (4) and (6)). Steps (1) and (3) are assumed to be at equilibrium. Calculations indicate that only diffusion steps have to be considered otherwise the rates would be constant, in contradiction with the experimental data. We have chosen to calculate the rate by taking into account the mixed diffusion of cations and anions.

In the discussion, we will assure that the adequation of the adjusted values of K_{AV} and K_{Ae} with these considerations gives self consistency to our process.

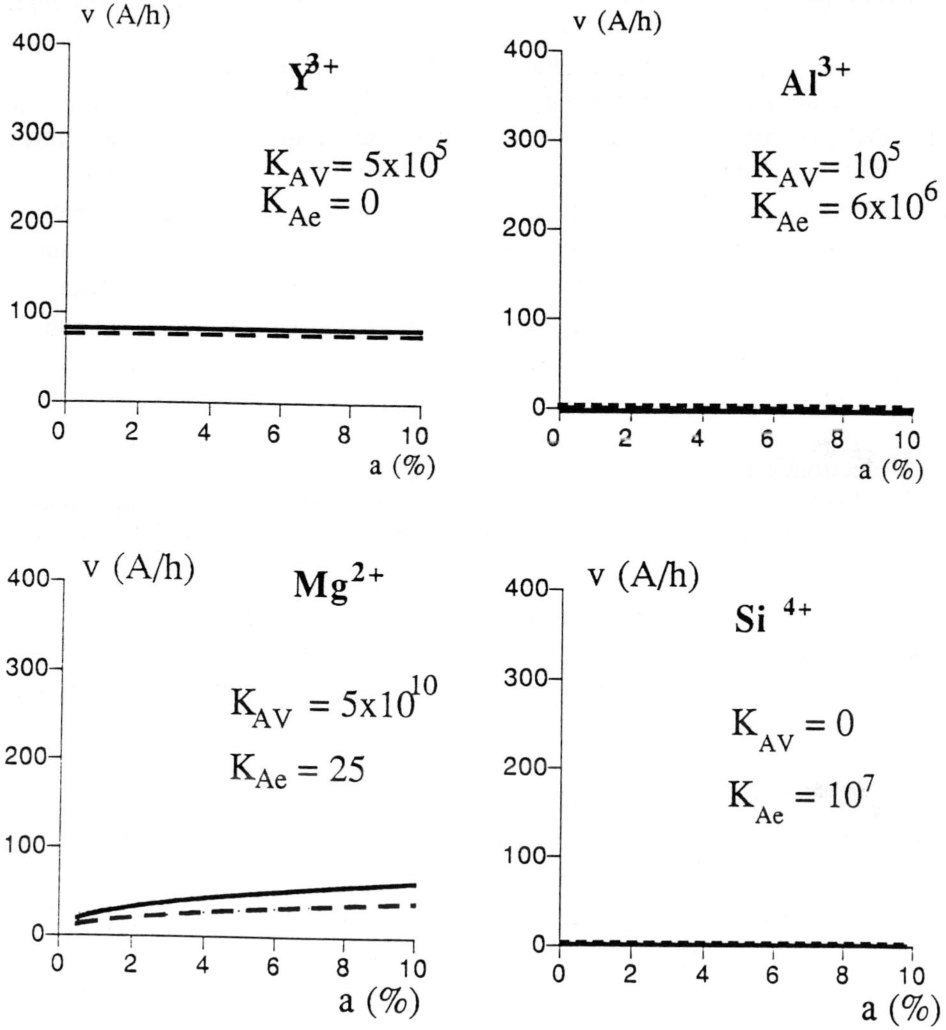

Figure 2: Theoretical variations of the rate of crystallite size increase in doped ceria at 943 K, with P_{O_2} = 13.33 kPa (---) and 0.13 kPa (—), versus the dopant concentration a (at %).

5. DISCUSSION

Theoretical rates of cristallite size increase versus additive content are displayed on figure 2. On each plot are reported the fitted values of the equilibrium constant K_{AV} and K_{Ae}. Figures 1 and 2 show the good agreement between experimental data and the model.

In the case of Mg^{2+}, the association constant K_{AV} must be rather high so that these cations are predominantly associated with the oxygen vacancies forming uncharged (Mg_{Ce},V_O).defects. For Mg^{2+} cations, the electronegativity is higher (1.3) than that of Ce^{4+} (1.12), so the rate has been calculated assuming that the concentration in oxygen vacancies is taken equal to the one of Mg^{2+} cations associated with electrons.

If we consider the trivalent cations, we may distinguish Y^{3+} and Al^{3+} due to their electronegativities which are respectively lower and higher than the one of Ce^{4+} ions (1.0 and 1.6, respectively). In addition, these two cations show a reasonnable tendency to form associations with oxygen vacancies. According to the low electronegativity of yttrium, the value of K_{Ae} is assumed to be zero. The best fit is obtained for Y^{3+} considered to be equally distributed in substituted and associated defects. In the case of Al^{3+}, the best fit is obtained for Al^{3+} associated for the same part with the oxygen vacancies and with the electrons.This explained the better stabilizing effect of Al^{3+} compared to Y^{3+}.

In the case of Si^{4+} ions, the electronegativity (1.9) is much higher than the one of Ce^{4+} ions (1.12), that indicates a strong tendency for association with electrons. As the substitution of cerium ions by Si^{4+} creates neutral defects, K_{AV} has been taken equal to zero. The rate has been calculated with the assumption that the concentration in oxygen vacancies is equal to the concentration of Si^{4+} associated with electrons.

Butler and col. [12] have computed the binding energies of the defects created by the association between some cations and oxygen vacancies. Our fitted values of K_{AV} are in good agreement with their results[11].

We have verified at higher temperature the validity of the model using the best surface area stabilizers shown in this study. The results are displayed in Table 1, where CeO_2(A) and CeO_2(B) are two different grades of ceria.

Table 1: Stabilization of ceria by Al^{3+} and Si^{4+} at 1073 and 1173 K.

Surface area	CeO_2 (A)	Al^{3+} doped CeO_2 (A)	CeO_2 (B)	Si^{4+} doped CeO_2
As is	135 m^2/g	138 m^2/g	200 m^2/g	185 m^2/g
1073 K (2 h)	15 m^2/g	36 m^2/g	50 m^2/g	75 m^2/g
1173 K (6 h)	9 m^2/g	17 m^2/g	29 m^2/g	56 m^2/g

6. CONCLUSION

According to this model, a foreign cation may exhibit :
- a strong tendency to form associates with oxygen vacancies (like Mg^{2+}), that can be evaluated from calculations of the binding energy of the associated defect [12],
- a strong tendency to form associates with Ce^{3+} ions (like Si^{4+}, Al^{3+}), that has been found to be related with the electronegativity,

in order to be strongly effective in improving the thermal stability of ceria.

REFERENCES

1 J. G. Nunan, H. J. Robota,M. J. Cohn and S. A. Bradley, J. Catal. 133 (1992) 309.

2 B. Harrison,A. F. Diwell and C. Halette, Platinium Metals Rev. 32 (1988) 73.

3 A. F. Diwell, R. R. Ragaram, H. A. Shaw and T. Truex, in "Catalysis and Automotive Pollution Control" (A. Crucq, ed.) [Elsevier, Amsterdam, 1991], p. 139.

4 B. K. Cho, J. Catal.131 (1991) 74.

5 G. N. Sauvion, J. Caillod and C. Gourlaouen, Patent of Rhöne-Poulenc, EP 0207857 ; T. Ohata, K. Tsuchitani and S. Kitaguchi, Patent of Nippon Shokubai Kagaku, JP-8890311; N. E. Ashley and J. S. Rieck, Patent of Grace W R and Co-Conn, US-484727.

6 M. Pijolat, M. Prin, M. Soustelle and P. Nortier, accepted in J. Chimie-Physique.

7 M. Prin, Doctorat Thesis, 1991, Saint-Etienne (France)

8 E. K. Chang and R. N. Blumenthal, J. Solid State Chem. 72 (1988) 330.

9 R. Gerhardt-Anderson and A. S. Nowick, Solid State Ionics 5 (1981) 547.

10 Wang Da Yu and A. S. Nowick, Solid State Ionics. 5 (1981) 551.

11 M. Pijolat, M. Prin, M. Soustelle, O. Touret.and P. Nortier, to be published in J. Chem. Soc. Farad. Trans.

12 V. Butler, C. R. A. Catlow, B. E. F. Fender and J. H. Harding, Solid State Ionics 8 (1983) 109.

A. Frennet and J.-M. Bastin (Eds.)
Catalysis and Automotive Pollution Control III
Studies in Surface Science and Catalysis, Vol. 96

RADIAL FLOW CONVERTER : NEW DEVELOPMENTS IN HIGH CELL DENSITY CATALYSTS.

F. Bonnefoy, F. Petitjean and P. Steenackers

Bosal Research
Dellestraat 20, 3560 Lummen, Belgium

ABSTRACT

The back-pressure, flow distribution and conversion efficiencies of two 1000 and 1600 Cpsi "Radial Flow Converters" (R.F.C) are is compared with those of a 400 Cpsi conventional axial flow converter. It is shown that the flow dynamical concept using radial flow through an original substrate layout, allows the use of high cell densities (up to 1600 Cpsi) without any back-pressure increase or even with important back-pressure reductions.

The high cell density substrates show a very homogeneous flow distribution over their total volume.

The design has shown the possibility of important savings on the PM usage without deterioration of the conversion efficiencies as measured during EEC test cycles.

1. INTRODUCTION

Automotive catalytic converters today generally make use of ceramic or metallic substrates which have cell densities of typically 400 cells/inch^2 (62 cells/cm^2).

Several authors [1], [2], have shown that, by increasing this cell density, higher specific conversion efficiencies can be obtained. This is mainly due to the fact that, in cells with smaller diameter, the mass transfer mechanism towards the catalytic active surface of the reactor is substantially improved. This results in an overall gain in catalytic reactivity which is so important that it becomes possible to reduce the precious metal contents of the reactor substantially without reducing converter efficiency rates.

Another advantage of higher cell densities is the possibility to reduce the total substrate volume. Smaller volumes will be helpful in many LEV and ULEV applications, in particular when converters need to be positioned closer to the engine compartment where space constraints often restrict volume.

However, until now, high cell density converters have had very limited applications. There are two main reasons for this : unacceptable back-pressure increase and technological limitations in the coating process.

A new concept is proposed here which enables to eliminate these two constraints. This approach uses a new type of flow dynamical layout and constructional principles for the converter substrate. It will be shown that such substrate layouts, using a radial flow arrangement, permit the use of very high cell densities without any increase of back-pressure. In fact, the concept often allows back-pressure decrease and it also assures a very even flow distribution over the total substrate volume.

2. Pressure Drop And Flow Distribution In A Converter Monolith

The gas flow in a "conventional" monolith is mainly laminar. The associated pressure drop can be calculated, as a first approximation, according to formula (1).

$$DP = C \,.\, N \,.L \,.\, Q \,/\, A \qquad (1)$$

Whereby : C : constant
N : cell density
L : length of substrate
Q : volume flow
A : monolith cross sectional area
ΔP : pressure drop over substrate

Equation (1) shows that an increase of cell density (N) will not cause pressure drop increase if it can be compensated by a reduction of the substrate length (L) and/or by an increase of the monolith cross sectional area (A).

Figures 1 and 2 show how a "Radial Flow" layout of the substrate will easily realize a substantial reduction of the substrate length L (a factor of 5 or more can be readily achieved) and, at the same time, will enable an increase of the substrate cross sectional area A (by a factor 2 or 3 e.g.).

This particular layout will assure an almost perfectly even flow distribution over the substrate.

Figure 1: New Radial Flow Catalytic Converter

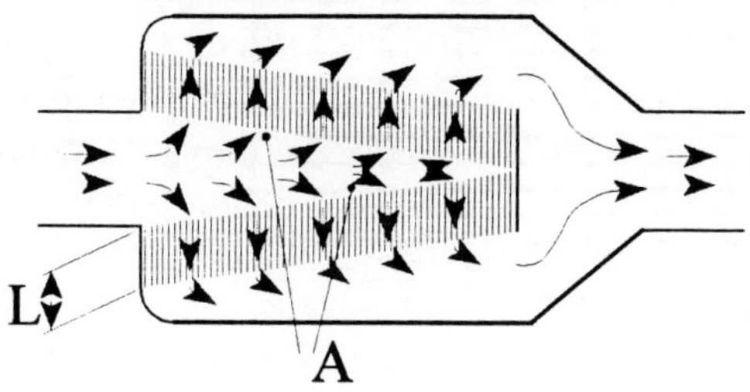

Figure 2: Conventional Catalytic Converter

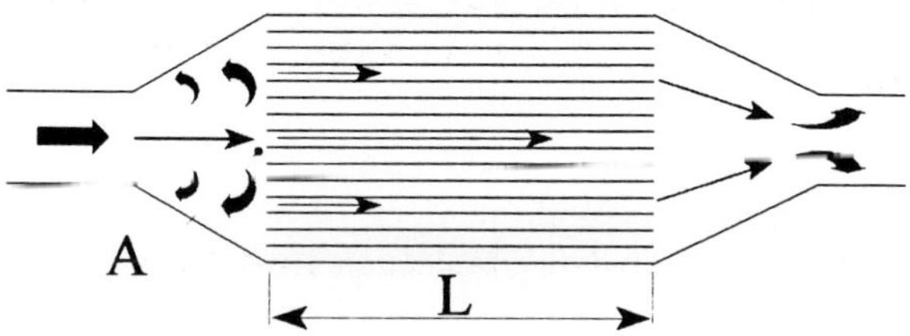

A: inlet cross sectional area
L: substrate length

Several authors, [3], [4], have shown that a non homogeneous flow distribution over the monolith cross sectional area can cause both a decline of conversion efficiency and a premature ageing of the catalyst wash coat due to localized poisoning and thermal degradation.

Figures 3 to 6 show flow dynamical computer simulations which allow us to compare the flow distributions in a 400 cells/inchs conventional ceramic monolith with the flow distribution in a "Radial Flow" 1600 cells/inch^2 metallic substrate. Also the pressure drop calculations are shown. It should be noted that in this example the substrate volumes are such that both substrates have the same total surface area (T.S.A.).

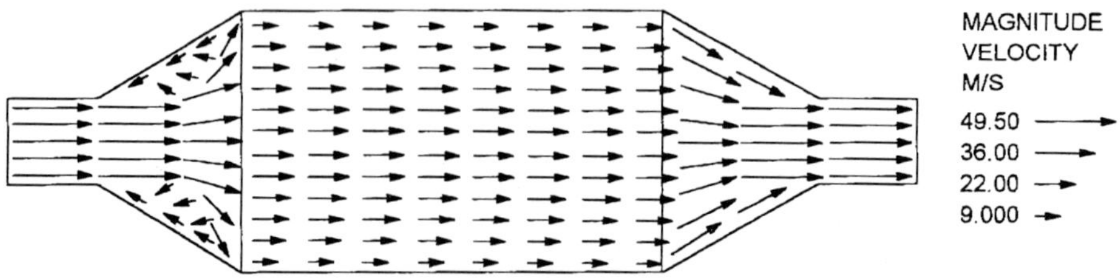

AXIAL FLOW CONVERTER
CELL DENSITY = 400 CPSI
SUBSTRATE VOLUME = 2.4 L

BOSAL RESEARCH

FIGURE 3

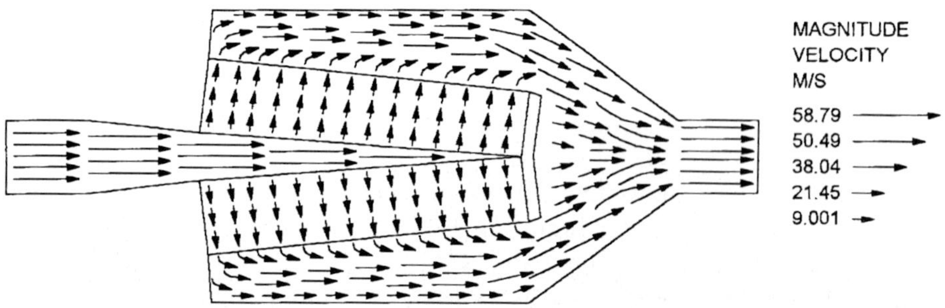

RADIAL FLOW CONVERTER
CELL DENSITY = 1600 CPSI
SUBSTRATE VOLUME = 0.9 L

PATENT APPLIED FOR IN VARIOUS COUNTRIES
BOSAL RESEARCH

FIGURE 4

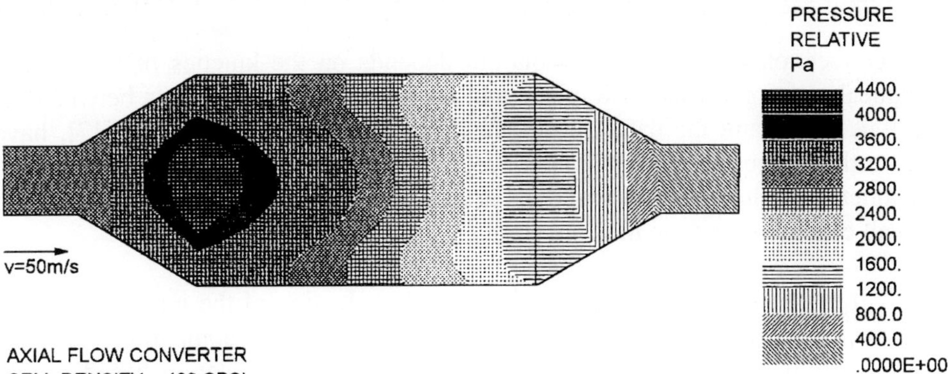

AXIAL FLOW CONVERTER
CELL DENSITY = 400 CPSI
SUBSTRATE VOLUME = 2.4 L

BOSAL RESEARCH

FIGURE 5

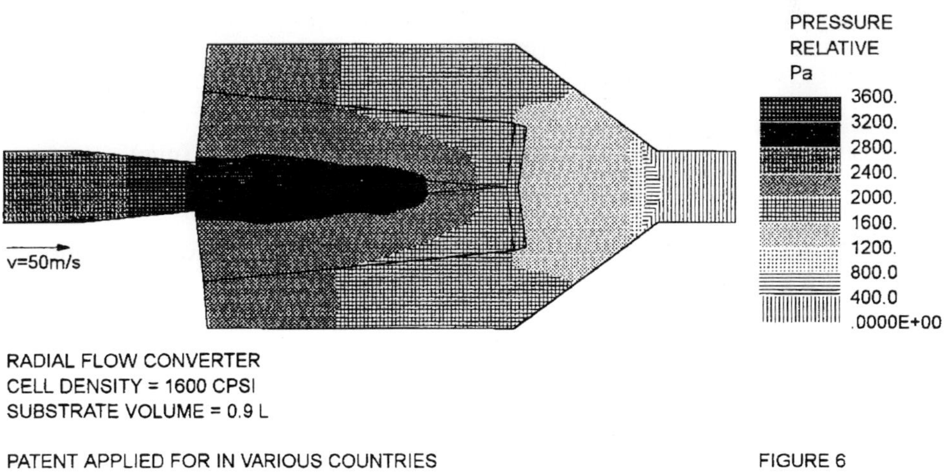

RADIAL FLOW CONVERTER
CELL DENSITY = 1600 CPSI
SUBSTRATE VOLUME = 0.9 L

PATENT APPLIED FOR IN VARIOUS COUNTRIES
BOSAL RESEARCH

FIGURE 6

3. MASS TRANSFER IMPROVEMENT IN HIGH CELL DENSITY SUBSTRATES

The conversion efficiency in a monolith depends on the kinetics of the various chemical reactions which take place and on the transfer of reactants between the gaseous phase and the catalyst surface. Several authors, [1], [5], [6], [7], have shown that, once the reactor temperature is well above the light-off temperature, the kinetics of the chemical reactions are such that the mass transfer controls the overall conversion rate in the converter.

It is possible to do a simple theoretical evaluation of the impact of higher cell densities on the mass transfer conditions. Following comparable analysis work by Lylykangas and al. [2], we hereby present a calculation based on the "film" model of Schweich and Leclerc [8].

The mass conservation equation reads :

$$v \cdot r \cdot Cp\,(dCe/dz) + (4.K_D/Dh) \cdot (Ce-Cs) = 0 \qquad (2)$$

with :

-	r	Gas density	(kg/m^3)
-	v	Gas speed	(m/s)
-	Dh	Hydraulic diameter	(m)
-	Cp	Specific heath	(j/Kg.K)
-	Ce	Concentration in the gas	(mol/m^3)
-	Cs	Concentration in the gas at the interface with the wash coat	(mol/m^3)
-	K_D	Mass transfer coefficient	(m/s)
-	z	Axial coordinate	(m)

The coefficient of the mass transfer K_D is obtained from the Sherwood number :

$$Sh = K_D.Dh/Dm \qquad (3)$$

HAWTHORN [9] proposes the following relation :

$$Sh = Sh_{lim}[1 + 0.095 \cdot Re \cdot Sc \cdot (Dh/z)]^{0.45} \qquad (4)$$

It is admitted [8] to put $Sh = Sh_{lim}$, if the following condition is respected :

$$z > 0.05\ Dh\ Re\ Sc \qquad (5)$$

This is the case for a standard monolith (400 Cpsi which operates maximally at Re=500) where the limiting Sherwood number is reached while z becomes longer than 18mm. If we suppose that this condition is also respected for the substrates with higher cell density, then the resolution of the equation (1) yields :

$$Ce(z) = C0 + (Cs-C0).(1 - e^{-z/l}) \quad (6)$$

$$l = (Dh/4) . (Re.Sc/Sh_{lim}) \quad (7)$$

whereby : C0 stands for the initial concentration in the gas phase. Ce(0) = C0

In the hypotheses of a mass transfer limitation of the conversion, the speed of the chemical reactions leads to a concentration of the reactants at the catalyst surface which is close to zero. If Cs equals zero, equation (6) becomes :

$$Ce(z) = C0.(e^{-z/l}) \quad (8)$$

This simple formula may give at least a qualitative estimation of the possible reduction in substrate volume at equivalent conversion efficiencies.

Table 1

Cell density (Cpsi)	Geometry of the duct	Hydraulic diameter (mm)	Theoretical volume reduction factor
400	Square	1	1
800	Sinusoidal	0.627	1.8
1200	Sinusoidal	0.478	3.2
1600	Sinusoidal	0.333	6.5

Table 1 was calculated from equation (8) using the following values for the relevant dimensionless coefficients :

Number of Schmidt = 0.7

Sherwood limiting number according to SHAH AND LONDON [10] :

$Sh_{lim} = 3.608$ square channel
$Sh_{lim} = 2.617$ sinusoidal channel

The results show that (within the limits of the theoretical assumptions) the increase of the cell densities will allow an important decrease of the substrate volumes. E.g. for equivalent inlet surfaces, the utilisation of a 1600 Cpsi substrate allows the monolith volume to be reduced by a factor as high as 6.5 (as compared with a standard 400 Cpsi substrate).

The following experimental results will show that indeed substantial reductions may be practically achieved.

4. Experimental Results

4.1 Emission test results

Figures 7 to 10 show the results of the emission tests executed according to the EEC/91/441 test cycle on a vehicle equiped with a 2.8 litre engine.

The figures compare the emission results obtained with 2 R.F.C. converters of respectively 1000 and 1600 cells/inch^2 and a substrate volume of 0.76 litres. The results are compared with those of a 2.4 litres conventional 400 cells/inch^2 converter. The total loading of the conventional converter was 3.4 g Pt/Rh (5/1) and the R.F.C.'s had a total loading of 1.7g Pt/Rh (5/1). The following table 2 resumes the data in terms of volume reduction factors and precious metal loading reduction factors. All converters had metal substrates and are in fresh conditions.

Table 2

	Cell density	Relative PM loading	Relative substrate volume
Cat A (conventional design)	400	100 %	100 %
Cat B (R.F.C.)	1000	50 %	31 %
Cat C (R.F.C.)	1600	50 %	31 %

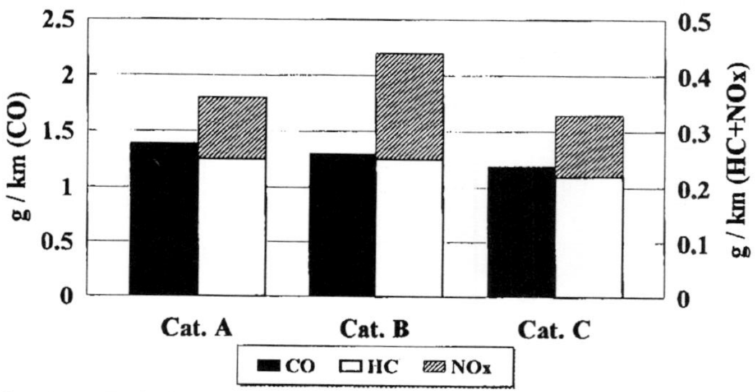

fig. 7 : *Emission test results UDC + EUDC*

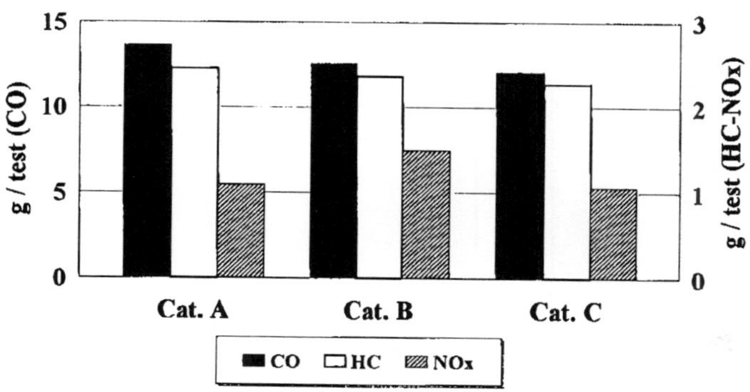

fig 8 : *Emission test results UDC phase 1*

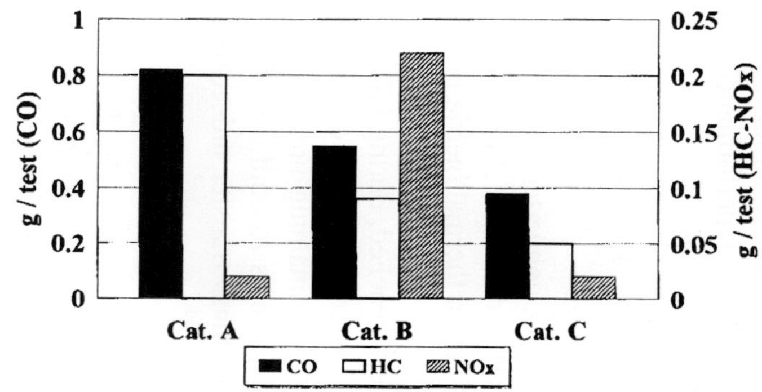

fig. 9 : *Emission test results UDC phases 2-3-4*

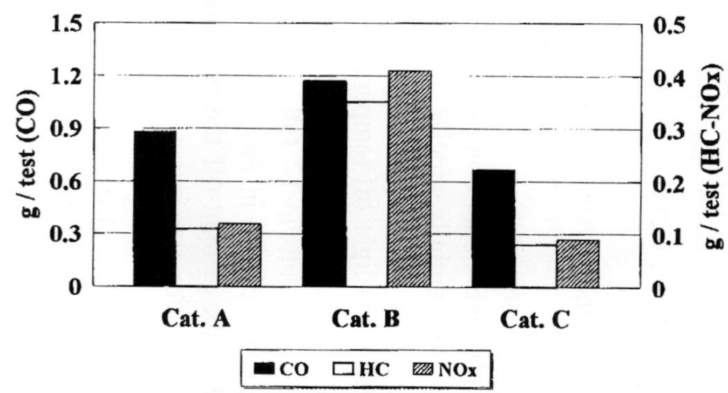

fig. 10: *Emission test results EUDC*

Figures 7 to 10 show detailed comparisons of the emissions during the EEC cycle. In particular, the comparison of emission performances during the various phases of the cycle are given separately.

The following interpretations can be made :

- Good performance is shown of both R.F.C. converters in comparison with the conventional converter at cold start conditions. This is mainly due to the substantially smaller substrate with good thermal inertia properties.

- After the light-off temperature has been reached, during UDC phase 2, 3 and 4 and during the EUDC phase, the increase of cell density from 1000 to 1600 cells/inch^2 (all other conditions remaining unchanged) shows a clear decrease in the emissions which seems to confirm the hypotheses that above light-off the phenomena of mass transfer indeed controls the conversion rate in a monolithic reactor.

- The substantial decrease in precious metal loading (by 50 %) is clearly counter- balanced by the improved mass transfer conditions. The excellent flow distribution over the substrate certainly helps further to assure optimal usage of all the P.M.'s present in the substrate.

4.2 Backpressure results

The backpressure of the total exhaust system equiped consecutively with converters A, B and C showed slightly better values for B and C.

Table 3 Backpressure of the total system (mbar)

RPM	Converter A	Converter B	Converter C
4500	250	225	235
5000	280	255	265
5800	345	340	345

4.3 Ageing

The standard axial flow converter A and the RFC B were submitted to an accelerated ageing of 50 hours at 850C (inlet gas temperature).

The following table shows the deterioration factors for the emissions of HC, CO and NO_x caused by the accelerated ageing :

Emissions (50 hours) = deterioration factor x Emissions (fresh catalyst)

Table 4

	Converter A	Converter B
Deterioration factor HC	1.55	1.32
Deterioration factor CO	1.66	1.57
Deterioration factor NO_x	2.22	1.79

The results in table 4 show that the two catalysts undergo a comparable deterioration of their performances. The decreased precious metal content on catalyst B does not result in premature activity loss. This is probably due to the very even flow distribution in the radial flow converter avoiding high temperature concentrations. More ageing tests are actually undertaken to confirm this further.

5. CONCLUSION

This study shows that the use of high cell density substrates, together with a specific substrate shape creating a radial flow through the substrate, allows a reduction of back-pressure while opening the possibility of a substantial reduction of PM usage without loss of conversion efficiency. Further work is ongoing to assure the long term durability properties of the concept.

REFERENCES

1 R.H. Heck, J. Wei and J.R. Katzer, Aiche journal (vol. 22, No. 3), May 1976.
2 M. Luoma, P. Lappi and R. Lylykangas, SAE paper nr. 930940 (1993).
3 J.S. Howitt and T.C. Sekella, SAE paper nr. 740244 (1974).
4 C.D. Lemme and W.R. Givens, SAE paper nr. 740243 (1974).
5 L.C. Young and B.A. Finlayson, Aiche journal (vol. 22, No. 2), March 1976.
6 L.L. Hegedus, Am. Chem. Soc. Dir. Petroleum. Chem., Prep. Symp. Catalytic Approaches to Environm. Control, Chicago Meeting, August 26-31, 1973, vol. 18, pp. 487-503.
7 M.I. Ryan, E.R. Becker and K. Zygourakis, SAE paper nr. 910610 (1991).
8 D. Schweich and J.P. Leclerc, Catalysis and Automotive Pollution Control II, pp. 437-463, A. Crucq (Editor), 1991 Elsevier Science Publishers B.V., Amsterdam.
9 R.D. Hawthorn, Aiche Symp. Ser., 70 (137), 428-438 (1974).
10 R.K. Shah and A.L. London, Laminar flow forced convection in ducts, Academic Press, New York, 1978.

A. Frennet and J.-M. Bastin (Eds.)
Catalysis and Automotive Pollution Control III
Studies in Surface Science and Catalysis, Vol. 96
1995 Elsevier Science B.V.

STABILIZATION OF RHODIUM/ALUMINA CATALYSTS BY SILICON EXCHANGE

R. W. McCabe, R. K. Usmen, G. W. Graham, W. L. H. Watkins and W. G. Rothschild

Ford Research Laboratory
Ford Motor Company
Dearborn, Michigan 48121

ABSTRACT

Modification of alumina by exchange of surface hydroxyl groups with tetraethyl orthosilicate was investigated as a method for preventing interactions between rhodium and alumina in high-temperature (1173 K) oxidizing environments. Temperature-programmed reduction experiments, together with catalytic reactivity measurements and BET surface area measurements were used to characterize both the silicon exchanged rhodium/alumina catalyst and reference catalysts comprised of rhodium/alumina (without silicon) and rhodium/zirconia admixed with blank alumina. Relative to the reference catalysts, the silicon exchanged catalyst showed less interaction between the Rh and alumina and better retention of BET surface area after high temperature steam-oxygen aging. The data suggest that dehydroxylation of alumina at elevated temperatures is an important factor in the interaction between rhodium and alumina. Exchange of surface hydroxyl groups with tetraethyl orthosilicate appears to stabilize those surface sites where rhodium and alumina would otherwise interact. On the other hand, silicon exchange results in a weaker interaction between Rh and the support and growth of larger Rh particles than obtained on the unmodified alumina.

1. INTRODUCTION

Rhodium interacts with γ-Al_2O_3 at temperatures above 873 K [1,2], sharply decreasing the amount of rhodium oxide that can be reduced at temperatures below 1073 K and lowering the activity for automotive exhaust reactions. A recent study from this laboratory [3] examined the effectiveness of La^{+3} for preventing interactions between Rh and γ-Al_2O_3. The La^{+3} largely blocks the interactions between Rh and γ-Al_2O_3, but does so by interacting with the Rh instead, apparently forming a mixed oxide with Rh under oxidizing conditions.The present study examines the effectiveness of a tetravalent cation, Si^{+4}, for preventing interactions between Rh and γ-Al_2O_3 in high temperature oxidizing environments. Recent studies have shown that exchanging surface hydroxyl species in γ-Al_2O_3 withsilicon compounds increases the stability of the γ-Al_2O_3 toward sintering [4], and inhibitsthe conversion of Rhx0 to Rh^{+1} [5,6]. The latter process involves the breakup of small metallic Rh particles into isolated Rh^{+1} species at isolated hydroxyl sites on the alumina surface [7]. Yates and coworkers [5,6] have shown that exchanging these hydroxyl sites with trimethylchlorosilane prevents the formation of Rh^{+1}. Our study was predicated on the assumption that the same hydroxyl sites may play a role in the high temperature oxidizing interactions between Rh and γ-Al_2O_3. Thus we have prepared silicon-exchanged Rh/γ-Al_2O_3 catalysts and examined their stability in both high temperature O_2 and H_2O-O_2 environments. The silicon-exchanged catalysts are compared to equivalently aged Rh/Al_2O_3 and Rh/ZrO_2+Al_2O_3 (physical mixture) reference catalysts,chosen as representative interacting (Al_2O_3) and non-interacting (ZrO_2) supports.

2. EXPRIMENTAL

2.1 Catalyst Preparation.

The silicon-exchanged catalysts were prepared by first exchanging tetraethyl orthosilicate, $Si(OC_2H_5)_4$ (Aldrich) with γ-Al_2O_3 (Degussa, aluminum oxide C) following the procedure of Beguin et al. [4]. This included calcination of the Si-exchanged catalysts in flowing oxygen at 773 K for 12 h as a final step. $Rh(NO_3)_3$.2H-2O was then added from aqueous solution by the incipient wetness technique, followed by drying at 373 K and calcination at 823 K for 4 h. Two concentrations were prepared, one containing 1.18 wt% Si and the other 1.75 wt% Si. Si concentrations were measured by energy-dispersive x-ray fluorescence spectroscopy and confirmed by electron microprobe analysis. Most of the data reported here were obtained with a 0.6wt%Rh/1.75wt%Si/γ-Al_2O_3 catalyst (designated Rh/Si/Al_2O_3). IR spectra obtained on 0, 1.18, and 1.75 wt% Si exchanged catalysts showed a progressive increase in the ratio of the 3730-3745 cm^{-1} hydroxyl stretching mode (associated with silica) to the 3685 cm^{-1}

mode (associated with the alumina), in keeping with the observations of Beguin et al.[4].

Two reference catalysts were prepared in addition to the silicon-exchanged catalyst: 0.6wt%Rh/γ-Al_2O_3 and a physical mixture of 1.46wt%Rh/ZrO_2 (41 wt%) and blank alumina (59 wt%) (both prepared by incipient wetness impregnation of $Rh(NO_3)_3 \cdot 2H_2O$ precursor from aqueous solution followed by drying at 373 K and calcination for 4 h at 823 K). The physical mixture was prepared with the same effective Rh loading (0.6 wt%) as for the other catalysts. Unless otherwise noted, the 0.6%Rh/1.75%Si/γ-$Al12_2O_3$ catalyst is referred to as Rh/Si/Al_2O_3, 0.6%Rh/γ-Al_2O_3 as Rh/Al_2O_3, and the 1.46%Rh/ZrO_2 (41 wt%)+ γ-Al_2O_3 (51 wt%) as Rh/ZrO_2+Al_2O_3.

2.2 Catalyst Aging.

All catalyst samples as prepared and calcined at 823 K are labeled "fresh". Some of the samples were oxidized in air at 1173 K for one hour and are referred to as "high-temperature oxidized". Others were oven-aged for 24 h or more at 1173 K in a laboratory retort oven under an atmosphere of steam and O_2. These samples are called "steam-O_2" aged catalysts for simplicity. In some instances, catalysts were "restored" after high temperature aging by reducing in flowing H_2 for 1 h at 1173 K and reoxidizing at 823 K prior to temperature-programmed reduction or activity measurements.

2.3 Temperature-programmed Reduction (TPR).

TPR experiments were conducted with an Altimira temperature-programmed system following procedures described previously [8]. Catalyst powder samples (100 to 300 mg) were heated from room temperature to 1073 K at a rate of 15o/min in a flowing stream (40 cc/min) of 8% H_2 in Ar. Prior to reduction, samples were given a standard in situ preoxidation at 773 K for 1 h. TPR traces reported here are baseline corrected (using a smooth curve subtraction procedure) and normalized to an equivalent sample size.

2.4 Catalytic Activity Measurements.

Catalytic activity was evaluated in an integral quartz flow reactor in steady-state experiments carried out at 823 K and a gas hourly space velocity (GHSV) of 60,000 (calculated at standard conditions of temperature (293K) and pressure (1 atm)). The feed was a simulated automotive exhaust stream containing 1.5% CO, 0.5% H_2, 1500 ppm HC (comprised of 1000 ppm propylene and 500 ppm propane), 1000 ppm NO, 0.6-1.4% O_2, 20 ppm SO_2, and N_2 balance. Conversions of NO, HC, and CO were measured as a function of the molar ratio of the reducing to oxidizing species, R, varied by changing the amount of O_2 in the feed while holding the other concentrations constant (R<1 = net oxidizing feed; R=1 = stoichiometric feed; R>1 = net reducing feed).

3. RESULTS

3.1 TPR Data

TPR traces for the Rh/Al_2O_3, $Rh/Si/Al_2O_3$, and $Rh/ZrO_2+Al_2O_3$ catalysts after various treatments are summarized in Figures 1-3, respectively. In each figure, the TPR trace for the fresh catalyst is shown at the top of the left-hand panel. Immediately underneath are the TPR traces for the high-temperature oxidized catalyst (left panel) and the 1173 K/24h steam-O_2 aged catalyst (right panel). The bottom traces in each figure are for the respective restored samples. In addition to the TPR traces, Table 1 summarizes the H_2 consumption for each sample obtained by integrating the area under the TPR traces and compares it to the theoretical H_2 consumption.

The fresh Rh/Al_2O_3 catalyst (Fig. 1) gave a single reduction peak near 370 K which accounted for reduction of all of the Rh in the sample from Rh_2O_3 to Rh metal (Table 1). After both high temperature oxidation and steam-O_2 aging, the low temperature reduction peak was replaced by broad, poorly defined reduction features extending from 380 K to about 550-600 K. A high temperature peak near 600 K was also observed in the steam-O_2 aged Rh/Al_2O_3 catalyst. Similar peaks were observed in other high-temperature treated catalysts (1173 K) in Figs. 1-3 and are attributed to the alumina support rather than the Rh (accordingly, their areas are not included in calculating the integrated H_2 uptakes in Table 1). The high temperature treated samples retained only about 10-25% of the Rh in a readily reducible form. High temperature reduction restored reducibility to all of the Rh in the case of the oxidized sample, but only to about two-thirds of the Rh in the case of the steam-O_2 aged sample.

TPR data for the $Rh/SiO_2/Al_2O_3$ catalyst are shown in Figure 2. In contrast to the fresh Rh/Al_2O_3 sample, less of the Rh was reduced in the low-temperature peak near 350 K, and more was reduced in a long tail that extended out to about 600K Only

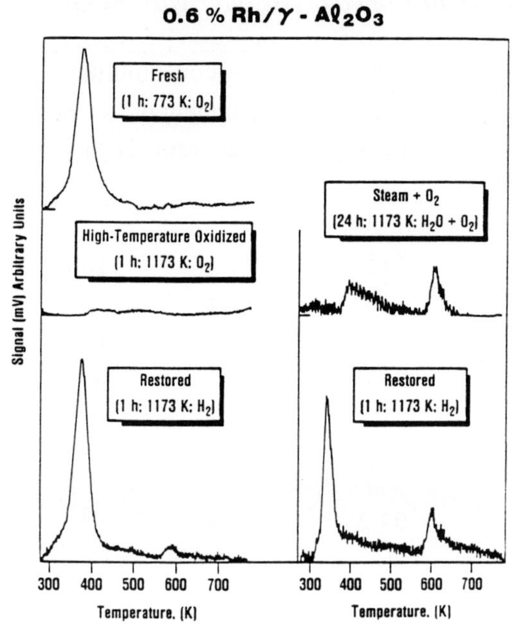

Figure 1. TPR spectra of Rh/Al_2O_3 catalysts after various pretreatments.

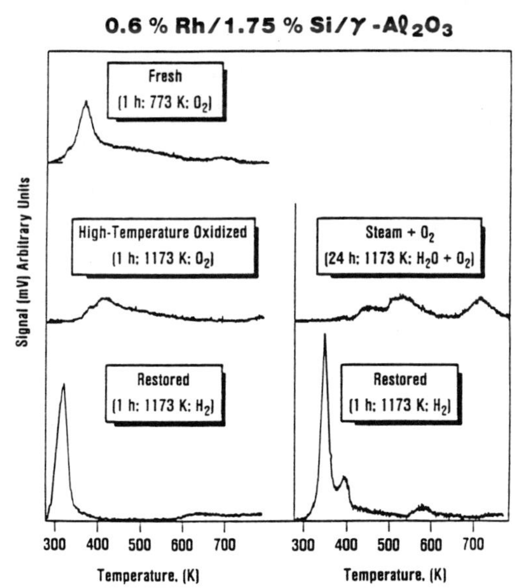

Figure 2. TPR spectra of $Rh/Si/Al_2O_3$ catalysts after various pretreatments.

Table 1. Characterization of catalysts after various aging treatments.

Sample	BET $m2/g$	H_2 consumption calculated (μmol H_2/g)	% of theoretical1 %	Maximum NO conversion
Rh/Al_2O_3				
Fresh	115	94.6	108	99
HTO_2		8.3	9	81
restored		92.2	105	99.5
O_2+H_2O	64.1	23.7	27	96
restored		58.5	67	
$Rh/Si/Al_2O_3$				
Fresh	93.2	67.6	77	98
HTO		39.0	44	90
restored		67.6	77	99
O_2+H_2O	86.4	42.0	48	97
restored		76.8	87	
$Rh/ZrO_2+Al_2O_3$				
Fresh	5.53	90.7	103	99.5
HTO		64.9	74	98
restored		92.6	105	99.5
O_2+H_2O	3.13	41.1	47	96
restored		50.8	58	

1The theoretical H_2 consumption is 87.8 μmol H_2/g based on $Rh_2O_3+3H_2=2Rh+3H_2O$.
2High temperature oxidized.
3Values are for 1.46%Rh/ZrO_2 only (i.e. without admixed Al_2O_3).

about 77% of the Rh underwent the equivalent of a three electron reduction, as evident from the H_2 consumption reported in Table 1. High temperature oxidation and steam-O_2 aging both resulted in loss of reducible Rh, but not as much as observed for the Rh/Al_2O_3 catalyst. Both the high temperature oxidized and steam-O_2 aged samples retained ca. 45% of the Rh in a readily reducible form whereas the corresponding aged samples without Si retained only 9 and 27%, respectively. Restoration of the Rh/Si/Al_2O_3 catalysts brought the amount of reducible Rh back to the fresh catalyst level in the case of the high-temperature oxidized sample, and to even higher levels in the steam-O_2 aged sample. Much of the high-temperature trailing edge in the reduction profile of the fresh sample was eliminated after restoring the high-temperature oxidized sample. The same was true of the steam-O_2 aged sample, although a new shoulder peak developed near 400 K that was not present in either the fresh catalyst or the restored high-temperature oxidized sample.

Figure 3 summarizes the TPR data for the physical mixture of Rh/ZrO_2+Al_2O_3. The fresh catalyst TPR trace was similar to that observed for the fresh Rh/Al_2O_3 catalyst except for a more pronounced shoulder near 400 K. High-temperature oxidation and steam-O_2 aging shifted the low temperature peak (at 350 K in the fresh sample) up to 400-450 K; not as much high temperature tailing beyond 600

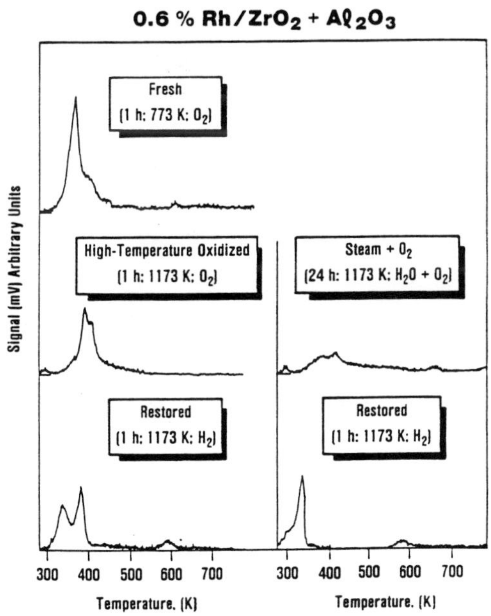

Figure 3. TPR spectra of Rh/ZrO_2+Al_2O_3 catalysts after various pretreatments.

K was observed as compared to either the Rh/Al_2O_3 or $Rh/Si/Al_2O_3$ catalysts. After restoration, both the high-temperature oxidized and steam-O_2 aged catalysts displayed a low temperature reduction peak near 330 K, while the high-temperature oxidized catalyst displayed an additional peak near 375 K.
The Rh/ZrO_2+Al_2O_3 catalyst retained a greater fraction of the Rh in a reducible form after high temperature oxidation than did the Rh/Al_2O_3 and $Rh/Si/Al_2O_3$ catalysts (see Table1). However, after steam-O_2 aging, the Zr- and Si-containing catalysts both yielded the same amount of reducible Rh, although most of the reduction occurred at higher temperatures in the Si- containing catalyst than in the Zr-containing catalyst. The high temperature oxidized Rh/ZrO_2+Al_2O_3 catalyst was completely restored after 1173 K reduction, but the corresponding steam-O_2 aged catalyst was restored to only the 58% level.

3.2 Catalytic Activity Measurements

Figure 4 shows steady-state activity measurements for the fresh Rh/Al_2O_3 (dashed curves/open symbols) and $Rh/Si/Al_2O_3$ (solid curves/solid symbols)

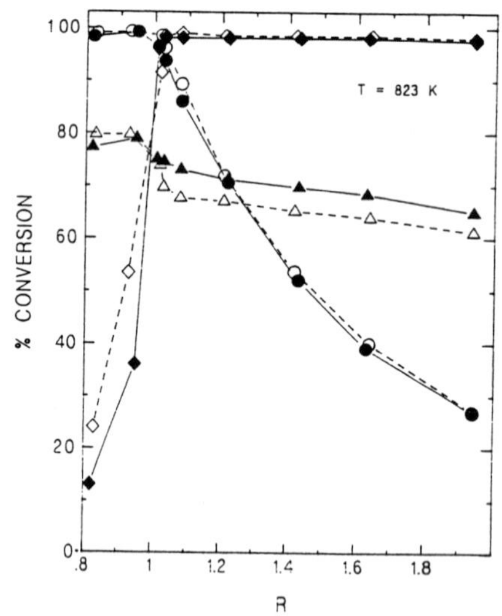

Figure 4. Activity comparison of fresh Rh/Al_2O_3 (dashed lines) and $Rh/Si/Al_2O_3$ (solid lines) catalysts at 823 K ((◊) NO, (o) CO, and (Δ) HC).

catalysts at 823 K. The CO conversion profiles were virtually identical for the two catalysts. The silica-containing catalyst showed 4-5 percentage points higher HC conversion under reducing conditions (R>1), but lower NO conversion under oxidizing conditions (R<1). Similar results were obtained with the zirconia-containing catalyst; the only significant difference between the Zr-containing catalyst and the Rh/Al_2O_3 catalyst was that the Zr-containing catalyst gave 5-10 points higher HC conversion at R>1 than observed for the Rh/Al_2O_3 catalyst.

Steady-state conversion measurements of the type shown in Figure 4 were obtained for all three catalysts after high-temperature oxidation, after the subsequent 1173 K H_2 restoration treatment, and after the extended steam-O_2 aging. Instead of presenting the full plots, however, we confine our attention to the NO conversion data here, since the primary role of Rh in the automotive

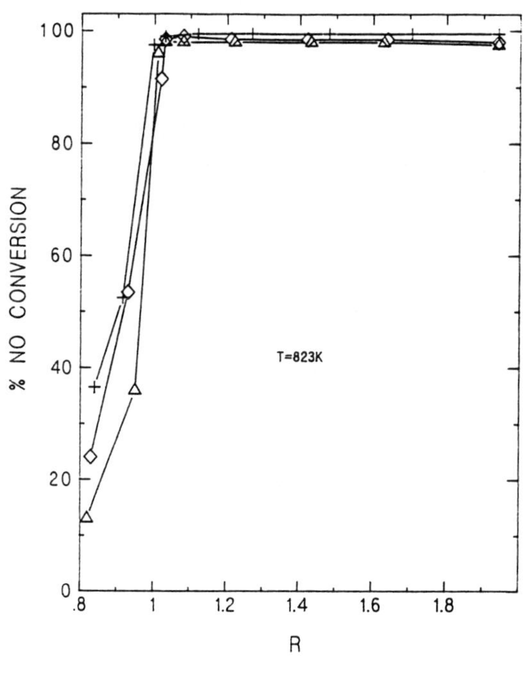

Figure 5a. NO conversion profiles of fresh Rh/Al_2O_3 (◊), $Rh/Si/Al_2O_3$ (Δ), and Rh/ZrO_2+Al_2O_3 (+) catalysts at 823 K.

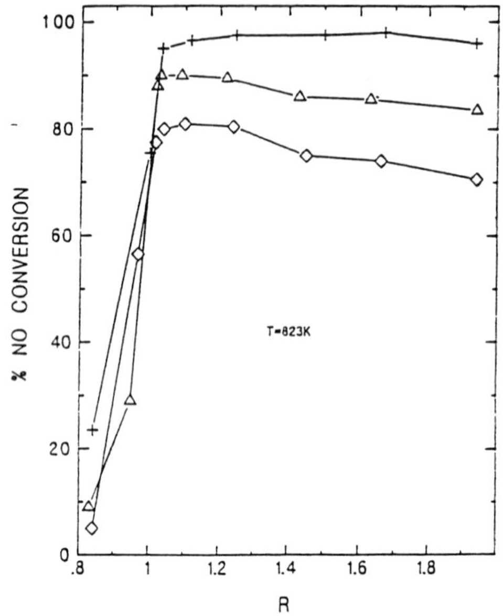

Figure 5b. NO conversion profiles for high temperature oxidized Rh/Al_2O_3 (◊),Rh/Si/Al_2O_3 (Δ), and Rh/ZrO_2+Al_2O_3 (+) samples at 823 K.

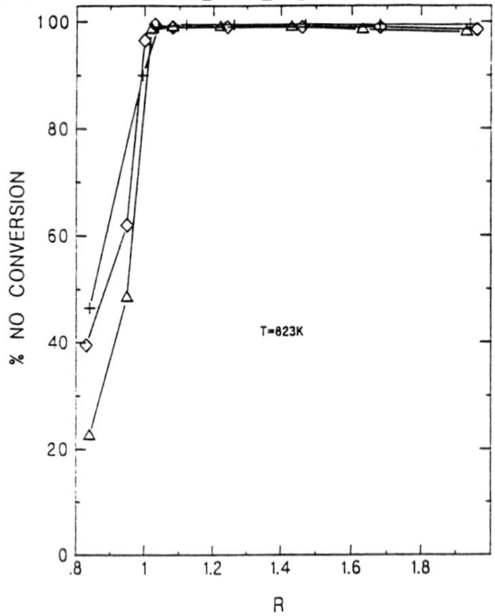

Figure 5c. NO conversion profiles for restored Rh/Al_2O_3 (◊),Rh/Si/Al_2O_3 (Δ), and Rh/ZrO_2+Al_2O_3 (+) samples at 823 K.

catalyst under warmed up conditions is to control NOx emissions. NO reduction activity was also affected more strongly than either HC or CO oxidation activity under the various treatments afforded these catalysts. Figure 5 compares the NO conversions of the three catalysts in the fresh state (Fig. 5a), after 1173 K oxidation (Fig. 5b), after restoration (Fig. 5c), and after extended steam-O_2 aging at 1173 K (Fig. 5d). Whereas the fresh catalysts all displayed rich-side NO conversions above 97% (Fig. 5a), much lower conversions were observed after high temperature oxidation (Fig. 5b). Specifically, the peak NO conversion for the Rh/Al_2O_3 catalyst was 81% while that of the Rh/Si/Al_2O_3 catalyst was 90%. The Rh/ZrO_2+Al_2O_3 catalyst was least affected, retaining a peak NO conversion of 98%. The peak NO conversions are summarized in Table 1.High temperature H_2 reduction restored all of the catalysts to their fresh NO conversion levels (compare Figs. 5a and 5c). Subsequent aging in steam-O_2 once again decreased NO conversions, but not nearly to the extent observed after high temperature oxidation (at least for the alumina-based catalysts). As evident from Figure 5d,

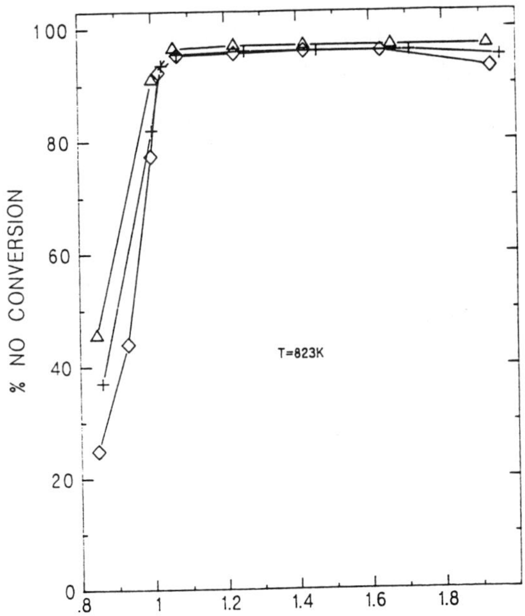

Figure 5d. NO conversion profiles for steam-O_2 aged Rh/Al_2O_3 (◊),Rh/Si/Al_2O_3 (Δ), and Rh/ZrO_2+Al_2O_3 (+) samples at 823 K.

NO conversion over the steam-O_2 aged silicon-exchanged catalyst was equivalent to or greater than that of either the equivalently treated Rh/Al_2O_3 or Rh/ZrO_2+Al_2O_3 catalysts over the full range of R examined. After steam-O_2 aging (Fig. 5d), the silicon-exchanged catalyst was the most effective of the three catalysts for converting NO under lean conditions (R<1). In contrast, it was the least effective of the three catalysts for lean NO conversion in the fresh, high-temperature oxidized, and restored forms.

4. DISCUSSION

Comparing the fresh Rh/Si/Al_2O_3 catalyst to the fresh Rh/Al_2O_3 and Rh/ZrO_2+Al_2O_3 reference catalysts reveals that the silicon exchanged catalyst has only about three-fourths the H_2 uptake of the reference catalysts and displays a long high temperature tail in the TPR trace that is not present in the reference catalysts. These differences suggest that silicon blocks or inhibits access to some of the Rh on the alumina catalyst, either by pore blocking or direct masking of Rh particles, although these effects are not apparent in the activity data.

Despite some loss of available Rh as prepared, the silicon-exchanged catalyst showed much higher retention of reducible Rh (44%) than the Rh/Al_2O_3 reference catalyst (9%) after high temperature oxidation. Thus the added silicon appeared to block the high temperature interaction of the Rh with the alumina support to a large extent, as also evident from the activity data where NO conversion, in particular, was significantly greater for the silicon exchanged catalyst (90% max) than for the Rh/Al_2O_3 reference catalyst (81% max). The Rh supported on zirconia showed both the best retention of reducible Rh (74%) and NO conversion (98% max), as expected for a non-interactive support. The return of all three catalysts to their fresh NO conversion levels and their fresh levels of total H_2 uptake in TPR after restoration indicated that no irreversible changes in the amount of accessible Rh were effected by the high temperature oxidation. Some differences were noted in the TPR traces and in the HC conversions of the restored catalysts compared to the fresh catalysts, and these differences appear to result from changes in particle size and/or morphology. The Rh/ZrO_2+Al_2O_3 catalyst, in particular, gave a set of TPR traces (fresh, high temperature oxidized, and restored) similar to those reported previously for Rh/SiO_2 catalysts [9]. In the Rh/SiO_2 study, the double peak in the TPR trace, such as that observed for the high temperature oxidized Rh/ZrO_2+Al_2O_3 catalyst following restoration, was attributed to a mixture of small, poorly crystallized Rh particles (i.e. high defect density), and large, highly crystalline Rh particles (i.e. low defect density). The long term steam-O_2 aging affected the catalysts much differently than the short term high temperature oxidation. The Rh/Al_2O_3 reference catalyst was less affected by the long term steam-O_2 aging than by the short term oxidation, as evidenced by its greater retention of reducible Rh (27% vs 9%) and higher peak NO conversion (96% vs 81%). These results suggest that dehydroxylation of the alumina may be a critical step in the high temperature interaction between Rh and Al_2O_3 under oxidizing conditions. The presence of steam in the long term aging would be expected to maintain a higher equilibrium concentration of hydroxyl groups on the surface of the alumina thus inhibiting the interaction with Rh to some extent.

The same trend was apparent in the Rh/Si/Al_2O_3 catalyst; the amount of reducible Rh was slightly greater after long term steam-O_2 aging compared to the short term hightemperature oxidation (48% vs 44%) and the NO conversion was also greater after long term steam-O_2 aging (97% vs 90%). The Rh/ZrO_2+Al_2O_3 catalyst, in contrast, showed less reducible Rh after steam-O_2 aging than after the high temperature oxidation (47%vs 74%) and roughly equivalent NO conversion (96% vs 98%). The decrease inreducible Rh following long term steam-O_2 aging may have resulted primarily from pore structure collapse and Rh occlusion, as suggested by the decrease in surface area of the 1.46% Rh/ZrO_2 catalyst (without admixed Al_2O_3) from a fresh value of 5.5 m^2/g to 3.1 m^2/g after long term steam-O_2 aging. Also, the amount of reducible Rh could only be restored to about 60% of the fresh catalyst level after the steam-O_2 aging, consistent with the retention of only about 60% of the fresh surface area.

The Rh/Si/Al_2O_3 catalyst retained 92% of its fresh surface area after extended steam-O_2 aging. In addition, high temperature H_2 restoration after the extended steam-O_2 aging resulted in 95% retention of the theoretical H_2 consumption as measured by TPR. This is greater than the H_2 consumption of the fresh catalyst (77% of theoretical), and indicates that high temperature steam-O_2 aging reverses some of the Rh masking present in the fresh catalyst. Moreover, in contrast to our previous experience with La addition to Rh/Al_2O_3 catalysts, the blocking of the interaction between Rh and the alumina was not accompanied by the formation of a compound between Rh and Si under high temperature oxidizing conditions. In the case of the Rh/La/Al_2O_3 catalyst, compound formation between the La and Rh following high temperature oxidation resulted in sharp degradation of NO conversion (59% peak) compared to the reference Rh/Al_2O_3 catalyst (81% peak). In contrast, the Si modified catalyst gave a peak NO conversion of 90% following the same treatment. Furthermore, even after extended hydrothermal aging, the Si exchanged Rh/Al_2O_3 catalyst shows advantages over both alumina and zirconia based reference catalysts in its retention of catalytic activity, Rh reducibility, and surface area. These advantages are due to a combination of (1) greater retention of BET surface area, and (2) blocking of deleterious interactions between Rh and Al_2O_3.

As a final note, it is somewhat surprising that, in Fig. 5d, the $Rh/Si/Al_2O_3$ and $Rh/ZrO_2+Al_2O_3$ catalysts do not show much higher activity than the reference Rh/Al_2O_3 catalyst despite the nearly factor of two greater fraction of reducible Rh retained by the Si-modified and Zr-supported catalysts. However, chemisorption measurements, carried out on the steam-O_2 aged catalysts, indicated only 2-3% dispersion for the Rh in the Si-modified catalyst and 1-2% dispersion in the Zr-supported catalyst. This contrasts with an apparent dispersion of 7% for the Rh in the reference Rh/Al_2O_3 catalyst after steam-O_2 aging. The combined TPR, reactivity, and chemisorption data suggest a trade-off in surface modifications of the type carried out here; the extent of Rh interaction with the support is decreased but at the expense of increased sintering of the Rh particles.

ACKNOWLEDGMENTS

We appreciate the assistance of F. W. Kunz, D. R. Liu, and Jong-Sim Park in carrying out XRF and microprobe analyses of the catalyst samples. We also appreciate helpful discussions with C. K. Narula on ion-exchange techniques.

REFERENCES

1 H.C. Yao, S. Japar, and M. Shelef, J. Catal., 50 (1977) 407.
2 C. Wong and R.W. McCabe, J. Catal., 119 (1989) 47 .
3 R.K. Usmen, R.W. McCabe, L.P. Haack, G.W. Graham, J.S. Hepburn, and W.L.H. Watkins, W. L. H., J. Catal., 134 (1992) 702.
4 B. Beguin, E. Garbowski, and M. Primet, J. Catal., 127 (1991) 595.
5 D.K. Paul, T.H. Ballinger, and J.T. Yates, Jr., J. Phys. Chem., 94 (1990) 4617.
6 D.K. Paul and J.T. Yates, Jr., J. Phys. Chem., 95 (1991) 1699.
7 T.H. Ballinger and J.T. Yates, Jr., J. Phys. Chem., 95 (1991) 1694; and references therein.
8 J.Z. Shyu and K. Otto, J. Catal., 115 (1989) 16.
9 C. Wong and R.W. McCabe, J. Catal., 107 (1987) 535.

A. Frennet and J.-M. Bastin (Eds.)
Catalysis and Automotive Pollution Control III
Studies in Surface Science and Catalysis, Vol. 96

STRUCTURAL, MORPHOLOGICAL AND SURFACE CHEMICAL FEATURES OF Al_2O_3 CATALYST SUPPORTS STABILIZED WITH CeO_2

C. Morterra[a], G. Magnacca[a], V. Bolis[a], G. Cerrato[a], M. Baricco[a], A. Giachello[b] and M. Fucale[b].

[a] *Department of Inorganic, Physical, and Materials Chemistry, University of Turin, via P. Giuria 7, I-10125 Turin, Italy.*

[b] *Centro Ricerche FIAT, Strada Torino 50, I-10043 Orbassano (Turin), Italy.*

ABSTRACT

Two families of CeO_2/Al_2O_3 (ACE) catalyst supports, containing 3 and 20 % wt CeO_2 respectively, have been examined and compared with the pure oxides CeO_2 and (transition-phase) Al_2O_3. Structural data indicate that, also in the presence of CeO_2 , the Al_2O_3 support undergoes all phase transitions at the usual temperatures. Electron micrographs show that particles morphology and crystals size are mainly imposed by the Al_2O_3 support. The porosimetric texture indicates a more pronounced mesoporosity in the ACE systems than in pure Al_2O_3, especially in the case of low CeO_2 loadings. FTIR spectra and volumetric/calorimetric data of the adsorption of CO at ≈300 K show that the surface Lewis acidity of aluminas is severely modified by the presence of CeO_2: the overall CO uptake is increased, there is the formation of one additional family of Lewis acid sites (Ce^{4+} ions), and the presence of CeO_2 either leaves unchanged or increases the concentration of the strongest Lewis centres of (transition-phase) aluminas (Al^{3+} in tetrahedral coordination).

1. INTRODUCTION

The addition of variable amounts of CeO_2 to alumina-supported three-way-conversion catalysts is now widely used. In fact the effectiveness of CeO_2 is well established, although its detailed role in activity enhancement is still a subject of competing hypotheses [1]. CeO_2 has been supposed to: (i) participate in the catalytic reactions, acting as an oxygen storage component; (ii) stabilize the dispersion of the noble metal and/or promote its catalytic activity; (iii) act as a phase stabilizing agent for the active γ-Al_2O_3 support.

The first two promoting roles of CeO_2 have been sometimes considered in some detail in recent years [1, 2], whereas the action towards the active alumina support has been usually explained only in terms of the different specific surface area of the various Al_2O_3 phases.

The present contribution is aimed at elucidating, by the use of different analytical techniques, the actual action exerted by the CeO_2 additive in respect of the alumina catalyst support.

2. EXPERIMENTAL

2.1. Materials

Pure and Ce-doped alumina samples were prepared starting from water dispersions (10%) of (pseudo)-boehmite (Disperal Alumina, Condea Chemie). Under continuous stirring, controlled amounts of Ce nitrate solution were added in order to obtain CeO_2:Al_2O_3 ratios of 3:100 (2.5 % wt) and 20:100 (16.6 % wt) in the final calcined material. The pH was adjusted with HNO_3 to a final value of 3. The suspensions were dried at 353 K (24 h), and then the crushed products underwent a first oven treatment at 773 K (3 h) to convert -AlOOH into -Al_2O_3.

Doped and undoped samples were fired in air (5 h) at three different temperatures (773, 1273, 1473 K), rehydrated by exposure to the atmosphere for several days, and used for IR measurements in the form of self-supporting pellets (15-25 mg cm^{-1}) or as loose powders for microcalorimetric measurements (ca. 0.5 g). All thermal treatments were carried out in vacuo for 2 h (residual pressure 10^{-5} Torr; 1 Torr $\approx$133.3 N m^{-2}).

As a reference material, a high purity microcrystalline CeO_2 sample was prepared from Ce nitrate [3], and fired in air at 773 K. Its surface area is 78 m^2 g^{-1}.

Samples are designated in the text by the following symbols: A (reference alumina), ACE3 and ACE20 (Ce-doped samples, CeO_2 : Al_2O_3 ratio of 2.5 and 16.6 % wt respectively). A subscript numeral T_1 represents the temperature (K) of the preliminary oven firing of the material. This symbol is often followed by a second numeral T_2, which represents the temperature (K) of vacuum activation undergone prior to adsorption and IR investigation.

2.2. Methods

BET surface areas were determined with N_2 at 77 K with an automatic apparatus Sorptomatic 1900, C. Erba. Porosimetric data were obtained using Kelvin equation [4].

Structural analysis was performed by X-ray diffraction (XRD). A conventional Bragg-Brentano geometry was used (Philips PW 1050), with Co K_α incident radiation.

(HR)TEM micrographs were obtained with a JEOL JEM 2000 EX apparatus (200 kV acceleration), equipped with a top-entry stage. The samples were deposited from isopropanol dispersions on Cu grids coated with a holey carbon film.

IR spectra were run at 2 cm^{-1} resolution at ≈300 K on a Bruker FTIR spectrometer (Mod. 113v), equipped with MCT detector. From all spectra of adsorbed CO, the contribution of the gas-phase was computer-subtracted interactively.

Heats of adsorption were measured at 303 K with a Tian-Calvet microcalorimeter (Setaram), connected to a vacuum/gas volumetric apparatus, that enabled the simultaneous determination of adsorbed amounts by the stepwise contact with subsequent doses of the adsorptive [5].

3. RESULTS AND DISCUSSION

- BET areas are summarized in Table 1. The presence of ceria turns out to have an appreciable detrimental effect on the surface area of alumina. This effect is frequently produced by the addition of lanthanide oxides and, in general, of oxides possessing high specific weight and a non porous texture. The effect is present for all Al_2O_3 phases, is higher the higher the thermal treatment, and becomes more important the higher the CeO_2 loading.
- A differential porosimetric distribution in the pore radius range 10-80 Å is shown in Figure1.

Table 1 BET data ($m^2\ g^{-1}$)

Samples	**Firing temperatures (K)**		
	773	**1273**	**1473**
A	186.5	120.0	4.5
ACE3	158.7	47.5	1.4
ACE20	133.7	23.0	2.1

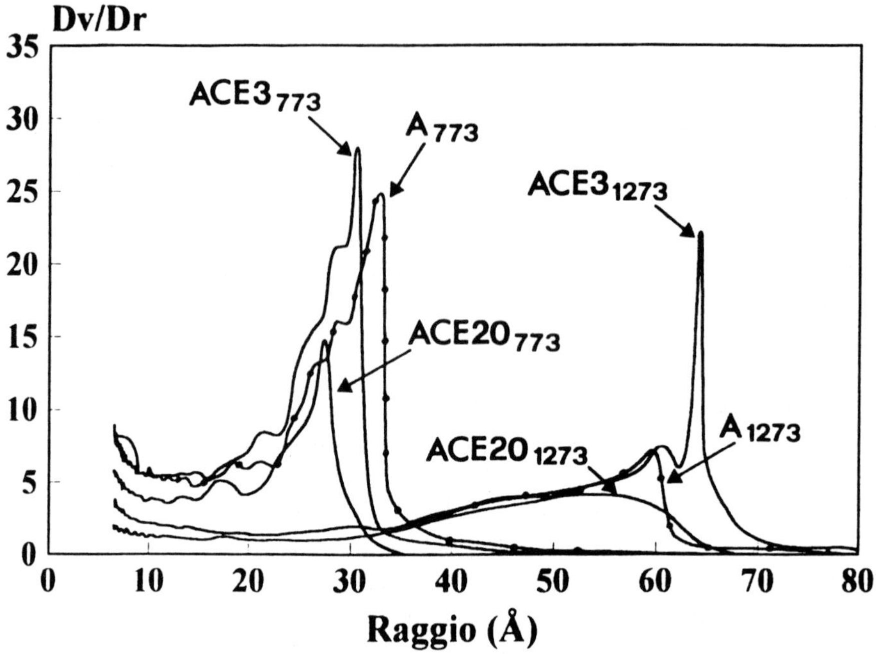

Figure 1. Differential porosimetric distribution of A and ACE pretreated at 773 and 1273 K.

Alumina fired at 773 K presents a sharp peak (≈30 Å). The peak grows and shifts to smaller sizes at low CeO_2 loadings, and then decreases at higher loadings. The small pore radius peak observed on A_{773} and ACE_{773} systems suggests the stabilization of crystallographically defective configurations, and thus of coordinatively unsaturated (**cus**) sites on the surface.

In A_{1273}, the main peak becomes broad and moves to about 60 Å: when the size of the particles increases, also the size of the pores is observed to increase. The addition of small amounts of ceria ($ACE3_{1273}$) tends to stabilize defective situations, and a pore peak persists at ≈ 65 Å, sharper and with higher intensity than on A_{1273}. Opposite to that, large percentages of CeO_2 ($ACE20_{1273}$) tend to reduce severely the number of pores and their average size, and the overall porosity becomes similar to that of the corresponding pure alumina.

• XRD patterns are reporter in Fig.2. Section A compares the materials fired at 773 K. In A_{773} (**a**), broad diffraction peaks of the γ-Al_2O_3 phase are observable.

In $ACE3_{773}$ (**b**), the pattern of γ-Al_2O_3 is almost unchanged, and the CeO_2 phase starts becoming evident (e.g. see the weak peak at $2\theta \approx 35^o$). In $ACE20_{773}$ (**c**), owing to the high atomic scattering factor of Ce, the CeO_2 phase becomes predominant, and the alumina phase is hardly distinguishable at all. The diffraction peaks of the CeO_2 phase in $ACE20_{773}$ are broader than in the case of a pure CeO_2 phase treated in the same conditions (**d**). No mixed phases can be observed, but there is basically the contribution of the two oxides (γ-Al_2O_3 and CeO_2).

Section B compares the materials treated at 1273 K. Pure alumina (**a**) appears now in the δ, θ-phase. At low CeO_2 loading, the ACE system (**b**) shows the prevailing features of the alumina phases, whereas at high CeO_2 loading (**c**) almost exclusively the features of CeO_2 can be observed. It is worth noting that, after thermal treatment at 1273 K, the $\gamma \rightarrow \delta,\theta$-$Al_2O_3$ phase transition has occurred regularly also in the presence of CeO_2. After firing at 1273 K, there is still no evidence of mixed phases. The breadth of CeO_2 diffraction peaks in $ACE20_{1273}$ appears now similar to that of pure CeO_2 treated at 773 K.

In section C, XRD patterns of materials treated at 1473 K are compared. Pure alumina (**a**) is now in the stable α-phase. Diffraction peaks of α-Al_2O_3 are also observable both in low (**b**) and high (**c**) CeO_2-loaded samples. No mixed phases are yet revealed by the XRD patterns. All phase transitions of Al_2O_3 seem thus to be unaffected by the presence of CeO_2 in the firing temperature range considered.

• <u>TEM data</u>. The reference CeO_2 773 sample presents flat particles, fairly large (5-10 nm diameter) and quite regular, in agreement with what suggested by the sharp XRD peaks. The regular fringe patterns (see fig. 3A) indicate a good crystalline order, with a probable preferential termination along the (111), (200), and (311) crystal planes.

It is known that the definition of the micrographs of transition aluminas is not very good, due to the high trasparency of the particles to the electron beam, whereas the situation is different for the α-phase (A_{1473}), as the thickness of the corundum crystallites causes very poor trasparency and quite dark images [6].

$ACE3_{773}$ (3B) and $ACE20_{773}$ (3C) consist of particles very similar to those of the parent A_{773} (see the inset in fig. 3B). In particular, the high porosity of the crystallites is still revealed by the dark grooves system, and the average size of the alumina particles in both ACE_{773} systems is about 10-20 nm. Moreover, some particles of CeO_2 are observed, caracterized by a sharper contrast, and exhibiting smaller sizes and the same preferential terminations described before for pure CeO_2. No overlayers and/or amorphous coatings can be observed around the particles, and phases other than CeO_2 and γ-Al_2O_3 are not detected.

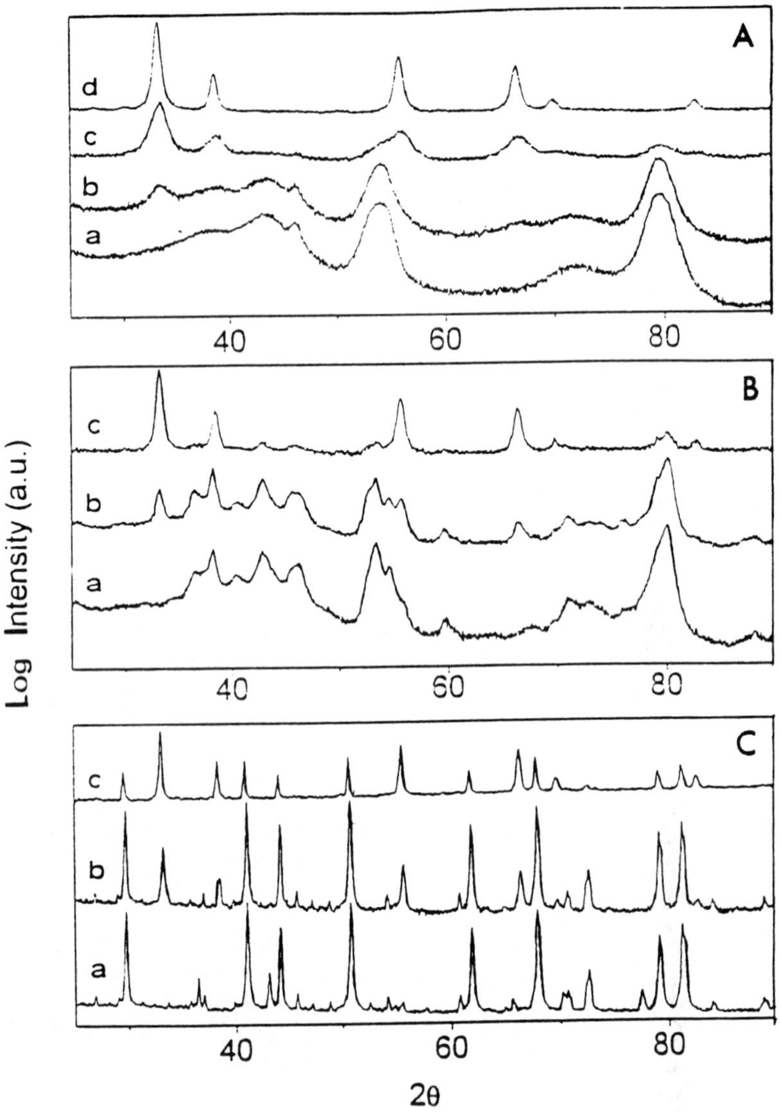

Figure 2. XRD patterns of samples pretreated at: 773 K (sect. ***A****), 1273 K (sect.* ***B****), and 1473 K (sect.* ***C****). Curves* ***a****: pure alumina; curves* ***b****: ACE3; curves* ***c****: ACE20; curve* ***d****:* CeO_2.

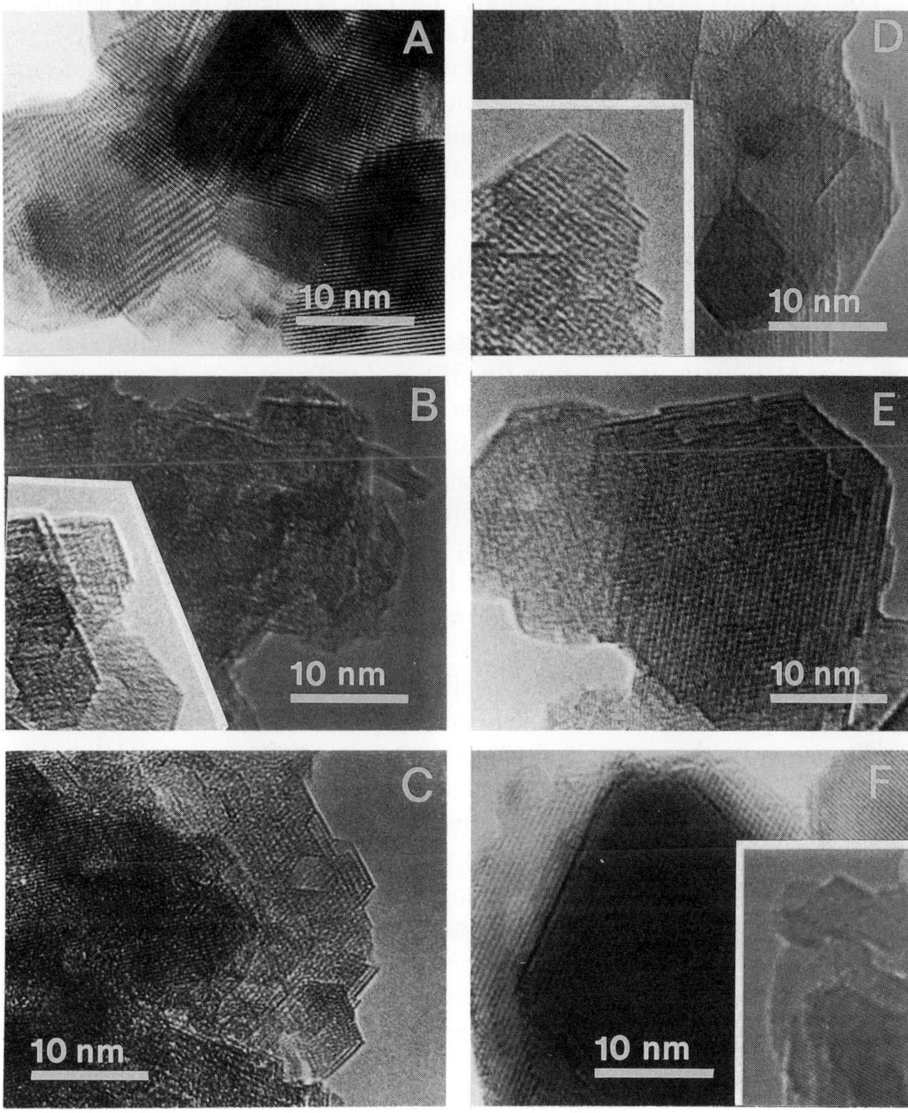

Figure 3. TEM images of some A and ACE samples. ***A***: CeO_2; ***B***: $ACE3_{773}$ (inset: A_{773}); ***C***: $ACE20_{773}$; ***D***: $ACE3_{1273}$ (inset: A_{1273}); ***E***: $ACE20_{1273}$; ***F***: $ACE20_{1473}$ (inset: A_{1473}).

For $ACE3_{1273}$ (3D) and $ACE20_{1273}$ (3E) the crystalline order is definitely higher than on the corresponding A_{1273}, (see the inset to fig. 3D). In $ACE3_{1273}$ the packing of the original sub-particles is very dense, the crystal terminations are still fairly irregular, and traces of the diffuse pore system of $ACE3_{773}$ are still evident. For $ACE20_{1273}$ the crystal terminations are more regular, the particles are definitely larger, and the fringe patterns are not assignable to definite crystalline planes of Al_2O_3 or CeO_2 without ambiguity. No new phases are observed also for
these samples, and there is no trace of amorphous overlayers around the crystallites.

The ACE3 and ACE20 systems pretreated at 1473 K are very similar (for this reason, only $ACE20_{1473}$ is shown in fig. 3F), and different from A_{1473} (see the inset to fig. 3F). The particles size is in the 60-150 nm range. The fringe patterns indicate that the α-Al_2O_3 phase is by far predominant, while there is some evidence for the presence of some residual δ,θ-Al_2O_3, of a CeO_2 phase ((111) and (200) planes), and of a phase $CeAlO_3$ ((012) or (202) planes)

TEM and XRD analysis give consitent results: the phase transitions and the bulk features of aluminas are not appreciably influenced by the presence of CeO_3, and the pure oxides seem to coexist separately, at least as long as alumina is in the spinel (transition) phases.

• <u>CO adsorption: FTIR data</u>. CO is a soft Lewis base, that adsorbs reversibly onto surface **cus** cations of d^0 metal oxides. It yields σ-coordinated adspecies, whose formation is spectroscopically witnessed by a ν_{CO} frequency higher than that of CO gas (2143 cm^{-1}). The polarization of adsorbed CO, monitored by the upwards shift of the ν_{CO} frequency, depends on several features of the adsorbing cation: chemical nature, actual charge, crystallographic configuration, and chemical environment. The shift of ν_{CO} upon adsorption is usually related to the energy of the acid-base interaction, whose strength can be measured as the adsorption heat.

For the sake of brevity, only the interaction between CO and alumina samples pretreated at 1273 K will be considered here. In fact, the samples fired at lower temperatures behave quite similarly, and exhibit minor (quantitative) differences that is not worthwhile describing in detail.

Figure 4 compares the spectral pattern of CO uptake at $\approx$ 300 K on
CeO_2 $_{773}773$ (**a**), $A_{1273}1073$ (**b**), $ACE3_{1273}1073$ (**c**), and $ACE20_{1273}1073$ (**d**). On CeO_2, CO uptake yields a couple of bands, centered at $\approx$ 2170 cm^{-1} (main band) and $\approx$ 2150 cm^{-1} (weak shoulder). Both bands are due to σ-coordination onto $Ce^{4+}{}_{cus}$ centres: on CO adsorption after reduction, both bands decrease and no new bands due to CO uptake onto $Ce^{3+}{}_{cus}$ centres are observed.

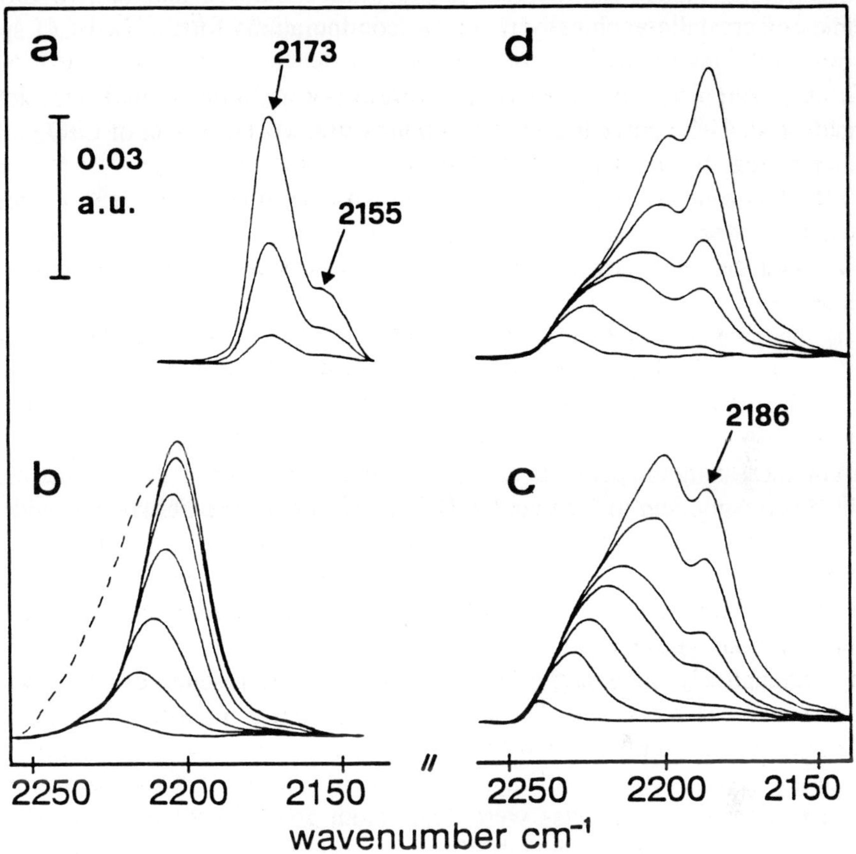

Figure 4. IR spectral patterns of CO adsorption at ≈300 K on: $CeO_{2\ 773}773$ (sect. ***a****); $A_{1273}1073$ (sect.* ***b****; dashed line: 140 Torr CO onto $A_{773}1073$); $ACE3_{1273}1073$ (sect.* ***c****); $ACE20_{1273}1073$ (sect.* ***d****). P_{CO} are in the 1.5×10^{-1} - 1.4×10^{2} Torr range.*

As described elsewhere [7,8], CO adsorbed onto δ,θ-alumina presents at least three families of bands whose spectral positions, in order of increasing acidic strength, are: ≈2200 cm^{-1} (most abundant), ≈2215 cm^{-1} (weak), and ≈2235 cm^{-1} (very weak; this component is much stronger in the γ-phase, as shown by the dashed trace in Fig. 4b). The high ν_{CO} frequencies confirm the formation of simple linear σ-coordinated complexes in all cases. The assignment is to **cus** Al^{3+} ions with tetrahedral coordination (Al^{IV}_{cus}). The Al^{IV}_{cus} sites are localized on regular crystal planes for the main band (≈2200 cm^{-1}), and in two different families of crystallographically defective configurations for the two CO species absorbing at higher ν [8].

On ACE samples, the spectrum of CO adsorbed at 300 K is quite complex and variable with CeO_2 percentage (see sections **c** and **d**). The effect of CeO_2 loading on alumina can be summarized as follows:

i) there is a new family of Lewis acidic sites, ascribable to $Ce^{4+}{}_{cus}$ centres, whose νCO frequency is at 2185-2190 cm^{-1} , i.e. ≈15 cm^{-1} higher than on pure CeO_2, indicating a perturbation of the CeO_2 network brought about by the interaction with the Al_2O_3 moiety;

ii) the presence of CeO_2 brings about the stabilization of the most acidic $Al^{IV}{}_{cus}$ sites (i.e. the sites yielding CO bands at ν > 2200 cm^{-1}), that are very few on the pure A_{1273} system. The effect is far more evident in the samples with lower CeO_2 loading;

iii) on increasing the percentage of CeO_2, there is a sharp decline of the number of sites corresponding to "regular" $Al^{IV}{}_{cus}$ sites, i.e. of the CO species with νCO at ≈2200 cm^{-1}

It is so deduced that CeO_2 perturbs the surface acidic features of the Al_2O_3 network, and the loading of CeO_2 involves mainly the aluminium sites present on regular faces of the crystallites. The other sites, crystallografically more defective and coordinatively more acidic, are stabilized by the presence of CeO_2, at least up to some CeO_2 coverage. These findings are consistent with the porosimetric and morphological data reported above.

• <u>CO adsorption: volumetric/calorimetric data</u>. Heats of CO adsorption and relevant adsorbed amounts were measured for the systems $A_{1273}1023$, $ACE3_{1273}1023$, and $ACE20_{1273}1023$.

The volumetric isotherms are reported in Figure 5a. In the whole interval of CO pressure examined, the adsorbed amounts per unit surface area are definitely larger on the ceria-doped specimens than on pure alumina, and the difference between doped and non-doped systems increases with increasing CO pressure. (For instance, at p_{CO} = 60 Torr, the uptake is 0.13 CO molecules per nm^2 (0.21 $\mu mol/m^2$) on pure alumina, whereas it is 0.31 and 0.40 CO molecules per nm^2 (0.51 and 0.66 $\mu mol/m^2$) in the case of ACE3 and ACE20 respectively).

These figures indicate that the overall population of sites sufficiently acidic to σ-coordinate CO at ambient temperature increases when the alumina system is loaded with increasing amounts of CeO_2 , that lead to the formation at the surface of (increasing amounts of) coordinatively unsaturated Ce^{4+} sites. This result could be somehow anticipated on the basis of the CO spectral patterns of Fig. 4, and gains here quantitative evidence.

As for the acidic strength of the sites that adsorb CO, it was measured as molar heat of adsorption, and is reported in Fig. 5b as a function of the adsorbed amounts. The heat of adsorption declines with increasing CO coverage for both pure and Ce-doped alumina, as typical of energetically heterogeneous surfaces.

The sites energy distributions of the two families of samples present some significant differences. The initial energy of interaction (q_0), corresponding to CO uptake onto the strongest sites, can be estimated by extrapolating to zero coverage the plots of adsorption

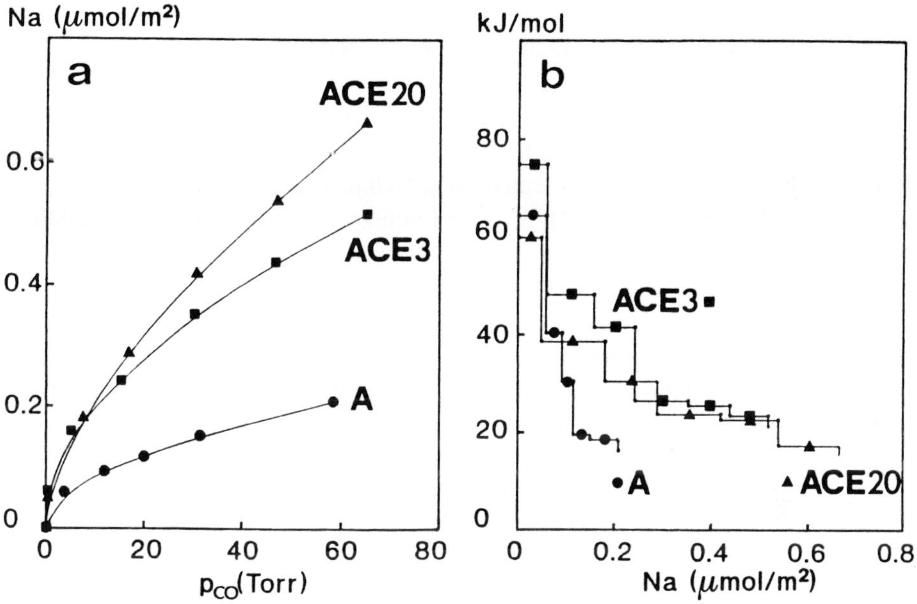

*Figure 5. Section **A**: CO adsorption isotherms at 303 K on $A_{1273}1073$ (●), $ACE3_{1273}1073$ (■) and $ACE20_{1273}1073$(▲); section **B**: partial molar heat of adsorption of CO at 303 K for the same samples as a function of adsorbed amount.*

heat vs. adsorbed amounts (the continuous plots derived from the block plots of Fig. 5b and yielding the extrapolated q_0 figures are not shown). The initial heat is quite high for ACE3 ($q_0 \approx$ 90 kJ/mol) with respect to pure alumina ($q_0 \approx$ 80 kJ/mol): this indicates that the presence of Ce^{4+} cations increases the Lewis acidity of the system, by increasing not only the overall population of sites but also the maximum acidic strength of a fraction of them. A high initial energy of interaction is consistent with the presence on ACE3 of a large population of CO adspecies absorbing at high frequency ($\nu_{CO} \approx 2235\ cm^{-1}$), as shown in Fig. 4c.

The heat of adsorption of CO on pure alumina declines quite steeply, and reaches a value of ≈20 kJ/mole when CO coverage is ≈ 0.13 CO molecules/nm^2. The heat values for the low-loading sample ACE3 decline quite slowly, and reach the value ≈20 kJ/mol for a coverage of ≈0.31 CO molecules/nm^2 This indicates the presence of a large proportion of sites, virtually absent on pure alumina, characterized by a medium-high acidic strength (heat values in the 50-30 kJ/mol interval), in agreement with the spectroscopic observation of CO adspecies with $\nu_{CO} \approx 2190\ cm^{-1}$). These sites correspond to the presence of abundant **cus** Ce^{4+} centres, located on flat regular planes of the microcrystals and active towards CO at ≈300 K.

As for the high-loading sample ACE20, the (extrapolated) initial energy of interaction is not much different from that estimated for pure alumina (q_0 ≈75 kJ/mole). This indicates (in agreement with what already monitored by the CO bands reported in Fig. 4d) that high CeO_2 loadings tend to suppress gradually the strong acidic sites, whose surface concentration was enhanced by low CeO_2 loadings. At medium CO coverages, also the high-loading ACE20 samples exhibit a strong contribution from abundant Lewis sites of medium-high strength. The latter are ascribed to **cus** Ce^{4+} centres, that gradually substituted **cus** Al^{3+} cations located in regular crystal planes of the Al_2O_3 network: this corresponds to the gradual formation of a surface mixed phase, that is not revealed by the bulk analytical techniques.

4. CONCLUSIONS

The present results confirm that the addition of CeO_2 does modify the surface properties of alumina, but indicate that the changes cannot be interpreted in terms of phase stabilization and/or of the preservation of a high surface area support. In fact, CeO_2 brings about a net overall decline of the surface area, and phase transitions of Al_2O_3 occur as on the pure system.

Even if the methods of structural and morphological analysis suggest that the gross features of the ACE preparations are those expected of a plain mixed system (at least up to the transition to the α-Al_2O_3 phase, catalytically inactive), surface physico-chemical analysis methods indicate that, on a microscopic scale, the situation is quite different. In fact, the two oxidic systems do affect one another appreciably. (**i**) The presence of the transition-phase alumina induces in the CeO_2 network smaller crystallites size and enhanced acidity of the surface

cus Ce^{4+} sites; **(ii)** The presence of CeO_2 induces in the alumina network an altered porous texture, an increased overall strong Lewis acidity (i.e., the acidity revealed by CO uptake at ≈300 K), the presence also in the high-temperature spinel phases of abundant amounts of the very strong Lewis acid sites that are probably important in the catalytic activity.

ACKNOWLEDGMENTS

This research was partly supported by the CNR, Progetto Finalizzato Materiali Speciali.

REFERENCES

1 J.G. Nunan, H.J. Robota, M.J. Cohn and S.A. Bradley, Catalysis and Automotive Pollution Control II, A. Crucq (Ed.), Elsevier, Amsterdam, 1991, p. 221.
2 A.F. Diwell, R.R. Rajaram, H.A. Shaw and T.J. Truex, Catalysis and Automotive Pollution Control II, A. Crucq (Ed.), Elsevier, Amsterdam, 1991, p. 139.
3 E. Abi-aad, R. Bechara, J. Grimblot and A. Aboukais, Chem. Mater., 5 (1993) 793.
4 S. Lowell and J.E. Shields, Powder Surface Area and Porosity, Chapman and Hall, London (2nd Edition), 1984.
5 V. Bolis, B. Fubini, E. Garrone and C. Morterra, J. Chem. Soc., Faraday Trans. 1, 85 (1989)
6 C Morterra, G. Magnacca, G. Cerrato, N. Del Favero, F. Filippi, C. V. Folonari, J. Chem. Soc. Faraday Trans., 89 (1993) 135.
7 L. Marchese, S. Bordiga, S. Coluccia, G. Martra, and A. Zecchina, J. Chem. Soc., Faraday Trans., 89 (1993) 3483.
8 C. Morterra, V. Bolis, G. Magnacca, Langmuir, in press.

A. Frennet and J.-M. Bastin (Eds.)
Catalysis and Automotive Pollution Control III
Studies in Surface Science and Catalysis, Vol. 96
1995 Elsevier Science B.V.

SUPPORT AND NEMCA INDUCED PROMOTIONAL EFFECTS ON THE ACTIVITY OF AUTOMOTIVE EXHAUST CATALYSTS

I.V. Yentekakis, C.A. Pliangos, V.G. Papadakis, X.E. Verykios and C.G. Vayenas

Department of Chemical Engineering
Institute of Chemical Engineering and High Temperature Chemical Processes
University of Patras, GR-26500 PATRAS - GREECE

ABSTRACT

Metal-support interactions can play an important role in catalytic oxidations and in the performance of three-way automotive exhaust catalysts. In this investigation the kinetics of CO and C_2H_4 oxidation and NO reduction by CO were studied on fifteen different model catalysts comprising combinations of three metals, Pt, Pd and Rh supported on five different supports, i.e. SiO_2, γ-Al_2O_3, ZrO_2 (8% Y_2O_3), TiO_2 and TiO_2 (4% WO_3). Significant support effects in turnover frequency were observed for many of these metal-support combinations for all three model reactions.

In a separate set of experiments the kinetics of C_2H_4 oxidation were investigated on polycrystalline Rh films interfaced with ZrO_2 (8 mol% Y_2O_3) solid electrolyte (YSZ) in a galvanic cell of the type

$$C_2H_4, O_2, Rh/YSZ/Pt, O_2.$$

It was found that by applying external potentials and thus supplying O^{2-} to the catalyst surface, up to 80-fold, increases in catalytic rate can be obtained (NEMCA effect). It was noted that the observed kinetic behavior upon increasing catalyst potential parallels qualitatively the observed increase in turnover frequency of Rh for C_2H_4 oxidation upon varying catalyst support.

1. INTRODUCTION

It has often been demonstrated in the literature that the effect of the support on the performance of metal catalysts can be very significant. The chemisorptive and catalytic properties of metal crystallites can be influenced by the nature of the carrier employed in catalyst formulation. This kind of metal-support interactions was first established by the early work of Schwab [1] and Solymosi [2] who attributed these phenomena to changes in the electronic state of the metal via electronic-type interactions with the carrier. In recent years, interactions of Group VIII metals with TiO_2 supports induced by high-temperature reduction were described under the concept of SMSI [3] and were shown to originate from migration of TiO_x species to the surface of the metal crystallites [4] and from an electronic factor operating concomitantly [5].

More recently it has been found that incorporation of altervalent cations into the crystal structure of semiconductive carriers can play an important role in defining the catalytic properties of the supported metal crystallites [6,7,9,10]. According to proposed explanations, these effects are due to electronic interactions at the metal-support interface caused by the different Fermi levels of the two solids in contact. For example, it was shown [6,7] that small Pt particles supported on TiO_2 doped with higher valence cations (Sb^{5+}, Ta^{5+}, W^{6+}) exhibit significantly reduced chemisorption capacity for H_2, O_2 and CO, because of the acceptance of charge transferred from the carrier to the Pt crystallites. However, the effect of higher valence doping of TiO_2 on the catalytic performance of Rh is opposite to that observed on Pt. Solymosi et al. [8] obtained enhanced activities for CO hydrogenation when Rh was supported on TiO_2 (W^{6+}). The same behavior was observed by Ioannides and Verykios [9,10] for CO, CO_2 and benzene hydrogenation, using Rh/TiO_2 (W^{6+}) catalysts.

Many investigators have found that the choice of support can affect the activity and stability of catalysts for several reacting systems. Vannice et al [11] observed that the type of the support can affect the behaviour of Pt catalysts for CO and CO_2 methanation, since higher activities were obtained when the metal was supported on TiO_2 rather than on Al_2O_3 or SiO_2. Metcalfe and Sundaresan [18] reported that Rh/YSZ is more active than Rh/Al_2O_3 for the CO/O_2 and CO/NO reactions and more recently Oh [13] found that a modified $Rh/CeO_2/Al_2O_3$ catalyst resulted in higher activities for NO reduction by CO than the Rh/Al_2O_3 catalyst. Different supports affect not only catalyst activity but also catalyst stability. For instance, Yao et al. [14] reported that the stability of Rh/ZrO_2 catalysts was considerably better than that of Rh/Al_2O_3, under H_2+NO reaction conditions.

The trends mentioned above can be important in the case of automotive emission-control catalysts. By using suitable supports, it is possible to achieve the same catalyst performance with smaller amounts of noble metals. The latter is crucial for Rh which is an expensive and scarce noble metal [15].

The use of solid electrolytes as active catalyst supports to induce the effect of Non-Faradaic Electrochemical Modification of Catalytic Activity (NEMCA) [16] or in situ controlled promotion [17] and alter the catalytic properties of metal catalysts has been described in detail previously [16-18]. A goal of the present work was to explore the possible relationship between the NEMCA effect [16-18] and the effect of dopant-induced-metal-support interactions [6,7,9,10].

2. EXPERIMENTAL

Materials: Rh, Pd and Pt catalysts were prepared by impregnation of the supports with a solution of $Rh(NO_3)\cdot 2H_2O$, $PdCl_2$, H_2PtCl_6 (Alfa Products) respectively, to yield 0.5 wt% metal loadings. The following powder supports were used: SiO_2 (Alltech Associates), TiO_2 (Degussa P25), ZrO_2 (8 mol% Y_2O_3) (Zirconia Sales) and γ-Al_2O_3 (Akzo Chemicals).The dopant material for TiO_2 supports was WO_3 (Alfa Products). The doped and undoped TiO_2 supports were thermally treated (5h, 900°C- rutile form) while the ZrO_2 (8% Y_2O_3) support had undergone thermal pretreatment at 1500°C during production.

Before any measurements, the catalysts were reduced in a quartz tube by He flow (80 cc/min) at 250°C for 1 h followed by H_2 flow (80 cc/min) at 400°C for1h.

Catalyst characterization: The characterization of the catalysts was carried out in a BET apparatus (Accusorb 2100E, Micromeritics) with H_2 chemisorption and N_2 and/or Ar physical adsorption experiments at 77 K.

Electrochemical measurements: The basic experimental setup is shown schematically on Fig. 1a. The metal working catalyst electrode is deposited on the surface of a ceramic solid electrolyte (i.e. Y_2O_3-stabilized-ZrO_2 (YSZ), an O^{2-} conductor). Details of catalyst, counter and reference electrode preparation and characterization have been reported in detail elsewhere [16]. The catalyst electrode is exposed to the reactive gas mixture (i.e. $C_2H_4+O_2$) in a continuous flow gradientless reactor. Under open-circuit conditions (I=0) it acts as a regular catalyst for the C_2H_4 oxidation reaction. The electrodes are connected with a galvanostat/potentiostat which is used to apply constant currents between the catalyst and the counter electrode or constant potentials between the catalyst and reference electrode. In this way O^{2-} ions are supplied from (or to) the solid electrolyte to (or from) the catalyst-electrode surface. The current is defined positive when anions are supplied to the catalyst electrode.

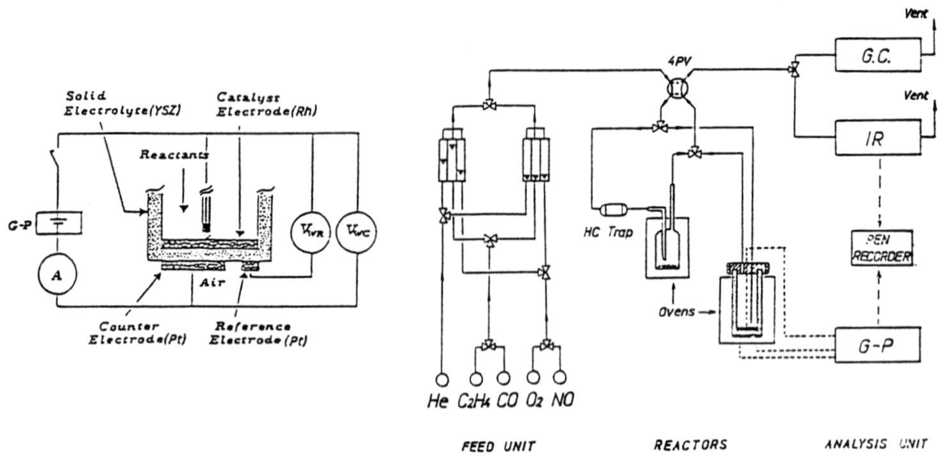

Fig. 1 a) The three electrode configuration used in the NEMCA experiments.
b) Experimental set-up for catalytic and NEMCA experiments; 4PV: Four-port-valve, GC: Gas chromatograph, IR: Infrared CO_2 Analyser, G-P: Galvanostat-potentiostat.

Apparatus: The catalysts were tested under reaction conditions in the experimental setup shown in Fig. 1b. Two types of reactors where used, one for the supported catalysts, and one for the NEMCA experiments. Experimental details for the latter have been reported elsewhere [16]. Both reactors have been shown to behave as CSTR.

Reaction systems: The reactions used for catalyst testing are the main ones which take place in a catalytic converter, i.e.: CO oxidation, C_2H_4 oxidation (as a model light HC) and NO reduction by CO. Experiments were conducted within the temperature range of 150-500°C.

3. Results and Discussion

3.1. Catalyst characterization

The results for all fifteen catalysts tested are summarized in Table 1. The H/M ratio (M: Rh, Pd, Pt) was calculated assuming a 1:1 H to metal stoichiometry.

Table 1: catalyst characterization

	DISPERSION [H/M] (*)			BET SURFACE AREA (**) (m^2/gr)	BET SURFACE AREA (***) (m^2/gr)
	Rh	Pt	Pd		
$ZrO_2(8\%Y_2O_3)$	0.44	0.44	0.37	1.55	5.2
TiO_2	0.10	0.07	0.33	Ø	0.13
$TiO_2(4\%WO_3)$	0.30	1.00	0.33	Ø	4.9
γ-Al_2O_3	1.00	0.68	0.61	177.1	Ø
SiO_2	0.65	0.54	0.65	261.7	Ø

(*) H_2 chemisorption at 298 K.
(**) N_2 physical adsorption at 77 K
(***) Ar physical adsorption at 77 K

3.2. CO oxidation

The kinetics of the CO oxidation reaction over all three noble metals exhibit a Langmuir-Hinshelwood type behavior, due to competitive adsorption of CO and oxygen, characterized by the appearance of rate maximum with increasing CO partial pressure. This is a typical behavior which has been described by many investigators for this reaction system [19,20]. The catalytic activity, presented as turnover number of CO_2 production, was measured by varying either the partial pressures of CO or O_2, or temperature, keeping the other parameters constant.

An example of comparative representation of catalytic activity under CO oxidation is given in Fig. 2. This figure shows the dependence of the rate (turnover number) on the partial pressure of CO and refers to 0.5 wt% Rh supported on different carriers. It can be observed that catalytic activity is enhanced when the metal is supported on the doped carriers, ZrO_2 (8% Y_2O_3) and TiO_2 (4% WO_3). The rates of CO consumption over Rh/ZrO_2 (8% Y_2O_3) and Rh/TiO_2 (4% WO_3) are significantly higher than those observed when Rh is supported on high surface area γ-Al_2O_3, which is the commercially used support. Furthermore, the beneficial effect of doping of the support is obvious, upon comparing the curves corresponding to the two TiO_2 supports (doped and undoped). A similar beneficial effect upon doping the TiO_2 support with W^{6+} cations was also observed in the case of Pd while the opposite effect was observed in the case of Pt.

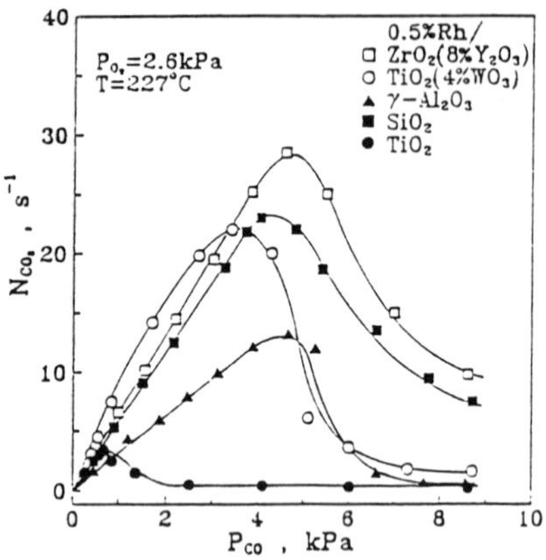

Fig.2 Activity of supported Rh catalysts under CO oxidation

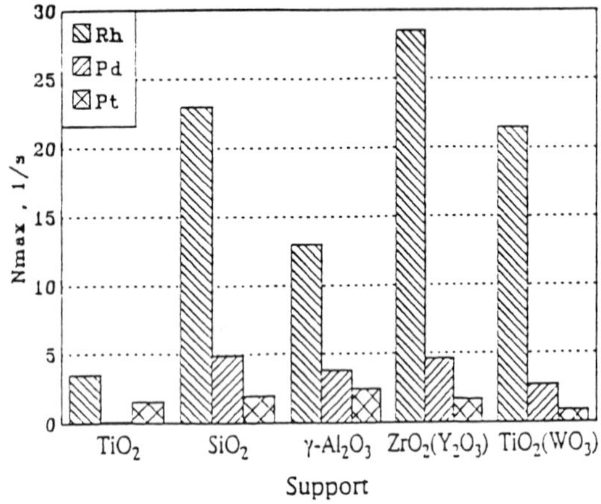

Fig. 3 Maximum measured turnover frequencies for CO oxidation (corresponding to the r maxima of r vs PCO) at T=227°C and PO2=2.6 kPa

A comparison of the intrinsic activity of all fifteen catalyst formulations, tested under CO oxidation reaction conditions, is given in the bar diagram of Fig.3 which shows the maximum measured turnover numbers at 227°C and oxygen partial pressure of 2.6 kPa. Rhodium is clearly a far superior CO oxidation catalyst as compared to Pt and Pd, in accordance with results of other investigators [21]. It is also observed in Fig. 3 that the activity of Rh for CO oxidation is very sensitive to the carrier employed for its dispersion. A weaker sensitivity is exhibited by Pd and Pt.

3.3. C_2H_4 oxidation

The kinetics of this reaction were also found to follow a Langmuir-Hinshelwood type behavior, with competitive adsorption of C_2H_4 and oxygen. A comparative kinetic diagram for all the supported Rh catalysts is given in Fig. 4. The dashed lines indicate an abrupt drop in the ethylene combustion rate which is probably due to surface oxide formation. The same behavior was also observed in the electrochemical promotion (NEMCA) experiments, as briefly discussed below and further described elsewhere [22].

The highest combustion activities were obtained over Rh/TiO_2(WO_3) and Rh/ZrO_2(Y_2O_3) catalysts. The beneficial effect of doping the TiO_2 carrier with W^{6+} cations is also illustrated in Fig. 4. The enhancement of the ethylene combustion activity of Rh when it is dispersed on W^{6+} doped TiO_2 is particularly pronounced at higher oxygen partial pressures (P_{O2} > 1 kPa).

Regarding the Pd and Pt catalysts supported on the same substrates, the following behavior was recorded: Higher activities were observed upon doping the TiO_2 supports with W^{6+} cations in the case of Pd catalysts, while the inverse behavior was observed in the case of Pt catalysts. Pd/ZrO_2 (8% Y_2O_3) is significantly better than Pd/γ-Al_2O_3. The Pt/ZrO_2 (8% Y_2O_3), and Pt/γ-Al_2O_3 exhibited the best activity among the Pt catalysts.

A comparative summary of these results concerning C_2H_4 oxidation is given in Fig. 5. It is worth noting the superior activity of Pd supported catalysts for this reaction, in agreement with results reported in the literature [23,24]. The best carriers for the different metals are $TiO_2(W^{6+})$ for Rh, $ZrO_2(Y_2O_3)$ for Pd and SiO_2 for Pt.

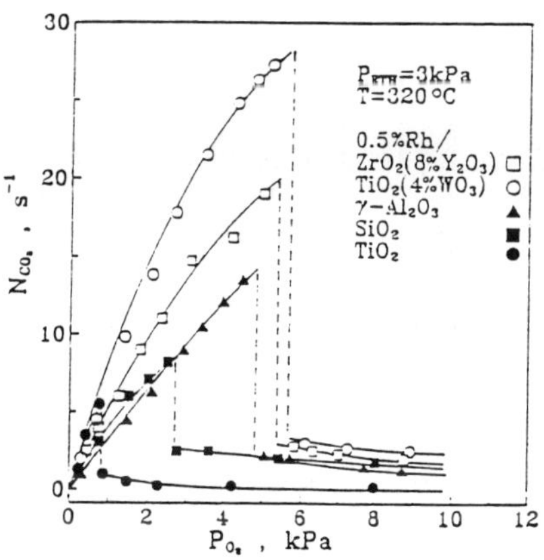

Fig. 4 Effect of P_{O2} and support on the rate of C_2H_4 oxidation on supported Rh

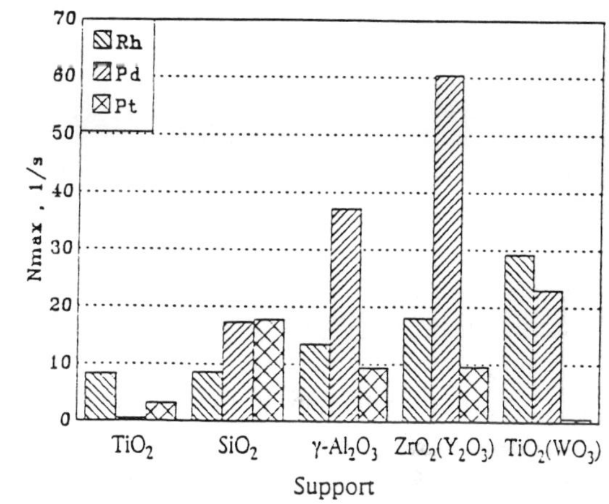

Fig. 5 Maximum measured turnover frequencies of C_2H_4 oxidation at T=320°C, $P_{C_2H_4}$=3 kPa.

3.4. CO-NO reaction

The removal of NO_x is of great interest for air pollution control and thus extended research has been carried out during the past fifteen years in the reduction of NO by CO. In the present study, the activity of Pt, Pd and Rh for NO reduction by CO was investigated and the influence of the carrier was examined. It was found that, in addition to CO_2 and N_2 formation, N_2O was also formed in appreciable quantities. The N_2O production rate was found to go through a maximum with increasing temperature, in the range of 320-350°C, due to decomposition of N_2O at higher reaction temperatures [25,26].

The influence of the carrier on NO reduction activity of Pt, Pd and Rh catalysts is illustrated in Fig. 6. It is apparent that, in agreement with results of earlier studies [21,27], among the three metals, Rh exhibits far superior NO reduction activity in the presence of CO. The activity of Rh is also sensitive to the support employed for its dispersion. Thus, Rh dispersed on W^{6+} -doped TiO_2 exhibits activity which is approximately 4 times higher than that of Rh on undoped TiO_2, while the activity of Rh on the other supports investigated falls in between.

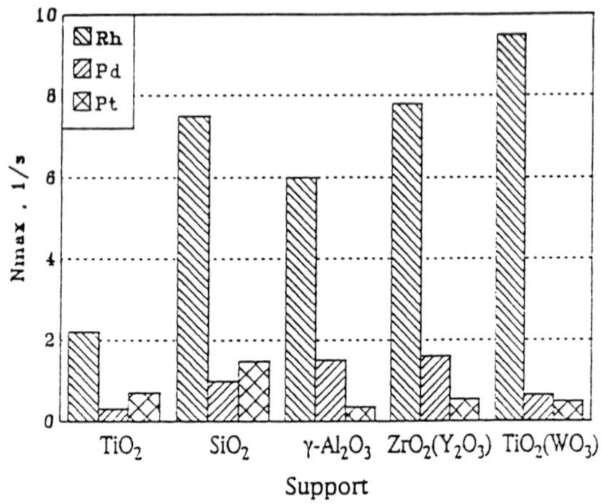

Fig. 6 Maximum measured turnover frequencies for the CO-NO reaction at $T=360^oC$, $P_{NO}=1$ kPa.

3.5. NEMCA studies for the system $C_2H_4/O_2/Rh$

The kinetics of C_2H_4 oxidation were also investigated on polycrystalline Rh films interfaced with ZrO_2 (8 mol% Y_2O_3), or YSZ solid electrolyte in a galvanic cell of the type: C_2H_4, O_2, Rh/YSZ/Pt, O_2. It was found that by applying external potentials and thus supplying O^{2-} to the catalyst surface, up to 80-fold increases in catalytic rate are obtained.

An example is shown in Fig. 7 which shows a typical galvanostatic transient, i.e., it depicts the transient effect of a constant applied current on the rate of C_2H_4 oxidation and on catalyst potential, V_{WR}. At the start of the experiment (t^30) the circuit is open (I=0) and the steady state catalytic rate value, r_o, is $1.8x10^{-8}$ g-at O/s. At t=0 the galvanostat is used to apply a constant current (I=400 μA) between the catalyst and the counter electrode. According to Faraday's Law, oxygen anions are supplied to the catalyst at a rate $G_O=I/2F=2.05x10^{-9}$ g-at O/s. This causes an 8800% increase in catalytic rate ($\Delta r=1.6x10^{-6}$ g-at O/s). The rate increase Δr is 770 times larger than the rate of oxygen anions G_O of supply

(Λ=770). This means that each O^{2-} supplied to the catalyst causes, at steady state, 770 chemisorbed oxygen atoms to react and form CO_2. In the present work the maximum observed value of the enhancement factor (Faradaic efficiency) Λ [16-18] was of the order of $5x10^4$.

The kinetic behavior upon varying catalyst potential is shown in Fig. 8. It is worth noting the similarities with Fig. 4. Increasing catalyst potential causes the same rate changes as variation of the catalyst support in the order TiO_2,

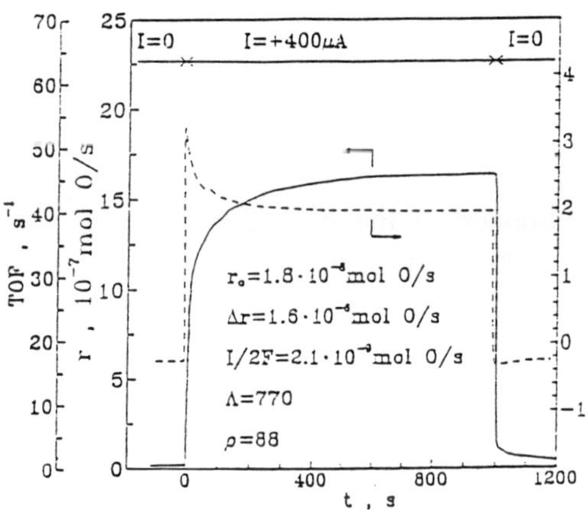

Fig 7 A galvanostatic transient during C_2H_4 oxidation on Rh catalyst. T=350°C, P_{O2}= 2 kPa, P_{C2H4}=5.5 kPa

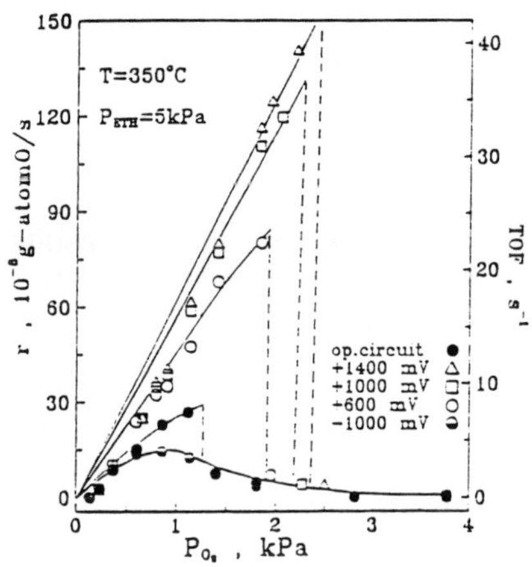

Fig. 8 Effect of P_{O2} and catalyst potential on the rate of C_2H_4 oxidation on Rh.

SiO_2, γ-Al_2O_3, $ZrO_2(8\%Y_2O_3)$, $TiO_2(4\%WO_3)$. Since the NEMCA effect is due to the change in catalyst work function eF with changing catalyst potential ($\Delta e\Phi = e\Delta V_{WR}$) [16,18] and to the concomitant back-spillover of promoting oxide ions onto the catalyst surface [16,28], it is likely that the above catalyst support sequence corresponds to increasing work function of the supported Rh crystallites, i.e. $e\Phi$ is lowest for Rh supported on TiO_2 and highest for Rh supported on $TiO_2(4\%WO_3)$.

4. CONCLUSIONS

It was found that the catalytic activity of three-way catalysts can be altered significantly by using different supports for each noble metal component. This may be due to electronic interactions between the metal and the support [9,10] and/or due to oxygen back spillover from the supports to the catalyst as in the NEMCA experiments [16-18,28].

The following sequence of catalytic activity was observed, with minor variations for each reaction:

Rh: TiO_2(4% WO_3)>ZrO_2 (8% Y_2O_3)> γ-Al_2O_3, SiO_2>TiO_2
Pd: ZrO_2(8% Y_2O_3)> γ-Al_2O_3>TiO_2(4% WO_3), SiO_2>TiO_2
Pt: SiO_2>ZrO_2(8% Y_2O_3), γ-Al_2O_3>TiO_2>TiO_2(4% WO_3)

Qualitatively similar results were obtained in the NEMCA study of C_2H_4 oxidation. It was shown that the catalytic rates and also the location of the abrupt break in the rate vs P_{O2} curves (most likely corresponding to the formation of a surface oxide) can be adjusted by means of the externally applied potential.

The observed strong similarities between the NEMCA effect [16-18] and the effect of dopant-induced-metal-support interactions [6,7,9,10] suggests that the origin of both effects may be similar. Surface spectroscopic and work function measurements could further elucidate this point.

REFERENCES

1 Schwab, G-M., in "Advances in Catalysis", Vol. 27, p. 1, Academic Press, Orlando, FL, 1978
2 Solymosi, F., Catal. Rev., 1 (1967) 233
3 Tauster, S.J., and Fung, S.C., J. Catal., 55 (1978) 29
4 Ko, C.S., and Gorte, R.J., J. Catal., 90 (1984) 59
5 Sadeghi, H.R., and Henrich, V.E., J. Catal., 109 (1988) 1
6 Akubuiro, E.C., and Verykios, X.E., J. Catal., 103 (1987) 320
7 Akubuiro, E.C., and Verykios, X.E., J. Catal., 113 (1988) 106
8 Solymosi, F., Tombacz, I., and Koszta, J., J. Catal., 95 (1985) 578
9 Ioannides, T., and Verykios, X.E., J. Catal., 140 (1993) 353; 143 (1993) 175; 145 (1994) 479
10 Ioannides, T., Verykios, X.E., Tsapatsis, M., and Economou, C., J. Catal., 145 (1994) 491
11 Vannice, M.A., Twu, C.G., and Moon, S.H., J. Catal., 79 (1983) 70

12 Metcalfe, I.S., and Sundaresan, S., AIChE Journal, Vol. 34(6) (1988) 1048
13 Oh, S.H., J. Catal., 124 (1990) 477
14 Yao, H.C., Stepien, H.K., and Gandhi, H.S., J. Catal., 61 (1980) 547
15 Kummer, J.T., J. Phys. Chem., 90 (1986) 4747
16 Vayenas, C.G., Bebelis, S., Yentekakis, I.V., and Lintz, H.-G., Catalysis Today, 11 (1992) 303-442
17 Yentekakis, I.V., Moggridge, G., C.G. Vayenas, and R.M. Lambert, J. Catal., 146 (1994) 292
18 Vayenas, C.G., Bebelis, S., and Ladas, S., Nature (London), 343 (1990) 625
19 Engel, J., and Ertl, G., in "Advance in Catalysis", Vol. 28 (1979) 1-78
20 Creighton, J., Tseng, F., White, J., and Turner, J., J. Phys. Chem., 85 (1981) 703
21 Kobylinski, T.P., and Taylor, B.W., J. Catal., 33 (1974) 376
22 Pliangos, C.A., Yentekakis, I.V., and Vayenas, C.G., J. Catal. (1994) submitted
23 Gordonna, G.W., Kosanovich, M., and Becker, E.R., Platinum Metals Rev., 33(2) (1989) 46
24 Hammerle, R.H., and Truex, T.J., S.A.E. Paper, No 760090 (1976)
25 McCabe, R.W., and Wong, C., J. Catal., 121 (1990) 422
26 Kim, Y., Schreiffels, J.A., and White, J.M., Surf. Sci., 114 (1982) 349
27 Hegedus, L.L., and Gumbleton, J.J., Chemtech, 10 (1980) 630
28 Ladas, S., Kennou, S., Bebelis, S., and Vayenas, C.G., J. Phys. Chem., 97(35) (1993) 8845

A. Frennet and J.-M. Bastin (Eds.)
Catalysis and Automotive Pollution Control III
Studies in Surface Science and Catalysis, Vol. 96
1995 Elsevier Science B.V.

SYNTHESIS AND STUDY OF HONEYCOMB MONOLITHIC CATALYSTS FOR CATALYTIC COMBUSTION

Z.R.Ismagilov, G.V.Chernykh, R.A.Shkrabina.

Boreskov Institute of catalysis, 5 Prospekt Ak. Lavrentieva, 630090 Novosibirsk 90, Russia

ABSTRACT

This paper concerns the synthesis of catalysts able to retain their high activity upon the catalytic combustion of hydrocarbons at temperatures up to 1100°C. SiO_2, La_2O_3, CeO_2 additives are used to privent alumina washcoat on cordierite monoliths from sintering. Catalysts containing Pt and Pd have been investigated. The catalysts with Pd on Al_2O_3+SiO_2 support have been found to be most active and stable at high temperature.

1. INTRODUCTION

Catalytic combustion as an alternative to conventional thermal combustion received considerable attention in the past decade [1-3]. It is known that the catalysts for high temperature combustion should possess the following properties:

high catalytic activity;

developed geometrical surface, low pressure drop, high thermal stability and high specific suface area of the support;

thermal stability of the active component and support under a long term catalyst operation at high temperatures.

Typically, monoliths made of various materials are used for supporting such catalysts [4]. A thin oxide layer (usually γ-Al_2O_3) covering the walls of monolith channels provides a high specific surface area. At high temperatures a phase transformation of γ-Al_2O_3 to α-Al_2O_3 occurs decreasing surface area resulting in catalyst activity reduction. Various additives such as La_2O_3, CeO_2,

ZrO_2, BaO, etc. introduced into alumina were found to increase its thermal activity [6,7].

Catalysts containing noble metals are known to be the most active catalysts for the complete combustion of hydrocarbons [8]. During the catalyst operation at temperatures higher than 800°C the active component starts to sinter and evaporate. Therefore, the active component should have a low vapour pressure, or it should be strongly bonded to the thermally stable support [9]. Pt and Pd are the most often used noble metals.

A better thermal stability performance of the Pd/Al_2O_3 system is explained by a strong interaction of PdO with Al_2O_3 surface. PtO interacts with alumina less strongly, therefore, Pt/Al_2O_3 system undergoes considerable sintering in air at temperatures exceeding 600°C [10]. PdO formation as well as the sintering of Pt particules decreases the catalyst activity.

Thus, the problem of catalyst desing can be divided into several subproblems:

1. synthesis of a thermally stable support including monolith washcoating with oxides, selection of modifying additives and their introduction into the oxide layer;
2. catalyst synthesis: deposition of the active component and the selection of their chemical composition improving its thermal stability.

2. EXPERIMENTAL

Cordierite monoliths 10×10×10mm wer used as the catalyst support. Its properties are listed in Table 1:

Table 1 Properties of monolith

Parameters	Value
Cell size, mm	1.2×1.2
Number of cells per cm^2	49
Wall thickness, mm	0.2
S_{BET}, m^2/g	0.1
Total pore volume, cm^3/g	0.819
Average pore radius (r), mkm	36.7
r_{1},mkm	1.43
r_{2},mkm	71.90

BET surface area was measured by argon thermal desorption. The pore structure was determined by the mercury intrusion porosimetry method (Porosizer-2300). Disperse suspensions prepared by mixing alumina hydroxide with pseudoboehmite stugture, HNO_3 and H_2O have been used for washcoating. La_2O_3 was introduced into the suspension from $La(NO_3)_3$ $6H_2O$, SiO_2- from silica sol, CeO_2 from Ce(NO3) $6H_2O$. cordierite monoliths were immersed in suspension for 5 min and, after being blown with air, dried at 60-80°C for 2h. to obtain high surface area of washcoat the impregnation was repeated twice. before the second immersion the monoliths were heated at 300°C for 2h.

To study the properties of the support with washcoat, the monoliths (after last impregnation and drying at 60-80°C) were calcined at 550°C and 1100°C for 3h.to avoid washcoat pore closure and burying of active components at high temperatures the support calcined at 1100°C was used for catalyst preparation. Pt and Pd were deposited from the aqueous solutions of H_2PtCl_6 and H_2PdCl_4via impregnation in excess solution followed by drying at 60-80°C. THE catalysts were calcined at 600°C and 1100°C. the content of the active component and additives was determined by X-ray fluorescence (VRA-20). Catalytic activity was studied using a flow-circulation technique and characterized by the rate of complete butane oxidation (W) at 5000C and a stationary butane concentration of 0.2vol.%. The inlet reaction mixture contained 0.5vol.% of C_4H_{10}, balance air.

Pt and Pd dispersion in the samples was determined by SAXS (KRM-1 chamber)using CuKα radiation, voltage 30 kV and filament current 20 mA. The distribution of additives and noble metals over the oxide film was studied using X-ray microprobes.

3.RESULTS

The data on washcoat preparation and thermal stability of the support are summarized in Table 2. The additive efficiency was defined as the ratio:

$$K= S_{BET}(1100°C/ S_{BET}(550°C°)$$

From this data it is evident that the introduction of additives allows the support surface area to increase considerably after high temperature treatement. It is well known that the additives differ in the mechanism of their interaction with alumina, but as seen from Table 2, the efficiency values of CeO_2,SiO_2 and La_2O_3 are rather close to each other. If the additive stabilizing effect is estimated as a

ratio between S_{BET} of pure alumina and that of doped alumina at 1100°C the following series of additive efficiency can be obtained:
SiO_2~CeO_2>La_2O_3.

This series is in good agreement with the reference data. Based on the results obtained and on the assumption that CeO_2 can be stabilize the active components (Pt and Pd) as well, the supports modified with this additive and SiO_2 were chosen for the catalyst preparation. As seen from Table 3 the catalysts calcined at 600°C and deposited on a support containing 5wt.% of CeO_2 are more active independently of the content and composition of the active component. The increased palladium loading leads to the catalyst activity enhancement for both supports

Table 2 Properties of washcoat and support thermal stability

N°	Washcoat	Al_2O_3 wt.%	S_{BET}, m^2/g 550°C	1100°C	K
1	Al_2O_3	8.1	25.0	0.7	0.03
2	Al_2O_3 + 5% CeO2	9.1	24.0	9.1	0.38
3	Al_2O_3+ 2% SiO_2	9.7	25.0	9.2	0.37
4.	Al_2O_3+ 2% La_2O_3	9.4	.23.9	8.3	0.35

Table 3 Activity of catalysts for C_4H_{10} oxidation

N°	Washcoat	Pt, wt.%	Pd,wt.%	W. 10^2,C_4H_{10}/gs 600°C	1100°C
1	Al_2O_3 + 5% CeO2	0.2	--	7.1	5.6
2	-- --	--	0.2	12.6	3.0
3					
4	Al_2O_3 + 2% SiO_2	0.2	--	9.1	2.6
5	-- --	--	0.2	10.7	10.7
6	-- --	--	0.5	11.7	10.7

For samples treated at 1100°C the following results were obtained. All catalysts on Al_2O_3 + CeO_2 support lose their activity. The samples containing 0.2wt.% of Pd (samples 2,3) are deactivated the most, and that containing 0.2wt.% ofPT (sample 1) - the least. Pd/ Al_2O_3 + SiO_2samples retain their high activity (samples 5,6).

in order to explain the effects observed, a more detailed study of supported samples cotaining CeO_2 additive was performed.

All samples were shown by SAXS to be polydispersed and to contain particules ranging from 50A to 180A in size. But no strong relationship between the active component particule size, the calcination temperature and the catalyst composition was observed. With increasing temperature to 1100°C polydispersive distribution is retained and the average size of active component particules tends to slightly increase. For some amples, motly with palladium, calcination at 1100°C leads tothe shift of SAXS particule size distribution curves to the interval 20-40 A. This formal increase of dispersion under calcination in air can be explained by significant growth of an oxide layer over sintered metal particules with simultaneous formation of an interphase boundary with the washcoat, thus, the metal particule size, observed by SAXS, is smaller.

It should be noted that due to the complexity of the catalyst studied, care must be taken in the interpretation of dispersion measurements made by SAXS.

It was found by X-ray microprobes (samples 1-3) that the thickness of the washcoat cover is around 50 mkm. CeO_2,PtandPd are distributed uniformaly with respect to Al_2O_3 in the samples calcined at 600°C. Calcined at 1100°C leads to noticeable redistribution of active components: platinum in the catalyst Pt/ Al_2O_3 + CeO_2 concentrates on the outer surface of the monolith wall to a higher extent than palladium inPd/ Al_2O_3 + CeO_2

4.CONCLUSIONS

The main findings of this work can be summarised as follows. The introduction of CeO_2 , SiO_2stabilizes the support surface area at high temperatures. The catalysts prepared on the thermally stable supports containing Pd on Al_2O_3 + SiO_2 retain their high activity after treatement at 1100°C.

REFERENCES

1 R. Prasad, L. A. Kennedy and E. Ruckenstein, Cat. Rev.-Sci. Eng., 26(1),1-58(1984)
2 H. Arai and M. Machida, Catal Today, 10, 81-94 (1991).
3 Z. R. Ismagilov and M.A. Kerzhentsev, Catal. Rev.-Sci. Eng., 32(1-2), 51-103(1990)
4 D.L. Trimm, Appl. Catal., 7,249-282(1983).
5 B.Beguin, E. Garbowski and M. Primet, J.Catal., 127, 595-604(1991).
6 M. Machida, K. Eguchi and H. Arai,J. Catal., 103, 385-393(1987).
7 M. Machida, K. Eguchi and H. Arai J; Catal., 120,377-386(1989).
8 Y. Morooka and A. Ozaki, J. Catal., 5, 116-124(1966).
9 M. F. M. Zwinkels, S.G. Jaras and P.G. Menon, Catal Rev.-Sci. Eng., 35(3), 319-358(1993).
10 M. A. Leeand E. Ruckenstein, Catal Rev.-Sci. Eng.,25,475-550(1983)

A. Frennet and J.-M. Bastin (Eds.)
Catalysis and Automotive Pollution Control III
Studies in Surface Science and Catalysis, Vol. 96

STRUCTURE AND CATALYTIC ACTIVITY OF MIXED OXIDES OF PEROVSKITE STRUCTURE

V. Mathieu - Deremince[1], J.B. Nagy and J.J. Verbist

Groupe de Chimie Physique, Facultés Universitaires Notre-Dame de la Paix, 61 Rue de Bruxelles, B-5000 Namur, Belgium

ABSTRACT

The $La_{1-x}Ce_xBO_3$ (B= Ti, Cr, Mn, Fe, Ni, Co) mixed oxides of perovskite structure present a catalytic activity for CO oxidation and NO reduction which increases with increasing Ce content. $LaCoO_3$ and $LaMnO_3$ are the more active compounds for both CO oxidation and NO reduction. The oxidation of CO by O_2 follows a suprafacial type mechanism where the adsorbed oxygen is the active species. The reduction of NO by CO is best explained by a redox mechanism.

Ce is only incorporated in a limited amount in the perovskite structure and a concomitant formation of CeO_2 is also observed. For a low Ce content, CeO_2 is highly dispersed on the surface of the catalyst particles, while larger particles are obtained at higher Ce content.

The presence of Ce leads to an increase in both CO oxidation and NO reduction activities. The CO oxidation is dependent on the dispersion of CeO_2 on the mixed oxides particles. On the other hand, the NO reduction does depend on the Ce content in the perovskite structure.

1. INTRODUCTION

Despite the fact that perovskite-type oxides have been suggested as substitutes for noble metals in automotive exhaust catalysis [1], relatively few studies were devoted to the synthesis, characterization and catalytic activity of these catalysts.

These compounds, of general formula ABO_3, generally include a lanthanide element A and a transition metal B. Partial substitution of A by Ce is also interesting, as Ce is a well known additive of exhaust catalyst [2]. Its oxygen storage ability is used to broaden the air/fuel ratio window, improving the activity of three way catalysts (CO, hydrocarbon oxidation, NO reduction).

---

[1] Present address : Solvay & Cie, Rue de Ransbeek 310, B-1120 Bruxelles, Belgium

It seems that the lattice oxygen plays a direct role in the oxidation of CO. Indeed, it was found that the catalytic activity is maximum, if the bond energy of the lattice oxygen is minimum [3]. Hence, the oxygen vacancy is also an important factor for the catalytic activity. The partial substitution of the A element can monitor the oxygen vacancy. If Ce(IV) is introduced in the perovskite structure A site vacancies are produced [4]. The partial substitution of B element by Pt (IV) increases greatly the CO oxidation activity of $La_{0.7}Pb_{0.3}MnO_3$ perovskite [5].

For the NO reduction by CO $AMnO_3$ perovskites were the most studied catalysts [6]. The importance of anion vacancies was emphasized for the good catalytic activity. The $ACoO_3$ was also found a good candidate for both CO + O_2 and CO + NO reactions [7]. Very recently perovskite-related oxides were also proposed as catalysts for the direct decomposition of NO into N_2 and O_2 [8, 9].

This paper deals with the characterization and catalytic activity of cerium substituted lanthanum perovskites $La_{1-x}Ce_xBO_3$ (B = Ti, Cr, Mn, Fe, Ni, Co),x ,varying from 0 to 0.6 [10].

2. EXPERIMENTAL

$La_{1-x}Ce_xBO_3$ compounds were synthesized by evaporation of a solution of the respective metal nitrates or oxalates and calcination of the precursors either at 1000°C or at 1200°C for 5 h [11].

X-ray powder diffraction patterns were recorded on a Philips PW 1730 diffractometer using Ni filtered Cu Ka radiation. XPS spectra were obtained on a Hewlett Packard 5950 A spectrometer with Al Ka radiation.

The catalysts were tested for the CO + O_2 and CO + NO reactions in a flow reactor system. Before the tests, the compounds (1.0 g) mixed with small glass beads (weight ratio 1:1) were pretreated overnight in N_2 at 400°C and then in a mixture of CO, O_2, N_2 and Ar or CO, NO, N_2, Ar, respectively for 30 min at the same temperature to attain steady - state conversion. The same flow rate (Q = 150 ml/min), the same mixture of reagents (20 % CO, 10% O_2, 70 % N_2 or 20% CO, 20% NO, 20% Ar, 40% N_2, respectively, in mol %) and the same reaction temperature (T = 100-400°C or 300-500°C, respectively) were used to compare the activity of the different catalysts.

For the determination of the kinetic schemes, the differential mode of the reactor was used (% conversion < 15%).

3. RESULTS AND DISCUSSION

3.1. X-ray Diffraction

Since no traces of $La_2 O_3$ or BO_x appear in the diffraction patterns of $LaBO_3$ (B = Cr, Mn, Fe, Co), the perovskite is completely formed at 1000°C in the systems not containing Ce. With B = Ni or Ti, additional phases appear which are $La_2 NiO_4$ or $La_2 Ti_2 O_7$, respectively. Moreover, the stoechiometry of the latter perovskite is $La_{0.7}TiO_3$.

The Ce introduced in the reagents is not totally incorporated in the perovskite lattice. The excess crystallizes as CeO_2 [4, 12]. The latter is influenced by the nature of metal B (Table 1).

For compositions x < 0,05, no trace of CeO_2 is detected. This can be due either to a total incorporation of Ce into the perovskite lattice or to a very highly dispersed (and/or amorphous) CeO_2 not detected by XRD. For higher Ce content, the relative intensity of CeO_2 increases with increasing x, the slope being in the following order as a function of B : Ni > Co > Mn > Cr > Fe > Ti.

Table 1 Relative heights of X-ray diffraction peaks characterizing the different catalysts calcined at 1000°C

	I_{CeO_2}/I_{LaBO_3} [a]					
x[b]	Ti	Cr	Mn	Fe	Co	Ni
0.01	-	-	-	0	0	0
0.03	-	-	0	0	0	0
0.05	0	0	0.06	0.04	0.12	0.13
0.07	-	-	-	0.08	0.17	0.21
0.10	0	0.06	0.14	0.08	0.22	0.33
0.15	-	-	-	0.17	0.26	0.28
0.20	0.05	0.22	0.31	0.21	0.35	0.58
0.40	0.13	0.73	0.77	0.46	0.88	1.20
0.60	-	-	1.37	0.98	1.41	1.79

a : I_{CeO_2} - d = 3.16 Å; $I_{La_{0.7}TiO_3}$ - d = 2.74 Å; I_{LaCrO_3} - d = 2.74 Å; I_{LaMnO_3} - d = 2.72 Å; I_{LaFeO_3} - d = 2.78 Å; I_{LaCoO_3} - d = 2.67 Å; I_{LaNiO_3} - d = 2.74 Å b :x is defined as $La_{1-x}Ce_xBO_3$

Table 2 Amount of adsorbed oxygen (in %) and $O_{lattice}$/La atomic ratios of perovskites calcined at 1000°C

Perovskites	O_{ads} (%)	$O_{lattice}$/La [a]
$La_{0.7}TiO_3$	35.9	4.67
$LaCrO_3$	42.5	3.04
$LaMnO_3$	49.2	2.53
$LaFeO_3$	61.9	1.75
$LaCoO_3$	84.9	0.59
$LaNiO_3$	71.0	1.09

a : Theoretical atomic ratios : 3.0, except for Ti(4.3)

The formation of extra perovskite lattice CeO_2 gives rise to three different results :

- appearance of A vacancy sites (Ø) following $La_{1-x}Ø_xBO_3 + xCeO_2$,
- migration of (excess) B ions outside the perovskite lattice to yield BOx : $La_{1-x}B_{1-x}O_{3-x} + xBO_x + xCeO_2$.
- the occurrence of both phenomena.

For low Ce content, the first process is predominant, while for higher Ce content BO_x phase also appears. The relative importance of the two processes is highly dependent on the nature of B. Neither TiO_2 nor Cr_2O_3 are detected for any initial x value. Only traces of Fe_2O_3 are shown for x = 0.40 and 0.60. Mn_2O_3 starts to appear for x = 0.20 and CoO for x = 0.10. NiO is detected for any x values studied.

3.2. X-ray photoelectron spectroscopy

The valence state of La and B ions in $LaBO_3$ perovskites is equal to (III) in all compounds without Ce, but for Ti, where Ti is in (IV) oxidation state, in agreement with the stoechiometry of $La_{0.7}TiO_3$ perovskite.

The atomic ratio computed form the B2p and La3d core peak intensities is close to the theoretical value of 1 in $LaBO_3$ for B=Cr, Mn, Fe and Co. For Ti, this ratio is equal to 1.44, corresponding to the above mentioned stoechiometry.

The XPS spectra of O1s are more interesting to characterize the non-stoechiometric composition of the perovskites. Indeed, their catalytic activity is dependent on the excess or default of oxygen in the lattice. The higher energy O1s peak (E_b = 530.5 eV with respect to E_b of carbon, 284.7 eV) is attributed to adsorbed oxygen and the lower energy peak (E_b = 529.0 eV) to lattice oxygen [13].

The amount of adsorbed oxygen increases as a function of the atomic number of B ions, exception for Ni (Table 2). From the comparison of the theoretical $O_{lattice}$/La ratio (equal to 3) with the experimental values, oxygen deficiency is detected for all B ions, excepting Ti and Cr (Table 2). The importance of oxygen deficiency increases with the atomic number of B ions. Hence, it seems that the oxygen is essentially adsorbed at anion vacancies.

The perovskites synthesized in presence of Ce do not exhibit any bond energy difference for La or B ions. On the other hand, despite of the complexity of the Ce XPS spectra, the Ce3d core peak intensities and the Ce(IV) satellite intensity allowed us to estimate the relative amount of Ce(III) and Ce(IV) on the surface [10, 11].

The amount of Ce(III) is considered as being incorporated in the perovskite lattice, Ce(III) replacing La(III) in $La_{1-x}Ce_xBO_3$. For x<0.20, the incorporation of Ce(III) decreases with increasing x values for all B ions (Figure 1).

The amount of Ce(III) is a function of the nature of B ions. The following sequence is obtained : Ti > Fe > Mn > Co > Cr. A rather high amount of Ce(III) accompanies Ti, while it is difficult to incorporate it in presence of Cr. Fe, Mn and Co represent intermediate cases.

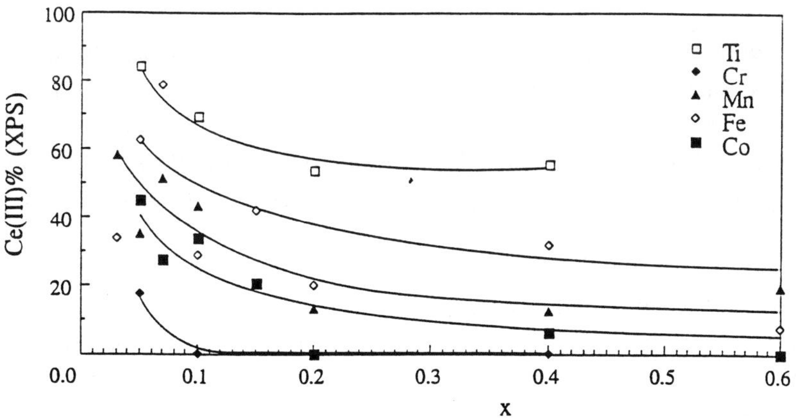

Figure 1. Variation of the amount of Ce(III) (in %) as a function of x in the perovskites $La_{1-x}Ce_xBO_3$.

Table 3 Atomic ratios Ce/La from XPS La3d and Ce3d peak intensities for the catalysts $La_{1-x}Ce_xBO_3$ calcined at 1000°C

x	Ce/La (synthesis)	Ce/La (XPS) Ti	Cr	Mn	Fe	Co
0.03	0.031	-	-	0.20	0.22	-
0.05	0.052	0.12	0.12	0.25	0.21	0.13
0.07	0.075	-	-	0.23	0.30	0.19
0.10	0.11	-	0.23	0.28	0.25	0.17
0.15	0.18	-	-	-	0.23	0.28
0.20	0.25	0.23	0.36	0.51	0.56	0.43
0.40	0.67	0.45	0.79	0.56	0.68	0.54
0.60	1.50	-	-	1.15	0.90	0.89

Following these values, two behaviours can clearly be distinguished. For x ^{2} 0.40, the surface atomic ratios are higher than the bulk values for all B ions, showing a surface excess Ce. Oppositely, for x > 0.40, due to the sintering of CeO_2 particles, the surface atomic ratios are lower than the corresponding bulk values. For low x values, the surface CeO_2 is highly dispersed and not detected by XRD (Table 1). For high x values, the sintering of CeO_2 particles is also confirmed by XRD.

Note, that the surface Ce(III)/La values are generally independent on the Ce/La ratio in the synthesis mixture. This amount of Ce is considered being incorporated in the perovskite lattice.

As a conclusion, it can be underlined that Ce(III) is incorporated in the lattice, while CeO_2 is either in a highly dispersed form or is found as larger particles.

The O1s bond energies are not influenced by the Ce substitution. Two types of oxygens - lattice and adsorbed - are also distinguished as for the pure perovskites. However, the quantitative analysis is more difficult in this case, because both perovskite and CeO_2 contain the two types of oxygen. The $O_{lattice}$/La ratios are all smaller than the corresponding theoretical values, indicating that all the catalysts are oxygen deficient.

For Fe, Co and Ni, the decrease of O_{ads} as a function of x shows the same variation as Ce(III) surface (Figure 2a). Hence, it can be concluded that the incorporation of Ce(III) in the lattice has a negative effect on the adsorption of oxygen.

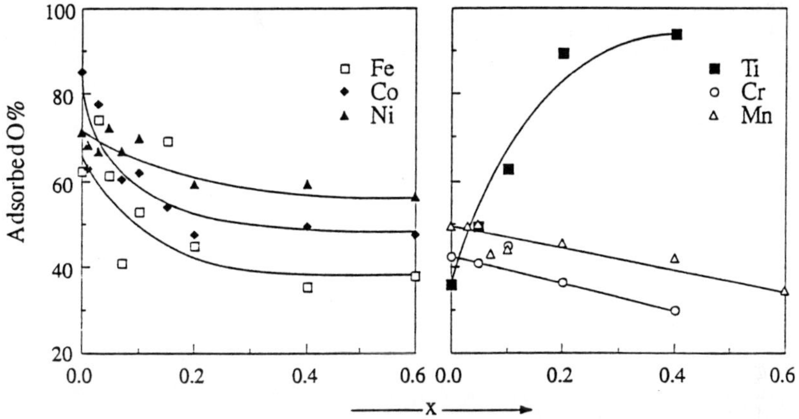

Figure 2. Variation of the amount of adsorbed oxygen (in %) as a function of x in the perovskites $La_{1-x}Ce_xBO_3$.

For Cr and Mn, a continuous decrease of O_{ads} is observed, while for Ti, O_{ads} increases greatly as a function of x (Figure 2b). The latter could be explained by the non-stoechiometry of the initial perovskite $La_{0.7}TiO_3$, by the high dispersion of the CeO_2 formed which also contribute to the amount of O_{ads}.

3.3. Catalytic activity

The pure perovskites are all more active for CO + O_2 than CO + NO reactions. The best catalysts for both reactions are $LaMnO_3$ and $LaCoO_3$. The activity of the different pervoskites for CO oxydation can be linked semiquantitatively with the ease of anionic vacancy formation in the lattice, described by the B-O bond energy (Figure 3).

The formal kinetics were determined only on the less active $LaCrO_3$ catalyst. The Langmuir-Hinshelwood model explains quite well the experimental rate equation :

$$R = kp_{CO}p_{O_2}^{0.5}$$

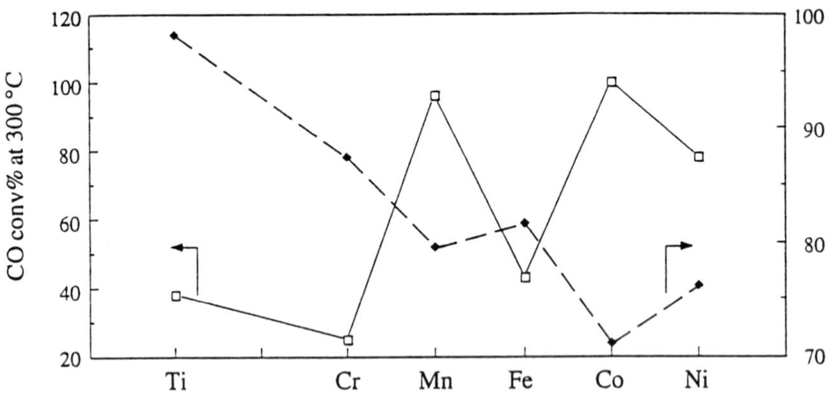

Figure 3. Comparison between the catalytic activity and the B-O bond energy in the perovskites $LaBO_3$ (B = Ti, Cr, Mn, Fe, Co, Ni).

No CO conversion was observed in absence of adsorbed oxygen, confirming the reaction mechanism involving both adsorbed species. The apparent energy of activation is equal to 12.2 kcal mol^{-1}.

The CO + NO reaction is better explained by a redox mechanism, where CO is oxidized by the catalyst which is regenerated by the reduction of NO. Moreover, a dissociative adsorption of NO better explains the experimental results on $LaCoO_3$ and $LaMnO_3$, while a molecular adsorption has to be supposed on $LaCrO_3$ adn $LaFeO_3$.

Simultaneous adsorption of CO and NO shows a competiton for the same active sites [14]. The latter are generally surface oxygens leading to the formation of nitrates, nitrosyls, dinitrosyls, and carbonates, carbonyls, respectively. On the other hand, the dissociative adsorption of NO leads to the formation of N_2 at higher temperature.

The following reaction scheme is adequate for the LaCoO3 and LaMnO3 catalysts :

CO + *	⇔¤ CO—*	(1)
NO + 2*	⇔¤ N—* + O—*	(2)
CO—* + O—*	⇔¤ CO_2 + 2*	(3)
N—*	⇔¤ $1/2N_2$ + *	(4)

Leading to

$$R = \frac{k p_{CO} p_{NO}^{\ 0.5}}{1 + k' p_{CO}}$$

The computation of the activation energy allows one to determine $Ea_3 + 0.5\ \Delta H_{NO}$: 14.4 kcal mol^{-1} for $LaMnO_3$ and 12.5 kcal mol^{-1} for $LaCoO_3$, respectively. Ea_3 is the activation energy for the CO oxidation (step 3) and ΔH_{NO} is the adsorption energy of NO in a dissociated form. The heat of adsorption of CO is determined to be equal to -7 kcal mol^{-1}, value to be compared to the literature value of - 6 kcal mol^{-1} [15].

The competition between CO and NO is detected in the reaction scheme CO + NO on $LaCrO_3$ and $LaFeO_3$. Indeed, a negative partial order equal to -1 is obtained for NO on both catalysts.

The reaction scheme allows one to explain the experimental partial orders and the rate equation

$$R = \frac{k p_{CO}}{1 + k' p_{NO}}$$

From the apparent activation energies, the heat of adsorption for NO adsorption is determined to be equal to - 9 kcal mol^{-1}.

The addition of Ce to the perovskites leads to different effects depending on the nature of B ions and on the relative amount of Ce. It has to be emphasized, that CeO_2 itself is also a good catalyst for CO oxidation with O_2 (97 mol % CO conversion at 300°C). This activity is equal to that of $LaMnO_3$, but it is inferior to the activity of $LaCoO_3$.

The following reaction scheme is adequate for the $LaCoO_3$ and $LaMnO_3$ catalysts :

CO + *	⇔¤ CO—*	(1)
NO + 2*	⇔¤ N—* + O—*	(2)
CO—* + O—*	⇔¤ CO_2 + 2*	(3)
N—*	⇔¤ $1/2N_2$ + *	(4)

Leading to

$$R = \frac{kp_{CO}p_{NO}^{\ 0.5}}{1 + k'p_{CO}}$$

The computation of the activation energy allows one to determine $Ea_3 + 0.5\ \Delta H_{NO}$: 14.4 kcal mol^{-1} for $LaMnO_3$ and 12.5 kcal mol^{-1} for $LaCoO_3$, respectively. Ea_3 is the activation energy for the CO oxidation (step 3) and ΔH_{NO} is the adsorption energy of NO in a dissociated form. The heat of adsorption of CO is determined to be equal to -7 kcal mol^{-1}, value to be compared to the literature value of - 6 kcal mol^{-1} [15].

The competition between CO and NO is detected in the reaction scheme CO + NO on $LaCrO_3$ and $LaFeO_3$. Indeed, a negative partial order equal to -1 is obtained for NO on both catalysts.

The reaction scheme allows one to explain the experimental partial orders and the rate equation

$$R = \frac{kp_{CO}}{1 + k'p_{NO}}$$

From the apparent activation energies, the heat of adsorption for NO adsorption is determined to be equal to - 9 kcal mol^{-1}.

The addition of Ce to the perovskites leads to different effects depending on the nature of B ions and on the relative amount of Ce. It has to be emphasized, that CeO_2 itself is also a good catalyst for CO oxidation with O_2 (97 mol % CO conversion at 300°C). This activity is equal to that of $LaMnO_3$, but it is inferior to the activity of $LaCoO_3$.

The activity of $LaCoO_3$ first increases as a function of x in $La_{1-x}Ce_xCoO_3$. This increase is explained by the beneficial effect of dispersed CeO_2 on the surface of the catalyst. However, the concomitant increase of Ce(III) incorporation into the perovskite lattice leading to A site vacancies could suggest that these cation vacancies could also have a positive effect on CO oxidation, as it was suggested earlier [16]. Note, that a mechanical mixture of CeO_2 (5 mol%) and $LaCoO_3$ is less active than the corresponding Ce containing $La_{0.95}Ce_{0.05}CoO_3$ catalyst. For higher Ce content, the activity decreases and is levelled off, showing that the larger CeO_2 particles are less active. Similar conclusions can be drawn for the Mn, Fe and Cr containing catalysts.

The activity of $La_{1-x}Ce_xTiO_3$ catalysts is systematically lower than that of $La_{0.7}TiO_3$. A first large decrease of the activity for x = 0.05 is followed by a slight increase for x > 0.05 due to the formation of CeO_2 on the surface.

Finally the $La_{1-x}Ce_xNiO_3$ catalysts are more active than $LaNiO_3$ for x > 0.2. For lower x values the influence of Ce is negligible.

CeO_2 is a relatively good catalyst for NO + CO reaction. Its activity (equal to 10 mol% CO conversion at 300°C) is higher than those of Ti, Cr, Mn, Fe and Ni perovskites (1.7 %, 2.5 %, 9.0 %, 5.9 % and 2.1 %, respectively). Only Co perovskite shows a higher activity (equal to 40.8 % conversion). Note, that the activities are always lower for the NO reduction than the CO oxidation.

For all perovskites the increase in activity as a function of x can be linked to the increase of Ce(III) content in the lattice. For Ti and Ni perovskites, after an initial increase of the activity a plateau is reached. For Fe and Cr perovskites, the activity is decreasing after the initial increase, showing a maximum in the activity vs x curves.

The Mn perovskite shows an initial small decrease of activity followed by a larger increase as a function of x. The behaviour of $La_{1-x}Ce_xCoO_3$ perovskites is more complex. A quite large initial decrease of activity is observed for x < 0.05. After that value a maximum is reached at x = 0.20 (activity = 65 mol % CO conversion at 300°C), followed by a decrease in activity (ca 40 mol% CO conversion at x = 0.60). For the initial addition of Ce, the activity decreases because the CeO_2 formed has a lower activity. At higher Ce content, larger CeO_2 crystallites are formed freeing the more active surface of the perovskite catalyst. Finally, the activity decreases because the amount of Ce(III) incorporated in the lattice decreases. Note, that for small x values, the influence of Ce(III) concentration is hidden by the effect of CeO_2 on the surface.

A kinetic study for the CO + NO reaction on $La_{0.95}Ce_{0.05}CoO_3$ revealed a decrease of the partial order with respect to NO : it is equal to 0.2, in comparison with the value of 0.5 obtained for the pure perovskite. Further studies are necessary to ascertain the specific role of both Ce(IV) and Ce(III) ions in the CO + NO reaction.

4. Conclusion

Ce(III) is incorporated in the perovskite lattice for low x values in $La_{1-x}Ce_xBO_3$ perovskites.The remaining Ce is in a highly dispersed CeO_2 form. For higher x values, larger CeO_2 crystals are formed. The different phases formed are active in both CO + O_2 and CO + NO reactions. The activity for the former reactions is always higher than that for CO + NO. The best perovskites are obtained with B = Mn and Co.

References

1 W.F. Libby, Science, 171 (1971) 499.
2 F. Le Normant, P. Bernhardt, L. Hilaire, K. Kili, G. Krill and G. Maire, Stud. Surf. Sci. Catal., 30 (1987) 221.
3 T. Shimizu, Chem. Lett., (1980) 1.
4 T. Nitadori, S. Kurihara and M. Misono, J. Catal., 98 (1986) 221.
5 D.W. Johnson, P.K. Gallagher, G.M. Wertheim and E.M. Vogel, J. Catal., 48 (1977) 87.
6 R.J.H. Voorhoeve, J.P. Rameika, L.E. Trimble, A.S. Cooper, F.S. Disalvo and P.K. Gallagher, J. Solid State Chem., 14 (1975) 395.
7 G.L. Bauerle, N.T. Thomas and K. Nobe, Chem. Eng. J., 4 (1972) 199.
8 Y. Teraoka, T. Harada, H. Furukawa and S. Kagawa, Guczi L. et al. (eds), New Frontiers in Catalysis, Stud. Surf. Sci. Catal., 75 (1993) 2649.
9 H. Shimada, S. Miyama and H. Kuroda, Chem. Lett. (1990) 1.
10 V. Deremince-Mathieu, H. Collette, J.B. Nagy, E.G. Derouane and J.J. Verbist, Inorg. Chim. Acta, 140 (1987) 41.
11 V. Deremince-Mathieu, PhD Thesis, FUNDP, Namur, 1989.
12 H. Arai, T. Yamada, K. Eguchi and T. Seyiama, Appl. Catal., 26 (1986) 265.
13 J.L. G. Fierro and L.G. Tejuca, Appl. Surf. Sci., 27 (1987) 453.
14 see e.g. M.A. Peña, J.M.D. Tascon, J.L.G. Fierro and L. Gonzalez Tejuca, J. Colloid Interface Sci., 119 (1987) 100 and references therein.
15 J.M.D. Tascon and L.G. Tejuca, React. Kinet. Catal. Lett., 15 (1980) 185.
16 R.J.H. Voorhoeve, in 'The Catalytic Chemistry of Nitrogen Oxides', J.C. Larson, ed., New York, 1975, p 215.

A. Frennet and J.-M. Bastin (Eds.)
Catalysis and Automotive Pollution Control III
Studies in Surface Science and Catalysis, Vol. 96

PREPARATION OF ALUMINA SUPPORTED CERIA. I : SELECTIVE MEASUREMENT OF THE SURFACE AREA OF CERIA AND FREE ALUMINA.

R. Fréty[a], P. J. Lévy[a], V. Perrichon[a], V. Pitchon[a], M. Primet[a], E. Rogemond[a], N. Essayem[a], M. Chevrier[b], C. Gauthier[b] and F. Mathis[b],

[a]*Laboratoire d'Application de la Chimie à l'Environnement, LACE/CNRS, Université Claude Bernard, Lyon I, 43 Bd. du 11 Novembre 1918, 69622 Villeurbanne Cedex, France.*

[b]*Régie Nationale des Usines RENAULT : Direction des Etudes Matériaux, 8-10 avenue Emile Zola, 92109 Boulogne-Billancourt Cedex, France. Centre de Lardy, 1 Allée Cornuel, 92510 Lardy, France. Direction de la Recherche, 9-11 Avenue du 18 Juin 1940, 92500 Rueil Malmaison, France.*

ABSTRACT

In a general study of the role played by each component of three-way catalysts, a selective measurement of alumina and ceria surfaces area has been developed on model ceria-alumina supports. Three CeO_2/Al_2O_3 solids (6.5, 13.4, 21 wt.% CeO_2) were prepared by grafting Ce acetylacetonate on the surface of an alumina and calcination at 673 K. They were characterized by BET and X-ray diffraction measurements. The hydrogen temperature-programmed reduction (TPR) resulted in profiles similar to those of unsupported cerias, thus allowing an estimation of the equivalent surface area of ceria alone from the hydrogen uptake at low temperature. By adsorbing CO_2 on the OH groups of the alumina surface, hydrogenocarbonates species were formed and the IR band at 1235 cm^{-1} was selected to determine quantitatively the unpertubed alumina surface. In that case, by comparison with the BET surface area, it was possible to deduce an other value of the ceria surface area. The two sets of values obtained by these independent methods were found in good agreement, thus giving a good support for the reliability of each method. They show that the unpertubed alumina surface decreases when increasing the cerium content. Only 20% of the original alumina surface is preserved in the support containing 21 wt.% CeO_2.

1. INTRODUCTION

Ceria and alumina, which form the intermediate porous layer (the washcoat) between the mechanical support and the supported metals and promotors, are important components in three-way catalysts (TWC) used for car exhaust gas cleaning. Although basic studies have been published on such systems [1], the interactions which can exist between alumina and ceria and which in turn may affect the interactions between the supported metal and the washcoat [2], are not fully understood. In particular, the multiple roles attributed to ceria, like stabilization of the alumina [3], of the supported metals [4], like storage and release of oxygen [2], are very probably dependent on preparation methods, activation and reaction conditions. Therefore, selective characterization of each oxide in ceria-alumina can be useful for a better understanding of the role played by both alumina and ceria in TWC. The present work describes two "chemical methods", the adsorption of carbon dioxide followed by IR spectroscopy and the temperature-programmed reduction (TPR), which lead to an estimation of the surface extent of alumina and ceria respectively.

2. EXPERIMENTAL

2.1. Preparation

Cerium (III) acetylacetonate in toluene solution has been grafted on the hydroxyl groups of a powdered form of a transition alumina support (SCM129/Rhône-Poulenc) in order to deposit ca. 3 Ce ions per nm^2 of alumina surface ; the population of OH groups in alumina has been estimated to be 10 OH/nm^2[5]. After reaction, the solid was dried *in vacuo* (373 K, 10 h) and calcined under flowing oxygen (0.5 K/min, 673 K, 4 h). It is refered thereafter to as Ce3Al. Solids with 6 and 9 Ce ions per nm^2 of alumina were prepared by repeating two times the previous procedure to give Ce6Al and Ce9Al. After each grafting, the solids were dried and calcined as before.

Unsupported ceria (HSA/Rhône-Poulenc) and pure alumina (SCM129, Rhône-Poulenc) were used as reference samples. These two oxides have specific surface areas of 127 and 107 m^2/g, respectively.

2.2. Characterization

Ceria-aluminas were studied by nitrogen adsorption at 77 K, in an automated volumetric set up built in the laboratory after a vacuum desorption at 773 K. All the BET measurements were refered to the initial mass of the sample.

The X-ray diffraction (XRD) patterns were obtained with a Siemens D5000 apparatus. Line broadening was used for the determination of the crystallite sizes.

FTIR spectroscopy (BRUKER IFS 110) study of the adsorption of carbon dioxide on the solids was used to determine the free alumina surface area. More precisely, the adsorption of CO_2 onto hydroxyl groups of the alumina produced characteristic bands of hydrogenocarbonate species [6,7]. The optical density of the band at 1235 cm^{-1}, (δ C-O-H bending mode), after taking into account the bands displayed by ceria, was measured and used to determine the free alumina surface. From this method and by comparison with total BET surface area, a ceria surface has been estimated. Ceria-aluminas, alumina and ceria were used as self supported wafers (0.01 to 0.03 g/cm^2); they were thermally treated up to 673 K, under oxygen and then under high vacuum, *in situ*, before adsorption of CO_2 at room temperature. Then, the catalysts were evacuated at 295, 373, 473 and 573 K, for 1 h. IR spectra were recorded at room temperature, after CO_2 adsorption and after each desorption temperature.

The TPR experiments were performed in a set up described previously [8]. A weight amount of ca.0.2 g was located on the porous disc of a quartz microreactor, pretreated under a flow of air up to 673 K, for 1 h, then outgassed under argon flow (773 K, 1.5 h) and cooled to room temperature before a mixture argon + 1 vol.% hydrogen was deviated onto the catalyst. The heating rate (20 K/min), the flow rate (20 ml/min) and the final reduction temperature (1073 K) were kept constant.

3. RESULTS

3.1. Preliminary observations

Table 1 gives the code of the samples, the ceria contents and the BET values. When increasing the cerium content, in the mixed solids, the BET surface area of the samples decreases but is always higher than that expected from the simple alumina contribution. For ceria-aluminas calcined at 673 K, XRD detected small particles of ceria. A mean crystallite size obtained from the (100) and (200) lines broadening gave values of 4.2, 5.5 and 6.0 nm, for the three samples Ce3Al, Ce6Al and Ce9Al, respectively. Therefore, the preparation method using cerium (III) acetylacetonate does not lead to big supported ceria particles. However, it is known [1] that multiple cerium species exist in ceria-alumina and XRD alone may be insufficient to fully describe the cerium dispersion. Considering the literature data showing linear correlation between the low temperature hydrogen consumption during TPR and the initial surface area of unsupported cerias [9-11], the TPR technique was applied to characterize the present ceria-aluminas samples.

3.2. Indirect determination of ceria surface area

A preliminary TPR-SM experiment was performed on Ce3Al without *in situ* pretreatment before reduction. In addition to the expected hydrogen consumption at temperatures above ca. 623 K, the MS analyses show that the catalyst releases carbon dioxide, in three different steps, two of them peaking before 673 K. Therefore, the catalysts have been flushed *in situ* at 673K-773 K to eliminate the majority of CO_x species, before cooling to room temperature and initiating the TPR run.

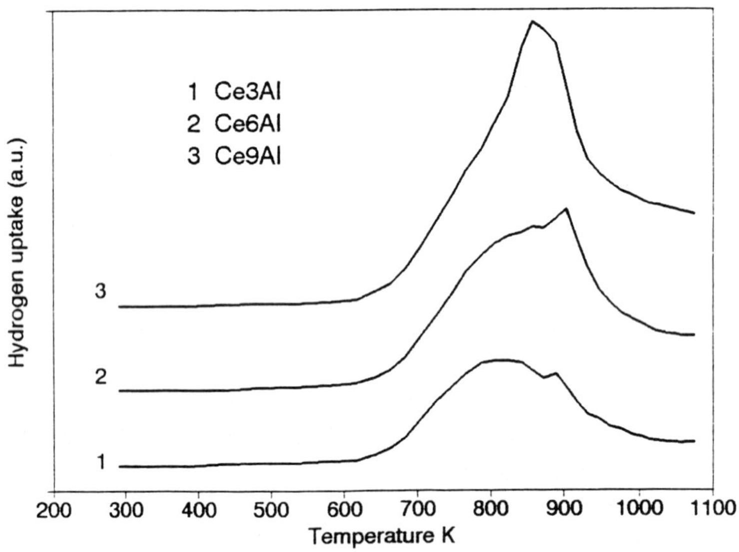

Figure 1 : TPR profiles corresponding to the ceria-alumina samples pretreated in air (673 K for 1 h) then under Ar (773 K for 1.5 h). $H_2/Ar = 1/99$*, heating rate : 20 K* min^{-1}.

The TPR profiles for Ce3Al, Ce6Al and Ce9Al, after *in situ* pretreatments, are shown in Figure 1. In previous experiments [9,11], TPR of unsupported ceria have shown the presence of at least two important hydrogen uptakes, in agreement with literature data [1,2,10], one at moderate temperature (600-870 K) attributed to the reaction of hydrogen with *capping* oxygen ions, and a second one at much higher temperature due to the bulk reduction of ceria.

The TPR profiles for supported ceria present two to three poorly resolved peaks, in the so-called low temperature domain (600-950 K), and a slowly decreasing consumption of hydrogen after stabilization of the temperature at 1073 K. The low temperature H_2 consumption phenomenon increases as the cerium content in the samples increases, but shows also that more than one chemical species is concerned. Since, the temperature for the beginning of this first phenomenon is rather constant, at ca. 600 K, the low temperature reduction peaks must always deal with the same type of cerium species.

Table 1: Alumina, ceria and ceria-aluminas : summary of the results obtained in the present study.

Code	CeO_2 wt%	S_{BET} m^2/g	OD_{1235} /g	SAl m^2/g cat.	SCe1 m^2/g cat.	SCe2 m^2/g
Al_2O_3	0	107	2.4	107*	0	0
Ce3Al	6.5	102	1.2	54	35	48
Ce6Al	13.4	101	0.8	36	55	65
Ce9Al	21.0	92	0.3	13	74	79
CeO_2	99.5	127	0	0*	127	127

* values used for a linear relationship between ceria and alumina (FTIR spectroscopy)

With the hypothesis that the hydrogen uptake at the lower temperature is characteristic of the reduction of Ce^{4+} at the surface of cerium species, and taking into account the linear relationship between the initial surface area of unsupported cerias and the hydrogen consumption during the low temperature uptake [11], it is possible to convert each hydrogen uptake at low temperature to an "equivalent initial surface area" of ceria (SCe1).

Table 1 reports the values of SCe1 deduced from TPR experiments, for the alumina-supported ceria and the reference samples.

By increasing the cerium content, the value of SCe1 increases significantly; for Ce9Al, the value of SCe1 is almost 60% that of the reference ceria, whereas its ceria content is only one fifth of the total solid. Thus, the dispersion of ceria is rather high, as well as the coverage of the original alumina support by cerium species.

3.3. Indirect determination of the specific surface area of alumina

In order to cross-check the preceding results with a second "chemical" method, experiments able to detect selectively the fraction of alumina surface uncovered by ceria were looked for. As other alumina-supported oxides [12-16], adsorption of carbon dioxide was considered. Preliminary volumetric results of the adsorption of carbon dioxide at room temperature, after calcination and outgassing under vacuum at 673 K, were not fully conclusive, as both ceria and alumina are able to chemisorb carbon dioxide; further, although the adsorption strength of carbon dioxide is much lower on alumina than on ceria, experiments for isolating selective contribution of ceria, after desorption of CO_2 from alumina at 373 K were also unsatisfactory, as some 20% of CO_2 adsorbed on ceria was also desorbed during evacuation.

Consequently, the IR spectroscopy of adsorbed CO_2 has been considered. IR spectra in the hydroxyl group range (3800-3400 cm^{-1}), for both pure alumina and ceria, are shown in Figures 2a and 2b, respectively. Curves 1 refer to samples calcined at 673 K and outgassed at the same temperature and then, cooled to room temperature. Curves 2 refer to the catalysts after adsorption of CO_2 at room temperature (pCO_2 = 9-15 torrs) and curves 3 to the samples after a further evacuation at room temperature for 1 h.

Curves 2 and 3 (Figure 2a) show by comparison with Curve 1 that the adsorption of CO_2 on alumina generates a band at 3610 cm^{-1}, stable after evacuation at room temperature. This band is attributed to the stretching mode of OH group in hydrogenocarbonate species [6,7], its intensity decreases after desorption at 373 K, for 1h, and the band is completely eliminated after desorption at 473 K, for 1 h.

As seen in Figure 2b (Curves 2 and 3), the adsorption of CO_2 on CeO_2 produces extremely complex spectra, both in presence of CO_2 and after evacuation of the gas phase. Furthermore, the band at 3618 cm^{-1} is very close to the position of the hydrogenocarbonate species in the case of alumina. These two reasons are sufficient to eliminate the possibility to obtain selective measurement of surface properties of alumina in this region of the IR spectrum.

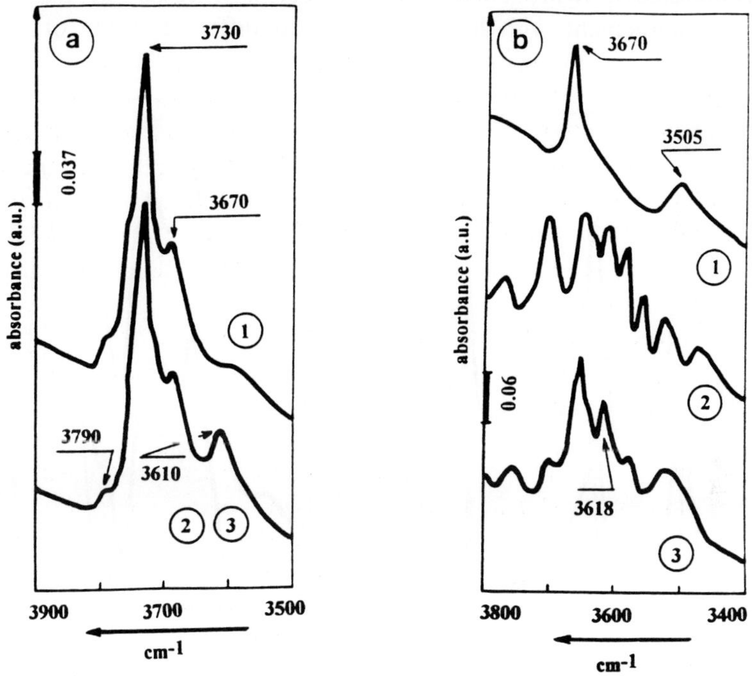

Figure 2 : FTIR spectra in the hydroxyl groups range for pure alumina (figure 2a) and pure ceria (figure 2b) :
1 : spectra after oxygen treatment at 673 K and evacuation at 673 K(alumina)
2 : introduction of CO_2 (10 torrs) at 298 K for 1h,
3 : desorption of CO_2 at 298 K for 1h.

Figures 3a and 3b present the IR spectra of CO_2 in the 1100-1900 cm^{-1} range, for alumina and ceria respectively. Curves 1 refer to the samples in the presence of a pressure of 10 torrs of CO_2 whereas curves 2 refer to the same solids after desorption at room temperature. In Figure 3a, a lot of absorptions are present; the vibrations at ca. 1860, 1820, 1775, 1705, 1437, 1320, and 1200 cm^{-1} show intensity stable or decreasing when the time of contact between CO_2 and the catalyst increases; these bands are also greatly decreased after evacuation at room temperature. The more intense vibrations at 1648, 1483 and 1235 cm^{-1} have an intensity growing with the contact time and are still clearly visible after evacuation at room temperature. All the preceding bands decrease sharply after desorption at 373 K, the stability of the bands at 1648, 1483 and 1235 cm^{-1} being however higher than that of the others. After evacuation at 473 K, the surface of the alumina is freed from all these adsorbed species. These three main bands are due to vibrations of hydrogenocarbonate species, whereas the others can be attributed, in agreement with literature data, to different forms of carbonates species [6,17,18].

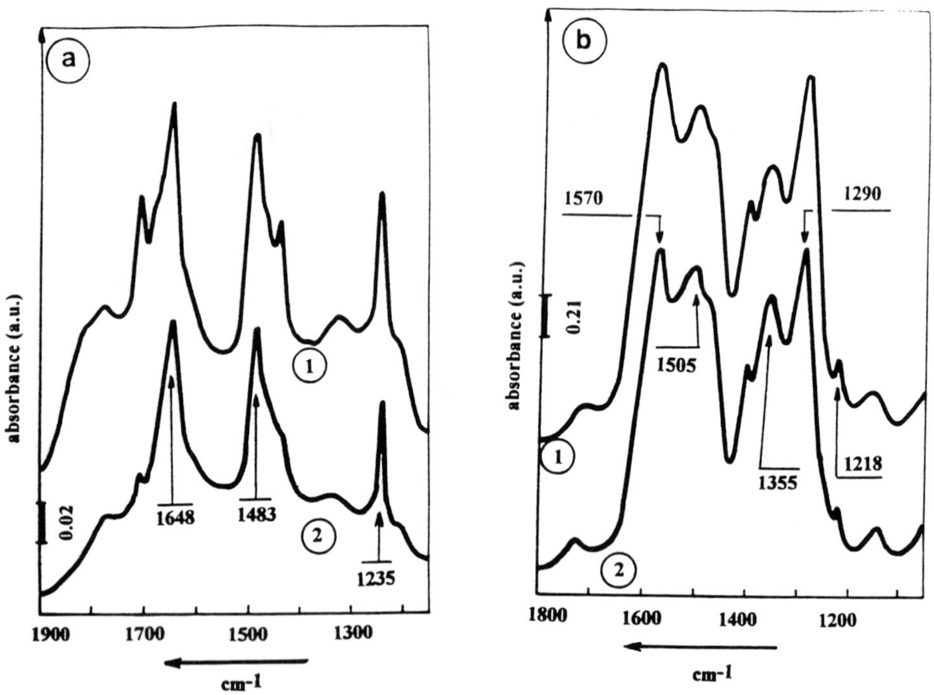

Figure 3 : FTIR spectra in the carbonate groups range for pure alumina (figure 3a) and pure ceria (figure 3b). Pretreatment identical to Figure 2 :
1 : introduction of CO_2 (10 torrs) at 298 K for 1h,
2 : desorption of CO_2 at 298 K for 1h.

The general aspect of the spectrum of CO_2 species adsorbed on CeO_2 is practically not modified by evacuation at room temperature (Figure 3b), excepted the small shifts of the bands at 1730, 1354 and 1140 cm^{-1}. All the bands at 1730, 1570, 1505, 1470 (shoulder), 1400, 1355, 1290, 1218, 1140, 1050 and 1040 (shoulder) cm^{-1} are attributed to carbonates and carboxylates species [19-23]. It is important to mention that the bands present at 1648 and 1235 cm^{-1} in the spectrum of CO_2 irreversibly adsorbed on alumina, appear at positions allowing their discrimination from the contribution of ceria. Therefore, they can be tentatively used to characterize free alumina surface in ceria-alumina samples.

Figure 4 shows the spectra of CO_2 irreversibly adsorbed at room temperature on the three ceria-alumina solids. All samples present a very wide absorbance in the region 1300-1700 cm^{-1}, in which dominate the features of alumina (Ce3Al and Ce6Al) or of ceria (Ce9Al). From Figure 4, it appears that only the band at 1235 cm^{-1} can be used with enough precision for the measurement of optical density. For all the other bands of the spectra, uncertainties in the position of the base line and strong shadowing of individual bands impede serious selective measurement.

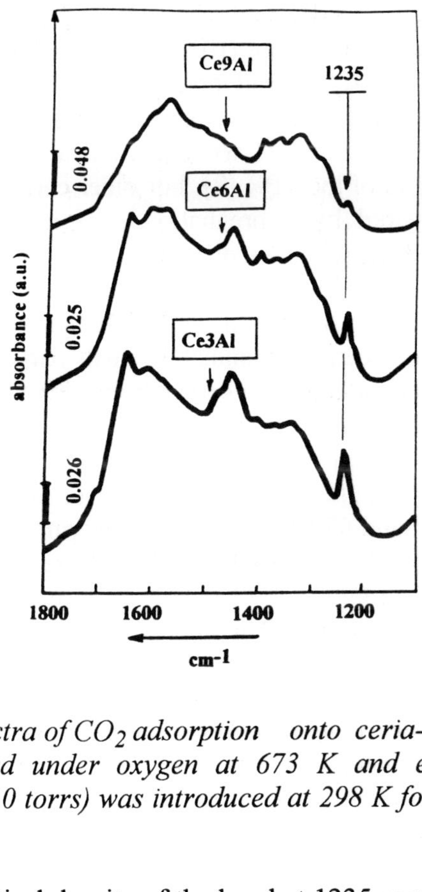

Figure 4 : FTIR spectra of CO_2 adsorption onto ceria-aluminas samples. The solids were calcined under oxygen at 673 K and evacuated at the same temperature. CO_2 (10 torrs) was introduced at 298 K for 1 h and evacuated for 1 h at 298 K.

The values of the optical density of the band at 1235 cm^{-1} expressed per gram of sample are also given in Table 1, and are converted in terms of not modified surface of alumina, SAl, using a linear relationship between the optical density of the 1235 cm^{-1} band for pure alumina and a value 0 for pure ceria. The difference between the total surface area of ceria-alumina (BET method) and SAl measured from the preceding data is noted SCe2; the values of SCe2 are also reported in Table 1. For the ceria-aluminas solids of the present study, the two sets of values SCe1 and SCe2, obtained by completely independent methods, are in relatively good agreement.

4. DISCUSSION

Ceria-aluminas have been studied in the past by different techniques, like TPR, Raman spectroscopy, XRD and XPS [1,2,10,24,25]. For ceria-alumina prepared by impregnation with cerium (III) nitrate and calcined for 16 h at 773-1073 K, TPR up to 1173 K showed poorly resolved TPR traces. The TPR profiles display one main TPR peak at low cerium loading and at least 3 hydrogen uptakes for cerium loading around 21 wt.% [2]. Apparently, all Ce^{4+} ions were reduced to Ce^{3+} at the end of the TPR [2]. For other ceria-aluminas containing 1 to 17 wt.% ceria, prepared by impregnation of a non microporous alumina (alumina C/Degussa, 100^2/g) with aqueous solution of cerium (III) nitrate, drying and calcination at 1073 K for 4 h, the presence of three different types of cerium species was advocated [1] : a species in close interaction with the alumina and refered to as "precursor of cerium aluminate", small particles of ceria with size lower than the XRD detection limit and ceria particles with size equal to or higher than 20 nm [1]. The former species would be prevalent at low ceria loadings, whereas the latter ones would exist mainly for highly loaded samples. In this second study, TPR results suggested that the precursor of the cerium aluminate is "partially" reduced at around 673 K and fully transformed to the aluminate at ca. 1000 K, whereas the small ceria particles and the large ceria particles would be reduced around 873 K and above 1073 K respectively. Later, with unsupported cerias of different specific surface areas, a reduction model was proposed, consisting in reducing first "capping" oxygen ions, during the low temperature peak, before reducing bulk ceria at higher temperatures [2,10]. Such a reduction model has been extended to other samples of unsupported cerias with larger surface areas, using both TPR and a selective quantification of Ce^{3+} through magnetic susceptibility measurements [9,11]. More precisely, a linear relationship between the initial specific surface area of ceria and the hydrogen consumed during the low temperature TPR peak, has been observed. A consumption 4.2 μmol H_2/m^2 of CeO_2, during the TPR, has been estimated [11].

The preceding results suggested that during the reduction of ceria, bulk reduction is very much slower than surface reduction, leading to a possible "selective measurement" of reducible species at the ceria surface.

In the present study, with the hypothesis that the density of the reducible oxygen ions at the surface of supported cerium species is similar to the one found in unsupported ceria samples, it was possible to transform the hydrogen uptake of the low temperature TPR phenomenon, during the reduction of ceria-alumina, in "initial equivalent ceria surface", SCe1. The values reported in Table 1 increase with the cerium loading, and seem consistent with a rather high coverage of alumina by cerium containing species, as expected from the grafting method used for the preparation. The high coverage of alumina by cerium species in the present samples is sustained by the comparison of the Ce/Al atomic ratios measured by XPS, in the present catalysts and in the ones described by Shyu et al. [1]. In our case, the atomic Ce/Al ratio was found close to 0.3, for the ceria-alumina with 21 wt.-% ceria, whereas Shyu et al. [1] measured only 0.03 in a catalyst containing some 20 wt.% ceria.

However, the specific surface area of ceria in ceria-alumina, as measured by TPR, is quite different from the values deduced from X-ray line broadening. This difference can be explained as the area calculated from XRD is 50% lower than the area calculated from TPR : XRD detects only ceria crystallites with a "sufficient" crystallite size, whereas TPR measures all reducible surface species, even those considered as $CeAlO_3$ precursor. In addition, with very small ceria particles or with phases prepared at rather low temperatures, numerous surface defects are probably able to interact with subsurface cerium species [11,26], and lead to hydrogen uptake higher than expected from a strict surface phenomenon. Anyway, considering that the low temperature TPR peak includes multiple cerium species, the difference between the cerium surface obtained by TPR and that deduced from XRD would represent the surface of cerium species in strong interaction with the alumina, a measure which could give deeper insight into the structure of ceria-aluminas.

The uncertainties discussed in the preceding section prompted us to look for another "chemical method" to estimate the surface of supported ceria. As for alumina-supported oxides [12-16], the adsorption of CO_2 has been considered, IR spectroscopy being prefered, with the aim of finding vibrations specific of CO_2 interacting either with ceria or with alumina. This method has been applied although very recent results, mainly with alumina supported molybdenum questionned the validity of CO_2 adsorption to measure the coverage of alumina by supported oxides [15,16]. In fact, these studies indicate that only a very limited fraction of the alumina OH groups are able to interact with CO_2 and that these OH are also those linked to the supported molybdenum species. However, if the use of total adsorption of CO_2 to estimate alumina coverage by supported oxides can be questionnable [15,16], the use of IR bands characteristic of HCO_3^- must give a reasonnable answer.

The results presented in Figures 2 to 4, confirmed the very complex situation of adsorbed CO_2 species at the surface of both alumina and ceria. However, the present results are in agreement with literature data [17-19,21]. With the hypothesis that neither the nature of the OH groups nor the formation of hydrogenocarbonate species on the alumina are modified by the presence of supported ceria (hypotheses well verified for Ce3Al and Ce6Al, in the OH region of IR spectra), it has been possible to calculate a free alumina surface in the mixed solids using the optical density (OD) of the band at 1235 cm^{-1}, since this band does not appear with pure ceria, and its frequency is not affected by the ceria presence. Although a more intense hydrogenocarbonate band, at 1645 cm^{-1} (symmetric C-O-H stretching), can also be used for this purpose, the problematic measure of its optical density made its use non reliable.

The results from CO_2 adsorption have been compared with those deduced from TPR. Considering the rather good agreement between the values reported in Table 1, each method seems to support the other.

Therefore, two different chemical methods allow to calculate selectively the surface area of alumina-supported cerium species.

It is important however to recall that for a true quantitative determination of the hydrogenocarbonate species by IR spectroscopy, the exact nature of the alumina and the thermal treatments which easily modify the hydroxyl population (type and number) have to be very well controlled. In the case of using TPR for estimating the coverage of alumina by ceria, it is necessary that all cerium species are initially in the Ce4+ oxidation state. Such a situation is probably not always warranted in samples calcined at a temperature high enough to form Ce^{3+}, in cerium aluminate phase. Further work is in progress to answer these questions.and to apply this methodology to industrial catalysts.

5. Conclusion

The aim of this work was to find a general method to discriminate ceria and alumina surface in ceria-alumina supports. Two methods were applied based either on the selective reduction of the ceria surface or the use of the 1235 cm^{-1} IR absorption band specific of the adsorption of CO_2 on the surface hydroxyl groups of alumina. The rather good agreement between the results obtained by the two methods supports the validity of the hypotheses made in each case. The main limitation concerns the possible modifications which may occur in the hydroxyl groups population of the alumina after various thermal and chemical treatments.

Acknowledgement

The technical assistance of Mr. J. Billy for infrared experiments is gratefully acknowledged.

References

1 J.Z. Shyu, W.H. Weber and H.S. Gandhi, J. Phys. Chem., 92 (1988) 4964.
2 H.C. Yao and Y.F. Yu Yao, J. Catal., 86 (1984) 254.
3 B. Harrison, A.F. Diwell and C. Hallett, Platinum Met. Rev., 32 (1988) 73.
4 J.C. Summers and S. Ausen, J. Catal., 58 (1979) 131.
5 A.V. Kiselev and V.I. Lygin (eds.), Infrared Spectra of Surface Compounds, J. Wiley and Sons, New York, (1975), p. 237.
6 J.B. Peri, J. Phys. Chem., 70 (1966) 3168.
7 C. Morterra, A. Zecchina, S. Coluccia and S. Chirino, J. Chem. Soc. Faraday Trans. I, 73 (1977) 1544.
8 F.M. Zotin, L. Tournayan, J. Varloud, V. Perrichon and R. Fréty, Appl. Catal., A General, 98 (1993) 99.
9 A. Laachir, V. Perrichon, A. Badri, J. Lamotte, E. Catherine, J.C. Lavalley, J. El Fallah, L. Hilaire, F. Le Normand, E. Quéméré, G.N. Sauvion and O. Touret, J. Chem. Soc., Faraday Trans. I, 87 (1991) 1601.
10 M.F.L. Johnson and J. Mooi, J. Catal., 103 (1987) 103, and 140 (1993) 612.
11 V. Perrichon, A. Laachir, O. Touret, G. Bergeret, R. Fréty and L. Tournayan, J. Chem. Soc., Faraday Trans. I, 99 (1994) under press.

12 K. Segawa and W Keith Hall, J. Catal., 77 (1982) 221.
13 Y. Okamoto and T. Imanaka, J. Phys. Chem., 92 (1988) 7102.
14 C. O'Young, C. Yang, S.J. DeCanio, M.S. Patel and D.M. Storm, J. Catal., 113 (1988) 307.
15 L. Gonzalez, J.L. Galavis, C. Scott, M.J. Perez Zurita and J. Goldwasser, J. Catal., 144 (1993) 636.
16 F.M. Mulcahy, K.D. Kozminski, J.M. Slike, F. Ciccone, S.J. Scierka, M.A. Eberhardt, M. Houalla and D.M. Hercules, J. Catal., 139 (1993) 688.
17 N.D. Parkins, J. Chem. Soc. (A), (1969) 410.
18 L.H. Little and C.H. Amberg, Canad. J. chem., 40 (1962) 1997.
19 C. Li, K. Domen, K. Maruya and T. Onishi, J. Amer. Chem. Soc.,111 (1989) 7683.
20 C. Li, Y. Sakata, T. Arai, K. Domen, K. Maruya and T. Onishi, J. Chem. Soc., Faraday Trans.I, 85 (1989) 929, and 85 (1989) 1451.
21 F. Bozon-Verduraz and A. Bensalem, J. Chem. Soc., Faraday Trans. I, 90 (1994) under press.
22 T. Jin, Y. Zhou, G.T. Mains and J.M. White, J. Phys. Chem., 91 (1987) 5931.
23 A. Badri, S. Lamotte, J.C. Lavalley, A. Laachir, V. Perrichon, O. Touret, G.N. Sauvion and E. Quémeré, Eur. J. Solid State, Inorg. Chem., 28 (1991) 445.
24 G.W. Graham, W.H. Weber, C.R. Peters and R. Usmen, J. Catal., 130 (1991) 310.
25 J.G. Nunan, H.J. Robota, M.J. Cohn and S.A. Bradley, J. Catal., 133 (1992) 309.
26 N. Kaufherr, L. Mendelovici and M. Steinberg, J. Less Common Metals, 107 (1985) 281.

A. Frennet and J.-M. Bastin (Eds.)
Catalysis and Automotive Pollution Control III
Studies in Surface Science and Catalysis, Vol. 96
1995 Elsevier Science B.V.

INFLUENCE OF THE NATURE OF THE METAL PRECURSOR SALT ON THE REDOX BEHAVIOUR OF CERIA IN Rh/CeO_2 CATALYSTS

S. Bernal[a], J.J. Calvino[a], G.A. Cifredo[a], J.M. Gatica[a], J.A. Pérez Omil[a], A. Laachir[b] and V. Perrichon[b]

[a]*Departamento de Ciencia de Materiales, Ingeniería Metalúrgica y Química Inorgánica. Facultad de Ciencias. Universidad de Cádiz. Apartado 40. Puerto Real. 11510 Cádiz. Spain.*
[b]*LACE/CNRS, Université Lyon I, IRC. 2, Avenue Albert Einstein. 69626 Villeurbanne Cedex. France.*

ABSTRACT

The hydrogen chemisorption on two Rh/CeO_2 catalysts prepared respectively from rhodium nitrate and rhodium chloride has been studied. The evolution of the ceria oxidation state with the series of treatments applied was monitored by means of a magnetic balance. Temperature Programmed Reduction and Oxidation (TPR/TPO) as well as Volumetric Adsorption techniques have also been used.

Upon treating with H_2, at 623 K, the Precursor/Support systems, the reduction level reached by ceria was 11.4% for the sample prepared from $Rh(NO_3)_3$ and much higher, 22.1%, for the chlorine containing catalyst. In the later case, the reduction degree varied very slightly with succesive evacuation/hydrogen adsorption cycles. This contrasts with the behaviour observed for the sample prepared from rhodium nitrate, for which the reduction process is to a much larger extent reversible. The results reported here for $RhCl_3/CeO_2$ are interpreted as due to the substitution of the ceria lattice oxygen ions by Cl^-, thus blocking the operation of both the direct and incorporation of Cl^- into the support. This induces a much higher irreversible reduction of ceria as well as the blocking of both the direct and back spillover processes responsible for the reversibility of the ceria reduction by hydrogen.

1. INTRODUCTION

It is generally acknowledged that the redox behaviour of ceria plays an important role in determining its singular properties as metal support and promoter [1,2]. We have recently shown that, in the presence of rhodium, ceria can chemisorb large amounts of hydrogen via a spillover process [3,4]. As deduced from magnetic measurements, this type of adsorption modifies the ceria oxidation state inducing a very strong increase of the concentration of paramagnetic Ce^{3+} ions [3-5]. Upon further evacuation, ceria becomes to a large extent reoxidized, thus indicating that there is an important reversible contribution to the final reduction degree reached by ceria [3-5]. The observations above were made on Rh/CeO_2 samples prepared from $Rh(NO_3)_3$; however, we noted in ref. [6] that the behaviour of catalysts prepared from $RhCl_3$ was significantly different. This prompted us to carry out the present study, the major objective of which was to investigate the influence of chlorine on the chemisorptive and redox properties of ceria. For this purpose, two Rh/CeO_2 catalysts prepared by impregnating the same ceria sample with either $Rh(NO_3)_3$ or $RhCl_3$ were investigated by means of a Faraday magnetic balance, Temperature Programmed Reduction/Oxidation (TPR/O) and Volumetric Adsorption techniques. Since chlorine containing noble metal precursors are often used in the preparation of TWC's and related catalytic systems, the present work would provide some clues helping us to understand the behaviour of this very important family of catalysts.

2. EXPERIMENTAL

The ceria sample used here was a medium-surface area sample (49 $m^2.g^{-1}$) prepared in the laboratory by calcining at 873 K, for 4 h, a cerium hydroxycarbonate precipitate obtained by addition of ammonium carbonate to a solution of $Ce(NO_3)_3$. To eliminate the residual carbonate, it was further treated with flowing H_2 at 773 K for 4h, and finally reoxidized in air at 773 K for 4 h. The surface area of the ceria sample did not significantly change throughout the whole series of treatments applied in the present work.

The Rh/CeO_2 catalysts were prepared by incipient wetness impregnation technique using an aqueous solution of either $RhCl_3$ or $Rh(NO_3)_3$. The precursor/support systems were dried in air at 383 K for 10 h, and stored in a dessicator with no further precautions. The metal loading of the two samples above, hereafter referred to as $Rh(Cl)/CeO_2$ and $Rh(N)/CeO_2$ respectively, was 3% by weight.

The Faraday microbalance used here has been described elsewhere [7]. Details about the procedure we have followed to determine the magnetic susceptibility of the Rh/CeO_2 catalysts are reported in refs. [3,8]. The experimental values, which were corrected from the traces of ferromagnetic impurities (6-12 ppm), allowed us to determine the content of paramagnetic Ce^{3+} ions, and therefore the percentage of Ce^{4+} ions reduced to the 3+ oxidation state [8]. The magnetic contribution of rhodium was considered to be negligeable (Maximum value to be expected: 3.10^{-8} emu CGS at 298 K, to be multiplied by 12.56 to obtain SI units in $m^3.g^{-1}$).

The experimental device used in Temperature Programmed Reduction/Oxidation (TPR/TPO) studies was similar to that described in ref. [9]. The quadrupole mass spectrometer used to analyse the evolved gases was a VG Spectralab SX-200 instrument interfaced to a PC-type microcomputer. The experiments were run under the following conditions: Flow rate of H_2 (O_2): 60 $cm^3 \cdot min^{-1}$; Heating rate: 10 $K.min^{-1}$.

Hydrogen volumetric adsorption measurements were carried out in a conventional high vacuum system equipped with a capacitance gauge, MKS Baratron, model 220 BHS. The time spent between succesive isotherm points was routinely 20 minutes. The final hydrogen pressure was 300 Torr.

3. RESULTS

3.1 Magnetic Susceptibility Study

Table 1 summarizes the results obtained from the magnetic susceptibility study carried out on both $Rh(Cl)/CeO_2$ and $Rh(N)/CeO_2$ catalysts. This study was aimed at investigating the evolution of the ceria redox state upon submitting the catalysts to a number of treatments including the initial reduction step at 623 K, several hydrogen adsorption/ evacuation cycles, a series of reoxidation experiments at increasing temperatures from 295 K to 773 K, and finally a new hydrogen reduction at 623 K of the reoxidized sample.

As deduced from Table 1, before the first hydrogen treatment, the $RhCl_3/CeO_2$ sample was slightly paramagnetic, thus indicating the presence in it of some Ce^{3+} ions (Percentage of Ceria Reduction: 0.4%). After the reduction treatment at 623 K, for 1h, in a flow of H_2 (60 $cm^3.min^{-1}$), at a heating rate of 10 $K.min^{-1}$, followed by cooling to 295 K, also in flowing H_2, the percentage of Ce^{3+} present in the catalyst was found to be equal to 22.1%. This value remained constant upon evacuation at 295 K; moreover, it diminished very slightly after pumping off the sample ($P<1.10^{-7}$ Torr), at 623 K: 20.7%. If the sample reduced and evacuated at 623 K is treated with H_2 (P_{H2}: 300 Torr), at

Table 1. Magnetic balance study of the redox processes occuring in a sample of $RhCl_3/CeO_2$ submitted to the treatments listed below. Comparison with a $Rh(NO_3)_3/CeO_2$ catalyst.

run	TREATMENT	ex chloride % Ce^{3+}	ex nitrate % Ce^{3+}
1	Vacuum-294 K- 20 h	0.4	1.4
2	H_2-623 K-cooling under H_2	22.1	11.4
3	Vacuum-298 K	22.0	11.0
4	Vacuum-623 K	20.7	5.0
5	H_2-294 K-20 h	20.6	10.1
6	H_2-523 K-cooling under H_2	21.7	11.7
7	H_2-623 K-cooling under H_2	-	11.7
8	Vacuum 773 K	20.9	4.1
9	H_2-294 K-18 h	21.1	9.8
10	H_2-523 K	22.0	11.7
11	Vacuum-623 K	20.9	(6.1)
12	Vacuum-773 K	-	4.2
13	H_2-773 K-cooling under H_2	22.9	14.3
14	Vacuum-294 K	22.7	13.0
15	Vacuum-773 K	20.4	6.5
16	H_2-294 K-17 h	20.6	8.4
17	H_2-523 K	21.8	12.9
18	Vacuum-294 K	21.8	12.5
19	Vacuum-773 K	20.5	6.1
20	O_2-294 K-20 h	12.8	1.6
21	O_2-473 K	(7.1)	(1.6)
22	O_2-623 K	(5.2)	1.7
23	O_2-773 K	0.7	1.9
24	Vacuum-773 K	1.7	1.5
25	H_2+Vacuum-623 K	11.4	
26	H_2-294 K-18 h	11.6	
27	H_2-523 K	12.5	

- The susceptibility was measured at 294 K, except for the values between parenthesis for which the measurement was performed at the actual temperature of the catalyst.
- Unless specified, all the treatments were carried out for 1 h.

295 K, for 20 h, and even at 523 K, for 1h, and cooled again to room temperature, no significant variation of the ceria reduction percentage could be

observed. Likewise, three succesive cycles of hydrogen treatment at 773 K followed by cooling to 295 K under hydrogen, and further evacuation at increasing temperatures up to 773 K, did not change the ceria oxidation state significantly.

After the series of hydrogen treatments/evacuations above, the Rh(Cl)/CeO_2 catalyst was further treated with O_2(P_{O2}: 40 Torr) at 295 K. This induced a fast partial oxidation of ceria, the reduction percentage of which decreasing from 20.5% to 12.8%. However, it was necessary to heat the sample up to 773 K, under O_2, to arrive at a reduction degree close to the initial one (1.7%). A further reduction treatment with H_2 at 623 K resulted in a reduction extent of 11.4%, a value much lower than that reached in the previous series of treatments.

The same experimental protocol was applied to the catalysts prepared from rhodium nitrate. The corresponding results are also included in Table 1 for comparison. The first hydrogen treatment at 623 K resulted in a ceria reduction degree of 11.4 %. This value is close to that expected for a fully surface reduced ceria. In accordance with refs. [10,11], for a ceria sample with a surface area of 49 $m^2.g^{-1}$ the reduction percentage would be 9.3 %. These results confirm what we have observed earlier on a high surface area ceria (120 $m^2.g^{-1}$) supported rhodium catalyst also prepared from $Rh(NO_3)_3$. For this catalyst the hydrogen treatment at 623 K followed by cooling to 295 K, also under H_2, leaded to the support reduction to an extent roughly corresponding to its complete surface reduction [4]. Furthermore, this reduction process was to a large extent reversible upon evacuation [4]. This latter observation is also in very good agreement with that found in the present work on a different ceria sample. In effect, according to Table 1, the ceria reduction degree in the Rh(N)/CeO_2 catalyst dropped to 5% upon evacuation at 623 K, and to a bit lower value, 4%, after pumping off at 773 K. This clearly indicates that the back spillover process is very important in the case of the catalyst prepared from rhodium nitrate. Also worth of noting, the ceria reduction degree is reversibly recovered when the evacuated catalyst is put again in contact with H_2. In accordance with Table 1 (run 9), the ceria re-reduction starts to be observed upon prolonged H_2 treatment (18 h), at 295 K. Finally, the behaviour against the reoxidation is also very different for the two catalysts investigated here. In effect, in contrast to that observed for the Rh(Cl)/CeO_2 sample, on Rh(N)/CeO_2 the oxygen treatment at 295 K leads to the recovery of the initial (unreduced) ceria oxidation state.

3.2 TPR/TPO-MS and Volumetric Chemisorption Studies

The thermal evolution of the Precursor/Support system in flowing hydrogen has been investigated with the help of the TPR-MS technique. By means of this technique we have monitored the gaseous products generated throughout the whole reduction process and, specifically, the evolution of chlorine containing species from the $RhCl_3/CeO_2$ sample. In accordance with our study, no signals corresponding to HCl (m/e: 36), Cl (m/e: 35) or even to Cl_2 (m/e: 70) could be observed up to temperatures as high as 1173 K, i.e. well above those applied in this work. This would indicate that the chlorine present in the starting sample remains trapped by the catalyst. To confirm this, a $Rh(Cl)/CeO_2$ sample reduced for 1h at 773 K, the highest reduction temperature used here, and flushed with inert gas at 773 K for 1h, was further studied by TPO-MS. Figure 1 accounts for the trace corresponding to m/e: 70 (Cl_2) signal. As presumed, the TPO-MS diagram in Fig. 1 confirms the presence of chlorine in the reduced catalyst. In accordance with Fig. 1, the evolution of Cl_2, which

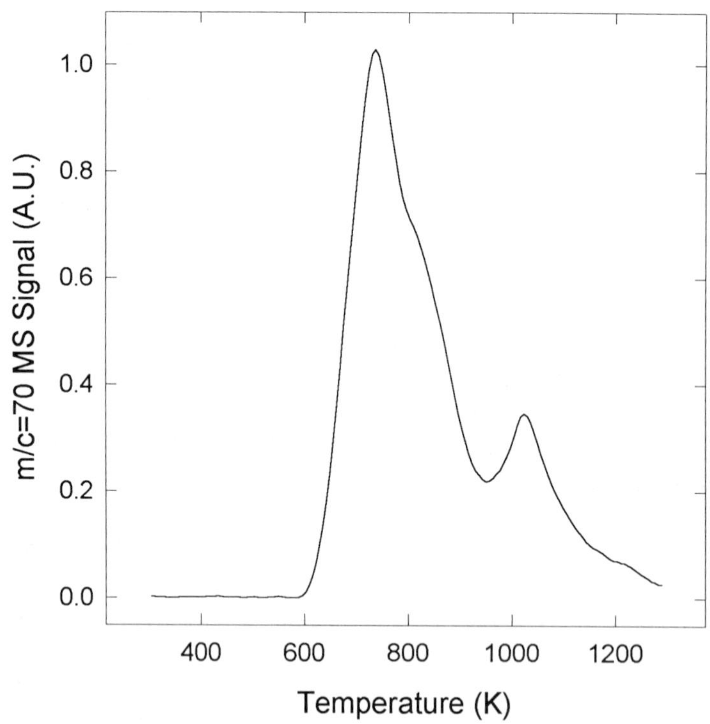

Fig 1.- OTP-MS study of a $Rh(Cl)/CeO_2$ catalyst reduced at 773 K. Trace for Cl_2 (m/e: 70)

starts to be observed at 600 K, takes place through at least two well resolved peaks at around 750 K and 1025 K. Also worth of noting, the relative intensity of the two peaks in Fig. 1 indicates, in good agreement with the magnetic susceptibility measurements, that a major part of the chlorine can be eliminated by treating the sample with oxygen at 773 K. However, the existence of the high temperature feature suggests that some residual Cl^- would be trapped in the catalyst after the reoxidation treatment at 773 K.

We have also studied the volumetric adsorption of H_2 on the Rh(Cl)/CeO_2 catalyst reduced and evacuated at either 623 K or 773 K. The adsorption studies were carried out at two different temperatures: 191 K and 295 K. In accordance with refs. [6,13], the H/Rh ratio determined from the low temperature isotherm can be considered as an estimate of the metal dispersion, and therefore, from the comparison of the H/Rh values determined at 191 K and 295 K the likely occurrence of spillover phenomena, at room temperature, can be established. For the catalyst reduced at 623 K, the H/Rh ratios were found to be 0.99, at 191 K, and 0.71 at 295 K. In the case of the low temperature adsorption, a second isotherm was recorded after pumping off the catalyst at 191 K, for 20 min. From this experiment, we determined the reversible contribution (H/Rh: 0.23) to the total hydrogen adsorption (H/Rh: 0.99). The difference between the two values above would give us an estimate of the irreversible chemisorption at 191 K, H/Rh: 0.76, a value quite close to that determined at 295 K (H/Rh: 0.71), and also very similar to those obtained in the same way for the catalyst reduced at 773 K: H/Rh: 0.73, at 191 K, H/Rh: 0.71 at 295 K. These results suggest that the rhodium dispersion is high and very similar at both reduction temperatures. It can also be deduced from this study, that the spillover contribution to the total amount of chemisorbed hydrogen is very small. This observation, which is in agreement with the results obtained from the magnetic balance study, contrasts with the behaviour deduced from Table 1 for the Rh(N)/CeO_2 catalyst, as well as with that reported in refs. [4,6] for a different Rh/CeO_2 catalyst, also prepared from $Rh(NO_3)_3$. In these latter cases, the catalysts reduced and evacuated at 623 K do exhibit hydrogen spillover at 295 K.

4. DISCUSSION

From the results reported in Table 1, we may conclude that the redox behaviour of ceria in Rh/CeO_2 catalysts depends very much on the nature of the metal precursor salt used. For the catalyst prepared from $RhCl_3$ the reduction level reached by ceria (22.1%) is much larger than that determined for the sample prepared from $Rh(NO_3)_3$ (11.4%). A second very important difference is the degree of reversibility of this reduction process. As deduced from Table 1, the Rh(N)/CeO_2 catalyst can be strongly reoxidized (Above 60% of the initial reduction degree) by simple evacuation. On the contrary, the Rh(Cl)/CeO_2 becomes reoxidized to an extent not larger than 5% of the total reduction level. In accordance with the results discussed in refs. [4], the reversible contribution to the ceria reduction degree can be associated to hydrogen chemisorbed on it that can be desorbed as H_2 via back spillover.

$$1/2\ H_2(g) + \mathrm{Rh(particle)} \rightleftharpoons \mathrm{H(ads)—Rh(particle)}$$

$$\mathrm{H(ads)—Rh(particle)} + —\underset{|}{Ce^{4+}}—\underset{|}{O^{2-}}— \rightleftharpoons \mathrm{Rh(particle)} + —\underset{|}{Ce^{3+}}—\underset{|}{\overset{H^{+}}{\overset{|}{O^{2-}}}}—$$

Then, we may conclude that the spillover phenomena are unimportant in the Rh(Cl)/CeO_2 catalyst. This conclusion is consistent with both the magnetic measurements and volumetric adsorption studies commented on above.

The irreversible reduction of ceria can be due to the existence of oxygen vacancies, to the substitution of lattice O^{2-} by Cl^- or even to both mechanisms operating simultaneously. The occurence of chlorine incorporation to ceria during the preparation of M/CeO_2 catalysts has already been shown by X-Ray Diffraction [14,15] and High Resolution Electron Microscopy [16]. Obviously, in the case of the Rh(N)/CeO_2 catalyst, the first mechanism is the only that can operate. Assuming that the evacuation treatment at 773 K completely eliminate the hydrogen chemisorbed on the Rh(N)/CeO_2 catalyst [5,13], the residual reduction degree determined from magnetic measurements after this treatment can be considered as an estimate of the concentration of oxygen vacancies in the support: 4.1% for the catalyst reduced at 623 K (Run 8 in Table 1), and 6.5% upon reduction at 773 K (Run 19 in Table 1).

Regarding the Rh(Cl)/CeO_2 catalyst, we propose that both oxygen vacancies and lattice Cl^-ions coexist. Furthermore, from the results of the reoxidation study reported in Table 1, we can make a quantitative estimate of their contributions to the total degree of irreversible reduction of ceria (20.9 %). In effect, the run 20 in Table 1 shows that the treatment with O_2, at 295 K, of the Rh(N)/CeO_2 catalyst leads to the almost complete reoxidation of ceria, i.e. this oxygen treatment can be used to determine the concentration of oxygen vacancies. This can also be concluded from the reoxidation study of bare ceria carried out with the help of a magnetic balance [8]. If it is so, the residual reduction degree determined for Rh(Cl)/CeO_2 after its reoxidation at 295 K (12.8 %, Run 20 in Table 1) would be assigned to the presence of Cl^- in the ceria lattice. This interpretation of the magnetic data is also supported by the

TPO-MS study in Fig.1, in accordance with which, no Cl_2 evolution can be observed below 600 K. Moreover, if the total amount of the chlorine present in the starting sample is trapped by ceria, the support reduction degree would be 15 %. This means that a major part of the Cl^-ions is retained by ceria, which, in turns, is consistent with the absence of any TPR-MS signal for chlorine containing species up to temperatures well above those used here.

From the discussion above, we may conclude that the contribution of the oxygen vacancies to the reduction degree of ceria in the Rh(Cl)/CeO_2 catalyst (8.1%) can be determined by substracting from the total irreversible reduction level (20.9%) the value associated to the presence of chlorine in the support lattice (12.8%). In accordance with this estimate, the concentration of support oxygen vacancies in Rh(Cl)/CeO_2 would be higher than that determined for Rh(N)/CeO_2. In this respect, it would be noted that, as discussed in refs. [4,17], in the absence of spillover phenomena, for instance for bare ceria, the contribution of oxygen vacancies to the total reduction level of ceria increases. This might well be the case for Rh(Cl)/CeO_2.

The TPO-MS diagram in Fig. 1 suggests that some residual chlorine ought to remain trapped by the Rh(Cl)/CeO_2 catalyst upon reoxidation at 773 K. This probably explains why the further reduction/evacuation at 623 K of the reoxidized catalyst leads to a ceria reduction degree, 11.4 % (Run 25 in Table 1), which though much lower than that observed after the initial reduction treatment (20.7 %), is still higher than the value determined for Rh(N)/CeO_2: 5 % (Run 4 in Table 1). These observations are however unconsistent with the low reduction degree reached by the Rh(Cl)/CeO_2 catalyst upon reoxidation at 773 K: 0.7 %. It can tentatively be suggested that some chlorine transfer from the support to the metal would occur during the reoxidation treatment at 773 K, the phenomenon being reversed upon reducing again the catalyst. Though some rather similar suggestions have already been made in the literature (18), the confirmation of such a proposal would deserve some additional studies.

In summary, the present work shows that the use of chlorine containing metal precursor salts has a dramatic effect on the chemisorptive and redox properties of ceria in Rh/CeO_2 catalysts. On the one hand, the usual hydrogen treatment aimed at the preparation of the final catalyst induces a much higher reduction level on ceria, when chlorine is present. This is interpreted as due to the almost stoichiometric incorporation of the chlorine present in the initial Precursor/Support system into the support ceria lattice, as well as to the creation of oxygen vacancies in the ceria lattice. On the other hand, the hydrogen chemisorption on the Rh(Cl)/CeO_2, becomes very small, the so-called reversible contribution to the total ceria reduction degree being almost completely supressed. This contrasts with the behaviour of the Rh(N)/CeO_2 catalyst, for which the reversible chemisorption of H_2 on the support represents a major contribution to the total reduction degree reached by ceria. Our results also show that an important part of the chlorine retained by ceria can be eliminated upon treating the catalyst with oxygen at 773 K; some Cl^- ions remain trapped by the catalyst. This residual chlorine would be responsible for the intermediate behaviour of the oxidized catalyst compared to that of Rh(Cl)/CeO_2 and Rh(N)/CeO_2.

ACKNOWLEDGEMENTS

This work has received financial support from the DGICYT (Project with Ref.: PB92-0483). We acknowledge Johnson-Matthey for a loan of precious metals. We also thank Mrs. C. Larese for her contribution to some measurements included in this work.

REFERENCES

1 R.K. Herz; Ind. Eng. Chem. Prod. Res. Dev. 20 (1981) 451.
2 T. Jin, T. Okuhara, G.J. Mains and J.M. White; J. Phys. Chem. 91 (1987) 3310.
3 S. Bernal, J.J. Calvino, G.A. Cifredo, J.M. Rodríguez-Izquierdo, V. Perrichon and A. Laachir; J. Catal. 137 (1992) 1.
4 S. Bernal, J.J. Calvino, G.A. Cifredo, J.M. Rodríguez-Izquierdo, V. Perrichon and A. Laachir; J. Chem. Soc. Chem. Comm. 460 (1992).
5 S. Bernal, J.J. Calvino, G.A. Cifredo, A. Laachir, V. Perrichon and J.M. Herrmann; Langmuir 10 (1994) 717.
6 S. Bernal, J.J. Calvino, M.A. Cauqui, G.A. Cifredo, A. Jobacho, J.M. Pintado and J.M. Rodríguez-Izquierdo; J. Phys. Chem. 97 (1993) 4118.
7 V. Perrichon and J.P. Candy; J. Catal. 89 (1984) 93.
8 A. Laachir, V. Perrichon, A. Badri, J. Lamotte, E. Catherine, J.C. Lavalley, J. El Fallah, L. Hilaire, F. LeNormand, E. Quéméré, G.N. Sauvion and O. Touret; J. Chem. Soc. Faraday Trans. I, 87 (1991) 1601.
9 S. Bernal, R. García and J.M. Rodríguez-Izquierdo; Thermochim. Acta 70 (1983) 249.
10 M.F.L. Johnson and J. Mooi; J. Catal. 103 (1978) 502 and 140 (1993) 612.
11 V. Perrichon, A. Laachir, O. Touret, G. Bergeret, R. Frety and L. Tournayan; J. Chem. Soc. Faraday Trans. I, 90 (5) (1994) 773.
12 S. Bernal, J.M. Gatica and C. Larese; Unpublished Results.
13 S. Bernal, J.J. Calvino, G.A. Cifredo, A. Jobacho and J.M. Rodríguez-Izquierdo; J. Rare Earths (Special Issue) 2 (1991) 838.
14 F. Le Normand, L. Hilaire, K. Kili, G. Krill and G. Maire; J. Phys. Chem. 92 (1988) 2561.
15 M. Wolcyrz and L. Kepinski; J. Solid State Chem. 99 (1992) 409.
16 S. Bernal, F.J. Botana, J.J. Calvino, G.A. Cifredo and J.A. Pérez Omil; Inst. Phys. Conf. Ser. 138 (Section 10) (1993) 485.
17 S. Bernal, J.J. Calvino, G.A. Cifredo, J.M. Gatica, J.A. Pérez Omil and J.M. Pintado; J. Chem. Soc. Faraday Trans. 89 (1993) 3499.
18 H. Lieske, G. Lietz, H. Spindler and J. Völter; J. Catal. 81 (1983) 8

A. Frennet and J.-M. Bastin (Eds.)
Catalysis and Automotive Pollution Control III
Studies in Surface Science and Catalysis, Vol. 96

THE PREPARATION OF THERMALLY STABLE WASHCOAT ALUMINAS FROM LOW COST GIBBSITE

Yafeng Huang [a], *N.W. Cant [b], J. Guerbois [a], D.L. Trimm [a] & A. Walpole [a]

[a] *Department of Chemical Engineering & Industrial Chemistry, University of New South Wales PO Box 1,KENSINGTON NSW 2033*
[b] *School of Chemistry, Macquarie University, NSW 2109*

ABSTRACT

Study of the formation and decomposition of aluminium sulphates has shown that low cost gibbsite can readily be converted to thermally stable washcoat alumina through the intermediate formation of the sulphate. Optimal reaction between hydroxidic aluminium starting materials and sulphuric acid occurred in the presence of water. A protective sulphate layer was formed on the surface of gibbsite on reaction with concentrated sulphuric acid which limited conversion. Higher conversion could be achieved by reaction with diluted acid. Conversion of the resultant aluminium sulphate to alumina was essentially complete on calcination at about 1000°C for 4 hours.

No differences in the product were observed on changing the starting material to pseudo-boehmite. The alumina produced from both starting materials was thermally stable with a surface area of ca 120 - 130m^2g^{-1}. The surface area dropped to ca 12-13 m^2g^{-1} on heating to 1200°C for 4 hours. Addition of 2% La_2O_3 or 4% BaO was found to stabilise the alumina, and samples were produced with surface areas of 70-100m^2g^{-1} after heating at 1200°C for 4 hours.

It was found to be necessary to grind the material to obtain a solution with solids loading suitable for washcoat preparation. Up to 67wt% suspension could be achieved at some cost in thermal stability.

1. INTRODUCTION

The use of catalysts to reduce emissions from automobile engines is now well established [1]. The active components of the catalysts consists of platinum and/or palladium together with rhodium. These precious metals are distributed through a washcoat which, in turn, is suspended on a pellet or a monolithic substrate.

The washcoat provides a high, stable surface area for dispersion of the precious metals [2,3]. The bulk of the washcoat consists of alumina, stabilised by the addition of small amounts of baria or lanthana. Thermal stabilisation is essential, since temperatures in the catalyst bed can rise to over 1000°C. The choice of alumina precursors (boehmite or gibbsite) and the preparation/stabilisation procedures are vital to the stability of the washcoat [2,3].

Washcoat alumina is generally produced from boehmite, which is relatively expensive but gives a more thermally stable product. The present studies were initiated in order to study the possibility of producing thermally stable alumina from gibbsite, via the intermediate production of sulphates. It is known that alumina can be produced by the dehydration and decomposition of aluminium sulphate [4] and it seemed possible that the conversion of cheaper precursors to the sulphate followed by decomposition to the oxide could offer some economic advantage in the production of washcoat material.

2. EXPERIMENTAL

Gibbsite powder, CHP-34OL, was supplied by the Sumitomo Chemical Co. The gibbsite contained >99.9% $Al_2(OH)_6$ with Na_2O impurity of 0.06%, and was of mean particle diameter 39.0 mm. Pseudo-boehmite (Pural 200) was obtained from Condea. Aluminium sulphate from Ajax Chemicals (AR grade) had a purity of >97.0% with less than 0.05% sodium (Na) impurity, and less than 0.01% insoluble matter. Concentrated sulfuric acid (98%) was supplied by Ajax. Some samples were doped by adding small amounts of $La(NO_3)_36H_2O$ (Aldrich Chemicals: 99.99% pure) or $Ba(NO_3)_2$ (Ajax: 99.5% pure)

2.1 Sulphation

Initial experiments attempted to react gibbsite with concentrated sulphuric acid. Conversions were low as a result of formation of a surface layer of aluminium sulphate which restricted further sulphation. Subsequent experiments were carried out with a fixed ratio of 98% sulphuric acid: gibbsite (1.92:1 - calculated to be the ratio required to convert 100% of gibbsite) in the presence of varying amounts of water. Conversion was found to be highest at an acid: water: gibbsite ratio of 1.92:2.02:1.

Subsequent sulphation was carried out using a standard procedure. 100g gibbsite were mixed with 200 ml distilled water and stirred. 194.5g of 98% H_2SO_4 were added slowly and with stirring. The temperature was controlled at about 110°C for 30 min, after which time the products were poured into a mould where total solidification occurred.

2.2 Doping

Several methods of doping were tested, using 2 mol% La_2O_3 (calculated with respect to final alumina) or 4 mol % BaO, both added as nitrates. No significant differences were observed between adding dopant to the solution of aluminium sulphate before cooling or to solid aluminium sulphate after drying (see below).

A general procedure was adopted in which a solution of the dopant nitrate was added to the sulphurisation beaker during the heating/stirring period and before cooling and dehydration.

2.3 Drying/calcination/grinding

The sulphate samples were broken up and dried at 150-200°C (5h) and 600°C (either 2 or 4h, depending on the mass of material).

Conversion of the sulphate to alumina was achieved by calcining in air at 1000°C for 4h. The reasons for the choice of these conditions are given below.

Grinding was carried out in a ball mill, using 50g alumina mixed with 250ml water. Both pure and doped alumina were ground.

2.4 Characterisation of products

Surface areas were measured using a Micromeritics 2200 high speed surface area analyser: reproducibility was ± 2%. X-ray diffraction patterns were obtained using a Rigaku powder diffractometer with CuKa radiation. Particle sizing was accomplished using a Malvern Mastersizer.

Water absorptivity was measured by adding demineralised water to 5g of dried alumina until the mixture just became fluid. The amount of water required was measured and the percentage solids calculated. Reproducibility was ± 1%.

3. Results and Discussion

Important properties for a washcoat alumina include surface area/porosity/thermal stability, particle size and coatability. The preparation of alumina from gibbsite via the sulphate was checked with respect to these properties.

The conversion of gibbsite to aluminium sulphate was measured in terms of the weight gain, on the assumption that, after dehydration, the product of the reaction was $Al_2(SO_4)_3$. The initial reaction resulted in the formation of a solid product containing large amounts of water of crystallisation, but this water could be removed during drying.

$$Al_2(OH)_6 + 3H_2SO_4 + 12H_2O = Al_2(SO_4)_3\ 18H_2O \quad \Delta H = -433\ kJ/mole$$

The importance of water to the sulphation was shown by the fact that <10% conversion of gibbsite was observed on passing gaseous sulphur dioxide or trioxide over the gibbsite, apparently as a result of the formation of a protective sulphate layer. Dissolution of some of the protective layer of sulphate accounts for more extensive conversion in aqueous solutions.

Decomposition of the sulphate to alumina was found to be extensive at ca 800°C (Figure 1). By 920°C, the majority of sulphate had been converted. These temperatures are somewhat higher than those reported by Nam & Gavalas [5] and this reflects the bulk decomposition required in the present studies.

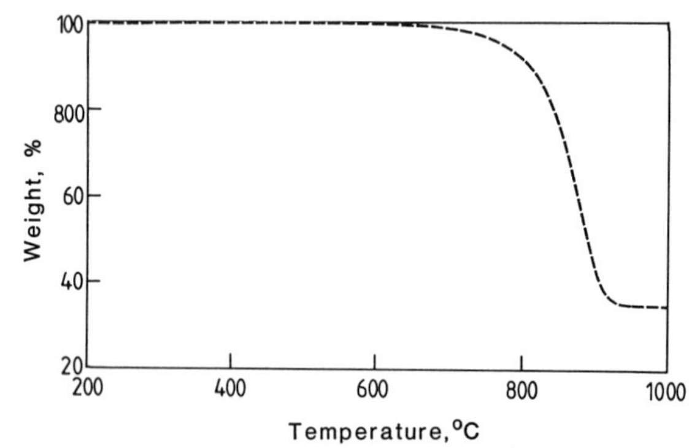

Figure 1 *Themogravimetric studies of the heating of a sample of aluminium sulphate dehydrated at 600°C.*

Using the weight changes from gibbsite to dehydrated sulphate and from sulphate to alumina, it was possible to obtain a measure of the conversion of gibbsite to sulphate and the effect of this on the properties of the alumina. A linear correlation was observed between conversion to sulphate and the surface area of the product alumina (Figure 2). At 100% conversion, aluminas of surface area of >120 m^2g^{-1} could be produced by calcination. The alumina derived from Ajax aluminium sulphate had a surface area between 120 and 130 m^2g^{-1}. X-ray diffraction showed that γ-alumina was the major product.

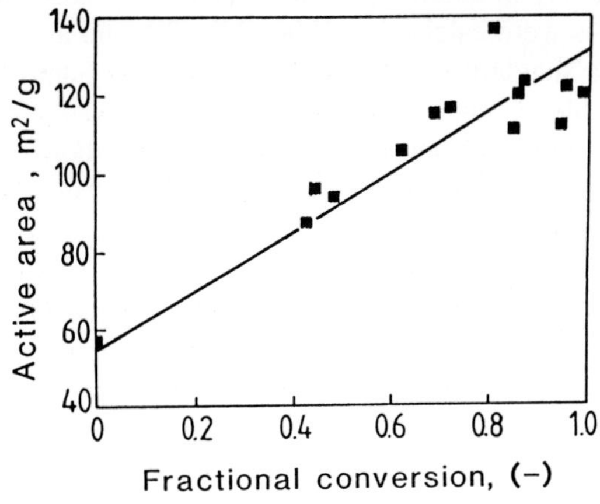

Figure 2 *The surface area of aluminas calcined at 1000°C for 2h. The samples contained aluminium sulphate produced from gibbsite. The abscissae shows the fractional conversion of the gibbsite.*

Similar experiments were carried out with pseudo-boehmite. The conversion was higher, as pseudo boehmite was more active than gibbsite (Table 1). The final product was indistinguishable from that produced from gibbsite as a starting material for the preparation of high surface area alumina.

Table 1 Surface Areas of Alumina Samples before and after Thermal Treatment

Sample Origin	Dopant	Calcined at 1000°C, 2h	Calcined at 1200°C, 4h
		$m^2\ g^{-1}$	$m^2\ g^{-1}$
Pseudo boehmite	zero	122	-
Aluminium sulphate	zero	120 - 123	12
Sumitomo gibbsite	zero	134	13.3
Gibbsite	zero	96.3	12-13
Gibbsite	2% La_2O_3	106.1	82.8
Gibbsite	4% BaO	121.2	103.6

The thermal stability of aluminas in the presence and absence of dopants was then examined. Samples were calcined at 1000°C for 2 h and at 1200°C for 4 h, the latter treatment being a standard washcoat test. The results, summarised in Table 1, show that thermally stable alumina can be produced from the sulphate in the presence of additives. At least in the presence of baria, the formation of $BaO.Al_2O_3$ and $BaO.6AlO_3$ was detected by X-ray diffraction [6].

Although these samples were thermally stable, it was found that the solids loading in suspension was comparatively low (Table 2). It was suspected that the particle size was too large and, as a result, ball mill grinding was initiated. In the dry state, the samples were found to cake badly, due to moisture adsorption. Wet grinding was found to lead to better suspension loadings (Table 2), but to somewhat lower thermal stability.

After 24 h grinding a mean particle size of 4.6 mm was observed.

Comparable tests were carried out using doped washcoat alumina. A final surface area of 30.7 m^2g^{-1} after calcination for 4 h at 1200°C was observed.

The results show that it is possible to produce a thermally stable alumina from low cost gibbsite through the intermediate formation of aluminium sulphate. It is necessary to add small amounts of dopant but a product suitable for application as a washcoat for car exhaust catalysts is produced.

Table 2 Effect of wet grinding alumina samples
Alumina Doped with 4 mole % BaO

Grinding time hours	Solids in Solution wt %	Surface Area after at 4h at 1200°C/4h
O	37.0	70.3
2	44.0	41.3
4	64.5	41.1
8	47.0	36.0
15	50.5	39.7
24	67.0	42.1
24	52.0	30.7

4. ACKNOWLEDGMENTS

The authors acknowledge, with gratitude, financial support from, Johnson Matthey Catalytic System Division. Some of this work was funded by a grant from the Australian Research Council.

5. REFERENCES

1 Taylor, K.C., Chem. Tech. Sept (1990) 551
2 Cooper, B.J. Evans, W.D.J. and Harrison, B., Ed A Crucq and A Frennet, Elsevier Amsterdam (1987) 117-141.
3 Cooper, B.J. Harrison, B., Shutt, E. & Lichtenstein, I. SAE Technical Paper 770367 (1977)
4 Udd, J.C., US Patent 3,265,464 (1966).
5 Nam, S.W. & Gavalas, G.R. Appl. Catal. 74 53 (1991).
6 Church, J.S. Cant, N.W. & Trimm, D.L. Appl. Catal. 101 105 (1993).

Gasoline Catalyst Technologies

A. Frennet and J.-M. Bastin (Eds.)
Catalysis and Automotive Pollution Control III
Studies in Surface Science and Catalysis, Vol. 96
1995 Elsevier Science B.V.

DEVELOPMENT OF IMPROVED Pd-ONLY AND Pd/Rh THREE-WAY CATALYSTS

B. H. Engler, D. Lindner, E. S. Lox, A. Schäfer-Sindlinger and K. Ostgathe

Degussa AG, Hanau, Germany

ABSTRACT

The results of a development program for Pd-only and Pd/Rh automotive emission control catalysts are discussed. A review of former experiences with Pd-containing systems, especially their poor tolerance against lead and sulfur poisoning is given. Model gas experiments were conducted with improved Pd-based catalysts. Not only the activity to convert simultaneously CO, NO_x and hydrocarbons (HC) was investigated, but also the formation of so-called secondary emissions such as N_2O or NH_3.

The catalyst development was supported by XPS-measurements paying special attention to the oxidation state, spread and the dispersion of Pd. Promising Pd and Pd/Rh-catalyst candidates were tested fresh and after engine aging on engine- and vehicle dynamometers. The results are compared to Pt/Rh-catalysts. For Pd-only catalysts the ratio of catalyst volume to vehicle mass seems to be one important factor to guarantee high conversion levels after aging. The Pd/Rh-technologies reported in this study showed similar activities on different vehicles and in different test cycles as Pt/Rh catalysts. With the new developed Pd/Rh-catalyst the European legislation for model year 1996 could be achieved thus representing an economical and technical alternative to Pt/Rh-catalysts. Major potential for further improving the Pd/Rh-technologies performance must be seen in an optimization work regarding fuel quality and A/F-management.

1. INTRODUCTION

The majority of modern gasoline fueled passenger cars is equipped with so-called closed-loop three-way catalysts to aftertreat their exhaust gases [1]. The purpose of this system is to convert simultaneously carbon monoxide, hydrocarbons and nitrogen oxides by means of a precious metal based heterogeneous catalyst, whereby the engine's air-to-fuel-ratio is controlled to obtain exhaust gas compositions that guarantee optimal conversions. For a long time Pt/Rh-containing converters were dominant, but there is a growing interest for using Pd in such catalysts for several reasons described in what follows:

First of all, economic factors have to be mentioned, because for the time being Pd is by far the cheapest precious metal that is applicable in automobile converters, see Figure 1.

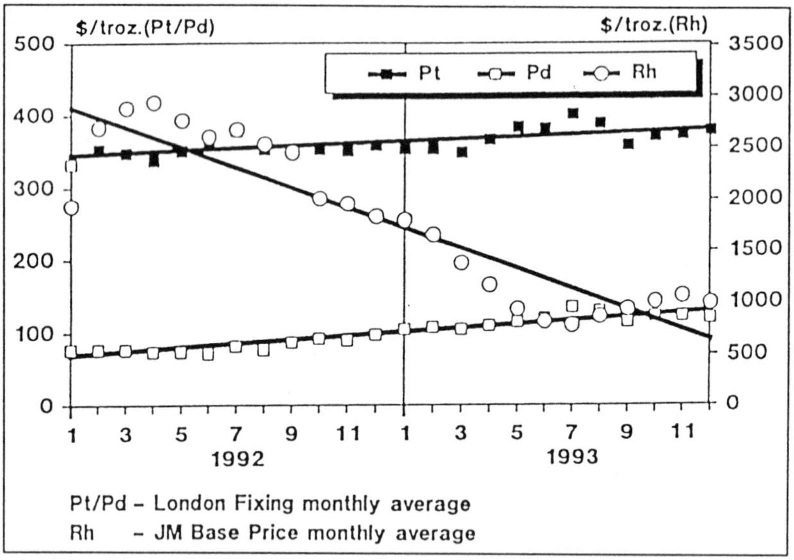

Figure 1: Pt, Pd and Rh monthly average prices in 1992 and 1993 [2].

But on a long term basis a major increase of the amount of Pd used in automobile converters might change the price situation completely. Indeed, presently only 9 % of the Pd available to the market is used in automotive emission control catalysts. In case of Pt this share is about 45 %, whereas the world supply of Pt and Pd is in the same order of magnitude, see Figure 2.

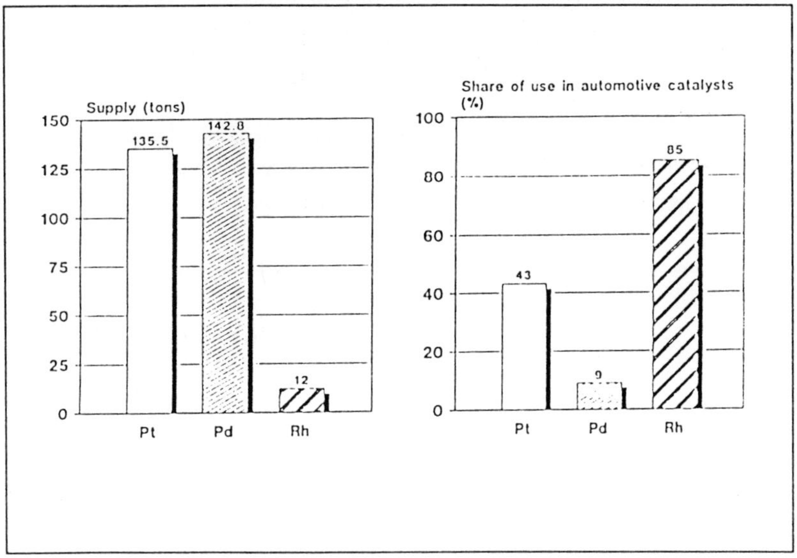

Figure 2: *World supply of Pt, Pd and Rh in 1992 and share of automotive catalyst application [2].*

Second, technical boundary conditions have been improved, such as the sulfur and lead content in the fuel. They limited in former times the use of Pd as precious metal component for high performance three-way catalysts [1]. Additional deposits of oil ashes caused major activity losses for Pd. Furthermore the A/F-window for Pd-containing catalysts was limited due to irreversible damage for the oxidation reactions and for the reduction of NO_x under rich conditions by SO_2. Besides this the literature reports that higher levels of secondary emissions like NH_3 and N_2O could be formed [3-7]. Next, Pd and Rh containing catalysts are more sensitive to lead poisoning than Pt-containing catalysts due to a bulk alloy or solid solution formation between Pd and lead. Several investigations have covered these findings [8, 9].

Due to a number of improvements in fuel quality, significantly lower lead and sulfur contents and improved engine management systems, Pd-containing catalysts gain more and more interest.

In the last 10 years the gasoline sulfur and lead content in Europe and the United States decreased significantly. Table 1 illustrates the situation [7, 11].

Table 1: Development of the fuel sulfur and lead content

Country	Sulfur content [ppm]	Country	Lead content [g/l]
USA (ASTM unleaded, 7-7-1986)[a]	1000	USA (1985/86)	0,013
Field USA	40 - 300	Field USA (1985/86) Today	0,29 0,13 0,026 ≤ 0,001
FRG (DIN 51607, August 1989)	1000	FRG (since 1989)	0,013
Field FRG (today)	50 - 450 ppm	Field FRG	≤0,003

a) EPA § 86.1213-87, fuel specifications

In Germany nowadays a span of sulfur levels between 50 and 450 ppm is detected in the field. The usual values range below 150 ppm. In Germany the lead content of unleaded fuel reaches values below 3 ppm.

To meet the more stringent emissions standards especially for Europe and the U. S. not only a tighter A/F-control around stoichiometry is implemented, but also in many cases secondary air is added to the exhaust gas to achieve a quicker light-off. Under such conditions Pd-containing converters also were reported to have advantages over state-of-the-art Pt/Rh-catalysts [7, 10, 12].

Summarizing there are both economic and technical reasons to explain the growing interest in Pd-containing catalytic converters to reduce automobile emissions. Main results of a long term development program are shown in this study.

2. EXPERIMENTAL

2.1 Catalysts

The catalyst samples were prepared by coating monolithic cordierite substrates with a cell density of 400 cpsi and a wall thickness of 6.5 mil with an aqueous slurry of different aluminum oxides containing certain stabilizers together with state-of-the-art as well as special stabilized oxygen storage components. The washcoat loadings ranged between 140 and 250 g/l catalyst. After drying and calcining the washcoated monoliths (700°C, 2 h, static air) the samples were impregnated with aqueous solutions of the desired amount and kind of precious metals, followed by drying, calcination and reduction steps. The precious metal loadings varied from 80 to approx. 1100 g/ft^3 for Pd, 40 to 50 g/ft^3 for 5Pt/1Rh, and 40 to 90 g/ft^3 for 5 Pd/1Rh. Parts of the catalysts were evaluated fresh and after different aging procedures in model gases (air, 4 h, 1050°C) or in real engine exhaust gases (fuel cut aging cycles, catalyst inlet temperatures between 870 and 950°C, 40 to 100 hours).

The model gas studies were performed with samples of 1 inch diameter and 3 inch length, for engine tests pieces of 1.5 inch diameter and 6 inch length were drilled out of larger samples. In the vehicle tests the ratio of catalyst volume to engine displacement ranged between 0.7 and 1.5.

2.2 Activity Test Procedures

2.2.1 Model Gas Tests

Different types of model gas tests were performed. Standard light-off and dynamic sweep test methods were applied for the activity comparison [13], furthermore the formation of secondary emissions such as NH_3 and N_2O was investigated in separate test cycles [14]. Parts of the tests were performed with SO_2-free and others with SO_2 containing gas mixtures.

2.2.2 Engine Tests

The engine performance tests were conducted on dynamometers equipped with EFI (Electronic Fuel Injection) gasoline engines using different types of light-off tests and A/F-scans. The space velocity was kept constant at 60.000 Nl/l/h. The exhaust gas composition for the light-off tests varied from about stoichiometry over slightly lean to a very lean composition, during the A/F-scans the perturbation frequence as well as the amplitude was changed. Details are given in [15, 16].

2.2.3 Vehicle Tests

The vehicles used in this study are developed for the current US or EU legislation. The engine displacements ranged from 1.3 to 3.2 liters (EFI engines using 4 to 8 cylinders). European and US driving cycles were applied, the fuel corresponded to current US and EU specifications.

2.3 Catalyst Characterization

Surface analysis investigations (XPS) were performed on a Leybold equipment already described in [16]. The fresh catalyst samples were stored under argon prior to the catalytic tests and the surface analysis. The aged catalyst samples were also handled under argon. The precious metal dispersion was determined for some of the catalysts by a pulsed CO chemisorption technique [16].

3. TEST RESULTS

As already reported the main drawbacks for the introduction of Pd-based three-way catalysts are their smaller A/F-window and their insufficient resistance against sulfur poisoning. To overcome these disadvantages a research program was conducted that first concentrated on a washcoat development using model gas activity tests for Pd-only catalysts. Based on these results engine tests followed by vehicle tests were performed with Pd/Rh- and partwise with Pd-only technologies.

3.1 Model Gas Test Results

In lean and rich model exhaust gases activity tests were performed with different Pd-only catalysts and compared to a Pt/Rh-reference system based on a $Al_2O_3/CeO_2/ZrO_2$ washcoat. Parts of the results are illustrated in Figure 3a and 3b.

Figure 3a gives the light-off temperatures for CO and HC, determined at a 50 % conversion of the pollutant. Under oxidizing as well as under reducing conditions a new developed stabilized washcoat containing modified Al_2O_3, special oxygen storage components and certain promoters showed by far the best light-off behaviour. Alumina-only based Pd-catalysts had approx. a 80°C higher light-off temperature.

A similar conclusion is reached from Figure 3b, where the dynamic conversion level for CO and NO_x are plotted. Here the Al_2O_3/Pd-catalysts fail totally, whereas the stabilized washcoat/Pd-system performs similar as compared to the Pt/Rh-based catalyst, even after aging. The high NO_x-conversions especially under rich conditions for the best Pd-only technology are remarkable.

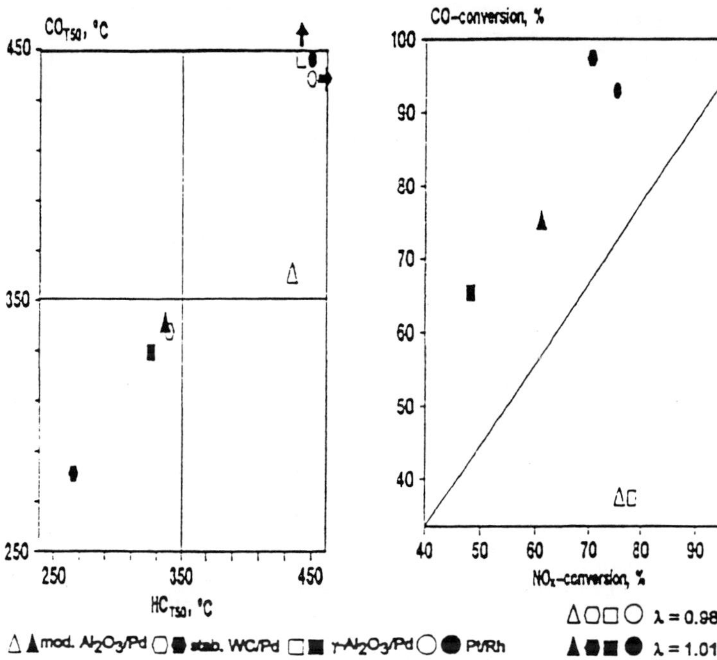

Figure 3: *Model Gas Test Results - Comparison of 5Pt/ 1Rh (40 g/ft^3) with Pd-only (80 g/ft^3, 4h, 1050 °C, air aged)*
a) light-off test, SV = 50.000 h^{-1}
b) A/F-scan, SV = 50.000 h^{-1}, T = 400 °C

Next the influence of SO_2 on the activity was investigated for the Pt/Rh- and the best Pd-only catalyst. Figure 4 reports the conversion rates for HC (Figure 4 a) and NO_x (Figure 4 b) for different temperatures in a rich gas atmosphere.

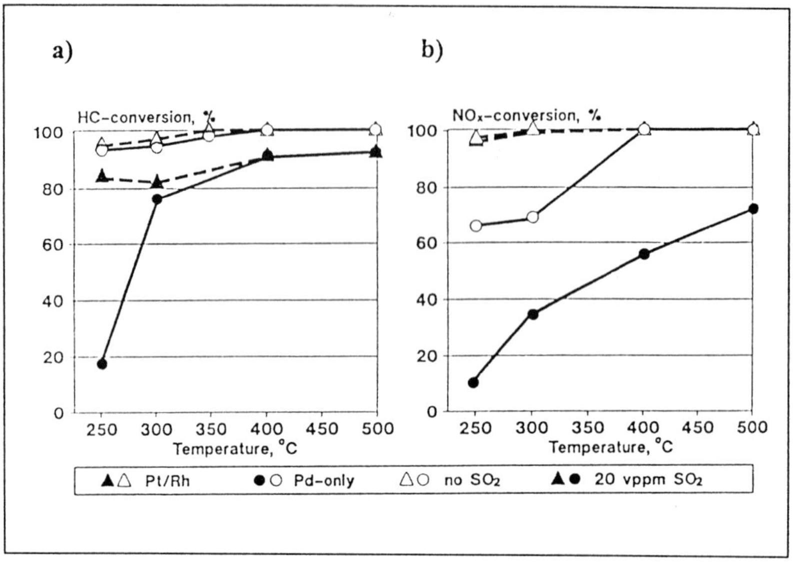

Figure 4: Model Gas Test Results - Comparison of 5Pt/1Rh (40 g/ft^3) with Pd-only (80 g/ft^3) catalysts; (SV = 50.000 h^{-1}, A/F = 14.45, fresh catalysts)

In the sulfur-free gas mixture the Pd-only catalyst already shows disadvantages for NO_x at temperatures below 350° C, and this becomes even more pronounced in the experiment with SO_2-containing gas. For HC differences are only obvious at temperatures below 300° C. But for sure these results indicate again the strong influence of sulfur on the performance of Pd-only catalysts.

Another interesting question that is raised regarding automotive catalytic converters is the formation of secondary emissions. In this study main emphasis was put on the nitrogen containing molecules NH_3 and N_2O, the latter of which becomes more and more important due to it's green house effect potential. First the NH_3-formation is compared for the Pt/Rh- and the stabilized Pd-system, see Figure 5.

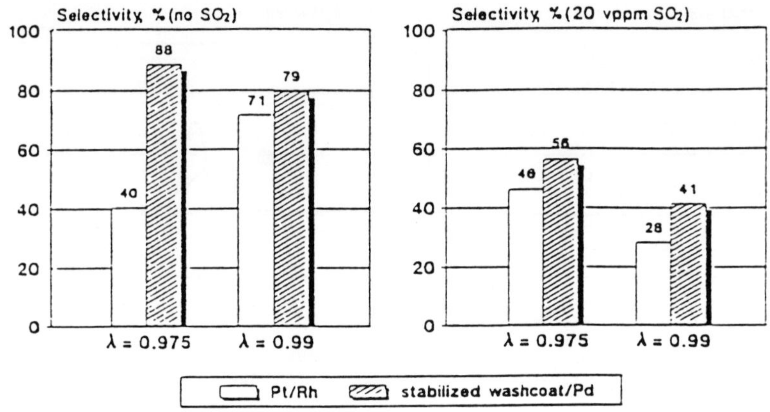

Figure 5: Model Gas Test Results - Comparison of 5Pt/1Rh (40g/ft³) with Pd-only (80g/ft³); Formation of NH_3 (T=300°C, SV=50.000h^{-1}, fresh catalysts);

Under the test conditions reported the selectivity S for the formation of NH_3, defined as

$$S_{NH_3} = \frac{n_{NH_3},out}{n_{NO_x},in - n_{NO_x},out} \qquad \text{(n = molar flow of the component)} \qquad (1)$$

is always more pronounced for the Pd-only catalyst. This is true for both SO_2-free and SO_2-containing gas mixtures and confirms data reported already in chapter 2. The influence of the temperature and the A/F-ratio on the selectivity is nearly negligible up to a temperature of approx. 400°C under rich conditions (A/F from 14.24 to 14.45). This might be of importance for vehicles operating rich during cold-start, where Pd-based converters might form larger amounts of NH_3 as compared to Pt/Rh. The conversion of NO_x was comparable in the SO_2 free gas mixture; in the presence of SO_2 the Pd-only system lost somewhat in performance.

The formation of N_2O for the same catalysts at A/F = 14.31 and A/F = 14.75 in sulfur-free gases is shown in Figure 6.

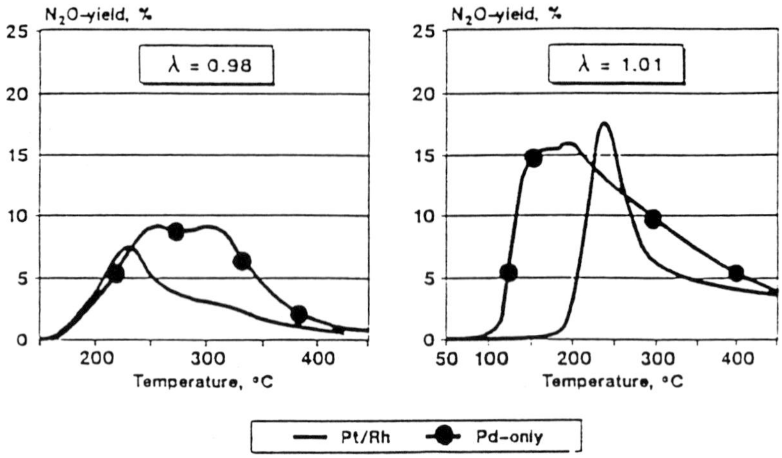

Figure 6: Model Gas Test Results - Comparison of 5Pt/1Rh (40g/ft^3) with Pd-only (80g/ft^3); Formation of N_2O (SV=50.000h^{-1}, fresh catalyst);

Again, for the Pd-only catalyst a higher overall formation of N_2O, in this case expressed as yield Y

$$Y_{N_2O} = \frac{n_{N_2O},out}{n_{NO_x},in} \cdot \frac{1}{2} \qquad \text{(n = molar flow of the component)} \qquad (2)$$

was determined especially under lean conditions at temperatures below 200°C and under rich conditions at temperatures in the light-off range at about 250 to 350°C. Generally it could be observed in the model gas test, that in the temperature range of catalyst light-off, approx. 10 to 15 % of the NO_x is converted into N_2O. In further experiments it will be checked wether these findings are confirmed under vehicle operation conditions and how dynamic vehicle operation conditions influence the catalyst behaviour.

Summarizing, promising activity improvements could be achieved in model gas experiments with Pd-only catalysts, even though some open questions regarding sulfur poisoning or the formation of secondary emissions have to answered.

3.2 Physicochemical Characterization

The oxidation state, i.e. the electron density of Pd particles is of significant interest. Recently de Vries et al. have investigated different lanthanum doped automotive catalysts and detected several different Pd species on the catalysts [17-21]. The higher electron density of the Pd particles has an influence on the chemisorption of hydrocarbon species.

The data of Table 2 give some literature reference binding energy values of different Pd species [17].

Table 2: Binding energy values

Pd species	Binding energy [eV]	Full width at half maximum [eV]
Pd me tal	334,9	1,1[a]
Pd black	335,0	1,1-1,5[b]
Pd	334,8 - 335,1	1,9[c]
PdO_{1-x}	335,6	1,6
PdO	336,3	2
PdO_2	337,4-338,3	2,5

[a] Pd foil; [b] XPS pass energy 50 eV, values depend on the particle size of the sample;
[c] values measured in the presence of La, Zr, etc.

In the present investigation different washcoat systems developed for Pd-catalysts were analyzed by XPS. For catalyst samples described in Table 3 PdO_2 species were not detected, but significant amounts of slightly oxidized Pd at 335,2 to 335,5 electron volts were found. Table 3 shows some results.

Table 3: XPS-binding energies of fresh Pd-only catalysts (full width at half maximum)

Catalyst sample	Binding Energy [eV]	Pd species
CeO_2/Al_2O_3	335,0 - 335,2	Pd/PdO_{1-x}
stabilized Al_2O_3	334,9 (1.9) 335,9 (2.0)	Pd/PdO_{1-x}
stabilized Washcoat	334,8 (1.9)	Pd

In all ceria containing catalysts cerium occurs mostly in the reduced oxidation state (Ce-III). $CeAlO_3$ was also detected.

The existence of a partially oxidized Pd species (PdO_{1-x}) can be assumed for the catalyst sample shown in Figure 7b due to the large width at half maximum of the peaks around 334,9 eV or the shift of the peaks up to 335,6 eV. This type of Pd on the catalyst surface can be interpreted as nonstoichiometric Palladium oxide i.e. PdO_{1-x} and is an efficient catalyst for the hydrocarbon oxidation [22]. After ageing in most of the cases stoichiometric Palladium oxide (PdO) was observed.

Electropositive elements lead to an increase of the surface concentration of Pd and tend to stabilize the dispersion at higher Pd loadings. This is shown in Figure 7a and b by comparing the intensities of the Pd 3d3/2 and Pd 3d5/2 peaks. The fresh catalyst sample in Figure 7a(stabilized washcoat) shows a lower surface concentration of Pd (0,16%-atom) compared to the catalyst sample shown in Figure 7b (stabilized alumina, 0,93%-atom).

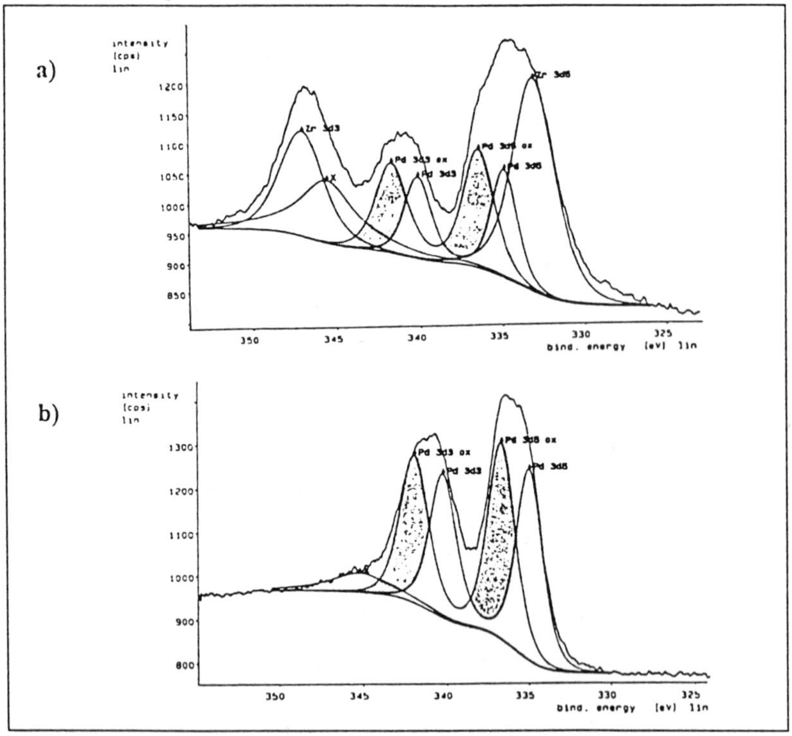

Figure 7: XPS-spectra a) stabilized washcoat
b) stabilized Al_2O_3

3.3 Engine and Vehicle Test Results

3.3.1 Pd-only catalysts

For the experiments described here, the stabilized washcoat system was used for the Pd-only catalysts. First the influence of the Pd-loading on catalyst light-off was investigated under both stoechiometric and lean conditions. The Pd-loading was varied over a very broad range from 3.5 to 40 g/l and the catalyst performance was compared to that of a reference Pt/Rh-catalyst.

Furthermore CO-chemisorption experiments were conducted with the Pd-catalysts. The results are summarized in Figure 8.

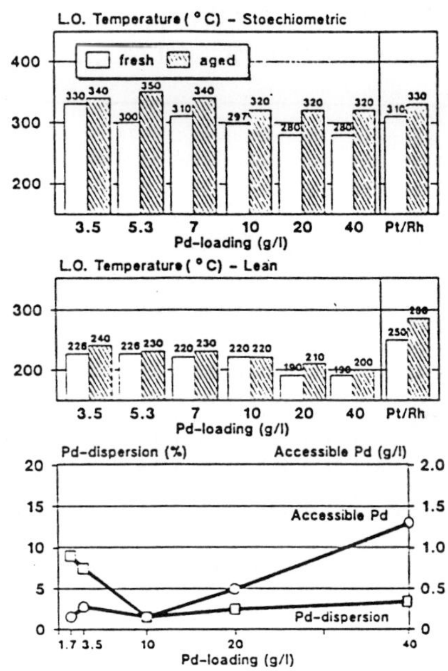

Figure 8: Temperatures for 50 % conversion of HC and Pd-dispersion as a function of the Pd-loading (aging = 100 h, 870 °C, fuel cut method);

For the fresh Pd-only catalysts under stoechiometric conditions a 50°C difference could be observed for the HC-light-off temperature comparing 3.5 to 40 g Pd/l. These differences more or less disappeared after aging and in all cases, i. e. fresh and aged, the Pt/Rh-system ranged in the midfield regarding HC-light off. But this result changed for the lean exhaust gas composition. For Pd-only a much lower light-off temperature was determined as compared to Pt/Rh and in all cases, fresh and aged the influence of an increasing Pd-loading could be observed, most probably due to the higher accessible Pd-surface that is obtained with higher loadings. The Pd-dispersion was nearly constant.

In the following vehicle test study catalysts with a 100 g Pd/ft^3 loading were used. For the three different vehicles chosen the ratio of catalyst volume to mass of the vehicle varied from 1.4 liter catalyst/ton vehicle (A) over 2.1 l/t (B) to 2.6 l/t (C). FTP 75 tests were conducted for the fresh and 50 h, 900°C, fuel cut aged systems. The conversion rates for CO, HC and NO_x are given in Figure 9.

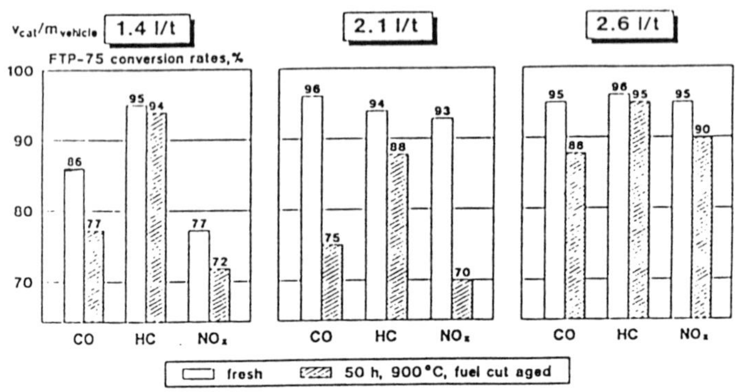

Figure 9: FTP 75 test results for Pd-only catalysts (100 g Pd/ft^3)

For the fresh samples especially for the vehicles B and C conversion rates of about 95 % could be achieved for all three components, but only for vehicle C similar activities were maintained after aging.

These results might indicate that for the use of Pd-only technologies in vehicles with tough emission control requirements a higher catalyst volume must be used as compared to Pt/Rh or Pd/Rh converters.

3.3.2 Pd/Rh-Catalysts

In the experiments described below a stabilized washcoat type impregnated with Pd and Rh was compared to an $Al_2O_3/CeO_2/ZrO_2$ - based Pt/Rh-catalyst. First some engine test results are given in Figure 10.

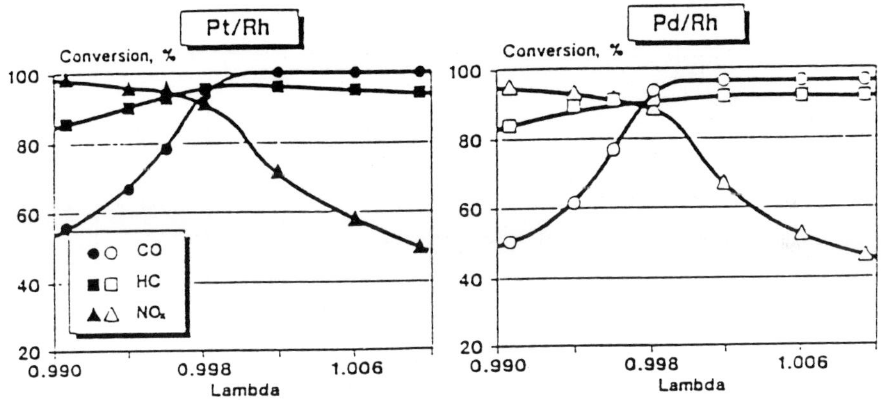

Figure 10: Engine sweep test comparison - Pt/Rh versus Pd/Rh (5/1, 40 g/ft^3: 50 h, 930 °C, fuel cut aged; T=450 °C, 1 Hz ± 1 A/F)

Comparing the aged Pd/Rh- with Pt/Rh-systems only little differences were observed, e. g. small deficits regarding NO_x around stoechiometry for Pd/Rh, but this is hardly influencing the A/F-window. Based on this promising data, this Pd/Rh-technology was investigated on different vehicles in different test cycles.

In the vehicle study first the influence of the vehicle's A/F-management on catalyst performance is investigated. The European MVEG-A test cycle was driven, the vehicle had a 6-cylinder 2.8 l engine. Figure 11 shows how a modification of the A/F-management influences exhaust gas temperature and raw emissions (Figure 11a) as well as the catalyst activity (Figure 11b) for Pt/Rh and Pd/Rh.

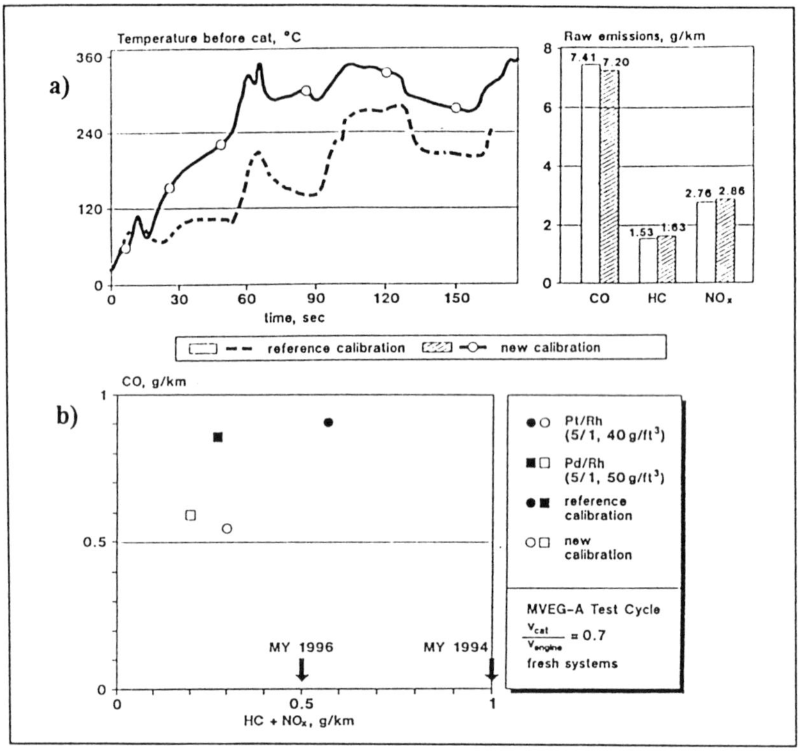

Figure 11: Influence of the A/F-management on a) exhaust gas temperature and raw emissionb) catalyst performance (MVEG-A test, V_{cat}/V_{engine}=0,7, fresh systems, 5Pt/1Rh, 40 g/ft^3, 5Pd/1Rh, 50 g/ft^3;)

From Figure 11 a it becomes clear that the new calibration mainly leads to a steeper temperature increase, whereas the raw emissions are hardly influenced. This gives in general improved emission levels due to a faster catalyst light-off, as shown in Figure 11b. The data clearly demonstrate at least equivalence for the Pt/Rh- and the Pd/Rh-technologies, the modelyear 1996 European standards could easily be met with the cheaper Pd/Rh-system with a relatively small catalyst volume.

Similar conclusions can be drawn from the results shown in Figure 12, where the same Pt/Rh- and Pd/Rh-catalysts were tested on two different vehicles in the MVEG-A and FTP75 test cycle and comparable activities were obained.

It should be put emphasize on the good CO bag 1 performance during the MVEG-A test cycle and some bag 3 HC advantages for Pd/Rh while the NO_x activity is comparable (see Figure 12a).

During the FTP 75 tests the good aging stability for the Pd/Rh technology especially for CO and NO_x is remarkable [Figure 12b].

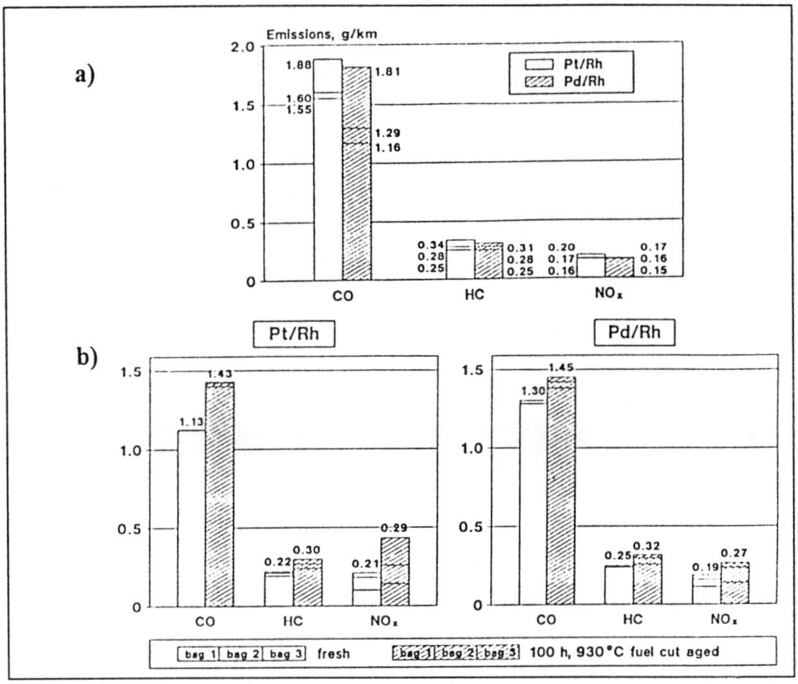

Figure 12: Vehicle test results - 5Pt/1Rh versus 5Pd/1Rh (50 g/ft³)
a) V_{cat}/V_{engine}=0.84, MVEG-A test cycle; 100 h,930°C fuel cut aged);
b) V_{cat}/V_{engine}=1, FTP75 test cycle; fresh and 100 h, 930°C, fuel cut aged, emissions in g/mile);

Summarizing these findings, Pd/Rh-technologies offer from the technical aspect as well as for cost reasons promising alternatives compared to the more expensive well-known Pt/Rh-technologies. This seems to be true already for vehicles that are approved for the current legislation and still as an outlook for future work , further improvement should be possible by optimizing A/F-management systems and/or the quality of the fuel.

4. CONCLUSIONS

From the experiments the following conclusions can be drawn for the Pd-only catalyst.

- Model gas test results

In SO_2-free gas mixtures new developed Pd-only catalysts performed equal to conventional Pt/Rh-systems. By addition of 200 vppm SO_2 the Pd-only catalyst lost activity. The Pd-only systems formed greater amounts of secondary emissions like NH_3 or N_2O than Pt/Rh-catalysts.

- Results of XPS-experiments

With new developed washcoats an improved Pd-dispersion could be achieved, also a stabilization of slightly oxidized PdO_{1-x} species was observed.

- Vehicle test results

Dependent on the ratio of catalyst volume to mass of the vehicle conversion rates exceeding 90 % for all pollutants were achieved after severe engine aging.

For the Pd/Rh-catalysts the following can be summarized:

- Engine test results

Comparable activities for fresh and aged Pd/Rh- and Pt/Rh-systems could be demonstrated.

- Vehicle test results

Comparing Pd/Rh- with Pt/Rh-technologies on different vehicles in European and FTP 75 - test cycles similar to lower emissions were achieved with the cheaper Pd/Rh-catalysts.

After severe aging model year 1996 legislation for the European Union could be met with Pd/Rh-systems.

5. ACKNOWLEDGEMENT

The authors wish to thank colleagues and co-workers for the valuable discussions and for the high quality experimental work. Especially Dr. P. Albers is thanked for carrying out and discussing the XPS-experiments. Ms. Eva Förnges and Ms. Bettina Schäfer are thanked for their dedication in editing the paper.

6.REFERENCES

1 K. C. Taylor, Automobile Catalytic Converters, Springer, Berlin, Heidelberg, New York, Tokyo (1984);
2 Annual Report Degussa AG (1993)
3 H.S. Ghandi and M. Shelef, Applied Catalysis, 77 (1991) 175;
4 D.R. Monroe, M.H. Krueger, D.D. Beck and M. D'Aniello, Jr., "Catalysis and Automotive Pollution Control II", A. Cruc, Ed., 593, Elsevier (1991);
5 D.D. Beck, D.R. Monroe and M.H. Krueger, SAE Paper No. 910844 (1991);
6 R.J. Farrauto and J.J. Mooney, SAE Paper No. 920557 (1992);
7 J.C. Summers, J.F. Skowron, W.B. Williamson and K.I. Mitchell, SAE Paper No. 920558 (1992);
8 H.S. Ghandi, W.B. Williamson, E.M. Logothetis, J. Tabock, C. Peters, M.D. Hurley and M. Shelef, Surface and Interface Analysis, 6 (1984) 149;
9 T. Mallat, Z. Bodnar, S. Szabo and J. Petro, Applied Catalysis 69 (1991) 85;
10 B.H. Engler, D. Lindner, E.S. Lox, K. Ostgathe, A. Schäfer-Sindlinger, W. Müller, SAE Paper 930386 (1993);
11 a) Automotive Engineering, 3(1994)59;
b) S. K. Hoekman, Environ. Sci. Technol., 26(1992)1206;
c) Bosch, Kraftfahrttechnisches Taschenbuch, 21. Edition, Düsseldorf, VDI Publishers;
d) K. C. Taylor, "Catalysis and Automotive Pollution Control I", A. Cruc, Ed. 97, Elsevier (1987);
e) W. B. Williamson, H. S. Ghandhi, M. E. Szepilka and A. Deakin SAE-Paper No. 852097 (1985);
f) Standard Specifications for Automotive Gasoline, ASTM D439, 1980 Annual Book of ASTM Standards, Part 23, November 1980;
12 D.J. Ball, SAE Paper No. 932765 (1993)
13 E.S. Lox, B.H. Engler, E. Koberstein, SAE Paper No. 910841 (1991);
14 B.H. Engler, E. Koberstein, D. Lindner, E.S. Lox, "Catalysis and Automotive Pollution Control II", A. Crucq, Ed., 641, Elsevier (1991)

15 B.H. Engler, E.S. Lox, K. Ostgathe, T. Ohata, K. Tsuchitani, S. Ichihara, H. Onoda, G. T. Garr, D. Psaras, SAE Paper No. 940928 (1994);
16 B.H. Engler, E. Koberstein, P. Schubert, Applied Catalysis, 48 (1989), 71;
17 K. S. Kim, A. F. Gossmann, N. Winograd, Anal. Chem. 46(1974)197;
18 T. E. Hoost, K. Otto, Applied Catalysis, 92(1992)39;
19 M. Shelef, L. P. Haack, J. E. de Vries, E. M. Logothetis, J. Catal. 137(1992)114;
20 K. Otto, L. P. Haack, J. E. de Vries, Applied Catalysis B: Environmental, 1(1992)1;
21 P. J. Schmitz, K. Otto, J. E. de Vries, Applied Catalysis A: General, 92(1992)59;
22 D. D. Beck, D. R. Monroe, C. L. DiMaggio, J. W. Sommers, SAE Paper No. 930084 (1993);

A. Frennet and J.-M. Bastin (Eds.)
Catalysis and Automotive Pollution Control III
Studies in Surface Science and Catalysis, Vol. 96

SMART Pd TWC TECHNOLOGY TO MEET STRINGENT STANDARDS

J. Dettling[a], Z. Hu[a], Y. K. Lui[a], R. Smaling[a] and C. Z. Wan[a]
A. Punke[b]

[a]*Environmental Catalyst R&D*
Engelhard Corp., Iselin, New Jersey 08830-0770 (U.S.A.)
[b]*Engelhard Technologies, GmBH & Co. OHG, 30625 Hannover, Germany*

ABSTRACT

This work highlights the use of model gas reactions to probe and elucidate the properties of complex palladium catalysts. Palladium loading in the catalyst is directly related to hydrocarbon and nitrogen oxide activity, while ceria, an oxygen storage component, improves carbon monoxide performance. An approach, Strategic Material Architecture at Reaction Temperatures (SMART), is discussed as it is applied to palladium automotive catalysts. It will be demonstrated that the catalyst design must incorporate features to achieve early light-off, high temperature performance, thermal resistance, sulfur poisoning tolerance and good TWC activity over a broad range of conditions. A highly integrated Pd catalyst washcoat design is shown to obtain high TWC performance typical of that seen in Rh-based catalysts. A correlation of model gas reaction data with dynamometer aged and vehicle evaluated catalysts will be demonstrated.

1. INTRODUCTION

California standards call for significantly reduced vehicle emissions between 1997 and 2003. To comply with the stricter hydrocarbon standards, emission design engineers are examining methods to cost effectively decrease the hydrocarbons early in Bag 1 of the FTP-75 cycle. Potentially expensive and complicated approaches, that include electrically heated monoliths, hydrocarbon traps, and burner or igniter systems are being evaluated against less costly alternatives like close coupled or manifold catalysts. Typically, the high temperature durability of the close coupled or manifold catalysts is being addressed by the use of palladium (Pd). This paper deals with the development of a high performance TWC palladium catalyst to meet this challenge.

To achieve high nitrogen oxide (NOx), carbon monoxide (CO), and hydrocarbon (HC) efficiencies from a palladium catalyst, H.S. Gandhi and co-workers [1,2] modified the precious metal function with other base metal oxides. Their work showed that the addition of MoO_3 improved the selectivity of Pd for converting NO_2 to N_2. Later, they found the conversion of saturated hydrocarbons could be improved with the same NOx selectivity when Pd was modified with WO_3 [3]. A.T. Bell and his group investigated the reducability of lanthanum oxide in the Pd-La catalyst system [4,5,6,]. This work suggested that strong metal support interactions, found by S. Tauster for TiO_2, were operative for other oxides [7]. H. Muraki et al. enhanced automotive palladium catalyst performance with lanthanum oxide. They found that, under reducing conditions, La increased the selectivity for NO reduction by H_2 and the catalytic activities for CO and propylene reactions by H_2O [8]. Not all of the early work considered the durability of the catalyst in vehicle application, nor the effects of sulfur. It should be noted that palladium interaction with lead is of less concern today, because misfueling of vehicles with catalytic converters is greatly reduced.

An approach, we find useful known as Strategic Material Architecture at Reaction Temperature (SMART), is practiced in designing the washcoat formulation of the Pd monolith catalyst. With the challenge to reduce development cycle time, model gas reactions are used to probe and elucidate the properties of complex palladium catalysts. To ensure a correlation between laboratory reactor and vehicle results, sulfur dioxide (SO_2) and water are present along with the normal reactive constituents. Component interaction studies aid in the design of a high performance TWC Pd catalyst. Engine dynamometer aging and vehicle evaluations confirm the validity of the laboratory bench results. This approach has resulted in a Pd TWC catalyst design that features early light-off with thermal and sulfur poison resistance over a broad range of conditions.

2. EXPERIMENTAL

2.1 Strategic Material Architecture at Reaction Temperatures (SMART)

Typical monolithic washcoats are composed of support materials to disperse the catalytic component and additives to stabilize and promote their function. The Strategic Material Architecture concept builds on the premise that each particular catalytic function within the washcoat is surrounded by its own optimum environment to enhance a particular reaction. Application of the washcoat to the monolith, followed by impregnation of the catalytic function, does not allow one to optimize the interaction between specific component materials. Segregated washcoat technology permits the deposition of the active constituents on to specific supports before application to the monolith. Catalytic particles of different composition can be blended within the same washcoat layer or physically separated. This process optimizes beneficial metal support interactions while avoiding the deleterious ones. To understand what is necessary to improve the Pd activity for CO, HC and NO_X activity above the current commercial products, stressful probe conditions are selected to observe performance differences. Equally important are the conditions for probing the washcoat constituents at reaction temperatures. The SMART approach looks at the Pd function in terms of CO and HC oxidation and NO_X reduction under typical operating conditions after high temperature exposure. For Pd to behave similarly to a Rh-based catalyst, improvements in CO and NO_X activity are required.

2.2 Effect of Ceria

The positive influence of ceria on HC, CO and NOx activities of precious metal catalysts is well documented. With increasing ceria loading, 3.53 g/l Pd catalysts display significant improvements in both CO and NO_X activity, as seen in Figure 1A. The model gas reactions were conducted at 550°C..

Figure 1A
Increasing Ceria Loading Improves Both CO & NOx Activity at 550 C

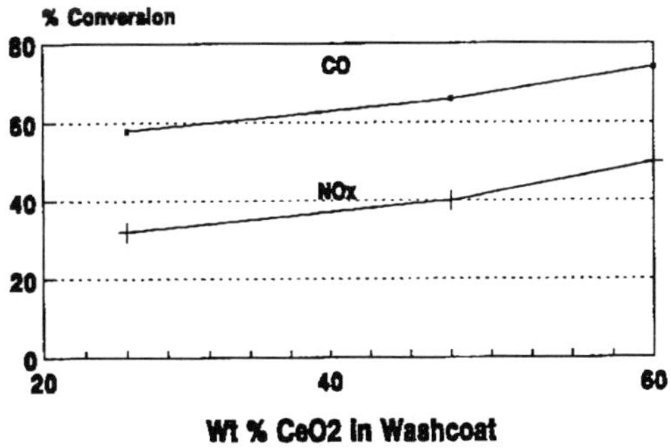

Aged 6 hours 1000°C @ Stoich. +/-0.2 A/F, Eval. 550°C, +/-0.1 A/F, 550,000 VHSV

Figure 1A. Increasing ceria loading improves both CO & NOx activity at 550°C.

This specific temperature mitigates the impact of sulfur and maximizes the influence of ceria. Similar catalysts were aged for 75 hours on a multi-mode dynamometer aging cycle in which the maximum bed temperature is 950°C. The FTP test data in Figure 1B were obtained on a vehicle equipped with a 3.8 l engine in which the 0.688 l catalyst sees a maximum bed temperature of 600-650°C. CO activity responds to the ceria loading, but there is less influence on NO_x performance. A global explanation is that sulfur adsorption on both the ceria and the Pd cause the discrepancy seen between the 550°C model gas data and the lower temperature vehicle results. Another consideration is that the rare earth oxides within the washcoat can give in situ strong metal support interactions with the Pd [9,10]. These interactions can alter the performance of the catalyst. Based upon the above data and the literature, it is proposed that above 500°C, ceria participates in the CO-NO_x reaction and below this temperature the Pd dominates the catalytic chemistry.

Figure 1B
In FTP Test, Ceria Effect Strong
ON CO & Weak on NOx Performance

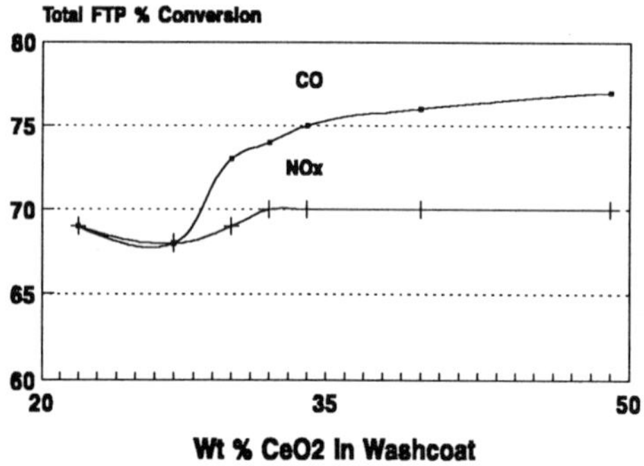

Figure 1B. In FTP test, ceria effect is strong on CO and weak on NO_x performance.

2.3 Influence of Sulfur

Two sets of conventional 3.53 g/l Pd catalysts were aged at 950°C for 12 hours and evaluated in laboratory equipment with and without sulfur in a typical automotive feedgas containing 0.75% CO, 0.2% H_2, 0.63% O_2, 16.3% CO_2, 10% H_20, 1400 ppm NO, 235 ppm C_3H_6, 235 ppm C_3H_8, 45 ppm SO_2 and N_2 balance. All evaluations are conducted employing a perturbation of ± 0.5 A/F @ 1

Hz @ stoichiometric set point and 50,000 VHSV space velocity. The light-off, represented as the temperature for 50% conversion, increase dramatically in the presence of 20 ppm SO_2 compared to the sulfur-free evaluation (Figure 2A). At 450°C and stoichiometric conditions, the catalysts as shown in Figure 2B give lower NO_x performance in the presence of SO_2. The more dramatic impact on NO_x activity suggests that sulfur is retarding the NO_x reaction by interfering with either CO or H_2. Therefore, all probe reaction studies contained SO_2 in the feedgas.

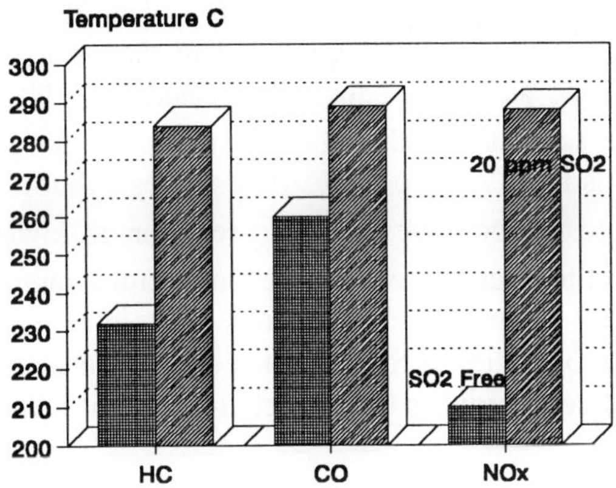

Aged 12 hours 950°C in gas mix without SO_2

Figure 2A. Sulfur has a strong influence on Pd light-off activity after aging.

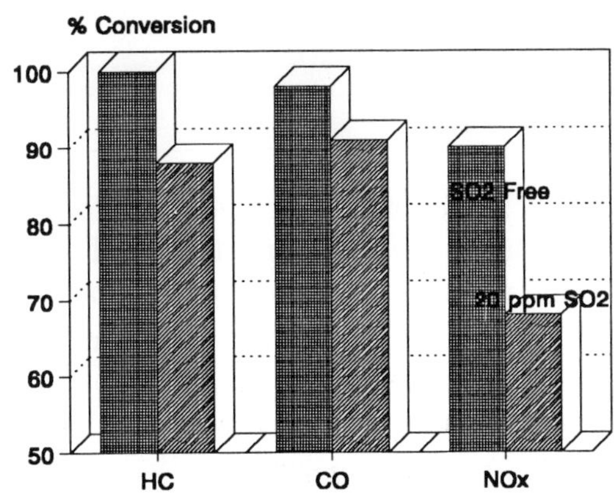

Aged 12 hours 950°C in gas mix without SO_2

Figure 2B. Sulfur effect seen under stoichiometric conditions at 450°C with Pd.

2.4 Washcoat Component Studies

In Table 1, the NO_x activity is given for a series of Pd catalysts aged for 12 hours at 950°C in a synthetic gas stream previously described. The evaluation is at 350°C, gas composition as stated above, and a space velocity of 50,000 hr-1. Doubling the Pd loading from 1.77 to 3.53 g/l increased the NO_x efficiency from 40% to 62%. This indicates that at lower temperatures the NO_x activity is directly dependent upon the relative concentration of Pd catalytic sites. To mitigate the loss of Pd sites due to high temperature oxidizing and reducing conditions, a stabilizer package was developed. In a 1.77 g/l Pd loaded catalyst, the addition of the stabilizer increased the NO_x efficiency from 40% to 66%. This improvement in performance is higher than that obtained with 3.53 g/l of Pd in the catalyst. With 3.53 g/l of Pd, the addition of the stabilizer package increases the NO_x efficiency to 72%. Interestingly, when ceria is added to the formulation (3.53 g/l Pd + stabilizer), the NO_x activity drops from 72% to 64%. This suggests that the Pd must be more metallic in nature to perform the CO-NO_x reaction at lower operating temperatures. For the catalyst to operate over a wide temperature range, Pd is segregated from ceria for low temperature operations and intricately associated with it for high temperature operation. ESCA studies on the washcoat constituents confirm that Pd devoid of ceria reduces to the metallic form at lower temperatures than if associated with it.

Table 1: Influence of Additives on Pd Low Temperature NO_x Activity

Pd Loading	Stabilizer	CeO_2	% NOx Conversion @ 350°C
1.77 g/l	No	No	40
3.53 g/l	No	No	62
1.77 g/l	Yes	No	66
3.53 g/l	Yes	No	72
3.53 g/l	Yes	Yes	64

Using the model gas reactor, the washcoat formulation is optimized for both low temperature (350°C) and high temperature (550°C) performance. Table 2 gives the CO and NO_x activity for the low and high temperature components of the washcoat. While each is optimized for a particular temperature regime, the combination of each component in a unique layered structure provides an overall synergism. A fully formulated monolithic catalyst, containing the optimized washcoat architecture, was compared to an existing Pd technology. Both 3.53 g/l Pd catalysts are heated in a tube furnace for 24 hours at 950°C in 10 % H_2O and the remainder N_2. They were evaluated in a typical synthetic automotive exhaust stream containing 20 ppm SO_2, perturbation of +/-0.5 A/F @ 1.0 Hz at 430°C and a stoiciometric set point. In Figure 3, improvements in HC, CO and NO_x are seen for the new formulation compared to the previous technology. The laboratory data gives evidence that a Pd catalyst with high activity levels is achievable after high temperature aging.

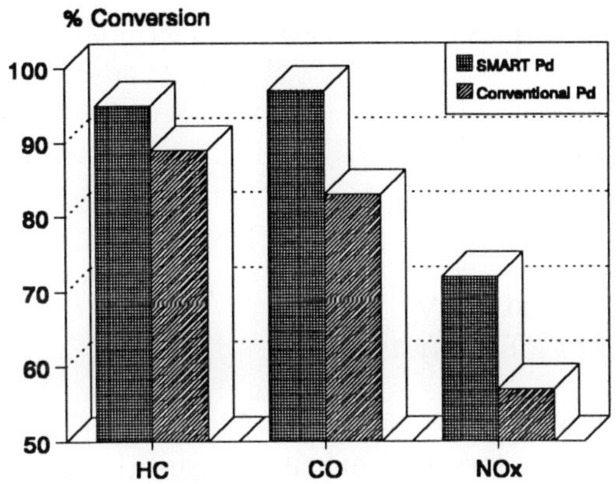

24 hours @ 950°C, 10% $H_2O + N_2$, Evaluation @ 430°C @ 50,000 VHSV

Figure 3. SMART Pd technology improved stability

Table 2: Optimized Pd Monolithic Washcoat Strategy

	% Conversion			
	350°C		550°C	
Washcoat Component	CO	NO_x	CO	NO_x
High Temperature Portion	60	31	68	47
Low Temperature Portion	69	53	62	33

Aging: +/-0.2 A/F @ Stoich. @ 1000°C for 6 hours
Evaluation: +/-0.1 A/F @ Stoich. 350,000 VHSV

2.5 Dynamometer Agings and Evaluations of SMART Pd Catalyst

Ammonia & H_2S Emissions: A catalyst containing 3.53 g/l of Pd is compared to a 1.41g/l 5Pt/1Rh catalyst for ammonia formation. Neither catalyst was aged. In Figure 4A, the ammonia formation is shown as a function of lambda at 475°C, +/- 0.5 A/F @ 1.0 Hz and 80,000 hr-1. While the Pd tends to make higher ammonia than the Pt/Rh catalyst under very rich A/F conditions, the reverse is true under stoichiometric operating conditions. In an emission control system with a stoichiometric set point, the Pd catalyst is equal to or more selective for NO_x

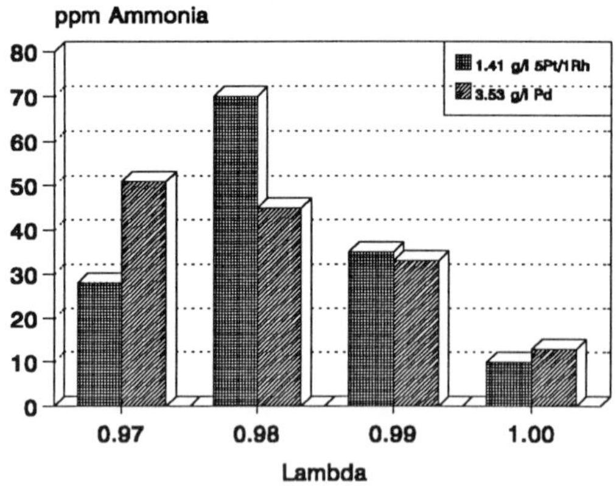

Sweep Test @ 475°C, +/-0.5 A/F @ 1.0 Hz, 80,000 VHSV
Figure 4A. NH_3 formation for Pd and Pt/Rh catalysts.

reduction to N_2 than a fresh Pt/Rh catalyst. In a separate test, the same catalysts were preconditioned at 590°C in a lean automotive exhaust stream (lambda = 1.02) for 30 minutes to store sulfur. They were then switched to a rich set point of lambda = 0.88 and measured for H_2S emissions at 530°C. From Figure 4B, it can be seen that the H_2S emission from the Pd-only catalyst is lower than that from a Pt/Rh catalyst containing a Ni scavenger. For comparison, a Pt/Rh catalyst without Ni is shown.

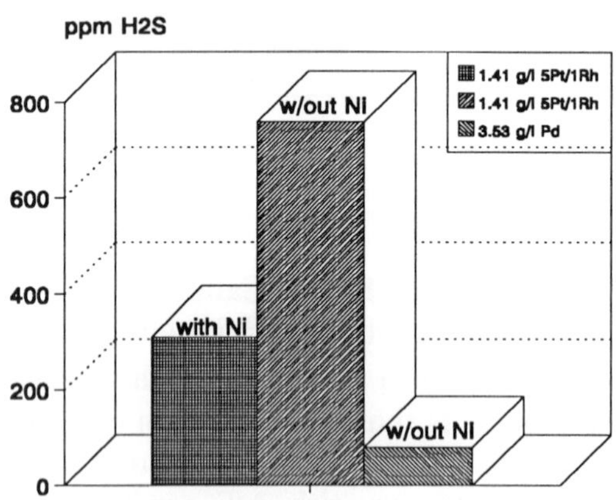

Figure 4B. H_2S low from SMART Pd technology.

<u>Operating Window</u>: A 3.53 g/l Pd catalyst optimized for high temperature exposure and a broad range of operating conditions was dynamometer aged for 95 hours. The cycle contains a high temperature mode in which the bed

temperature reaches 920°C for 68% of the cycle. There is a poisoning and fuel cut portion within the cycle. In Figure 5, the Pd catalyst is compared to a 1.41 g/l 5Pt/1Rh after the above aging in an evaluation at 450°C, 64,000 hr-1, a frequency of 1.1-1.5 Hz, and an amplitude of +/- 0.03. The non-Rh Pd technology has a profile similar to the Pt/Rh catalyst over the lambda scan window.

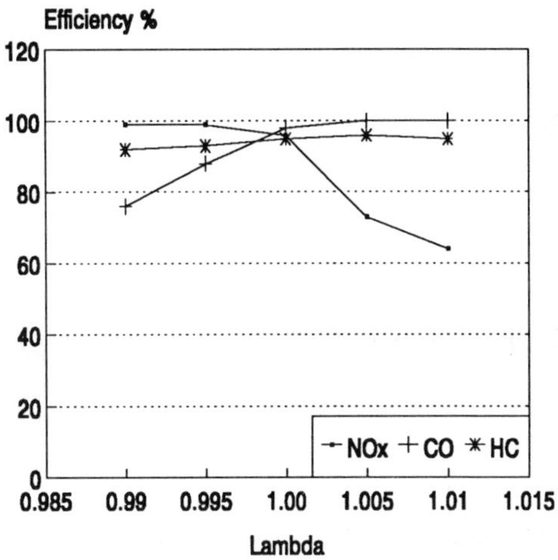

Figure 5A SMART Pd technology sweep curve

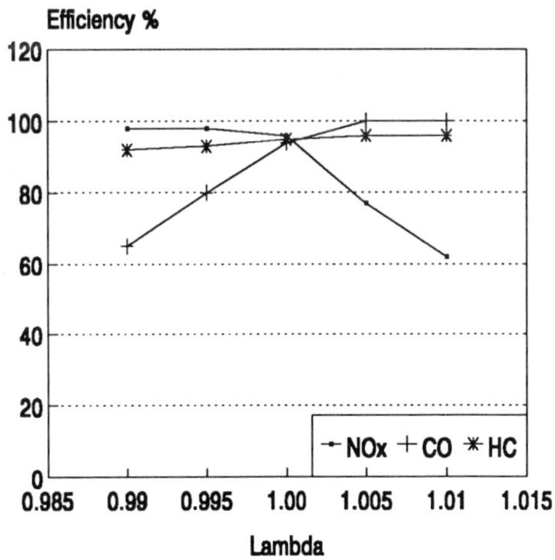

. *Figure 5B. Pt/Rh catalyst sweep curve.*

Two 3.53 g/l Pd catalysts were aged on a cycle for 75 hours in which the inlet temperature is 850°C for 16% of the time. A stoichiometric evaluation

(Figure 6) at 400°C compares the CO-NO_x crossover performance of a standard technology catalyst(10) to that of a high performance Pd catalyst. There is no drop-off in performance for the Pd catalyst when going from tight control (+/-0.1 A/F @ 1.0 Hz) to loose control (+/-0.5 A/F @ 0.5 Hz). Conversely, the standard reference catalyst activity is lower and drops further when going to the loose control.

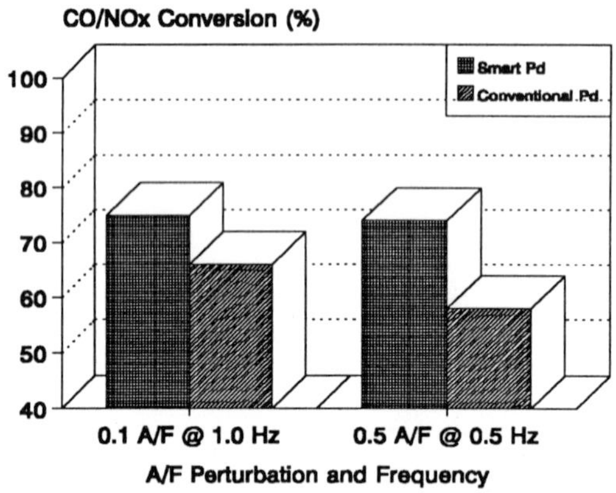

Aged 75 hours on Multi Mode Dyno Cycle

Figure 6. Superior CO/NO_x crossover with SMART Pd.

Vehicle Performance: In an attempt to meet the strict LEV/ULEV standards, a preliminary close coupled and underfloor catalyst system was evaluated. The close coupled catalyst was a 0.4 liter unit containing 10.6 g/l of Pd. The underfloor converter contains 2 units with a combined volume of 2.67 liters and a Pd loading of 3.53 g/l. The system was aged on a multi-step dynamometer cycle for 55 hours in which the close couple converter sees a maximum bed temperature of 950°C and the underfloor catalyst 800°C. The FTP test evaluations are conducted on a 1993 vehicle with a 2.2l DOHC engine.

Table 3: SMART Pd Technology System Testing

	g/mile		
	NMOG	CO	NO_x
Fresh	0.04 (97%)	0.46 (95%)	0.23 (95%) 0.14[1]
Aged	0.06 (95%)	0.60 (94%)	0.33 (93%) 0.17[1]

[1] Adjusted to 2.75 g/mile baseline.

The baseline emissions on the vehicle were 1.5 g/mile HC, 9.6 g/mile CO and 4.5 g/mile Nox. While the HC and CO emissions appear to be normal, our experience indicates that a 2.5 - 3.0 g/mile NO_x value is more indicative of vehicles of this displacement. In Table 3, the fresh and aged performance is given for the system. As seen, a SMART Pd-only system appears very stable with this aging history.
While the NMOG value is below the 0.075 g/mile LEV standard and the CO is significantly below the 1.7 g/mile ULEV, the NOx emission is above LEV/ULEV standard on this vehicle. If the NOx value is factored for a more typical baseline, it is anticipated that the vehicle meets the 0.20 g/mile ULEV standard. Without being able to control the engine management system, it is not possible to demonstrate the full advantage of the SMART Pd technology on a commercial vehicle.

3. Summary

Smart Material Architecture at Reaction Temperatures (SMART) is a methodology for optimizing individual washcoat components within a Pd catalyst to give a technology having Pt/Rh attributes. Laboratory reactor studies, coupled with characterization work, indicate that Pd having a more metallic character gives higher NO_x efficiency at lower temperatures. The Pd associated with ceria is effective for oxidation reactions and NO_x chemistry at the higher temperatures. By integrating both of these features into one washcoat, a Pd technology is obtained that gives high TWC performance.

Engine and vehicle evaluation results indicate that SMART Pd technology achieves performances equivalent in three-way conversion, and superior in ammonia and hydrogen sulfide formation, to those of Pt/Rh catalysts. FTP testing on a 1993 vehicle suggests that LEV NMOG and ULEV CO and NOx are possibilities with SMART Pd technology.

REFERENCES

1 Gandhi, H.S.; Watkins, W.L.; Stepien, H.K.; U.S. Patent 4,192,779 (1980).
2 Gandhi, H.S.; Yao, H.C.; Stepien, H.K.; ACS Symp. Ser. 1982, 178, 143.
3 Adams, K.M.; Gandhi, H.S.; Ind. Eng. Chem. Prod. Res. Dev. 1983, 22, 207.
4 Fleisch, T.H.; Bell, A.T. J. Catal. 1984, 87, 398.
5 Hicks, R.F.; Yen, Q.J.; Bell, A.T.; J. Catal. 1984, 89, 498.
6 Rieck, J.S.; Bell, A.T.; J. Catal. 1985, 96, 88.
7 Tauster, S.J.; ACS, Accounts of Chem. Res., 1987, 20, 389.
8 Muraki, H.; Shinjoy, H.; Sobukawa, H.; Yokota, K.; Fujitani, Y.; Ind. Eng. Chem. Prod. Res. Dev. 1986, 25, 202.
9 Sudhakar, C.; Vannice, M.A.; Appl. Cat. 1986, 14, 47.
10 Wan, C. Z. and Dettling, J. C.; U. S. Pat. 4,624,940 (1986)

A. Frennet and J.-M. Bastin (Eds.)
Catalysis and Automotive Pollution Control III
Studies in Surface Science and Catalysis, Vol. 96

A PALLADIUM FRONT BRICK STUDY

Douglas J. Ball[a] and Etienne Jacque[b]

[a]*AC Rochester, 1300 N. Dort Hwy., Flint, Michigan 48556, USA*
[b]*AC Rochester, ACG Technical Center of Luxembourg, L-4940 Bascharage, G.D. de Luxembourg*

ABSTRACT

With the recent adoption of stricter vehicle emission regulations both in Europe and in the United States, it has become necessary for emission engineers to evaluate various strategies to significantly reduce HC, CO and NO_x tailpipe emissions. Emission engineers are pursuing electrically heated converters, HC adsorbers, close-coupled converters, fuel heated converters, warm-up converters and advanced catalyst technologies to meet these regulations. The HC and NO_x regulations are the most challenging.

With regard to advanced catalyst technologies, recent developments in palladium (Pd) catalyst technologies have demonstrated superior HC and adequate NO_x performance compared to production platinum/rhodium (Pt/Rh) catalyst technologies. This paper investigates the application of advanced Pd catalyst technologies in converters that contain a Pd front brick and a Pt/Rh rear brick.

Two studies are presented. The first study evaluates the effects of Pd loading, Pd catalyst technology, Pd front brick catalyst volume and air/fuel control when a Pd catalyst is used in front of a Pt/Rh catalyst. Converters having 2.7 liters of catalyst volume were dynamometer aged and evaluated on two engines using the US FTP The second study not only investigated the effects of Pd loading and Pd catalyst technology of the front brick, but also investigated the effects of Pt/Rh loading and catalyst technology of the rear brick. Converters having 1.8 liters of catalyst volume were dynamometer aged and evaluated on a 1.6L European engine using the revised ECE + EUDC driving cycle. The effect of improved fuel injection on converter performance was also evaluated.

Within the scope of the experiments it was determined that Pd catalyst technologies have been developed that can significantly reduce emissions. Pd 3-Way converters (no Pt/Rh) should only be applied to vehicles with good to excellent air/fuel control. Pd front brick, Pt/Rh rear brick converters are appropriate for many vehicle applications. Significant reductions in HC and CO emissions can be achieved without increasing NO_x emissions. However, the choice of Pd catalyst technology and loading are important. High Pd loadings and active catalyst Pd catalyst technologies can be too selective for HC and CO, making the reduction of NO_x difficult. Increasing the Pt/Rh loading and using a high temperature Pt/Rh washcoat can reduce NO_x emissions.

1 Introduction

The stringent emission regulations of the near future have forced emission engineers to develop strategies to improve converter light-off performance. One strategy is to move the converter or a portion of the catalyst to the exhaust manifold. Earlier studies by Ball [1,2] and Summers et al. [3] showed that close coupled and small Pd containing warm-up converters can be very effective at improving cold start HC and CO performance on the US FTP. However, not all vehicle manufacturers can afford the cost or space to package a close-coupled or even a small warm-up converter on all their applications. For these applications, it would be desirable to have an underfloor converter that would have the light-off performance of a close- coupled or a warm-up converter.

This paper applies information from earlier "Warm-up - Underfloor Converter" studies [1,2,3] and investigates the use of a Pd front brick, Pt/Rh rear brick underfloor converter. This paper consists of two studies. These studies will investigate the effects on vehicle emissions of 1) Pd front catalyst volume, 2) Pd loading, 3) Pd catalyst technology, 4) rear brick Pt/Rh loading, 5) rear brick Pt/Rh washcoat technology and 6) air/fuel control.

2 Experimental

2.1 Scope

The first study, "US FTP Emission Study", investigates the effects of Pd loading, Pd catalyst technology and Pd front brick volume on US FTP emissions using aged 2.7L converters. The second study, "The European ECE + EUDC Emission Study", was performed to verify results of the first study and also to investigate the effects of varying the rear Pt/Rh brick catalyst technology and precious metal loading. The revised European ECE + EUDC was used to evaluate the performance of aged 1.8L two brick converters.

3 US FTP Emission Study

3.1 Converters

Table 1 describes the converter matrix used in the first study. 17 converters were built having a catalyst volume of 2.7 liters and using 400 cpsi ceramic substrates with a frontal area of 88 cm^2. Pd front catalyst volume varied from 0 (all Pt/Rh catalyst) to 100% (all Pd catalyst). Pd loading varied between 100 and 300 g/ft3. Two different types of Pd catalyst technology were also investigated.

The "Warm-up" Pd catalyst technology was developed for small converters located close to the exhaust manifold. The "3-Way" Pd catalyst technology was developed primarily for underfloor 3-way applications. The performance of these converters was compared to a baseline Pt/Rh only converter. This Pt/Rh converter used one of the first "High Tech." ceria containing washcoats developed in the early 1980's. The precious metal loading is 14/1 (Pt/Rh) @ 25 g/ft3. This catalyst technology and Pt/Rh concentration was also used as the rear catalyst technology for all of the converters in the study.

Table 1. 2.7L Converter Matrix

Parameter	Levels
Pd Front Catalyst Volume	0,8.3%,25%,50%,100%
Pd Loading	100, 300 g/ft3
Pd C atalyst Technology	Warm-up, 3-Way

3.2 Aging

All of the converters were dynamometer aged to estimate about 80,000 kilometers of field aging for low temperature underfloor applications [4]. See reference 4 for more detail on the aging schedule.

3.3 Performance Testing

After aging, the converters were evaluated in random order using an auto-driver FTP dynamometer stand[1,2]. The stand consisted of an engine,

Table 2. FTP Engines

	2.3L	3.8L
Cylinders	4	6
EGR	None	Yes
Air Meter	None	Yes
EPA Power (Kilowatts)	7.93	9.28
Inertia (kilograms)	1364	1761
Manifold to Manifold Distance (cm)		89
Manifold to Converter Distance (cm)	69	64
Average Engine-out (g/mile)		
HC (Std Dev.)	1.94 (0.049)	1.84 (0.041)
CO	8.89 (0.206)	11.82 (0.143)
NO_X	2.56 (0.077)	1.61 (0.072)
Average FTP Lambda	1.006	1.0006

transmission and an electric dynamometer. A computer controlled the electric dynamometer and the engine to simulate US FTP. In addition, two modal benches were used to measure engine-out and tailpipe emissions. A 1991 four cylinder 2.3L and a prototype V-6 cylinder 1993 3.8L were used to evaluate the FTP emissions of these converters. Due to the reproducibility of the FTP dynamometer stand only one FTP per converter was generally required. A few converters required additional FTP evaluations. A detailed description of the engines is found in Table 2.

Upon comparing the vehicles, the 1993 3.8L engine has better air/fuel control than the 1991 2.3L engine. The 3.8L responds better to engine transients and the engine is calibrated closer to stoichiometry.

3.4 US FTP Emission Study Results

Figures 1 and 2 present the FTP results in graphical form for the 2.3L and 3.8L engines, respectively. Each figure consists of four graphs representing HC, CO, NO_X and HC + NO_X emissions. The lines in each graph present the emission results for each combination of Pd catalyst loading and technology as the Pd catalyst volume is varied form 0% to 100%. The common point on each of these lines is the all Pt/Rh converter.

3.5 2.3L FTP Results

The top graph in Figure 1 presents the HC emission results. HC emissions are reduced between 33 and 50% when the Pd front catalyst volume is increased from 0 - 25%. Also, for each of the Pd loading and catalyst technology combinations, HC emissions continue to decrease as Pd catalyst volume and Pd loading are increased. At similar Pd catalyst volumes, the Pd 3-Way catalyst technology performs better than the Warm-up catalyst technology. The best converter for HC emissions was the 100% Pd converter using the Pd 3-Way catalyst technology at 300 g/ft3. The tailpipe emissions were 0.09 g/mile, a 62% reduction from the baseline HC emissions.

The next graph in Figure 1 presents the CO emissions results. Similar trends are seen. Again, significant reductions are seen in CO emissions as Pd catalyst volume is increased from 0 - 25%. More Pd catalyst volume and loading further decreases CO emissions.At similar Pd loadings and catalyst volumes, the Pd 3-Way catalyst technology performs better than the Warm-up technology. It is interesting to note that at similar Pd front catalyst volume, the Pd 3-Way at 100 g/ft3 converter performs similarly to the Pd Warm-up at 300 g/ft3.

The next graph in Figure 1 presents the NO_X emission results. These results are very interesting. Slight to no improvement in NO_X emissions are observed when increasing Pd front catalyst volume from 0 - 25% for each of the Pd loading and catalyst technology combinations. However, NO_X emissions increase as Pd catalyst front catalyst volume is increased from 25 - 100%. The worst converter is the 100% Pd converter with the 3-Way catalyst technology at 300 g/ft3. This converter was best for HC and CO emissions. Also, the best 100% Pd converter used the 3-Way catalyst technology with 100 g/ft3. Here it is observed that less Pd loading is better for NO_X emissions when using this Pd 3-Way catalyst technology.

The bottom graph of Figure 1 presents the HC + NO_X emission results. For each of the Pd loading and Pd catalyst technology combinations, the sum of the HC + NO_X emissions decreased as Pd catalyst volume was increased from 0 to 25%. However, as Pd catalyst volume is increased from 25 to 100%, HC + NO_X emissions increased for three of the four Pd loading and catalyst combinations. The Pd 3-Way at 100 g/ft3 did not exhibit this same trend. At 25% Pd catalyst volume it appears the performance of the converters is insensitive to Pd catalyst loading and technology.

3.6 3.8L FTP Results

The top graph in Figure 2 presents the HC emission results for the 3.8L engine. These results are similar to the 2.3L results. HC FTP emissions can be significantly reduced by increasing Pd front brick volume, increasing Pd loading and using the 3-Way Pd catalyst technology. Once again, the best converter is the Pd only converter with the 3-Way Pd catalyst technology at 300 g/ft3. HC emissions are reduced to 0.09 g/mile, a 61% reduction from the 0.23 g/mile emissions level of the all Pt/Rh converter.

The next graph in Figure 2 presents the CO emission results. Again, a similar trend is seen. CO emissions are reduced as the Pd catalyst volume and loading are increased and the Pd 3Way technology performs better than the Pd Warm-up technology. Also in general, the differences in performance of converters at the same Pd catalyst volumes are small, less than 0.1 g/mile.

The next graph in Figure 2 presents the NO_X emission results. NO_X emissions are significantly reduced as Pd catalyst volume is increased from 0 to 25%. Slight or no improvement is seen as Pd catalyst volume are increased from 25 to 100%. Also, at the 50 and 100% Pd catalyst volumes the NO_X emissions are not a function of catalyst technology but a function of Pd catalyst loading. A Pd loading of 300 g/ft3 performs better than 100 g/ft3.

The bottom graph in Figure 2 presents the HC + NO_X emissions results. As expected, the HC + NO_X emissions are significantly reduced as Pd catalyst volume is increased from 0 25%. Further reduction in emissions is observed as the Pd catalyst volume is increased from 25 to 100% At the higher Pd catalyst volumes, higher Pd loadings give lower emissions levels.

Figure 1. 2.3L FTP Results

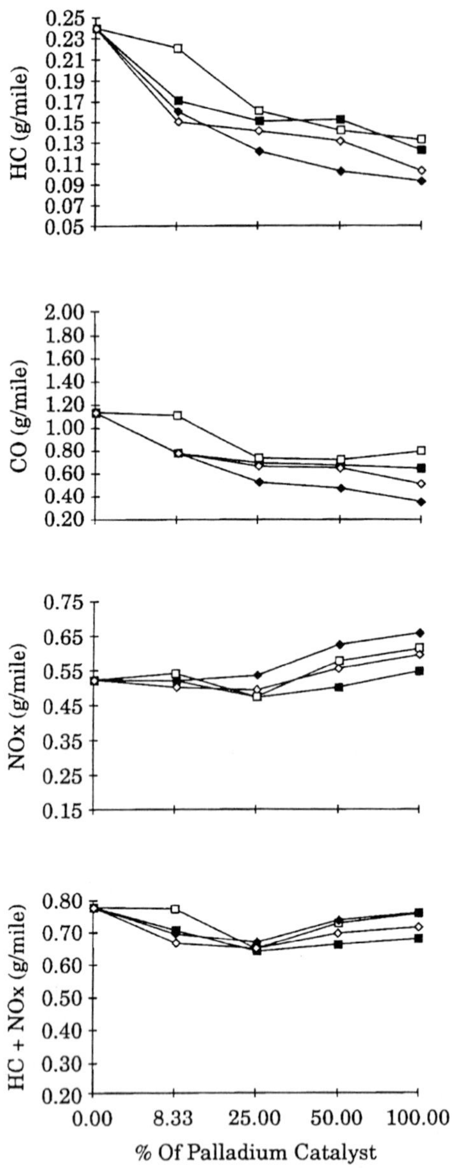

Figure 2. 3.8L FTP Results

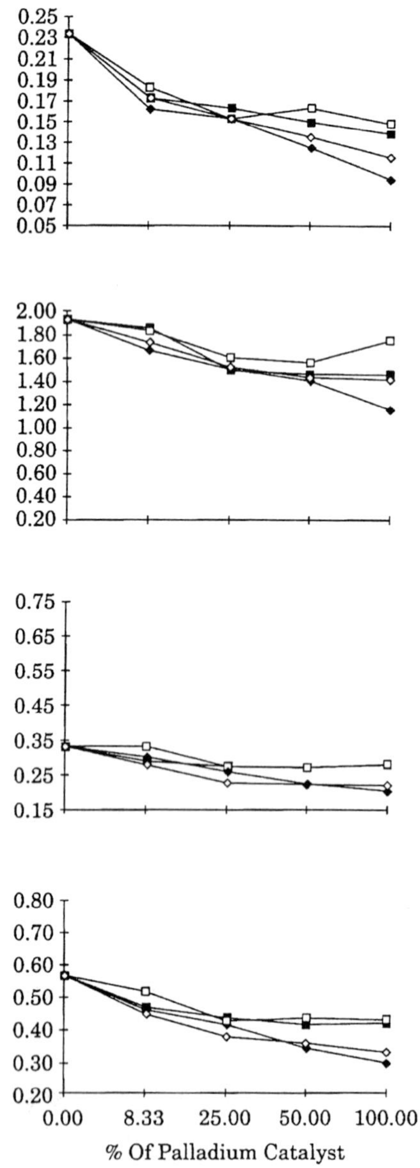

Legend

■ Pd 3-Way @ 100 g/ft3
□ Pd Warm-up @ 100 g/ft3
◆ Pd 3-Way @ 300 g/ft3
◇ Pd Warm-up @ 300 g/ft3

The Pd 3-Way and Warm-up catalyst technologies performed similarly at the same Pd catalyst volumes.

3.7 Effects of the Air/Fuel Control Systems

The emission results in Figures 1 and 2 allow a comparison of air/fuel control systems of these two engines. Remember that the 3.8L has better air/fuel control and operates closer to stoichiometry than the 2.3L engine. See Table 1. Comparing the HC and CO FTP results from Figures 1 and 2 it is observed that for both engines HC and CO emissions are reduced as Pd catalyst volume and loading are increased.

The comparison of the NO_X emissions is more interesting. On the 2.3L engine the best converter for HC and CO is worst for NO_X. This result implies that NO_X emissions are affected by the selectivity of the converter to oxidize HC and CO, especially at Pd catalyst volumes greater than 25%. In other words, the more active a converter is for HC and CO the worst it performs for NO_X. This result is not surprising because 1) a converter must have significant quantities of reductants like HC and CO to reduce NO_X and 2) lean calibration aids in the oxidation of HC and CO, thereby further reducing their concentrations for NO_X reduction. Conservatively, within the scope of this experiment, the 2.3L engine should use a Pd front catalyst volume of about 25% to achieve the significant reductions in HC and CO emissions without increasing NO_X emissions. Also, at a Pd catalyst volume of 25% the most cost effective Pd catalyst technology and loading can be used.

With 3.8L engine, which has higher engine-out CO and better air/fuel control about stoichiometry, no sacrifice in NO_X emissions (See Figure 2.) is observed as Pd catalyst volume was increased from 25 to 100%. The selection of Pd technology does not appear to have a large effect on emissions, but higher Pd loadings give lower HC, CO and NO_X emissions.

4. European ECE + EUDC Emission Study

4.1 Converters

Table 3 describes the 12 different converters built for this study. All of the converters had a catalyst volume of 1.8 liters and used 400 cpsi ceramic substrates with a frontal area of 88 cm^2. Each converter was built with two catalyst bricks having an equal volume of 0.9L. This experiment varied both the Pd front brick technology and loading and the Pt/Rh rear brick catalyst technology and loading. The Pd and the Std Pt/Rh washcoat technologies are the same ones described in the first experiment. The "HT" Pt/Rh washcoat technology is designed to improve CO and NO_X emission durability for high temperature applications. Pd loadings were varied from 150 to 300 g/ft3. Pt/Rh loading were varied between 40 g/ft3 @ 7/1 to 50 g/ft3 @ 5/1. The last two converters in Table 3 are all Pd and all Pt/Rh, respectively. These converters were built for baseline comparisons.

4.2 Aging

Two converters of each type were dynamometer aged to estimate about 80,000 kilometers of field aging for high temperature European underfloor applications.

4.3 Performance Testing

After aging the converters were evaluated using the revised ECE + EUDC driving cycle on a 1993 1.6L MPFI 16 valve vehicle. Emission runs were performed on this vehicle with production and selected fuel injectors. The selected injectors were specifically picked to minimize injector-to-injector flow variability. A detailed description of this vehicle is found in Table 4.
This 1.6L has relatively cool exhaust and achieved converter light-off with the Pd Front Brick converters about 175 seconds into the ECE portion of the driving schedule. Converter ligh-toff is defined when the front brick catalyst bed temperature reaches 350°C.

Table 3. 1.8L Converter Descriptions

	Front Brick		Rear Brick	
Converter Designation	Technology	Loading (g/ft3)	Technology	Loading (g/ft3)
P1,R1	Pd Warm-up	150	Pt/Rh Std	40 @ 7/1
P1,R2	Pd Warm-up	150	Pt/Rh Std	50 @ 5/1
P1,R3	Pd Warm-up	150	Pt/Rh HT	50 @ 5/1
P2, R1	Pd Warm-up	300	Pt/Rh Std	40@7/1
P2,R2	Pd Warm-up	300	Pt/Rh Std	50 @ 5/1
P2,R3	Pd Warm-up	300	Pt/Rh HT	50 @ 5/1
P3,R1	Pd 3-Way	150	Pt/Rh Std	40 @ 7/1
P3,R2	Pd 3-Way	150	Pt/Rh Std	50 @ 5/1
P3,R3	Pd 3-Way	150	Pt/Rh HT	50 @ 5/1
P4,R3	Pd 3-Way	300	Pt/Rh HT	50 @ 5/1
P4'P4	Pd 3-Way	300	Pd 3-Way	300
R3,R3	Pt/Rh HT	50 @ 5/1	Pt/Rh HT	50 @ 5/1

Table 4. 1.6L Vehicle Description

Cylinders	4	
EGR	No	
Air Meter	Yes	
Power (Kilowatts)	4.9	
Inertia (kilograms)	1139	
Manifold to Converter Dist. (cm)	120 (Dual takedown, 4-2-1)	
Average Engine-out (g/km)	Production Injectors	Selected Injectors
HC (Std Dev)	1.37 (0.067)	1.44 (0.049)
CO	5.81 (0.143)	5.34 (0.105)
NO_X	1.92 (0.046)	1.94 (0.042)
Average Lambda	1.013	1.011
Certified	US83	

4.4 European ECE + EUDC Emission Results

Figures 3 and 4 present the ECE + EUDC emission results for the 1.6L vehicle with production and improved injectors, respectively. Each converter type was evaluated at least three times on each vehicle. Stacked bar charts are used to show the emission contributions of the ECE and EUDC portions of the driving cycle. Again, each figure consists of four graphs representing HC, CO, NO_X and HC + NO_X emissions. The graphs are organized in groups of similar Pd front brick loading and technology. This is done so that the impact of front and rear brick catalysts can be easily seen. The converter designations in Table 4 are used as legends to describe the converters in Figures 3 and 4.

4.5 1.6L Emission Results, Production Injectors

A subset of nine converters described in Table 3 was evaluated on the 1.6L vehicle with production fuel injectors. The top graph of Figure 3 presents the HC emission results. 90% of the HC emissions are generated during the ECE driving schedule. No significant differences in HC emissions are seen with the Pd washcoat technologies when comparing converters with the same rear Pt/Rh catalysts. Higher Pd loadings appear to reduce HC emissions. Higher Pt/ Rh loadings in the rear brick do not appear to have an impact on HC emissions. The impact of the "HT" Pt/Rh washcoat technology is unclear. The "Std" Pt/Rh washcoat technology may perform better.

The next graph in Figure 3 presents the CO emission results. Here both the ECE and EUDC contribute significant emissions. However, there is considerable variability in the CO measurements. The impact of front and rear catalyst on CO emissions is difficult to determine. The next graph in Figure 3 presents the NO_X

Figure 3 1.6L ECE + EUDC Results
Production Injectors

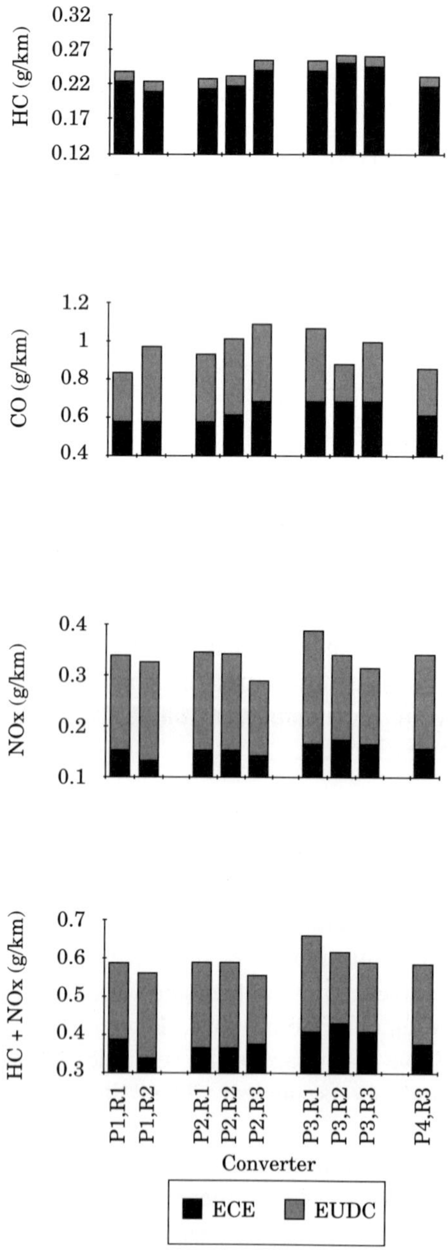

Figure 4 1.6L ECE + EUDC Results
Selected Injectors

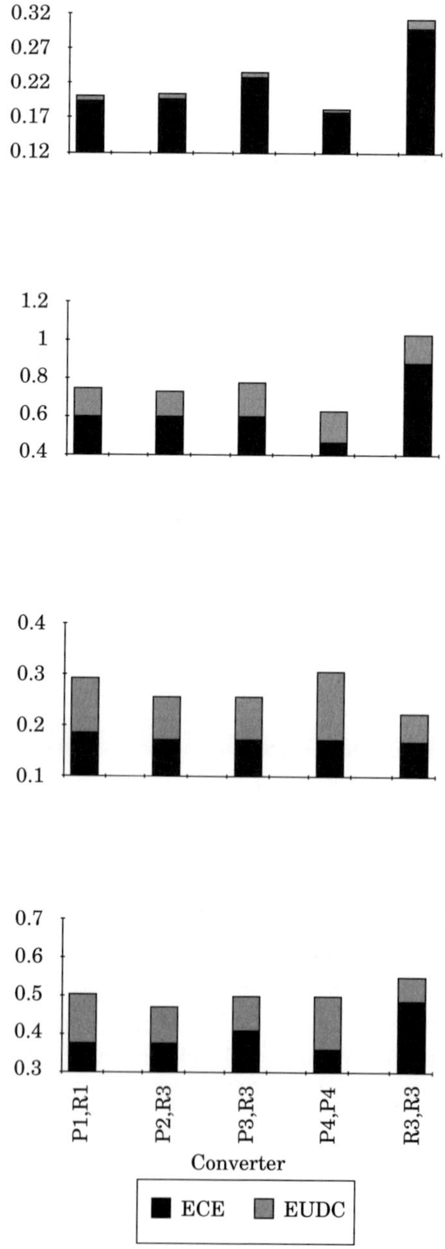

emissions results. These results show that adding Pt and Rh to the rear brick does improve NO_X emissions. The Pt/Rh "HT" washcoat technology appears to perform better than the "Std" technology. NO_X emissions were affected by the Pd front brick catalyst technology and loading. Lower Pd loadings and the Pd "Warm-up" catalyst technology tended to have better NO_X performance. It is interesting to note that the worse converters for HC are the best for NO_X.

The bottom graph of Figure 3 presents the HC + NO_X emission results. No big differences are seen between the converters, since the best converters for HC are the worse for NO_X. The selection of the best converter for this application should be based on cost.

4.6 1.6L Emission Results, Selected Injectors

Five converters described in Table 3 were evaluated on the 1.6L vehicle with the selected fuel injectors. This subset of converters include a Pd only converter (P4'P4) and an all Pt/Rh converter (R3,R3). The top graph in Figure 4 presents HC emissions results. The Pd only converter with the Pd 3-Way catalyst technology @ 300 g/ft3 (P4'P4) has the lowest HC emissions and the Pt/Rh converter (R3,R3) has the highest emissions during the ECE portion of the driving schedule. The HC emissions of the Pd front brick converters are slightly higher than the Pd only converter (P4'P4).

The next graph in Figure 4 presents the CO emission results. Similar results are found. The all Pd converter (P4'P4) had the lowest CO emissions and the all Pt/Rh converter (R3,R3) has the highest emissions during the ECE portion of the driving schedule. Again, the CO emissions of the Pd front brick converters are slightly higher than the Pd only converter (P4'P4).

The next graph in Figure 4 presents the NO_X emission results. Not surprisingly, the all Pt/ Rh converter (R3,R3) had the lowest NO_X emissions and that all Pd converter had the highest NO_X emissions. Again the best HC and CO performing converter has the worst NO_X emissions. This result is consistent with data presented earlier in Figure 1. The best two Pd front brick converters are P2,R3 and P3,R3.

The bottom graph of Figure 4 presents the HC + NO_X emission results. Converter, P2,R3 has the lowest HC + NO_X emissions and the Pt/Rh only converter has the highest. The other Pd front brick converters have similar emissions between converters, P2,R3 and the Pt/Rh only converter (R3,R3).

4.7 Effects of Fuel Injection

The effects of fuel injection quality on converter performance can be seen in Figures 3 and 4 since three Pd front brick converters, P1,R1, P2,R3, and P3,R3 were evaluated with both sets of injectors. All three of the converter realized

reductions in HC, CO and NO_x emissions with the selected converters. For converter 'P2,R3', HC, CO, NO_x and HC + NO_x emissions were reduced by about 0.04, 0.4, 0.07 and 0.11 g/km, respectively. Improvements in HC tailpipe emissions were largest during the ECE portion ofthe driving schedule, whereas, improvements in CO and NO_x emissions were primarily seen during the EUDC portion of the driving schedule.
Improving the air/fuel mixture distribution with selected injectors has shown to be an effective way to significantly reduce HC, CO and NO_x tailpipe emissions.

5. CONCLUSIONS

Within the scope of these experiments:

1) A 3-Way Pd converter (No Pt/Rh) can be used on specific vehicles with good to excellent air/ fuel control (and sufficient quantities of CO), to significantly reduce HC and CO emissions.
NO_x emission reductions were less dramatic.

2) A Pd front brick, Pt/Rh rear brick converter is appropriate for many vehicles. Such converters combine the excellent HC and CO light-off properties of Pd catalysts and the NO_x reduction properties of Pt/Rh catalysts. However, care must be taken in choosing the appropriate Pd front brick converter for a particular application, for high Pd loadings and active technologies can be too selective for HC and CO making the reduction of NO_x difficult. NO_x emissions of these converters can be improved by increasing the Pt/Rh in the rear brick and using a high temperature Pt/Rh washcoat. These Pd front brick converters can be used as a tool by emission engineers to address application specific emission problems.

ACKNOWLEDGEMENTS

The authors wish to thank J Okenka, D. Trytko, M. Zielinski, T. Kudza, C. Cole, C. Maly and J Kupe for their assistance.

REFERENCES

1 Ball,D.J "Distribution of Warm-up and Underfloor Catalyst Volumes", SAE Paper #922338, '92.

2 Ball,D.J "A Warm-up and Underfloor Converter Parametric Study", SAE Paper #932765,'93.

3 Summers,J.C., Skowron,J.F., and Miller,M.J., "Use of Light-Off Catalysts to Meet California LEV/ULEV Standards", SAE Paper #930386, 1993.

4 Sims,G., "Catalyst Performance Study Using Taguchi Methods", SAE Paper #881589, 1998.

A. Frennet and J.-M. Bastin (Eds.)
Catalysis and Automotive Pollution Control III
Studies in Surface Science and Catalysis, Vol. 96

NON-NOBLE METAL ENVIRONMENTAL CATALYSTS: SYNTHESIS, CHARACTERIZATION AND CATALYTIC ACTIVITY

Philip G. Harrison Nicholas C. Lloyd and Wan Azelee

Department of Chemistry, University of Nottingham, University Park, Nottingham NG7 2RD (U.K.)

ABSTRACT

The preparation, characterization, and catalytic activity of Cr(VI)- and Cu(II)-doped tin(IV) oxide catalysts are described. The catalysts, particularly the mixed Sn-Cu-Cr-O catalysts, exhibit comparable activity to conventional platinum catalysts for CO and hydrocarbon oxidation.

1 INTRODUCTION

The control of noxious emissions resulting from either the combustion of fossil fuels or from other industrial activities is one of the most immediate and compelling problems faced by nearly every country in the world. Environmental problems caused by two sources, automobile exhaust emissions and flue emissions from coal and oil fired power stations, have received much publicity over recent decades.

Emission problem arising from both automobile and stationary industrial internal combustion engine give rise to similar pollutants: carbon monoxide (CO), hydrocarbons (HC's), and nitrogen oxides (NO_x). For automobiles, these are now the subject of ever increasingly stringent legislation controlling the maximum permitted levels of emissions of each substance [1,2].

Platinum group catalysts currently represent the state-of-the-art in internal combustion engine emission technology. Current usage of rhodium and platinum in these catalysts exceeds the Rh/Pt mine ratio, and hence it is essential to reduce consumption in order to conserve the limited noble metal supply. It would be very beneficial, therefore, to reduce dependence on noble metals for catalytic converter usage and to seek viable alternative catalytic materials.

The driving force for the development of non-platinum exhaust emission catalysts is the price, strategic importance and low availability of the platinum group metals. Our studies have shown that catalysts based on tin(IV) oxide (SnO_2) promoted with chromium and/or copper (Cr-SnO_2 and Cu-Cr-SnO_2 catalysts) which exhibit excellent three-way catalytic activity - activity which is comparable to that shown by noble metals dispersed on alumina. This family of materials offers tremendous promise as cheap and efficient catalyst systems for the catalytic conversion of noxious emissions from a variety of sources. In this paper we describe three aspects of this family of environmental catalysts: (1) the synthesis, (2) their characterisation, and (3) their catalytic activity.

Knowledge of the solid-state chemistry of catalysts is invaluable to an understanding of their catalytic behaviour. Characterisation of this type of mixed oxide presents a major problem and is notoriously difficult. No single technique can yield anything but a very small amount of information, and a complete picture can only be gained by using a combination of methods. We report data obtained from a number of techniques including gas adsorption, FT-Raman, EPR and ESEEM (electron spin echo envelope modulation), XRD, and electron microscopy and electron diffraction.

2 CATALYST PREPARATION

The catalysts have been prepared by a combination of coprecipitation and impregnation techniques, and also by sol-gel methods. Although coprecipitation constitutes one of the most widely applied methods for the preparation of oxide catalysts, more recently it has become apparent that sol-gel techniques offer several advantages including (a) greater control of catalyst stoichiometry and homogeneity, (b) more efficient and intimate particulate mixing at the nanometer level, (c) greater thermal stability towards deleterious solid state processes such as sintering, segregation of metal components to grain boundaries, and phase separation and (d) greater control of catalyst dispersion on inert supports.

Tin(IV) Oxide Gel and Sol-gel:

Tin(IV) oxide gel was obtained by the ammonia precipitation method. Conversion to a stable sol-gel modification was effected by peptisation using choline.

Chromium(VI)- and Copper(II)-doped Tin(IV) Oxide

To a suspension of tin(IV) oxide gel in triply distilled water was added a solution of chromium(VI) oxide and the solution stirred for 24h. The resulting yellow mixture was filtered and the yellow solid air dried at 60°C for 24h. If desired, the powdery material was then washed with distilled water until the washings were colourless. The resulting material was again air dried. Copper(II)-doped tin(IV) oxide was either prepared in a similar manner using a solution of copper(II) nitrate, or by coprecipitation from an aqueous solution containing tin(IV) chloride and copper(II) nitrate in the required ratio. Both chromiun(VI)- and copper(II)-doped catalysts were also prepared by treatment of tin(VI) oxide sol-gel using aqueous solutions of chromium(VI) oxide or copper(II) nitrate. Mixed Sn-Cu-Cr-O catalysts were obtained by the sol-gel route.

3 FUNDAMENTAL CHEMISTRY UNDERLYING CATALYST PREPARATION

3.1 The Nature of Tin(IV) Oxide Sols: Photon Correlation Spectroscopy

The effect of added chromium(VI) oxide and copper(II) nitrate on the aggregation of a choline-stabilised tin(IV) oxide sol is illustrated in Figure 1. The choline-stabilised tin(IV) oxide sol exhibits an average particle size of *ca.* 700nm. However, addition of copper(II) nitrate to the tin(IV) oxide sol results in a decrease in the mean particle size. At a Cu:Sn ratio of 0.001 the mean particle size is 220nm rising steadily to a value of 400nm at a ratio of 0.02. In contrast, addition of even small amounts of chromium(VI) oxide to the sol has a dramatic effect. With a Cr:Sn ratio of 0.001, the mean particle size increases to 1250nm. Increasing the Cr:Sn ratio further results in a steady increase in the mean particles size until at a Cr:Sn ratio of 0.025, the mean particle size is 1850nm. At high metal:Sn ratios (>0.025 the tin sol is destabilised).

3.2 The Sorption of Chromate Species Onto Tin(IV) Oxide Gel

Exposure of tin(IV) oxide gel to aqueous solutions of CrO_3 result in the sorption of chromium species onto the oxide giving orange powders after filtering and drying in air. The amount of chromium species sorbed onto the oxide is dependent on several variables including the concentration of the CrO_3 solution, the temperature, the time of exposure, and the separation and washing procedures adopted. Figure 2 illustrates the effect of CrO_3 solution concentration on the loading achieved after stirring the slurry for 16 hours at ambient temperature. For CrO_3 solution concentrations in excess of *ca.* 0.1M followed by filtering and drying but no washing, the loading achieved is linearly proportional to the concentration up to the high concentration studied (1M). Increased time of exposure results in an increased loading, whereas washing the catalyst dramatically decreases the loading thus indicating the weak nature of the adsorption.

FT-Raman spectra in the range 700-1100 cm^{-1} (the ν(Cr-O) region) of three Sn-Cr-O catalysts with Cr:Sn atom ratios of 1:0.12, 1:0.23 and 1:0.38 are illustrated in Figure 3. All three spectra are similar in form exhibiting two principal maxima together with several shoulders. However, the positions of the peaks maxima shift with the Cr:Sn ratio. At the lowest chromium loading the spectrum exhibits maxima at 886 and 946 cm^{-1}, the former corresponds to the presence of CrO_4^- ions whilst the latter indicates the presence of $Cr_2O_7^{2-}$ anions (Table 1, assignments are by consistent with those made previously[3-5]. Other bands due to these

Table 1. Assignment of FT-Raman bands for chromate anions (cm^{-1})[a].

CrO_4^-	$Cr_2O_7^{2-}$	$Cr_3O_{10}^{2-}$	Assignment
		987	$\nu_{as}(CrO_2)$
		956	$\nu_s(CrO_2)$
886	942		$\nu_{as}(CrO_4)/(CrO_3)$
848	904	904	$\nu_s(CrO_4)/(CrO_3)$
		844	δ_{as}(Cr'OCr")

(a) Refs. 3-5.

species are present as shoulder features, and it is also probable that the shoulder at high wavenumber is due to a small amount of $Cr_3O_{10}^{2-}$ ion.

As the chromium loading is increased, however, the peak maxima shift to 890/901 and 952 cm^{-1} for ratio 1:0.23 and 901 and 959 cm^{-1} when the ratio is 1:0.38. In both cases pronounced shoulder features are present both to higher and lower wavenumber. This observed shift to higher wavenumber is readily rationalised by an increased concentration of $Cr_2O_7^{2-}$ and $Cr_3O_{10}^{2-}$ anions in these catalysts, and CrO_4^- ions are present only in small amounts. Higher polychromate species cannot be excluded at high ratios.

3.3 The Physical Nature of the Sn-Cr-O Catalysts and the Effect of Calcination

Nitrogen adsorption isotherms for the freshly prepared Cr(VI)/SnO_2 catalyst and after calcination for 24h. at various temperatures up to 1000°C are shown in Figure 4. Numerical data are collected in Table 2.

Table 2. Nitrogen adsorption data calculated by the BET and α_S methods.

Calcination	BET Data		α Data		V_p /cc g^{-1}	d / Å
Temp./°C	A_S /m^2 g^{-1}	C	A_S /m^2 g^{-1}	V_{mic} /cc g^{-1}		
60	114	167	125	0.007	0.057	18
300	96	425	114	0.008	0.053	19
600	58	16	59	-	0.059	40
1000	0.4	6	1	-	0.011	42

The specific surface area (A_S) of the uncalcined Cr(VI)/SnO_2 catalyst (BET, 114 m^2 g^{-1}; α_S, 125 m^2 g^{-1}) is considerably lower than that of tin(IV) oxide gel itself (185 m^2 g^{-1}). The specific surface area of the neat oxide decreases steadily with increase in temperature to a value of 40 m^2 g^{-1} after calcination at 1000°C. In contrast, that of the Cr(VI)/SnO_2 catalyst decreases by *ca.* 15% to *ca.* 100 m^2 g^{-1} after calcination at 300°C, by *ca.* 50% to 58 m^2 g^{-1} after calcination at 600°C, and falls dramatically to nearly zero after calcination after calcination at 1000°C.

The BET isotherm plots for the Cr(VI)/SnO_2 catalyst (Figure 4) show that the isotherm for uncalcined catalyst is typical for adsorption onto a microporous solid (type 1 isotherm). This form of isotherm is retained after calcination at 300°C, but after calcination at higher temperatures the isotherm changed drastically in form. Calcination at 1000°C resulted in an isotherm characteristic of a type III non-porous solid. Whilst after calcination at 600°C the material exhibited intermediate behaviour and was mesoporous.

The powder XRD spectra of freshly prepared (dried at 60°) samples of the Sn-Cr-O catalysts exhibit only broad diffuse lines characteristic of nanoparticulate tin(IV) oxide. These peaks are seen to gradually sharpen on calcination, but are still relatively broad after calcination at 600°. Only after calcination at high temperatures (*ca.* 1000°) is a characteristically sharp spectrum due to crystalline SnO_2 obtained. When the concentration of chromium in the catalyst is low (*eg* for a Sn:Cr ratio of 0.015), the chromium in the catalyst remains amorphous and no chromium-containing phase can be detected. Only at high ratios (*eg* 1:0.13) and at high calcination temperatures (1000°) does a chromium-containing phase, Cr_2O_3, separate (XRD 2theta values at 24.5, 36.2, 41.56, 50.2, 63.5).

Transmission electron microscopy of the Sn-Cr-O catalyst corroborates these observations. For example, the TEM micrograph of the catalyst with a Sn:Cr ratio of 0.015 at 600° shows only the presence of small crystallites of tin(IV) oxide and no chromium-containing phase although chromium is detected by EDXa analysis. We deduce therefore that the chromium is present in an amorphous surface layer of an as yet unknown composition on the crystallites of the tin(IV) oxide.

3.4 Preparation of Sn-Cu-O Catalysts - Coprecipitation or Sorption?

A key issue in the activity of these catalysts concerns the specific role of the Cu and Cr active sites which are expected to be located on the surface of the catalyst particles. Since both Cu(II) and Cr(III) are paramagnetic, EPR spectroscopy can be used to identify the nature of the heterometallic species within the catalyst matrix after different thermal and chemical treatments. However, a detailed picture of the local environment of the transition metal center cannot be obtained by conventional continuous wave EPR spectroscopy alone. These can, nevertheless, be obtained by pulsed EPR methods, namely the electron spin echo envelope modulation (ESEEM) technique.

Two types of catalyst, prepared by (1) impregnation of SnO_2-gel by Cu^{2+}, and (2) coprecipitation have been examined. Both have the same composition of a Sn:Cu^{2+} ratio of 1250:1.

Although EPR is very sensitive to the surroundings of the paramagnetic cation, it cannot provide directly information concerning the fine interactions with the framework. Such interactions can, however, be measured by ESEEM. This technique is particularly useful for the measurement of weak hyperfine interactions. The modulation frequencies are the NMR frequencies of the coupled nuclei. In disordered systems, the modulation frequencies are essentially the Larmor frequencies of the coupled nuclei, which serve to identify the coupled nuclei. The modulation depth can be related to the distance between the electron spin and the coupled nuclei, and to their number. The ESEEM measurements described here were carried out at 4 K using an operating frequency of 9.1 GHz. Both two-pulse and three-pulse sequences were employed.

The EPR spectra recorded at 120K of the SnO_2-Cu^{2+} catalyst sample prepared by impregnation (a) as prepared and (b) after calcination at 600° show that a single Cu^{2+} species with a cylindrical environment is present in each case although with differing $\boldsymbol{g}$ and $\boldsymbol{A}$ values. At intermediate calcination temperatures both species exist. When a compositionally identical sample prepared by coprecipitation is examined, both species are observed in the freshly prepared sample even prior to calcination.

Three peaks are observed in FT-ESEEM spectra of the SnO_2-Cu^{2+} catalyst sample prepared by impregnation, one at the Larmor frequency for hydrogen at *ca.* 13 MHz together with two peaks at the first overtone value. These spectra are very similar to the spectra of the aquo $[Cu(H_2O)_6]^{2+}$ cation found in aqueous solution indicating that this is the species initially sorbed on to the tin(IV) oxide gel most probably *via* hydrogen bonding. In the FT-ESEEM spectra for the same sample calcined at 300°C. The three features due to echoes from hydrogen still persist, but in addition new frequencies at the Larmor frequency of tin-119 at *ca.* 5 MHz and overtones appear. This indicates that, unlike for the uncalcined sample where no echoes from tin are observed and hence the copper is at a distance >5Å from any tin atoms, the result of calcination is to incorporate the copper ions into the tin oxide lattice. Calcination of the same sample at 600°C essentially removes all the initial $[Cu(H_2O)_6]^{2+}$ sorbate, and only the lattice-bound copper species is present. The

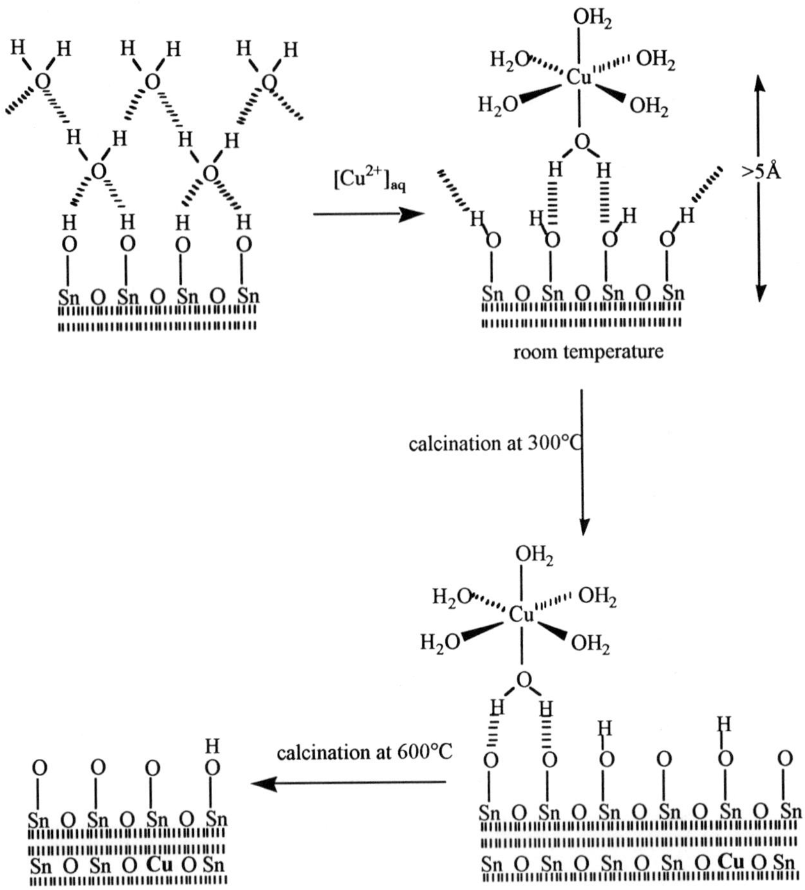
[Cu2+]aq
room temperature
>5Å
calcination at 300°C
calcination at 600°C
Sn O Sn O Cu O Sn

Scheme 1

processes are illustrated schematically in Scheme 1. No evidence for any intermediate covalently bound surface copper species could be detected.

In contrast, FT-ESEEM spectra of the second sample, that prepared by coprecipitation, exhibit both hydrogen and tin-119 frequencies for the freshly prepared sample. The conclusion, therefore, is that preparation of SnO_2-Cu^{2+} catalyst samples by coprecipitation results in the incorporation of copper into the tin(IV) oxide lattice as well as sorption of $[Cu(H_2O)_6]^{2+}$ cations on to the oxide particle surface.

4 Catalytic Activity

Catalytic activity studies were carried out using a typical catalytic microreactor. Catalyst sample sizes were 0.5g for CO oxidation studies and 2.0g for n-propane oxidation studies. Flow rates of the gas mixtures, which varied from stiochiometric to oxygen-rich, were in the range 90-100 ml min^{-1}. Table 3 summarises the activity of selected Sn-Cu-O, Sn-

Table 3. Temperatures necessary for the complete removal of CO and n-propane over Cu-Cr-Sn-O catalysts together with specific activities

Catalyst	T100(CO) /°C[a]	T100(C3H8) /°C[b]	Spec. Act. (CO)[c]	Spec. Act. (C_3H_8)[c]
SnO_2	360	525	0.06	0
71Sn 29Cu-O	150	400	0.53	0.82×10^{-3}
77Sn-23Cr-O	250	280	0.072	8.4×10^{-3}
62Sn-19Cu-19Cr	175	310	0.43	5.9×10^{-3}
Pt/Al_2O_3	200	400	0.064	1.6×10^{-3}

(a) Temperature required for the complete removal of CO.
(b) Temperature required for the complete removal of C_3H_8.
(c) Moles converted/g catalyst/h at 300°C at complete conversion.

Cr(VI)-O and Sn-Cu-Cr(VI)-O catalysts for the oxidation of CO and n-propane compared with tin(IV) oxide itself and a comparable alumina-supported platinum catalyst. Activity towards oxidation of CO of the catalysts is good if copper is present with specific activities which are approximately ten times that of a comparable alumina-supported Pt catalyst. Light-off temperatures tend to be lower than 100°, with complete conversion occurring below *ca.* 150°. In contrast, catalysts containing chromium(VI) enhance the conversion of propane, and the Sn-Cu-O catalysts performed less well than either the Sn-Cr(VI)-O or Sn-Cu-Cr(VI)-O catalysts. Again, specific activities compare favourably with the alumina-supported Pt catalyst. The best performance is exhibited by the mixed Sn-Cu-Cr(VI)-O catalysts which are efficient for the removal of both CO and propane. Figures 5 and 6 illustrate the comparative performances of the tin(IV) oxide catalysts together with data for commercial copper chromite for both the conversion of CO and propane.

REFERENCES

1 EC Directive Dir. 88/76/EEC, December 1987. See also Directives Dir. 88/436/EEC, 16 June 1988, and Dir. 89/458/EEC, 18 July 1989.
2 EC Communication COM (89) 662, 2nd February 1990.
3 G. Michel and R. Machiroux, J. Raman Spectrosc., 1983, 14, 22; 1986, 17, 79.
4 F. Gonzalez-Vilchez and W.P. Griffith, J. Chem. Soc., Dalton Trans., 1972, 1417.
5 M.A. Vuurman, Doctoral Dissertation, University of Amsterdam, 1992.

ACKNOWLEDGEMENTS:

We thank the Science and Engineering Research Council and the Government of Malaysia for support. Dr. Daniella Goldfarb and Mr Khalid Matar for the EPR and ESEEM measurements, and Dr. Carole C. Harrison for assistance with the electron microscopy.

Diesel Catalyst Technologies

A. Frennet and J.-M. Bastin (Eds.)
Catalysis and Automotive Pollution Control III
Studies in Surface Science and Catalysis, Vol. 96
1995 Elsevier Science B.V.

CATALYTIC OXIDATION OF DIESEL PARTICULATES WITH BASE METAL OXIDES

K.E. Voss, J.K. Lampert, R.J. Farrauto, G.W. Rice, A. Punke and R. Krohn

Engelhard Corporation, 101 Wood Ave., Iselin, NJ 08830-0770, USA and Engelhard Technologies GmBH, Misburger Strasse 81b, Hannover, Germany

ABSTRACT

Reduction of emissions from diesel engines using novel flow-through oxidation catalysts containing base metal oxides has been demonstrated. The soluble organic fraction (SOF) of the particulates can be converted and gas phase hydrocarbons can be removed via the proper choice of oxides. Low levels of platinum on the catalysts also compliment HC and CO conversion without leading to "sulfate-make". Catalyst performance is shown for transient and steady state engine testing relative to emissions standards for U.S. truck and European auto applications. Laboratory studies using PYRAN™ thermal chromatography and engine storage & release test results provide insight of the mechanisms of removal and conversion of SOF and HC's by the catalysts.

1. INTRODUCTION

The exhaust emissions from a diesel engine are composed of three phases: solids, liquids and gases. The combination of solids and liquids comprise the so called particulates, or total particulate matter (TPM), of the emissions. These particulates in turn are composed of: dry carbon (soot), liquid hydrocarbons and "sulfate".

The dry carbon comes from the incomplete combustion of the fuel and because it is principly a solid, a flow-through type catalyst typically has little effect on this fraction of the particulates.

The main liquids present in the particulates are a combination of lighter unburned fuel and lubricating oil, e.g. swept from the cylinder walls. Together these are the so called soluble organic fraction (SOF) or volatile organic fraction (VOF) which are in the form of discrete aerosols and/or are adsorbed on the dry carbon particles [1]. It is this liquid fraction of the particulates which a flow-through catalyst can remove and oxidize. When diesel fuel is burned in the engine most of the sulfur it contains is oxidized to SO_2 which is emitted in the gas phase. However, a portion of the sulphur is oxidized completely to SO_3 which in turn combines with the moisture in the exhaust and becomes condensed H_2SO_4 or "sulfate" which adds to the particulate emissions. Because of this an effective diesel catalyst must minimize any additional oxidation of SO_2 to SO_3.

The gas phase emissions of interest are composed primarily of hydrocarbons, carbon monoxide, nitrogen oxides and sulfur. Diesel emissions are clearly more complex than those from a spark ignited gasoline engine, and thus their reduction by catalytic treatment is much more complicated, requiring new technology. Control of diesel emissions is being addressed worldwide. In the U.S., Europe and Japan most of the trucks and buses operate with diesel fueled engines. Each vehicle or engine must meet specific standards as measured by standardized emissions tests which reflect the duty cycle anticipated for the particular engine. Within the U.S. a significant reduction in allowable TPM emissions for medium and heavy duty trucks has been required for MY 1994 with a standard of 0.10 g/hp-hr, down from 0.25 g/hp-hr prior to 1994 (see TABLE 1).

Table 1U.S. Heavy/Medium Duty Diesel Emissions Standards

(HD U.S. Transient Test Cycle)
Mass Emissions (g/HP-hr)

Year	HC	CO	NOx	TPM
1991	**1.3**	**15.5**	**5.0**	**0.25**
1994	**1.3**	**15.5**	**5.0**	**0.10**

This reduction for the most part can not be achieved via engine modifications alone making exhaust aftertreatment required. Generally, in the U.S. medium and heavy duty engines meet the standards for gas phase hydrocarbon, carbon monoxide and nitrogen oxides for MY 1994.

The emissions test is conducted using the heavy duty U.S. transient (FTP) cycle which reflects a continuous measure of emissions during various speed and loads reflecting urban and expressway driving [2]. The transient test itself and different engines with different power ratings, result in a variety of exhaust temperature and flow conditions. Thus, the catalyst must remove and convert the particulates over a broad range of conditions.

A newly developed catalyst, containing a proprietary, non-toxic base metal oxide as the active material for combusting the SOF portion of the particulates, is now commercialized in the U.S. to meet these standards [3].

In Europe, diesel passenger cars must meet standards reflecting both urban and high speed driving conditions. The standards are given in TABLE 2 below.

Table 2European Diesel Passenger Car Emissions Standards

(European Transient Test Cycle)
Mass Emissions (g/km)

	Timing	HC + NOx	CO	TPM
EURO I	**Current**	**0.97**	**2.72**	**0.14**
EURO II	**1996**	**IDI 0.70**	**1.00**	**IDI 0.08**
		DI 0.90		**DI 0.10**
EURO III	**1999**	**0.50**	**0.50**	**0.04**

The current standards are scheduled to become stricter in 1996 (EURO II) and as proposed again in 1999 (EURO III). In addition to particulates, the catalyst is also required to reduce gas phase HC and CO emissions.

The European Transient Test (Cycle A) evaluates emissions on the basis of low speed urban (ECE) and high speed (EUDC) driving segments [2]. For the overall test cycle the weighting for the ECE segment (avg. 19 km/hr) is approximately twice that of the EUDC segment (avg. 63 km/hr). Although catalyst inlet temperatures for the EUDC segment of the test can run between 200° and 550°C, they are typically quite low for the ECE segment (100-250°C). As a result an effective catalyst must operate over a broad range of temperatures and low temperature removal of emissions and catalyst light-off are significant challenges. Especially challenging is the requirement for CO and HC conversion during the low temperature ECE segment. Increasing the precious metal content will reduce the temperature for gas phase oxidation. However, this adds significantly to the cost of the catalyst since precious metals are expensive and furthermore, can be highly active towards the oxidation of SO_2 to SO_3 during the higher temperature EUDC portion. Consequently, alternative materials, which use a minimum of precious metals, would be more desirable.

The work reported in this paper describes the results of engine and vehicle tests of diesel oxidation catalysts containing base metal oxides for the combustion of the SOF portion of the particulate emissions and of a new generation of diesel oxidation catalysts specifically formulated for European passenger car applications. This new family of catalysts is based on the technology developed for the U.S. truck catalyst but also contains other proprietary non-toxic, inexpensive base metal oxide components which enhance the reduction of hydrocarbons at low temperatures without producing excessive amounts of sulphate. This is accomplished with relatively small amounts (< 30 g/ft^3) of precious metals.

A laboratory method simulating the adsorption and subsequent catalytic combustion of the soluble organic fraction (SOF) on these newly developed diesel oxidation catalysts, is also described in this paper. This technique, based on thermal chromatography, allows characterization of the initiation of catalytic oxidation and analysis of the products giving insight into the mechanism of operation.

2. MEDIUM DUTY DIESEL ENGINE TEST RESULTS

We have reported previously [3, 4] on the performance of diesel oxidation catalysts containing base metal oxides for SOF combustion and low levels of platinum for gas phase HC and CO reduction without production of "sulfate". This type of catalyst, designated in this paper as Catalyst "C", for the U.S. diesel truck application has given emissions reduction performance in the Heavy Duty Transient Test (FTP) as shown in TABLE 3. For this test the platinum loading of Catalyst "C" was 0.5 g/ft^3 or approximately two orders of magnitude lower than a conventional three way catalyst for SI gasoline applications. The tests were run using a MY 1991 5.9 liter DI/TCI engine rated at 190 HP which produced a maximum exhaust temperature of 305°C in the FTP test. The catalyst was coated onto a 9" dia.x6" long ceramic substrate (6.25 liters) having 400 cells/in^2. Low sulfur fuel (0.05 wt% S) was used for catalyst evaluations. Preconditioning and aging of the catalysts were done on an engine using fuel with a sulfur content of 0.3 wt% S.

Table 3Catalyst "C" - U.S. HD Transient (FTP) Emissions Test Results

	Mass Emissions (g/HP-hr)			
Emission Component	**Engine-Out Engine-Out**	**Catalyst"C" (Fresh)***	**Fresh % Removal**	**1000 hr Aged % Removal**
1. Total				
Particulates	**0.172**	**0.116**	**33**	**33**
a. SOF	**0.061**	**0.026**	**58**	**53**
b. Sulfate	**0.004**	**0.002**	**55**	**23**
2. Gas Phase				
a. HC	**0.299**	**0.188**	**37**	**29**
b. CO	**1.49**	**1.11**	**26**	**9**

*** Fresh = Preconditioned on the engine for 24 hrs.**

As can be seen Catalyst "C" gives a high level of SOF removal (58%) on this engine which contributes to a removal of 33% of the total particulates (TPM). Furthermore, the particulate removal performance is stable over extended periods of engine aging (1000 hrs). The MY 1991 engine meets the pre-1994 TPM and gas phase emissions standards without a catalyst. Catalyst "C" gives a substantial reduction in the SOF emissions. However, to meet the 1994 emissions targets further reductions in the "Dry Soot + Other" component of the particulates through new engine technology was also needed. This has been accomplished, and through joint catalyst development and engine development the TPM targets have been met for U.S. truck applications.

Inspite of the extremely low platinum loading level of this catalyst and the low engine exhaust temperatures in the test, it gave relatively good conversion of gas phase HC and CO (37% and 26%, respectively) in the fresh state. Gas phase activity was decreased; however, after extended engine aging.

U.S. emissions standards (LEV & ULEV) in the late 1990's will require catalysts with improved HC removal and probably at least some NOx reduction activity, as well. Emissions standards for diesel autos in Europe will require improved HC removal to meet EURO II levels for HC + NO_x. Catalyst "C" has been reformulated with additional oxide components and varying precious metal loadings to address these needs. The new type of catalyst, designated Catalyst "D" in this paper, has been found to exhibit improved HC removal, especially at low exhaust temperatures.

Steady state engine tests were conducted using a MY 1991 5.9 liter DI/TCI diesel engine rated at 230 HP running on low sulfur fuel (0.05 wt% S). The steady state test modes were chosen from the European R-49 Test Procedure which gave different catalyst inlet temperatures ranging from ca. 120 to 550°C. Tests comparing Catalyst "C" and Catalyst "D" were run. Each catalyst was coated onto a 9" dia. x 6" long ceramic substrate with cell spacing of 400 cpsi.

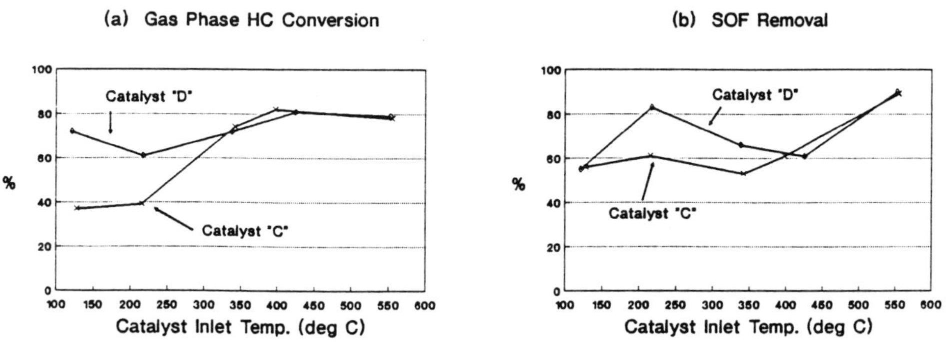

Figure 1. - Steady State Engine Performance of Catalysts "C" and "D"

For these tests both Catalyst "C" and Catalyst "D" had platinum loading levels of ca. 2 g/ft^3. The test results for gas phase hydrocarbon conversion and SOF removal and conversion are shown in Figure 1a and 1b, respectively.

As can be seen the reformulated Catalyst "D" exhibited substantially better removal of gas phase HC's at catalyst inlet temperatures in the 120-200°C range. At temperatures of ca. 300°C and above both catalysts gave comparable HC conversion. Thus Catalyst "D" gave a high level of HC removal and conversion (>60%) over the entire exhaust temperature range and with a very low platinum loading level. This improved HC removal is the type needed to address the coming European diesel auto and future U.S. diesel truck and bus emissions requirements. Catalyst "D" also exhibited slightly better overall SOF conversion than did Catalyst "C".

3. EUROPEAN DIESEL PASSENGER CAR TRANSIENT TEST RESULTS

The improved HC removal and conversion of Catalyst "D" was demonstrated further for the diesel auto application through European Transient Testing using a chassis dynamometer. The test vehicle had a 2.5 liter DI/TCI engine with EGR and was run on low sulfur fuel (0.05 wt% S). In the tests the engine gave catalyst inlet temperatures in the range 100-200°C for the urban driving (ECE) segment and in the range 200-355°C for the high speed (EUDC) driving segment. Overall this represented a cool running engine and thus a challenge for catalytic HC emission control. The raw emissions from the engine are given in Table 4.

Table 4.Emissions from 2.5 Liter DI/TCI+EGR Diesel Auto Engine

(European Transient Test Cycle)

	Mass Emissions (g/km)			
Test Segment	**HC**	**NOx**	**CO**	**TPM**
Urban (ECE)	**0.80**	**0.56**	**3.10**	**0.143**
High Speed (EUDC)	**0.11**	**0.40**	**0.44**	**0.113**
Overall (Cycle A)	**0.36**	**0.46**	**1.41**	**0.124**

The catalysts were coated onto 5.66" Dia. x 6" Long (2.47 liters) ceramic substrates with a cell spacing of 400 cpsi. Both Catalyst "C" and Catalyst "D" were evaluated with different platinum loading levels ranging from 2.0 to 45 g/ft^3. The test results (% removal) for HC and total particulate (TPM) performance as a function of Pt loading are shown in Figure 2a through 2d.

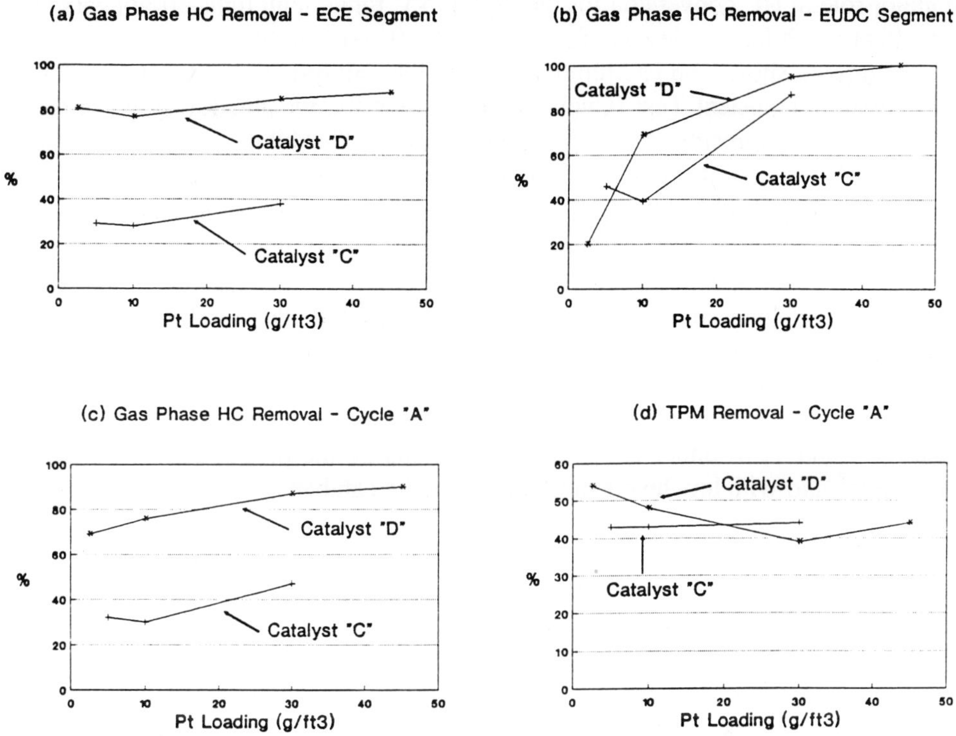

Figure 2. - European Transient Test Results for Catalyst "C" and Catalyst "D" - Performance for Gas Phase HC and TPM Removal

Figure 2a shows the gas phase HC removal performance for the low temperature (ECE) urban driving segment of the test. The reformulated Catalyst "D" exhibits ca. 80-90% HC removal with a only a slight increase as a function of increasing Pt loading. Catalyst "C" gives lower HC removal levels, between ca. 20% and 40%, again with only a slight increase with increasing Pt loading. The relatively flat behavior of HC removal with increased Pt loading in the low temperature range (100-200°C) of the ECE segment of the test indicates that HC adsorption is playing a major role in performance and light-off of the Pt function has not yet been fully attained.

Figure 2b gives the HC removal performance in the hotter high speed (EUDC) segment of the test. This shows a clear dependence on Pt loading for both Catalyst "C" and Catalyst "D"and light-off of the Pt appears to be playing the major role in HC conversion for the 200-355°C range. Full light-off appears to be in effect for Pt loading levels of just over 10 g/ft^3 for Catalyst "D".

Figure 2c shows the HC removal results for Catalyst "C" and Catalyst "D" for the overall Transient Test (Cycle "A") which is the weighted-average of the ECE and EUDC segments. As can be seen Catalyst "D" exhibits very high overall HC removal because of its excellent low temperature performance. Even at the lowest Pt loading level Catalyst "D" gives HC removal levels to meet the EURO II HC+NO_X standards with a comfortable margin and comes quite close to the proposed EURO III standards.

Figure 2d gives the particulate (TPM) removal of the catalysts over the Transient Test Cycle and it can be seen that high levels of performance (> 40% removal) are attained, especially for the lower Pt loading levels. Overall, these results show that it is possible to achieve good removal and conversion of gas phase HC's and TPM under demanding conditions by the proper choice of base metal oxides and with low levels of platinum on the catalyst.

4. STORAGE AND RELEASE EFFECTS OF EMMISIONS ON DIESEL OXIDATION CATALYSTS

The exhaust temperatures of diesel powered vehicles are quite low at idle and at low speeds and loads encountered in urban driving. To provide effective emissions reductions in normal use, as well as, in tests such as the U.S. FTP and European Transient (Cycle "A"), the catalyst needs to remove SOF and gas phase HC's at temperatures that are many times below precious metal light-off. We have found that the catalyst washcoat which has both storage and oxidation activity will provide the required low temperature removal of SOF and HC's and then at higher temperatures together with the precious metal function convert and oxidized the collected organics. In order to evaluate catalysts for this type of behavior a test was developed to exaggerate the characteristics of the two segments of the European Transient (Cycle "A"). Testing was conducted using a 5.9 liter DI/TCI diesel engine rated at 175 HP and running on low sulfur fuel (0.05 wt% S). For gas phase emissions the test involved an extended low load, low temperature (153°C) storage period (1 hr.) followed by a rapid ramp to 70% load in only 2 min., a hold for 8 min. and then a step to 100% load and a hold for 8 min. Concentrations of gas phase HC's and CO, at both the catalyst inlet and outlet, along with exhaust flow rates were measured at 1-2 second intervals and thus gas phase emissions could be integrated and totalized over the storage and ramp segments of the test. The results of this test for two samples of Catalyst "D", one with a Pt loading of 2.5 g/ft^3 a one with 10 g/ft^3 Pt, are given in Table 5.

This table gives the test segment, its time interval and the associated maximum catalyst inlet temperature. Also shown are the corresponding % removal of HC and CO. The ramp segment (18 min. total) is broken down into each of the first 5 one min. intervals and the last 13 min. interval. Also shown is the % removal of HC and CO totalized for the overall storage + ramp segments of the test.

Table 5 Storage and Combustion of Gas Phase HC's by Catalyst "D"

(Engine Storage and Ramp Test)

			% Removal			
	Time	*Max. Inlet*	($2.5 g/ft^3$)		($10 g/ft^3$)	
Test Segment	**Interval**	**Temp.(C)**	**HC**	**CO**	**HC**	**CO**
Storage	**50min.**	**153**	**62**	**-4**	**65**	**17**
Ramp	**1min.**	**178**	**57**	**-2**	**58**	**15**
	1min.	**255**	**-47**	**-16**	**-34**	**46**
	1min.	**307**	**-214**	**-51**	**-24**	**65**
	1min.	**330**	**-23**	**25**	**63**	**82**
	1min.	**344**	**50**	**77**	**70**	**84**
	13min.	**455**	**60-81**	**84-96**	**74-77**	**85-93**
Storage+Ramp	**68min.**		**52**	**-5**	**61**	**19**

These data indicate that the washcoat and platinum performed different functions over the different temperature ranges of the test. Platinum loading level (2.5 vs 10 g/ft^3) had little effect on HC removal below ca. 180°C where HC-storage by the washcoat occurred. Neither was there a significant effect of loading level on HC and CO conversion above ca. 350°C which was well beyond the Pt light-off point for both catalysts. The higher Pt loading; however, did provide significant improvements during the intermediate temperatures in the ramp interval by decreasing HC release and giving a lower CO light-off temperature. It is believed that the results indicate that organic material stored by the catalyst masks the CO conversion sites, so that CO light-off during the ramp occurs only after the sites are burned clean. The lower CO light-off with the higher Pt loading is attributed to more rapid burning of stored HC's. The low temperature storage followed by the rapid ramping to high load represents an extreme but challenging condition for a catalyst and as can be seen some release of gas phase HC's can occur during the ramp segment. However, this can be at least partially alleviated with proper choice of Pt loading level. Furthermore, over the entire storage and ramp cycle the net HC removal remains a net positive 52% or 61%, depending on Pt loading level.

For particulate emissions tests simultaneous particulate samples were collected upstream and downstream of the catalyst using two mini-dilution tunnels. In separate tests the time of the low temperature storage segment was varied from 30 to 120 min. and after each storage the ramp to high load was run. The results for inlet versus outlet levels of total particulates and their breakdown into components is shown in Figure 3 as a function of storage time.

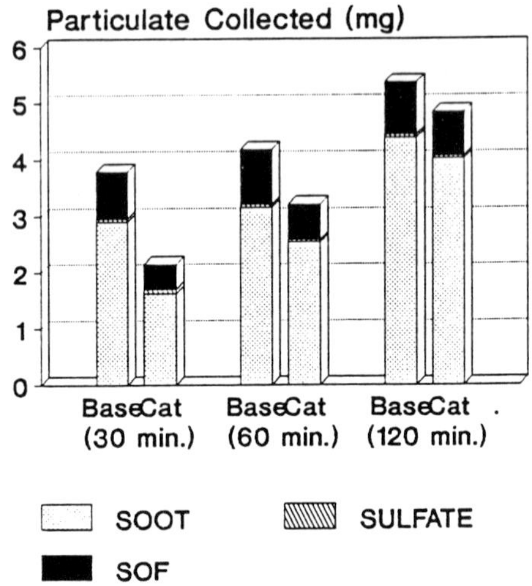

Figure 3 Diesel Engine Storage and Conversion of Particulates

As can be seen for each of the storage times (30, 60 and 120 min.) followed by a single rapid ramp to 100% load, there is no release of particulate material from the catalyst outlet relative to the catalyst inlet. In fact catalyst-out levels of particulates are less than catalyst-inlet. Although the gradual increase in baseline particulate levels with longer storage times suggests some holdup of material in the mini-tunnels, the catalyst-out SOF levels were always less than the inlet levels indicating that the catalyst is active for SOF removal even under these difficult conditions. The distinction between HC and SOF behavior in the tests is attributed to molecular weight differences in these species. The proportion of lighter, gas phase HC released or burned varied with Pt loading (and also with ramp rate). However, the higher molecular weight SOF was held and converted by the washcoat under the most demanding conditions used, and with little or no Pt effect. It is also possible that at least a portion of the gas phase HC's released during the rapid ramp to high load might have come from the breakdown of some of the liquid SOF into lighter gas phase HC-fractions by the catalyst.

5. LABORATORY STUDIES OF MODEL SOF ADSORPTION AND COMBUSTION BY DIESEL OXIDATION CATALYSTS

In order to understand the processes occurring during conversion of the soluble organic fraction (SOF) of the particulate emissions by the diesel oxidation catalyst, a laboratory method has been devised that simulates the dynamic adsorption of the SOF by the catalyst and also analyzes the combustion products generated when the SOF is converted. Commercial diesel lubricating oil and diesel fuel were used as models for actual lube and fuel derived SOF. Using a Pyran™-thermal chromatograph, shown schematically in Figure 4, lubricating oil (or diesel fuel) was impregnated on a quartz frit and placed in a quartz sample holder upstream of a small (0.25cm^3) sample of catalyst coated honeycomb substrate (400 cells/in^2).

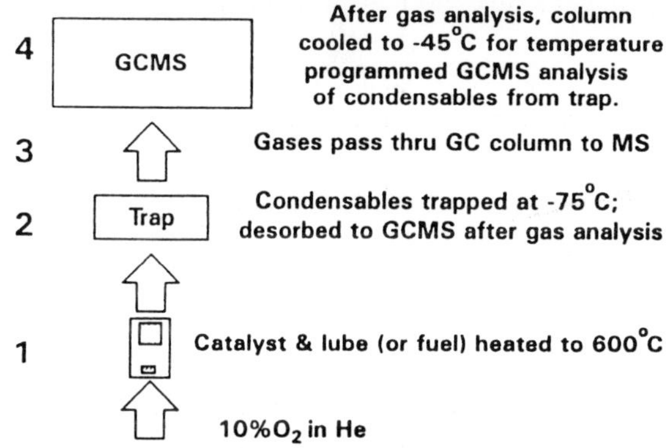

Figure 4. - Schematic of Pyran™ Thermal Chromatograph System

The lube oil/catalyst assembly was then heated from ambient to 600°C at 10°C/min. in a gas stream containing 10% O_2 in He at a flow rate of 40mL/min. As the temperature increased, the lube oil (or fuel) was volatilizes into the gas stream and swept through the catalyst sample. The products exiting the catalyst were then analyzed. Condensable combustion products were trapped cyrogenicaly and the non-condensables were analyzed in real time by a mass spectrometer. Subsequently the condensables were thermally desorbed for analysis by a gas chromatograph-mass spectrometer (GCMS). In separate experiments the by-products exiting the catalyst were analyzed for total hydrocarbons using a flame ionization detector (FID). This allowed determination of the volatilization curves for lube (or fuel) and the effects with or without the catalyst present. The ratio (wt/wt) of lube oil or diesel fuel to catalyst on the honeycomb substrate for the experiments was 0.04 and 0.01, respectively. In diesel exhaust, the SOF to wash coat ratio (wt/wt) is on the order of 0.001.

The thermal desorption-GCMS analysis shown in Figure 5 gives a comparison of (a.) chromatograms of SOF desorbed from particulates collected from the exhaust of a diesel engine with (b.) the chromatograms of diesel lube oil and diesel fuel. Superimposed on the chromatograms for calibration are the positions of the corresponding GC retention times for a normal aliphatic hydrocarbon simulated distillation series with different molecular weights.

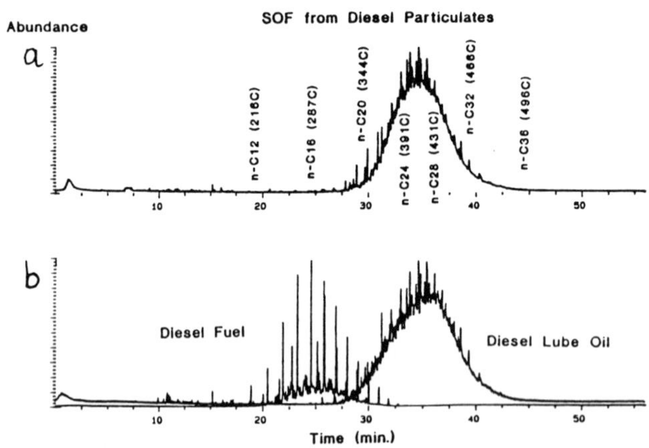

Figure 5. - Thermal Desorption-GCMS Analysis of SOF From Diesel Particulates Compared With Diesel Lube Oil and Diesel Fuel

As can be seen (b.) diesel fuel is composed of fairly distinct components in a range comparable to normal aliphatic hydrocarbons having 12- to 20-carbons. Diesel lube oil, on the other hand, is predominantly a continuum of many compounds with boiling points comparable to normal aliphatic hydrocarbons mainly in the range between 18- and 36-carbon atoms long. The SOF (a.) from particulate emissions collected on a diesel engine, run at low speed and high load, was found to be composed primarily of diesel lube with a small contribution from the higher boiling range of diesel fuel.

Figure 6. again shows the chromatogram of the SOF desorbed from particulates collected from the engine exhaust (upper curve) compared with the chromatogram of SOF of approximately the same quantity of particulates after the engine exhaust had passed through Catalyst "C" (lower curve).

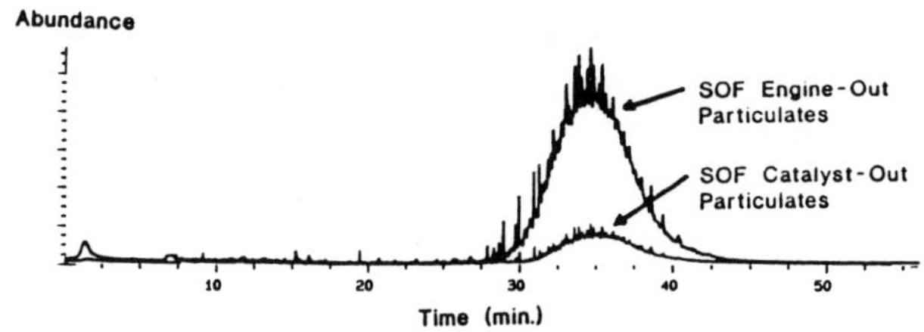

Figure 6. - Pyran™ Analysis of SOF from Diesel Exhaust Particulates (Engine-Out vs Treated with Catalyst "C")

As can be seen treatment of the exhaust by Catalyst "C" results in a dramatic reduction in the SOF of the particulate emissions.

The FID traces in Figure 7 compare the total hydrocarbon traces for diesel lube oil as a function of temperature and in a gas stream of either helium (nonoxidizing) or 10% O_2 in helium (lean). In the bottom curves it can be seen that with no catalyst present diesel lube vaporizes into either gas stream between 100 and 200°C, well below its median simulated

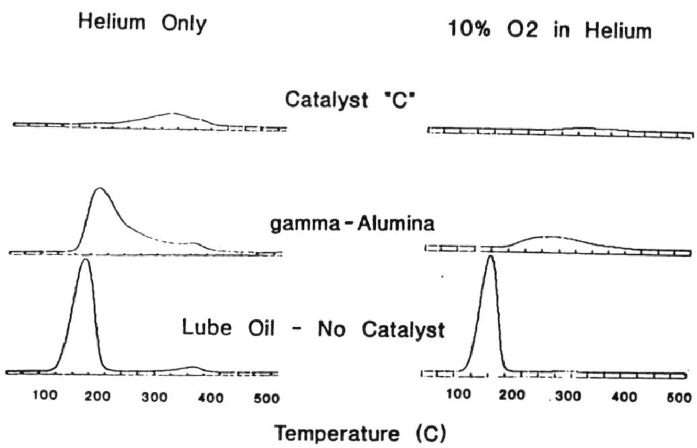

Figure 7. - FID Traces of Lube Oil With and Without Catalysts Present

distillation boiling point of 430°C. With Catalyst "C" present (top curves) and in the absence of oxygen, lube vapors are adsorbed by the catalyst and no hydrocarbons are released below 200°C. Those that are released are of a lesser amount. In the "lean" gas stream virtually no hydrocarbons pass through Catalyst "C" at any temperature. By comparison a honeycomb substrate coated with high surface area gamma-alumina does not adsorb and hold lube oil very effectively in the nonoxidizing gas stream (middle curves). In the "lean" gas stream gamma-alumina adsorbs and holds the lube oil better, but not as well as Catalyst "C".

Volatilization and adsorption of lube oil by the catalysts is followed by catalytic oxidation to CO_2 at higher temperatures as is shown in Figure 8. In the Pyran™-GCMS experiments, CO_2 is the only non-condensable oxidation product measured from the combustion of lube oil in the "lean" gas stream by either Catalyst "C" or the gamma-alumina. The onset of CO_2 evolution with Catalyst "C" occurs at approximately 180°C, which is just above the temperature range for the lube oil volatilization curve which is also shown. Lube conversion to CO_2 is virtually complete with Catalyst "C" as determined by CO_2 mass balance (95-100%) and the absence of detectable condensable species.

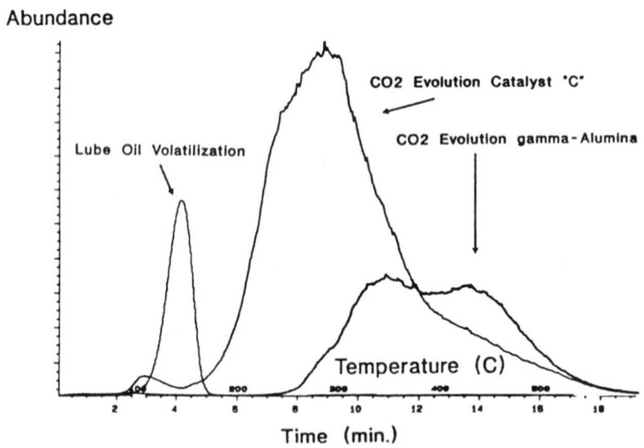

Figure 8. - Conversion of Lube Oil to CO_2 by Catalysts

Gamma-alumina by comparison begins to combust lube oil to CO_2 at approximately 300°C, which is well above the temperature of lube oil volatilization the combustion onset temperature for Catalyst "C". Furthermore, gamma-alumina typically converted only 40 to 50% of the lube to CO_2.

Diesel fuel was found to volatilize approximately 50°C lower than lube oil. This was accompanied by a corresponding shift to lower combustion temperatures with Catalyst "C" (Figure. 9). Fuel conversion with Catalyst "C"was typically 80% by mass balance, with CO_2 the only non-condensable product. In contrast to the onset of fuel combustion shifting to lower temperatures over Catalyst "C", the onset of fuel combustion over alumina was at higher temperature and occurred at almost the same temperature as did lube oil combustion.

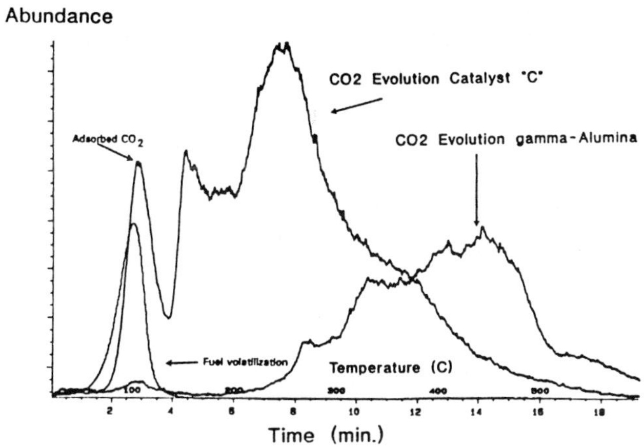

Figure 9. The Volatilization of Diesel Fuel and Its Combustion by Catalysts

Non-burned fuel hydrocarbons were observed in the product stream with both Catalyst "C" and gamma-alumina. However, the molecular weight distributions were quite different (Figure 10). Catalyst "C" combusts all the heavier fuel hydrocarbons, with only the lightest fuel species bypassing the catalyst, while the entire molecular weight spectrum of the fuel is observed in the non-burned hydrocarbons from the gamma-alumina catalyst.

These data suggest that while both Catalyst "C" and gamma-alumina adsorb SOF components at low temperatures and burn them as the temperature increases, each catalyst's mechanism of adsorption and subsequent combustion is quite different. With Catalyst "C", the light-off temperature is an apparent function of the volatility of the SOF component. Catalyst light off occurs when the temperature is high enough for the adsorbed component to desorb from the catalyst surface: 150°C for the fuel component, and 200°C for the lube. The observed difference in light-off temperature between the fuel and lube for Catalyst "C" is due to the difference in the lube and fuel volatility.

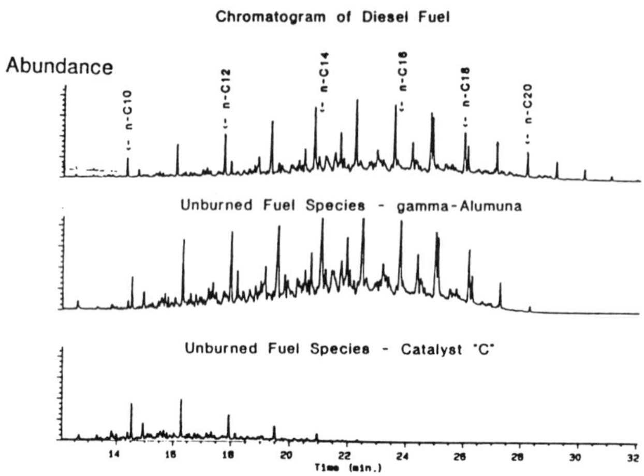

Figure 10. - Chromatograms of Diesel Fuel and Unburned Fuel Fractions from Catalyst "C" and Gamma-Alumina

However, the catalyst is not hot enough to burn the fuel's lowest molecular weight fraction in these experiments. These components are not adsorbed by Catalyst "C", and pass over the catalyst at temperatures below the catalyst light off temperature. The contrast with gamma-alumina is striking. Gamma-alumina lights off at 300°C, 100°C higher than does Catalyst "C", and at approximately the same temperature for both the diesel fuel and lube, independent of the components' volatility. This suggests that the alumina is forming similar intermediate species on its surface from the adsorbed lube or fuel SOF components. Gamma-alumina has lower conversions for both the fuel and lube than does Catalyst "C" because the alumina catalyst adsorbs less of each SOF component and the adsorbed species desorb from the alumina catalyst surface before the light-off temperature is reached.

6. SUMMARY AND CONCLUSIONS

Diesel oxidation catalyst technology using base metal oxides in the washcoat for removal and conversion of particulate SOF and gas phase HC's has been demonstrated in steady state and transient engine tests.

These active washcoats have been found to adsorb emissions at low exhaust temperatures and oxidize them at higher temperatures in the engine duty cycle. Precious metal in the form of low loading levels of platinum can also be incorporated into the catalysts for additional gas phase HC activity and for CO conversion. By proper choice of catalyst oxides in the washcoat the technology can be tailored for current U.S. truck applications which require particulate emission reduction or by modification include improved gas phase HC removal and conversion, especially at low exhaust temperatures. This gives catalyst technology to address the needs for European diesel auto emissions standards. Engine storage and release experiments show that adsorption of emissions (HC, SOF) can occur on a catalyst under extended low load, low temperature conditions. Under subsequent rapid ramping to high load and temperature in a matter of minutes reemission of gas phase HC's can be controlled fairly effectively with as little as ca. 10 g/ft^3 platinum. Storage and release tests show that the catalyst is very effective for controlling SOF emissions and catalyst-out SOF is consistently lower than engine-out baseline levels. A laboratory test was developed to investigate the mechanisms of adsorption and conversion of SOF on diesel catalysts. This test used diesel lube oil and diesel fuel as models for SOF. Test results have shown that SOF components are adsorbed effectively by the catalyst at temperatures below ca. 200°C and then fully oxidized by the catalyst washcoat at temperatures above about 200°C. This behavior is consistent with what has been seen in engine tests.

RERERENCES

1 P. Zelenka, W. Kriegler, P. Herzog and W.P. Cartellieri, "Ways Toward the Clean Heavy-Duty Diesel," SAE No. 900602 (1990).
2 J.S. McArragher, et al.,"Motor Vehicle Emission Regulations and Fuel Specifications - 1992 Update," CONCAWE, Brussels (1992) 111.
3 R.J. Farrauto, K.E. Voss and R.M. Heck, SAE No. 932720.
4 K.E. Voss, B.O. Yavuz, C. Hirt and R.J. Farrauto, SAE 940239.

A. Frennet and J.-M. Bastin (Eds.)
Catalysis and Automotive Pollution Control III
Studies in Surface Science and Catalysis, Vol. 96

PERFORMANCE OF OXIDATION CATALYSTS FOR HEAVY-DUTY DIESEL ENGINES

H. J. Stein[a], G. Hüthwohl[b] and G. Lepperhoff[b]

[a]*Mercedes-Benz AG, Dept. EN/MGA, D-70322 Stuttgart, Mail Code T 352, Germany*

[b]*FEV Motorentechnik GmbH & Co. KG, Neuenhofstrasse 181, D-52078 Aachen, Germany*

ABSTRACT

A variety of catalyst substrates, washcoats, and formulations were screened on a Mercedes OM 366 LA direct injection, turbocharged-intercooled engine in order to identify the optimal engine/catalyst combination for US applications. The engine was operated at five steady state modes, and on the US transient and European 13-mode cycles.

The carbon monoxide (CO) and hydrocarbons (HC) light-off temperatures were lowest for catalysts with a high platinum loading and an active washcoat, slightly higher for catalysts with low platinum loadings, and considerably higher for the palladium catalyst. The particulate (PM) emission at light load operation with an exhaust gas temperature about 423 K was reduced by 75 % to 85 % with all catalysts due to the high conversion rate of the PM soluble organic fraction. The PM efficiency still increased by 5 % at medium load operation, and then started to decrease around a temperature of 550 K due to the beginning of sulfate formation. Therefore, TWC technology cannot be used for Diesel engines. The sulfate formation could be minimized by a higher space velocity, an optimized washcoat, and a lower platinum loading, or palladium loading, respectively.

On the US transient cycle, PM was reduced by 20 % to 35 % with all catalysts tested. Among the two catalysts with 35 % efficiency, catalyst D which produced less sulfate at high temperatures was selected as first choice. Deterioration of catalyst D over 2200 hours was 8 % for PM, 13 % for CO and 21 % for HC.

On the 13-mode cycle, PM increased by 100 % to 350 % due to a considerable sulfate formation at the highly weighted high power modes. Since the SOF fraction of the particulates is less pronounced in this cycle, the potential of an oxidation catalyst for PM removal is limited.

1. Introduction

Stringent emission standards for heavy-duty Diesel engines will require in the near future significant reductions of the oxides of nitrogen (NO_x) and particulate (PM) emissions. To a high degree, modification of the engine and the introduction of a low sulfur Diesel fuel will contribute to reaching this goal. For some applications, however, the use of aftertreatment technologies is necessary for passing the 1994 PM standard in the USA with a sufficient safety margin. Since particulate filtering systems proved to be too complicated and expensive for the OM 366 LA engine, flow through oxidation catalysts were considered as a viable solution [1]. Such systems have already been successfully introduced for passenger car Diesel engines, and transfer of the technology to truck engines seemed to be a feasible way. The OM 366 LA is a turbocharged and intercooled direct injection six cylinder in-line engine with a swept volume of 5,9 dm^3, and is offered for the US market in four power versions of 127, 142, 157 and 172 kW. In this program, a 172 kW prototype engine that did not yet meet the 1994 PM emission limits was used to select the optimal converter configuration.

1.1. Particulate composition

Diesel particulates (PM) are collected on a teflon-coated glass fiber filter after dilution of the raw exhaust gas with clean ambient air in a dilution tunnel to a temperature at or below 325 K [2]. The dilution process changes the particle characteristics significantly. In the raw exhaust gas at high exhaust gas temperatures, the particles mainly consist of soot (carbon), the so called insoluble fraction (Insol). Through the dilution and cooling process, part of the high boiling exhaust hydrocarbons from the fuel and lubricating oil of the engine will condense and/or adsorb onto the soot forming the Soluble Organic Fraction (SOF) [3]. Additionally, sulfates (SO_4) out of the fuel sulfur will condense onto the carbon core. Depending on the water content of the particulate weighing chamber, the sulfate is usually associated with 1,3 g to 2 g water per gram of sulfate [4].

A typical PM profile of the OM 366 LA on the US transient cycle is shown in Figure 1. Oxidation catalysts are able to oxidize HC, CO and the SOF portion of the particulates, but will nearly not affect their Insol portion nor the NO_x emission. The efficiency depends on the catalyst formulation and on the converter size. A major problem associated with the use of oxidation catalysts for Diesel engines is the possible oxidation of SO_2. Depending on the catalyst efficiency, this can considerably increase the sulfate portion of the particulates, and thus the total PM [5].

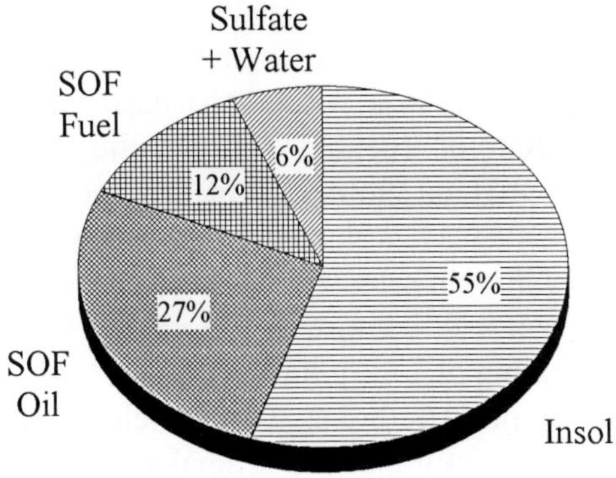

Figure 1 Typical particulate composition of the OM 366 LA on the US transient cycle

1.2. Catalyst selection

The catalysts evaluated in this program are listed in Table 1. Both ceramic and metallic monolith substrates with a diameter of 266 mm were used. The lengths of the substrates were 152 mm and 76 mm resulting in a catalyst volume of 8,6 dm^3 and 4,3 dm^3, respectively. The precious metal catalyst formulation comprised Platinum (Pt) and Palladium (Pd) with various loadings. Catalyst A* was only tested for a comparison of the ceramic vs. metallic substrates using identical catalyst formulation and converter size. Before running the test program, all catalysts were aged 50 hours at high temperature conditions.

Table 1 List of catalysts

Catalyst	Material	Cell density Cells/cm^2	Coating Metal	g/dm^3	Space Velocity
A	Ceramic	46	Platinum (Pt)	1,77	95140
A*	Metallic	23	Platinum (Pt)	1,77	95140
B	Ceramic	46	Platinum (Pt)	1,77	190280
C	Ceramic	62	Platinum (Pt)	0,35	95140
D	Ceramic	62	Platinum (Pt)	0,07	95140
E	Metallic	23	Palladium (Pd)	1.77	95140
F	Ceramic	62	Platinum (Pt)	0,02	95140

1.3 Test program

First, the catalysts were screened on five engine modes, as shown in Table 2. The catalyst screening was performed with three basic viewpoints. First, the influence of the substrate material (ceramic or metallic) was evaluated using catalysts A and A*, since it affected the design of the test program. Second, the influence of the space velocity was evaluated by comparing catalysts A and B with identical loadings, but different size. Third, optimized catalyst coatings were specifically tested for PM removal, comprising a reduced Pt content (0.35, 0.07 and 0,02 g/dm³) and a Pd coating of 1,77 g/dm³.

The result of the screening procedure was used to select a number of catalysts for further testing on the two most important official exhaust test cycles: the US transient cycle and the European 13-mode cycle. This part of the program consisted of verifying the conclusions from the screening procedure on these cycles. On the transient cycle, the engine is mainly operated under light or medium load conditions, whereas on the 13-mode cycle high load operation is predominant [6]. PM was measured in a full flow dilution tunnel, and was analyzed for SOF and sulfate.

Table 2 Engine modes for catalyst screening

Mode	Engine speed		Engine		Catalyst temperature	
	s^{-1}	min^{-1}	Torque Nm	Power kW	K	°C
1	15	910	90	8,6	423	150
2	26	1560	83	13,6	458	185
3	26	1560	208	34,0	548	275
4	26	1560	415	67,8	673	400
5	26	1560	625	102,1	753	480

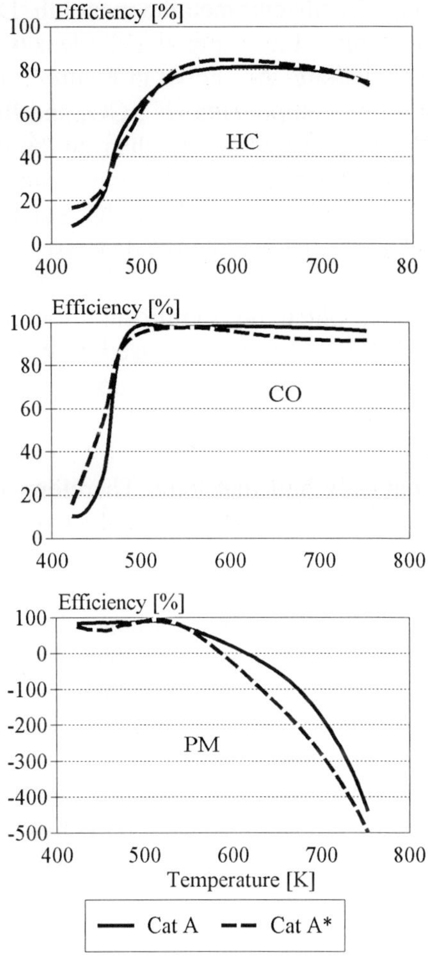

Figure2 Influence of the reactor material on the catalyst efficiency

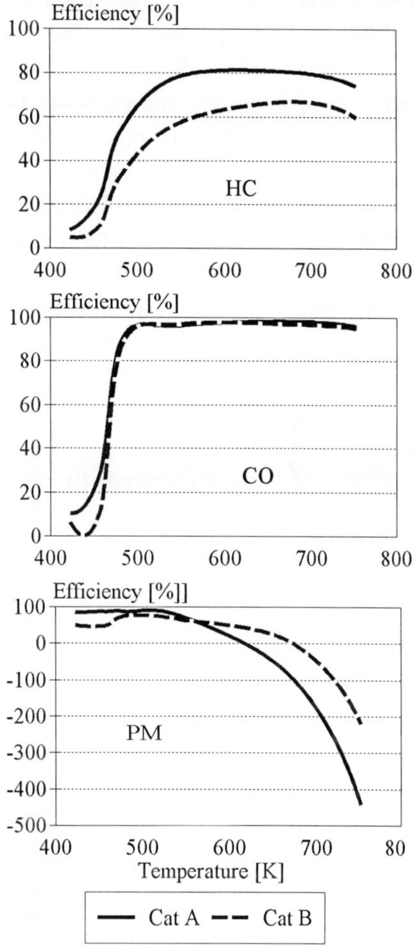

Figure3 Influence of the space velocity on the catalyst efficiency

2. EVALUATION OF CATALYST PERFORMANCE

2.1. Catalyst substrate

Catalyst A was prepared on a 46 cells/cm² ceramic monolith. It is based upon a washcoat optimized for Diesel engine operation, but with the usual Pt content of 1,77 g/dm³. Catalyst A* had a washcoat and a precious metal kind and loading identical to catalyst A, but was prepared on a 23 cells/cm² metallic monolith. Due to the slotted structure of the metallic substrate, the nominal cell density is equivalent to the cell density of the ceramic substrate. As shown in Figure 2, the catalyst efficiencies are very similar for the three components HC, CO, and PM. The efficiency for the gaseous components HC and CO is less than 20 % at a temperature of 420 K, then rapidly increases with increasing temperature reaching the maximum of 98 % at 500 K for CO and of 80 % at 550 K for HC. For PM, the trend is opposite: the PM removal of 95 % at low temperatures decreases with increasing temperature, and is then followed by an increase of PM at high temperatures due to the formation of sulfate. Since no significant difference between the two substrates was found, a ceramic substrate was used in most of the following program because of the cost advantage in mass production.

2.2. Space velocity

Catalyst B was identical to catalyst A, but only half that long. Therefore, the space velocity was twice as high, and the reaction time on the catalyst surface reduced. The results are shown in Figure 3. While the space velocity has only a minor influence on the CO efficiency at low temperatures only, HC and PM efficiencies are highly affected over the whole temperature range. The HC efficiency is 10 % to 20 % lower with the higher space velocity . For PM, the higher space velocity leads to a lower efficiency at low temperatures due to a reduced SOF removal. However, at high temperatures there is less PM increase due to a reduced sulfate formation. In total, reduction of the catalyst size is not satisfactory: on one hand, PM formation at high temperatures is still considerable, on the other hand the poor SOF efficiency is not sufficient for effective PM removal, and the poor HC efficiency might cause partial oxidation of hydrocarbons which often coincides with odor formation.

2.3. Precious metal loading

So far, it was found that the substrate had only a minor influence on the catalyst efficiency, and that reduction of the catalyst size was disadvantageous, in general. As a consequence, the precious metal loading was changed: the Pt content was reduced to 0,35 g/dm³ (catalyst C) and to 0,07 g/dm³ (catalyst D). In another formulation, Pd was used instead of Pt (catalyst E), since Pd catalysts are less likely to form sulfates by direct oxidation [5].

For the gaseous emissions HC and CO, the results are shown in Figure 4. The HC efficiency of the two Pt formulations is similar to that of the highly active catalyst A (see Figure 2). It reaches the maximum around 550 K, but with a better low temperature activity of catalyst C. The Pd catalyst has a poor HC conversion at low temperatures, and reaches the maximum conversion rate not before 700 K.

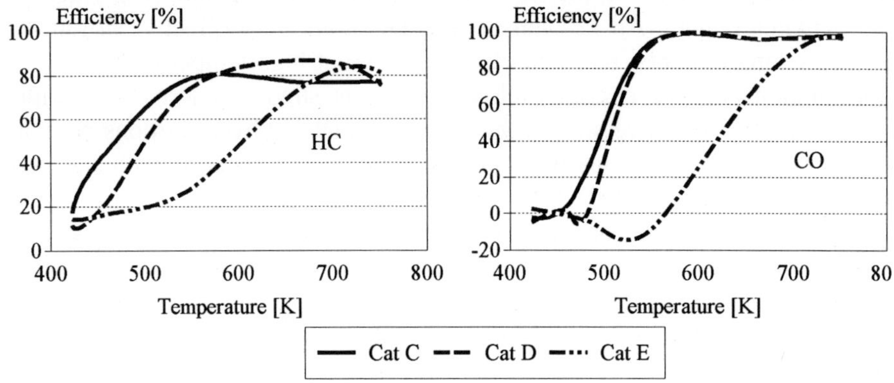

Figure 4 Influence of the precious metal loading on the gaseous emissions efficiency

Above 700 K, the activity of the three catalysts is very similar. These findings also apply to the CO conversion, but with a lesser difference between the two Pt formulations over the whole temperature range.

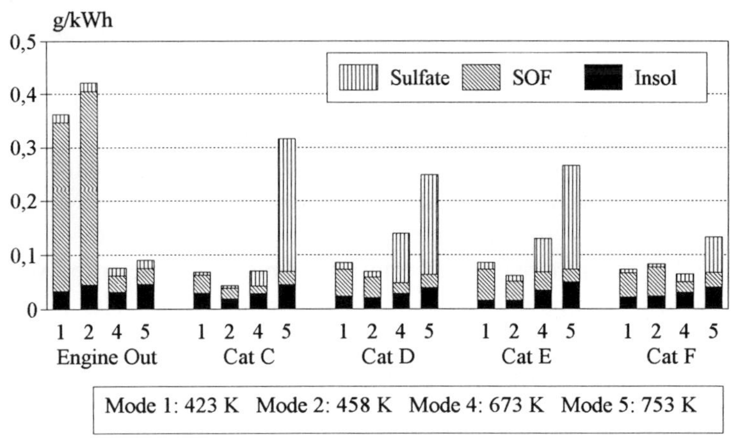

Figure 5 Influence of the precious metal loading on the PM level and on the particulate composition

For PM, the particulate composition is crucial in assessing the catalyst efficiency. Figure 5 exhibits the PM emission with and without catalysts at the engine modes 1, 2, 4 and 5 and broken into the respective Insol, SOF and sulfate/water fractions. An additional formulation (catalyst F) was considered in this comparison. Catalyst F is identical to catalyst D, but with a further reduced

Pt content of 0,02 g/dm³. At the two low temperature modes, the SOF fraction of the particulates is high. All Pt catalysts (C, D, F) are very effective in removing SOF at low temperatures, but much less effective at high temperatures. With the Pd formulation (E), the overall SOF efficiency is slightly lower. This counter-oriented trend of the conversion of the particulate (SOF) hydrocarbons compared to the gaseous (HC) hydrocarbons (see Figure 4) is also seen with the highly active catalyst A, and points to a different reaction mechanism [7].

The gaseous hydrocarbons are oxidized directly through contact with the catalytically active surface, and the efficiency therefore increases with increasing temperature. With all catalysts, some storage of the insoluble carbon fraction was observed at the modes 1 and 2. At low temperatures, a large portion of the SOF is adsorbed onto this accumulated soot layer leading to sticky particles which remain on the washcoat. With increasing temperature, the SOF hydrocarbons are oxidized leaving back dried particles. The majority of these particles is blown out with increasing exhaust gas flow. However, a small part of the soot is oxidized in the catalyst. At high temperatures, the particulates pass through the converter nearly unchanged. No adsorption of SOF can take place, and the SOF efficiency decreases.

At the high temperature mode 5, sulfate formation is considerable with all formulations. The most sulfate production is observed with the highest Pt content of catalyst C, the least with the lowest Pt content of catalyst F. Sulfate formation of Catalyst D with 0,07 g/dm³ Pt is similar to that of catalyst E with 1,77 g/dm³ Pd. From that it can be concluded that there is a direct relationship between sulfate formation and precious metal loading and formulation.

These results clearly point to the strategy of finding the optimal catalyst selection. SOF removal, the catalyst's only way of reducing PM, is largely independent of precious metal content or choice. Contrary to that, sulfate formation, which in turn increases the total PM emission, is retarded with lower Pt loadings or with Pd catalysts. The gaseous components HC and CO are effectively oxidized at low temperatures only with high Pt loadings. Since Diesel engines have inherent low HC and CO emissions, a catalyst is not required for passing even the strictest emission limits. Overall, the best solution is therefore a catalyst with a low Pt loading. This catalyst combines a high SOF removal potential with the low sulfate formation potential usually associated with a Pd catalyst, without compromising too much the efficiency of the gaseous components, like the Pd catalyst.

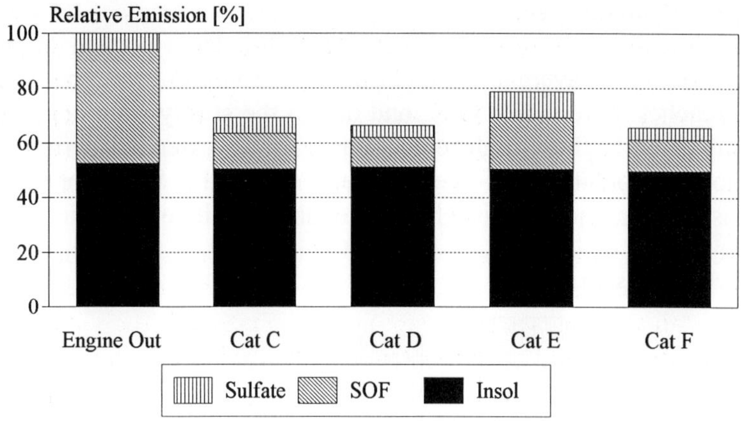

Figure 6 PM emissions on the US transient cycle

3. CATALYST VERIFICATION

3.1. US transient cycle

The US transient cycle is light to medium load cycle where the engine is operated mainly in the idle and rated speed regimes. The average load factor accounts for only 20 %. The maximum temperature during the cycle does not exceed 670 K, and the average temperature is around 490 K. Therefore, the potential for unwanted sulfate formation is limited.

In Figure 6, the PM engine out level of the prototype engine is set to 100 %. With all catalysts, total PM is reduced between 20 % and 35 % due to a high SOF conversion. The Pt content does not significantly influence the SOF efficiency, but the efficiency is lower with the Pd catalyst E. As expected, no sulfate formation is observed. In total, an oxidation catalyst is an effective means for reducing the PM emission to a level that is well below the 1994 PM emission standards.

Based on the above results, catalyst D was selected as first choice. Compared to catalyst C, the high temperature behaviour, i.e. sulfate formation is better due to the lower Pt content, although this is basically not needed on the US transient cycle. Furthermore, the lower Pt content is also connected with a lower catalyst price. The very low Pt content of catalyst F was not considered necessary for the US engine application. Such a low Pt loading is also critical with regard to catalyst ageing, a very important factor in the certification process of engines for the US market.

3.2. European 13-mode cycle

The 13-mode cycle is a steady state cycle with a large portion of full load engine operation. The average load factor is about 45 %. The maximum temperature reaches more than 850 K, and due to the steady state conditions the catalyst is subjected to these high temperatures for an extended period of time. Compared to the conditions of the transient cycle, the SOF portion of the particulates is smaller, and there is a high potential for sulfate formation.

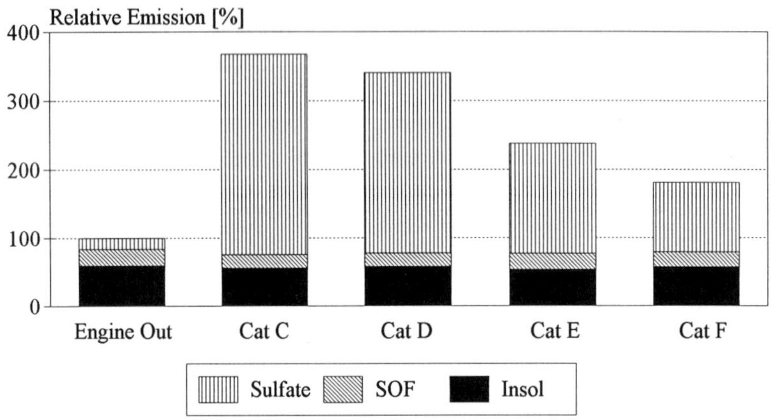

Figure 7 PM emission on the 13-mode cycle

Figure 7 shows that the PM level with catalyst is higher by a factor of 2 to 3,5 than the engine out level. Even with the very low Pt content of catalyst F, the Euro 2 PM standard is by far exceeded. This is caused by a dramatic increase of the sulfate fraction. The SOF fraction is lower than on the US transient cycle, and due to the high temperatures SOF efficiency is poor. Thus, it can be concluded that a catalyst for European applications will be smaller in size (higher space velocity will reduce sulfate formation) and very limited in its overall efficiency on the 13-mode cycle. Such a catalyst was not tested in this program, since the investigation was only aimed to developing a US version.

4. DURABILITY TESTING

Catalyst D was tested in durability program over a total time of 1600 hours (= 640 durability cycles) which corresponds to 213.000 km of field operation. The durability cycle consisted of four different full load modes at 2300, 1380, 950, and 2600 min^{-1}. The first mode was run for one hour, the other modes for 0,5 hours, each. The emissions were measured every 400 hours, and the four measuring points subjected to a linear regression analysis. The regression line was then extrapolated to the 2200 hours endpoint of the durability run. This endpoint corresponds to 296.000 km of field operation, which is defined in the regulations as the "useful life" of the OM 366 LA engine. By definition, the deterioration factor is the difference between the starting point and the endpoint of the regression line. For the certification of an engine, the emissions values must be multiplied with the individual deterioration factors. From that it is clear that low deterioration factors are very important.

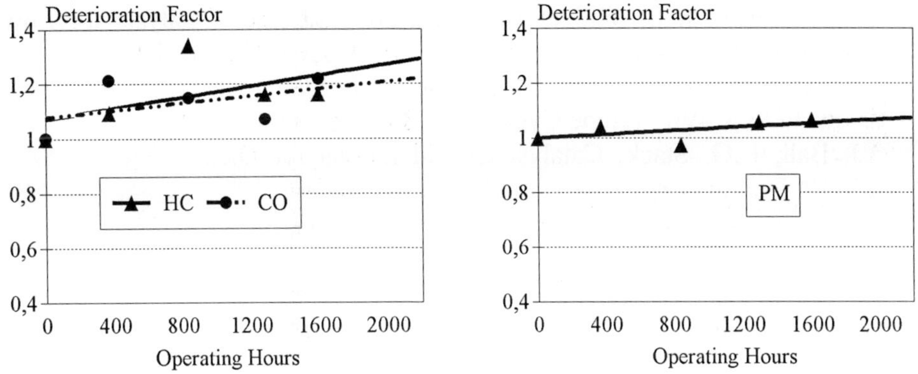

Figure 8   Influence of catalyst ageing on the HC, CO and PM emissions

The results are shown in Figure 8. The zero hour emissions values are set to 1, and the other values are related to the zero hour values. The deterioration factors for the gaseous components HC and CO are 1,21 and 1,13, respectively. Since both emissions are far below the respective emission standards, these values are acceptable, especially with regard to the low Pt loading. The PM emission is nearly constant over the whole durability run which results in a deterioration factor of 1,08. This value is low enough for the engine to comply with the PM standard. It also demonstrates that advanced catalyst technology, i.e. low precious metal loadings can be applied to heavy duty Diesel engines over their useful life period. Apart from the catalyst evaluation in this program, a slight decrease of the NO_X emission was observed in the durability run of the engine.

REFERENCES

1 J. Widdershoven, F. Pischinger, G. Lepperhoff et al., Possibilities of Particle Reduction for Diesel Engines, SAE 860013, Detroit, 1986

2 Code of Federal Regulations, Title 40, Part 86, Subpart N, Emission Regulations for New Gasoline- and Diesel-Fueled Heavy Duty Engines

3 S. Reichel, F. Pischinger, G. Lepperhoff, Influence on Particles in Diluted Diesel Engine Exhaust Gas, SAE 831333, Detroit, 1983

4 J.C. Wall, S.A. Shimpi, M.L. Yu, Fuel Sulfur Reduction for Control of Diesel Particulate Emissions, SAE 872139, Detroit, 1987

5 5 M.G. Henk, W.B. Williamson, R.G. Silver, Diesel Catalysts for Low Particulate and Low Sulfate Emissions, SAE 920368, Detroit, 1992

6 G.M. Cornetti, K. Klein, G.J. Fränkle, H.J. Stein, US Transient Cycle Versus ECE R 49 13-Mode Cycle, SAE 880715, Detroit, 1988

7 D.J. Ball, R.G. Stack, Catalyst Considerations for Diesel Engines, SAE 902110, Detroit, 1990

A. Frennet and J.-M. Bastin (Eds.)
Catalysis and Automotive Pollution Control III
Studies in Surface Science and Catalysis, Vol. 96
1995 Elsevier Science B.V.

CATALYTIC REDUCTION OF NITROGEN OXIDES IN DIESEL EXHAUST GAS

B. H. Engler, J. Leyrer, E. S. Lox and K. Ostgathe

Degussa AG, Hanau, Germany

ABSTRACT

This paper reports first results of research and development work to achieve nitrogen oxide reduction under lean diesel exhaust gas conditions. Much attention is paid to the influence of operation conditions on catalyst performance. A major part of the paper deals with the influence of the hydrocarbon component, the hydrocarbon concentration and the HC/NO ratio on the activity of a special developed platinum based catalyst. Other aspects discussed are a spectroscopic characterization and a selectivity study. A hypothesis of a "dual-site" reaction mechanism for NO_x-reduction in lean diesel exhaust gas precious metal based catalyst is established. Finally, first promising results on the performance of the catalyst system in a vehicle dynamometer test are given.

1. INTRODUCTION

Due to current and proposed legislation in Europe and the United States there is a growing need for low-emission diesel passenger cars and trucks. Main emphasis is put on diesel engines with limited emission of nitrogen oxides and particulate matter as well as gaseous hydrocarbons and carbon monoxide. Various catalytic aftertreatment techniques are available to reduce those pollutants from the exhaust gas of on-road diesel vehicles, which are the flow-through diesel oxidation catalyst [1-4] and the particulate trap [5-6]. Whereas the diesel oxidation catalyst is a well accepted technology to limit particulate, hydrocarbon and carbon monoxide emissions in the exhaust gas of passenger cars in Europe and of light and heavy duty trucks in the US, the catalytic diesel trap is still lacking a reliable regeneration technology. A major challenge left is the removal of NO_x from the oxygen rich exhaust gas of diesel engines. One possibility is the use of an exhaust gas recirculation system (EGR). But it is well known that EGR systems, while lowering the tailpipe NO_X emission, could increase the particulate emission [7]. Also the feasibility of the use of catalysts is discussed in the literature. Unfortunately, conventional precious metal containing three-way type catalysts do not convert nitrogen oxides under excess of oxygen [8]. With the selective catalytic reduction process by using ammonia as reducing agent, NO_X emission can be reduced under diesel engine exhaust gas conditions [9], but the use of nitrogen containing substances like ammonia or urea, makes this process difficult to apply to on-road vehicles, especially passenger cars.

Several laboratory investigations have been reported on the catalytic removal of nitrogen oxides from oxygen-rich exhaust gases without the addition of nitrogen containing substances. Direct nitric oxide decomposition would be the preferred method, but the catalytic activity of the best current technology is very limited under typical diesel exhaust gas conditions [10]. Hence, there is much interest in the reduction of nitrogen oxides by adding CO [11], hydrocarbons [12-17] or alcohols [18] to the exhaust gas stream.

In a recent paper [19] first results were described of an investigation to achieve NO_X reduction in diesel exhaust gas by using special coated, zeolite type catalyst. The infuence of the type of gaseous HC-components, HC-concentration, sulfur dioxide, carbon monoxide, water and oxygen content as well as space velocity on catalytic activity was discussed.

This paper describes further investigations on this subject. Different parameters affecting the performance and particularities to the use of this system in modern diesel engines are discussed.

2. Selectivity versus space time analysis of integral reactor conversion data - a tool to identify reaction paths

In the literature several reaction mechanisms are discussed for the reduction of nitrogen oxides by hydrocarbons under excess of oxygen with precious metal free catalysts [13,16-20]. For precious metal containing catalysts only a few reference data exist. The majority of the investigations use the measurement of reaction intermediates by spectroscopic techniques to determine the reaction path [20]. Another way , used in this study is the use of a selectivity analysis [21-22]. For a given reaction the conversion and selectivity of the reactants can be calculated by following equations:

Reaction: A → B (1)

Conversion:

$$X_A = \frac{F_{A^\circ} - F_A}{F_{A^\circ}} = \frac{\text{Moles of A disappeared}}{\text{Moles of A fed}} \quad (2)$$

$$X_B = \frac{F_B}{F_{A^\circ}} = \frac{\text{Moles of B formed}}{\text{Moles of A fed}} \quad (3)$$

with

X_A = Conversion of reactant A
X_B = Conversion of product B
F_A and F_B = Molar flow of reactant A and product B [mol/s]
F_{A° = Molar flow of reactant A at reactor inlet [mol/s]

Selectivity:

$$S_B = \frac{X_B}{X_A} = \frac{\text{Moles of A converted to B}}{\text{Moles of A converted}} \quad (4)$$

with S_B = Selectivity for product B

From these data thc initial selectivity S_{B° can be determined by the following equation:

$$S_{B^\circ} = \lim_{\left(\frac{V}{F_{A^\circ}}\right) \to 0} \left[\frac{1}{X_A} \cdot \int_0^{\infty} r_B \cdot d\left(\frac{V}{F_{A^\circ}}\right)\right] \quad (5)$$

with:

S_{B° = Initial selectivity for product B
r_B = Reaction rate of B [mol/kg · s]
V = Catalyst volume [l]
V/F_{A° = Space time [l· s/mol] = Residence time of reactant A
= Inverse space velocity

This integration is best carried out graphically by plotting S_B against V/F_{A°. From these data the identification of the reaction paths is possible by calculation of initial selectivity for different reaction products. The interpretation is summarized in Table 1.

Table 1: Theory and interpretation of the selectivity study

Reaction	Initial Selectivity	First derivation
$A \rightarrow B$	$S_{B^\circ} = 1$	$\dfrac{dS_B}{d\left(\dfrac{V}{F_{A^\circ}}\right)} = 0$
A → B, A → C (parallel)	$S_{B^\circ} + S_{C^\circ} = 1$	$\dfrac{dS_B}{d\left(\dfrac{V}{F_{A^\circ}}\right)} \neq 0$
$A \rightarrow B \rightarrow C$	$S_{B^\circ} = 1$ $S_{C^\circ} = 0$	$\dfrac{dS_B}{d\left(\dfrac{V}{F_{A^\circ}}\right)} < 0$ $\lim_{\left(\frac{V}{F_{A^\circ}}\right) \rightarrow \infty} S_C = 1$

3. EXPERIMENTAL

3.1 Catalyst

The monolithic samples were prepared by coating cordierite honeycombs (cell density 62 cells/cm^2; wall thickness 20 μm) with an aqueous slurry of the desired washcoat oxide. After drying and subsequent calcination at 530°C for 2 h in air, the washcoated supports were impregnated with an aqueous solution of the platinum salt, dried and activated in hydrogen for 2h at 530°C. The total platinum loading was fixed at 50 g/ft^3 Pt .

For model gas experiments samples with a diameter of 2.5 cm and a length of 7.6 cm were used. The vehicle test results were obtained with full size substrates with a diameter of 14.4 cm and a length of 15.2 cm.

3.2 Performance Tests

The majority of the performance tests were done with an integral model gas reactor, described in a recent paper [2]. It consists of a gas mixing section, a reactor section and an analytical section. For simulation of typical diesel exhaust gas hydrocarbons several selected liquid HC-components were introduced in the exhaust gas stream by means of an HPLC-pump (Shimazu LC9A) and using an stainless steel evaporator (T>180°C). The model gas compositions used in this study are given in <u>Table 2</u>.

Table 2: Model gas compositions and test procedures

	Test 1	Test 2	Test 3	Test 4	Test 5
CO [ppm]	350	350	350	350	350
H_2 [ppm]	117	117	117	117	117
HC	3200	3200	800-12800	3200	3200
[ppmC_1]	270	270	270	100-800	270
NO [ppm]	25	25	25	25	25
SO_2 [ppm]	6	6	6	6	6
O_2 [vol.-%]	10.7	10.7	10.7	10.7	10.7
CO_2 [vol.-%]	10	10	10	10	10
H_2O [vol.-%]	balance	balance	balance	balance	balance
N_2 [vol.-%]					
Space Velocity [Nl/l/h]	50.000	50.000	50,000	50.000	20.000-100.000
Temperature range [°C]	125-500 500-125	125-500	125-500	125-500	125-500

Test 1: HC-component: n-Hexadecane ($C_{16}H_{34}$)

Test 2: HC-component: Methane (CH_4), Ethane (C_2H_6), Propane (C_3H_8), Butane (C_4H_{10}), n-Heptane (C_7H_{16}), Iso-octane [= 2,2,4 Trimethyl-pentane] (C_8H_{18}), n-Hexadecane ($C_{16}H_{34}$), Ethylene (C_2H_4), Propylene (C_3H_6), 1-Butylene (C_4H_8), 1-Octene (C_8H_{16}), Methanol (CH_3OH), Ethanol (C_2H_5OH), 1-Propanol (C_3H_7OH), 1-Butanol (C_4H_9OH), Toluene (C_7H_8), 1-Methylnapthalene ($C_{11}H_{10}$), ortho-, meta-, para-Xylene (C_8H_{10})

Test 3-5: HC-component: n-Hexadecane ($C_{16}H_{34}$)

The liquid hydrocarbons n-hexadecane (cetane number = 100) and 1-methylnapthalene (cetane number =0) were chosen as model components to simulate diesel fuel, the hydrocarbons iso-octane (octane number = 100) and n-heptane (octane number = 0) as model components to simulate gasoline, respectively.

On-line analysis was performed for the hydrocarbons by a flame ioniziation detector, for CO, N_2O and CO_2 by NDIR-detectors, for NO, NO_2 and NO_x by a chemuluminescence detector and for O_2 by a paramagnetic detector. Except for the O_2 detector, all sample lines were heated to > 120°C to avoid condensation of the exhaust gas components. The amount of N_2 formed by the reaction was calculated from the mass balance. A detailed list of the on-line analyzers used is given in Table 3.

Table 3: List of on-line analysis methods

Component	Detector	Type
HC	FID	Ratfisch RS 55 HC
CO	NDIR	Leybold Binos 100
CO_2	NDIR	Leybold Binos 100
N_2O	NDIR	Leybold Binos 100
NO_x, NO, NO_2	Chemiluminescence	Tecan CLD 700 EL
O_2	Paramagnetic	Siemens Oximat

Final test results were performed on a chassis dynamometer according to the European MVEG-A test procedure. The passenger car used had a inertia weight of 1500 kg and was equipped with a 2.5 l DI/TC four cylinder diesel engine.

3.3 Aging Procedure

All samples were measured in fresh or diesel engine aged conditions. The stationary diesel engine bench was equipped with a 1.9 l IDI/TC diesel engine described in a recent paper [1]. Table 4 describes the 50h aging cycle used in this

Table 4: Diesel bench aging cycle for NO_x-catalyst (1.9 l IDI/TC diesel engine;diesel fuel S-content: 0.15 wt.-% S; lubricating oil phosphorous content: 0.1 wt.-% P; duration 50h) [19]

Step	Engine Speed [rpm]	Load [Nm]	Duration [min]	Converter Inlet Temperature [°C]
1	2000	50	18	250
2	3000	115	6	465
3	4000	10	6	290
4	4000	110	30	620
5	Repeat step 1-4			

study. This cycle was chosen to give a first indication of the long term stability of the catalyst. The aging cycle includes low temperature conditions as well as high temperature phases. The diesel fuel used contained 0.15 wt.-% sulfur, the lubricating oil had 0.1 wt.-% phosphorous. The cetane number, density and all other specifications correspond to the German standard DIN 51601 for automotive diesel fuel [22].

3.4 Spectroscopic Characterization Methods

In fresh conditions the catalyst was analyzed by diffuse reflectance infrared Fourier transform (DRIFT-) spectroscopy to determine the NO adsorption state. The DRIFT spectra were measured in the wavelength range of 1500-2300 cm^{-1} on a Bruker IFS 88 spectrometer equipped with a Spectra-Tech in situ cell. The spectra resolution was 2 cm^{-1}. The spectra were recorded in situ at 225°C in flowing 2% NO in Helium (50 cm^3/min) after preconditioning for 0.5 h at 400°C in hydrogen (50 cm^3/min).

4. Results and Discussion- Model Gas Reactor Test Results and Spectroscopic Characterization Data

4.1 Influence of Exhaust Gas Temperature

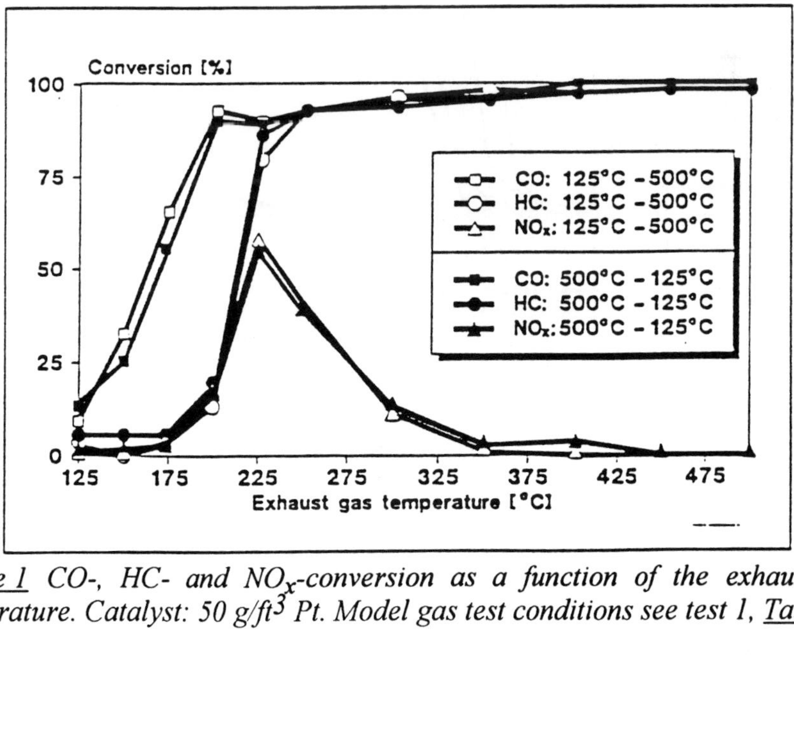

Figure 1 CO-, HC- and NO_x-conversion as a function of the exhaust gas temperature. Catalyst: 50 g/ft^3 Pt. Model gas test conditions see test 1, Table 2

Under realistic diesel engine operation conditions the exhaust gas temperature will switch from low to high temperature and reverse. For this reason NO_X-conversion efficiency was tested in the model gas reactor by increasing the exhaust gas temperature from 125°C to 500°C and from 500°C down to 125°C (test 1, Table 2). The NO_X-, HC- and CO-conversion efficiency obtained are shown in Figure 1 The results reveal clearly that the catalytic efficiency in this area is independent from temperature decrease or increase.

4.2 Influence of Exhaust Gas HC-Components

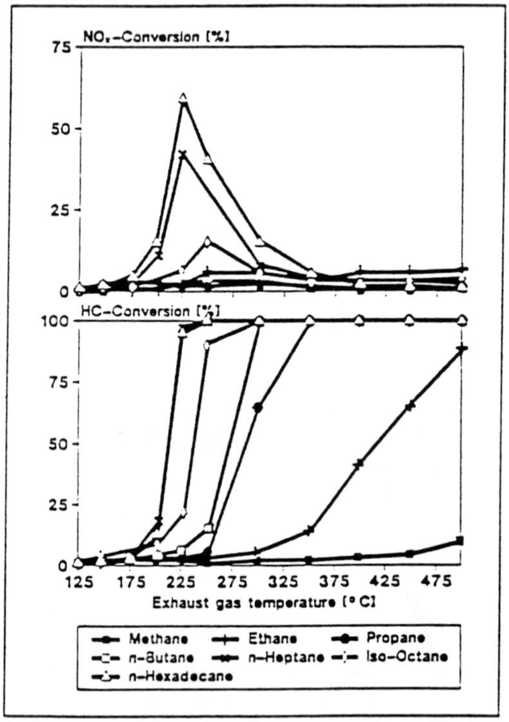

Figure 2A: HC- and NO_X-conversion as a function of the exhaust gas temperature for different paraffinic hydrocarbons. Catalyst: 50 g/ft^3 Pt, fresh. Model gas test condition see test 2, Table 2

To evaluate the influence of the exhaust gas HC-components the catalyst sample was tested in the fresh condition with various olefinic, paraffinic, alcoholic and aromatic hydrocarbons. The concentration of the various components was fixed to 3200 ppmC_1 (test 2, Table 2). The model gas test results for HC- and NO_x-conversion efficiency are shown in Figure 2A-D The graphs reveal clearly the influence of the C-number per molecule and the nature of the HC-component on catalyst efficiency. Paraffinic hydrocarbons, see Figure 2A, show NO_X-conversion only at higher C-numbers per molecule. Olefinic hydrocarbons, see Figure 2B, reveal NO_X-conversion also at low C-numbers per molecule. Alcoholic hydrocarbons, see Figure 2C, show the highest overall activity for NO_X-reduction under the applied conditions. Aromatic hydrocarbons Figure 2D show NO_X-conversion efficiency depending on the reactivity of the molecule.

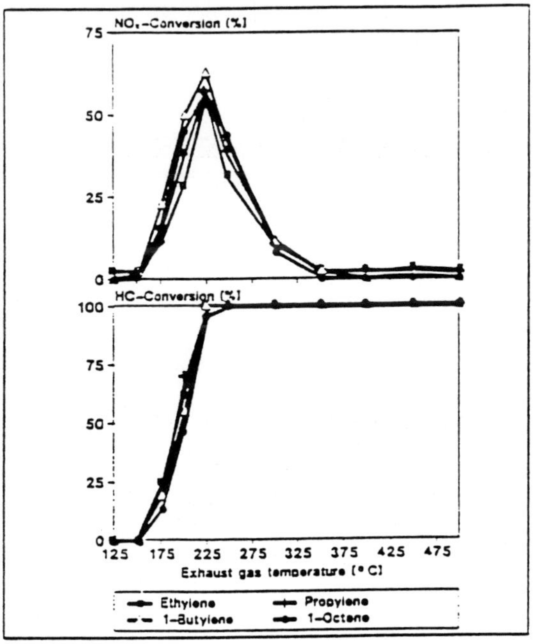

Figure 2B: HC- and NO_X-conversion as a function of the exhaust gas temperature for different olefinic hydrocarbons. Catalyst: 50 g/ft^3 Pt, fresh. Model gas test condition see test 2, Table 2.

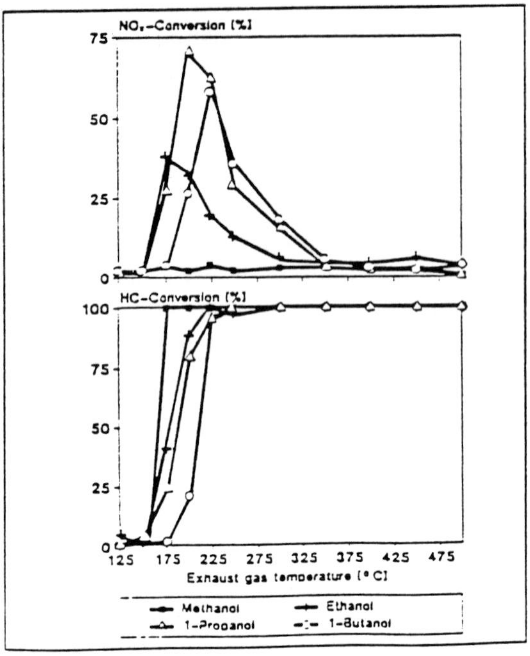

Figure 2C:

HC- and NO_x-conversion as a function of the exhaust gas temperature for different alcoholic hydrocarbons. Catalyst: 50 g/ft^3 Pt, fresh.
Model gas test condition see test 2, Table 2.

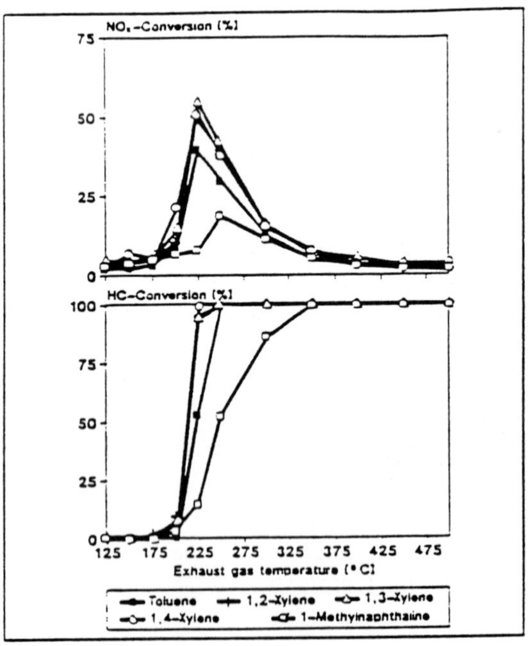

Figure 2D:

HC- and NO_x-conversion as a function of the exhaust gas temperature for different aromatic hydrocarbons. Catalyst: 50 g/ft^3 Pt, fresh.
Model gas test condition see test 2, Table 2

In *Figure 3* the NO_x-conversion efficiency, measured at 225°C, is plotted versus the C-number per molecule for all hydrocarbon components tested. As mentioned before, the data clearly reveal that alcoholic and olefinic hydrocarbons exhibit excellent activity for NO_x-reduction, but also that n-hexadecane, a component already present in diesel fuel, is a suitable reactant to reach a high NO_x-conversion activity level.

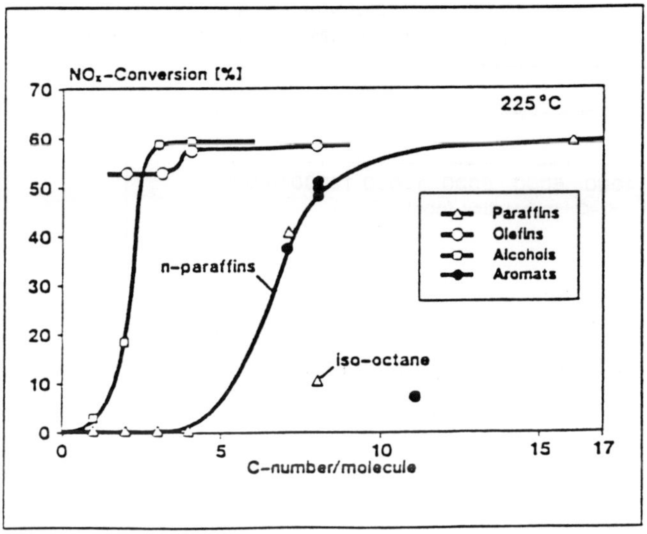

Figure 3: NO_x-conversion as a function of the C-number per molecule for different hydrocarbon species. HC-concen-tration 3200ppmC_1. Catalyst: 50 g/ft^3 Pt, fresh.Model gas test condition, see test 2, Table 2.

4.3 Influence of the Exhaust Gas Hydrocarbon Concentration

To evaluate the influence of the exhaust gas hydrocarbon concentration, the Pt based catalyst was tested in fresh condition at a hydrocarbon content of 800-12800 ppmC_1. *Figure 4* shows the NOx-conversion efficiency for exhaust gas temperatures of 225°C, 250°C and 300°C (test 3, *Table 2*). N-hexadecane was chosen as hydrocarbon component . At HC-concentrations above 3200 ppmC_1 of n-hexadecane the NO_x-conversion efficiency is independent of the amount of hydrocarbons introduced. At lower concentration the maximum of NO_x conversion was shifted to slightly higher temperatures and decreased. This results show that certain levels of hydrocarbons are necessary to reach acceptable NO_x-conversion efficiencies. On the other hand the data reveal that high HC-concentrations have no poisoning effect on the catalyst.

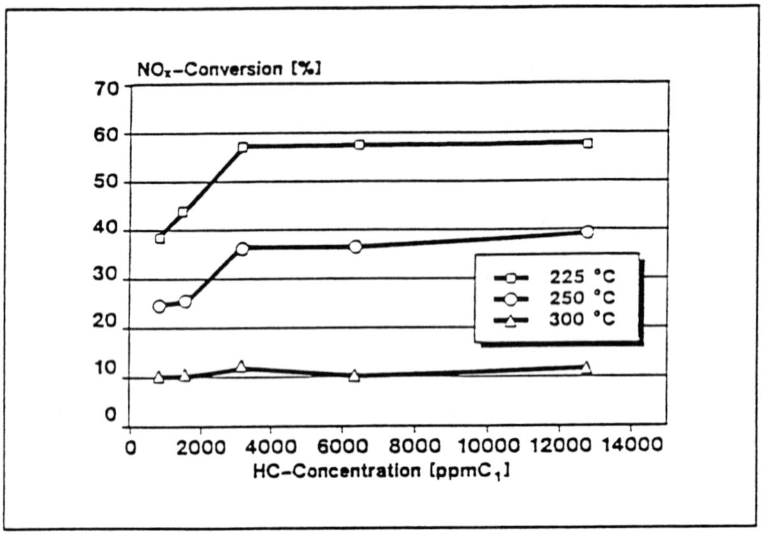

Figure 4: *NO_x-conversion for different hydrocarbon concentrations (n-hexadecane) as a function of the exhaust gas temperature. Catalyst: 50 g/ft^3 Pt, fresh. Model gas test condition see test 3, Table 2.*

4.4 Influence of the Exhaust Gas Nitrogen Oxide Concentration

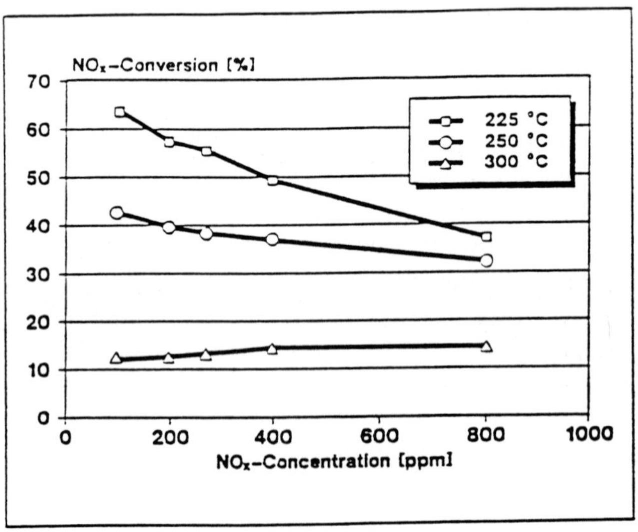

Figure 5: *NOx-conversion for different NO concentrations as a function of the exhaust gas temperature. HC-concentration 3200 ppmC$_1$. Catalyst: 50 g/ft^{3}Pt, fresh. Model gas test condition see test 4, Table 2.*

The effect of the exhaust gas NO concentration on the NO_x-conversion has been studied in the model gas reactor by varying the NO content in the range of 100-80 ppm NO (test 4, *Table 2*). The hydrocarbon concentration (n-hexadecane) was fixed to 3200 ppmC_1. The results obtained are shown in *Figure 5*. It can be demonstrated that the NO concentration had a significant effect on NO_x conversion efficiency at an exhaust gas temperature of 225°C. The effect ist negligible at an exhaust gas temperature above 250°C.

4.5 Spectroscopic Characterization

The DRIFT spectrum of *Figure 6* shows a peak at 1787 cm^{-1} at a temperature of 225°C. The band is characteristic for a linearly adsorbed NO on platinum [24]. This spectrum gives an indication that the adsorption of NO on Pt sites is a first crucial step for the reduction of nitrogen oxides in the exhaust gas stream.

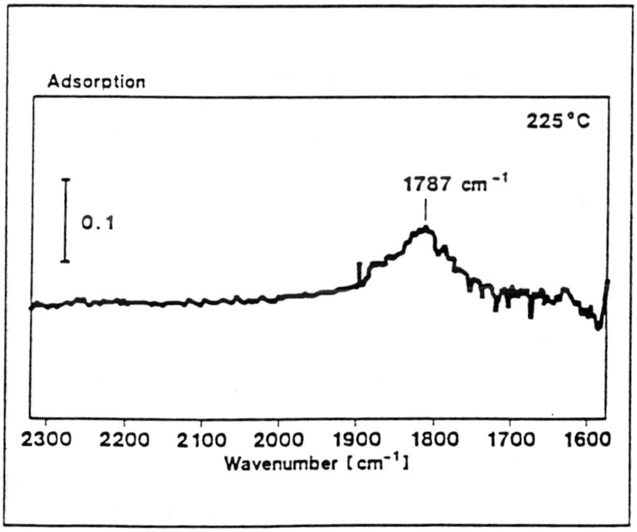

Figure 6: Diffuse reflectance infrared Fourier transform (DRIFT-) spectrum of the platinum based catalyst measured at 225°C in flowing 2% NO in helium (50 cm³/min)

5. RESULTS AND DISCUSSION - SELECTIVITY STUDY AND HYPOTHESIS OF A REACTION MECHANISM

The term "space velocity" has established itself as a measure for the exhaust gas flow referred to the catalyst volume. By this definition it is an indication for the residence time of exhaust gas molecules within the catalyst. As mentioned before the inverse residence time plotted versus the selectivity of the reactants gives a tool for the identification of reaction paths. Therefore, the conversion efficiency over the catalyst was measured at 225°C by varying the space velocity (test 5, Table 2). From these data the selectivities $S(N_2)$, $S(NO_2)$ and $S(N_2O)$ were calculated by using the equations given in chapter 2 and plotted in Figure 7.

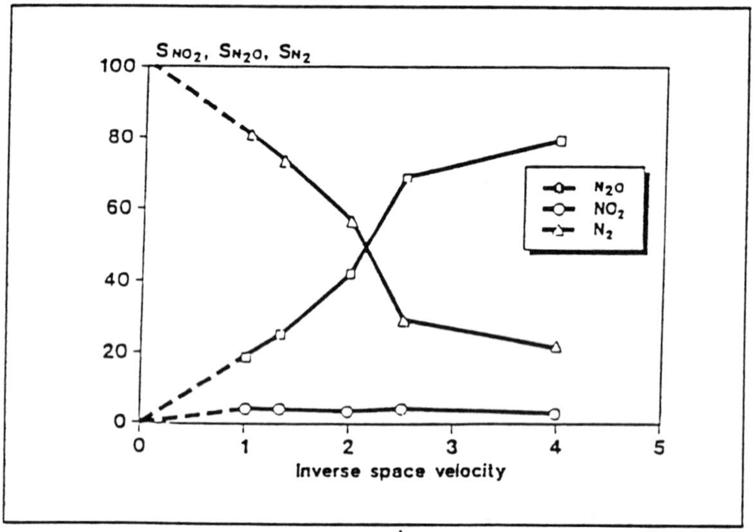

Figure 7: Initial selectivity for N_2, N_2O and NO_2 as a function of the residence time or inverse space velocity. Catalyst: 50 g/ft^3 Pt, fresh. Model gas test conditions see test 5, Table 2.

By consideration of the discussed initial selectivity the diagram characteristics and the spectroscopic characterization data can be interpreted in the following way:

- NO_2 and N_2 are most probably primary products of the reaction
- N_2O is most probably a secondary product of the reaction

These reported observations can be rationalized by the following sequence of surface (elementary) steps:

$$NO + [S_1] \xrightarrow{k_1} NO - [S_1] \qquad (I)$$

$$NO - [S_1] + 1/2\, O_2 \xrightarrow{k_2} NO_2 - [S_1] \qquad (II)$$

$$NO_2\text{-}[S_1] \xrightarrow{k_3} NO_2 + [S_1] \qquad (IIA)$$

$$C_xH_y + zO_2 + [S_2] \xrightarrow{k_4} C_xH_y\, O_{2z} - [S_2] \qquad (III)$$

$$NO_2 - [S_1] + C_xH_y\, O_{2z}\text{-}[S_2] \xrightarrow{k_5} 1/2N_2O + xCO_2 + yH_2O + [S_1] + [S_2] \qquad (IVA)$$

$$NO - [S_1] + C_xH_y\, O_{2z} - [S_2] \xrightarrow{k_6} 1/2N_2 + xCO_2 + yH_2O + [S_1] + [S_2] \qquad (IVB)$$

Figure 8 gives a graphical evaluation of the same sequence.

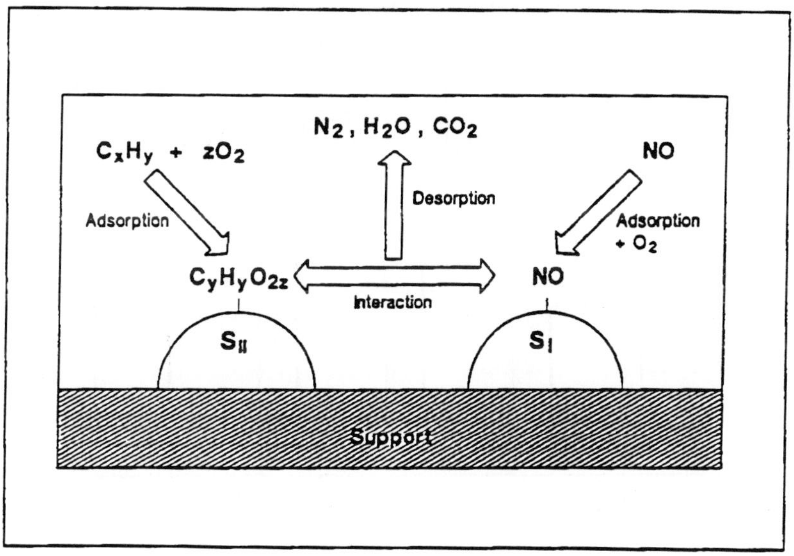

Figure 8: Proposed "dual-site" reaction mechanism for NOx-reduction in lean diesel exhaust gas on precious metal containing catalysts

Further research and development work it is necessary to proof this hypothesis. Especially the identification of the adsorbed hydrocarbon species is a major task for this work.

6. RESULTS AND DISCUSSION - VEHICULE EVALUATION

In preliminary tests the catalyst efficiency was tested on a vehicle dynamometer in the European MVEG-A test cycle. During the first 430 sec. of the test cycle no hydrocarbons were added to the exhaust gas stream. In this phase of the cycle the catalyst inlet temperature is in the average lower than 150°C and therefore the exhaust gas temperature does not reach the light-off temperature of the catalyst. After 430 sec. till the end (1220 sec.) a constant level of 800 ppmC_1 of gaseous hydrocarbons was added to the engine exhaust gas stream in front of the catalyst. During this phases of the cycle the exhaust gas temperature is in the range of 200°C to 450°C. As hydrocarbon components a butylene/butane (2:1) gas mixture was choosen for technical reasons. Figure 9 shows the NO_x-, HC- and CO- and particulate conversion rates in the MVEG-A test cycle. The data were obtained after 50h engine bench aging.

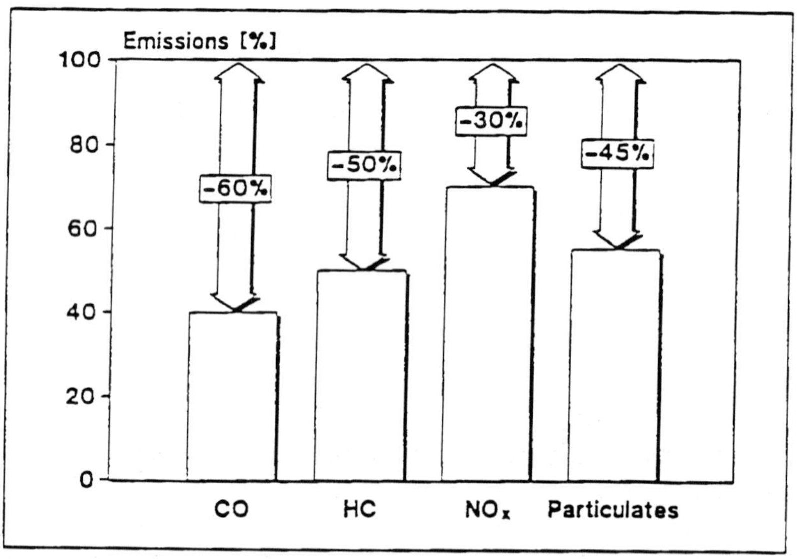

Figure 9: CO-, HC-, NO_x- and particulate conversion rates measured in the European MVEG-A test cycle. Catalyst: 50 g/ft^3 Pt. Aging : 50 h diesel engine bench (see Table 4)

7. Conclusions

It is obvious that a further reduction of the tailpipe emission from diesel vehicles by the use of heterogeneous catalysts is a challenging task for both the catalyst and the engine manufacturers. With regard to the catalyst development, the investigations presented in this paper have shown that the first steps are done, but also that additional work is necessary. From this study the following conclusions can be drawn:

- A reduction of NO_x emission in diesel exhaust by injection of hydrocarbons upstream of a catalyst is possible.
- The model gas experiments have shown that the catalyst activity is influenced by thc hydrocarbon concentration, the hydrocarbon component and the C-number per molecule of the hydrocarbon used .
- Alcoholic and olefinic hydrocarbons show the highest overal activity for NO_x reduction under these conditions
- n-Hexadecane, a hydrocarbon component already existing in diesel fuel, shows high an efficiency for NO_x-reduction.
- By a selectivity study and spectroscopic characterization data it was possible to propose a "dual-site" reaction mechanism for NO_x-reduction in lean diesel exhaust gas
- Finally, it was demonstrated in a vehicle dynamometer test according to the European MVEG-A test cycle that it is possible to achieve NO_x reduction of approximatelly 30%, combined with CO-, HC- and particulate conversion levels of about 50%, 60% and 45%, respectively.

Acknowledgement

The authors wish to thank all co-workers and colleagues for fruitful discussions. Special thanks to Prof. Dr. H. Knözinger and N. Schlensog for carrying out the DRIFT measurements and the extremely helpful discussions. Mr. Blumrich, Mr. Fischer and Mr. Emge for the engine vehicle experiments, Mr. Loesche for the model gas experiments, Mrs. Laber, Mrs. Gulde and Mrs. Bintz for sample preparation and Mrs. Förnges for carefully typing and preparing the manuscript.

REFERENCES

1 R. Beckmann, W. Engeler, E. Mueller, B.H. Engler, J. Leyrer, E.S. Lox and K. Ostgathe: "A New Generation of Diesel Oxidation Catalysts", SAE Technical Paper Series 922330 (1992)

2 P. Zelenka, E.S. Lox and K. Ostgathe: "Reduction of Diesel Exhaust Emissions by Using Oxidation Catalysts", SAE Technical Paper Series 902111 (1990)

3 E.S. Lox, B.H. Engler and E. Koberstein: "Diesel Emission Control", Proc. 2nd Intern. Symp. CAPOC II, Brussels, p. 291 (1990)

4 J. Leyrer, E.S. Lox, K. Ostgathe, B.H. Engler and E. Koberstein: Proc. Intern Seminar "Worldwide Engine Emission Standards and How to Meet Them", IMechE, London, p. 163 (1990)

5 R.W. Horrocks: "Particulate Control Systems for Diesel Engines", IMechE, London, England, Paper C349/87, p. 319 (1987)

6 F. Pischinger, G. Lepperhoff, U. Pfeiffer, K. Egger and G. Hüthwohl: "Modular Trap and Regeneration System for Buses, Trucks and Other Applications", SAE Technical Paper Series 900325 (1990)

7 J.R. Needham, D.M. Doyle, S.A. Faulkner and H.D. Freeman: "Technology for 1994", SAE Technical Paper Series 891949 (1989)

8 P. Öser and H. Völker: "Optimization of Catalyst Systems with Emphasis on Precious Metal Usage", SAE Technical Paper Series 872096 (1987)

9 S. Blumrich and B.H. Engler: "The DESONOX / REDOX-Process for Fuel Gas Cleaning. A Simultaneous Fuel Gas Purification Process for the Simultaneous Removal of NO_x and SO_2 respectively CO and HC", 1st Conf. Environmental Industrial Catalysis, Louvain-La-Neuve, Belgium, p. 263 (1992)

10 M. Shelef: "On the Mechanism of Nitric Oxide Decomposition over Cu-ZSM-5", Catal. Letters, 15, 305 (1992)

11 G.L. Bauerle, G.R. Service and U. Nobe: "Ind. Eng. Chem. Prod. Res. Develop., 11, 54 (1972)

12 W. Held, A. König, T. Richter and L. Puppe: "Catalytic NO_x-Reduction in Non-Oxidizing Atmosphere", SAE Technical Paper Series 900496 (1990)

13 T. Inui, S. Kojo, M. Shibaka, T. Yoshida and S. Iwamoto: "NO Decomposition on Cu-Incorporated A-Zeolites Under the Reaction Conditions of Excess Oxygen with a Small Amount of Hydrocarbons", Zeolite Chemistry and Catalysis, p. 355 (1991)

14 J.O. Petruchi, G. Sill and W.K. Hall: "Studies of the Selective Reduction of Nitric Oxide by Hydrocarbons", Appl. Catal. B:Environmental, 2, 303 (1993)

15 C.N. Montreuil and M. Shelef: "Selective Reduction of Nitric Oxide Over Cu-ZSM-5 Zeolite by Water-Soluble Oxygen Containing Organic Compounds", Appl. Catal. B:Environmental, 1, L1-L8 (1992)
16 M. Iwamoto and H. Hamada: "Removal of Nitrogen Monoxide from Exhaust Gases Through Novel Catalytic Processes", Catal. Today , 10, 57 (1991)
17 M. Iwamoto, N. Mizuno and H. Yahiro: "Selective Reduction of NO and Hydrocarbon in Oxidizing Atmosphere", Proc. 1st Joint Workshop of Catalytic Science and Technology (JECAT), Tokyo, Japan, p. 199 (1991)
18 H. Hamada, Y. Kintaichi, T. Yoshinari, M. Tabea, M. Sasaki and T. Ito: "Performance of Solid Acid Type Catalyst for the Selective Reduction of Nitrogen Oxides by Hydrocarbons and Alcohols, 1st Conf. Environmental Industrial Catalysis, Louvain-La-Neuve, Belgium, p. 219 (1992)
19 B.H. Engler, J. Leyrer, E.S. Lox and K. Ostgathe: "Catalytic Reduction of NO_x with Hydrocarbons Under Lean Diesel Exhaust Gas Conditions", SAE Technical Paper Series 930735 (1993)
20 Z. Chajar, M. Primet, H. Praliaud, M. Chevrier, C. Gauthier, F. Mathis and E.S. Lox: "Influence of the Copper Dispersion on the Selective Reduction of Nitric Oxide over Cu/Al_2O_3 Catalysts: Nature of the Active Sites", Proc. 3rd Intern. Symp. CAPOC III, Brussels (1994)
21 O.H. Hougen and K.W. Watson: "Chemical Process Principles", J. Wiley, New York (1947)
22 G.F.Froment and K.B. Bishoff: "Chemical Reactor Analysis and Design", J. Wiley, New York (1979)
23 German Specifications of Automotive Diesel Fuel, DIN 51601 (1986)
24 B.A. Morrow, J.P. Chevrier and I.E. Moran: " An Infrared Study of the Adsorption of NO on Silica-Supported Platinum over a Wide Temperature Range", J. Catal., 91, 208 (1985)

A. Frennet and J.-M. Bastin (Eds.)
Catalysis and Automotive Pollution Control III
Studies in Surface Science and Catalysis, Vol. 96

CATALYTIC OXIDATION OF DIESEL SOOT: CATALYST DEVELOPMENT

John P.A. Neeft, Olaf P. van Pruissen, Michiel Makkee and Jacob A. Moulijn*

Delft University of Technology, Section Industrial Catalysis
P.O. Box 5045, 2600 GA Delft, The Netherlands

ABSTRACT

A number of metal oxides was screened upon catalytic activity for soot oxidation by means of TGA/DSC. Several metal oxides appeared to be active soot oxidation catalysts. Contact between catalyst and soot was found to play a major role in this solid-solid-gas reaction; varying this contact, activities for several catalysts ranged from active to hardly any activity. It is further tentatively suggested that contact of soot, deposited on catalytic coated particulate filters, is poor, which has major implications for the development of soot oxidation catalysts under diesel operation conditions.

1 INTRODUCTION

After the successful introduction of the three way catalyst for otto engines in the 80's, attention has been focused on the emission characteristics of diesel engines during the last decades. It was found that NO_x and soot (better: particulates) are the main contributors from diesel engines to environmental pollution and health hazards. Therefore, NO_x and particulate standards have tightened, and will tighten in years to come, which resulted in the launching of large research projects worldwide.

At Delft University of Technology, in 1990 the project 'Catalytic removal of soot and NO_x from diesel exhaust gases' was initiated. Within the soot part of the project, the attention is being focused on the development of a catalytic soot filter as an add-

* The authors would like to acknowledge financial support for the DUT-project *'Catalytic removal of soot and No_x from diesel exhaust gases'* by the Dutch Ministry of Housing and the Environment (VROM).

on device for diesel engines, as well as on the chemical processes taking place at a smaller scale when soot is combusted over a catalyst.

Catalytic oxidation reactions of carbonaceous materials have been extensively studied. Carbons, chars, and graphites have been oxidized catalytically under carefully controlled steady state conditions in order to accurately determine reaction rates. Often, high purity carbonaceous materials were used in combination with low catalyst and oxygen concentrations. In others studies, ignition of carbonaceous materials was studied under less well controlled conditions.

The reaction rate of catalyzed carbon oxidation reactions depends on a number of intrinsic parameters:

- The carbonaceous material used. Its reactivity depends on the hydrogen-to-carbon ratio, the surface area, the ash content, and the amount of adsorbed components, in particular hydrocarbons.
- The catalyst used, including its preparation.
- The catalyst-to-carbon ratio.
- The contact between carbon and catalyst, as will be discussed shortly.
- Pretreatment of the sample. For example, the amount of adsorbed hydrocarbons on soot is known to decrease as a function of time; an initial heat-up in inert gases leads to desorption of hydrocarbons and it may have other effects, as a change in contact between soot and catalyst, or a reaction between soot and metal oxide, reducing the catalyst.
- Oxygen partial pressure. The soot oxidation rate was found to depend linearly on the oxygen partial pressure [1,2].

The reaction rate can be measured in various ways. In this study, Thermogravimetrical Analysis (TGA) in combination with Differential Scanning Calometry (DSC) was used. Also this technique and the experimental method can influence the observed reaction rate. The following parameters are of importance:

- The sample mass, the gas flow rate, and the specific geometry of the thermobalance. These parameters influence heat and mass transport to and from the sample and determine when, at high carbon oxidation rates, heat production will pass a threshold value after which the carbon will ignite.
- The temperature history of the sample. Peak temperatures (temperatures at highest mass loss rate) shift to lower temperatures as heating rates decrease.

In literature, not always sufficient attention is paid to these parameters. In catalytic carbon and graphite studies, as reviewed by McKee [3], experimental conditions are usually thoroughly controlled and described. In soot oxidation studies for diesel aftertreatment purposes (e.g. [1,4-7]), experimental conditions are often less carefully reported. As also the intrinsic parameters in these two types of studies differ (low oxygen and catalyst concentrations in the former type of experiments versus high concentrations in the latter; often very small, nanometer-scale catalyst particles in the former experiments versus large, micron-scale particles in the latter), results from one type of study cannot be used in the other type of study without due consideration.

The objective of this study is to investigate the influence of above mentioned intrinsic parameters of catalyst and soot on the catalytic combustion of soot. In order to be able to study these parameters, first the activity of different soot oxidation catalysts has to be defined properly. The screening of catalyst materials, as presented in this paper, aims to fulfil this need.

A second objective of this paper is to assess the influence of -according to our views- one of the most important parameters apart from the type of catalyst used; the contact between soot and catalyst. So far, little attention has been paid to this parameter in soot oxidation literature.

2 EXPERIMENTAL

As the composition of diesel particulates (e.g. fraction of adsorbed hydrocarbons) depends upon many motor characteristics as engine load, speed, and various temperatures, it is difficult to collect batches of soot with constant properties. Therefore, we choose to work with printex-U (a flame soot supplied by Degussa AG) as a model soot. Properties of this model soot and diesel particulates collected from a one cylinder direct injected diesel engine (Yanmar L90E diesel generator set) are listed in Table 1.

Table 1: Some properties of Printex-U (model soot) and diesel particulates

	Printex-U	Diesel particulates
Elementary analysis (wt%)		
C	92.2	90.1
H	0.6	2.4
wt% volatiles*	6	55±10 (at 0% load) 24±8 (at 75% load) 10±5 (at 100% load)
Surface area (BET)	100 m^2/g	60 - 85 m^2/g
Elementary particle size**	25 ± 3 nm	20±3 nm (at 75% load)

*: desorption as measured by TGA weight loss in N_2 to 960 K.

In the catalyst screening experiments, powder-form catalysts were used, which were milled and sieved. The fraction smaller than 125 mm was used.

Activities of soot oxidation catalysts were determined using an STA 1500H thermobalance (combined TGA and DSC). A heating rate of 10 K/min and a flow rate of 50 ml/min 21 vol% O_2 in N_2 were used. Thermal runaways were observed when using too large soot samples or samples that were not diluted, as is demonstrated in Figure 1. These thermal runaways ('ignitions') could be avoided using small, diluted samples. Silicon carbide (SiC) was found to be a proper, inert material having a high heat capacity as well as a high thermal conductivity. Employing sample sizes of 2 mg soot and 4 mg catalyst, diluted with about 55 mg SiC, thermal runaways as well as strong heat-up of the samples during soot oxidation were avoided and smooth and reproducible combustion peaks were observed. The maximum of these peaks named combustion temperature (T_{comb}), determined by the heat flux (DSC) signal (which coincides within 10 Kelvin with the peak of the slightly noisy derivative mass signal), was used as a first indication for catalyst activity. Combustion temperatures could be reproduced within typically 5 K for the same catalyst batch.

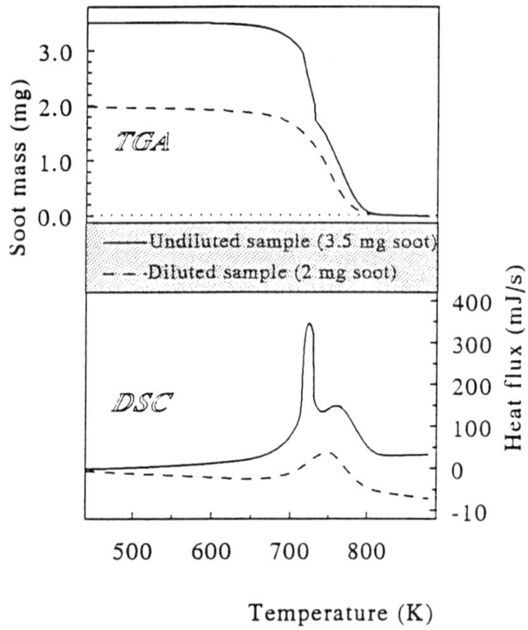

Figure 1 Example of thermal runaway (sharp peak in upper DSC curve). A simultaneous sharp temperature increase is observed (discontinuation in TGA-curve).

In order to investigate the influence of contact between soot and catalyst on soot combustion temperature, two methods were employed to prepare soot catalyst mixtures: firstly milling a mixture of the soot and catalysts, using a mechanical agate mill; this type of contact is referred to as tight contact, and, secondly, just physical contact by shaking soot and catalyst in a sample bottle, which will be referred to as loose contact. In all experiments a catalyst-to-soot ratio of 2 (on weight basis) was used.

For evaluation purposes, particulates from a one cylinder diesel engine were collected on catalyst powder as well as on small segments of catalyst impregnated wall flow monolith (EX-47 from Corning) to be able to assess the degree of contact when collecting soot on filters coated with a catalyst. These segments (diameter of 12.5 mm and length of 20 mm) were impregnated with $Co(NO_3)_2$- or $Cu(NO_3)_2$-solutions, or washcoated with γ-Al_2O_3 particles which had in a previous step been coated with cobalt by wet impregnation of $Co(NO_3)_2$. The segments were calcined at 675 K, then the alternate channels were closed using an alumina sol based glue, and subsequently the segments were dried and calcined at 1000 K.

3 RESULTS AND DISCUSSION

Screening of catalysts

During the catalyst screening tests, it became readily clear that contact between soot and catalyst played a very important role in the catalytic oxidation of soot. Among the catalysts showing activity in tight contact with soot, some catalysts showed also activity in loose contact. Others, however, did hardly or did not. Two examples are shown in Figure 1 (Co_3O_4) and Figure 3 (MoO_3).

The activity for MoO_3 in loose contact mode is still considerable, while Co_3O_4 has hardly any activity in loose contact mode. As it is not clear which contact is important in soot oxidation in diesel applications, both combustion temperatures were measured for a series of catalysts showing activity in tight contact mode. Results are summarized in Figure 1.

For a number of catalysts, two different oxides have been tested. For lead, no significant difference was found in combustion temperature starting from either oxides PbO and PbO_2. For copper, a higher combustion temperature was found for cuprous oxide ($T_{comb, CuO}$ = 797 K) than for cupric oxide ($T_{comb, CuO}$ = 763 K). Silver oxide (Ag_2O) decomposes between 490 and 530 K, so that the actual catalyst was probably metallic Ag. For Co_3O_4, quite a large range of combustion temperatures was found using different batches of Co_3O_4: from 664 to 762 K. The largest number of data was found close to 685 K. The exact reason for this lack of reproducibility is not yet understood. For Cr_2O_3, initially lower combustion temperatures were found as those presented in Figure 1. Small amounts of CrO_3, which was used as the precursor salt for Cr_2O_3, were thought to cause this irreproducibility. CrO_3 is more active than Cr_2O_3, but it decomposes to Cr_2O_3 upon reaction with soot.

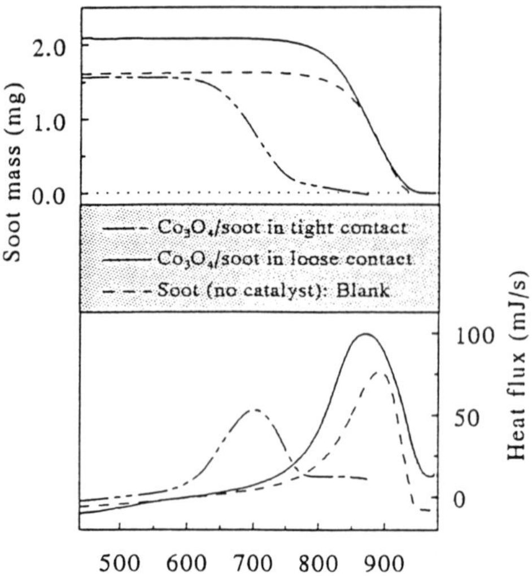

Figure 2 Tight and loose contact mode soot oxidation: Co_3O_4

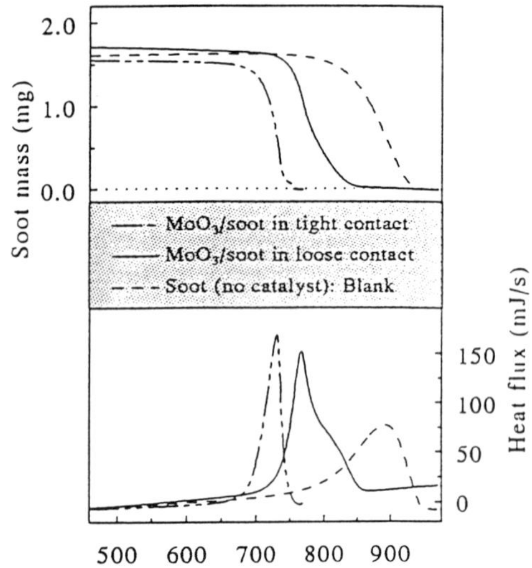

Figure 3 Tight and loose contact mode soot oxidation: MoO_3

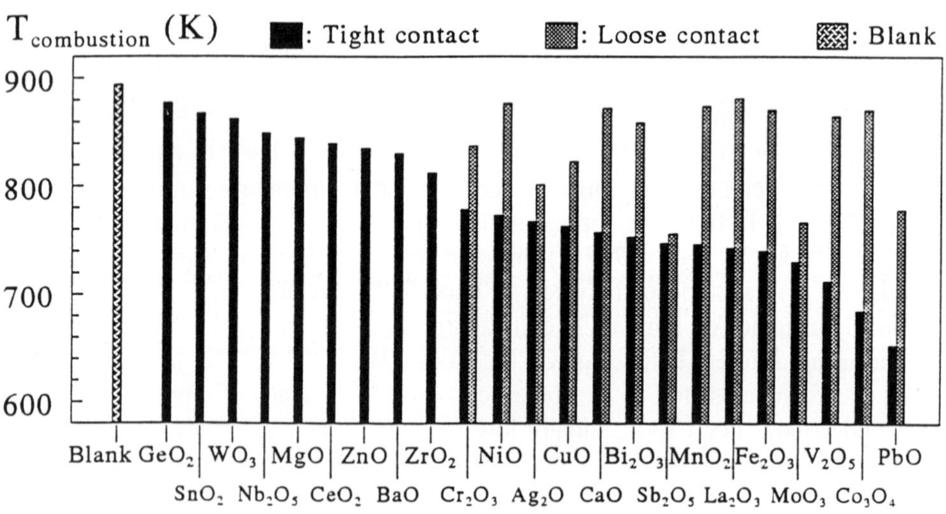

Figure4 Model soot combustion temperatures in tight and loose contact mode

A classification of the catalysts screened is shown in Table 2.

Table 2: Classification of catalysts according to tight and loose contact mode activity.

Tight contact:		
Hardly or no activity (T_{comb} > 800 K):	Moderate activity (800 K > T_{comb} > 700 K)	High activity (T_{comb} < 700 K)
GeO_2, SnO_2, WO_3, Nb_2O_5, MgO, CeO_2, ZnO, BaO, ZrO_2	Cr_2O_3, NiO, Ag_2O, CuO, CaO, Bi_2O_3, Sb_2O_5, MnO_2, La_2O_3, Fe_2O_3, MoO_3, V_2O_5	Co_3O_4, PbO
Loose contact		
Hardly or no activity (T_{comb} > 850 K):	Moderate activity (850 K > T_{comb} > 700 K):	
NiO, CaO, Bi_2O_3, MnO_2, La_2O_3, Fe_2O_3, V_2O_5, Co_3O_4	Cr_2O_3, CuO, Ag_2O, PbO, MoO_3, Sb_2O_5	

As can be seen in Table 2, only Cr_2O_3, CuO, Ag_2O, PbO, MoO_3, and Sb_2O_5 show high or moderate activity in tight as well as in loose contact mode. The other catalysts found to be active in tight contact mode, showed no activity in loose contact mode.

Oxidizing graphite using low catalyst concentrations, Amariglio and Duval [8], Heintz and Parker [9], and Magne and Duval [10] found high activities for the oxides of Pb, Mn, Ag, Cu, V, Au, Ba, Cd, and Ni. McKee [11], determining ignition temperatures of graphite-catalyst mixtures, found several metal or metal oxide catalyst to be active (in order of decreasing activity): Pb, V, Mn, Co, Cr, Cu, Mo, Ag, Cd, Fe, Pt, Ni, Rh, Ir, Ru, Pd, Ce, Zn, W, Hg, and Sn. McKee further reported the trioxides of As, Sb, and Bi to be moderately active catalysts [3]. These data are more or less consistent with our ranking of catalysts; almost all the active and moderately active catalyst in tight contact mode, as shown in Table 2, are also reported active in above mentioned studies [3,8-11].

Alkali compounds are known to be active carbon oxidation catalysts (as reviewed in [3]). Results for Li, K, Na, and Cs are not reported here as problems were encountered, preparing soot-alkali hydroxide mixtures, following our tight and loose contact mode procedure.

Precious metals were not examined in this study. These metals are too expensive to be used in large amounts in particulate filters. Using low amounts of catalyst, as in flow-through oxidation catalysts, we expect the contact between precious metal and soot to become a problem. Besides, sulfur dioxide oxidation over precious metal catalysts can form a large problem under practical conditions [12].

Contact between soot and catalyst

The difference between soot oxidation in loose and tight contact with catalyst is remarkable for some catalysts; using Co_3O_4, V_2O_5, Fe_2O_3, La_2O_3, MnO_2, and NiO, high or moderate activities are found in tight contact mode while in loose contact with soot hardly any activity is found. Other catalysts; PbO, MoO_3, Sb_2O_5, CuO, Ag_2O, and Cr_2O_3, exhibit also activity in loose contact mode, although this activity is lower than the activity in tight contact mode (see Figure 1).

In literature, the activity of graphite oxidation catalysts has been correlated to the mobility of the catalyst, as expressed by the Tamman temperature of the metal or metal oxide, being the temperature at which the lattice atoms of the compound become mobile. This Tamman temperature was found to be roughly half the melting temperature (T_m) in Kelvin [13]. Besides, some metal oxides exert a measurable vapour pressure in the temperature window of this study, e.g. MoO_3 and Sb_2O_5 [14]. This mobility does not provide an explanation for our data: It would explain the high loose contact mode activities of PbO (T_m = 1159 K), MoO_3 (decomposes, MoO_2: T_m = 1068 K), Sb_2O_5 (decomposes via Sb_2O_4 to Sb_2O_3 with T_m = 929 K), possibly CuO (T_m = 1599 K, Cu: T_m = 1356 K) and Ag_2O (decomposes, Ag: T_m = 1235 K), but it fails on vanadium and bismuth, also having low melting temperatures (V_2O_5: T_m = 963 K, Bi_2O_3: T_m = 1098 or 1133; phase-depending) and on chromium, which is moderately active in loose contact mode although it has a high melting point (Cr_2O_3: T_m= 2266 K) [15]. In screening bulk solids it appears that the sample history is quite important [16], e.g. commercial PbO samples contain a large amount of carbonates, which decompose upon heating at high temperature.

The observation of low activities for active catalysts, when the soot-catalyst contact is poor, might have important implications for catalysts used in diesel exhaust applications. In order to assess the contact between soot collected on a filter coated with a catalyst, two other experiments were performed. In the first experiment, diesel particulates were collected on large 250 mm Gelman Sciences Type A/E filters, of which a well defined part was covered with catalyst powder. Subsequently, the soot was combusted in a thermobalance using standard conditions as flow rate and heating rate. Results for a Co_3O_4 catalyst are shown in Figure 3. In this figure, the combustion temperatures are indicated in the DSC patterns by their temperatures: 668, 765, and 881 Kelvin for tight contact with hydrocarbon-desorbed soot, tight contact, and collection on catalyst powder, respectively. Furthermore, in the DSC patterns two sharp soot ignitions are visible in the tight contact curve for both the straight contact as well as the tight contact after desorption of hydrocarbons. In the tight contact mode curve also two small hydrocarbon combustion peaks can be seen. From this figure it becomes clear that diesel soot, collected on catalyst powder, combusts at a temperature comparable to loose contact mode combustion of Printex-U and Co_3O_4 (see Figure 1). The same soot, collected on a filter without catalyst, shows a higher activity when milled with a Co_3O_4 catalyst. This activity, however, is lower than the tight contact mode activity in Figure 1. Probably the adsorbed hydrocarbons cause

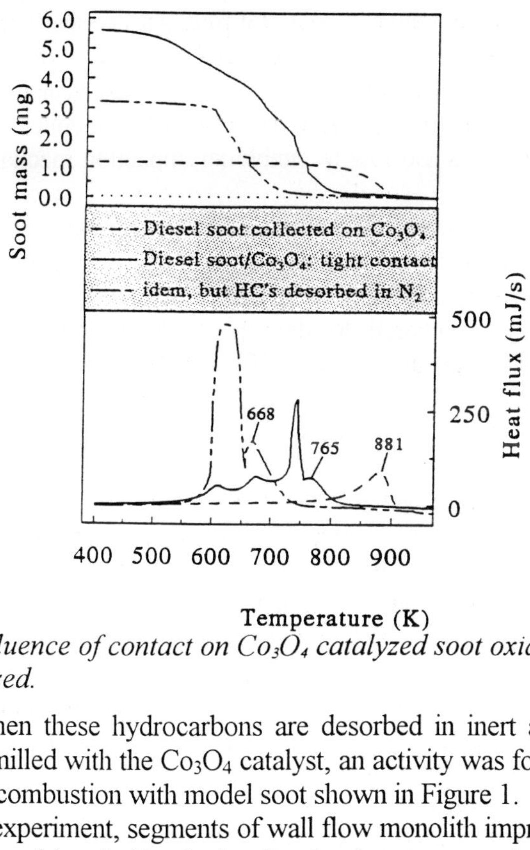

Figure 5 Influence of contact on Co_3O_4 catalyzed soot oxidation. Diesel soot (75% load) is used.

this effect; when these hydrocarbons are desorbed in inert atmosphere before the diesel soot is milled with the Co_3O_4 catalyst, an activity was found similar to the tight contact mode combustion with model soot shown in Figure 1.

In a second experiment, segments of wall flow monolith impregnated with Co_3O_4 or CuO, were placed in a holder in the diesel exhaust gas, and diesel exhaust gas was sucked through the monolith by means of a gas pump. In this way, particulates were deposited on the segments of wall flow monolith. The soot was succeedingly burned in a thermobalance. The results of these tests were that, using Co_3O_4 or CuO catalysts, no large catalytic activities could be found: although soot combustion started at about 100 K lower than was observed for blank wall flow monoliths without a catalytic coating (about 700 K versus 800 K), the catalyst impregnated segments of monolith showed a similar soot combustion temperature as the segments of monolith without a catalyst.

From these tests it seems that, using catalysts being not or moderately active in loose contact mode (see Table 2), catalytic coatings on a particulate filter are not able to catalyze soot oxidation. The contact between soot and catalyst is probably poor under these circumstances and resembles the loose contact. This contact is apparently too poor to significantly enhance the soot oxidation rates. Our observation was made for a wall flow monolith, which is a surface filter (particulates are deposited on top of the filter surface). The contact between soot and filter surface and between soot and catalyst is not expected to be better in other particulate filters. In all these filters, including in filters acting on a deep filtration mechanism (particulates are deposited throughout the filter structure), the driving force for particulate collection is diffusion. Density of soot particulates is too low to enable an impaction mechanism to play a role in particulate filtration.

In literature, some supports for this poor contact between soot and catalyst are given. Precious metals are reported to give only a minor decrease of soot ignition temperatures [5,17]. Base metals showed a higher activity (up to 150 K [5,18]), which is, however, low compared to the up to 300 K reduction in ignition temperature observed using fuel additives [19,20].

Löwe and Mendoza-Frohn [21] examined a model soot similar to the soot used in this study and brought it in contact with a Pt/Co/La catalyst in a number of ways. They come to the same observations as reported in this paper when they compared soot, impregnated with catalyst, with soot, pressed or sprinkled upon catalytic layers. Their best results were obtained when the soot was impregnated with a catalyst solution. They argue that contact, instead of the amount of catalyst, causes higher soot oxidation rates. Moreover, our experiments show that, when using intrinsically active catalysts (catalysts active in tight contact mode), the degree of incapability of enhancing soot oxidation rates is dependent on the properties of the catalyst itself. Some catalysts can significantly increase soot oxidation rates when contact between catalyst and soot is poor, whereas for other catalysts no increase was observed.

Currently we are further investigating the contact between soot and catalyst, including possibilities for optimization of this contact. A number of methods to establish a better contact seems to be feasible. In literature, some groups have reported on the use of small inorganic fibres or whiskers in order to 'blow up' the washcoat and catalyst, within typical diesel particulate dimensions (typically 0.2 - 1 μm). Although very low soot ignition temperatures were published (Saito [22]), we have not been able to duplicate these results. A second possibility to ameliorate contact is 'bringing the soot to the catalyst'; some kind of mechanical means to establish tighter contact. For a number of reasons (limited strength of filters, unfavourable mechanical constructions in diesel exhaust gases) this option seems to be irrelevant. A third and more feasible option seems to be 'bringing the catalyst to the soot', which we already discussed explaining the loose contact mode activities. Additional experiments indicate that other, modified catalyst formulations are very active soot oxidation catalysts, which is caused by their improved mobility. However, this mobility has a drawback: Deactivation rates of the catalyst are highly increased and catalyst deactivation becomes a very important catalyst design parameter. It may be clear that these types of catalysts are still in an explorative state. We know of no literature data, reporting on studies trying to improve catalyst mobility. Our data will be published in another paper. Fourth and finally, an extensively studied option is the use of organometallic fuel additives, in which an improved contact between soot and catalyst seems likely. Upon combustion, the organometallic species are decomposed, resulting in metal oxide or sulfate particles dispersed in the soot particulates. Only a few data has been collected upon dispersion of the catalyst and the contact between catalyst and soot (Otto and coworkers showed a good dispersion of a lead catalyst in soot particulates [23]). From ignition temperature measurements and field tests using wall flow monoliths in combination with fuel additives, it seems that the activity of these catalysts (including catalysts not active in loose contact as cerium oxide and manganese oxide [24,25]) is quite high. This indicates that the contact between soot and catalyst is better than loose contact.

CONCLUSIONS

A large number of metal oxides shows activity for the soot oxidation reaction, in case of good contact between soot and catalyst. The most active catalyst found in this study are, in order of decreasing activity, PbO, Co_3O_4, V_2O_5, MoO_3, Fe_2O_3, La_2O_3, MnO_2, Sb_2O_5, Bi_2O_3, CuO, Ag_2O, NiO, and Cr_2O_3.

A number of ways was employed to make contact between soot and catalyst. It was shown that the contact between soot and catalyst is of prime importance for the reactivity of the catalyzed soot oxidation. Several catalysts were found to be unable enhancing the soot oxidation rate when the contact between soot and catalyst is poor. This explains why catalyst coated particulate filters have been found quite inactive under practical conditions. For catalytic systems which seem to be able to decrease soot combustion temperatures a possible explanation can be probably found in catalyst mobility.

As methods, bringing soot and catalyst into contact, greatly differ (or are not even mentioned) in different studies, the activities of catalysts as described in carbon, soot or graphite oxidation studies may only be compared qualitatively.

REFERENCES

1 Ciambelli, P.; Parrella, P. and Vaccaro, S., Kinetics of soot oxidation on potassium-copper-vanadium catalyst, In: Catalysis and Automotive Pollution Control II, (Ed: Crucq, A.) Elsevier Science Publishers, Amsterdam, 323-35 (1991).
2 Lin, C. and Friedlander, S.K., Soot oxidation in fibrous filters. 2. Effects of temperature, oxygen partial pressure, and sodium additives, Langmuir 4(4), 898-903 (1988).
3 McKee, D.W., The catalyzed gasification reactions of carbon, In: Chemistry and physics of carbon, Vol. 16, (Eds: Walker, P.L.J. and Thrower, P.A.) Marcel Dekker, New York, 1-118 (1981).
4 Ahlström, A.F. and Odenbrand, C.U.I., Catalytic combustion of soot deposits from diesel engines, Appl. Catal. 60(1), 143-56 (1990).
5 Niura, Y.; Ohkubo, K. and Yagi, K., Study on catalytic regeneration of ceramic diesel particulate filter, SAE Paper 860290 (1986).
6 Hillenbrand, L.J. and Trayser, D.A., A concept for catalyzed ignition of diesel soot, SAE Paper 811236 (1981).
7 Watabe, Y.; Irako, K.; Miyajima, T.; Yoshimoto, T. and Murakami, Y., "Trapless Trap" - A catalytic combustion system of diesel particulates using ceramic foam, SAE Paper 830082 (1983).
8 Amariglio, H. and Duval, X., Etude de la combustion catalytique du graphite, Carbon 4, 323-32 (1966).
9 Heintz, E.A. and Parker, W.E., Catalytic effect of major impurities on graphite oxidation, Carbon 4, 473-82 (1966).
10 Magne, P. and Duval, X., Comparaison des effets catalytiques de divers métaux dans les réactions graphite-oxygène et graphite-protoxyde d'azote, Bull.Soc.Chim.France A5, 1593-7 (1971).
11 McKee, D.W., Metal oxides as catalysts for the oxidation of graphite, Carbon 8(5), 623-35 (1970).
12 Ball, D.J. and Stack, R.G., Catalysts for diesel powered vehicles, In: Catalysis and Automotive Pollution Control II, (Ed: Crucq, A.) Elsevier Science Publishers, Amsterdam, 337-51 (1991).
13 Baker, R.T.K., In situ electron microscopy studies of catalyst particle behavior, Catal.Rev.-Sci.Eng. 19(2), 161-209 (1979).
14 Knacke, O.; Kubaschewski, O. and Hesselmann, K., Thermochemical properties of inorganic substances, Vol. I and II., Springer-Verlag, Berlin, 2412 pages (1991).
15 Weast, R.C. and Astle, M.J., Handbook of chemistry and physics, 62nd ed., CRC Press, Inc., Boca Raton (Florida) (1981).

16 Arnoldy, P.; Heijkant, J.A.M.van den; Bok, G.D.de and Moulijn, J.A., Temperature-programmed sulfiding of MoO_3/Al_2O_3 catalysts, J. Catal. 92, 35-55 (1985).

17 Koberstein, E.; Pletka, H.-D. and Völker, H., Catalytically activated diesel exhaust filters - engine test methods and results, SAE Paper 830081 (1983).

18 Pattas, K.N.; Stamatellos, A.M.; Patsatzis, N.A.; Kikidis, P.S.; Aidarinis, J.K. and Samaras, Z.C., Forced regeneration by exhaust gas throttling of the ceramic diesel particulate trap, SAE Paper 860293 (1986).

19 Daly, D.T.; McKinnon, D.L.; Martin, J.R. and Pavlich, D.A., A diesel particulate regeneration system using a copper fuel additive, SAE Paper 930131 (1993).

20 Rao, V.D.; White, J.E.; Wade, W.R.; Aimone, M.G. and Cikanek, H.A., Advanced techniques for thermal and catalytic diesel particulate trap regeneration, SAE Paper 850014 (1985).

21 Löwe, A. and Mendoza-Frohn, C., Zum Problem der Dieselrub-Verbrennung auf einem katalysatorbeschichteten Filter - der Kontakt zwischen Katalysator und Feststoff, Chem.-Ing.-Tech. 62(9), 759-62 (1990).

22 Saito, K.; Ueda, K. and Ikeda, Y., Exhaust gas cleaning catalyst and process for production thereof, European patent application 0211233, Filed 1-7-86, Issued 25-10-89.

23 Otto, K.; Lehman, C.; Bartosiewicz, L. and Shelef, M., Carbon oxidation catalyzed by lead, Carbon 20(3), 243-51 (1982).

24 Pattas, K.N. and Michalopoulou, C.C., Catalytic activity in the regeneration of the ceramic diesel particulate trap, SAE Paper 920362 (1992).

25 Wiedemann, B.; Doerges, U.; Engeler, W. and Poettner, B., Application of particulate traps and fuel additives for reduction of exhaust emissions, SAE Paper 840078 (1984).

A. Frennet and J.-M. Bastin (Eds.)
Catalysis and Automotive Pollution Control III
Studies in Surface Science and Catalysis, Vol. 96

CATALYTIC COMBUSTION OF DIESEL SOOT ON PEROVSKITE TYPE OXIDES

W. Sri Rahayu, L. Monceaux, B. Taouk and P. Courtine

Département de Génie Chimique, Université de Technologie de Compiègne, B.P. 649, 60206 Compiègne Cédex, France

ABSTRACT

The combustion of soot particulates is studied on perovskite type oxide catalysts having the general formula : $La_{0.8}Sr_{0.2}Mn_{1-x-y}B'_x\phi_yO_3$ (B' = Pt, Ru or Pd, x < 0.01, and F represents a deficiency in B site, y < 0.09). In order to determine the relative activity of various catalytic compositions, TGA and DTA experiments are performed on a mixture of catalyst and soot under air flow. The ignition temperature (T_i), the temperature of the maximum of combustion rate (T_m) and the soot conversion are determined. Soot is simulated either by a mixture of carbon black, fuel and lube oil or by carbon black alone. After a screening, the best catalytic results are obtained for two compositions, i.e. $La_{0.8}Sr_{0.2}Mn_{0.999}Pd_{0.001}O_3$ and $La_{0.8}Sr_{0.2}Mn_{0.91}\phi_{0.09}O_3$. In both cases, the combustion temperature (T_m) of carbon black is found to be lowered from 650°C to about 440°C. The influence of several parameters on the catalytic properties of these latter formulations is studied : catalyst preparation method, specific area, ageing, carbon black/catalyst ratio.

1.INTRODUCTION

Compared to gasoline engines, diesel powered automobiles emissions are low in carbon monoxide but, unfortunately, they contain a higher proportion of NO_x and soot particulates which have a serious impact on urban air quality and human health. Therefore, legislation makes it a duty to significantly reduce these emissions [1]. Soot particulates may be separated from the exhaust stream by filtration through a porous trap [2], but this device causes the increase of exhaust back pressure and therefore affects fuel economy and vehicle performance, consequently traps must be periodically regenerated by oxidizing the collected particulates. Several methods are used in order to regenerate traps. These methods can be divided into catalytic and non-catalytic ones. In the first case, fuel additives [3, 4] and catalytic trap coatings are necessary [5-13] whereas in the non-catalytic case, the trap is electrically heated or with the help of a burner. As fuel additives may cause engine disorders and external heating may lead to the trap destruction, then, it seems that catalytic trap coatings could be a satisfying alternative. Different kinds of catalysts have already been studied. Research works in this field proceeded through noble metals and transition metals or alkaline oxides based catalysts. Pt and Pd, though showing some activity [5-7], cause serious problems, as they are poisoned by sulphur. It has been shown that catalytic performance and durability are improved by using noble metals doped praseodymium, neodymium and samarium oxides [13].

Comparison between oxides of copper, manganese, chromium and vanadium shows that V_2O_5 is one of the most active catalysts for the combustion of diesel soot [10]. Catalyst made up of V_2O_5 and CuO with a molar fraction 0.9 in vanadium leads to an enhancement of the activity [11]. Another screening study on oxides of potassium, vanadium, copper, manganese and cobalt shows that the systems (K-V-O) and (K-Cu-O) can reduce the ignition temperature of carbon black of about 260°C [12]. It is to be noticed that no study about the stability of these compounds is reported.

Then, the purpose of this study was to investigate the possibility of lowering of combustion temperature of diesel soot particulates by using perovskite type oxides (ABO_3) as catalysts. These phases are known to be good catalysts for total oxidation of hydrocarbons and CO. They are thermally and chemically resistant. It is possible to introduce in their structure little amount of noble metals or other ions [14, 15]. The basic formula of the studied phases is $La_{0.2}Sr_{0.8}MnO_3$ doped or not by platinum, palladium and ruthenium. To evaluate the catalytic activity of catalysts, thermogravimetric (TGA) and differential thermal analysis (DTA) are used.

2. EXPERIMENTAL PROCEDURE

2.1. Preparation of perovskite catalysts

The studied catalysts have the following general formula :
$La_{0.2}Sr_{0.8}Mn_{1-x-y}B'_x\phi_yO_{3-\lambda}$ where B' is a noble metal (Pt, Pd and Ru), F cation vacancies and l refers to the oxygen non stoichiometry. The perovskite phases are obtained by decomposition of equimolar mixtures of the corresponding metal nitrates (La, Sr and Mn) and noble metal (Pt, Pd and Ru) chlorides.

The precursors are synthesized by a sol-gel process. This method that was developed and accorded to Baythoun et al. [16], consists in dissolving, gelating, drying and activation steps. The salts are dissolved, in either water (hereafter first method), or ethylene glycol as solvent (second method) with citric acid.

i) In the first case, a solution of mixed salts is evaporated to dryness at 80°C and the gel obtained is dehydrated at 60-70°C for 5 hours and calcined with a heating rate of 5°C/h for 6 hours at 600°C under air.

ii) In the second case, the mixtures are stirred while heating around 80-110°C until the sol is obtained. The as-prepared gels are dried and calcined by heating rate of 5°C/min., under air, for 6 hours up to 600°C for crystallization.

2.2. Preparation of simulated diesel soot

The simulated diesel soot is synthesized by a mechanically mixing of carbon black (70%), diesel fuel (15%) and lube oil (15%). The carbon black used (Regal 660 supplied by the Cabot France Co) has a specific area of 112 m^2/gr., and the particle size is around 0.24 μm in diameter.

2.3. Activity test

The combustion of diesel soot is conducted using TG-DTA Setaram 92. Sample to be analyzed is prepared by mixing catalysts with simulated diesel soot or carbon black alone and grinding in an agate mortar. Then it is loaded in a platinum crucible on initial weight around 66 mg. The experimental conditions are 6°C/min of heating rate from 20°C to 600°C, 1.5 l/h of oxidizing atmosphere flow. The computer system allows to draw both TGA and DTA curves as shown in Figure 1. TGA curve gives the conversion of diesel soot and the ignition temperature T_i. On the other hand, DTA curve provides the maximal rate of oxidation corresponding to the top of the exothermic signal (T_m).

2.4. Characterization of samples

The specific areas of samples are measured using a triple point BET method of the nitrogen adsorption at 77K on a Quantasorb Jr. surface area analyser.

The X-ray diffraction (XRD) analysis of the samples is carried out with an X-ray powder diffractometer using CuKa radiation and curved sensitive detector.

Temperature-programmed desorption (TPD) is carried out under vacuum (~10^{-6}mbar). Prior to each run the sample is treated in air stream at 600°C and then cooled to room temperature. Then, the temperature of the sample is raised at a constant rate of 10°C/min. under vacuum. Oxygen, CO and CO_2 are detected using a quadrupole mass spectrometer (QTMD Carlo Erba).

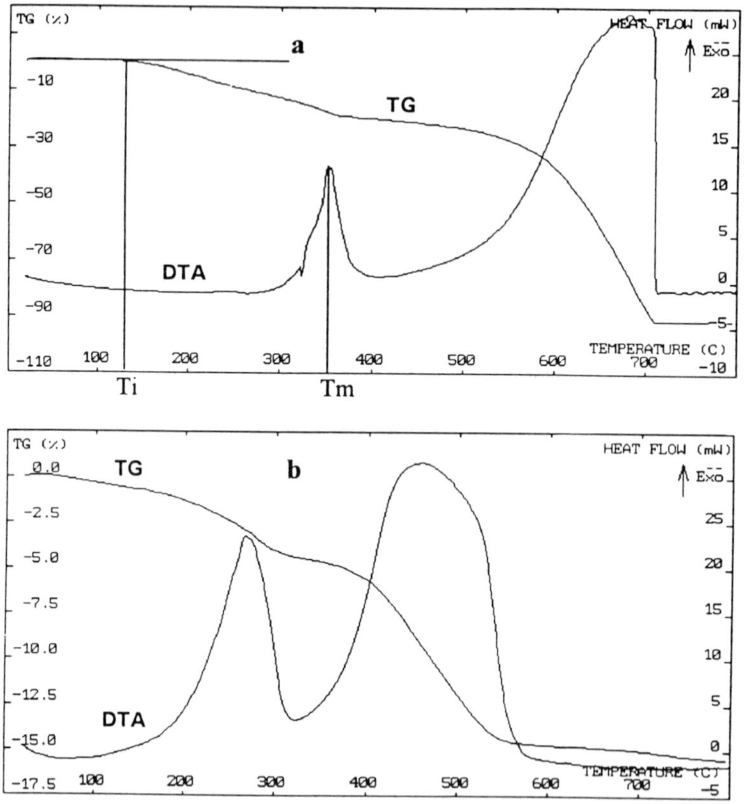

Figure 1. Thermal analysis (TG and DTA) of :
a) simulated soot ; b) mixture of LMPds catalyst and soot (15 wt %)

3. Results

3.1. Catalytic Test

Reactivity of simulated soot : Results obtained with the simulated soot without catalyst are shown in Figure 1.a. Two steps of weight loss are observed. In the temperature range 150°C to 400°C a 25 % weight loss is assigned to oxidation of hydrocarbons (fuel and oil). In the range 500°C-700°C a second loss corresponds to the oxidation of carbon black. The DTA curve shows two exothermic peaks at 350°C and 670°C associated respectively with hydrocarbons and carbon black oxidation. With dry carbon black, no modification on ignition temperature of the particulates is observed.

Screening : TG and DTA experiments are carried out with mixtures of soot and catalysts. In Table 1, ignition temperature T_i and maximum rate of combustion T_m are noted and related to temperatures obtained for soot combustion without catalyst. An important catalytic effect is observed in all cases, and the maximum effect is reached for the sample containing palladium (in stoichiometric substitution). From comparison of the various combustion temperatures, it is possible to deduce the following order of catalytic activity : "LMs" < "LMPts" < "LMPdn" < "LMRun" < "LMRus" < "LMn" < "LMPds".
These results show that catalytic activity is influenced by two parameters : the presence of the substituted noble metal and the non stoichiometry.

Soot/catalyst ratio : A series of experiments with different [soot/catalyst] ratios is carried out with stoichiometric substituted Pd doped perovskite as catalyst. Results show that the combustion temperature increases when the [soot/catalyst] ratio is higher than 15 wt%. This is probably due to the lack of contact between catalyst surface and all the soot particulates beyond this point.

Catalyst ageing and specific area : Simulation tests on Pd containing catalyst in the presence of 5 wt % of soot are conducted in a furnace in the same conditions as for TG-DTA, from room temperature up to 800°C during 10 minutes, then an aliquot part of the sample is taken off, mixed with soot and tested in TG-DTA, whereas the remaining of the sample is kept in the furnace for further runs.

The results (Table 2) show that catalytic reactivity is weakly influenced by ageing test despite the decrease of surface areas, the stabilization of which is reached after the third run. (Ethylene-glycol preparation gives the highest specific area and activity.)

Isothermal test : Heating in TG-DTA up to the chosen temperature is performed under pure nitrogen which is then replaced by air for combustion. The sample is then heated at higher temperatures until the total combustion of soot is achieved. These experiments evidence that the combustion rate of soot increases with temperature, and that the carbon black can be totally and rapidly eliminated at a temperature higher than 450°C, whereas the liquid hydrocarbon part of the simulated soot is completely oxidized at 200°C.

Table 1. Screening test results. Catalysts are tested with 15 wt % soot

Sample	Symbol	T_{calc} (°C)	1st DTA peak % loss	1st DTA peak T_m	2nd DTA peak % loss	2nd DTA peak T_m
Hydrocarbon			64.0	305	90.2	350
Carbon Black	(CB)		-	-	59.0	675
Stimulated Soot			18.2	350	73.7	670
$La_{0.8}Sr_{0.2}MnO_{3+\lambda}$	LMs	600	25.0	300	56.7	480
$La_{0.8}Sr_{0.2}Mn_{0.91}F_{0.09}O_{3+\lambda}$	LMn	600	20.0	275	60.2	460
$La_{0.8}Sr_{0.2}Mn_{0.999}Pt_{0.001}O_{3+\lambda}$	LMPts	600	20.0	295	56.7	485
$La_{0.8}Sr_{0.2}Mn_{0.9}Pt_{0.008}F_{0.092}O_{3+\lambda}$	LMPtn	600	23.3	305	56.7	460
$La_{0.8}Sr_{0.2}Mn_{0.9}Ru_{0.1}O_{3+\lambda}$	LMRus	900	16.7* 20.0	340 355	76.7	560
$La_{0.8}Sr_{0.2}Mn_{0.9}Ru_{0.1}O_{3+\lambda}$	LMRus	600	18.3* 25.9	305 345	63.3	490
$La_{0.8}Sr_{0.2}Mn_{0.9}Ru_{0.008}F_{0.092}O_{3+\lambda}$	LMRun	600	18.3* 23.3	280 302	63.3	475
$La_{0.8}Sr_{0.2}Mn_{0.999}Pd_{0.001}O_{3+\lambda}$	LMPds	900	16.7* 23.3	325 352	60.2	540
$La_{0.8}Sr_{0.2}Mn_{0.9}Pd_{0.008}F_{0.092}O_{3+\lambda}$	LMPdn	900	16.7	307	50.0* 80.0	525 605
$La_{0.8}Sr_{0.2}Mn_{0.999}Pd_{0.001}O_{3+\lambda}$	LMPds	600	20.0	260	60.0	450
$La_{0.8}Sr_{0.2}Mn_{0.9}Pd_{0.008}F_{0.092}O_{3+\lambda}$	LMPdn	600	18.3	290	56.7	490

T_{calc} : temperature of calcination , *: two DTA peaks.
"s" indicated on symbol means a stoichiometric formula, and "n" is non stoichiometry

3.2. Characterization

3.2.1. XRD

Carbon black : The XRD pattern (Figure 2) shows four wide peaks due to poorly crystallized graphite. These peaks correspond to the stacks of parallel hexagonal layer planes.

Catalysts : XRD patterns indicate the existence of one phase which crystallizes in rhombohedral structure with the following parameters :
$a \approx 7.75$Å and $\alpha \approx 90.27°$ ($R\bar{3}m$).

Table 2 : Preparation method, surface area and ageing influence on the catalytic activity of the catalyst containing Pd "LMPds" calcined at 600°C. The soot to catalyst ratio is 5 wt %.

Preparation method	Test No	spec. area m^2/gr.	1st DTA peak*		2nd DTA peak**	
			T_m °C	% wt loss	T_m °C	% wt loss
Ethylene-glycol solution	1	23.4	225	28	445	70
	2	17.78	230	28	480	62
	3	16.17	-	-	-	-
	4	15.67	260	22	455	75
Aqueous solution	1	16.45	250	30	460 490	65 85
	2	14.78	260	16	500 600	60 81
	3	9.54	-	-	-	-
	4	8.94	265	20	470	62

* : hydrocarbons oxidation ; ** : carbon black oxidation.

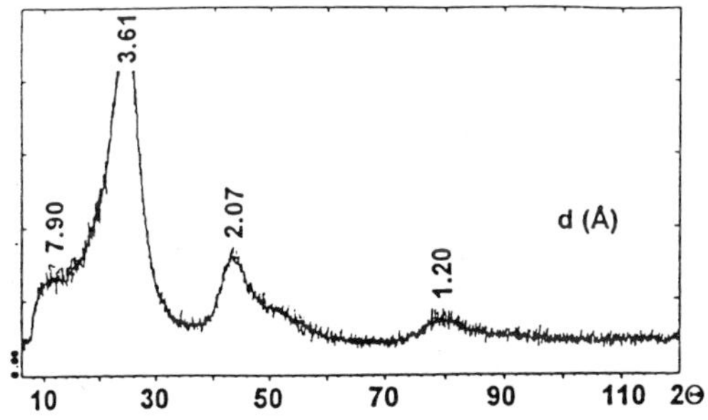

Figure 2. XRD pattern of carbon black

3.2.2. TPD

Several TPD analysis are carried out concerning carbon black, Pd containing catalyst and mixture of both of them.

In the first sample (Figure 3.a), desorption of CO and CO_2 was observed. CO_2 desorbs at low temperature (200-400°C) while CO desorption begins at 200°C to become important at 550°C. CO and CO_2 arise from surface carbon-oxygen complexes. The TPD curves of catalyst (Figure 3.b) show two important desorbed species : CO_2 and O_2. CO_2 certainly results from the decomposition of residual carbonated components remaining during the preparation. O_2 may correspond to adsorbed and absorbed oxygen species on the perovskite structure [17]. When the catalyst is mixed with the carbon black, CO_2 and CO desorptions are observed while oxygen desorption disappears (Figure 3.c).

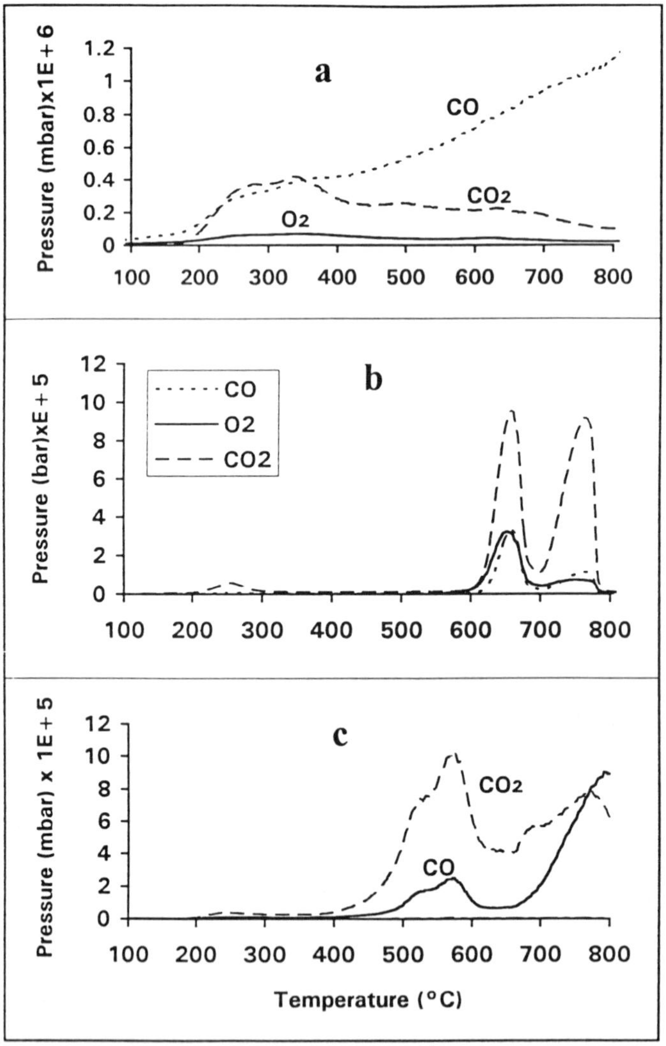

Figure 3. TPD profiles of O_2, CO and CO_2 from : a) carbon black ; b) $La_{0.8}Sr_{0.2}Mn_{0.999}Pd_{0.001}O_{3+\lambda}$; c) mixture of catalyst and carbon black

4. Discussion

i) Reactivity of carbon black alone

According to the XRD pattern (Figure 2), the carbon black Regal 660 has a surface characteristic similar to graphite structure. Recent investigations in scanning tunnelling microscopy [26, 27] have shown that this type of carbon black is actually constituted by the superposition of the overlapping of many graphite planes, rolled up in an "icospiral" structure, whose exposed edges take the form of "fish scales". This led to models of soot formation. Figure 4 also gives our example of HRTEM characteristic image, similar to that obtained by Bourrat [28]. On the particulate surface, carbon-oxygen complexes, which are previously formed, decompose and give rise to a formation of CO_2 at low temperature and CO at higher temperature [18, 19]. TPD measurements on carbon black sample confirmed these phenomena (Figure 3.a).

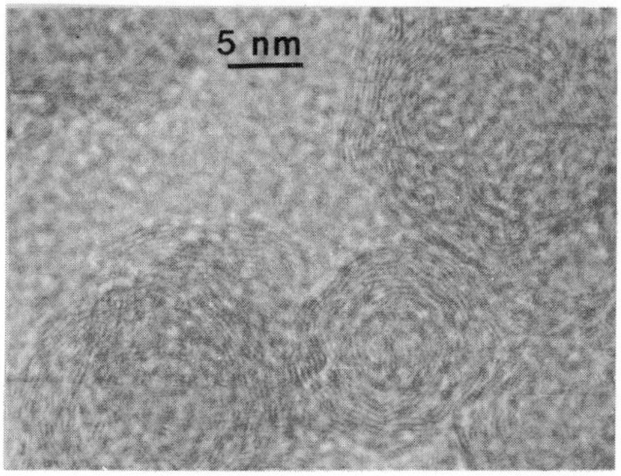

Figure 4. HRTEM image of the carbon black used as a reactant

The mechanism of carbon oxidation is very difficult to investigate for several reasons : i) the high exothermicity of the reaction provides a difference of temperature between the solid surface and the gas ; ii) the existence of several varying parameters during reaction due to mass transfer effects (pore structure and particulate size modifications, swelling, cenosphere formation).

In spite of these problems, many authors agree with the fact that the mechanism of the $C\text{-}O_2$ reaction is the same for the different kinds of carbon (graphite, char, coal, carbon black, etc.) However, the reactivity varies with the structure, the morphology, the specific areas and the surface complexes.

Generally, the mechanism comprises three steps : firstly, oxygen adsorption at the edges described above with C-O complexes formation, secondly, complex decomposition, CO_2, CO formation and finally new active sites liberation.

ii) Reactivity of carbon black on catalyst

When the reaction is performed in presence of a catalyst, the mechanism is more complicated. In the case of perovskite type oxide, the catalytic effect is due to transition ion in B site. It is known that the octahedral environment of the B ions splits the d-orbitals into two levels ; the lower (t_{2g}) level contains orbitals that are less repulsed by negative point charges (oxygen) than those in the higher (e_g) level. At the surface, $d_z{}^2$ orbital is the lowest e_g level. In regard to CO oxidation, it is observed that the maximum activity is reached in both cases for an occupation of the e_g levels of less than one electron, the t_{2g} levels being half-filled or totally filled [14, 15, 20, 21]. In our case, the B cation is Mn^{3+} and Mn^{4+}. The electronic configuration is respectively $3d^4$ ($t_{2g}{}^3$, $e_g{}^1$) and $3d^3$ ($t_{2g}{}^3$, $e_g{}^0$). In both configurations, Mn cation is in the maximum space interval of activity.

Concerning CO/O_2 reaction on perovskite oxide, the "suprafacial" mechanism is assumed, as well as for the carbon black/oxygen reaction. Carbon adsorption on the catalyst surface could be made through the C-C bond or the C-O surface complexes, assuming that C is bonded to the Mn ion with donation of carbon lone pair into the empty $3d_z{}^2$ orbital to form s bond accompanied by back donation of the t_{2g} electrons of Mn ion to anti-bonding π-orbital of C-O or C-C. Moreover, the mechanism begins by simultaneous adsorption of carbon and oxygen, the interaction between adsorbed species causes the CO_2 formation, the desorption of which releases the catalytic active sites.

Finally, the screening investigation shows :

1) that the best performance is obtained with "LMn" and "LMPds". The enhancement of the activity is due to the non-stoichiometry of Mn cation in the former case and to Pd substitution in the latter. The Mn^{4+}/Mn^{3+} ratio is in turn 0.57 for LMs, 1.0 for LMPds and 1.2 for LMn. Therefore, the improvement of activity may be correlated, first of all, to the Mn^{4+}/Mn^{3+} ratio, particularly when noble metals are absent. In this way, Vrieland suggested that Mn^{4+} and Mn^{3+} do not act as individual surface ions, but form a part of a large group which acts as either an electron donor or acceptor [22].

2) As far as precious metal dopes are concerned, Pd is the most active for the carbon oxidation. Inside the perovskite matrix its oxidation state is Pd^{2+} as it was reported that the platinum in the same kind of structure is in the form of dissolved tetravalent ion [23].

The role of palladium can be due to its known catalytic property :

- its affinity to the π-allylic bond favours the formation of activated surface complexes with the carbon black ;

- in both Pd^{2+} and Pd^{0} forms, it is assumed that the palladium is a soft Pearson acid. Thus, it preferably reacts with a soft Pearson base such as the >C=C< bonds present in the carbon black. Among the four noble metals studied, Pt, Pd, Rh and Ir for the catalytic oxidation of graphite single crystals [24], the catalytic activity is indeed in the following order : Pd >> Pt > Ir > Rh.

CONCLUSION

In conclusion our results have shown that :

- the perovskite type oxides ($La_{0.8}Sr_{0.2}Mn_{1-x-y}B'_{x}\phi_{y}O_{3}$) are highly active in the catalytic combustion of diesel soot at temperature above 320°C (T_i) ;

- the catalytic activity is significantly influenced by the noble metal substitution (Pd, Pt, Ru) in B site and by the non-stoichiometry. The best performance is obtained by two samples ; $La_{0.8}Sr_{0.2}Mn_{0.999}Pd_{0.001}O_{3}$ and $La_{0.8}Sr_{0.2}Mn_{0.91}\phi_{0.09}O_{3}$.

- the enhancement of the activity is due to : i) the catalytic properties of Pd ions in the former catalyst and ii) the increase of Mn^{4+}/Mn^{3+} ratio in the latter.

REFERENCES

1 D.J. Ball and R.B. Stack, CAPOC II, Ed. A. Crucq, Elsevier, Amsterdam (1991) p. 337.
2 F. Indra, SAE Paper No. 885151 (1988).
3 V.D. Rao, J.B. White, W.R. Wade, M.G. Aimone and H.A. Cikanek, SAE Paper No. 850014 (1985).
4 M.R. Montierth, SAE Paper No. 840072 (1984).
5 R. Domesle, E. Koberstein, H.D. Pletka and H. Voelker, US Patent No.4,515,78 (1985).
6 B.J. Cooper and J.E. Thoss, SAE Paper No. 890404 (1989).
7. D.J. Ball and R.G. Stack, SAE Paper No. 902110 (1990).
8 Y. Watabe, K. Irako, T. Miyajima and T. Yoshimoto, SAE Paper No. 830083 (1983).
9 J. Widdershoven, F. Pischinger, G. Lepperhoff, K.P. Schicki, J. Strutz and S. Stahlhut, SAE Paper No. 860013 (1986).
10 A.F. Ahlström and C.U.I. Obenbrand, Appl. Catal., 60 (1990) 143.
11 A.F. Ahlström and C.U.I. Obenbrand, Appl. Catal., 60 (1990) 157.

12 P. Ciambelli, P. Parrella and S. Vaccaro, CAPOC II, Ed. A. Crucq, Elsevier, Amsterdam (1991) p. 323.
13 H. Makoto and S. Koichi, European Patent No 0397411 A_2 (1990).
14 R.J.H. Voorhoeve, Adv. Materials in Catalysis, Eds. J.J. Burton and R.L.Garten, Academic Press, London (1977).
15 L.G. Tejuca, J.L.G. Fierro and J.M.D. Tascon, Adv. in Catalysis, 36 (1989) 237.
16 M.S.G. Baythoun and F.R. Sale, J. Mater. Sci., 17 (1982) 2757.
17 Y. Teraoka, M. Yoshimatsu, N. Yamazoe and T. Seiyama, Chem. Lett., (1984) 893.
18 D.L. Trimm, Catalysis-Special Report of Royal Society of Chemistry, No. 4, London (1980).
19 N.M. Laurendeau, Prog. Energy Combust., 4 (1978) 221.
20 R.J.H. Voorhoeve, J.P. Remeika and L.E. Trimble, Ann. N.Y. Acad. Sci., 272 (1976) 3.
21 J.M.D. Tascon and L.G. Tejuca, Reac. Kinet.Catal.Lett., 15 (1980) 185.
22 E.G. Vrieland, J. Catal., 32 (1974) 415.
23 D.W. Johnson Jr., P.K. Gallagher, G.K. Werthem and E.M. Vogel, J. Catal., 8 (1977) 87.
24 R.T.K. Baker and R.D. Sherwood, J. Catal., 61 (1980) 378.
25 H.W. Kroto, A.W. Allaf and A.P. Balm, Chem. Rev., 91 (1991) 1213.
26 J. Lahaye and G. Prado, Soot in combustion and its toxiproperties, NATO Series VI, Plenum (1983).
27 J.B. Donnet and E. Custodero, Bull. Soc. Chim. Fr., 131 (1994) 115-117.
28 X. Bourat, Extended Abstracts 21th Biennal Conf. Carbon, ACS, 229 (1993).

ACKNOWLEDGEMENTS

The authors are indebted to Peugeot SA and STTS for financial support, and particularly thank Dr. Belot, Dr. Le Borgne and Mr. Foucaud for their contribution and helpful discussions.

Lean NO_x Catalyst Technologies

A. Frennet and J.-M. Bastin (Eds.)
Catalysis and Automotive Pollution Control III
Studies in Surface Science and Catalysis, Vol. 96

THE MECHANISM OF THE LEAN NO_X REACTION OVER Pt-BASED CATALYSTS

GP Ansell[a], SE Golunski[a], JW Hayes[a], AP Walker[a] and R Burch[b] and PJ Millington[b]

[a] *Johnson Matthey Technology Centre, Sonning Common, Reading RG4 9NH, UK*

[b] *University of Reading, Whiteknights, Reading RG6 2AD, UK*

ABSTRACT

It is now well established that Cu/ZSM-5, the first generation lean NO_x catalyst, is not suitable for widespread application on vehicles because it can undergo rapid and irreversible deactivation under real operating conditions. Pt-based catalysts offer an active and stable alternative to Cu/ZSM-5.

Steady state reactor experiments have demonstrated that Pt/Al_2O_3 catalysts are active for the lean NO_x reaction at temperatures as low as 200°C. Both N_2 and N_2O are generated by such catalysts. While there is no simple correlation between NO reduction activity and Pt surface area, there is certainly a very good inverse correlation between the maximum NO reduction activity and the temperature. The most effective catalysts for NO reduction are those which are active at the lowest temperatures.

Temporal Analysis of Products (TAP) has been used to obtain an in-depth mechanistic understanding of the lean NO_x reaction over a Pt/Al_2O_3 catalyst. The predominant mechanism for selective NO reduction involves the reduction of oxidised Pt sites by the hydrocarbon, followed by the decomposition of NO on these reduced Pt sites. A detailed model of the lean NO_x reaction over Pt/Al_2O_3 is presented which is capable of explaining all the results obtained in this work, as well as those reported in the literature.

1. INTRODUCTION

The emission of NO_x from both stationary and automotive sources has serious environmental implications. Ammonia can be used to reduce NO_x from stationary sources, but such an approach cannot easily be implemented on motor vehicles. The three-way automotive catalyst has been highly successful in controlling exhaust emissions from conventional petrol engines which operate under closed-loop conditions at a gas feed composition which is close to stoichiometric. However, the exhaust from diesel and from lean burn engines is substantially lean (ie net oxidising), containing over 5% oxygen. To reduce NO_x under these conditions, a new family of automotive catalyst must be developed.

Copper ion-exchanged into ZSM-5 zeolite was the first material to exhibit highly promising activity for the removal of NO under strongly oxidising conditions. Held et al [1,2] and Iwamoto et al [3] discovered that this Cu/ZSM-5 material could utilise small quantities of hydrocarbon to reduce the NO_x efficiently in such an environment.

Since these initial reports, a large number of investigations have been carried out, aimed at understanding which of the characteristics of the Cu/ZSM-5 catalyst are responsible for its high lean NO_x activity. Both the structure of the material and the nature of the reaction mechanism have been studied extensively. This paper will briefly discuss the salient features of the Cu/ZSM-5 catalyst, and will demonstrate that the Cu/ZSM-5 material deactivates rapidly under typical operating conditions. This observation, amongst others, necessitates the search for alternative lean NO_x catalysts.

One such alternative system comprises platinum-group metals supported on metal oxides. A number of groups have recently reported that such materials are active for NO_x reduction under strongly oxidising conditions [4-6]. However, none of these studies have addressed the fundamental questions pertaining to the mechanism of the selective reduction of NO on such catalysts, or to the nature of the catalytically active surface. In this paper, we describe the main features of the lean NO_x reaction over a series of Pt/Al_2O_3 catalysts. We also present an extensive study of the mechanism of this reaction over Pt-based catalysts.

2. EXPERIMENTAL

5% Cu/ZSM-5 was prepared by ion-exchanging the H-form of the zeolite (Si/Al = 50) using an aqueous solution of copper (II) ethanoate. The catalyst was dried (110°C, 16h), before being activated by calcination at 500°C under static air. The Pt/Al_2O_3 catalysts were prepared by incipient wetness impregnation using standard Pt precursors. After impregnation and drying, the catalysts were calcined in air at 500°C for 16 hours.

The Cu/ZSM-5 microreactor experiments were performed by heating a powdered sample (particle diameter <150mm) from 150 to 500°C at 10°C min^{-1} under a simulated exhaust gas. The gas mixture comprised 300ppm NO, 300ppm CO, 800ppm C_3H_6, 5% O_2, 10% H_2O, 13% CO_2 in N_2, and was flowed over the sample at 120 dm^3 h^{-1} g^{-1}. Nitric oxide conversion was monitored by a chemiluminescent analyser (Analysis Automation), while propene conversion was measured using a gas chromatograph (Perkin Elmer 8700, fitted with FID). The activity measurements on the Pt-based catalysts were carried out using a simplified gas mixture, containing 500ppm NO, 1000ppm C_3H_6, and 5% O_2 in nitrogen/helium. CO_2, N_2O, and C_3H_6 were monitored using a Perkin Elmer Sigma 4B gas chromatograph with a TCD, while NO and total NO_x (NO + NO_2) were analysed using a Signal NOX Analyser Series 4000 chemiluminescence detector.

Detailed investigation of the reaction mechanism was performed using a TAP reactor (Autoclave Engineers). Both the TAP apparatus and technique have been described in detail by the inventors [7,8], and only those aspects most relevant to this study will be included here. The system comprises four main features: (i) a valve assembly which permits the introduction of either a very narrow gas pulse or a continuous flow, (ii) a catalytic microreactor, (iii) a high vacuum system, (iv) a quadrupole mass spectrometer.

The continuous gas flow mode of operation allows one to evaluate the steady state performance of the catalyst and to pre-treat the catalyst prior to either pulsing experiments or temperature programmed desorption experiments. Two fast-acting pulsing valves are also incorporated into the valve assembly. These deliver gas pulses containing between 10^{13} and 10^{20} molecules with a full width at half height (FWHH) of ca. 200ms. These fast-acting valves can be used either singly or in tandem. Pump/probe experiments can be used to obtain unique mechanistic information, including reaction intermediate lifetimes. Within the pump/probe mode, the catalyst is initially treated with a reactant pulse from the first valve, and is later "probed" by a pulse of a different reactant from the second valve. Within this study, pump/probe experiments involving CO - NO ("pump species" - "probe species"), H - NO, C_3H_6 - NO, CO/O_2 - NO, and C_3H_6/O_2 - NO were performed. The mean gas pressure during the time that the "pump" species was in the reactor was typically 1 Torr (=133.3 Pa).

The time interval, dt, between the firing of the first and second pulsing valves is set by the operator and was varied from 0.01s to 10s in this study.

During most of the studies described here, a C_3H_6:O_2 ratio of 1:3 was employed. However, experiments were also carried out at other ratios ranging from 1:3 to 1:20. In each case the salient features of the reaction were found to be the same. The 1:3 data sets are preferred here bacause they gave rise to the highest nitrogen production yield within the following NO gas pulse. Therefore, the signal-to-noise characteristics of the data were optimum at this ratio, meaning that small changes in the yield would be readily seen and would be seen to be statistically significant.

Carbon monoxide and nitrogen both occur at mass 28. In some cases these species were differentiated by using isotopically labelled ^{15}NO as a reactant (which gave rise to $^{15}N_2$ at mass 30). In other cases they were differentiated by monitoring mass 12 - CO gives rise to a response at 12 amu, while N_2 does not. Carbon dioxide and nitrous oxide (both mass 44) were differentiated using the same procedures.

3. RESULTS AND DISCUSSION

3.1. Activity and Durability of Cu/ZSM-5

Figure 1 shows the nitric oxide and propene conversion traces obtained when the Cu/ZSM-5 catalyst was heated under the lean exhaust mixture. The NO conversion peaks at around 425°C, before falling away as the temperature is increased further. In addition, the propene conversion trace rises gradually as the temperature is increased; this type of light-off behaviour is typical of base metal containing systems. One of the problems associated with the Cu/ZSM-5 catalyst is its inability to combust the hydrocarbon completely - substantial amounts of carbon monoxide are generated during the light-off ramp, onsetting at ca. 300°C and peaking at 350°C [9]. However, in a practical application, this CO release could be dealt with by incorporating an oxidation catalyst downstream of the Cu/ZSM-5 catalyst, to oxidise the CO up to CO_2.

A far more serious limitation of the Cu/ZSM-5 catalyst system is illustrated by the data in Figure 2. These data were generated at Ford Motor Co. using a Johnson Matthey catalyst. Here we see that the durability of the catalyst is extremely poor under typical operating conditions. This lack of stability has also been reported by other workers [eg 10,11], and the cause of the deactivation has been postulated to be associated with both the migration and sintering of the Cu entities [11], and with the steam-induced dealumination of the parent zeolite [12]. However, whatever the cause of this deactivation, it is clear that unless the Cu/ZSM-5 catalyst can be greatly stabilised, it cannot be regarded as a serious contender for a lean NO_x catalyst under real operating conditions. (Since the catalyst deactivates fairly rapidly at 485°C, it is reasonable to expect that its deactivation will be accelerated by any transient increases in temperature on the vehicle - it is not unrealistic to say that these transients could, on a lean burn vehicle, take the catalyst temperature above 800°C. Therefore, a huge increase in the stability of the Cu/ZSM-5 catalyst would be necessary to turn it into a commercially realistic lean NO_x catalyst.)

3.2. Performance Characteristics of Pt/Al_2O_3 Catalysts

It is clear, therefore, that we need to investigate alternative lean NO_x catalysts. One such family of catalysts comprises the platinum group metals supported on metal oxides. Promising results from such systems have already appeared in the literature [5]. Here we restrict our attention to a series of Pt/Al_2O_3 catalysts of various loadings and prepared from various Pt precursors.

Figure 3 plots the steady state NO conversion against temperature for a series of Pt/Al_2O_3 catalysts of various Pt loadings. It is clear that as the peak activity moves to higher temperature so the maximum NO conversion decreases. This can be seen more clearly in Figure 4, where the maximum conversion of all of the catalysts

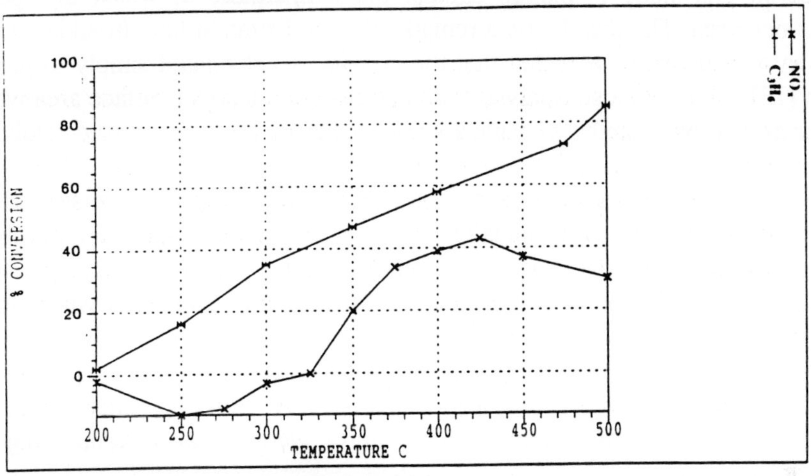

Fig.1 The NO and C_3H_6 conversion profiles obtained over Cu/ZSM-5 during temperature programed reaction under a full gas mix.

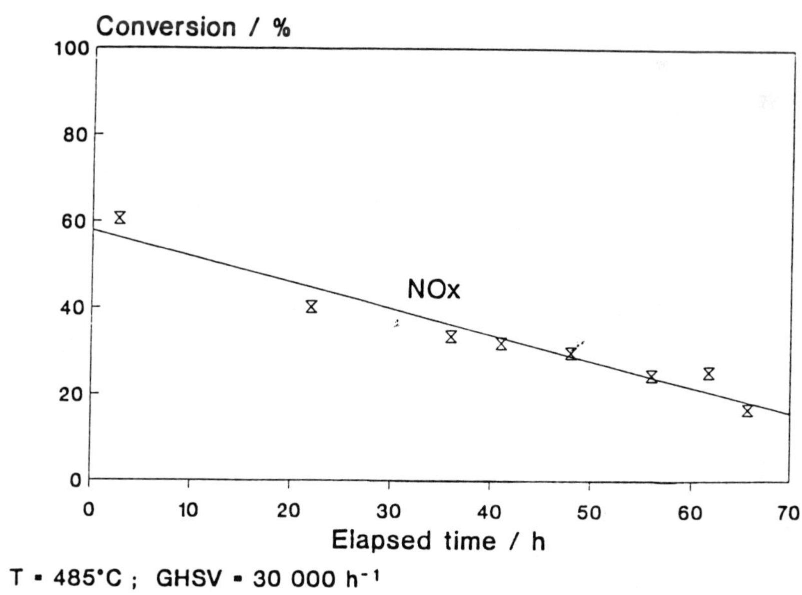

Fig.2 The long-term conversion performance of a Cu/ZSM-5 monolith sample. Data reproduced by kind permission of Dr H.S. Gandhi of Ford Motor Co.

investigated in this study is plotted against the temperature at which this peak conversion occurred. The data lie on a remarkably good straight line. In addition, Pt surface area measurements revealed that the activity is not related simply to metal dispersion [13], some of those catalysts with extremely high metal surface area were rather poor as lean NO_x catalysts, while some of the more poorly dispersed catalysts were highly active.

For all of the NO conversion traces shown in Figure 3, the propene conversion takes off as the NO conversion rises initially. Then, as the peak temperature for NO conversion is approached, the propene conversion reaches 100%. This coincidence of hydrocarbon light-off and NO_x conversion is a general phenomenon within lean NO_x catalysis.

The conversion of NO is not completely selective to N_2, some N_2O is also produced. Other authors have reported similar observations [4,5]. However, those catalysts which had NO conversions at higher temperatures were found to exhibit higher selectivity to N_2.

Li and Armor [14] have reported that supported Pt catalysts are poor N_2O decomposition catalysts. When we passed an $N_2O/O_2/C_3H_6$ mixture over the catalysts we found that less than 5% of the N_2O is decomposed to N_2 at temperatures below 450°C. Therefore, N_2O is not an intermediate on the way to N_2 under lean NO_x conditions.

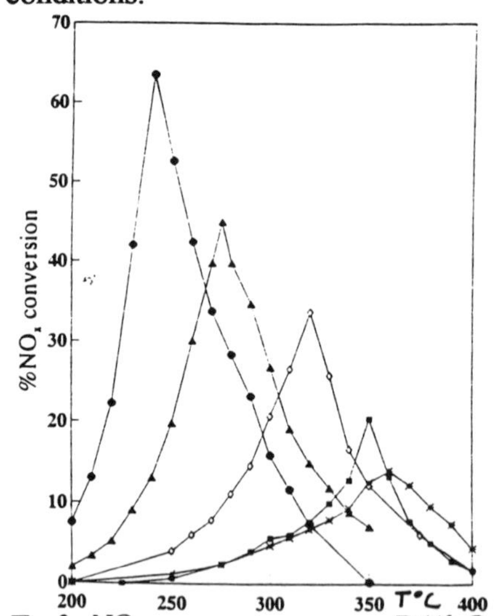

Fig.3 NO_x conversions over Pt/Al_2O_3 catalists as a function of temperature and Pt loading (✱0.1%Pt, ■0.25% Pt, ◇0.5% Pt, ●2% Pt)

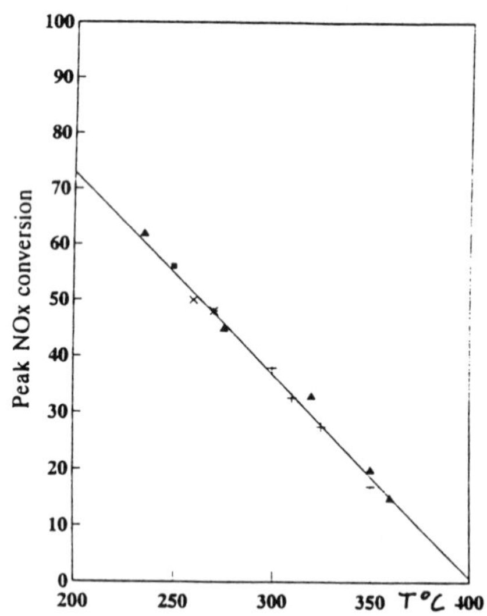

Fig.4 Variation of peak NOx conversion with peak conversion temperature for a range of Pt catalysts. The different symbols refer to catalysts made using different Pt precursors.

3.3. Mechanistic Studies over a 1% Pt/Al_2O_3 Catalyst

These microreactor experiments have clearly demonstrated that Pt-based catalysts are promising lean NO_x catalysts. The experiments have also provided one or two insights into the underlying mechanism. In order to obtain a full mechanistic understanding, a series of experiments were carried out on the TAP reactor using a 1% Pt/Al_2O_3 catalyst.

3.3.1. NO decomposition

First of all, the interaction of NO with the Pt/Al_2O_3 catalyst was studied. These experiments [13] revealed that reduced Pt/Al_2O_3 was capable of dissociating the NO molecule. Within these pulsing experiments, the nitrogen atoms rapidly recombined and desorbed to generate gas phase molecular nitrogen, while the oxygen atoms remained on the surface of the Pt (even at 400°C) and inhibited further NO decomposition. Indeed, the oxidised Pt/Al_2O_3 catalyst was not capable of decomposing NO under our conditions. These observations are in agreement with the results of earlier studies [15,16].

3.3.2. CO-NO and H_2-NO reactions

The CO-NO and H_2-NO reactions were investigated using the pump/probe capability of the TAP reactor. The results from the two studies were essentially equivalent [13], so only the CO-NO results will be discussed in any detail here. In an initial set of experiments, the reductant was pulsed over the (initially oxidised) catalyst at 400°C, followed one second later by a pulse of NO. Figures 5 and 6 show the responses at 28 amu and 44 amu during such an experiment. Several experiments were also performed to probe for the existence of a direct reaction between CO (or H_2) and NO. These were conducted by starting with a time interval (dt) between the CO (or H_2) pulse and the following NO pulse of 0.01s. The nitrogen yield was recorded. Then, dt was gradually increased from 0.01s to 10s. In each case, an identical N_2 yield was obtained, indicating that no direct reaction takes place between an incoming NO molecule and an adsorbed CO (or H) species. All of these results are completely consistent with the following reaction model:

$$2\ Pt + NO \text{ ----> } Pt\text{-}N + Pt\text{-}O \qquad (1)$$
$$2\ Pt\text{-}N \text{ ----> } 2\ Pt + N_2 \qquad (2)$$

followed by:

$$Pt\text{-}O + CO\ (H_2) \text{ ----> } Pt + CO_2\ (H_2O) \qquad (3)$$

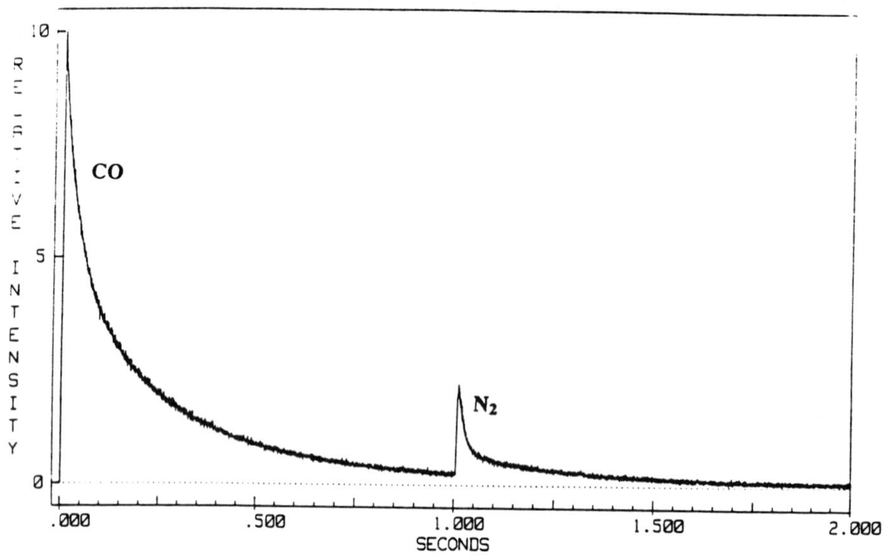

Fig.5 28 amu trace for the CO-NO reaction over oxidised Pt/Al_2O_3 at 400°C

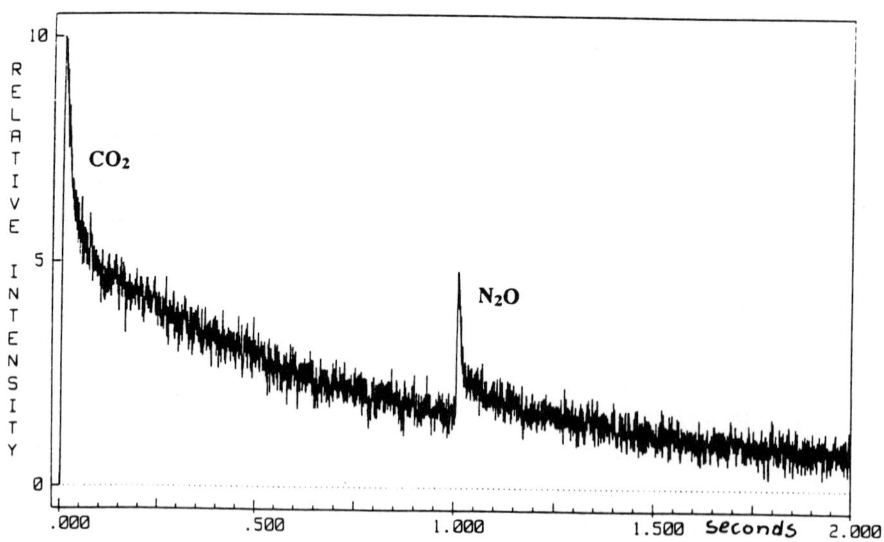

Fig.6 44 amu trace for the CO-NO reaction over oxidised Pt/ Al_2O_3 at 400°C

Other workers have postulated this sequence of reactions under conditions where the partial pressure of the reactants is low [17,18]. It should be noted that low levels of N_2O were generated during the NO pulse (see Figure 6), but the dominant N-containing product was N_2 under these conditions.

In a further set of experiments, the pulsing pressure of CO or H_2 (whichever was being used as the reductant) was increased in a gradual manner while the NO pulsing pressure was kept constant. In each case, NO was pulsed one second after the reductant pulse. The concentration of each of the reductants was increased until further increases in the size of the reductant pulse did not lead to any increase in the N_2 yield generated within the NO pulse. The corresponding yields of N_2 (shown in traces B and C of Figure 7) represent the maximum possible yield under conditions in which the reaction proceeds according to equations (1) to (3) above. The significance of these results is discussed below.

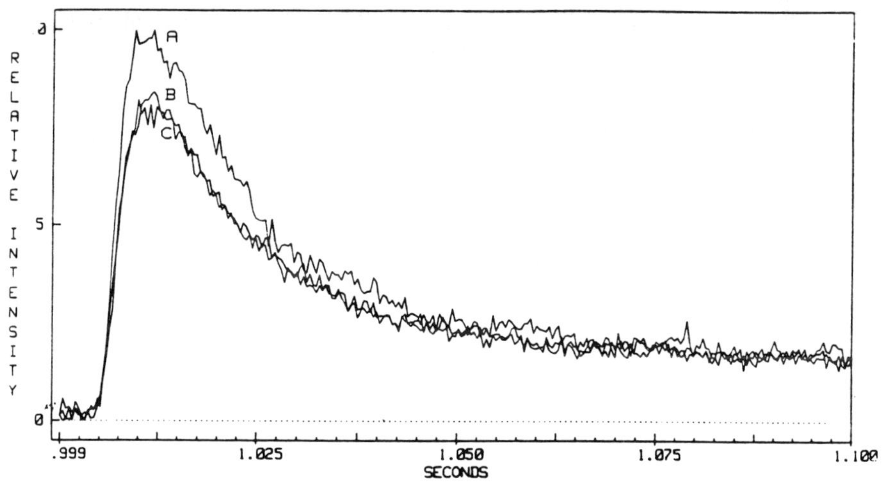

Fig.7 A comparison of the relative efficiences of C_3H_6/O_2, CO, and H_2 as reductants of NO; N_2 yield for (A) C_3H_6/O_2, (B) CO, (C) H_2

3.3.3. C_3H_6/O_2-NO reaction

The next series of experiments was designed to provide a model of the lean NO_x reaction. Therefore, the three principal components of the reaction were pulsed, C_3H_6 (representing hydrocarbons), NO and O_2. First of all, experiments similar to those described above for the CO-NO and H_2-NO reactions were performed. Trace A in Figure 7 shows the N_2 yield obtained from an experiment in which a C_3H_6/O_2 (1:3) feed was pulsed one second before the NO. It is clear that the vast majority of the N_2

generated within this experiment can be accounted for by a mechanism in which the hydrocarbon species reduces (or "cleans-off") the Pt surface, and the NO then decomposes over this reduced surface. Such a mechanism would be analogous to that outlined in equations (1) to (3) above. In addition, pump/probe experiments revealed that the N_2 yield within the NO pulse was independent of the value of the time interval, dt, between the C_3H_6/O_2 pulse and the NO pulse. This observation parallels those made in the CO-NO and H_2-NO experiments described above, and reinforces the suggestion that mechanistic similarities exist between the three reactions. However, Figure 7 does reveal that the N_2 yield in the C_3H_6/O_2 experiment is significantly higher than that in either the CO or H_2 experiments. It is possible that a second, minor reaction mechanism is responsible for this higher yield in the C_3H_6/O_2 experiment. Indeed, a small amount of CO_2 was released during the NO pulse in this experiment; this contrasts with the behaviour within the experiments in which CO and H_2 were used as reductants, where no CO_2 was observed within the NO pulse. Therefore, it seems likely that there are two mechanisms occurring when a C_3H_6/O_2 feed is used as the reductant - the first (predominant) sequence is analogous to that outlined in equations (1) to (3) above, while the second (minor) mechanism involves the reaction between a deposited carbon-based species and an NO (or NO_2) species. It is this latter mechanism which gives rise to the low level of CO_2 observed within the NO pulse in the C_3H_6/O_2-NO experiment.

Activation energy measurements for the lean NO_x reaction over Pt-based catalysts support the "HC clean-off, NO decomposition" model described above.

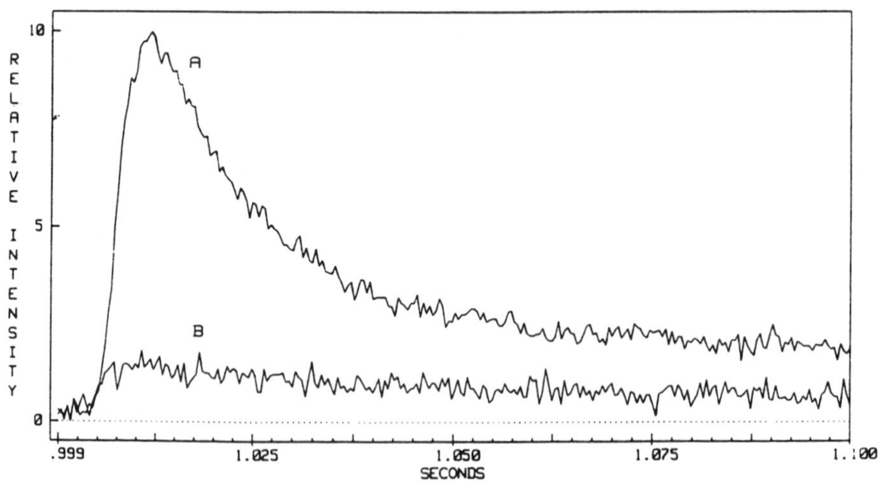

Fig.8 Effect of O_2 on the CO/NO reaction; N_2 yield for (A) CO-NO, (B) CO/ O_2(1/1)-NO

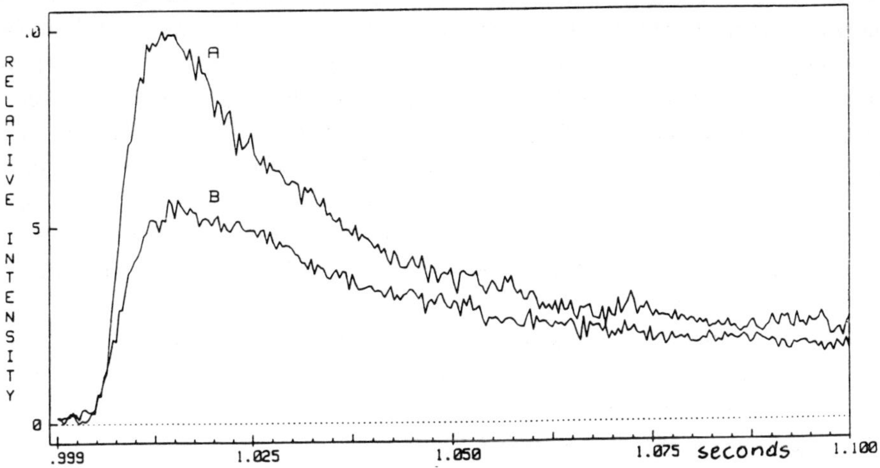

Fig.9 Effect of $P(O_2)$ on the C_3H_6/O_2-NO reaction; N_2 yield for (A) C_3H_6/O_2 (1/3), (B) C_3H_6/O_2 (1/9)

The measured activation energy value of 69 kJ/mol [19] falls within the range from 55 to 85 kJ/mol which other workers have reported for the dissociation of NO over Pt [20,21].

Therefore, it can be seen that propene can effectively reduce NO even when O_2 is present. In contrast, Figure 8 reveals that the addition of O_2 into the CO feed has a devastating effect on the N_2 yield within the following NO pulse. The nitrogen yield is virtually zero when a CO/O_2 (1/1) gas mix is used. Figure 9 demonstrates that the further addition of O_2 to the C_3H_6/O_2 gas feed has a much smaller effect on the resulting N_2 yield. Note that the reductant-containing pulses in trace B of Figures 8 and 9 have exactly the same net-oxidising power ($CO/O_2 = 1/1$, $C_3H_6/O_2 = 1/9$).

So, it is clear that propene is an effective reductant of NO in the presence of oxygen, while CO is an effective reductant only in the absence of oxygen. These observations parallel those made within microreactor experiments, where propene is found to be an effective lean NO_x reductant over Pt-based catalysts, while CO is not [22].

3.3.4. The mechanism

Figure 10 illustrates the individual steps in the lean NO_x reaction mechanism over Pt-based catalysts. Within this mechanism the hydrocarbon reduces a patch of Pt atoms from Pt-O to Pt metal. NO dissociation then takes place on these reduced Pt sites. This is the mechanism which predominates over Pt-based catalysts.

Once NO dissociation has been initiated, there are two ways in which the N atoms generated by the dissociation can be removed. At low temperatures, we expect that most of the NO on the Pt surface will be present as molecular NO. As the temperature rises, so does the rate of NO dissociation [18, and refs therein] and the rate of NO desorption [13]. So, at low temperatures, where the surface will accommodate a low concentration of N adatoms and a high concentration of molecular NO, a significant reaction product is expected to be N_2O, formed by the reaction of adsorbed N adatoms with adsorbed molecular NO. At higher temperatures, where the rate of NO dissociation is higher (leading to a higher concentration of N adatoms) and the rate of NO desorption is higher (leading to a lower surface concentration of molecular NO), it is expected that N_2, formed by N-N recombination, will be the predominant product. This explains why Pt catalysts generate both N_2 and N_2O. It also explains why those catalysts which have peak NO conversion at higher temperature exhibit higher selectivity to N_2 rather than N_2O. In addition, the importance of NO dissociation in the lean NO_x mechanism is reinforced by the activation energy data mentioned above.

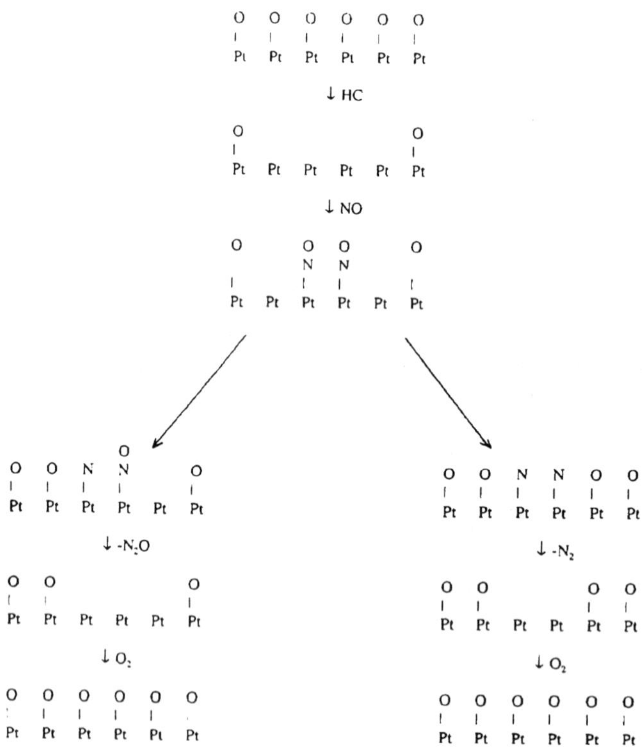

Fig.10 The mechanism of NO reduction by hydrocarbons over Pt-based catalysts.

We have demonstrated that propene is an effective lean NO_x reductant, while CO is not. This observation can be rationalised in terms of the size of the reduced patch of Pt atoms generated by the combustion of one C_3H_6 molecule and one CO molecule. Each propene molecule can remove 9 oxygen atoms from the Pt surface, generating a large patch of reduced Pt atoms on which NO decomposition can readily occur. (Note that the decomposition of each NO molecule requires the existence of at least two adjacent, reduced Pt sites.) In contrast, however, each CO molecule can only remove one oxygen atom from the Pt surface, which may allow an NO molecule to adsorb, but will not lead to the NO decomposition required for N_2 generation.

Clearly, once the propene molecule has generated a reduced patch of Pt atoms, there will be a competition between NO adsorption/ dissociation and the re-oxidation of the sites by molecular oxygen dissociation. The experimental evidence strongly suggests that as the temperature is increased we move from a region in which NO adsorption/ dissociation is favoured (low temperature), to one in which re-oxidation by O_2 is preferred (high temperature). Such a hypothesis explains why the peak NO activity decreases as the temperature of peak NO conversion increases.

CONCLUSIONS

The data generated within this study indicate that the predominant mechanism by which lean NO_x conversion occurs over Pt-based catalysts comprises two basic steps: reduction of the Pt sites by the hydrocarbon, followed by the decomposition of NO (to generate N_2 and adsorbed O species) over these reduced Pt sites. We believe that this mechanism can account for all of the results presented here, and those so far published in the literature. The mechanism leads us to predict that it will be difficult to produce a platinum-based catalyst that will exhibit significant lean NO_x activity at temperatures above about 350°C.

ACKNOWLEDGEMENTS

We are grateful to Dr H S Gandhi for allowing us to use Figure 2, and to Johnson Matthey plc for permission to publish this work.

REFERENCES

1 W. Held and A. Koenig, Ger. Offen, DE 3 642 018 (1987) to Volkswagen AG.
2 W. Held, A. Koenig, T. Richter and L. Puppe, SAE Paper No. 900496 (1990).
3 M. Iwamoto, H. Yahiro, S. Shundo, Y. Yu-u and N. Mizuno, Shokubai (Catalyst), 33 (1990) 430.
4 G. Zhang, T. Yamaguchi, H. Kawakami and T. Suzuki, Appl. Catal. B:Env., 1 (1992) L15.
5 A. Obuchi, A. Ohi, M. Nakamura, A. Ogata, K. Mizuno and H. Ohuchi, Appl. Catal. B:Env., 2 (1993) 71.
6 M. Sasaki, H. Hamada, Y. Kintachi and T. Ito, Catal. Lett., 15 (1992) 297.
7 J.R. Ebner and J.T. Gleaves, (Monsanto Co.) European Patent Application, EP 0266334A3 (1988).
8 J.T. Gleaves, J.R. Ebner and T.C. Keuchler, Catal. Rev.-Sci. Eng., 30 (1988) 49.
9 C.J. Bennett, P.S. Bennett, S.E. Golunski, J.W. Hayes and A.P. Walker, Appl. Catal. A, 86 (1992) L1.
10 D.R. Monroe, C.L. DiMaggio, D.D. Beck and F.A. Matekunos, SAE Paper No. 930737, (1993).
11 K.C.C. Kharas, H.J. Robota and D.J Liu, Appl. Catal. B:Env., 2 (1993) 225.
12 R.A. Grinsted, H.W. Jen, C.N. Montreuil, M.J. Rokosz and M. Shelef, Zeolites, 13(8) (1993) 602.
13 R. Burch, P.J. Millington and A.P. Walker, Appl. Catal. B:Env., accepted for publication.
14 Y. Li and J.N. Armor, Appl. Catal. B:Env., 1 (1992) L21.
15 S. Pancharatnam, K.J. Lim and D.M. Mason, Chem. Eng. Sci., 30 (1975) 781.
16 K.J. Lim, D.G. Loffler and M. Boudart, J. Catal., 100 (1986) 158.
17 B.A. Banse, D.T. Wickham and B.E. Koel, J. Catal., 119 (1989) 238.
18 R.I. Masel, Catal. Rev.-Sci. Eng., 28 (1986) 335.
19 C.J. Bennett and P.S. Bennett, unpublished results.
20 M. Shelef, K. Otto and H.S. Gandhi, Atm. Environ., 3 (1968) 107.
21 K. Otto and H.C. Yao, J. Catal., 66 (1980) 229.
22 G.P. Ansell and J.W. Hayes, unpublished results.

A. Frennet and J.-M. Bastin (Eds.)
Catalysis and Automotive Pollution Control III
Studies in Surface Science and Catalysis, Vol. 96

INFLUENCE OF THE COPPER DISPERSION ON THE SELECTIVE REDUCTION OF NITRIC OXIDE OVER Cu/Al_2O_3 : NATURE OF THE ACTIVE SITES.

Z. Chajar[a], M. Primet[a], H. Praliaud[a], M. Chevrier[b], C. Gauthier[c] and F. Mathis[d].

[a] *Laboratoire d'Application de la Chimie à l'Environnement, Université Claude Bernard LYON I, 43 bd du 11 Novembre 1918, 69622 Villeurbanne Cedex, France*

[b] *Renault Automobiles, Direction Etudes Matériaux, 92109 Boulogne Billancourt Cedex, France.*

[c] *Renault Automobiles, Centre de Lardy, 1 allée Cornuel, 92510 Lardy, France.*

[d] *Renault Automobiles, Direction de la Recherche, 9-11 Avenue du 18 Juin 1940, 92500 Rueil Malmaison, France.*

ABSTRACT

Cu/Al_2O_3 solids with various Cu loadings, between 0.3 and 6.4 wt %, are used for the reduction of NO by propane in the absence and in the presence of oxygen (up to 10 vol. %) in the 423 - 773 K temperature range. At a given temperature and for high Cu loadings, the introduction of oxygen induces a decrease in the activity in nitrogen formation. For low Cu loadings the activity increases with the oxygen content in the 1-2 vol. % range, then slightly decreases for higher oxygen amounts. The nature of the Cu species accessible to CO and NO is determined by infrared spectroscopy. High Cu loadings favor the formation of bulk oxides at the surface of the support whereas low Cu loadings favor the formation of isolated Cu species. The selective reduction of NO is thus related to the presence of these isolated copper species easily reduced and reoxidized.

1. INTRODUCTION

Iwamoto et al. [1, 2, 3] and Held et al. [4] independently reported that the selective catalytic reduction of NO by hydrocarbons, such as ethene, propene, propane, is possible in an oxidizing atmosphere on copper ion-exchanged ZSM-5 zeolites at temperatures as low as 473 - 673 K. Addition of oxygen to the reactants mixture is necessary to achieve the selective reduction of NO at high conversion levels. Afterwards many catalysts, zeolitic and non-zeolitic, has been found to be also active : proton-exchanged zeolites [5], cerium [6] or iron [7] or gallium [8] ion-exchanged zeolithe, alumina, silica and silica-alumina [9, 10, 11]. For non-zeolitic solids, the selective reduction of NO by hydrocarbons in the presence of oxygen has been considered to occur over solid acid catalysts such as alumina, silica-alumina, titania, zirconia [10] and sulfate promoted metal oxides such as titania, zirconia, ferric oxide [12].

In the case of alumina (or other simple oxides), the reaction occurs at high temperatures and under low space velocity conditions. The activity was found to be improved by the addition of platinum group metals [13, 14] and of transition metal oxides [15], especially copper [16, 17, 18]. For alumina-supported copper oxide catalysts a maximum effect has been found by the addition of 0.3 wt % Cu and it has been considered that, for higher copper contents, the formation of cupric oxide would give a solid selective for the oxidation of the hydrocarbon by oxygen [16]. In the case of alumina-supported Cu-Cs oxide catalysts the formation of an isocyanate species has been evidenced by exposition to mixtures "nitric oxide/oxygen/propene (or acetylene)" but not with propane [18, 19]. In fact the mechanism of the reaction and the nature of the active sites are still unknown.

In this paper we report a study of the activity of Cu/Al_2O_3 solids with various copper contents towards the reduction of NO by propane in the presence of oxygen. The nature of the superficial copper species is determined by infrared spectroscopy of adsorbed CO and NO. The results are compared to the catalytic activity.

2. EXPERIMENTAL

2.1. Preparation of the catalysts

Four Cu/Al_2O_3 solids with various copper loadings (0.3, 1.7, 3.2 and 6.4 wt. Cu %) were prepared by impregnation of an alumina from Degussa (BET area 100 $m^2\ g^{-1}$) with aqueous solutions of copper nitrate. The solids were thus dried at 383 K and calcined under oxygen at 773 K (heating rate 1 K min^{-1}) overnight.

The DRX patterns were characteristic of the gamma alumina support and the CuO phase was only detected for the solid containing 6.4 wt % copper, but not for lower copper amounts.

2.2. Catalytic activity measurements

Catalytic activity measurements were carried out with a fixed bed flow reactor. The reactor was a quartz tube. The flow rates were adjusted using Brooks mass flow controller units. The composition of the effluents was analyzed by gas chromatography using a dual CTR1 column from Alltech (porapak for CO_2, N_2O, molecular sieve for O_2, N_2, CO) with a thermal conductivity detector and a porapak Q column with a flame ionisation detector for hydrocarbons. The mixture was analyzed every 13 minutes. NO and NO_2 amounts were measured continuously on-line by means of Rosemount Infrared Analyzers. Helium was used as carrier gas as well as diluent gas.

100 mg of catalyst were charged with 400 mg of diluent (inactive low surface area α Al_2O_3) and the total flow rate (reactants in He) was 10 $dm^3 \cdot h^{-1}$ (gas hourly space velocity of 50 000 h^{-1} for the catalyst considering a density of ca. 0.5 g cm^{-3}).

The experiments were carried out into two steps :

- the calcined solids were contacted at room temperature with a C_3H_8 (2000 vpm) - NO (2000 vpm) mixture free of oxygen (reducing mixture). The reaction temperature was first increased from 298 to 773 K with a ramp of 4 K min^{-1} and then decreased from 773 to 298 K.
- Oxygen was thus admitted at a given temperature, and for each temperature, the oxygen content was progressively increased in the 0.2 - 10 vol. % range.

NO may be converted into N_2O, N_2, NO_2. In fact, substantial amounts of NO_2 were formed because of the gas phase reaction at room temeperature between NO and oxygen in the pipes of the apparatus, before the reactor and in the exit lines from the reactor, as already noticed [20, 21]. In the present study, the catalytic activity only concerns the conversion of NO into nitrogen.

2.3. Infrared spectroscopy of CO and NO adsorptions

The nature of the superficial copper species accessible to CO and NO was deduced from infrared spectroscopy measurements.

The samples were pressed in order to obtain thin discs of known weight between 15 and 40 mg. The discs were placed in a sample holder made of quartz and introduced in a cell that allowed "in situ" treatments. Prior to IR measurements the samples were heated at 773 K for 1 h under oxygen (80 Torr), then degassed at 773 K for 1 h. After cooling to 298 K the background spectrum was recorded, then carbon monoxide (around 50 Torr) or nitrogen monoxide (around 10 Torr) was introduced at 298 K, the contact time varying between 0 and 20 h. The gaseous phase was evacuated at the same temperature between 10 s and 3 h.

Infrared absorbance spectra were recorded at room temperature on a FTIR spectrometer (I.F.S. 110 from BRUKER) with a resolution of 4 cm^{-1}.

3. RESULTS AND DISCUSSION

3.1. Activity in the absence of oxygen (C_3H_8-NO-He mixture)

First of all, it has been checked that the conversion does not depend on the particle size in the 20 - 500 μm range.

In the absence of oxygen, the NO conversion into N_2 increases with the temperature and the copper content but remains weak. It begins at around 623 K with the solid containing 6.4 wt. % Cu. It reaches 2, 11, 36 and 40 % for the solids containing 0.3, 1.7, 3.2 and 6.4 wt. % Cu respectively. The activity expressed per gram of introduced copper is approximatively independent of the copper content (between 130 * 10^7 and 250 * 10^7 moles NO converted into N_2 s^{-1} g^{-1} Cu). A small activation under the reactants mixture has been observed and the values here reported are obtained during the decrease in temperature, from 773 to 298 K.

CO_2 and N_2 are the only products detected. The C_3H_8 conversion corresponds to the NO reduction according to the reaction :

$$C_3H_8 + 10\ NO \rightarrow 5\ N_2 + 3\ CO_2 + 4\ H_2O \qquad \text{(reaction 1).}$$

For comparison a Cu-ZSM5 solid prepared by impregnation of a ZSM5 zeolite (Si/Al = 18.9) by an aqueous solution of copper nitrate and containing 2.6 wt. % Cu is able to convert, at 773 K, 20 % of NO into N_2, value corresponding to a rate of 173 * 10^7 mol. s^{-1} g^{-1} Cu (22).

Let us remark that, in the absence of oxygen, the reduction of NO by C_3H_8 is weaker than the reduction of NO by CO. As an example, with a mixture NO (2000 vpm) - CO (2000 vpm) - He, for the 6.4 wt % Cu solid, the NO conversion reaches 100 % at 746 K and the light-off temperature is 553 K during the decrease in temperature, between 773 and 298 K. Propane is a less efficient reducing agent than carbon monoxide in spite of the more reducing character of the mixture (ratio NO/CO = 1 and ratio NO/10 C_3H_8 = 0.1).

3.2. Activity in the presence of O_2 (C_3H_8-NO-O_2-He mixture)

3.2.1. NO conversion into N_2

The effect of the oxygen addition on the NO conversion into N_2 at 773 K is shown on the figure 1. This temperature has been selected in order to obtain a good activity without oxygen. Two behaviours are observed according to the copper contents.

For the solids containing 3.2 and 6.4 wt. % Cu the activity decreases as soon as oxygen is introduced and decreases continuously for higher oxygen contents.

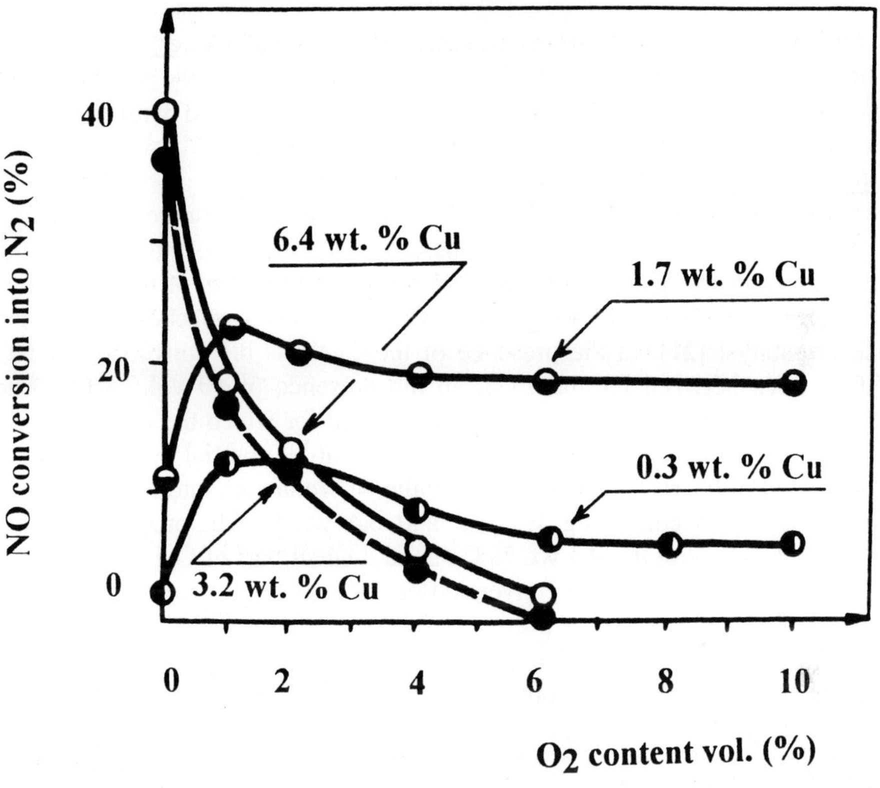

Figure 1. Conversion of NO into nitrogen at 773 K as a function of the oxygen content for the three Cu/Al_2O_3 catalysts

On the contrary, for the solids containing 0.3 and 1.7 wt. % Cu the NO conversion into N_2 is enhanced by oxygen addition. It goes through a maximum in the 1-2 vol. % O_2 range, but remains relatively important even with a large excess of oxygen. With 10 vol. % O_2 and with the 1.7 and 0.3 wt % Cu solids, the NO conversion is equal to 19 and 7 % respectively. This enhancement is lower than the one observed with Cu-ZSM5 catalysts [22]. Let us recall that, at 623 K, for a Cu-ZSM5 solid prepared by impregnation and containing 2.6 wt. %

Cu, the NO conversion into N_2 reached 78 % in the presence of 0.5 vol. % O_2 and is equal to 25 % in the presence of 10 vol. % O_2.

It may be noticed that, with CO as a reducing agent, the introduction of oxygen leads to a drastic decrease in NO conversion whatever the copper content.

3.2.2. Activity for the N_2 formation expressed per gramm of copper.

The activities expressed per mole of NO converted into N_2 per gramm of copper and per second are reported in the figure 2. The solid with the lower copper content is the more active one.

3.2.3. Main products and C_3H_8 conversion

The main products are N_2, NO_2 and CO_2., N_2O was never detected.

The conversion of NO into NO_2 increases significantly with the oxygen content and is performed by homogeneous reaction in the pipes of the apparatus. Let us recall that with 10 vol. % O_2 this conversion into NO_2 reaches 25 % in the absence of catalyst [21]. In the presence of the catalysts the conversion of NO into NO_2 varies between 25 and 17 % in the presence of 10 vol. % O_2. The lowest value is obtained with the 1.7 wt. % Cu solid for which the conversion of NO into N_2 is the highest. For this solid, the residual NO partial pressure, in the pipes at the exit of the reactor, is lowered by the NO reduction into N_2.

With the solids containing 1.7, 3.2 and 6.4 wt. % Cu, propane is fully oxidized into CO_2. With the 0.3 wt. % Cu solid a small part of C_3H_8 is oxidized into CO at 773. For instance the conversion of C_3H_8 into CO reaches 4 % for an overall conversion of 19 % in the presence of 10 vol. % O_2.

The C_3H_8 conversion continuously increases with the temperature and the oxygen concentration. Propane conversion and NO reduction begin at the same temperature. However this C_3H_8 conversion exceeds the conversion which would be due to the reaction with NO (reaction 1). For instance in the presence of 10 vol. % O_2, with the 0.3 and 1.7 wt. % Cu solids, when the conversion of NO into N_2 is equal to 7 and 19 %, the C_3H_8 overall conversion reaches 19 and 60 %, respectively, showing the importance of the following reaction :

$$C_3H_8 + 5\ O_2 \quad \rightarrow \quad 3\ CO_2 + 4\ H_2O \text{ (reaction 2).}$$

For the 6.4 wt % Cu solid the C_3H_8 conversion reaches 89 %.

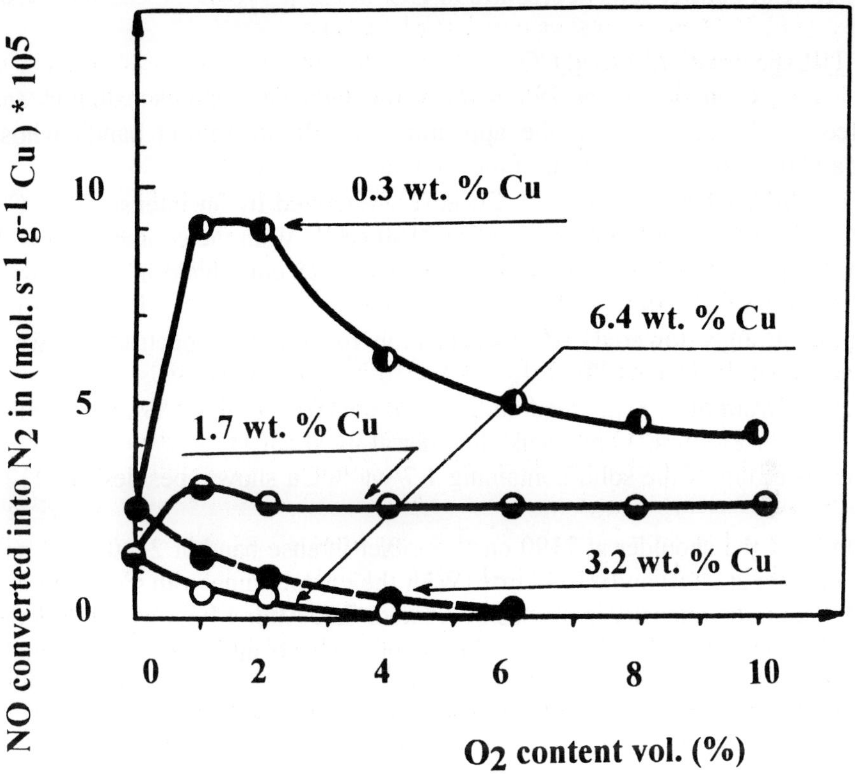

Figure 2. Rate of reduction of NO into N_2 (mol. s^{-1} g^{-1} Cu) at 773 K as a function of the oxygen content for the three Cu/Al_2O_3 catalysts.

The C_3H_8 conversion is never complete. However the maximum in NO conversion observed when the oxygen content increases is not due to a deactivation of the catalyst because additional runs with low oxygen contents lead to NO conversion similar to those of the first run. The decrease may be due to the influence of the propane partial pressure.

3.3. Superficial copper ions accessible to CO

3.3.1. FTIR spectra of adsorbed CO

The admission of CO at 298 K onto the Cu/Al_2O_3 solids calcined and evacuated at 773 K results in the appearance of IR absorption bands whose number and positions are function of the Cu loading (figure 3).

The solid containing 6.4 wt % Cu is characterized by an intense IR ν CO band at 2125 cm^{-1} and a weak shoulder at 2190 cm^{-1}. With the contact time with CO (between 0 and 20 hours) the intensity of the 2125 cm^{-1} band increases by a factor 1.3. Upon evacuation at 298 K the species show a poor stability since the intensity is strongly lowered. After evacuation for 1 h the spectrum shows a unique and weak band at 2120 cm^{-1}.

The spectrum of CO adsorbed on the solid containing 3.2 wt % Cu is similar (intense band at 2125 cm^{-1} and weak shoulder at 2190 cm^{-1}).

The spectrum of the solid containing 1.7 wt % Cu shows, besides the 2125 cm^{-1}
νCO band and the shoulder at 2190 cm^{-1} another intense band at 2110 cm^{-1} and weak shoulders at 2160 and 2165 cm^{-1}. With the contact time with CO, between 0 and 20 hours, the intensity increases by a factor 1.7. After 1 hour evacuation the spectrum shows only the 2120 and 2110 bands whose intensities are strongly decreased.

The spectrum of the 0.3 wt % Cu is more complex showing the presence of an intense band at 2160 cm^{-1} besides the bands at 2110, 2125, 2135 cm^{-1} and the shoulder at 2200 cm^{-1}. The 2110 cm^{-1} band is the more intense one. With the contact time with CO the intensity increases by a factor 1.1. Upon evacuation at 298 K the intensity of all the bands is strongly decreased. After 1 hour evacuation the spectrum shows bands at 2145, 2155 and 2110 cm^{-1}; the 2110 cm^{-1} band remains the more intense one.

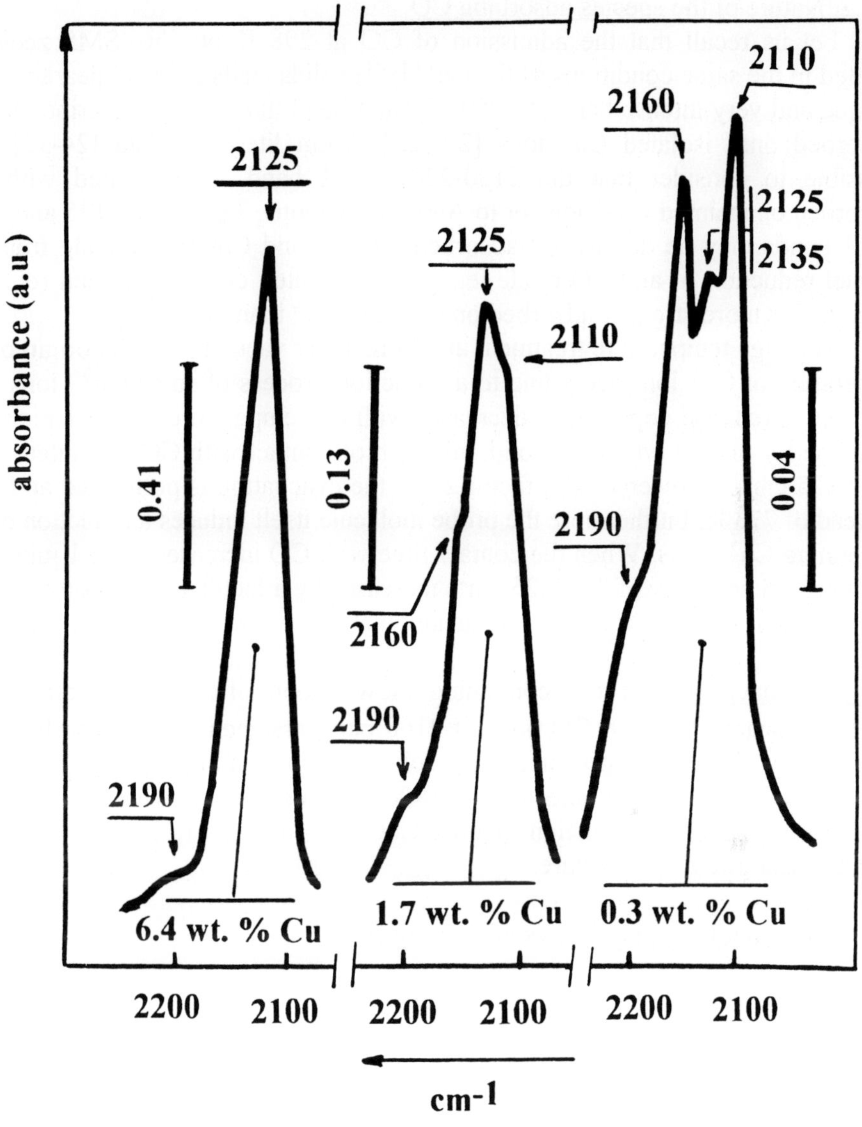

Figure 3. FTIR spectra of CO adsorbed on Cu/Al_2O_3 samples (previously calcined under O_2 and evacuated at 773 K). 50 Torr CO pressure, 1 h at 298 K.
6.4 wt. % Cu : 38 mg pellet
1.7 wt. % Cu : 32 mg pellet
0.3 wt. % Cu : 39 mg pellet

3.3.2. Nature of the species adsorbing CO

Let us recall that the admission of CO at 298 K on Cu-ZSM5 zeolithes treated in the same conditions as the Cu/Al_2O_3 solids leads to the appearance of a unique and very intense band at 2150-60 cm^{-1} band that we have ascribed to CO adsorbed onto isolated Cu^+ ions [22, 23]. From literature data [24-32] it is possible to consider that the 2190-2200 cm^{-1} band is associated with CO adsorbed on isolated Cu^{2+} ions or to Al^{3+} surface ions. The 2135, 2125 and 2110 cm^{-1} bands may be due to CO adsorbed on Cu^+ and Cu^0 ions arising from the partiel reduction of an CuO oxide ie., on non-isolated copper Cu^+ and (or) Cu^0 ions. CO is more strongly adsorbed on Cu^0 and Cu^+ than on Cu^{++}.

The high temperature treatment at 773 K under vacuum is at the origin of the generation of Cu^+ ions according to a reduction process of some Cu^{2+} ions. The number of reduced copper ions decreases with the temperature of evacuation. For instance for the 6.4 wt. % Cu solid, after 1 h of contact with CO, the intensity of the 2125 cm^{-1} is lowered by a factor 2.1 if the evacuation is performed at 423 K instead of 773 K. Furthermore the probe molecule itself induces a reduction of the remaining Cu^{2+} ions. When the contact time with CO increases from 1 minute to 20 hours the intensity of the 2125 cm^{-1} increases by a factor 1.35 after evacuation at 773 K and by a factor 1.8 after evacuation at 423 K.

3.3.3. In conclusion the solid containing a low amount of copper is characterized by the appearance of a υ CO band at 2160 cm^{-1} assigned to CO adsorbed onto isolated Cu^+ ions. The spectrum of the solid containing a high amount of copper, for which CuO has been detected by DRX, is characteristic of CO adsorbed on Cu^+ arising from the reduction of bulk CuO. CO appears to be able to reduce Cu^{++} ions at room temperature.

3.4. Superficial copper species adsorbing NO

3.4.1. FTIR of adsorbed NO

The admision of NO on the solids calcined and evacuated at 773 K results in the appearance, besides the bands of gaseous NO (1875 cm^{-1}), of gaseous N_2O (2215 and 2240 cm^{-1}) and of nitrates and nitrites on the support (1480-1665 cm^{-1} range) of two absorption doublets, a intense one (1890-95 and 1870-60 cm^{-1} bands) and a weak one (1765-60 and 1740 cm^{-1} bands) due to NO adsorbed onto Cu^{n+} ions.

Figure 4 shows the spectra of NO adsorbed at 298 K on the two solids 6.4 and 0.3 wt. % Cu calcined and evacuated at 773 K, after substraction of the gas phase.

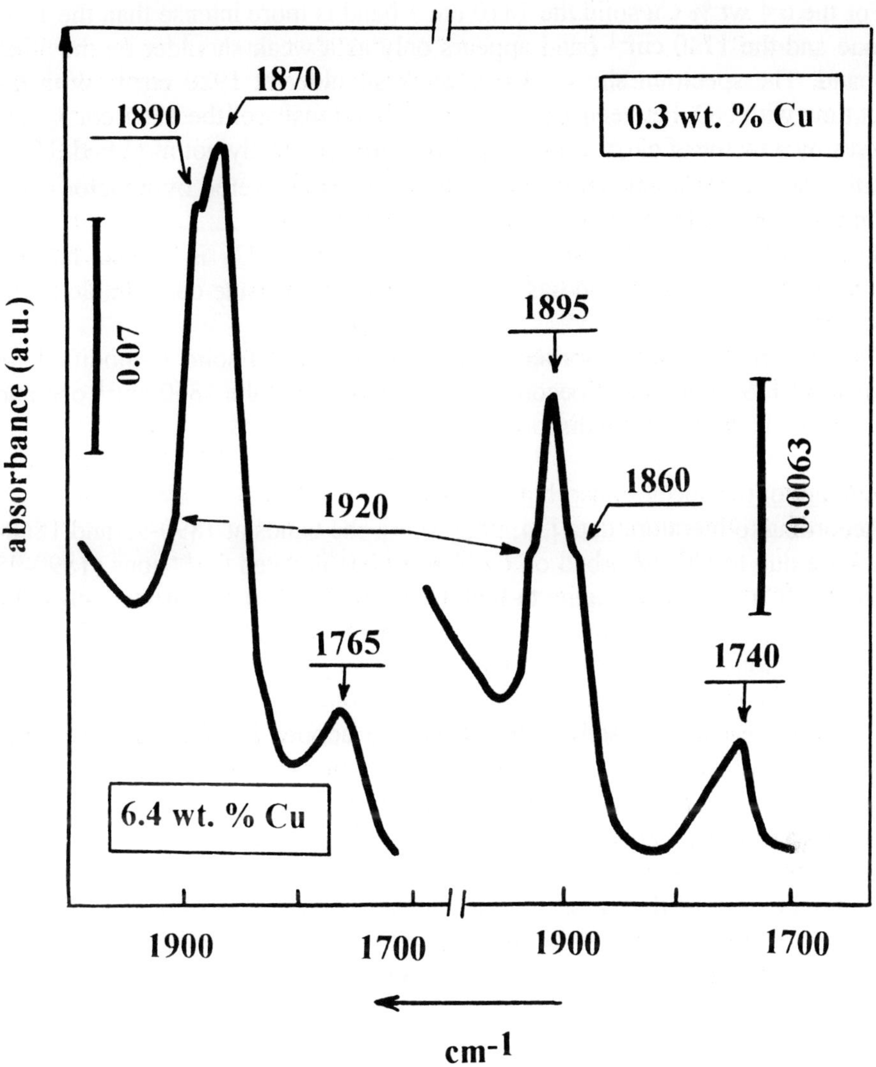

Figure 4. FTIR spectra of NO adsorbed on Cu/Al_2O_3 samples (previously calcined under O_2 and evacuated at 773 K). 10 Torr NO at 298 K, the gas phase has been substracted 6.4 wt. % Cu : 30 mg pellet 0.3 wt. % Cu : 24 mg pellet

For the 6.4 wt % Cu solid the 1870 cm^{-1} band is more intense than the 1890 cm^{-1} one and the 1740 cm^{-1} band appears only as a weak shoulder on the 1765 cm^{-1} band. The spectrum shows also a weak shoulder at 1920 cm^{-1}. With the contact time with NO, between 1 mn and 1 h, the intensity of the 1765 cm^{-1} band decreases by a factor of ca. 5. The other bands are apparently not modified. Upon evacuation at 298 K the intensities of all the bands are lowered by a factor ≅ 10. NO is not strongly held on the Cu^{n+} ions.

The spectrum of NO adsorbed on the 1.7 wt. % Cu is similar but the intensity of the 1740 cm^{-1} band has increased and has become comparable to the intensity of the 1760 cm^{-1} band.

The spectrum of NO adsorbed on the 0.3 wt. % Cu solid is modified. In particular the 1895 cm^{-1} band becomes more intense than the 1860 cm^{-1} one and the 1760 cm^{-1} band pratically disappears.

3.4.2. Nature of the species adsorbing NO

According to literature data [24, 28, 33-36], the bands at 1890-95 and 1860-70 cm^{-1} are due to NO adsorbed on Cu^{++} ions : on isolated Cu^{++} ions (1890-95 cm^{-1}) and on Cu^{++} ions belonging to bulk CuO (1860-70 cm^{-1}) respectively. The bands at 1760-65 and 1740 cm^{-1} are characteristic of NO adsorbed onto Cu^{+} ions : isolated Cu^{+} ions (1740 cm^{-1}) and Cu^{+} ions belonging to bulk Cu_2O (1760-65 cm^{-1}).

From the results exposed above, when the copper content decreases, the intensities of the bands due to isolated copper species (NO on Cu^{++} at 1890-95 cm^{-1} and NO on Cu^{+} at 1740 cm^{-1}) become higher than the intensities of the bands assigned to the oxides (CuO at 1860-70 cm^{-1} and Cu_2O at 1760-65 cm^{-1}). This phenomenon is especially marked with the 0.3 wt % Cu solid.

It may be noticed that, at 298 K and whatever the copper content, the NO probe molecule oxidizes the Cu^{+} reduced species created during the evacuation at 773 K. In effect when the contact time with NO increases from 1 min to 20 h, the intensity of the 1890 cm^{-1} band increases whereas the intensity of the 1760 cm^{-1} band decreases. All the species possess a very poor stability upon evacuation at 298 K. NO adsorbed on Cu^{++} belonging to CuO is the more stable species.

3.4.3. In conclusion the presence of isolated copper species is favoured by low copper amounts. The NO adsorption corroborates the results deduced from the CO adsorption. NO is able to reoxidize Cu^{+} at room temperature.

4. Conclusion

The infrared spectroscopy of adsorbed CO and NO has allowed us to determine the nature of the copper species at the surface of the alumina support.

High copper loadings favor the formation of bulk oxides on this surface. At the same time with high copper contents the introduction of oxygen induces a decrease in the reduction of NO into nitrogen by propane. The presence of bulk copper oxides leads to the formation of a catalyst very active for the oxidation of the hydrocarbon molecule by oxygen. Because of the decrease in the reductant partial pressure NO cannot be converted into nitrogen.

Low copper loadings favors the formation of isolated copper ions less active in the oxidation of propane by gaseous oxygen. In addition, the activity toward the NO reduction into nitrogen enhanced by the presence of oxygen. The selective reduction of NO is thus related to the presence of isolated copper ions at the surface of the support.

Because of the presence of various copper containing species in the studied catalysts, it is not possible to quantify the amounts of copper belonging to every species : isolated Cu^{2+} of Cu^{+} ions, bulk oxides (CuO and Cu_20).

Furthermore the oxidation state of copper is easily changed, Cu^{++} being reduced into Cu^{+} by the CO probe molecule and Cu^{+} being oxidized by the NO probe molecule even at 298 K.

Acknowledgements

The authors are very grateful to Mr. J. Billy for his assistance in FTIR experiments.

References

1 M. Iwamoto, Proc. of the Meeting on Catalytic Technology for Removal of Nitrogen Monoxide, Tokyo, January. 1990 p. 17, cited in reference 3.
2 M. Iwamoto, H. Yahiro, S. Shundo, Y. Yu-u and N. Mizuno, Appl. Catal., 69 (1991) L15.
3 S. Sato, Y. Yu-u, H. Yahiro, N. Mizuno and M. Iwamoto, Appl. Catal., 70 (1991) L1.
4 W. Held, A. Koenig, T. Richter and L. Puppe, SAE Paper 900496, 1990.
5 H. Hamada, Y. Kintaichi, M. Sasaki, T. Ito and M. Tabata, Appl. Catal., 64 (1990) L1.
6 M. Misono and K. Kodo, Chem. Lett., (1991) 1001.

7 S. Sato, H. Hirabayashi, H. Yahuro, N. Mizuno and M. Iwamoto, Catal. Lett, 12 (1992) 193.
8 K. Yogo, M. Ihara, I. Terasaki and E. Kikuchi, Appl. Catal. B, 2 (1993) L1.
9 Y. Kintaichi, H. Hamada, M. Tabata, M. Sasaki and T. Ito, Catal. Lett., 6 (1990) 239.
10 M. Iwamoto and H. Hamada, Catal. Today, 10 (1991) 57.
11 M. Sasaki, H. Hamada, Y. Kintaichi and T. Ito, Catal. Lett., 15 (1992) 297.
12 H. Hamada, Y. Kintaichi, M. Tabata, M. Sasaki and T. Ito, Chem. Lett., 1991 p. 2179.
13 A. Obuchi, A. Ohi, M. Nakamura, A. Ogata, K. Mizuno and H. Ohuchi, Appl. Catal., B, 2 (1993) 71.
14 G. Zhang, T. Yamaguchi, H. Kawakami and T. Suzuki, Appl. Catal., B, 1
15 H. Hamada, Y. Kintaichi, M. Sasaki, T. Ito and M. Tabata, Appl. Catal., 75 (1991) L1.
16 Y. Torikai, H. Yahiro, N. Mizuno and M. Iwamoto, Catal. Lett., 9 (1991) 91.
17 S. Sumiya, G. Muratratsu, N. Matsumura, K. Yoshida and R. Schenk, SAE paper 920853, 1992.
18 Y. Ukisu, S. Sato, A. Abe and K. Yoshida, Appl. Catal., B, 2 (1993) 147.
19 Y. Ukisu, S. Sato, G. Muramatsu, K. Yoshida. Catal. Lett., 16 (1992) 11.
20 Y. Li and W.K. Hall, J. Phys. Chem., 94 (1990) 6145.
21 Z.Chajar, M. Primet, H. Praliaud, M. Chevrier, C. Gauthier and F. Mathis, Catal. Lett., submitted.
22 Z. Chajar, M. Primet, H. Praliaud, M. Chevrier, C. Gauthier and F. Mathis, Appl. Catal. B, submitted.
23 J. Sarkany and W.M.H. Sachtler, Zeolithe, 14 (1994) 7.
24 J.W. London and A.T. Bell, J. Catal., 31 (1973) 32.
25 G.J. Millar, C.H. Rochester and K.C. Waugh, J. Chem. Soc., Farad.Trans., 87 (1991) 1467 and 88 (1992) 1477.
26 J. Sárkány, J. d'Itri and W.M.H. Sachtler, Catal. Letters, 16 (1992) 241.
27 G. Busca, J. Molec. Catal., 43 (1987) 225.
28 Y. Fu, Y. Tian and P. Lin, J. Catal., 132 (1991) 85.
29 K.P. de Jong, J.W. Geus and J. Joziasse, J. Catal., 65 (1980) 437.
30 M.A. Kohler, N.W. Cant, M.S. Waintwright and D.L. Trimm, J. Catal., 117 (1989) 188.
31 D.B. Clarke, I. Suziki and A.T. Bell, J. Catal., 142 (1993) 27.
32 R. Hierl, H. Knözinger and H.P. Urbach, J. Catal., 69 (1981) 475.
33 M. Iwamoto, H. Yahiro, N. Mizuno, W.X. Zhang, Y. Mine, H. Furukawa and S. Kagawa, J. Phys. Chem., 96 (1992) 9360.
34 J. Valyon and W.K. Hall, J. Phys. Chem., 97 (1993) 1204.
35 E. Giamello, D. Murphy, G. Magnacca, C. Morterra, Y. Shioya, T. Nomura and M. Anpo, J. Catal., 136 (1992) 510.
36 G. Spoto, S. Bordiga, D. Scarano and A. Zecchina, Catal. Lett., 13 (1992) 39.

A. Frennet and J.-M. Bastin (Eds.)
Catalysis and Automotive Pollution Control III
Studies in Surface Science and Catalysis, Vol. 96

LEAN NOx REDUCTION ON Cu-NaY AND Cu-HZSM5 ZEOLITES AT THE SPARK IGNITION ENGINE EXHAUST

P. Ciambelli[a], P. Corbo[b], M. Gambino[b], V. Indovina[c], G. Moretti[c] and M.C. Campa[c]

[a] *Dipartimento di Ingegneria Chimica e Alimentare, Università di Salerno, 84084 Fisciano (SA), Italy*
[b] *Istituto Motori del CNR, Via G. Marconi 8, 80125 Napoli, Italy*
[c] *Centro Studio del CNR su "Struttura e Attività Catalitica di Sistemi di Ossidi" (SACSO), c/o Dipartimento di Chimica, Università La Sapienza, P.le A. Moro 5, 00185 Roma, Italy*

ABSTRACT

The lean NO_x reduction on the Cu-NaY and Cu-HZSM5 catalysts has been investigated at the spark ignition engine exhaust. Catalysts prepared by exchange method at different Cu loadings were characterized by XRD, XPS, ESR and redox treatments with CO and O_2. The results show similar performances of Cu-HZSM5 and Cu-NaY. Reaction rates, normalized with respect to the total copper, resulted five times higher for Cu-HZSM5 compared to Cu-NaY. The nature of sites active for the lean NO_x reduction is compared with that for the NO decomposition. It is concluded that different sites are operated for the two reactions and that a much higher fraction of copper is involved in the lean NO_x reduction compared to the NO decomposition.

1. INTRODUCTION

Due to the combined requirement for improving environmental protection and fuel efficiency, the development of lean-burn and Diesel engines for automobiles is linked to the availability of catalysts able of removing NO_x from oxygen-rich exhaust gas. Copper-zeolite has been proposed as a potential catalyst for NO_x control in lean exhaust gas, selectively converting NO_x to N_2 in the presence of hydrocarbons and excess O_2 [1-2]. Most of the results reported have been obtained with laboratory reactors in experimental conditions rather different from those of engine exhaust gas. The effects of zeolite type and exchange level, of hydrocarbon type, and of experimental conditions of reaction have been investigated. The system Cu-HZSM5 seems to be superior to other Cu-zeolites towards the selective reduction of NO_x with hydrocarbons, but a comparison of performances of different zeolites in real conditions is not available [2-6].

In this paper the catalytic properties of copper-exchanged ZSM-5 and Y zeolite for the selective reduction of NO_x at engine exhaust gas have been investigated and compared to those previously found in the decomposition of NO on the same catalysts [7]. The main results from NO decomposition have shown that (i) the catalytic activity of Cu-HZSM5 is much higher compared to that of Cu-NaY and (ii) the high catalytic activity of Cu-HZSM5 is due to the very last fraction of the copper exchanged in the zeolite framework (20 % of total copper at most) immediately below 100 % exchange.

In the present paper the dependence of the activity for the lean NO_x reduction on the nature of zeolite and copper content is examined in real conditions. The features of the copper active sites for the NO decomposition and NO_x reduction with hydrocarbons are discussed.

2. EXPERIMENTAL

2.1 Catalyst preparation

The zeolites used as starting materials were HZSM5 (Enichem Anic) and NaY (Union Carbide). Copper was introduced into the zeolites by the ion-exchange method. The exchange was performed at room temperature (HZSM5)

Table1.Catalysts and their main features

Catalyst	Si/Al atomic ratio	Cu wt%	% of exchange
NaY	2.4	0	0
Cu-NaY	2.4	2.49	24
Cu-NaY	2.4	7.66	76
HZSM5	17	0	0
Cu-HZSM5	17	0.92	31
Cu-HZSM5	17	3.28	115
Cu-HZSM5	17	3.74	130
Cu-HZSM5	17	5.64	203

or at 70°C (NaY) employing 0.1 M aqueous solution of cupric acetate. Copper content was determined by atomic absorption (Varian SpectrAA-30) and expressed both as Cu wt% on a wet basis and as % of exchange (100% exchange level corresponds to 1 Cu^{2+} per 2 Al atoms). The samples and their main features are reported in Table 1. Further details about the preparation procedure will be reported in a separate paper [8].

2.2 Catalyst characterization

X-ray powder diffraction patterns of samples were obtained with a Philips automated PW 1729 diffractometer. Scans were taken with a 2ϑ size of 0.01° using Cu-Kα radiation (Ni-filtered).

Diffuse Reflectance spectra (DRS) of samples were recorded using a Cary 5 spectrometer with a diffuse reflectance accessory in the wavelength range 200-2500 nm, covering UV, visible and near-infrared regions.

X-ray photoelectron and X-ray excited Auger spectra were obtained using Mg Kα radiation ($h\nu$ = 1253.6 eV) with a Leybold-Heraeus LHS-10 spectrometer operating at constant transmission energy (E_0 = 50 eV). The spectra were recorded at room temperature and at low X-ray fluxes (12 KV, 10 mA) to minimize X-ray-induced reduction of Cu^{2+} species. The surface compositions of samples were obtained on the basis of the peak area intensities of the Cu ($2p_{3/2}$) and Si (2p) emissions using the sensitivity factor method [9]. The Fermi level of the samples were determined using the binding energy values of Si (2p) (102.3 eV for Y zeolites and 102.9 eV for ZSM5 zeolites) and the adventitious carbon with a binding energy of C (1s) fixed at 284.8 eV [8].

Redox experiments and ESR determination of Cu^{2+} were performed with a circulation all-glass apparatus equipped with a magnetically driven pump. The sample (0.2 to 1.0 g) was placed in a silica reactor equipped with a side ESR tube. All the samples before the redox cycles were treated in O_2 at 773 K. The redox cycles consisted of (i) heating in He flow at 823 K for 2h, followed by evacuation at 773 K and heating in O_2 at 773 K; (ii) evacuation at RT followed by reduction with CO at 773 K; (iii) evacuation at 773 K followed by a second treatment with O_2 at 773 K. During the treatments (i) to (iii), the pressure of O_2 or CO was monitored with a pressure transducer (MKS Baratron, sensitivity 1 Pa) until a nearly constant pressure was reached. All these measurements allowed the variation of the average oxidation number of copper to be followed. The acquisition or loss of electrons are expressed as e/Cu (number of electrons/total number of Cu atoms). At the end of treatments (i) to (iii), ESR spectra of Cu^{2+} species were recorded at RT. ESR measurements were carried out on a Varian E-9 spectrometer equipped with an on line computer. Absolute concentrations of

Cu^{2+} species were obtained from the integrated area by using a Varian strong pitch with linear spin density, $3 \cdot 10^{15}$ spin cm^{-1}, as a standard.

2.3 Catalytic tests apparatus

In Figure 1 the experimental apparatus used for the evaluation of catalytic activity is shown.

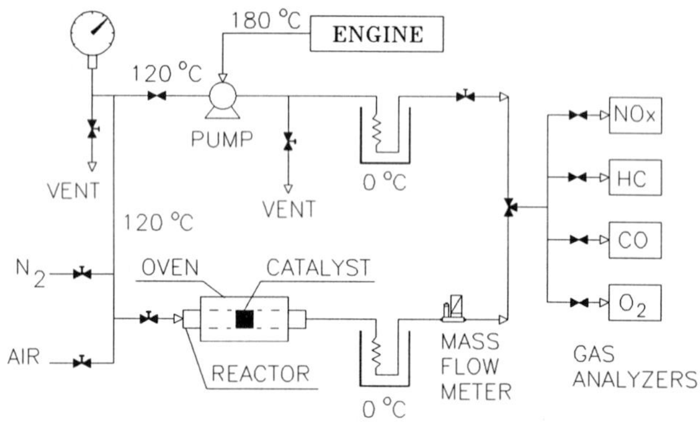

Figure 1. Experimental apparatus used for catalytic activity tests.

The engine, spark ignition four cylinders, 1350 cm^3 displacement, is installed on test bench and coupled to an electric dynanometer. A fraction of the exhaust gas (from 0.2 to 1.0 Nl/min) was collected by a probe heated at 180°C and sent to the reactor by a pump heated at 120°C. The reactor, a stainless steel tube of 10 mm i.d., was inserted in a tubular electric oven, driven by a temperature programmer. The catalyst was loaded as powder (200-400 μm), selecting the quantity in order to realize space velocitiy values comparable to those typical of the commercial catalytic converters. The pollutants (NO_x, HC, CO) and the oxygen concentrations were measured by on line analyzers: chemiluminescence for NO_x, flame ionization for HC, NDIR for CO and electrochemical for O_2. For all catalytic tests the engine operated in the following conditions: 3000 rpm, 20 kW, air/fuel mass ratio = 17. Commercial unleaded gasoline was used as fuel. The selected engine operating conditions corresponded to the following exhaust gas volumetric composition, as calculated from air and fuel flow rate fed to the engine: CO_2=11%, H_2O=12%, O_2=3%, N_2=74%. The NO_x, HC, CO and O_2 average concentrations as measured at the engine exhaust were: NO_x=1200 ppm, HC= 400 ppm, CO=1300 ppm, O_2=4%. The NO/NOx ratio resulted 0.9, the balance being NO_2. Programmed temperature tests were carried out in the range

25-600°C (25°/min). The conversions were determined measuring the pollutant concentrations before and after the catalyst.

3. Results

3.1 Catalyst characterization

All the fresh copper-zeolite samples presented the typical XRD pattern of the parent zeolites.

DRS spectra of the fresh samples show that the d-d transitions lie in the range expected for Cu^{2+} in octahedral environment of O-containing ligands (range 10000-15000 cm^{-1}) [10]. O^{2-} to Cu^{2+} charge transfer transitions occur in the range 39000-44000 cm^{-1} as one broad and intense band. Cu-NaY and Cu-HZSM5 zeolites treated under O_2 at 500 °C show d-d transitions in the same energy range observed for the fresh samples. No signal due to the presence of segregated CuO phase is detected in all sample, except in overchanged Cu-HZSM5 containing 5.64 Cu wt%.

XPS spectra of the fresh samples evidence a reduction process of the Cu^{2+} species under X-rays irradiation. The reduction stops to Cu^{+} ions. The water molecules held by the Cu^{2+} ions in the fresh samples seem to play a role in the process which is much faster in the Cu-HZSM5 catalysts. This result is in agreement with the work reported by Jirka and Bosacek [11] and suggests a better stabilization of Cu^{+} species in ZSM5 zeolites in comparison to Y zeolites. The Auger parameter of Cu^{+} species, defined as $\alpha'_{Cu} = E_b\ (2p_{3/2}) + E_k$ ($L_3M_{45}M_{45}, {}^1G$), both in Y and ZSM5 zeolites, is close to 1846.5 eV. This value is about 1.5 eV lower than the value reported for Cu_2O and can be explained as a consequence of: (i) differences in the coordination number of the core-ionized Cu^{+} ions; and (ii) the small polarizability of the oxide anion in the zeolite framework [8, 11].

The surface composition of the fresh samples with exchange levels less than 100% is similar to that obtained by chemical analysis whereas for the over-exchanged HZSM-5 sample XPS shows a surface enrichment in copper species. Moreover the copper seems to be more dispersed across the zeolite crystal as a result of dehydration. Further details and results of the XPS characterization will be reported elsewhere [8].

The (e/Cu) values measured from the redox cycles with He-CO-O_2 on Cu-HZSM5 and Cu-NaY samples are reported in Table 2 together with the percent of the total copper (analytical) detected by ESR as Cu^{2+} after the various treatments.

The ESR results are considered first. Samples treated in He at 823 K or heated with O_2 at 773 K exhibit an intense, broad and poorly resolved ESR spectrum of Cu^{2+}. With increasing copper content, the fraction of total copper detected by ESR as Cu^{2+} decreases monotonically. In particular, in samples Cu-HZSM5 containing 0.92, 3.28, 3.74 and 5.64 Cu wt%, the corresponding fractions of Cu^{2+} detected by ESR are 82, 48, 36 and 19%, respectively; in samples Cu-NaY containing 2.49 and 7.66 Cu wt%, the corresponding fractions of Cu^{2+} are 100 and 39%, respectively. In both zeolites, the decreasing of the copper fraction which is detected by ESR is accounted for by the increasing of dipolar interaction between neighbouring Cu^{2+} ions. After reduction with CO at 773 K, the intensity of the signals markedly decreases and the ESR spectra are much better resolved, particularly so in Cu-HZSM5, where three different types of Cu^{2+} species can be recognised. The relevant spin-hamiltonian parameters of these species are $g_{\|}$=2.31, $g_{\perp}$=2.06, $A_{\|}$=158 Gauss $A_{\perp}$=20 Gauss; $g_{\|}$=2.30, $g_{\perp}$ =2.05, $A_{\|}$=160 Gauss, $A_{\perp}$=18 Gauss; and $g_{\|}$=2.27, $g_{\perp}$=2.03, $A_{\|}$=168 Gauss, $A_{\perp}$=18 Gauss, respectively. The ESR parameters of the three species are in very good agreement with those reported by Giamello et al. [12] for Cu^{2+} in Cu-ZSM5.

Table 2. The extent of reduction (e/Cu) of Cu-HZSM5 and Cu-NaY catalysts after various redox treatments with O_2 or CO.

Catalyst(a)	He 823 K(b)		CO 773 K(c)		O_2 773 K(c)	
	$(e/Cu)_{O_2}$	$\%Cu^{2+}$(e)	$(e/Cu)_{CO_2}$	$\%Cu^{2+}$(e)	$(e/Cu)_{O_2}$	$\%Cu^{2+}$(e)
CuZ-0.92	0.00	82	0.5	33	0.4	67
CuZ-3.28	0.20	48	0.8	8	0.8	48
CuZ-3.74	0.20	36	1.0	5		
CuZ-5.64	0.20	19	1.6	3		
CuY-2.49	0.02	100	0.7	75	0.7	100
CuY-7.66	0.03	39	0.5	19	0.4	36

(a) CuZ = Cu-HZSM5, the figure after CuZ and CuY specifies the Cu wt.%
(b) Samples heated in He at 823 K for 2 h and exposed thereafter to O_2 at 773 K. The extent of reduction, $(e/Cu)_{O_2}$, was determined from the amount of oxygen consumed.
(c) Samples heated in O_2 at 773 K for 0.5 h, evacuated at RT and reduced thereafter with CO at 773 K. The extent of reduction, $(e/Cu)_{CO_2}$, was determined from the amount of CO_2 produced.

(d) After the reduction treatment, as described in (c), samples were treated in O_2 at 773 K. The values $(e/Cu)_{O_2}$ were determined from the amount of O_2 consumed.
(e) The percent of the analytical copper detected as Cu^{2+} by ESR.

As regards the oxidation state of copper after reduction, in all Cu-HZSM5 samples the e/Cu values (0.5 to 1) measured after the treatment with CO are consistent with the reduction of Cu^{2+} to Cu^{+}. In the case of the CuZ-5.64 sample, the value e/Cu = 1.6 suggests that a fraction of copper, and in particular that exceeding the 100% exchange, is segregated as CuO. The reduction of CuO to Cu metal accounts for the high e/Cu value measured with the excessively exchanged Cu-HZSM5 sample.

In Table 2 a reasonable agreement is observed when $(e/Cu)_{CO_2}$ values measured from the reduction with CO and $(e/Cu)_{O_2}$ values measured from the subsequent oxidation with O_2 are compared. This finding shows a satisfactory reversibility of both Cu-HZSM5 and Cu-NaY systems.

3.2 Catalytic performance

In Figures 2 and 3 the results of programmed temperature tests carried out at the engine exhaust are reported in terms of pollutant percentage conversion. For all catalytic tests the apparent contact time, expressed as the ratio of catalyst weight to volumetric flow rate (W/F value), was 0.15 gs/cm^3.

No NO_x conversion was obtained in all the range of reaction temperatures investigated either on HZSM5 or NaY, whereas HC conversion by partial oxidation to CO is indicated by the increased CO concentration at the reactor outlet with respect to that in the exhaust.

A maximum value for the NO_x conversion (34%) in the range 200-550°C was obtained on the Cu-HZSM5 catalyst containing 3.28 wt% of copper at 400°C (Figure 2). This maximum coincides with nearly complete conversion (higher than 90%) of both CO and HC. Comparable results were obtained for the CuY catalyst containing 7.66 wt% of copper (Figure 3), which exhibits a NO_x maximum conversion of 29% at 400°C. In comparison to the above ZSM-5 based catalyst the HC conversion was slightly lower, whereas CO conversion is detected only after 400°C.

In the same Figures the effect of copper content on the catalytic activity can be observed for both Cu-HZSM5 and Cu-NaY catalysts. In fact, a decrease of NO_x conversion from 34 to 25% at 400°C was observed when the copper content of Cu-HZSM5 was lowered from 3.28 to 0.92 wt%. Similarly, a maximum NO_x conversion of only 12% was obtained for Cu-NaY containing 2.49 wt% of copper. Adsorption effects do not allow significant differences in hydrocarbon

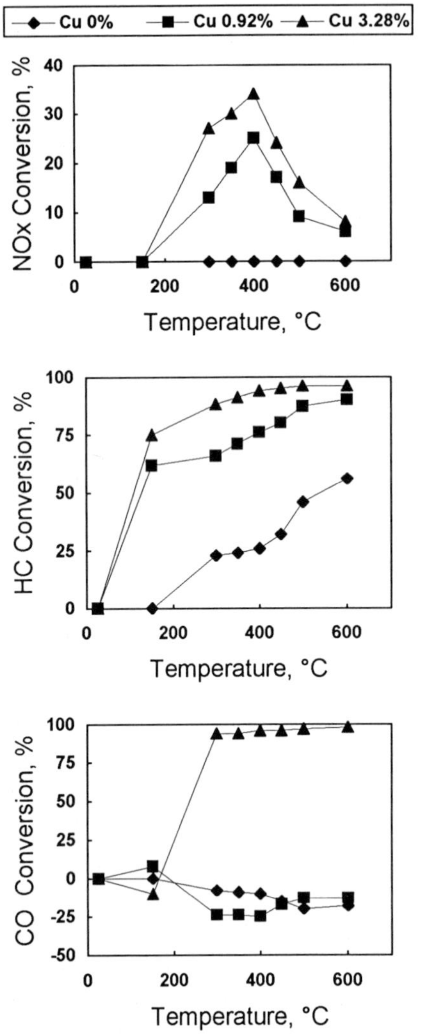

Figure 2. NO_x, HC and CO conversion profiles obtained over Cu-HZSM5 during temperature programmed reaction at W/F=0.15 gs/cm^3. For gas feed composition see Experimental (2.3).

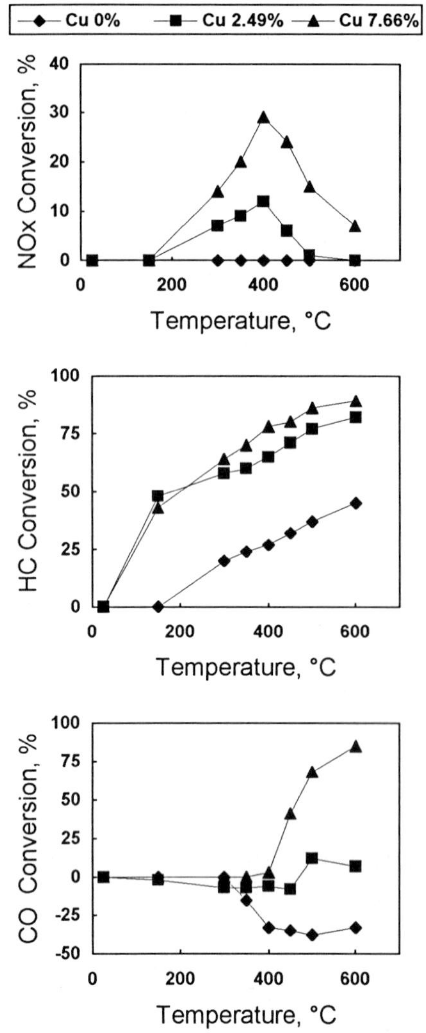

Figure 3. NO_x, HC and CO conversion profiles obtained over Cu-NaY during temperature programmed reaction at W/F=0.15 gs/cm^3. For gas feed composition see Experimental (2.3).

conversion to be observed at lower temperatures, but data at 400°C show that lowering zeolite copper affects also the hydrocarbons conversion which decreases of 19% for both Cu-HZSM5 and Cu-NaY. Moreover, a remarkable effect of copper content on CO conversions is evident for the two zeolites. In particular, in the absence of copper the CO concentration at the outlet was higher than at the inlet reactor, due to the partial oxidation of hydrocarbons with oxygen. The presence of copper promotes the conversion of CO to CO_2, up to reaching nearly total conversion at copper loadings corresponding to about 100% exchange level, but at lower temperatures over Cu-HZSM5 with respect to Cu-NaY.

The catalytic activity of the over-exchanged Cu-HZSM5 sample (203% exchange level), evaluated at W/F = 0.05 gs/cm^3, resulted only slightly lower with respect to Cu-HZSM5 containing 3.28 wt% of copper.

4. Discussion

At variance of the laboratory data reported in [3, 6, 13], in the experimental conditions investigated by us for the selective reduction of NO_x with the hydrocarbons at the engine exhaust gas neither HZSM5 nor NaY zeolite are active in converting NO_x. Copper loading gives comparable activity to both zeolites for the selective reduction of NO_x (about 30% conversion at 400°C). The decrease of NO_x conversion for temperatures higher than 400°C, observed for all catalysts tested, can be attributed to the competition between NO_x reduction by hydrocarbons and HC oxidation by oxygen, the latter becoming the predominant reaction when the temperature increases.

It is well known that Cu-HZSM5 zeolites are much better catalysts for the direct decomposition of NO than other catalysts based on different zeolites. However, the lack of large differences in the catalytic performances of the two Cu-loaded zeolites investigated must be related to the specific features of the selective reduction of NO_x with hydrocarbons, especially in the real conditions of the engine exhaust gas. Moreover the activity exhibited by the H form of ZSM5 and Y zeolites in the selective reduction of NOx [6, 13] requires the role of the transition metal to be understood, in comparison to that played in the direct decomposition of NO [2, 6, 14].

The analysis of the data in Table 2 shows that not all Cu sites are equivalent in their reducibility. Schematically, three different behaviours can be envisaged. A portion of Cu is easily reduced with He at 823 K, a second portion requires a treatment with CO at 773 K, a third one is unaffected by the reduction with CO, being still stable after the latter treatment, on ESR evidence. It is remarkable that only Cu-HZSM5 samples with a copper content in the vicinity of 100% exchange contain Cu^{2+} species which are reduced by the treatment with He at 823 K. The maximum fraction of Cu^{2+} which can be reduced with He is roughly 20% of total copper. In all samples, after reduction with CO at 773 K, ESR spectra show the presence of a fraction of Cu^{2+} which is not reduced by this treatment. In particular, 0.2 to 0.3 Cu wt% and 1.5 to 1.9 Cu wt% are the amounts of Cu^{2+} not reduced with CO in ZSM5 and Cu-NaY, respectively. Are these features in some

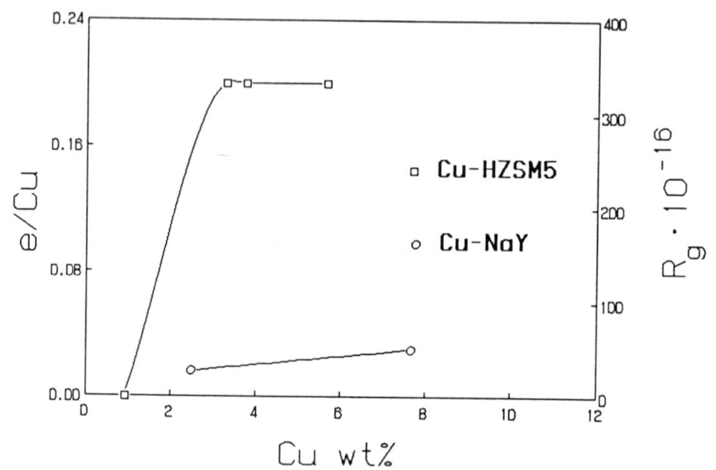

Figure 4. A correlation between the molar fraction of Cu^{2+} reduced in the treatment with He at 823 K (□, Cu-HZSM5 and O, Cu-NaY) and the catalytic activity at 773 K for the NO decomposition (full line and dotted line refer to Cu-HZSM5 and Cu-NaY respectively). Catalytic activity data (R_g, NO molecules converted to N_2/s g, scale on the right) are taken from ref. [7].

way correlated to the catalytic activity in the selective reduction of NO_x and in the direct decomposition of NO?

In a previous investigation of NO decomposition it was found that the first addition of copper to ZSM-5, up to about 80% exchange level, leads to catalysts with rather low activity. Conversely, the activity of Cu-HZSM5 catalysts increases by roughly 100-fold, when the extent of copper exchange increases from 80 to 100%. At any Cu-exchange level, up to 100%, the catalytic activity of Cu-NaY is substantially lower compared to Cu-HZSM5. All these facts showed that not all Cu sites are equivalent in their decomposition activity and that 20% is an upper limit for the fraction of Cu which is active in Cu-HZSM5 catalysts [7].

Considering further that a redox mechanism has been generally proposed for the decomposition of NO, the copper sites active for this reaction are only those endowed with the suitable redox properties. In view of the striking coincidence between the values reported above as upper limit for the active fraction of copper in Cu-HZSM5 on one side and the active fraction of Cu which is reduced during the treatment in He at 823 K on the other side (20% in both cases) we propose that the active copper is that reduced to Cu^+ during the treatment with He.

This hypothesis is here substantiated by the correlation presented in Figure 4, where the molar fraction of the Cu^{2+} reduced in the treatment with He at 823 K (data taken from Table 2) and the catalytic activity on Cu-HZSM5 and Cu-NaY (data from ref. [7]) are reported as function of the Cu wt%.

In the case of the selective reduction with hydrocarbons the effect of copper content on the rate of NO_x reduction has been evaluated from the low conversion data obtained at 300°C. The rates increase with copper content (Table 3) and

Table 3.Catalytic activity (R_g, NO_x molecules/s g and R_{Cu}, NO_x molecules/s Cu-atom) at 573 K and W/F=0.15 gs/cm³ for the NO_x reduction on Cu-HZSM5 and Cu-NaY.

Catalyst(a)	$Rg \cdot 10^{-16}$	$R_{Cu} \cdot 10^{6}$
CuZ-0.92	3.5	6.3
CuZ-3.28	7.2	3.7
CuZ-5.64	9.0	2.6
CuY-2.49	1.6	1.1
CuY-7.66	2.3	0.5

(a) CuZ=Cu-HZSM5, the figure after CuZ and CuY specifies the Cu wt%.

have similar values for both series of catalysts. The rates of NO_x reduction normalised with respect to the total copper content are nearly independent of the Cu content, showing a slight decrease by increasing the Cu loading. The values for Cu-HZSM5 catalysts are only slightly higher than for Cu-NaY.

This different behaviour with respect to the direct decomposition could be still related to a redox mechanism if one considers the comparable fraction of Cu reduced by the treatment with CO instead of the fraction reduced by He in Cu-HZSM5 and Cu-NaY. (Table 2). The alternative explanation is that a redox cation is not essential for the reduction of NOx with hydrocarbons [6, 14]. The marked dependence of the CO conversion on the type of zeolite and on the copper content (Figures 2 and 3) indicates that the end product of hydrocarbons oxidation (CO or CO_2) is strictly related to the redox cation.

5. Conclusions

The comparison of catalytic activity in the selective lean NO_x reduction at the exhaust of a spark ignition engine has shown that similar performances in terms of NO_x conversion can be obtained on Cu-loaded ZSM-5 and Y zeolites at 400°C. Reaction rates of NOx reduction evaluated at low temperature (300°C) and normalised with respect to copper atom resulted about five times higher for Cu-HZSM5 than for Cu-NaY.

This behaviour is much different from that found on the same catalysts for the direct decomposition of NO, whose rate is strongly enhanced on Cu-HZSM5 in the range 80-100% of exchange level, with respect to Cu-exchanged Y zeolite.

It has been proposed that the copper active for the direct decomposition of NO on Cu-HZSM5 is the last fraction of exchanged copper, which is reduced by He treatment at 823 K. A redox mechanism involving a much higher fraction of the exchanged copper could explain the lack of differences in the performances of Cu-HZSM5 and Cu-NaY zeolites. Moreover the conversions of CO on the two zeolites are somewhat different, suggesting that the redox Cu cation should play a significant role in determining the end product of hydrocarbons oxidation.

Acknowledgements.

The authors acknowledge Dr. G. Petrini of Enichem Anic for kindly supplying H-ZSM5 zeolite sample and Mr. G. Minelli for the chemical analysis.

References

1 W. Held, A. Konig, T. Richter and L. Puppe, SAE Paper 900496 (1990).
2 M. Iwamoto and H. Hamada, Catal. Today, 10 (1991) 57.
3 M. Iwamoto, H. Furukawa and S. Kagawa, in: New Developments in Zeolite Science and Technology, Proc. 7th Int. Zeolite Congr., Y. Murakami, A. Ijima and J. W. Ward (eds.), Kodansha, Tokyo, 1986, p. 943.
4 M. Iwamoto, N. Mizuno and H. Yahiro, in : New Frontiers in Catalysis, Proc. 10th Intern. Congr. on Catalysis, L. Guczi, F. Solymosi and P. Tètènyi (eds.), Akadèmiai Kiado, Budapest, 1993 p. 1285.

5 B. K. Cho, J. Catal., 142 (1993) 418.
6 J. O. Petunchi, G. Sill and W. K. Hall, Appl. Catal. B, 2 (1993) 303.
7 M. C. Campa, V. Indovina, G. Minelli, G. Moretti, I. Pettiti, P. Porta and A. Riccio, Catal. Lett., 23 (1994) 141.
8 G. Moretti et al., to be published.
9 C. D. Wagner, L. E. Devis, M. V. Zeller, J. A. Taylor, R. H. Raymond and L. H. Gale, Surf. Interf. Anal., 3 (1981) 211.
10 R. A. Schoonheydt, Catal. Rev.-Sci. Eng., 35 (1993) 129.
11 I. Jirka and V. Bosacek, Zeolites, 11 (1991) 77.
12 E. Giamello, D. Murphy, G. Magnacca, C. Morterra, Y. Shioya, T. Nomura and M. Anpo, J. Catal., 136 (1992) 510.
13 H. Hamada, Y. Kintaichi, M. Sasaki, T. Ito and M. Tabata, Appl. Catal. 64 (1990) L1.
14 M. Shelef, Catal. Lett., 15 (1992) 305.

A. Frennet and J.-M. Bastin (Eds.)
Catalysis and Automotive Pollution Control III
Studies in Surface Science and Catalysis, Vol. 96

OXIDATION STATE OF COPPER DURING THE REDUCTION OF NO_X WITH PROPANE ON H-Cu-ZSM-5 IN EXCESS OXYGEN.

T. Pieplu, F. Poignant, A. Vallet, J. Saussey, J.C.Lavalley & J.Mabilon *[+]

Laboratoire catalyse & spectrochimie,CNRS URA 0414, ISMRA-Université, Bd M[al] Juin, 14050 Caen Cedex, France
**Institut Francais du Pétrole, B.P. 311, 92506 Rueil-Malmaison France.*

ABSTRACT

NO (reactant) and CO (product) molecules were used as IR probes of the copper oxidation state in H-Cu-ZSM-5 catalysts. CO adsorption is specific to Cu^{I} sites. Its characteristic band at 2158 cm^{-1} provides quantitative results on integrating its molar extinction-coefficient (ε) NO decomposes oxidizing Cu^{I} to Cu^{II}. Propane and oxygen in a special IR reactor cell always yielded chemisorbed CO. Use of ε indicates the NO influence on Cu state.

1.INTRODUCTION

The removal of NO in exhaust gases of automobile equipped with gasoline engines has been sucessfully performed by using the so-called three-way catalyst system. However, under conditions of substantial excess O_2, the rate of reduction decreases markedly on this kind of catalytic system [1]. Therefore, de-NOx from the exhaust gases of diesel engines, which involves a substantial excess O_2, is still an unresolved problem. Recently, it has been reported that NO can be

[+] This work was supported by a Brite Euram project

reduced to N_2 with hydrocarbons as reducing agents even in the presence of excess O_2 using solid catalysts such as Cu-zeolites [2]. To improve the process which is not yet of practical interest, basic research on the copper electronic state in working conditions is needed.

In this paper we report the results obtained by FT-IR spectroscopy on a H-Cu-ZSM-5 catalyst containing 2.5 wt% Cu. We first test in static conditions, whether reactants such as NO or products like CO are able to discriminate between Cu^{I} and Cu^{II} oxidation states. Then the results are applied under working conditions using a special IR cell acting as a reactor.

2. EXPERIMENTAL

The H-Cu-ZSM-5 catalyst was prepared by cationic exchange with $Cu(NO_3)_2$ solution. From the following data Si/Al=16.8 and 2.48 weight % of copper we deduce the amount of aluminium: 5.39 Al/unit cell e.g. $H_{5.4-2n}\,Cu_n^{(II)}Al_{5.4}\,Si_{90.6}\,O_{192}$ and then the amount of copper: 2.31 Cu/unit cell. Thus, the sample was labelled Cu-ZSM-5-17-85 i.e. the standard identification used in all publications.

For IR measurements, the sample was pressed into self-supporting discs (~10 mg.cm^{-2}). The spectra of the adsorbed species were obtained by subtracting the spectrum of the wafer from the spectrum obtained after adsorption.

Static experiments

The wafer was placed in a cell, where it underwent all activation and adsorptive treatments in a strictly *in situ* configuration. Spectra were recorded at room temperature (hereafter denoted by r.t.) at a resolution of 4 cm^{-1} on a Nicolet 5SX FT-IR spectrometer.

Prior to spectral recording, samples were oxidised or reduced, according to the following procedures :

(1) Oxidised sample (hereafter referred to as H-Cu-ZSM-5(Ox)) : First the sample was treated twice as follows : oxygen (100 Torr) was admitted to the sample at 773 K for 1/2 h and evacuated for 1/4 h. Then, oxygen (100 Torr) was admitted to the sample at 773 K for 2 h. The sample was cooled to r.t. in the oxygen atmosphere, and O_2 was removed by evacuation at r.t.. Evacuation of O_2 at higher temperatures would result in a partial reduction of copper.

(2) Evacuated sample (hereafter referred to as H-Cu-ZSM-5(Evac)) : After O_2 treatment at 773 K as before, the sample was evacuated at 773 K overnight and then cooled to r.t. under a dynamic vacuum.

Dynamic experiments

The cell design has already been reported [3]. It is installed in a continuous flow system built directly on the FT-IR spectrometer (Nicolet Magna 750) to minimize the dead volume. The cell can be considered as a gas flow reactor with a dead volume less than 0.2 cm^3. Those capabilities permit reactivity experiments, combined with rapid IR spectroscopic measurements of surface species to be made in a manner which has not been possible in standard IR cell. The GC collection was performed with a Delsi chromatograph on line. Gases were also analysed by FT-IR in a short spacial cell, also directly connected to the IR cell-reactor, with a Nicolet 5SX spectrometer. NOx reduction was measured as the ratio $C\% = (NO° - (NO + NO_2)) \cdot 100/NO°$

Prior to spectral recording the sample was treated, according to the following procedure : the catalyst was heated progressively to 623 K (2.5 $K.min^{-1}$) under O_2 (5 %) in He and maintained for 3 hours in these conditions (calcination). It was then exposed (under flow: 25 $cm^3.min^{-1}$; GHSV = 50000 h^{-1}) to different previously analysed gas mixtures.

3. RESULTS AND DISCUSSION

3.1 Static experiments

The spectra of H-Cu-ZSM-5(Evac) and (Ox) are presented in Figure 1. Both spectra show ν(OH) bands at 3740 and 3610 cm^{-1} corresponding to silanol and acidic Si-OH-Al framework groups. Moreover, overtones and combination bands of the framework vibrations are situated at 2000, 1850 and 1620 cm^{-1}. Moreover, the spectrum of H-Cu-ZSM-5(Ox) shows a broad band between 3750 and 3200 cm^{-1} and a sharper one at 1620 cm^{-1} whereas the intensity of the 3610 cm^{-1} band is higher for H-Cu-ZSM-5(Evac). Subtracted spectra clearly evidence two absorptions near 3750-3200 and 1620 cm^{-1}. Owing to the fact that the 3610 cm^{-1} OH band intensity is weaker after oxidation, the broad band in the ν(OH) range presents a window at this wavenumber. These features are explained by taking into account the reversibility of the copper oxidation state change according to the pretreatment or the nature of the surrounding atmosphere. The concomitant equilibrium of charges when reducing the sample would be assured by the water coordinated on Cu^{II} which gives rise to H^+ ions increasing the Brönsted acidity under reduction conditions (these protons provide the charge equilibrium). Thus, study of the ν(OH) bands intensity can be used as a means to determine the variation of the copper oxidation state.

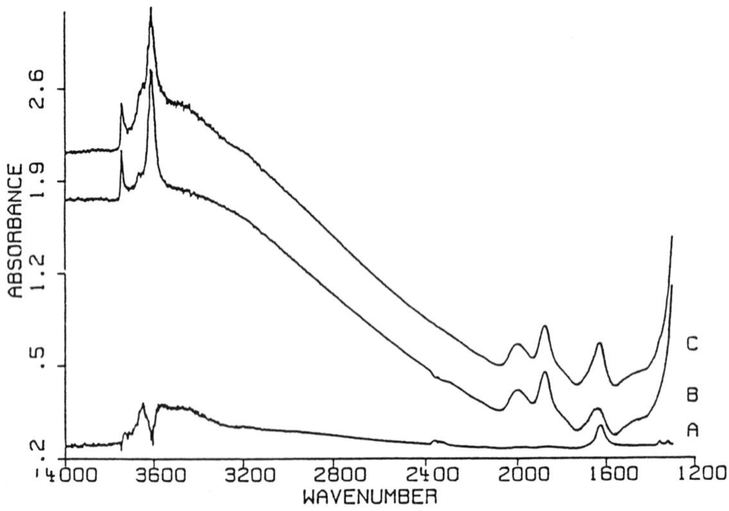

Fig. 1 - Spectra of H-Cu-ZSM-5(Evac) (B) and (Ox) (C) ; (A) subtraction of the two spectra (C - B).

3.1.1 CO adsorption

OnH-Cu-ZSM-5(Evac) successive introduction of CO portions at r.t. leads first to a ν(CO) band at 2158cm^{-1} characterising CO adsorption on Cu^{I} sites [4]. A very weak band is also noted at 2110 cm^{-1} ; it corresponds to the same species with ^{13}CO contained in natural abundance. As in the case of the parent sample, no CO adsorption on Si-OH-Al groups is observed at r.t.. When the equilibrium pressure reaches 1 Torr, another band is observed at 2177 cm^{-1} A shoulder is also noted at 2151 cm^{-1} (Figure 3a). They both increase in unison (Figure 2) at the expense of the 2158 cm^{-1} band. They could be due to CO adsorption on Cu^{II} sites [5].

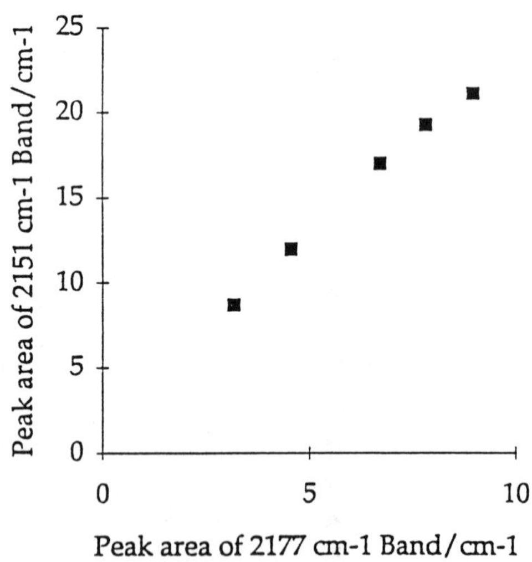

Fig. 2 - Correlation of peak area between 2177 and 2151 cm^{-1} bands over H-Cu-ZSM-5

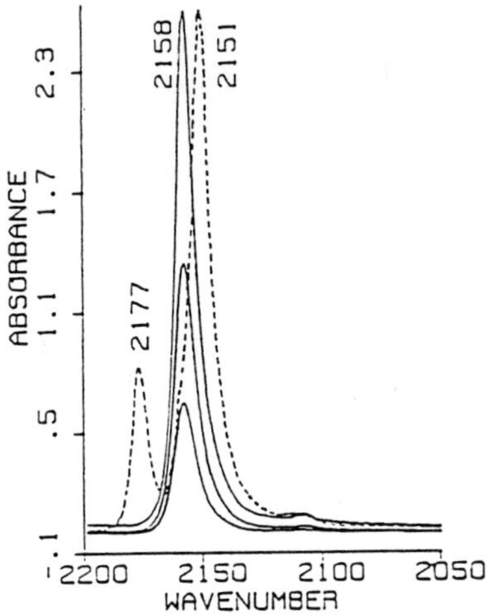

Fig. 3a - CO adsorption at r.t. on H-Cu-ZSM-5(Evac). The two first portions 1 & 2 μmol, dashed - 100 Torr, and after evacuation at r.t.

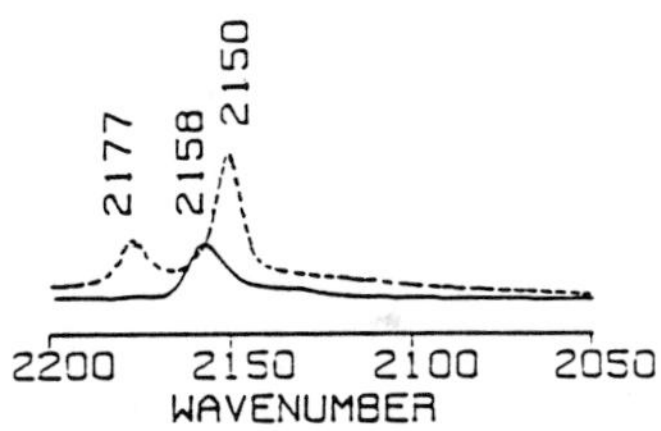

Fig. 3b - CO adsorption at r. t. on H-Cu-ZSM-5(Ox). Plain- 0.4 Torr ; dashed- 100 Torr.

To check the assignments, CO has been adsorbed on H-Cu-ZSM-5(Ox). Figure 3b shows that CO adsorption is in such a case very low : only a weak band at 2158 cm^{-1} is observed when first portions are introduced whereas the two bands (2177, 2150 cm^{-1}) appear under CO pressure. Their weak intensity discards the possibility of CO adsorption on Cu^{II} sites. We prefer to assign the 2177, 2151 cm^{-1} couple to gemdicarbonyl species adsorbed on Cu^{I} sites [6]. It seems from the literature, that $Cu^{I}(CO)_2$ species have never been reported before. However, gem-dicarbonyl species have been observed on oxidized transition metals like Ni^{+} [7] and Pd^{+}[8].

It has been seen that the 2177, 2151 cm^{-1} couple disappears on evacuation at r.t., restoring the 2158 cm^{-1} band corresponding to irreversibly chemisorbed species. In all cases (especially with H-Cu-ZSM-5(Evac)), for the first portions of CO introduced, it has been checked that all CO is totally chemisorbed. Moreover, there is a linear variation between the intensity of the 2158 cm^{-1} band and the amount of CO introduced (Figure 4). This allows us to deduce the integrated molar extinction coefficient :

$\bar{\varepsilon}_{CO}$ = 13.5 cm.μmol-1

It appears that CO is able to characterise the presence and the number of Cu^{I} sites. By contrast, it does not adsorb on Cu^{II} sites. The weak bands observed on H-Cu-ZSM-5(Ox) are due to a few (about 5 % of the total copper content) Cu^{I} residual sites persisting even after oxidation. Such a result is in agreement with [9] reporting that Cu^{II} sites are very easily reducible, as shown by results obtained under pure oxygen or O_2 + He mixture. As for the H-Cu-ZSM-5(Evac) catalyst, from the intensity of the 2158 cm^{-1} band observed after the introduction of a amount of CO followed by evacuation at r.t., it appears that CO detects only 1.4 Cu^{+}/unit cell, e.g. 60 % of the total copper amount. This confirms literature results [2] reporting that CO does not detect all the copper sites. Note that if Cu° were present, a ν(CO) band below 2100 cm^{-1} should appear [10].

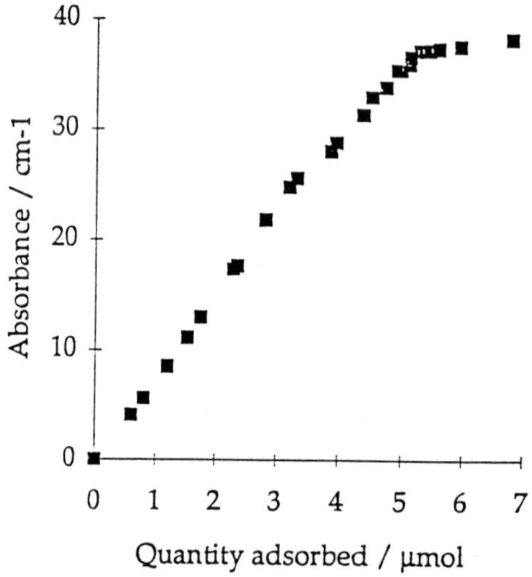

Fig. 4 - Determination of the integrated molar extinction coefficient of ν(CO) band at 2158 cm^{-1}.

3.1.2 NO adsorption

Nitric oxide does not adsorb on the parent H-ZSM-5 sample.

Introduction of a small amount of NO on H-Cu-ZSM-5(Evac) leads mainly to one band at 1811 cm^{-1} due to Cu^{+}-NO species. Increasing the NO amount leads to a more complex spectrum showing 4 main bands : 1811 cm^{-1} (Cu^{+}-NO), 1826 and 1733 cm^{-1} ($Cu^{+}(NO)_2$) and 1907 cm^{-1} (Cu^{2+}-NO) [11] (Figure 5a). At r.t. the intensity of the latter increases with time, at the expense of the others, indicating the oxidation of some Cu^{I} sites into Cu^{II} sites.

Introduction of a small amount of NO on H-Cu-ZSM-5(Ox) mainly leads to the 1907 cm^{-1} band ; however a small one is observed at 1811 cm^{-1} showing that not all the copper sites were completely oxidised like in the case of CO adsorption. Note that a spectrum similar to that observed on H-Cu-ZSM-5(Ox) is obtained when a large amount of NO is introduced on H-Cu-ZSM-5(Evac) and remained in contact with the catalyst for 1 h at r.t..

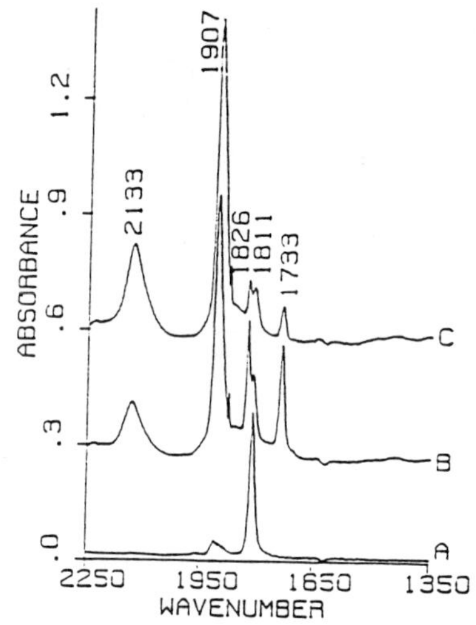

Fig.5a - NO adsorption at r.t. on H-Cu-ZSM-5(Evac). A- 0.1 Torr, B-100 Torr, C - 100 Torr after 3 hours

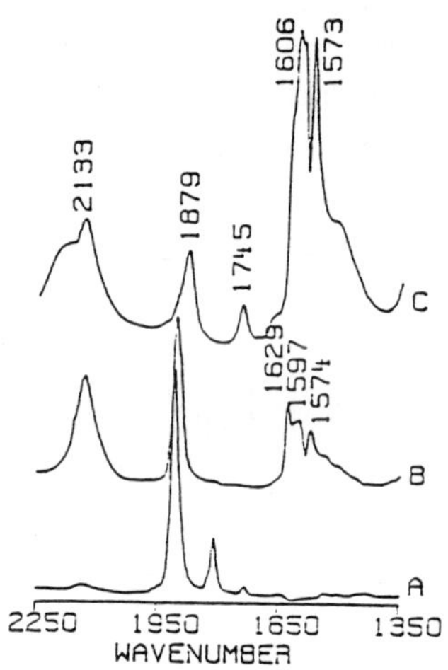

Fig. 5b - NO adsorption followed by O_2 at r.t. on H-Cu-ZSM-5(Ox). A- 2 Torr NO, B-plus 4 Torr O_3, C - after 3 hours

Whatever the oxidation state of the catalyst two other bands are noted at 2232 (weak) and 2133 cm^{-1} for large amounts of NO introduced. They are assigned to N_2O and NO_2^+ chemisorbed species, respectively [12].

When O_2 is introduced on a small amount of NO adsorbed on the oxidized sample (Figure 5b) other bands appear at 1629, 1597 and 1574 cm^{-1}. Hall et al. assigned the first one to NO_2 adsorbed species [13], whereas the other two would be characteristic of nitrate species [14].

Although NO is able to differentiate Cu^I and Cu^{II} sites, complex spectra are often observed (Cf figure 5b, spectrum C) due to the formation of N_2O and NO_2. Moreover, even at r.t., NO oxidizes Cu^I sites into Cu^{II} sites and therefore cannot be used as a probe molecule for the characterisation of the copper site oxidation state.

3.2 Dynamic experiments

The method has been applied to study the NO reduction (2000 ppm) by propane (2000 ppm) at 623 K in the presence of an excess O_2 (5 %) under helium flow. Before the study of this complex reaction, preliminary experiments have been performed using only two reactants.

3.3.1 Propane + oxygen

Under flow, in steady conditions, the propane conversion is 60 %. Only CO_2 and H_2O are detected by gas chromatography.

The spectrum of the species formed on the catalyst is presented in Figure 6. Water formation explains the broad band between 3700 and 3000 cm^{-1}. CO is chemisorbed on Cu^{I} sites as shown by the sharp band at 2156 cm^{-1}. From its intensity, one can estimate that the amount of Cu^{I} is 2.7 % of the total copper amount. The strong bands between 1625 and 1400 cm^{-1} correspond to carboxylate species.

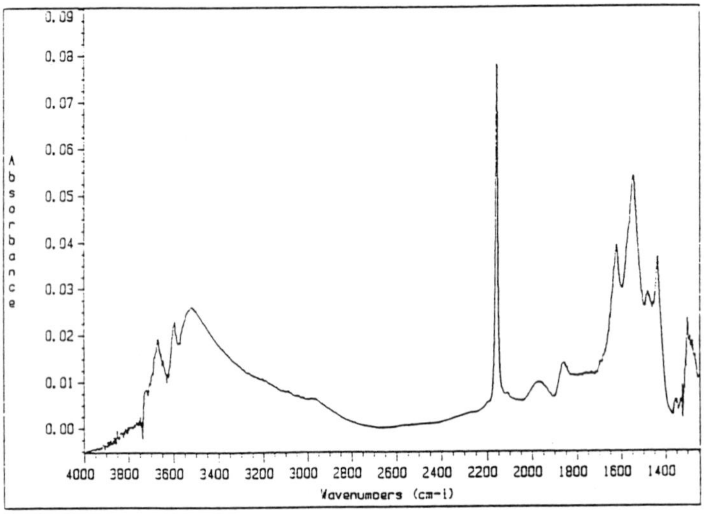

Fig. 6- IR spectrum of species formed on H-Cu-ZSM-5 under $C_3H_8 + O_2$ stream at 350°C.

3.3.2 Propane + nitric oxide

After 30 min of time-on-flow, the propane conversion is 4.4 % and that of NO is 3 %. The products formed are CO_2 and N_2O as observed by IR spectroscopy. No H_2O nor NO_2 has been detected as reaction products. As soon as the C_3H_8 + NO mixture is introduced, subtracted spectra (Figure 7) show a negative absorbance between 3700 and 3500 cm^{-1} with a positive sharp band at 3610 cm^{-1} indicating a reduction of some Cu^{II} sites into Cu^{I} sites. Although CO is not detected as a product, the sharp band at 2158 cm^{-1} indicates that a small amount of CO is formed and remains chemisorbed on Cu^{I} sites. This confirms the catalyst reduction. Weak bands are observed in the ν(CH) range. The high wavenumber of some of them (3060 cm^{-1}) characterises the formation of ethylenic species. Another band is noted at 2047 cm^{-1} and tentatively assigned to fulminate species [15]. With time on-stream the propane oxidation and NO_x reduction decrease. A new species is formed, characterised by a band at 2248 cm^{-1}, assigned to isocyanate species [16]. The intensity of ethylenic ν(CH) bands and those at 2248 and 2047 cm^{-1} increase with time-on-flow whereas that at 2158 cm^{-1} decreases. Since the spectrum in the ν(OH) range does not change, the intensity decrease of the ν(CO) band at 2158 cm^{-1} does not correspond to a partial reoxidation of the catalyst but rather to partial poisoning of Cu^{I} sites by more stable species.

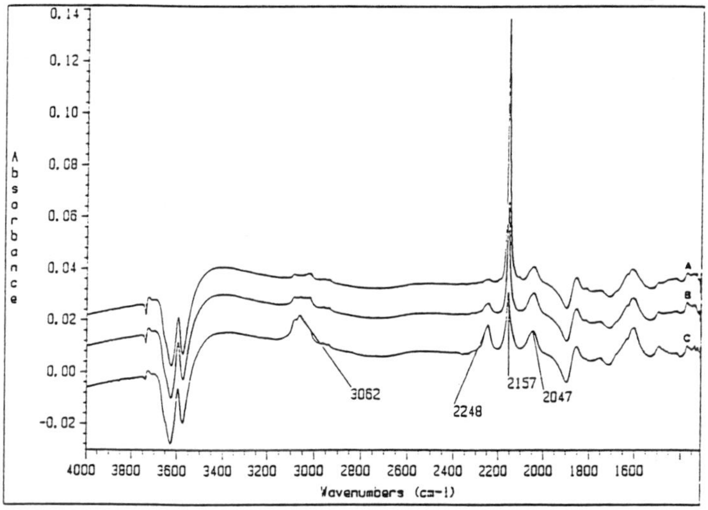

Fig. 7 - IR spectra of species formed on H-Cu-ZSM-5 under NO + C_3H_8 stream at 350°C. A) 7 min (C = 8%) ; B) 15 min (C = 4%) ; C) 42 min (C = 3%)

3.3.3 Propane + oxygen + nitric oxide

NO addition in propane and oxygen flow increases up to 68 % propane oxidation whereas NO reduction reaches 45 %. This result shows a positive effect of NO in propane oxidation in agreement with recent work on H-ZSM-5 reporting that addition of small amounts of NOx drastically increases the propane conversion [17]. This suggests a direct reaction between propane and NOx. One of the O_2 roles would be to eliminate strongly chemisorbed species evidenced under propane and NO flow, without O_2 ; such species could poison active copper sites. No CO is detected in the products but, again, a band is observed at 2158 cm^{-1} (Figure 8), characteristic of CO adsorption on Cu^{I} sites. Its intensity is weaker than that observed under flow without NO. This weaker intensity, the study of the ν(OH) bands indicating no catalyst reduction and the increase of propane combustion are in favour of a more oxidized copper state.

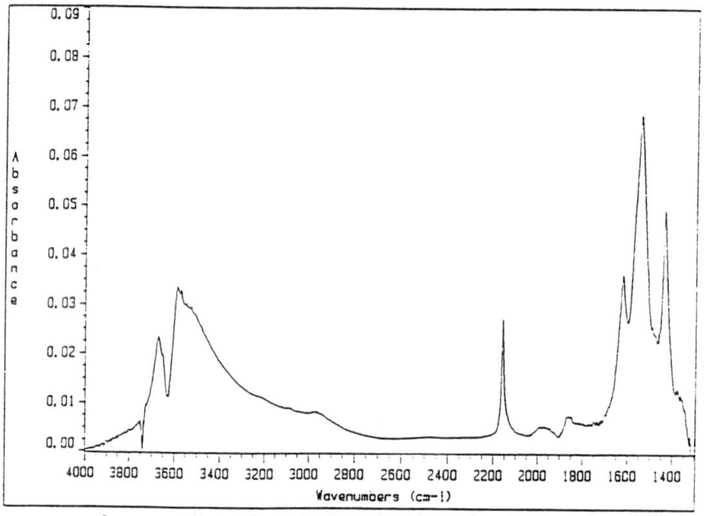

Fig. 8 - IR spectrum of species formed on H-Cu-ZSM-5 under $C_3H_8 + O_2 + NO$ stream at 350°C.

4. Conclusion

IR spectra obtained under flow conditions always present a band at 2158 cm^{-1} due to CO formation and adsorption on Cu^{I} sites. Its weaker intensity when introducing NO in the propane + oxygen flow shows that Cu^{II} sites are predominant in reaction conditions. Since the NO conversion in the presence of propane without oxygen is very low, one of the oxygen effects would be to maintain the catalyst in a high oxidation state, favouring the reaction. Another role would be to avoid the surface poisoning. Our method with gas flow reactor combined with rapid IR spectroscopic measurements of surface species permits detection of some species adsorbed on the working catalyst (carboxylate, isocyanate). Future experiments would determine whether some of these species could be intermediates in the NOx reduction by propane.

REFERENCES

1 Held, W. ; König, A. ; Richter, T. ; Puppe, L. SAE Paper 900496
2 Iwamoto, M. ; Mizuno, N. ; Yahiro, H. Sekiyu Gakkaishi 1991, 34, 375.
3 Joly, J.F. ; Zanier-Szydlowski, N. ; Colin,S. ; Raatz, F. ; Saussey, J. Lavalley, J.C. Catal. Today, 1991, 9, 31.
4 Huang, Y. Y. J. Catal. 1973, 30, 187.
5 Fu, Y. ; Tian, Y. ; Lin, P. J. Catal. 1991, 132, 85.
6 Lavalley, J.C. et al, to be published.
7 Kermarec, M. ; Olivier, D. ; Richard, M. ; Che, M. J. Phys. Chem. 1982, 86, 2818.
8 Lokhov, Y. A. ; Davydov, A. A. Kinet. Katal. 1980, 21, 1523.
9 Hall, W. K. ; Li, Y. J. Catal. 1991, 129, 202.
10 Ghiotti, G. ; Boccuzi, F. ; Chiorino, A. Stud. Surf. Sci. Catal. 1985, 21, 235.
11 Giamello, E. ; Murphy, D. ; Magnacca, C. ; Shioya, Y. ; Nomura, T. ; Anpo, M. J. Catal. 1992, 136, 510.
12 Chao, C.C. ; Lunsford, J.H. J. Amer. Chem. Soc 1971, 93, 71.
13 Valyon, J. ; Hall, W. K. J. Phys. Chem. 1993, 97, 1204.
14 London, J. W. ; Bell, A. T. J. Catal. 1973, 31, 32.
15 Nakamoto, K. Infrared and Raman spectra of inorganic and coordination compounds ; 4th Ed., Wiley Interscience Publication ; 1986, 289.
16 Ukisu, Y. ; Sato, S. ; Muratmasu, G. ; Yoshida, K. Catal. Lett. 1991, 11, 177.
17 Sasaki, M. ; Hamada, H. ; Kintaichi, Y. ; Ito, T. Catal. Lett. 1992, 15, 297.

A. Frennet and J.-M. Bastin (Eds.)
Catalysis and Automotive Pollution Control III
Studies in Surface Science and Catalysis, Vol. 96

NO DECOMPOSITION OVER PARTIALLY REDUCED METALLIZED CeO_2 CONTAINING CATALYSTS.

Gangavarapu Ranga Rao[a], Paolo Fornasiero[b], Jan Kašpar[b], Sergio Meriani[c], Roberta Di Monte[c] and Mauro Graziani[b]

[a]*International Centre for Science and High Technology, Padriciano 99, 34100 Trieste (Italy);* [b]*Dipartimento di Scienze Chimiche, Università di Trieste, Via Giorgieri 1, 34127 Trieste (Italy);* [c]*Dipartimento di Ingegneria dei Materiali e Chimica Applicata, Università di Trieste, Via Valerio 2, 34127 Trieste (Italy).*

ABSTRACT

M/CeO_2-ZrO_2 (M=Rh,Pt) solid solutions are investigated as catalysts for the reduction of NO by CO. It is shown that incorporation of CeO_2 into a ZrO_2 framework strongly promotes the reduction of the metallized support. The reduced support efficiently decomposes NO suggesting a direct participation of the support in the NO conversion.

1.INTRODUCTION

NO removal by reduction with CO is a key step in the catalytic conversion of automotive exhaust. Rhodium is extensively added to the commercial catalysts to promote this reaction and the high NO conversions and selectivity towards N_2 formation which are observed over rhodium catalysts compared to other noble metals, are generally associated with its ability to dissociatively chemisorb NO [1]. The effectiveness of cerium oxide in improving the three-way catalyst performances is well established and it is ascribed to the following effects of the promoter: i) stabilization of metal dispersion [2] and of alumina support [3], ii) promotion of the water-gas shift reaction [4] and iii) enhanced oxygen storage and release by shifting between CeO_2 under oxidizing conditions and Ce_2O_3 under reducing conditions respectively [5]. Recently, the oxygen vacancies associated with reduced ceria in the proximity of noble metal particles have been suggested as promoting sites for NO and CO conversions [6-7].

One of the major concern of the actual three-way catalysts is their thermal stability since, due to the close-coupled locations of the converter, the temperature of the catalyst may exceed 1200 K in the driving conditions. Upon ageing at such high temperatures, deactivation of the catalyst may occur due to sintering of metal particles, formation of irreducible rhodium species and collapse of the Al_2O_3 surface area. In addition, the Oxygen Storage Capacity (OSC) of the CeO_2 promoter strongly decreases upon thermal ageing, due to the growth of CeO_2 crystallites and/or formation of $CeAlO_3$ [8]. It was shown that addition of ZrO_2 to the washcoat significantly improved the thermal stability of the catalyst by hindering the sinterization of the CeO_2 particles [9]. This suggests that a close interaction between CeO_2-ZrO_2 may thermally stabilize the CeO_2 promoter. Incorporation of CeO_2 into a solid solution with ZrO_2 might provide a suitable way to improve the thermal resistance of the CeO_2 component due to the ceramic nature of these ceramic alloys. Furthermore, in the solid solution, the undesirable decline of the OSC due to fixation of the $Ce^{3+} \Leftrightarrow Ce^{4+}$ redox couple in the 3+ state ($CeAlO_3$, $Ce_2(SO_4)_3$) should be limited. On the basis of these considerations we decided to investigate a number of CeO_2-ZrO_2 ceramic alloys as supports for noble metal catalysts for the reduction of NO by CO. Here we report clear evidence for NO decomposition occurring over reduced ceria containing catalysts, suggesting a possible new role of both the metal and ceria promoter in the three-way catalysts under transient conditions. Temperature Programmed Reduction (TPR) experiments show that incorporation of ZrO_2 into a solid solution with CeO_2 strongly enhances the reduction of the metallized samples which occurs in the bulk of the support at fairly low temperatures. As a consequence, a strong dependence of the reduction behaviour upon the structural properties of the solid solution is also observed.

2. EXPERIMENTAL

CeO_2-ZrO_2 solid solutions with CeO_2 content ranging from 10% to 90% were prepared by firing mixtures of the oxides at 1873 K for 1h (surface areas ≈ 0.5-1.5 m^2g^{-1}). Powder x-ray diffraction analysis (Siemens Kristalloflex Mod.F Instrument, Ni-filtered CuK_a) showed presence of pure fluorite type structure for CeO_2 molar contents greater than 60%. For lower CeO_2 contents, tetragonal TZ° and TZ' phases [10] were detected. Supports were impregnated with $RhCl_3{\cdot}3H_2O$ or $Na_2PtCl_6{\cdot}nH_2O$ to incipient wetness, then the catalysts were dried at 393 K overnight and finally calcined at 723 K for 5 h. Temperature Programmed Reduction was carried out in a conventional system at a heating rate of 10 K min^{-1}. H_2 consumption in the TPR was estimated by using CuO as a standard. Oxygen Storage Capacity was measured by pulse technique in the same equipment used for the TPR experiments. The samples were reduced in H_2/Ar mixture (10 K min^{-1} to 700 K and then held for 2 h). H_2 was then desorbed in Ar flow (20 ml min^{-1}) at 700 K for 2h. The oxygen uptake was measured by injecting pulses of O_2 (0.092 ml) into the flow of Ar passing over the catalyst (0.04-0.05 g) until the breakthrough point was attained. Catalytic experiments were carried out in differential conditions using a U-shaped glass micro reactor (NO (2%) in He, total flow 15 ml min^{-1} or NO (1%) and CO (3%) in He, total flow 30 ml min^{-1}, GHSV = 12500-50000 h^{-1}). The effluents of the reactor (NO, N_2, N_2O, CO and CO_2) were analyzed gaschromatographically.

3. RESULTS AND DISCUSSION

3.1 X.R.D. and Temperature Programmed Reduction Characterization of the Samples

According to the cerium content, the ceria-zirconia ceramic alloys exist in three different phases, namely monoclinic, tetragonal and cubic [11]. Below 1300 K the monoclinic and cubic phases appear to be thermodynamically stable, however, when the ceramic method is employed for the synthesis of the solid solution, metastable tetragonal phase is easily formed in a wide compositional interval and it is fairly stable at ordinary temperatures [10c]. In our synthesis we have used a non quenching cooling rate which produces two phases of tetragonal symmetry referred as TZ° and TZ'. These phases exist pure in the compositional range 5-20% CeO_2 and 40-60% CeO_2 respectively, while at CeO_2 content 20-40% a mixture of TZ° and TZ' is obtained. The former phase is characterized by a larger ortogonality (c/a = 1.018) compared to the TZ' phase (c/a ≈ 1.010) [10b]. As shown in Fig.1, powder X.R.D. spectra confirmed formation of solid solutions in the present samples. The d(111) linearly decreases in the cubic region with decreasing CeO_2 content due to the smaller Zr^{4+} ionic radius (0.80 Å) compared to Ce^{4+} (0.97 Å). In the range of composition 30-40% of CeO_2, the constant values of (hkl) spacing indicate that a two phase region is present as expected on the basis of the phase diagram. Consistently, both $Ce_{0.3}Zr_{0.7}O_2$ and $Ce_{0.4}Zr_{0.6}O_2$ are mixtures of respectively TZ° (77 and 22%), TZ' (17 and 67%) and cubic (6 and 11%). All other samples present a single phase composition. Reduction behaviour of all the samples was investigated by means of Temperature Programmed Reduction. TPR profiles of the $Ce_{0.6}Zr_{0.4}O_2$ sample as well as Rh and Pt/$Ce_{0.6}Zr_{0.4}O_2$ samples are compared in Fig.2 with those carried out on pure CeO_2 and Rh/CeO_2 samples.

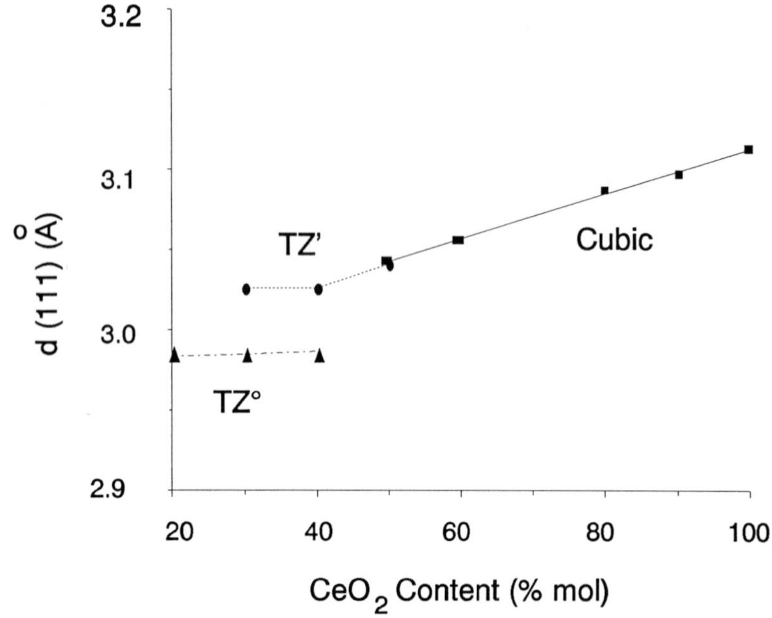

Figure 1. (111) spacing as determined by powder X.R.D. of CeO_2-ZrO_2 solid solutions.

They show a strong promotion of the reducibility of the $Ce_{0.6}Zr_{0.4}O_2$ solid solution in the presence of noble metals as shown by the appearance of a strong reduction feature between 500 and 700 K in the TPR profiles of both supported Rh and Pt/$Ce_{0.6}Zr_{0.4}O_2$ catalysts. This feature is absent in the other samples and it is associated with a partial bulk reduction of the support. By using the Mooi and Johnson model [12], 0.5 ml H_2 g^{-1} can be estimated for surface reduction of a 1.0 m^2g^{-1} surface area CeO_2 sample and 0.7 ml H_2 g^{-1} are estimated for the reduction of the Rh_2O_3 precursor. A consumption of about 1.3 ml H_2 g^{-1} is measured for the peak at around 350 K (Fig.2, traces 2,4) which is consistent with occurrence of Rh_2O_3 and surface reduction. In the range of temperatures 500-700 K, a consumption of 9.5 ml H_2 g^{-1} is estimated for the reduction process depicted in Fig.2, trace 4, which gives a bulk composition of Rh/$Ce_{0.6}Zr_{0.4}O_{1.9}$. Therefore the support reduction can be associated with the formation of anionic oxygen vacancies in the bulk of solid solution due to the reducible Ce^{4+} cation. This is supported by magnetic susceptibility measurements which showed close coincidence between the degree of reduction as estimated from TPR and that estimated according to [13] from this technique by considering that only Ce^{3+} contributes to the magnetic moment. The absence of the feature centered at 700 K in the unmetallized samples (Fig.2, trace 3) indicate that support reduction is promoted by metal particles which spill H_2 over the support. The importance of H_2 spill-over in facilitating the reduction of CeO_2 has long been recognized. Generally speaking, CeO_2 with a surface area higher than 20-30 $m^2\ g^{-1}$, shows presence of two reduction peaks approximately at 770 K and 1100 K which have been associated respectively with surface and bulk reduction [5]. In the presence of supported metal, the latter feature shifts to lower temperatures and splits into several peaks. At variance with this, TPR profile of low surface CeO_2 samples is almost unaffected by the presence of supported metal particles.

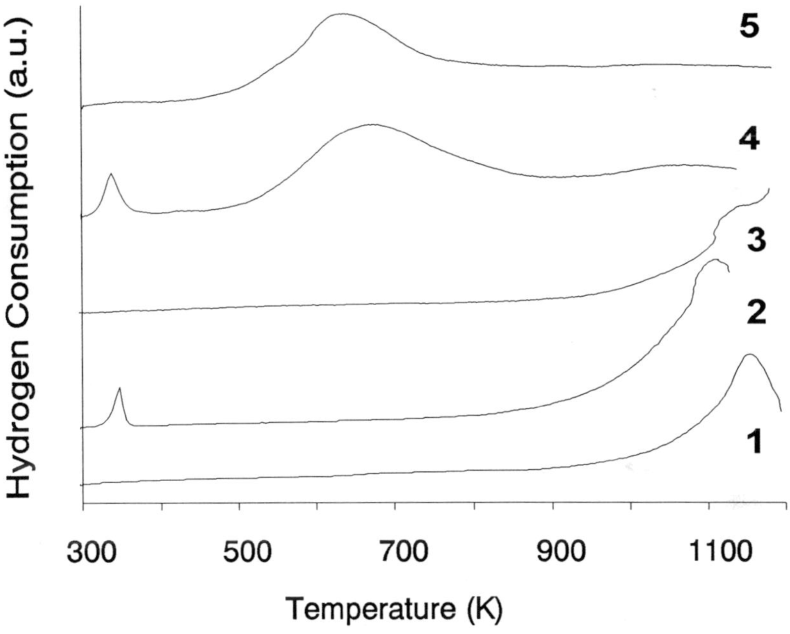

Figure 2. TPR profiles of (1) CeO_2, (2) 0.5%Rh/CeO_2, (3) $CeO_{0.6}Zr_{0.4}O_2$, (4) 0.5%Rh/$CeO_{0.6}Zr_{0.4}O_2$ and (5) 0.5%Pt/$CeO_{0.6}Zr_{0.4}O_2$.

The observation that bulk reduction is promoted upon incorporation of CeO_2 into a solid solution with ZrO_2 prompted us to investigate the dependence of the reduction behaviour upon structural properties of the samples. Fig.3 shows the TPR profile of a series of Rh/CeO_2-ZrO_2 samples with amount of CeO_2 between 20% and 100%. It appears clearly that formation of a solid CeO_2-ZrO_2 solution even at relatively low ZrO_2 contents strongly promotes bulk reduction of the support, as new feature -hereinafter indicated as LT (Low Temperature) peak- below 1050 K is observed. Moreover, in the cubic phase, the maximum of the peak broadly shifts towards lower temperatures upon decreasing the CeO_2, while the reverse is observed in the tetragonal region. The relative magnitude of the LT peak centered at 600-840 K and that of the one centered at 1050-1250 K - hereinafter indicated as HT (High Temperature) peak- is also affected by the amount of CeO_2 in the sample. On decreasing CeO_2 content, the HT peak becomes smaller and broader, and it is almost nihil in the tetragonal samples. Reduction of CeO_2 is suggested to proceed initially via a surface reduction and then on increasing the temperature bulk reduction occurs through a lattice oxygen diffusion limited process [13]. Quantitative estimation of the degree of reduction showed that after LT peak, a bulk composition of $Rh/Ce_mZr_{1-m}O_x$ ($0.2 \leq m \leq 0.9$) with $1.87 \leq x \leq 1.92$ is obtained. In the fluorite oxides, for defect

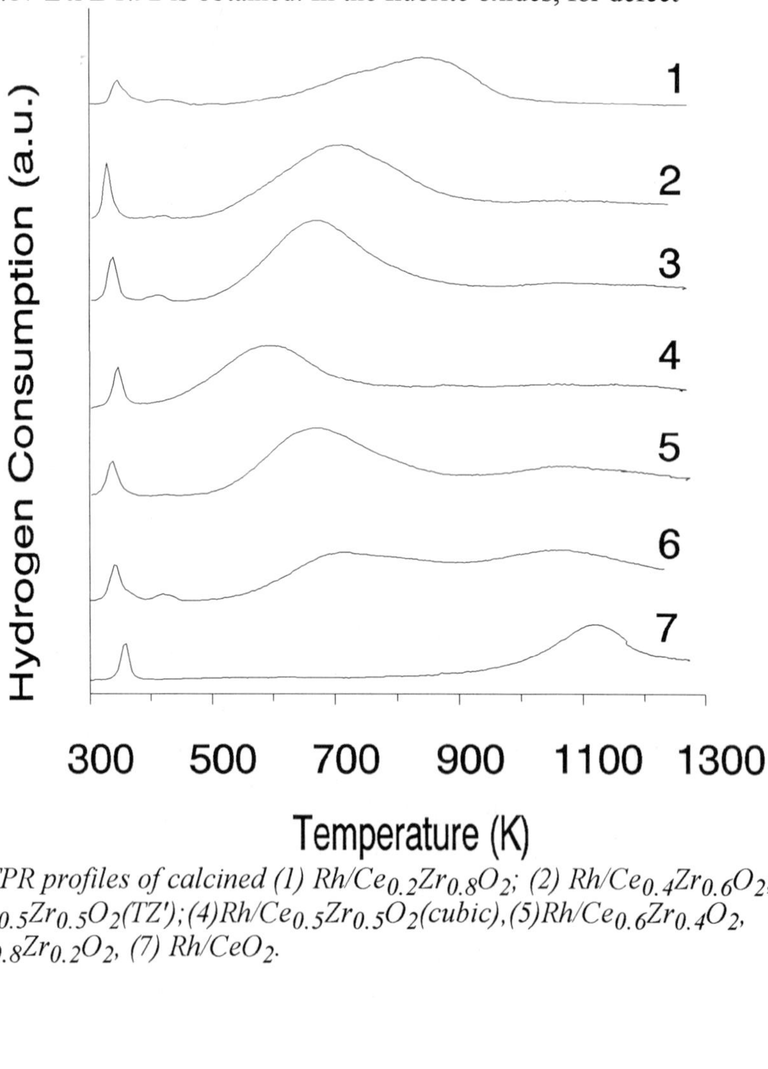

Figure 3.TPR profiles of calcined (1) $Rh/Ce_{0.2}Zr_{0.8}O_2$; (2) $Rh/Ce_{0.4}Zr_{0.6}O_2$; (3) $Rh/Ce_{0.5}Zr_{0.5}O_2$(TZ');(4)$Rh/Ce_{0.5}Zr_{0.5}O_2$(cubic),(5)$Rh/Ce_{0.6}Zr_{0.4}O_2$, (6) $Rh/Ce_{0.8}Zr_{0.2}O_2$, (7) Rh/CeO_2.

concentration above 2-x=0.08, the activation energy for the mass transport increases with increasing defect concentrations [14a]. This suggests that the defects which form in the course of the reduction might clusterize [14], which makes the oxygen transport more difficult, and this is responsible for the appearance of the HT peak. A full discussion of the reduction and reoxidation behaviour is reported elsewhere [15].

There is another very important point which should be noted: the occurrence of a bulk reduction also promotes the Oxygen Storage Capacity of the present samples as measured by oxygen uptake in comparison to Rh/CeO_2 catalysts irrespectively of their surface area. For example, after a reduction at 700 K, a Rh/$Ce_{0.5}Zr_{0.5}O_2$ sample of surface area 1.5 m^2g^{-1} showed, at the same temperature, an oxygen uptake of 9.0 ml O_2 g^{-1} compared to 4.8 ml O_2 g^{-1} measured on a Rh/CeO_2 sample (surface area 130 m^2g^{-1}) [15]. In summary, the observed reduction behaviour shows that incorporation of ceria into a solid solution with zirconia strongly favours formation of reduced Ce^{3+} species, and there is an optimum range of composition (CeO_2 = 40-60%) where the highest degree of support reduction and lowest reduction temperatures are observed. Finally, the 60 K shift of the LT peak toward lower temperature observed over the cubic Rh/$Ce_{0.5}Zr_{0.5}O_2$ sample in comparison with the tetragonal one (Fig.3, traces 3,4) is an indication that in the cubic structure the reduction process is favoured due to the higher oxygen mobility in the bulk.

3.2 Catalytic Activity Measurements

In order to investigate the effects of the nature of the support on the catalytic properties of the present catalysts, selected catalytic runs were carried out in a flow reactor at 473 K. The effects of the initial degree of reduction of the support were also analyzed and therefore prior to the catalytic run, the catalysts were reduced both at 473 K and 673 K for 2h. As indicated from the results of the TPR experiments, after a reduction at 473 K, no significant reduction of the bulk of the support is obtained while after a reduction at 673 K, a significant fraction of the bulk of the support is reduced. Consistently, separate TPR experiments showed that after an isothermal reduction at 473 K, the magnitude of the LT peak appears unaffected, while after an isothermal (2h) reduction at 673 K, the LT feature is no long observed. Notably, in the latter experiment carried out over the Rh/$Ce_{0.6}Zr_{0.4}O_2$ sample, after the isothermal reduction we measured a degree of reduction very close to that obtained after the LT peak.

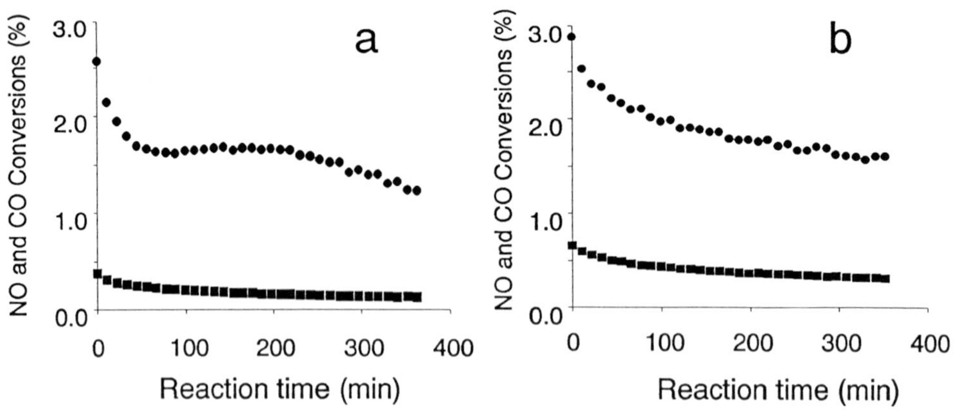

Figure 4. NO (•) and CO (■) conversions over Rh/$Ce_{0.6}Zr_{0.4}O_2$ prereduced respectively at 673 K (a) and 473 K (b).

Typical CO and NO conversion profiles as a function of time on stream at 473 K are reported in Fig.4b. At this temperatures N_2O is the major nitrogen containing product. For all the catalysts employed we observe an initial partial deactivation of the catalysts to reach a stationary state after approximately 5-6 h which is consistent with previous studies [16]. Recently, we observed over a Rh/Al_2O_3 catalyst that at a reaction temperature of 473 K in the presence of NO and CO, oxidative disruption of the supported rhodium particles to give isolated Rh^I sites is favoured while higher temperatures favour reductive agglomeration [17]. This suggests that upon steaming at 473 K, rhodium particles might be disrupted decreasing the efficiency of the catalysts as a result of smaller particle size.

The effects of steaming, however, appear affected by the reducing pretreatment. As shown by NO conversion over the Rh/$Ce_{0.6}Zr_{0.4}O_2$ catalyst prereduced at 673 K (Fig.4a), while the CO conversion shows only a slight decrease with time on stream, the NO conversion show an unusual behaviour showing a rapid initial deactivation which is followed by a slight increases of the activity starting at about 80 min of reaction and peaking at 200 min. Then, the NO conversion slowly decreases to the steady state value after approximately 6-8 hours of reaction. No such unusual behaviour was observed over the Rh/$Ce_{0.6}Zr_{0.4}O_2$ catalyst prereduced at 473 K (Fig.4b). The influence of the pretreatment temperature on the NO conversion becomes more apparent when the reaction rates measured in steady state conditions are reported in an Arrhenius fashion (Fig.5).

importantly, an apparent activation energy of 67-68 kJ mol^{-1} was measured in this experiment over both $Rh/Ce_{0.6}Zr_{0.4}O_2$ and $Ce_{0.6}Zr_{0.4}O_2$.

There is a further point of interest in these experiments. Surface oxygen vacancies, produced by a reductive treatment of CeO_2 have been suggested as promoting sites for NO and CO conversions. Recently, we observed that CO_2 hydrogenation over Rh/CeO_2 catalysts is promoted by a high temperature reduction and the observed transient rate enhancement was associated with formation of oxygen vacancies located at the periphery of the metal particles which in turn facilitate CO_2 dissociation [19]. Generally speaking, all these investigation point out the important role of surface oxygen vacancies [3,6,7,19] or surface oxygen nests [21] in determining the catalytic behavior of M/CeO_2 systems. Recent evidence, however, shows that even long range effects involving bulk CeO_2 lattice oxygen may play important role in these phenomena. CO TPD from a Rh/CeO_2 catalyst showed that migration of lattice oxygen from CeO_2 to the metal occurs [22]. The oxygen storage capacity was increased by formation of a La_2O_3-CeO_2 solid solution in Pt, Rh/La_2O_3-CeO_2/Al_2O_3 catalysts which was attributed to formation of lattice oxygen vacancies due to La^{3+} incorporation [23]. Maire et al [24] recently observed an enhancement of rate of CO oxidation in the presence of prereduced Pt/CeO_2 catalysts which was attributed to the presence of bulk oxygen vacancies. Participation of lattice oxygen has been recognized in several oxidation reactions while the present results suggest that the bulk oxygen vacancies may play an important role of in promoting the rate of the oxygen exchange reaction also in the reduction.

4. CONCLUSIONS

Metallized CeO_2-ZrO_2 appear to be potentially interesting materials for automotive catalysts as the incorporation of CeO_2 into the solid solution strongly promotes the reducibility of the support which occurs in the bulk at moderate temperatures. Consequently, these materials exhibit a high Oxygen Storage Capacity notwithstanding the low surface area of the present samples. The present observations also point out a new mechanism for the role of ceria in the reduction of nitrogen oxide attributable to the presence of redox couple $Ce^{3+} \Leftrightarrow Ce^{4+}$, which, at least in the low temperature regimes, effectively dissociates NO. In view of these results, an alternative explanation may be suggested for the increased conversion of a wet exhaust compared to dry conditions [4]. In the presence of water, both CeO_2 and the supported metal efficiently catalyze the water gas shift reaction producing H_2 which then may be spilled-over the support by the supported metal. The Ce^{3+} sites thus produced, efficiently decompose NO and are reoxidized to Ce^{4+}.

ACKNOWLEDGMENTS.

Ministero dell'Università e della Ricerca Scientifica (Roma) is acknowledged for financial support.

REFERENCES

1 K.C.Taylor, in "Catalysis-Science and Technology", (J.R.Anderson and M.Boudart, eds.), Springer-Verlag, Berlin, 1984, vol.5.
2 J.C.Summers and S.A.Ausen, J.Catal. 58 (1979) 131.
3 B.Harrison, A.F.Diwell, and C.Hallett, Plat. Met. Rev. 32 (1988) 73.
4 G.Kim, Ind.Eng.Chem.Prod.Res.Dev. 21 (1982) 267.
5 H.C.Yao and Y.F.Yu Yao, J.Catal. 86 (1984) 254.
6 J.G.Nunan, H.J.Robota, M.J.Cohn, and S.A.Bradley, J.Catal. 133 (1992) 309.
7 a) A.F.Diwell, R.R.Rajaram, H.A.Shaw, and T. J.Truex, in "Catalysis and Automotive Pollution Control II", Stud.Surf.Sci.Catal. (Crucq, A. ed.), Vol. 71, pp.139-152, Elsevier, Amsterdam, 1991; b) G.Munuera, A.Fernandez, and A.R.Gonzalez-Elipe, ibidem, pp.207-219.
8 J.Z.Shyu, W.H.Weber and H.S.Gandhi, J.Phys..Chem. 92 (1988) 73.
9 M.A.Harkonen, E.Aitta, A.Lahti, M.Luoma, T.Maunula, SAE 910846 (1991).
10 a) S.Meriani, Mat.Sci.Eng. 71 (1985) 369; b) S.Meriani and G.Spinolo, Powder Diffraction 2 (1987) 255; c) S.Meriani, Mater.Sci.Eng. A109 (1989) 121.
11 A.E.McHale, Phase Diagrams for Ceramists, Annual 1991, p.20.
12 M.F.L.Johnson and J.Mooi, J.Catal. 103 (1987) 502; erratum: J.Catal. 140 (1993) 612.
13 A.Laachir, V.Perrichon, A.Badri, J.Lamotte, E.Catherine, J.C.Lavalley, J.El Fallah, L.Hilaire, F.le Normand, E.Quemere, N.S.Sauvion and O.Touret, J.Chem.Soc. Faraday Trans., 87 (1991) 1601.
14 a) J.A.Killner and B.C.H.Steele, in "Non Stoichiometric Oxides" (O.T.Sorensen, ed.), ch.5, Academic Press, New York (U.S.A.), 1981; b) H.Matzke, ibidem ch.4; c) C.R.A.Catlow, ibidem ch.2.

15 P.Fornasiero, R.di Monte, G.Ranga Rao, J.Kašpar, S.Meriani, A.Trovarelli and M. Graziani, J.Catal. 151 (1995) 168.
16 W.C.Hecker and A.T.Bell, J.Catal. 84 (1983) 200.
17 J.Kašpar, C.de Leitenburg, P.Fornasiero, A.Trovarelli, and M.Graziani, J.Catal. 146 (1994) 136.
18 K.Otsuka, M.Hatano, and A.Morikawa, J.Catal. 79 (1983) 493.
19 A.Trovarelli, G.Dolcetti, C.de Leitenburg, J.Kaspar, P.Finetti, and A.Santoni, J.Chem.Soc., Faraday Trans. 88 (1992) 1311; A.Trovarelli, G.Dolcetti, C.de Leitenburg, J.Kaspar,, J., in "New Frontiers in Catalysis", Stud.Surf.Sci.Catal. (L.Guczi, F.Solymosi, and P.Tetenyi, eds.), Vol.75, pp.2781-2784, Elsevier, Amsterdam, 1993.
20 R.K.Hertz, Ind.Eng.Chem.Prod.Res.Dev. 20 (1981) 451.
21 M.G.Sanchez, and J.L.Gazquez, J.Catal. 104 (1987) 120.
22 G.S.Zafiris and J.Gorte, J.Catal. 139 (1993) 561.
23 T.Miki, T.Ogawa, M.Haneda, N.Kakuta, A.Ueno, S.Tateishi, S.Matsuura, and M.Sato, J.Phys.Chem. 94 (1990) 644.
24 C.Serre, F.Garin, G.Belot, and G.Maire, J.Catal. 141 (1993) 9.

A. Frennet and J.-M. Bastin (Eds.)
Catalysis and Automotive Pollution Control III
Studies in Surface Science and Catalysis, Vol. 96

SELECTIVE CATALYTIC REDUCTION OF NO_X IN DIESEL EXHAUST GASES WITH NH_3 OVER Ce & Cu MORDENITE AND $V_2O_5/TiO_2/WO_3$ TYPE CATALYSTS: CAN Ce SOLVE THE NH_3 SLIP PROBLEM?

R.J. Hultermans[a], E. Ito[a], Á. József[b], P.M. Lugt[a] and C.M. van den Bleek[a]

[a]*Department of Chemical Technology and Material Science, Delft University of Technology PO Box 5045, 2628 BL, The Netherlands*
[b]*Department of Chemical Technology, Technical University of Budapest Budafoki ut 8, 1521 Budapest, Hungary*

ABSTRACT

The performance of two zeolite type catalysts on their NO_x-SCR activity and their SO_2 oxidative activity is compared to a vanadium type catalyst. Home made Ce&Cu mordenite and a commercially available $V_2O_5/TiO_2/WO_3$ type catalyst have been tested for NO_x reduction with ammonia in model and diesel exhaust gases. The Cu mordenite catalyst showed good performance from 300 to 550°C. However, SO_3 formation was detected starting at 400°C and selectivity loss caused by oxidation of NH_3 to NO starting at 450°C. The vanadium type catalyst showed the same behaviour as Cu mordenite; selectivity decline above 450°C and SO_3 formation above 400°C. However, the Ce mordenite catalyst showed high temperature (>400°C) activity without oxidation of SO_2 to SO_3 and NH_3 to NO; a large excess of NH_3 resulted only in a limited NH_3 slip.

1. INTRODUCTION

In the last few decades a lot of attention has been paid to diesel engine emission reduction. Particulary soot and NO_x emissions have been the subject of extensive research trying to obtain reductions by engine and fuel modifications and various aftertreatment devices. This paper is aimed at the development of an aftertreatment device for NO_x control: selective catalytic reduction of NO_x using an injection of an aqueous urea solution. This technique stems originally from the field of power plant NO_x control and as such is proven to be quite effective [1]. Because of safety reasons and ease of use, urea is preferred over NH_3 [2-4]. The main problem in applying this technique to non-stationary engines (HD trucks) results from the fact that the exhaust gases vary widely and rapidly in temperature as well as in NO_x content. This means that the catalyst must be active in a wide temperature range and one must be very well able to control the amount of urea injected

The temperature of diesel engine exhaust gas varies somewhere between 100 - 600 °C. Catalysts so far used in power plants can be used from about 300 - 450°C. The lower limit is determined by ammonia sulfate formation, the upper limit by NH_3 oxidation and catalyst stability. The limit at low temperature is less severe when using low sulfur fuel. Furthermore, most diesel engines don't produce much NO_x at low loads and the relative importance of such low loads/low temperature conditions is further diminished by the weighting criteria used in standard road tests, e.g. the 13-mode test procedure [5]. An additional phenomenon at high temperatures is the formation of SO_3, thereby increasing the particulate emission. Essentially a SCR aftertreatment system should posses the following characteristics: 1) SCR activity, also at high temperatures, 2) no SO_2 oxidation activity and 3) no ammonia slip.

1.1.Temperature window

A lower temperature limit will be set at 300°C. Using urea injection at lower temperatures, polymerisation products can be formed. Furthermore, at these temperatures ammonium sulfate formation will block most SCR activity. The temperature range in which the catalyst will have to work is consequently from about 300°C to 550°C.

Standard vanadium type SCR catalysts have an upper temperature limit of about 450°C. Above this temperature conversion drops rapidly as a result of ammonia oxidation and at even higher temperatures deactivation follows. For developing new catalysts zeolite type catalysts were chosen; new developments showed a very high SCR activity for a combination of CuH-MOR & H-MOR in a temperature window of 200 - 600°C [6].

1.2.SO_2 oxidation

In this paper much attention is paid to an effect which is typical for diesel engine exhaust; the implications of the presence of sulfur in diesel fuel with regard to using catalysts. These implication can roughly be divided into two parts: the effect on the particulate emission and the effect on SCR activity. The sulfur in diesel fuel is mainly converted to SO_2. An increase in particulate emission would be caused by catalytic oxidation of SO_2 to SO_3 over the SCR catalyst. Since SO_3 forms aerosols, this is being detected as particulates. The second effect is deactivation of the SCR catalyst by poisoning (p.e. ammonium sulfate formation).

Standard diesel fuel contains up to 0.3 wt % sulfur. In the near future this will be limited to 0.05 wt % [5]. This sulfur is converted mainly to SO_2 (more then 95 %), and the remainder leaves the engine as SO_3, H_2SO_4 mist, or some other oxide form. About 1-3 percent is collected during the EPA transient test cycle as H_2SO_4 mist [7].

All oxidation catalysts tend to catalyse the oxidation of SO_2 to SO_3. Especially vanadium and platina type catalysts are well known SO_2 oxidation catalysts. For a commercial or newly developed SCR catalyst, this is undesirable. Because the emission of particulates is increased, an estimation is needed of the particulate emission caused by SO_3 formation [7,8]:

$$\textit{particulates} = 0.1 * \%\,\textit{Sulphur} * \%\ SO_2\ \textit{conversion}$$

It can be calculated that for a 5 percent SO_2 oxidation rate using a low sulfur fuel (0.05 wt %) the resulting additional emission of particulates would be 0.03 g/kWh . Since the near future emission limit is 0.15 g/kWh it is clear that SO_2 conversion should be low. The exact amount depends of course on the weighting factors applied for the high load points in the testing procedures.

In the literature [8] a related source of SO_3 emission is described: sulfur storage. This is the effect that stored sulfur on a catalyst support is released during a temperature transient. When oxidation catalysts (Pt on g-Al_2O_3 or SiO_2) are subjected to aging (typically 300 hours) the amount of particulates during an EPA transient cycle increases drastically. The way to handle this problem is using a support that doesn't store sulfur (SiO_2).

1.3. Ammonia slip

Ammonia slip typically occurs when overstoichiometric amounts of urea/ammonia are injected. Due to the ammonia storage capacity of the catalyst this is not emitted directly so if an appropriate control action is taken, ammonia slip can be avoided. However, up to today an inexpensive and fast NO_x and/or ammonia sensor is not available to accommodate such a control action. Solutions have been proposed in the use of engine maps for predicting the NO_x output [9] and/or an additional oxidation catalyst for avoiding ammonia slip [3]. The latter is detrimental to the sulfate emissions and as most of the oxidized ammonia will leave the system as NO, the overall NO_x removal efficiency will be lowered. Therefore, it would be convenient if a possible overdose of ammonia will leave the system in the form of nitrogen (N_2). This would solve the SCR control problem.

2. EXPERIMENTAL

For initial catalyst screening purposes a standard laboratory flow setup was used. Typical test gas compositions for the catalyst activity tests were 1000 ppm NO and NH_3, 5% O_2 and balance Argon. An initial screening on SO_2 oxidizing capacity was performed. The Cu-MOR and Ce-MOR were first pressed and then crushed again to yield a particle diameter range of 0.3-0.7 mm. Catalyst preparation and characterization is documented by Ito et al [10]. Normally 0.5 gram of catalyst was used with a flow of 500 Nml/min; this corresponds with a space velocity of about 35 000 h^{-1}. A VG Prima 600 type Mass Spectrometer was used to measure all nitrogen containing species and SO_2.

Further testing was performed with a somewhat larger apparatus (30 times larger; 15 grams of catalyst and 15 Nl/min). With this setup it was also possible to use diesel exhaust gases. Gas mixtures consisting of NO, NH_3, O_2, H_2O and SO_2 in balance Nitrogen or air were used. The diesel exhaust gas composition used was 7 % H_2O, 7 % CO_2, 50 ppm SO_2, 10 % O_2, 570 ppm NO_x, balance nitrogen [11] with 600 ppm NH_3 added for the reaction. The diesel exhaust gases were filtered to remove any particulates. Two home made metal exchanged zeolite type catalysts were tested: Cu-MOR and Ce-MOR. A $V_2O_5/TiO_2/WO_3$ catalyst, the Degussa AG DN 32 washcoated monolith (300 cpsi), was fitted into the reactor which had an internal diameter of 1.95 cm. Somewhat more of the DN 32 catalyst, 24.8 versus 15.0 grams, was used to obtain the same space velocity.

The NO_x concentration was measured with an Eco Physics chemiluminescence NO_x analyser. Ammonia was measured with a Siemens MIPAN (MIcrowave Process ANalyser). In case of dry model gases the disappearance of SO_2 was measured with a Siemens Ultramat 5E NDIR (Non- Dispersive Infra-Red) analyser. In case of diesel gases SO_3 was measured with a Severn Science Instruments automatic titration unit and the SO_2 concentration was calculated on basis of the fuel sulfur content.

Two diesel fuels have been used. The first one was a certified fuel, Haltermann CEC RF-03-A84, containing 0.14 wt % sulfur. The second fuel was a low sulfur fuel, obtained from van Gelder Aardolie B.V., with less than 0.001 wt % sulfur. In the case that certified fuel was used, the MIPAN ammonia analyser was equipped with an SO_2 scrubber.

3. RESULTS AND DISCUSSION

3.1. Screening results of Cu & Ce-MOR

The reduction of NO and the oxidation of SO_2 as a function of temperature using Cu-MOR and Ce-MOR was studied (Figure 1 & 2). The activity of Cu-MOR for the SCR reaction synthetic gases is very high; 100 percent conversion is already reached at 200°C. Increasing temperature leads to a conversion drop which is probably caused by ammonia oxidation (the ammonia conversion was 100 percent). The Ce-MOR reaches this conversion level at 300°C. A decline in NO conversion is found which is also attributed to ammonia oxidation. Behaviour of other, comparable, zeolite type catalysts in model gases are reported by Ito et al [10].

SO_2 oxidation occurs for Cu-MOR at about 400°C and leads at higher temperatures to excessive sulfate production. The Ce-MOR catalyst however, only shows a little amount of oxidation at 600°C. It should be noted that for practical purposes, especially in the low temperature range, it is possible to use far less Cu-MOR to reach the same conversion as the Ce catalyst.

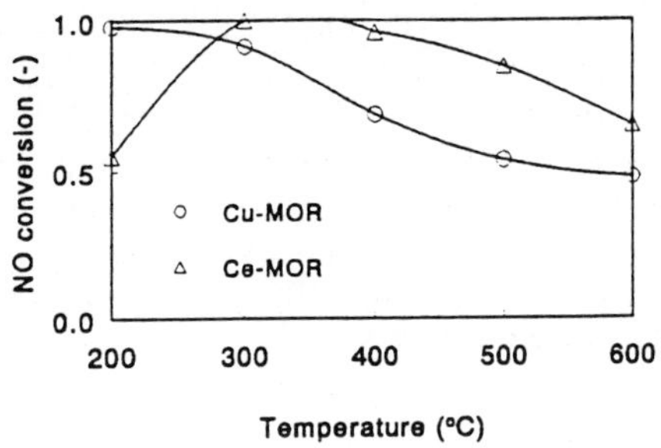

Figure 1: Activity of 0.5 grams of Cu- & Ce-MOR in 1000 ppm NO&NH_3, 5% O_2 and 500 Nml/min balance Argon.

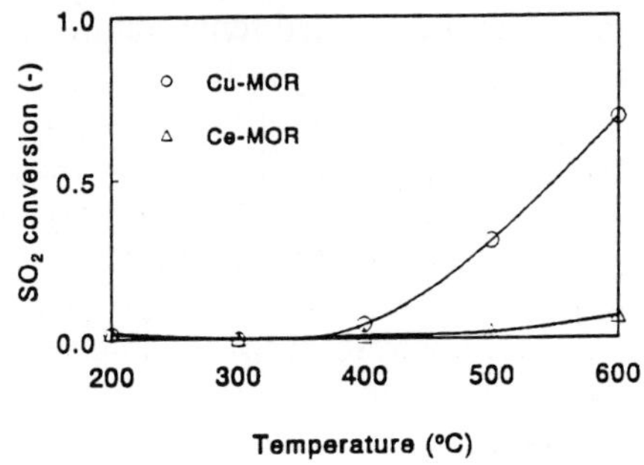

Figure 2: Activity of 0.5 grams Cu- & Ce-MOR in 500 ppm SO_2, 5 % O_2, 500 Nml/min balance Argon.

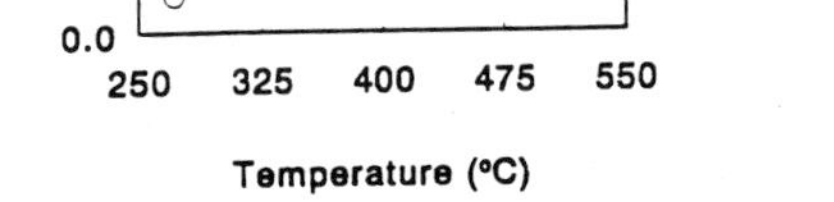

Figure 3: SCR activity of Cu- & Ce-MOR and DN 32 in diesel exhaust using diesel fuel with 0.14 wt % sulfur (sv 35 000 H^{-1}).

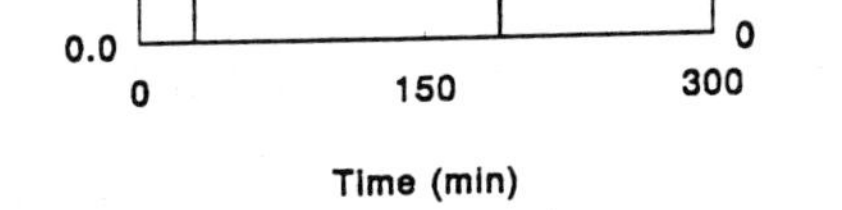

Figure 4: Influence of SO_2 on the performance of Ce-MOR in model gas containing 500 ppm NO & NH_3, 7% H_2O, 10% O_2 and balance N_2 (sv 35 000 h^{-1}).

3.2.Catalyst testing with diesel exhaust

The next step was testing these catalysts with diesel exhaust gases. It was expected that activities would be lower because of the presence of water and sulfur compounds. This effect is shown in Figure 3, where, compared to the results from Figure 1, the onset of the SCR reaction shifts at least 100°C towards the high temperature region. For the Cu-MOR catalyst enough activity was retained in the desired temperature range, the Ce-MOR catalyst did lose some of its activity in the range of 300 - 400 °C. This could be attributed to a combination of reversible deactivation of water and an irreversible deactivation of sulfur compounds (Figure 4). Partial regeneration was possible by heating up the catalyst, so the sulfur compounds can be desorbed. As already stated by Ham (11) ammonium sulfates can be found in Cu-MOR pores at temperatures up to 400°C. The Cu-MOR did also show some deactivation at temperatures below 300°C but because of the much higher activity a high conversion level is retained.

In Figure 3 the $DeNO_x$ performance of the vanadium catalyst is also given. It should be noted that comparison with the zeolite type catalysts is somewhat complicated because the vanadium catalyst is already put on a monolith carrier. However, what can be seen is the characteristic behaviour of a vanadium catalyst: it starts to work well above 300°C (also because of ammonium sulfate formation) and shows an activity decline at 450°C because of ammonia oxidation.

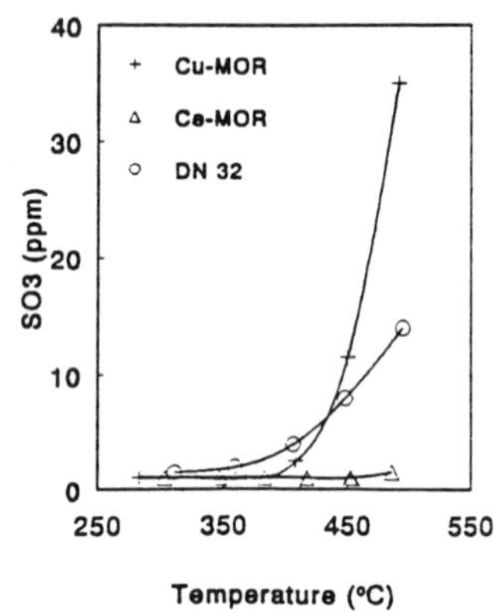

Figure 5: SO_3 production of DN 32, Ce-MOR and Cu-MOR in diesel exhaust using 0.14 wt % sulphur containing diesel fuel.

The sulfate production of this type of catalysts, Figure 5, follows the same trend as depicted in Figure 2; at about 400°C the vanadium type catalyst and Cu-MOR start producing sulfates whereas Ce-MOR does not produce any sulfates. One might expect no ammonium sulfate formation when the catalyst does not oxidize SO_2. However, the deactivation in a diesel exhaust or in model gases occurs at a much lower temperature than when SO_2 oxidation is observed.

Summarizing, a Ce-MOR catalyst is under development which does show a high activity over a wide temperature range. It should however be noted that, especially in presence of SO_2, in diesel exhaust gases this temperature window is decreased significantly starting at about 400°C. However, an important advantage is that no sulfates are formed. For practical purposes in diesel engine exhausts it will be necessary to use this catalyst in series with for example copper mordenite catalyst as already proposed by Medros [6]. Without physical separation and/or careful optimization with regard to catalyst activity, some sulfates will be produced. Furthermore, since most zeolites tend to absorb large amounts of SO_2, in practise some temperature transient related desorption effects can be expected.

3.3. Ammonia slip

The main technical problem for application of SCR in transient operation lies in the control of the amount of ammonia/urea injected as explained before. Consequently, attention must be paid to the ammonia oxidative capacity of the catalysts. As can be seen in Figure 6 and 7 Ce-MOR does not produce any NO during SCR operation or ammonia oxidation whereas the vanadium type catalyst (and also the Cu-MOR) start to produce NO at higher temperatures.

The question is wether it is possible to use an overstoichiometric amount of ammonia to reduce the NO_X. The idea is that superfluous ammonia will be oxidized to nitrogen rather than NO. This would facilitate the control problem considerably and possibly make an additional oxidation catalyst superfluous. As can be seen in Figure 8 this works in a diesel exhaust gas at high temperatures. Of course, the amount of overdosing is not unlimited because the onset of ammonia oxidation may also shift to a higher temperature in case of diesel exhaust. Furthermore, chances of ammonia slip caused by desorption due to temperature transients will increase because more ammonia will be stored on the catalyst.

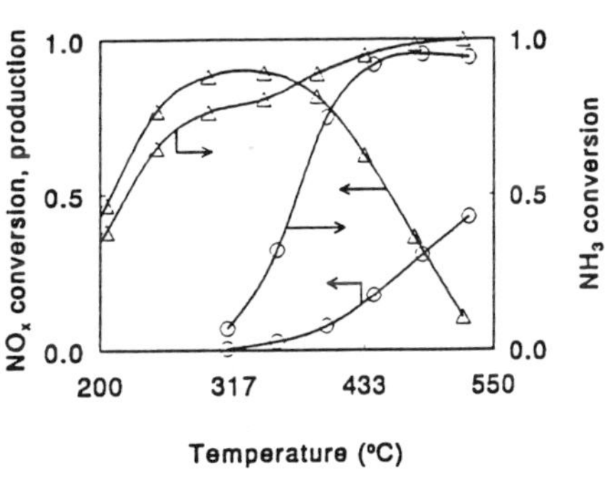

Figure 6: NO and NH3 conversions during SCR operation (Δ), NH3 conversion and NO formation during ammonia oxidation (o) for DN32. The mixture used was 500 ppm NH3 (& NO) in air, sv 35 000 h^{-1}.

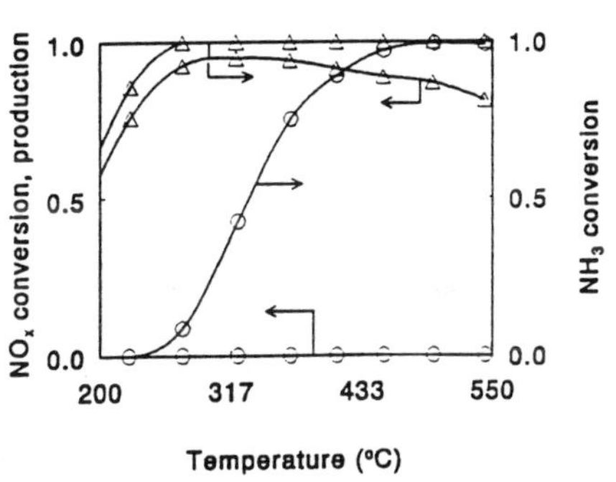

Figure 7: NO and NH3 conversions during SCR operation (Δ), NH3 conversion and NO formation during ammonia oxidation (o) for Ce-MOR. The mixture used was 500 ppm NH3 (& NO) in air, sv 35 000 h^{-1}.

4. Concluding Remarks

It has been shown that a metal exchanged mordenite (Ce) can have a high SCR activity without oxidizing SO_2 to SO_3 and NH_3 to NO. The main problem to be solved for diesel exhaust is the activity in the temperature region of 300-400°C. This problem is less when using low sulfur fuels. Until then, the Ce catalyst can only be used in high temperature applications or for diesel exhaust in combination with other compounds (i.e. copper).

The problems with the control of the urea injection can be avoided when the dosing needs not to be very precise.

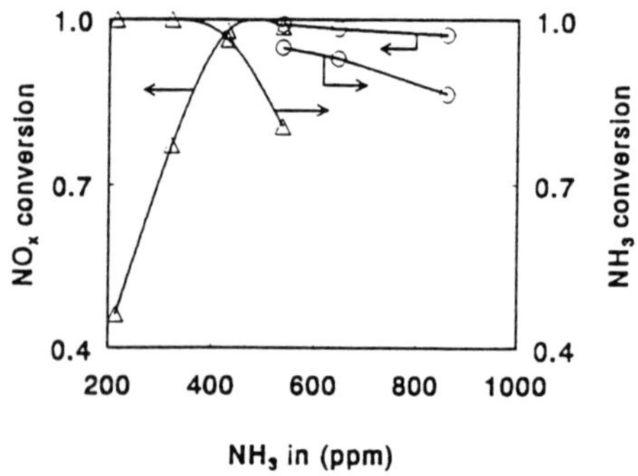

Figure8: Behaviour of Ce-MOR in diesel exhaust using low sulfur fuel as a function of NH_3 concentration, 440°C (Δ) and 530°C(o).

The advantage of the Ce catalyst is that ammonia oxidation forms only nitrogen so relieving the control considerably. Until now, ammonia slip is oxidized (in most cases to NO) by placing an oxidation catalyst behind the SCR system.

Acknowledgement

The authors would like to thank the Dutch Ministry of Housing and Environment (VROM) for their financial support.

REFERENCES

1 H. Bosch and F. Janssen, Catal. Today, 2 (1988) 369.
2 C.M. van den Bleek, P.J. van den Berg and A.G. Montfoort, 466, Second European Conference on Environmental Technology, Amsterdam, The Netherlands, 1987.
3 H.T. Hug, A. Mayer and A. Hartenstein, SAE Paper 930363.
4 A.W. Wypkema, PhD thesis, Delft, The Netherlands, 1991.
5 CONCAWE, Motor vehicle emission regulations and fuel specifications - 1992 update. ,rep. no. 2/92, Brussels, Belgium, 1992.
6 F.G. Medros, J.W. Elridge, and J.R. Kittrel, Ind. Eng. Chem. Res. 28 (1989) 1171.
7 K.J. Springer, Journal of Engineering for Gas Turbines and Power, 111 (1989) 361.
8 D.J. Ball and R.G. Stack, 337, Catalysis and Automotive Pollution Control II, Amsterdam, The Netherlands, 1991.
9 J. Walker and B.K. Speronello, SAE 921673.
10 E. Ito, R.J. Hultermans, P.M. Lugt, M.H.W. Burgers, H. van Bekkum and C.M. van den Bleek, this conference, Catalysis and Automotive Pollution Control III.
11 D.M. Heaton, TNO-MEL report 733030013, Delft, the Netherlands.
12 S.W. Ham, H. Choi, I.S. Nam and Y.G. Kin, Catal. Today, 11 (1992) 611.

A. Frennet and J.-M. Bastin (Eds.)
Catalysis and Automotive Pollution Control III
Studies in Surface Science and Catalysis, Vol. 96

REACTION STUDY OF DIESEL EXHAUST GASES OVER COPPER OXIDE BASED MODEL CATALYSTS

C.M.Pradier[§], H.Vikström, and J.Paul

KTH/Royal Institute of Technology, Physics III, S–100 44 Stockholm, Sweden
§ for correspondance

ABSTRACT

Nitric oxide reduction by hydrocarbons over oxidized copper samples has been studied by multiplexed mass–spectrometry in an all quartz reactor operated in static mode. At 600°C the reaction proceeds exclusively via consumption of HC and NO. NO is not consumed in the presence of dioxygen but preoxidation of the copper sample is necessary to facilitate high conversion rates. Infrared spectra obtained in situ at 45- incidence angle verify the conversion and show that the bulk of the active catalyst consists of cuprous oxide. A combination of IR– and mass–spectroscopy has proven to be a successful way to identify additional products and intermediates in the gasphase.

1.INTRODUCTION

The catalytic conversion of NO is a subject of great importance due to the environmental impact of the gas. New legistration, which puts stern limits on the emission from automobiles, has recently been proposed in the United States as well as in a number of European countries. This forms a new incentive to find economically attractive and efficient catalysts. The platinum group metals are extensively used in commercial catalysts for Otto engines but one major drawback with these metals is their high cost. Another shortcoming of these conventional three– way catalysts is their limited ability to operate under the oxidizing conditions of lean burn gasoline engines and diesel engines. Various materials containing copper have shown activity for NO reduction under these conditions but sulphur poisoning and deactivation by moisture limit their use [1]. Our approach is based on model studies where we isolate and characterize different parts of the overall reaction scheme. The present work concentrates on NO/HC reactions under reducing conditions. With our ongoing project we will achieve a better understanding of the mechanisms involved on the metal and the metal oxide surfaces. This knowledge will be applied to tests under oxidizing condtions, mimicing the exhaust gas composition, with a higher yield and an improved process as the result. The materials aspects of the active catalyst are discussed elsewhere [2]. A new deposition method for copper oxide with alumina as the carrier relates to our model studies of the unsupported catalyst and has proven to be superior to conventional impregnation methods [3].

2. EXPERIMENTAL

2.1. Mass–spectroscopy

A commercial quadrupole mass spectrometer (Balzer 421) was used to obtain information about gas conversion and consumption during the reaction. The sample was a 50 cm^2 disc of high purity copper (Goodfellow). Different gas mixtures, containing oxygen, nitric oxide, and iso– butane, were tested in order to find optimal conditions for NO conversion. The sample and gas mixture (total pressure 5–10 torr) were heated in an all quartz batch reactor by an external oven. This arrangement mimics realistic reactor conditions with a gas–temperature close to the catalyst´s temperature, a condition rarely fullfilled in surface science experiments. A small amount of gas was continously removed through a pipe to the mass spectrometer. The activity of the empty reactor, including thermal decomposition and combustion, has been studied and found to be negligible. Kinetics were evaluated from mass spectrometric data without considering the thermalization factor.

2.2. Infrared Spectroscopy

Specular reflection Fourier transform infrared spectroscopy, performed with the beam striking the sample at 45- from the surface normal, was also used for in situ studies of the catalytic reaction. This technique makes it possible to observe changes in the gasphase composition [4–5]. The broad band reflectivity of the overlayer can be used to obtain the composition of a thin oxide film [6]. Infrared spectra have been obtained from polycrystalline samples and from Cu(110). The former samples are akin to the ones used for mass spectrometric studies. Distinguished from those studies our FTIR work was performed with a reaction vessel at room temperature. The sample was in this case resistively heated by tantalum wires. This combination separates the gas temperature from the temperature of the catalyst. The temperature of the impinging molecules is a complicated function of the sample temperature and the pressure in the pressure range $>10^{-3}$ torr. At this pressure the mean free path of molecules is shorter than the distance between the sample and the walls of a typical ultra high vacuum system integrated with a catalytic reactor.

3. Results and Discussion

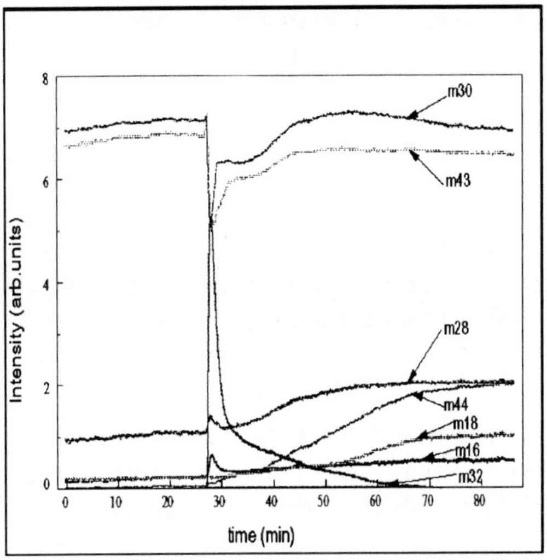

Fig. 1. Multiplexed mass–spectra from a sampling device in the batch reactor. The period to the left is characterized by preoxidation of the sample and consumption of oxygen and the period to the right by a simultaneous decrease of NO and isobutane and increase of nitrogen, water, and carbon dioxide. The reaction temperature was 600°C.

Figure 1 shows the gas–composition as a function of time during the preoxidation period and during typical batch mode conversion of NO and iso–butane to nitrogen, water, and carbon dioxide. Small quantities of other products: carbon monoxide, other nitrogen oxides, and hydrocarbons, were also detected. These data were obtained from the all quartz reactor with a polycrystalline copper sample. The initial total pressure was 10 torr with a O_2:NO:iso–C_4H_{10} ratio of 1:5:10. Preoxidation occurs during the transitory stage of the reaction. This stage is characterized by the rapid consumption of oxygen and by the growth of a cuprous oxide overlayer [2]. NO and HC conversion start during this stage but leads to incomplete oxidation of HC to CO and H_2O. The bulk structure of the active catalyst is further investigated by X–ray diffraction following an extensive reaction cycle [2].

The transitory stage is followed by a stationary phase with a constant and higher rate of NO conversion. We measured the rate of conversion during this stationary stage. These data give a yield of 2 NO molecules converted per second and adsorption site. This value should be taken as a nominal number rather than an absolute value due to the lack of an independent evaluation method for the number of available adsorption sites. The above number was calculated from the number of copper atoms in a closely packed planar metal surface.

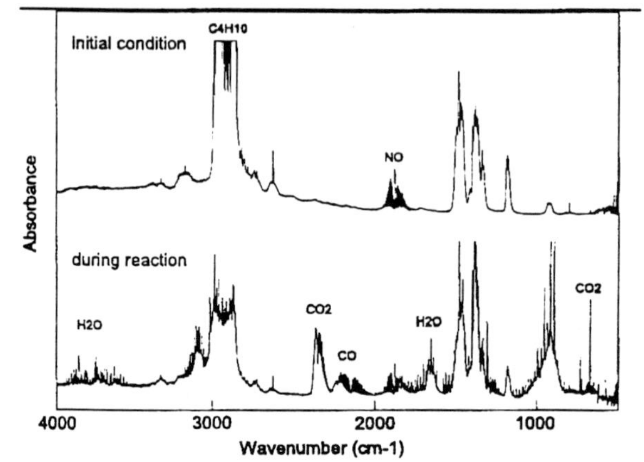

Fig. 2. Infrared spectra obtained by reflection off the sample in the reactor vessel, before and during the reaction. The incidence angle was 45° and the intensity was measured in the specular direction.

The qualitative conversion was confirmed by infrared spectroscopy (Figure 2). IR– spectra show the formation of gasphase intermediates in addition to the main products: N_2, CO_2, and H_2O. These intermediates were ethylene, propene, and methane. The high incidence angle, 45-, is suitable to follow changes is the oxidized overlayer rather than to detect surface intermediates [2].

The optimum conditions for rapid conversion are sensitive to the temperature and to the sequence of gases introduced into the reactor, in addition to the relative pressures. Too low temperatures and too oxidizing conditions following the preoxidation period result in continous oxide growth [2]. Other erronous conditions result in carbon build–up and passivation or run– away oxidation with extensive internal generation of heat [2]. On the other hand, considerably lower conversion temperatures with maintained high yields have been observed by in situ FTIR as a result of a gradual decrease of the externally supplied heat, once the converter is operating.

The above quoted high yields at lower temperatures may correspond to the enhanced yields at constant temperature, observed after regeneration cycles. Regeneration cycles by calcination were applied at regular time intervals for rich blends. The partial pressure of NO remains constant when the ratio O_2:NO is higher than 1:30. Each regeneration period is characterized by a reoxidation period with no NO consumption, during which oxidation of the copper sample or of surface residues occur. Removal of surface residues, mainly carbon, can be diagnosed by increased amounts of CO and CO_2. As the oxygen is consumed, the NO reduction rate again approaches the same level as during previous cycles. The decrease of the NO partial pressure sometimes shows an oscillatory behavior followed by a very rapid decrease.

The above observations in combination with unavoidable changes in the gas composition, due to the rapid rates and a limited vessel volume, give that the optimum conditions are limited to a narrow temperature and gas–composition window. Nevertheless, the high yields observed under the present optimal conditions and the likelihood of stable operation at lower temperatures are most encouraging. These yields are not out of range with conventional catalysts based on rare metals and compare favourably to complex oxides and zeolites.

4.CONCLUSIONS

Copper, modified by oxygen, was shown to be an active catalyst for reducing NO in the presence of a hydrocarbon. It is a realistic alternative to noble metals. Oxygen is necessary during a preliminary step of the reaction. Subsequently, the reduction of NO is effective on an oxide surface without oxygen in the gas phase. Kinetics results i.e. the rate of NO consumption and the N_2/CO ratio in the products are strongly dependent upon the temperature of preoxidation of the catalyst. This result suggests that the reaction is sensitive to the oxidation state of copper in the superficial layers.

This original experimental approach led us to the conclusion that in situ FTIR, combined with Mass Spectrometry, is the appropriate analysis technique for i) products, ii) intermediates, and iii) catalytically active surfaces.

ACKNOWLEDGEMENT

We acknowledge the joyful visit of John Robbins to our laboratory. He boosted our progress in this project.

REFERENCES

1 Ertl G., Hierl R., Knözinger H., Thiele N., and Urbach H.P., Appl.Surf.Sci.5(1980)49.
2 Pradier C.M., Robbins J.L., Liang K., Hoffmann F.M., and Paul J., (manuscript)
3 Pradier C.M., Hall R.B., Myers G., Öhman L.O., and Paul J., (manuscript)
4 Greenler R.G., Hahn R.R., and Schwartz J.P., J.Cat.23(1971)42.
5 London J.W. and Bell A.T., J.Cat.31(1973)96.
6 Wood B.J., Wise H., and Yolles R.S., J.Cat.15(1969)362.

A. Frennet and J.-M. Bastin (Eds.)
Catalysis and Automotive Pollution Control III
Studies in Surface Science and Catalysis, Vol. 96

SELECTIVE REDUCTION OF NO_X WITH AMMONIA OVER CERIUM EXCHANGED ZEOLITE CATALYSTS: TOWARDS A SOLUTION FOR AN AMMONIA SLIP PROBLEM

E. Ito, R.J. Hultermans, P.M. Lugt, M.H.W Burgers, H. van Bekkum and C.M. van den Bleek[1]

Delft University of Technology, Department of Chemical Technology and Material Sciences, Julianalaan 136, 2628 BL Delft (The Netherlands)

ABSTRACT

Cerium-exchanged ZSM-5 and mordenite showed a high NO conversion (> 70 %) and a high selectivity to N_2 (> 97 %) at 300 - 600 °C for NO reduction with ammonia in the presence of oxygen. Ammonia was found to be oxidized by oxygen over these cerium zeolite catalysts exclusively towards N_2 without production of N_2O and NO. NO reduction with up to 30 % excess of ammonia exhibited a high NO conversion and complete conversion of NH_3 at 300 - 500 °C at a gas space velocity of 12,000 h^{-1}. This offers a possible solution for the ammonia slip problem in a selective catalytic reduction (SCR) system with NH_3. Strong NH_3 adsorption up to 600 °C and a high amount of adsorbed reactive NO species in associated with the redox property of cerium (Ce^{III}/Ce^{IV}) are assumed to be responsible for the high NO reduction activity of cerium exchanged zeolite catalysts.

1. INTRODUCTION

The removal of NO_x from exhaust gases is an urgent issue, and environmental regulations are becoming more stringent than ever. The selective catalytic reduction (SCR) of NO_x with ammonia is the technique, which was extensively applied in stationary NO_x sources from the 1970s, while the successful development of a three-way catalyst made it possible to reduce NO_x in gasoline automobile exhaust.

---

[1]To whom correspondence should be addressed

Nowadays an intense focus is being placed on the development of a catalyst applicable in oxygen-rich conditions, aiming at cleaning of diesel exhaust or lean-burn exhaust gas. After a promising report on Cu-ZSM-5 [1, 2], a variety of zeolite catalysts has been investigated for NO_x reduction with hydrocarbons as reductants: Cu-ZSM-5 [1-6], Fe-ZSM-5 [7], Ce-ZSM-5 [8] and H^+ zeolites [9] mostly with ethene, propane or propene; and recently Ga-ZSM-5 [10] and Co-ZSM-5 [11] were reported to be able to reduce NO_x with methane as a reductant.

Ammonia-type reductants, which include ammonia, urea, cyanuric acid and different ammonium salts, have been applied so far mainly for stationary sources. The apparent preference for a hydrocarbon reductant for a mobile $deNO_x$ system is mainly based on the following reasons: (i) Hydrocarbons are always present in a combustion exhaust; (ii) Equipping mobile sources with an ammonia tank is considered unsafe; (iii) Ammonia-type reductants require a stringent injection control to prevent ammonia slip (< 5 ppm required [12]); (iv) Ammonia is found to be converted to NO and N_2O at higher temperatures (> 400 °C) over various SCR catalysts like V_2O_5 [12 - 14], WO_3 [13] and Cr_2O_3 [15]; (v) SO_2 present in a combustion exhaust is often converted to SO_3 to produce acidic particulates [13].

Nevertheless, ammonia-type reductants are generally much more effective for NO reduction than hydrocarbons, as seen in a comparison between different literature data at a given gas space velocity (GHSV) [16 - 18]. This is mainly ascribed to an intrinsic high reactivity of ammonia towards NO_x species to produce a N°N bond as a direct consequence. In contrast, in the case of hydrocarbon reductants, two NO molecules should be eventually combined to produce N_2 with an intervention of the reductant hydrocarbon. For instance, Iwamoto *et al.* [19] and Ukisu *et al.* [20] have proposed isocyanate as an intermediate formed from a hydrocarbon and NO over a copper-based catalyst, while Yasuda *et al.* have suggested organic nitro and nitroso compounds as intermediates over Ce-ZSM-5 [21]. These species are hereafter expected to react with NO [19], or NO_2 formed in a separate activation of NO with O_2 [21], finally leading to N_2 production. Such a complicated reaction route with plural steps is clearly subjective to many side reactions. A loss of reductant by combustion with oxygen is probably the largest side reactions in a hydrocarbon system, and it drastically lowers the reductant utilization for NO reduction. In contrast, ammonia is relatively unreactive towards oxygen.

Focusing on the intrinsic effectiveness of ammonia-type reductants, we have been investigating the application of a urea solution for the NO reduction in diesel exhaust. Since urea decomposes into two reductant molecules (ammonia and/or cyanic acid) above 300 °C in the presence of water, it can serve as a safe reductant reservoir. A high performance of urea solutions in NO reduction has already been reported in detail elsewhere [16, 22]. The other disadvantages remained to be solved are (iii) the ammonia slip, (iv) unfavorable ammonia oxidation to N_2O and NO, and

(v) a high SO_2 oxidation. Ammonia slip can be prevented in principle by a precise control of reductant injection to meet a maximum NO_x conversion and a minimum ammonia slip. However, this is rather difficult, in particular, under the non-stationary operation conditions of vehicles, and the present process control is not at a stage of application yet. An alternative solution is the development of a catalyst, over which NO is converted with ammonia efficiently towards N_2, and an excessively injected ammonia is converted selectively to nitrogen. Such a catalyst system enables the present control system applicable without concerns about an ammonia slip and undesirable production of NO or N_2O from ammonia. In addition, such a catalyst is desired not to promote SO_2 oxidation.

We have recently reported that cerium-exchanged mordenite (CeNa-MOR) is a highly active and selective catalyst for NO reduction with ammonia in oxygen-rich conditions [23]. We have further found that it oxidizes SO_2 to a negligible extent [24, 25], and that it appears to fulfill the aforementioned requirements for an ideal NH_3-SCR catalyst. In the present paper, we will report on these aspects of cerium-exchanged zeolite catalysts. Furthermore, its surface interaction with two reactants, NO and NH_3, is examined with the temperature programmed desorption (TPD) technique.

2. EXPERIMENTAL

2.1 Catalyst preparation

Na-MOR (SiO_2/Al_2O_3 = 13.1, PQ Zeolite), Na-ZSM-5 (SiO_2/Al_2O_3 = 40, Uetikon) and Na-Y (SiO_2/Al_2O_3 = 5.2, Akzo) were obtained commercially. Cation exchange was carried out with 8 - 25 g of zeolite in one of following metal salt solutions (1.5 l) at a given temperature for 5 - 20 h: 4.0 - 35 mM copper(II) acetate or 4.0 - 35 mM cobalt(II) acetate at room temperature, 3.9 mM iron(II) sulfate at 70 °C, 4.0 - 8.3 mM cerium(III) acetate or 2.0 mM lanthanum chloride at 70 - 100 °C. Exchange stoichiometries (Na^+ and M^{n+}) were confirmed on a charge equivalent basis for these samples; these were denoted as CuNa-ZSM-5(95), CeNa-MOR(71) *etc.*, the number in parentheses refers to the percentage of ion exchange. Partly H^+ exchanged cerium samples were prepared by an ion-exchange in 2.7 - 8.3 mM cerium(III) sulfate solution at 50 - 100 °C. They were denoted as CeNaH-ZSM-5(17) and CeNaH-MOR(41), in which H^+ sites were present at 67 % and 51 % (exchange stoichiometry), respectively. H-MOR was prepared from fully exchanged NH_4^+-MOR followed by an air calcination at 550 °C for 24 - 48 h. Exchanged samples were checked with X-ray powder diffraction for crystallinity, and elemental analysis was carried out with ICP-AES (Inductively Coupled Plasma Atomic Emission Spectroscopy) and AAS (Atomic Absorption Spectroscopy). A commercial

V_2O_5/Al_2O_3 catalyst (DN110, 10 wt% V_2O_5, Rhône-Poulenc) was used for a comparison.

2.2 Catalytic reactions

Catalytic reactions were carried out using conventional fixed bed reactors; a bench-top reactor BTRS-900 (Autoclave Engineers) or a home-made automated system. The reactant gases, NO (1024 ppm or 1 % in argon), NH_3 (952 ppm or 1 % in argon) and O_2 (20 % in argon) were commercially obtained (Scott specialty gases) and used without further purifications. These gases were applied under the following reaction conditions: NO (465 - 1000 ppm), NH_3 (465 - 1000 ppm) and O_2 (1 or 5 %) in argon at a total flow rate of 400 - 500 ml/min. 0.5 - 1.0 g of a pelletized catalyst (d = 0.5 mm on average) was used in a quartz tube reactor. These test conditions give gas space hourly velocities (GHSV) between 12,000 and 30,000 h^{-1} based on the apparent bulk density of zeolite (0.5 g/cm^3). After a catalyst pretreatment under argon at 200 - 300 °C for 2 - 3 h, reactions were started, and the effluent gas was monitored using a mass spectrometer (VG Prima 600, VG Analysis) for NO (m=30), NH_3 (m=17), N_2 (m=28), N_2O (m=44), NO_2 (m=46) and H_2O (m=18), or by a combination of two methods: chemiluminescence NO_x analysis (Signal 4000; Signal) and liquid-phase NH_3 analysis (AutoAnalyser; Technikon).

2.3 NH_3-TPD (Temperature programmed desorption)

NH_3-TPD was performed over CeNa-, LaNa-, H-, and Na-MOR with 0.1 g of catalyst powder in a quartz tube reactor surrounded by a heater, and the effluent gas was monitored by a conventional thermoconductivity (TCD) detector. Catalysts were first pretreated in nitrogen from room temperature up to 600 °C at a heating rate of 5 °C/min, followed by dwelling at 600 °C for 5 - 10 h. Samples were cooled to 100 °C under a N_2 flow, and a pure NH_3 (99.9%, Air Products) flow was admitted on the catalyst for 0.5 h. After flushing nitrogen until obtaining a stable zero signal, TPD was started at a constant temperature increase of 5 °C/min up to 600 °C under nitrogen, and desorbed ammonia was monitored.

2.4 NO-TPD (Temperature programmed desorption)

NO-TPD was carried out with CeNa-, LaNa-, H-, and Na-MOR in a similar manner as described in the NH_3-TPD section except for a maximum desorption temperature of 560 °C. The effluent gas was monitored by a mass spectrometer (MAT44; Varian) for NO (m=30), NO_2 (m=46) and N_2O (m=44) at a scan rate of 15 scans/min for each species.

3. RESULTS

3.1 NO reduction with ammonia in the presence of oxygen

NO reduction with ammonia was performed in the presence of oxygen over a variety of transition metals (Cu, Co, Fe and Ce) exchanged in ZSM-5 (Fig. 1). Iron and cerium turned out to be highly active at temperatures above 300 °C. Iron was, however, not further investigated, considering its undesirable high SO_2 oxidation feature [26]. As shown in Fig. 2, cerium-exchanged mordenite (Ce 5.66 wt%) was found to be more active than Ce-ZSM-5 (Ce 3.09 wt%), whereas CeNa-Y (Ce 9.67 wt%) showed only a poor activity in spite of its highest cerium content. As shown in Table 1, partly H^+ exchanged cerium-mordenite and -ZSM-5 were effective as well with much smaller cerium contents, suggesting a catalytic role of Brønsted acid sites.

A comparison between cerium-, copper-mordenite and a commercial vanadium oxide catalyst illustrates declining activities of copper and vanadium oxide in the higher temperature range (Table 1) and, an undesirable N_2O formation (Fig. 3).

In contrast, there was no N_2O production observed over cerium-exchanged mordenite.

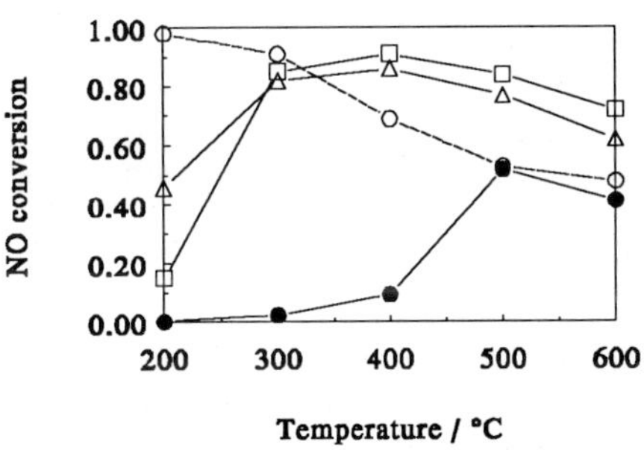

Figure 1. NO reduction with ammonia over metal-exchanged ZSM-5: NO 1000 ppm, NH_3 1000 ppm and O_2 5 % at GHSV of 30,000 h^{-1}. (□) CeNa-ZSM-5(95), (○) CuNa-ZSM-5(96), (●) CoNa-ZSM-5(69) and (△) FeNa-ZSM-5(96).

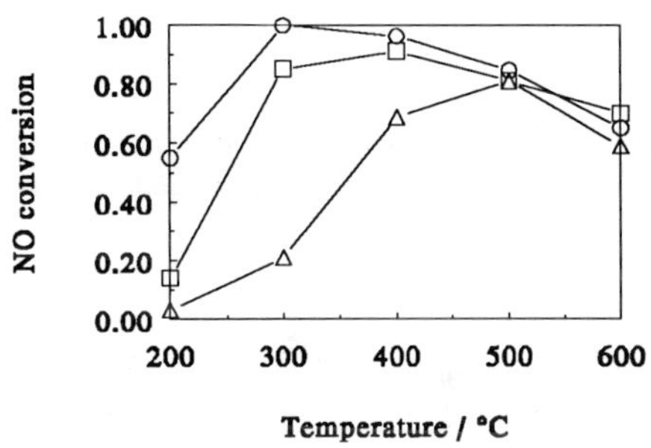

Figure 2. NO reduction with ammonia over different cerium exchanged zeolites NO 1000 ppm, NH_3 1000 ppm and 5% O_2 at GHSV of 30,000 h^{-1}. (○) CeNa-MOR(70), (□) CeNa-ZSM-5(95) and (△) CeNa-Y(65).

Table 1 NO reduction with ammonia in the presence of oxygen: NO 1000 ppm, NH_3 1000 ppm and O_2 5% at GHSV of 30,000 h^{-1}.

Catalyst	Metal content (wt%)	NO conversion (%) 200 °C	300 °C	400 °C	500 °C	600 °C
CeNaH-ZSM-5(17)	0.37	2	67	88	87	73
CeNa-ZSM-5(95)	3.09	13	84	90	82	68
CeNaH-MOR(41)	3.23	61	100	95	92	81
CeNa-MOR(70)	5.66	55	100	96	84	64
H-MOR	-	3	32	71	77	46
CuNa-MOR(69)	3.74	97	95	78	58	34
V_2O_5/Al_2O_3	5.60	53	100	85	19	0

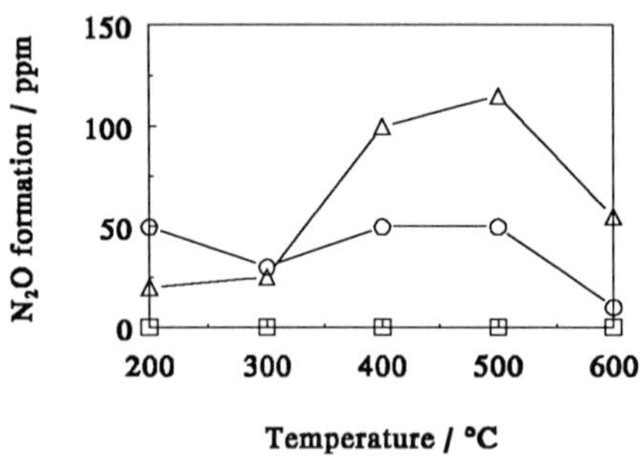

Fig. 3 N_2O formation in NO reduction with ammonia in the presence of oxygen: NO 1000 ppm, NH_3 1000 ppm and 5 % O_2 at GHSV of 30,000 h^{-1}. (o) CuNa-MOR(69), (□) CeNaH- MOR(41) and (Δ) V_2O_5/Al_2O_3

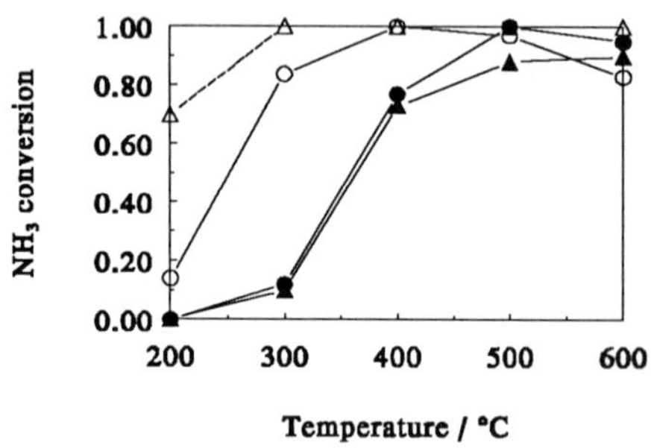

Fig. 4 NH_3 oxidation with oxygen: NH_3 1000 ppm and O_2 5 % at GHSV 30,000 h^{-1} over (●) CeNa- ZSM-5(95) and (▲) CeNaH-MOR(41), compared with ammonia conversions in NO reduction (see legend in Table 1 for reaction conditions) over (o) CeNa-ZSM-5(95) and (Δ) CeNaH-MOR(41).

3.2. Ammonia oxidation with oxygen compared to the NO reduction with ammonia

Oxidation of ammonia with oxygen was carried out with cerium exchanged mordenite and ZSM-5 (Fig. 4). To our surprise, no production of N_2O or NO was observed over both catalysts in this reaction temperature range. Compared with the NH_3 conversions under the NO reduction conditions, it is evident that the conversion of ammonia is much enhanced in the presence of NO. It is a rather different trend than that observed in NO reduction with hydrocarbons, where an onset temperature of reductant combustion coincides with an onset temperature of NO reduction [3, 27].

3.3. NO reduction with an excess amount of ammonia

The NO reduction with ammonia in the presence of oxygen was performed over CeNa-MOR(58) at three NH_3 to NO feed ratios, *i.e.*, NH_3/NO = 1, 1.3 or 1.6. As shown in Table 2, the maximum NO conversion (87 %) observed at equivalent amounts of NH_3 and NO was found to improve to a complete conversion using a 30 % excess of ammonia at 300 - 500 °C. Moreover, there was no ammonia observed in the product gas at this excess ratio above 200 °C. In other words, there was no ammonia slip even at a 30 % ammonia excess condition above 200 °C. At the NH_3/NO ratio of 1.6, NO conversion remained at 100 %, though NH_3 conversion was not complete under these conditions. It should be noted that at the NH_3/NO ratio of 1.6, the converted ammonia (0.63 × 572 ppm) at 200 °C is calculated to be 1.26 times more than the converted NO amount (0.81 × 357 ppm), and 1.33 times more at 300 °C (Table 2). These results suggest that under the present reaction conditions, ammonia can be introduced in excess of up to 30 % to obtain complete NO conversion without an ammonia slip above 200 °C.

Table 2 NO reduction in the presence of oxygen over CeNa-MOR(58) under excess ammonia conditions: a feed NO/NH_3 ratio (NO, NH_3) = 1.0 (465 ppm, 465 ppm), 1.3 (406 ppm, 526 ppm) or 1.6 (357 ppm, 572 ppm) in the presence of 1 % O_2 at GHSV 12,000 h^{-1}.

NH_3/NO ratio	NO (NH_3) conversion (%) 200 °C	300 °C	400 °C	500 °C
1.0	81(>97)	83(>97)	86(93)	87(n.d.)
1.3	82(n.d.)	100(>97)	100(n.d.)	100(>97)
1.6	81(63)	100(81)	100(n.d.)	100(n.d.)

n.d.: not determined

3.4 NH_3-TPD of CeNa-, LaNa-, H- and Na-MOR

CeNa-MOR(58) and LaNa-MOR(57) exhibited a strong NH_3 adsorption, showing its desorption continuing even at the final temperature of 600 °C. Low temperature NH_3 desorption (200 - 300 °C) observed with all four mordenite samples is ascribed to weakly adsorbed ammonia on the mordenite matrix. Over CeNa- and LaNa-MOR, another peak was observed at 400 - 500 °C, which was attributed to NH_3 coordinated to metal ions (cerium or lanthanum). A high temperature desorption ($\geq$ 600 °C) was observed over CeNa-, LaNa- and H-MOR. It might be associated with Brønsted acid sites, which can be expected in lanthanide-exchanged zeolite as reported for La-Y [28] and Ce-Y [29]. The strong ammonia adsorption on CeNa-MOR explains a zero reaction order in NH_3 observed for NO reduction in the presence of oxygen at 300 °C [26].

3.5 NO-TPD of CeNa-, LaNa-, H- and Na-MOR

NO-TPD was carried out with CeNa-MOR(58), LaNa-MOR(57) and H-MOR, and desorption of NO and N_2O was observed (Figs. 5(a) and 5(b)). A separate experiment with Na-MOR exhibited only a negligible amount of desorption products. As shown in Fig. 5(a), CeNa-MOR(58) exhibited two NO desorption peaks: one at 170 °C and another large peak at 290 °C. Since LaNa-MOR(57) and H-MOR showed the peak at 170 °C as well, it is probably related to Bronsted acid sites. CeNa-MOR(58) exhibited a substantial amount of N_2O desorption (Fig. 5(b)). N_2O is a weak adsorbate, and reported to desorb below 200 °C from copper-exchanged zeolite [30]. Therefore, N_2O observed above 200 °C is assumed to result from a reaction of NO species on the catalyst surface. The large N_2O peak observed with CeNa-MOR(58) at around 300 °C appears to correspond to a large NO desorption peak at 290 °C. It indicates that NO desorption at 290 °C is associated with N_2O formation observed at 300 °C. N_2O formation on bulk cerium oxide was reported by Niwa *et al.* [31]. NO_2 was not found during desorption with any of these mordenites.

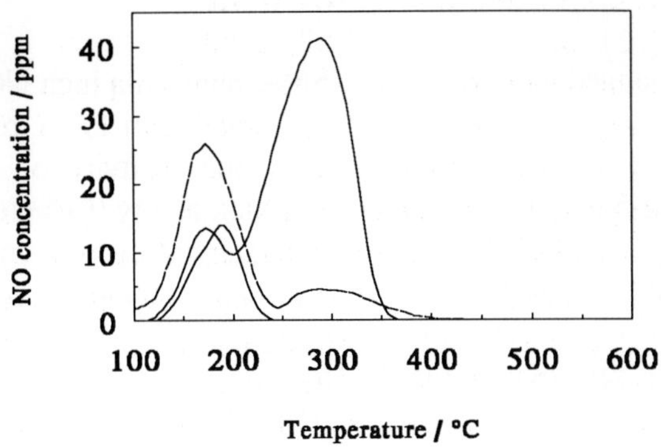

Fig. 5(a) NO desorption during temperature programmed desorption (TPD) of NO observed with

(—) CeNa-MOR(58), (- -) LaNa-MOR(57) or (···) H-MOR. After pretreating a sample under argon for 5 - 6 h at 560 °C, NO adsorption was carried out at 100 °C under NO (1024 ppm) flow at 100 ml/min for 2 -2.5 h, and consecutively, switched over to argon flow. After obtaining a stable zero signal, the temperature was raised at 5 °C / min.

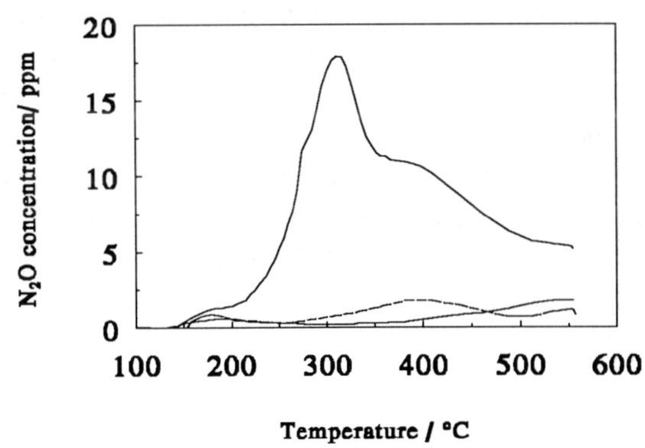

Fig. 5(b) N_2O desorption during the temperature programmed desorption (TPD) of NO observed with

(—) CeNa-MOR(58), (--) LaNa-MOR(57) or (···) H-MOR. For experimental conditions, see legend in Fig. 5(a).

4. DISCUSSION

Cerium-exchanged mordenite and ZSM-5 showed a high NO reduction ability at 200 - 600 °C, which is actually the essential part of a real diesel exhaust temperature window (200 - 700 °C). This high activity of cerium-exchanged mordenite and ZSM-5 may be ascribed to its redox ability (Ce^{III}/Ce^{IV}) and to a strong oxidizing property of Ce^{IV} [23]. Ce^{III} ions are oxidized by oxygen to Ce^{IV}, and Ce^{IV} activates NO to NO_2 (at 300 °C) or NO^+ (500°C) efficiently promoting a further catalytic cycle.

NO reduction selectivity towards nitrogen was considerably high over cerium catalysts. During our experiments, we have noticed that there was no N_2O production over cerium catalysts for reactions *in the presence of oxygen*, such as NO reduction in the presence of oxygen and NO oxidation with O_2 [23] N_2O was observed during NO-TPD under an argon atmosphere. Over cerium-exchanged zeolite, the oxidation of NO towards NO_2 (a NO-O_2 reaction) [23] might be far more advantageous than the disproportionation of NO (a NO-NO reaction) towards N_2O [31] in the presence of oxygen. However, this does not rule out a possible N_2O formation from a NH_3-NO reaction [32]. Besides, it should be noted that cerium does not possess *d*-valence electrons. This may distinguish cerium from transition metals in its coordination behavior in catalysis.

The exclusive selectivity towards nitrogen in the NH_3 oxidation with oxygen is striking, considering the fact that NO or N_2O production is observed with many other catalysts [13]. A reaction pathway of ammonia oxidation over cerium zeolite remains to be elucidated. The comparison of ammonia conversions in its oxidation and in the NO reduction illustrates a characteristic of the NO reduction with ammonia over cerium zeolite. At lower temperatures, NH_3 appears to be activated more easily under NO reduction conditions than with oxygen only. This picture of reductant activation differs form that assumed for hydrocarbons, where the activation of hydrocarbons is regarded as an important factor for NO reduction [3, 27]. In a previous paper, we have proposed the intermediacy of NO_2 in the NO reduction with ammonia at lower temperatures (*e.g.* 300 °C) [23]. This intermediate NO_2 can probably oxidize ammonia more efficiently than oxygen.

At a higher temperature ($\geq$ 400 °C), ammonia oxidation with oxygen becomes substantial, and a certain reaction competition is expected between NO reduction and ammonia oxidation. The utilization of ammonia for NO reduction was high (> 80 %) up to 500 °C, and starts declining at 600 °C (Fig. 1).

There are two other specific features observed in the catalysis of cerium zeolites. The first is the poor activity of CeNa-Y. A similar low activity of metal-exchanged Y compared to ZSM-5 is reported in the NO decomposition with CuNa-Y [33], and in the NO reduction with methane over Co-Y [34]. Sepúlveda-Escribano *et al.* have attributed this inferiority of Y to a lower stability of active copper species, *i.e.* Cu^+ [33], whereas Li *et al.* have suggested electronic influences from the specific zeolite environment [34]. Other properties of zeolite Y can also be considered. The first is the presence of large cages and a high framework charge density, which allows the formation of CeO_x clusters [29]. It should decrease the number of free coordination sites eventually leading to a low utilization of cerium ions. Such an oxide formation is more difficult to take place in a high silica zeolite as mordenite or ZSM-5, where cation sites are more remote from each other. Secondly, we mention the known preference of rare earth metal ions for positions in hexagonal prisms and sodalite cages [35] providing a low accessibility to reactants.

The second specific feature is the high performance of partly H^+ exchanged cerium zeolites. As shown in Table 1, CeNaH-MOR and CeNaH-ZSM-5 exhibited higher activities than CeNa-MOR and CeNa-ZSM-5, respectively, in spite of a much lower cerium content. In addition, fully exchanged H-MOR becomes substantially active only above 300 °C. The activity of partly H^+-exchanged cerium samples can not be explained by proportional contributions from cerium and H^+ sites, indicating an ensemble effect of cerium and H^+ sites. It should be noted that H^+ sites were assumed to catalyze NO reduction through formation of NO_2 as a reaction intermediate [36], while cerium was found to produce NO_2 efficiently from NO and O_2 [23]. Thus, H^+ sites perform probably more efficiently in cooperation with cerium sites, over which a large amount of intermediate NO_2 is produced.

NO-TPD results (Figs. 5(a) and 5(b)) indicate a specific interaction of NO with cerium-exchanged mordenite. First, the amount of NO desorbed from cerium is much higher than that observed on LaNa-MOR and H-MOR. NO desorption at around 290 °C was particularly large with cerium. The second feature is the reactivity of adsorbed NO on cerium. From LaNa- and H-MOR, NO was observed to desorb reversibly without substantial formation of N_2O, while NO was found to desorb as both NO and N_2O from cerium mordenite. Since, N_2O observed above 200 °C is ascribed to a reaction of adsorbed NO [30], the high N_2O formation up to the final temperature (560 °C) over cerium suggests a high reactivity of NO on the cerium zeolite surface. The large N_2O peak at 300 °C is apparently resulted from the disproportionation of NO, which originates from the NO desorbed at 290 °C. With respect to this specifically high N_2O formation over cerium, the redox couple (Ce^{III}/Ce^{IV}) may be responsible according to:

$$2\,NO + 2\,Ce^{III} + 2\,H^+ \rightarrow 2\,Ce^{IV} + H_2O + N_2O \qquad (1)$$

Although NO-TPD is informative on the specific interactions between NO and cerium, it probably does not fully reflect the behaviour of cerium under SCR conditions. In particular, the presence of oxygen may change the behaviour of these species drastically, considering the observed shifted trend in catalysis in the presence of oxygen as discussed above. Further TPD studies using both NO and O_2 may provide a more clear picture of the behaviour of these species under NO reduction conditions.

In conclusion, the observed complete conversions of NH_3 and NO under excess ammonia conditions (> 200 °C) indicates great potential of cerium zeolite. With ammonia applied in excess over NO, a stoichiometric amount of NH_3 converts NO completely, and the excess NH_3 will be simply converted to N_2 by oxygen. With a certain shortage of NH_3, a slightly lower NO conversion may be obtained, but ammonia will be exhausted in the NO reduction anyway. This flexibility in ammonia feed concentration makes an approximate reductant injection control applicable in practice. Under the present reaction conditions, an excess of ammonia up to 30 % is maximally allowed without ammonia slip. However, the effects of the reaction conditions, *e.g.*, the space velocity, the presence of water or SO_2, should be further examined to estimate such a "maximum ammonia excess value" in practice.

ACKNOWLEDGMENT

The authors would like to thank the Dutch Ministry of Housing and Environment (VROM) for their financial support.

REFERENCES

1 S. Sato, Y. Yu-u, H. Yahiro, N. Mizuno and M. Iwamoto, *Appl. Catal.* 70 (1991) L1.
2 C.N. Montreuil and M. Shelef, *Appl. Catal. B*, 1 (1992) L1.
3 B.K. Cho, *J. Catal.*, 142 (1993) 418.
4 J.O. Petunchi and W.K. Hall, *Appl. Catal. B*, 2 (1993) L17.
5 K.C.C. Kaharas, H.J. Robota and D.J. Liu, *Appl. Catal. B*, 2 (1993) 225.
6 G.P. Ansell, A.F. Diwell, S.F. Golunski, J.W. Hayes, R.R. Rajaram, T.J. Truex and A.P. Walker, *Appl. Catal. B*, 2 (1993) 81.
7 E. Kikuchi, K. Yogo, S. Tanaka and M. Abe, *Chem. Lett.* (1991) 1063.
8 M. Misono and K. Kondo, *Chem. Lett.* (1991) 1101.
9 H. Hamada, Y. Kintaichi, M. Sasaki and T. Ito, *Appl. Catal.* 70 (1991) L15.
10 K. Yogo, M. Ihara, I. Terasaki and E. Kikuchi, *Appl. Catal. B*, 2 (1993) L1,
11 Y. Li and J.N. Armor, *Appl. Catal. B*, 2 (1993) 239.
12 J.W. Byrne, J.M. Chen and B.K. Speronell, *Catalysis Today* (1992) 33.
13 H. Bosch and F.Janssen, *Catalysis Today*, 2 (1987) 436.

14 U. Ozkan, Y. Cai, M.W. Kumthekar and L. Zhang, *J. Catal.*, 142 (1993) 182.
15 H.E. Curry-Hyde and A. Baiker, *Appl. Catal. A*, 90 (1992) 183.
16 W. Held, A. König, T. Richter and L. Puppe, *SAE* 900496.
17 F.G. Medros, J.W. Eldridge and J.R. Kittrell, *Ind. Eng. Chem. Res.* 28 (1989) 1171.
18 S. Sato, Y. Yu-u, H. Yahiro, N. Mizuno and M. Iwamoto, *Appl. Catal.* 70 (1991) L1.
19 H. Yahiro, Y. Yu-u, H. Takeda, N. Mizuno and M. Iwamoto, *Shokubai*, 35 (1993)130.
20 Y. Ukisu, S. Sato, A. Abe and K. Yoshida, *Appl. Catal. B*, 2 (1993) 147.
21 H. Yasuda, T. Miyamoto, C. Yokoyama and M. Misono, *Shokubai*, 35 (1993) 386.
22 C.M. van den Bleek and A.W. Wypkema, The D.U.T. NO_x/Urea Process phase 3, NOVEM report contractnr. 42.33-005.30 (1990).
23 E. Ito, R.J. Hultermans, P.M. Lugt, M.H.W. Burgers, M.S. Rigutto, H. van Bekkum and C.M. van den Bleek, *Appl. Catal. B*, (1994) in press.
24 E. Ito, C.M. van den Bleek, H. van Bekkum, J.C. Jansen, R.J. Hultermans and P.M. Lugt, Delft University of Technology, Dutch Patent Appl., NL 93.02288 (1993).
25 R.J. Hultermans, E. Ito, Á. József, P.M. Lugt and C.M. van den Bleek, in the present proceedings of the third international symposium of Catalysis and Automotive Pollution Control III (CAPoC3), Brussels, April 20 - 22, 1994.
26 E. Ito, R.J. Hultermans, P.M. Lugt, M.H.W. Burgers, H. van Bekkum and C.M. van den Bleek, paper in preparation.
27 J.L. d'Itri and W.M.H. Sachtler, *Catal. Lett.*, 15 (1992) 289.
28 W. Klading, *J. Phys. Chem.*, 80 (1978) 262.
29 E.F.T Lee and L.V.C. Rees, *Zeolites*, 7 (1987) 446.
30 Y. Li and J.N. Armor, *Appl. Catal.*, 76 (1991) L1.
31 M. Niwa, Y. Furukawa and Y. Murakami, *J. Coll. Int. Sci.*, 86 (1982) 260.
32 L. Singoredjo, R. Korver, F. Kapteijn and J. Moulijn, *Appl. Catal. B*, 1 (1992) 297.
33 A. Sepúlveda-Escribano, C. Márquez-Alvarez, I. Rodoígez-Ramos, A. Guerrero-Ruiz and J.L.G.Fierro, *Catalysis Today*, 17 (1993) 167.
34 Y. Li and J.N. Armor, *Appl. Catal. B*, 2 (1993) 239.
35 W.J. Mortier (Editor), "Compilation of extraframework sites in zeolite", Butterworth Scientific Limited, London.
36 P.M. Hirsch, *Environmental Progress*, 1 (1982) 24.

A. Frennet and J.-M. Bastin (Eds.)
Catalysis and Automotive Pollution Control III
Studies in Surface Science and Catalysis, Vol. 96

SELECTIVE REDUCTION OF NO OVER COPPER-CONTAINING MODIFIED ZEOLITES

J. Halász, J. Varga, G. Schöbel, I. Kiricsi, K. Hernádi, I. Hannus, K. Varga, and P. Fejes

Applied Chemistry Department, József Attila University
Rerrich tér 1, H-6720 Szeged, Hungary

ABSTRACT

The most efficient method for NO removal from stationary and mobile sources is catalytic reduction with ammonia, hydrocarbons, CO or H_2. Modified zeolites are active catalysts in these processes. For Cu-ZSM-5 especially high activity and stability have been reported. In this work the properties of copper-containing ZSM-5 zeolites prepared by wet or solid state ion-exchange have been investigated. The Brönsted acidity of the Cu^{2+}-exchanged samples was much lower than that of the parent zeolites, and they had high activity in selective reduction with ammonia, propene or propane. A comparison of Cu-ZSM-5 activity in the decomposition of NO and in the reaction of NO with propene or propane revealed that the hydrocarbons as well as the nitrogen oxides play important role in the performance of NO reduction catalysis.

1. INTRODUCTION

The removal of nitrogen oxides from combustion and industrial exhaust remains an important problem which has been studied extensively. Combustional modifications (controlling burner stoichiometry and lowering flame temperature) have led to methods that are both cost-effective and energy-efficient, however, these methods by themselves cannot achieve the reduction of NO_x to levels required in new regulations [1]. The wet methods (and the adsorption techniques) for NO_x control, which can be used for stationary source emissions only, are expensive to operate and additionally, these methods have serious problems with adsorbent treatment and disposal [2].

Selective non catalytic reduction (SNCR) with NH_3 is limited to industrial boilers in consequence of the relatively narrow temperature range for the reaction. Selective catalytic reduction (SCR) by ammonia has high efficiency and it can be used for many stationary sources, especially for nitric acid plants [1], and it is based on the catalytic pairing of nitrogen atoms, one from nitric oxide, one from ammonia. This method, however, is unsuitable for small sources and vehicles. As far as automotive emission is concerned nonselective catalytic reduction (NSCR) by hydrocarbons, CO and H_2 from the exhaust stream has been reported over various catalysts recently [1,3,4].

The discovery by Iwamoto et al. in 1981 that copper-exchanged zeolite Y possesses high and stable activity for the direct decomposition of NO was the breakthrough which strongly suggested that a practical NO decomposition catalyst could be developed on zeolite basis [5]. It has been shown that catalysts based on copper-exchanged ZSM-5 are also very active for direct decomposition and selective reduction is also feasible with hydrocarbons, especially with propene [6].

The activity of copper-exchanged zeolite catalysts in direct decomposition was influenced by (i) the degree of copper ion-exchange [7] (no decomposition was observed in the absence of copper) and (ii) the Si/Al ratio of the zeolite [8], (iii) the preparation method of the parent Na-ZSM-5 [9]. The effect of admixed dioxygen and water vapour to the feed stream has been reported too [10].

For catalytic reduction processes, in which hydrocarbons were used as reducing agents, copper exchanged zeolites showed the best performance, for example for propene containing feed compositions [11], since close to 100% NO conversion could be attained at as low as 623 K. An acidic form of iron-silicate showed high activity and stability for reduction of NO with propene, but this catalyst was very sensitive to the presence of SO_2 in the feed [12]. Metallosilicates having ZSM-5 structure [13,14] seem to be a new class of catalysts for these reactions.

The objective of present paper has been to study the role of Cu-containing modified ZSM-5 type zeolites in the reactions of nitrogen monoxide.

2. EXPERIMENTAL

Nitric oxide decomposition and reduction by ammonia, propene or propane were carried out on H-ZSM-5, Cu^{2+} ion-exchanged X-type and ZSM-5 zeolites.

2.1. Catalysts

The Na-form of zeolites were synthesized by usual methods. Modification of ZSM-5 zeolites was carried out by two different methods: (i) conventional ion exchange in Cu^{2+}-acetate solution, and (ii) heat treatment of the physical mixtures of $CuCl_2$ and H-forms of zeolites at 873 K (solid-state ion exchange).

In liquid phase the copper-ion exchange was carried out by stirring about 25 g sample in 0.05 M Cu-acetate solution at room temperature for 24 hours, the exchanged sample was washed, dried and calcined at 773 K for 5 hours.

The solid-state ion-exchange was carried out by baking the physical mixture of H-ZSM-5 and $CuCl_2$ at 873 K for 6 hours.

2.2. Catalyst characterization

Composition of samples and copper ion-exchange level were determined by X-ray fluorescence (XRF) analysis.

The catalysts were characterized by X-ray diffractometry (XRD) and infrared (IR) spectroscopy. Acidity of the catalysts was tested by pyridine adsorption monitored by IR spectroscopy. Self-supported wafers pressed from zeolite powder (thickness 15 mg cm^{-2}) were placed in the sample holder and outgassed at 770 K in vacuum (final vacuum was better than 10^{-3} Pa) for 2 hours followed by cooling to room temperature where the spectrum of the activated zeolites were registered. 1.33 kPa pyridine was adsorbed at 473 K for 1 hour followed by evacuation at the same temperature for 1 h. For calculating the concentration of acid sites extinction coefficients available in the literature were used [15].

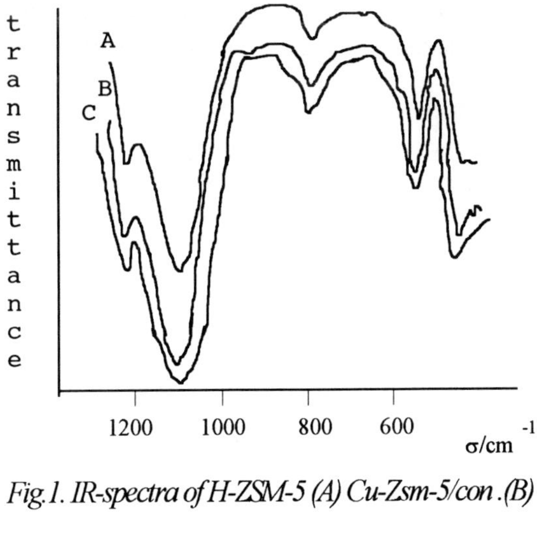

Fig.1. IR-spectra of H-ZSM-5 (A) Cu-Zsm-5/con.(B) and Cu-ZSM-5/sol. (C)

2.3. Reaction studies

The NO reactions (decomposition and reaction with propene or propane) were carried out in a recirculatory batch reactor with mass spectrometric analysis (details see in [2]). The catalyst sample (0.5 g) was activated in dioxygen at 723 K for 4 hours before each measurement. The gas-phase concentrations of reactants and products were measured by mass spectrometry. Mass numbers 41 of propene, 43 of propane, 46 of nitrogen dioxide and 30 of NO were used for analysis.

The SCR reaction (with NH_3) was studied in a fixed-bed flow reactor with gas chromatographic (GC) analysis. The inlet gas composition was 10% NO + 10% NH_3 in nitrogen; the feed rate was 120 cm^3/min (GHSV = 1500) at atmospheric pressure. Before each run, 5 g of the catalyst was pretreated in flowing air at 773 K for 2 h in tubular quartz reactor. Reactant and product analysis was performed by thermal conductivity detector using Porapack-Q column.

3. Results and Discussion

3.1. Catalyst characterization

Table 1 shows the Si/Al ratio, Cu content and the acidities of catalysts used.

Table 1. Properties of catalysts used

Sample	Si/Al ratio	Cu content mass %	Acidity/ μmol g^{-1}	
			Brönsted	Lewis
Cu-ZSM-5/con.	13.8	2.3	0.21	0.588
Cu-ZSM-5/sol.	14.5	2.7	0.027	0.289

3.1.1. Characterization by IR spectroscopy

Catalysts prepared were characterized by IR spectroscopy using the KBr matrix method generally applied to check the changes in the framework vibration region (400-1300 cm^{-1}). As it can be seen in Fig. 1, non of the conventional exchange method (B) and of the solid state exchange (C) resulted in observable structural changes.

3.1.2. Characterization by XRD

Each modified catalyst sample preserved the crystal structure characteristic of zeolite ZSM-5. No reflexions due to separate copper oxide phases could be detected for the solid-state exchanged material.

3.1.3. Acidity measurements

Upon adsorption of pyridine IR bands at 1540 and 1450 cm^{-1} appear characteristic for Brönsted and Lewis acid sites, respectively. The band at 1490 cm^{-1} is due to the combination vibration of both types of acid sites. On pyridine adsorption the OH band typical of Brönsted acidic hydroxyl groups disappeared while those of the terminal SiOH groups only decreased in intensity (see Fig. 2). These observations reflect the different acidity of these OH groups. The parent Na-ZSM-5 sample possessed no Brönsted acidity detectable by this method.

3.2. Catalytic measurements

Selective reduction of NO was investigated in the absence and presence of propene, propane or ammonia.

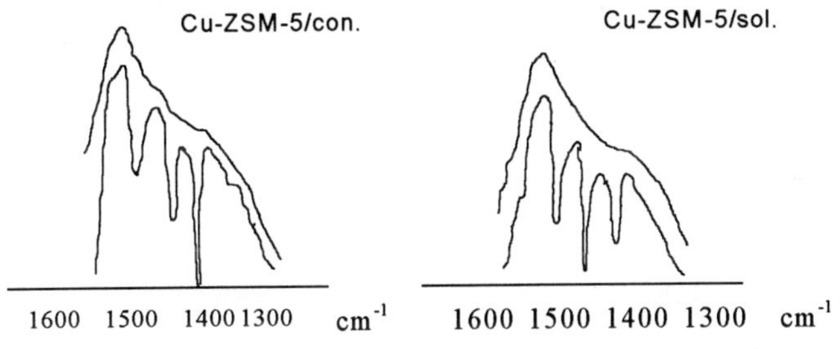

Fig.2. Determination of acidity of Cu-ZSM-5 by pyridine adsorption

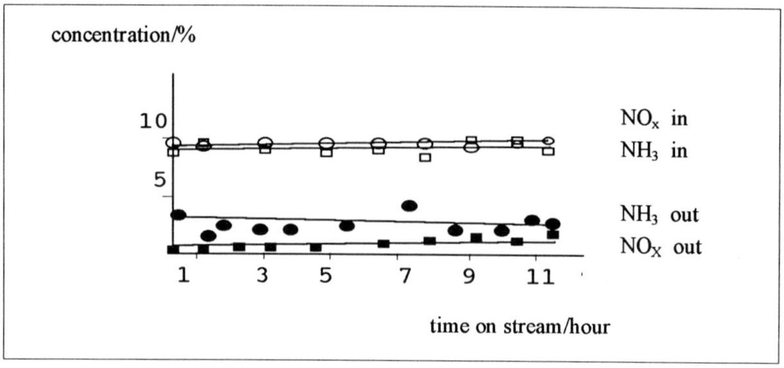

Fig.3. Selective reduction of NO over Cu-ZSM-5 zeolite at 573 K in fixed bed flow reactor.

3.2.1. Selective reduction with ammonia

In the selective reduction of NO with ammonia the Cu-ZSM-5 prepared by conventional ion exchange proved to be an effective catalyst, as can be seen in Fig. 3, where the results of reaction at 573 K is presented. The NO conversion to N_2 was higher than 95% throughout the run. In case of H-ZSM-5 only 15% conversion for NO and 12% for NH_3 was observed due to the high Brönsted acidity of this catalyst. The activity of Cu-X zeolite was much lower than that of Cu-ZSM-5 at the same temperature (57% conversion for NO and 60% conversion for ammonia).

3.2.2. NO decomposition

In the absence of any reducing agent NO over Cu-ZSM-5 prepared by wet method transformed to N_2 and NO_2 with molar ratio of 1:2 in a closed recirculatory reactor system (Fig. 4). This transformation has taken place with measurable rate above 573 K.

3.2.3. Catalytic reduction with propene

In the reduction proceeding in the presence of propene, formation of N_2, H_2O and CO_2 was detected over the copper containing catalyst.

Kinetic curves of NO consumption (Fig. 5) showed that the reaction was very slow at 473 K, fast at 593 K and above. The propene consumption (adsorption) was so fast that its disappearance could not be followed under these experimental conditions. The measured kinetic curves are depicted in Fig 6.

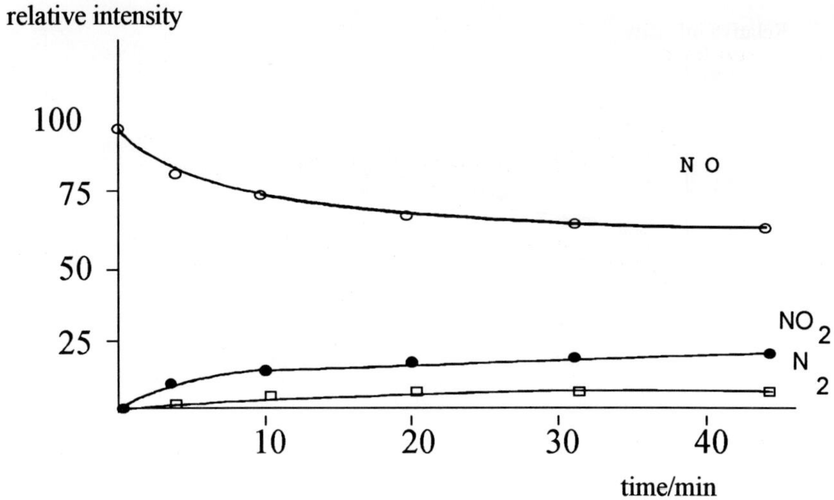

Fig.4. Decompositionof NO over Cu-ZSM-5 at 623K

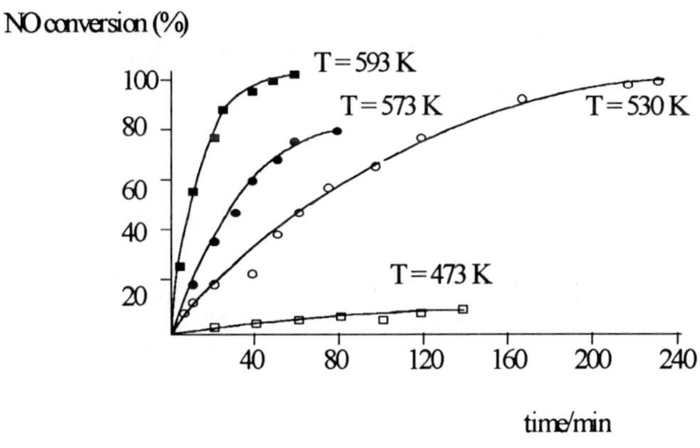

Fig.5. Temperature dependence of NO transformaton over Cu-ZSM-5/con. in the presence of propene.

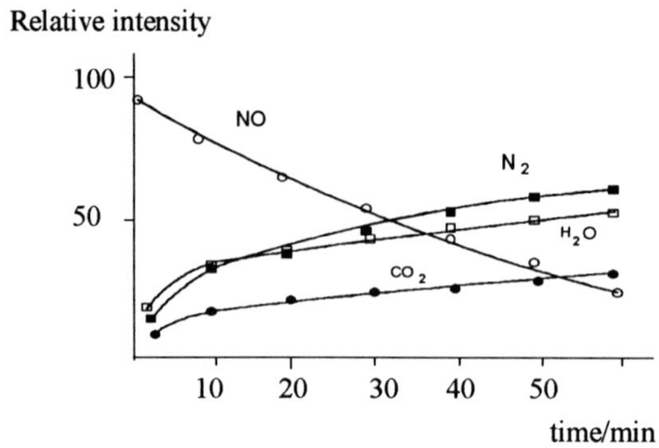

Fig.6. Kinetic curves of NO transformation over Cu-ZSM-5/con. in the presence of propene.

3.2.4. Catalytic reduction with propane

As Fig 7. shows completely different temperature range was required for NO reduction with propane. Here, the consumption of NO was easily measurable even at 823 K. The kinetic curves reveal that below 673 K only very slow reduction occurred.

As far as the transformation of hydrocarbons is concerned, complete consumption was observed for propene over 600 K, but it was not the case for propane. Fig 8. shows the kinetic curves measured for a typical reaction parameter set.

The NO conversion over Cu-ZSM-5 prepared by solid-state ion-exchange was much slower than that of over catalyst prepared by conventional method (comparable activity was detected at the temperature of 100 K higher); this behaviour could be attributed to the very low Brönsted acidity of this sample.

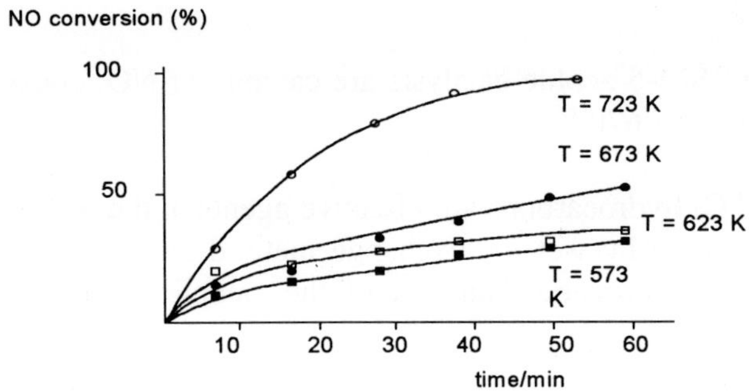

Fig.7. Temperature dependence of NO transformation over Cu-ZSM-5/con. catalyst in the presence of propane

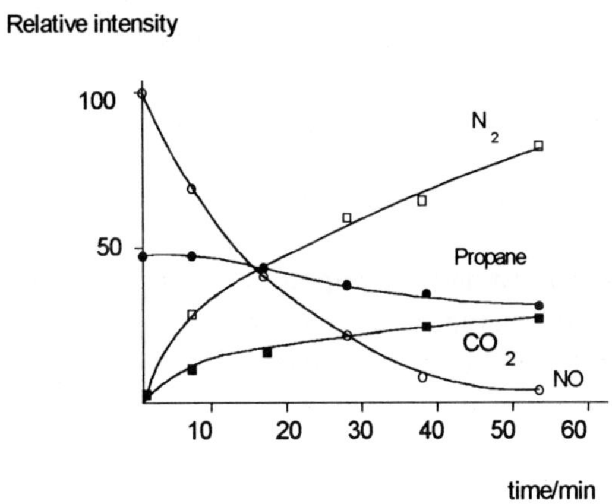

Fig. 8. Kinetic curves of NO transformation over CU-ZSM-5/con. in the presence of propane.

4. CONCLUSIONS

(i) copper containing ZSM-5 zeolite catalysts are capable of NO_x conversion up to 90% at temperatures over 670 K;

(ii) both ammonia and C_3 hydrocarbons are effective agents in the NO_x reduction, however, the catalyst made by wet ion-exchange method is more active in the reactions with propene or propane than that of the sample prepared by solid exchange;

(iii) the role of Brönsted acidic sites is to promote the formation of carbonaceous deposits which can be act as active centres in this process [16];

(iv) the effect of oxygen addition was dependent on the method of ion-exchange and the O_2/NO pressure ratio.

ACKNOWLEDGEMENT

The financial support of the National Science Foundation of Hungary (OTKA No. T 007601 and 1182/90) is gratefully acknowledged.

REFERENCES

1 H. Bosch and F. Janssen, Catal. Today, 2, (1988) 369.
2 I. Hannus, J. Halász, I. Kiricsi, Gy. Schöbel, Gy. Tasi and P. Fejes, Acta Phys. et Chem. (Szeged) 35, (1989) 3.
3 G. Centi, S. Perathoner, Y. Shioya and M. Anpo, Res. Chem. Intermed., 17, (1992) 125.
4 M. Iwamoto, in "Future Opportunities in Catalytic and Separation Technology, Stud. Surf.-Sci. Catal., 54, (1990) 121.
5 M. Iwamoto, H. Furukawa and S. Kagawa, Stud. Surf. Sci. Catal., 27, (1986) 943.
6 M. Iwamoto, H. Hamada, Catal. Today, 10, (1991) 57.
7 M. Iwamoto, S. Yokoo, K. Saaki, S. Kagawa, JCS Faraday Trans.I., 77, (1981) 1629.
7 M. Iwamoto, H. Yahiro, Y. Mine, S. Kagawa, Chem. Lett., (1989) 213.
8 Y. Li, J. N. Armor, Appl. Catal., 76, (1991) L1.
9 V. P. Shirakar, A. Clearfield, Zeolites, 9, (1989) 363.
10 M. Iwamoto, H. Yahiro, K. Tanda, Stud. Surf.-Sci. Catal., 44, (1989) 219.

11 B. K. Cho, J. Catal., 142, (1993) 418.
12 E. Kikuchi, K. Yogo, S. Tanaka, M. Abe, Chem. Lett., (1991) 1063.
13 K. Yogo, M. Ihara, I. Terasaki and E. Kikuchi, Appl. Catal. B2, (1993) L1.
14 K. Yogo, M. Ihara, I. Terasaki and E. Kikuchi, Catal. Lett., 17, (1993) 303.
15 J. Datka, J., Catal., 102, (1986) 43.
16 G.P. Ansell, A.F. Diwell, S.E. Golunski, J.W. Hayes, R.R. Rajaram, T.J. Truex and A.P. Walker, Appl. Catal. B2 (1993) 81.

A. Frennet and J.-M. Bastin (Eds.)
Catalysis and Automotive Pollution Control III
Studies in Surface Science and Catalysis, Vol. 96

ZEOLITE CATALYST FOR THE PURIFICATION OF AUTOMOTIVE EXHAUST GASES

Delia Florea, Lucia Georgescu, Floreta Constantinescu, Dumitru Mãnoiu, Dinu Gaber, Mihaela Comãnescu.

Research Institute for Petroleum Processing and Petrochemistry-ICERP-Ploiesti.
B-dul Republicii no. 291 A - Romania

INTRODUCTION

The latest literature data include a new catalytic process for nitrogen oxides (NO_x) selective reduction with hydrocarbons in oxidizing atmosphere on heterogeneous catalysts such as alumina [1], H - form zeolites [2] and metal supported zeolites [3, 4]. This process breaks the widely accepted concept that ammonia is the only selective reductant for NO_x in the presence of oxygen and the hydrocarbons are not useful for the selective reduction. The new process of selective reduction is remarkably improved in the presence of oxygen or even SO_2 and can be used for the purification of exhaust gases from engines (Diesel or gasoline type).

Generally, copper modified ZSM - 5 zeolite (Cu-ZSM-5) has been reported to have a high activity in reduction of nitrogen oxides (NO_x) by hydrocarbons under net oxidizing conditions. Recently, gallium ion exchanged ZSM-5 catalyst showed extremely high selectivity for the reduction of nitric oxide by ethene in the presence of excess oxygen [5].

Based on the above mentioned data the paper shows the results of research performed for a Cu-ZSM-5 catalyst production to the purpose of purification of vehicle engines exhaust gases. The influence of the ion exchange level in ZSM-5 upon the catalytic performances is also studied. The catalysts used were Cu-ZSM-5 granules or Cu-ZSM-5 coating on a ceramic monolith. The trials of correlation between the catalysts performances and the state of copper in catalysts prepared is also showed.

EXPERIMENTAL

Na-ZSM-5 having a molar SiO_2 / Al_2O_3 ratio of 42 was supplied by the factory of catalyst Vega. Copper ion exchanged ZSM-5 was prepared by the ion exchange of ammonium form ZSM-5 using an aqucous solution of copper acetate 0,06 M. Copper ion exchanged ZSM-5 catalysts with an excess loading of copper ions were prepared according to the method reported by Ywamoto et al [6]. The catalysts were produced by binding the Cu-ZSM-5 with alumina and nitric acid 10% and their formation as granules. Than, the catalysts were dried at 120°C and calcinated at 500°C. The amount of copper in zeolite was determined by atomic absorption spectroscopy.

The catalysts test samples having a cylindrical ceramic monolithic substrate (258 cells/in^2, 35 mm in diameter x 120 mm in length) were also produced. To prepare the catalysts, a cylindrical ceramic monolithic substrate was coated with a washcoat consisting of zeolite (69%) and silica (31%). The ZSM-5 had a silica alumina ratio 43,5. The copper ions were introduced in ZSM-5 by ion exchange with an aqueous solution of copper acetate - 0.06M. The ion exchange was performed after the coating of the monolithic substrate. The amount of zeolite on the monolithic substrate was 16-19%, and the copper content in the samples was 0.4 - 0.6 % corresponding to an exchange level of copper (> 100%).

As a reference, a monolithic oxidic catalyst type $Cu_xCo_{3-x}O_4$ (x=1) was prepared by coating the monolithic substrate with a washcoat consisting of alumina and cobalt - copper oxides. The cobalt and copper content in this catalyst was 33 g/l - Co and 18 g/l - Cu, corresponding to 4.2% metal content in catalyst.

The performances of the catalyst prepared were estimated by model reaction tests which consist in NO_x (NO + NO_2) reaction with methane and methane oxidation, performed in a laboratory test unit coupled with gas cromatografs. 3.5 g of the catalyst crushed in 1-1.6 mm was packed in a stainless steel tube of 16 mm inner diameter. The desired reaction gases were prepared by means of the mixing gas pumps type 4 Wösthof OHG - Bochum (Germany). The gases were routed through a by-pass line for analysis and finally routed through the catalyst. The reactant mixture contained 0.16 % NO_2, 1% methane and 20% oxygen in nitrogen with a total flow rate of 1630 cm^3/min. ($w/F = 0.13\ g \cdot s \cdot cm^{-3}$) in the test of NO_2 reduction with methane, and 1.5% CH_4 in air with a total flow rate of 830 cm^3/min. ($w/F = 0.25\ g\ .\ s\ .\ cm^{-3}$) for test of methane oxidation. The effluent reactor was analyzed by gas chromatografs equipped with a molecular sieve 5A column (for nitrogen and carbon monoxide) and a Porapak Q column (for nitric oxides, carbon dioxide and methane). The formation of nitrous oxide was hardly detected and the activity of the selective reduction was evaluated in terms of the conversion of nitric oxide to nitrogen.

The activity of the catalysts samples having a cylindrical monolithic substrate were determined only in the model reaction test of methane oxidation, performed in a laboratory test unit coupled with gas cromatografs, equipped with a special reactor for testing monolithic samples.

The state of copper in the prepared catalysts was studied by ESR spectroscopic technique and by thermal analysis in flow hydrogen medium. Experimental measurements were performed with a spectrometer ART-6 in X frequency bands analysis (u = 9010 MHz) and a thermoanalitical instrument SETARAM.

RESULTS AND DISCUSSION

In fig.1 is presented conversion of nitric oxides and methane versus concentration of oxygen. From the above figure is possible to see that nitric oxides conversion remain practically unchanged at a increasing of oxygen amount from 4% to 20%. The methane conversion have a increase at increasing of oxygen concentration. At a concentration of 20% oxygen the conversion of methane reaches 100%.

Taking in consideration these experimental dates our work was carried out at a 20% oxygen concentration with the purpose to see how is the evolution of NO_x conversion in case when hydrocarbons from reaction mixture are converted just before nitric oxides consumption.

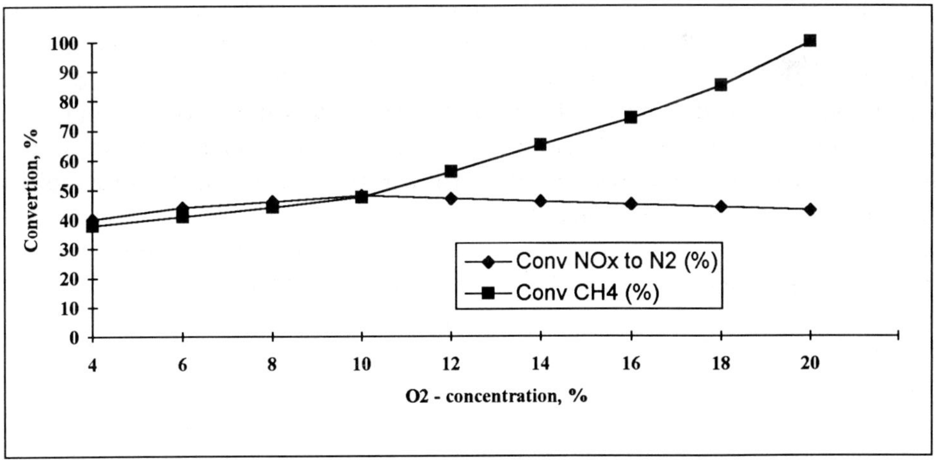

Fig. 1. Effect of oxygen concentration on the reduction of NO_x with methane and the methane oxidation over Cu-ZSM-5 (126) at 450°C (Reaction conditions: NO_x: 0.16%, CH_4: 1%, w/F=0.13g·s·cm^{-3})

Table 1 shows activity of catalysts obtained from ZSM-5 zeolite modified with copper at which the ion exchange level of the copper varied from 55% to 126% in the reduction reaction of NO_x with methane. The catalysts with a 100-126% degree of ion exchange of the copper have been found to show a higher conversion of the nitrogen oxide to nitrogen, especially at lower temperatures (300-400°C).

The activity of zeolite type catalysts is compared with the activity of an oxide catalyst supported on alumina, type $Cu_xCo_{3-x}O_4$ where x=1. Although the literature data report a good activity in the selective reduction reaction of nitrogen monoxide with propane [7] for a number of Cu/Al_2O_3 and Co/Al_2O_3 catalysts (especially for those prepared from acetates and calcinated at temperatures of 600°C or higher) and for alumina itself, the catalyst $Cu_xCo_{3-x}O_4$ with 8% cobalt and 4% copper has no activity in the reduction reaction of nitrogen monoxide with methane. We mention that this catalyst was prepared from cobalt and copper nitrates and calcined at 500°C.

Table 1 also includes methane total conversion in the reduction reaction of NO_x with methane and selectivity to carbon dioxide. The ability of Cu-ZSM-5 catalysts for hydrocarbon oxidation is generally known. Watchind the data presented in table 1, the activity of zeolite type catalyst in methane oxidation reaction is found to be comparable with the activity of the oxide catalyst even higher for the catalysts resulted from zeolite in which the copper degree of ion exchange is 124-126%. Such catalyst start converting the methane (as well as NO_x) at lower temperatures (300-400°C). A high selectivity for methane oxidation ca be remarked for all the tested catalysts. At the temperature of 500°C, methane conversion reaches values of 100% while NO_x conversion to nitrogen reaches values of only 50-60% for most tested catalysts. However NO_x conversion rapidly increase to 100%, within a narrow temperature range (500-550°C) after methane conversion reached 100%. Such behaviour could give us certain indications upon the reaction mechanism. From among the reaction schemes which have been proposed for the conversion of NO_x to nitrogen in the presence of hydrocarbons and oxygen excess, the one that suggest the coke deposited on catalyst surface as an active intermediate in NO_x reduction reaction, seems to be supported by our results [8]. Although the methane oxidized entirely, the deposits existing on catalyst surface play a reducing role and reduce NO_x to nitrogen when this reaction is no more in competition with methane oxidation reaction.

Zeolite catalyst performances were also checked in a direct manner in the methane oxidation reaction, a model reaction which tested the spinel oxide type catalysts prepared by us for hydrocarbons oxidation to the purpose of purifying engines exhaust gases. The results are presented in Table 2. From among the

Table 1. Reduction of nitric oxides (NO_x) over granular catalysts
Reaction conditions: NO_x - 0.16%, CH_4 - 1%, O_2 - 20%   w/F = 0.13 g · s · cm^{-3}

Catalyst	Metal content (%)	Conversion of NO_x to N_2 (%)				Conversion of CH_4 (%)				Selectivity to CO_2 (%)			
		300°C	400°C	500°C	550°C	300°C	400°C	500°C	550°C	300°C	400°C	500°C	550°C
$Cu_xCo_{3-x}O_4$	8 - Co, 4 - Cu	-	-	-	-	-	2	100	100	-	11	88	93
Cu - ZSM - 5 (55)	0.84 - Cu	-	-	47	100	-	11	100	100	-	8	77	93
Cu - ZSM - 5 (108)	1.856 - Cu	-	-	63	100	-	-	63	100	-	22	97	100
Cu - ZSM - 5 (126)	1.9 - Cu	28	38	61	100	11	20	100	100	2	15	85	97
Cu - ZSM - 5 (124)	2.3 - Cu	-	26	57	100	9	26	100	100	-	18	80	85

Values in parentheses for the expression of the catalysts represent the level of cation exchange.

zeolite type catalysts, those whose copper degree of exchange in zeolite is > 100% achieve higher methane conversions. Such results are in agreement with those obtained in the reduction reaction of nitrogen monoxide with methane and confirm the fact that those catalysts with high activity on NO_x conversion to nitrogen have also a high activity on methane oxidation.

Table 2. Catalytic activities of granules catalysts in reaction of methane oxidation.
Reaction conditions: CH - 1.5%, O_2 - 20% in N_2 w/F = 0.25 g · s · cm^{-3}

Catalyst	Conversion of CH_4 (%)		Selectivity to CO_2 (%)	
	400°C	500°C	400°C	500°C
$Cu_xCo_{3-x}O_4$ (8% Co, 4% Cu)	10	52	13	53
Cu-ZSM-5 (55)	11	42	5	29
Cu-ZSM-5 (108)	15	43	6	43
Cu-ZSM-5 (126)	14	57	9	56
Cu-ZSM-5 (124)	16	55	9	66

The ESR studies performed on the catalysts whose activity data are here in above presented show the existence of copper absorbed in the lattice under two forms which according to the literature data would be Cu^{2+} and Cu^{+}.

The samples of Cu-ZSM-5 (126) and Cu-ZSM-5 (124) show higher absorbed spins into the lattice as Cu-ZSM (55) sample. These results are in good agreement with the higher catalytic activity of these samples.

Reduction in the flow system with hydrogen at different temperatures (from room temperature to 900°C) give a significant difference between the samples. Fig. 2 shows that for the Cu-ZSM-5 (55) sample with an ion exchange level under 100%, the reduction of copper occur in a single step at a higher temperature (810°C). This fact demonstrated that at a lower copper content, this exists in a single form according to this method. For the samples with higher copper content the reduction of this cation occur in two steps. The first step takes place between 520°C and 605°C with a maximum at 565°C, the second step was found between 760 and 960°C with a pick at 845°C. An exploration of this fact might be that copper exists in two forms, in good agreement with the published literature which shows that for the zeolites with large amounts of copper, this cation exists as Cu^{2+} and Cu^{+}.

The ESR spectra for reduced samples with hydrogen have a different aspect. In the case of Cu-ZSM-5 (124) and Cu-ZSM-5 (126) the signal is lower than the signal of Cu-ZSM-5 (55) sample. This fact shows a large amount of copper which is possible to be reduced in the samples with the higher copper content. If the

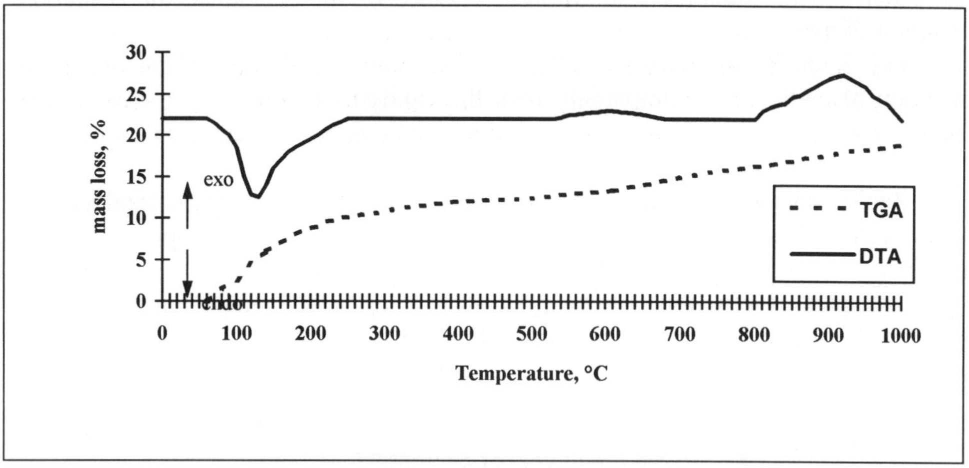

-a-

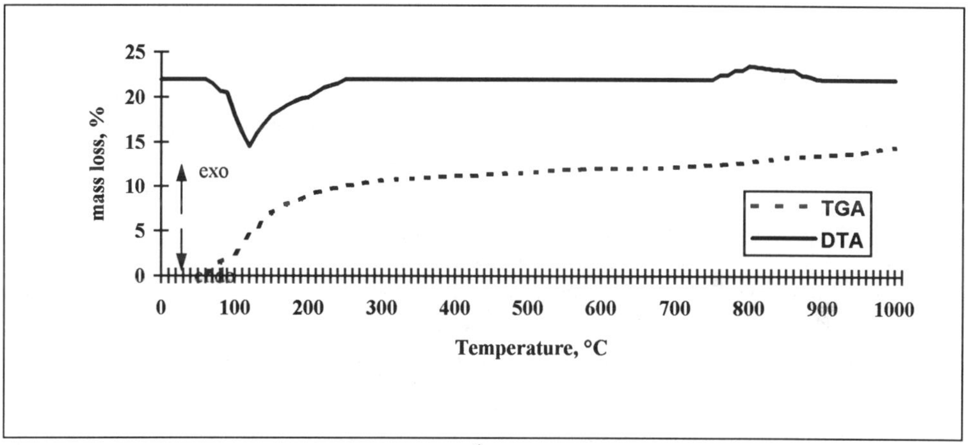

-b-

Fig. 2 TGA / DTA curves of :
a-CuZSM-5 (124)
b-CuZSM-5 (55)

reduced samples remain in air for a few hours spin number increases but below toinitial value. A redox cycle $Cu^0 \Leftrightarrow Cu^+ \Leftrightarrow Cu^{2+}$ in zeolite, quated by the literature data [9] could justify such modification.

The results obtained with granulated catalyst were checked with monolithic catalyst. The samples with a copper content of 0.6 wt % and 0.4 wt %, with an exchange level of copper over 100% and with nearly the same amount of zeolite deposited on the monolithic substrate (16-19wt%) were tested in the oxidation reaction of methane.

Fig. 3 shows the catalytic activity of monolithic samples in the oxidation reaction of methane in comparison with the catalytic activity of the oxidic type sample $Cu_xCo_{3-x}O_4$ (4.2) (x=1). This latter sample is a representative catalyst for hydrocarbons oxidation.

Zeolite catalysts samples show a lower activity than the oxidic catalyst, but the Cu-ZSM-5 sample with the copper content of 0.6 wt % approaches this catalyst.

The plot of reaction selectivity to carbon dioxide versus temperature is presented in fig. 4 and shows the same behaviour as that of the studies samples.

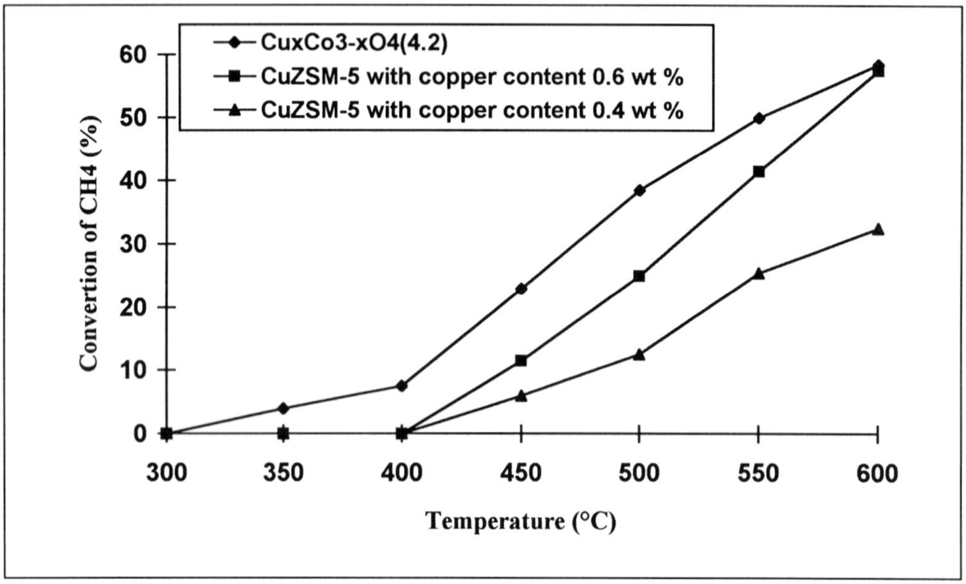

Fig. 3 Variation in the convertion of methane as a function of reaction temperature.
(Reaction conditions: 0.4% CH_4, 20% O_2 in N_2
Sample volum - 147 cm^3, space velocity - 10000 h^{-1})

In the case of monolithic catalysts, the ESR spectra present the same aspect as the granular samples. The copper is absorbed in the lattice under two forms which according to the literatura data would be Cu^{2+} and Cu^{+}. The number of spins is larger in the case of Cu-ZSM-5 sample with the copper content of 0.6 wt % than in the case of Cu-ZSM-5 with the copper content of 0.4 wt %. This fact is also in good agreement with the catalytic activity, in a similar maner as in the case of granular catalysts.

The reduction study with hydrogen of the monolithic catalysts under the same reduction conditions as for granular samples shows that in this case there is only a single reduction step at high temperature (910°C). After reduction a specific ESR signal appears for copper only in the Cu-ZSM-5 sample with the copper content of 0.6 wt %. This fact indicates that the amount of unreducible copper is large in this catalyst.

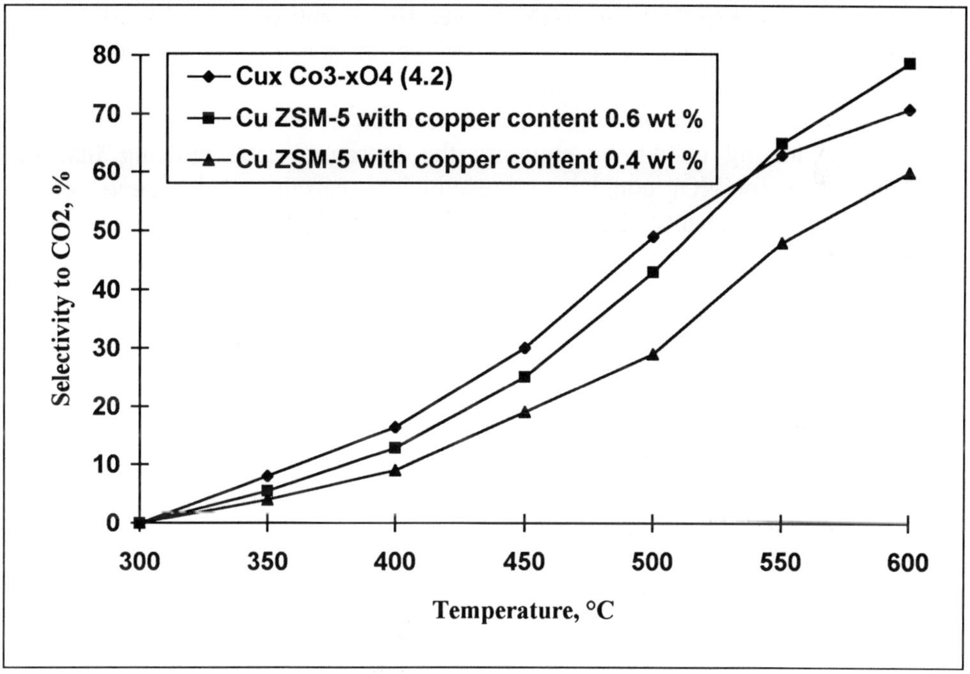

Fig. 4. Variation in the selectivity to carbon dioxide as a function of reaction temperature (Reaction conditions: 0.4% CH_4, 20% O_2 in N_2 sample volume - 147 cm^3, space velocity - 10000 h^{-1})

According to the literature it is very important to have copper too in the catalyst, which is difficult to reduce in order to carry out NO_x reduction with hydrocarbons in the presence of oxygen.

CONCLUSION

The results presented in this paper have confirmed the fact that the zeolite type catalysts which are highly active in methane oxidation reaction, are also highly active in NO_x reduction reaction.

They also have tried to explain copper state in the zeolite type catalysts used in the reduction reaction of NO_x with hydrocarbons in the presence of a greater amount of oxygen in the reaction medium, therefore under distinctly oxidizing conditions. They evidenced the importance of copper excess in such catalysts (the level of copper ion exchange in ZSM-5 over 100%) and confirmed the literature data which show that copper is present on the surface of these catalysts under two forms (Cu^{2+}, Cu^{+}) [6, 9].

Catalysts behaviour in the reduction reaction of NO_x with methane under distinctly oxidizing conditions supports the literature data showing that the reducing agent for NO_x could be represented by certain species deposited on catalyst surface (probably coke) which make possible NO_x reduction even after the complete oxidation of the methane from the reaction medium [8].

REFERENCES

1 Y. Kintaichi, H. Hamada, M. Tabata, M. Sasaki and T. Ito, Catal. Lett., 6 (1990) 239.

2 H. Hamada, Y. Kintaichi, M. Sasaki, T. Ito and M. Tabata, Appl. Catal. 64 (1990) L.

3 S. Sato, Y. Yu-u, H. Yahiro, N. Mizuno and M. Iwamoto, Appl. Catal. 70 (1991) L.

4 M. Iwamoto and H. Hamada, Catalysis Today 10, 57-71-(1991).

5 K. Yogo, M. Ihara, I. Terasaki and E. Kikuki - Appl. Catal. - B. Environmental, 2 (1993) L_1 - L_5.

6 M. Iwamoto, H Iahiro, S. Shundo, Y. Yu-u and N. Mizuno, Appl. Catal. 69 (1991) L_{15} - L_{19}.

7 H. Hamata, Y. Kintaichi, M. Sasaki and T. Ito., Appl. Catal. 75 (1991) L_1 - L_8.

8 G. P. Ansell, A.F. Diwell, S. E. Golunski, Y. W. Hayes, R. R. Rajaram, T. J. Truex and A. P. Walker, Appl. Catal. B, Environmental,2 (1993) 81 - 100.

9 R. Burch and P. J. Millington, Appl. Catal. - B Environmental, 2 (1993) 101 - 116.

A. Frennet and J.-M. Bastin (Eds.)
Catalysis and Automotive Pollution Control III
Studies in Surface Science and Catalysis, Vol. 96
1995 Elsevier Science B.V.

STUDIES OF SELECTIVE NO REDUCTION BY CH_4 AND CH_3OH OVER Co AND Cu EXCHANGED MORDENITE

J. Vassallo, M. Lezcano, E. Miró and J. Petunchi

Instituto de Investigaciones en Catálisis y Petroquímica - INCAPE - (UNL,FIQ - CONICET) - Santiago del Estero 2829 - 3000 - Santa Fe, Argentina

ABSTRACT

A systematic study of the selective reduction of NO using CH_4 and CH_3OH as reducing agents is presented. CoMordenite proved to be active and selective for both reactants while CuMordenite was so only when CH_3OH was used. The redox properties of the solids could explain that behavior. Interestingly, the NO to N_2 conversion started after the CH_3OH conversion reached 100%. Experiments made without oxygen in the feed stream showed a strong deactivation of the catalysts when methanol was the reducing agent, probably due to carbon deposits on the solid surface. The same effect was observed by transient experiments when oxygen was removed. Results showed that CH_3OH could be an oxygenated intermediate on the selective reduction of NO_x with CH_4 over CoMordenite.

1. INTRODUCTION

More efficient engine performance can be achieved by using higher air-fuel ratios than those permitted by the "window" required for present three-way catalyts [1]. Thus, the development of a suitable NO_x reduction catalyst would be highly desirable. Earlier reports on the use of hydrocarbons as selective reducing agents for NO in the presence of excess oxygen [2,3] have led to an increasing interest in these reactions because of their obvious practical importance. Although the general characteristics of these reactions are by now well known, the reaction mechanism is not completely understood and several suggestions have been advanced.

A redox mechanism has been proposed by Burch and Millington [4], while other authors have related active sites with the formation of carbon residues [5]. The possibility of a bifunctional redox-acid mechanism has been reported in [6] as well as the preferential NO reaction with O_2 to form NO_2 as the initial stage of the reaction [7]. However, one of the suggestions which have deserved more attention proposes the existence of an oxygenated intermediate, which would be responsible for NO reduction [8].

In the present work, this topic is further investigated through a systematic study of the selective reduction of NO by CH_4 and CH_3OH over metal zeolites. Accordingly, the role of the partial oxidation hydrocarbon in the reaction mechanism of the selective NO reduction is considered.

2. EXPERIMENTAL

The catalysts employed, Cu and Co Mordenite (hereinafter CuM and CoM) were prepared by ionic exchange of a LZM5 Mordenite provided by Linde (Si/Al 5.0 ratio), and aqueous solutions of CuAc (0.012 M) and $CoNO_3$ (0.025 M). The exchange was performed at ambient temperature, using Mordenite/solution ratios of 1 g/lt and 3 g/lt, respectively. In all cases, the preparation was performed during 24 hours and pH = 5. The solids obtained by filtration were dried on a stove at 120°C, and afterwards calcined in O_2 at 500°C. The amount of exchanged metal was determined by atomic absorption, 54.5% of the C.E.C. (cation exchange capacity) corresponding to CuM, and 25.0% to CoM (5.9 wt% and 2.4 wt%, respectively).

The kinetic experiences were carried out in a quartz tubular reactor (1.3 cm o.d. and 50 cm in length) under steady state conditions. The feed composed of He, NO and HC (whether CH_4 or CH_3OH) was introduced to the reactor after mixing in a manifold. O_2 was incorporated to the gaseous mixture close to the catalytic bed (about 2 cm above this one) with the purpose of minimizing reactions in the gas phase, especially when CH_3OH was used as reducing gas. In order to study the reaction $CH_3OH+NO+O_2$ in the gas phase, experiments were performed replacing the catalyst bed by the same volume of quartz chips.

In all experiences the composition of the feed consisted in NO (1000 ppm), CH_4 or CH_3OH (1000 ppm), O_2 (1%) and He as diluting gas. CH_3OH was introduced to the manifold by means of a saturator at 0°C. The total stream employed was 150 cm^3/min and the catalyst load was 0.5 g in every case (GHSV 6500h^{-1}).

The gaseous effluents were analyzed by gas chromatography (SHIMADZU GC-8A) employing a 5A zeolite to separate N_2, O_2, NO, CO , CH_4, and chromosorb to separate CO_2, CH_3OH and N_2O. The catalytic activity for the selective reduction was evaluated in terms of the N_2 production as $C_{NO} = 2N_2/NO$ and for the oxidation of CH_3OH and CH_4 as $C_{CH} = CO_x/HC$ (CO and CO_2 were the only oxidation products). C_j conversions were evaluated upwards and downwards with temperature to test results.

NO selective reduction (HC + NO + O_2) and oxidation (HC + O_2) experiences were performed. Within the text, the former will be referred to as "reaction A" and the latter as "reaction B", HC being equal to either CH_4 or CH_3OH, as it corresponds.

3. Results and Discussion

3.1. Methane as reducing agent

In the process of nitric oxide selective reduction, two main reactions are involved: NO_x to N_2 reduction, and the oxidation of the hydrocarbon with oxygen. When the hydrocarbon employed is methane, the copper exchanged in ZSM5 is ineffective to reduce NO_x since CH_4 preferably reacts with oxygen due to the great oxidating activity of the extralattice oxygens in CuZSM5 [9].

Figure 1.a shows the results obtained for reactions (A) (CH_4 + O_2 + NO) and (B) (CH_4 + O_2) when CuMordenite is used as catalyst. In agreement with what has been reported for CuZSM5 [2], no reduction of NO to N_2 is observed, the methane conversions being similar in both (A) and (B) reactions.

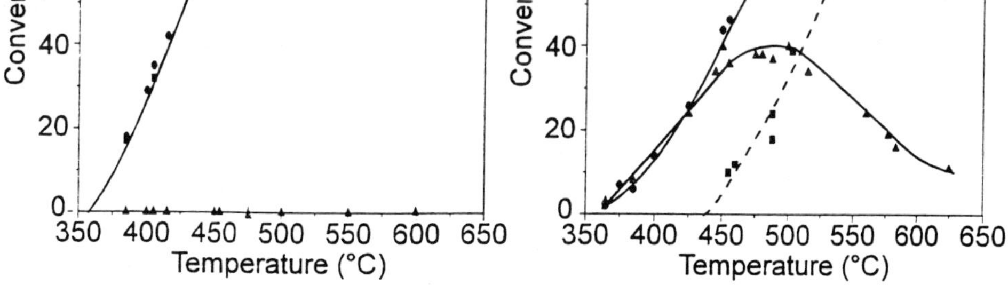

Figure 1.a: CH_4 + O_2 + NO and CH_4 + O_2 reactions over CuMordenite

Figure 1.b: CH_4 + O_2 + NO and CH_4 + O_2 reactions over CoMordenite

—●— Methane conversion (CH_4 + O_2 + NO), ---■--- Methane conversion (CH_4 + O_2), —▲— NO conversion

Instead, when CoMordenite is used as catalyst (Figure 1.b) it is clearly seen that this one is active for NO reduction; besides, the methane conversion in reaction (A) occurs at lower temperatures than in reaction (B). The temperature at which NO reduction begins to appear in Figure 1.b coincides with the beginning of CH_4 conversion in reaction (A). This behavior leads us to think of two stages for the selective reduction of NO. In the first one, at low temperatures, CH_4 oxidation promoted by the presence of NO_x starts. The second one consists in the direct oxidation of CH_4 with O_2, and as a consequence of its consumption, the conversion of NO to N_2 decreases. These two stages are consistent with the experimental observations usually reported in the literature: the maximum in the conversion of NO to N_2, and the promotion of HC oxidation due to the presence of NO.

However, the first stage mentioned above, does not simply consist in the reduction of NO with HC, since by performing a mass balance for the CH_4 consumed, it is deduced that there exists a catalytic effect of NO for the oxidation of HC with O_2. This results from the observation of the fact that the amount of HC reacting with O_2 in reaction (A) is greater than the amount of the same HC reaction with O_2 in reaction (B), be it NO or NO_2 the reactant in the mass balance. This comparison was performed in the low conversions zones of Figure 1, calculating the amount of CH_4 reacting with O_2 in reaction (A) as: CH_4 moles totally converted minus CH_4 reacted with NO_x.

This same effect was observed when the hydrocarbon employed was ethylene [10]. The exception occurred when CuMordenite and CH_4 were used (Figure 1.a), in which case NO does not react and both CH_4 conversion curves coincide.

These results may be explained through the existence of an oxygenated hydrocarbon as reaction intermediate, as proposed by Sasaki et al. [8]. According to this mechanism, HC would be partially oxidated at low temperatures (to methanol or formaldehyde if the hydrocarbon is methane), this compound being easily oxidated by NO_x or O_2. The possibility of the catalytic effect of NO in this partial oxidation stage is supported by the studies of McConkey and Wilkinson [11] who investigated the obtention of HCHO from CH_4 and O_2 using NO as catalyst in homogeneous phase. The said authors propose a mechanism according to which NO_2 dehydrogenates CH_4, producing methyl radicals which can be partially oxidated to formaldehyde. Sequentially, HCHO oxidates to CO_x.

As already pointed out, the CuMordenite catalyst is ineffective to reduce NO with CH_4 in the presence of oxygen, different from the CoMordenite behavior. This difference is consistent with the different redox capacity of these solids. Miró et al. [12] reported that the copper exchanged in mordenite can be totally reduced to Cu(I) when the reducing agent is carbon monoxide, and to Cu(0) when the reducing agent is hydrogen, at temperatures lower than 300°C in both cases. On the other hand, cobalt exchanged with mordenite is very difficult of being reduced. Miró and

Petunchi [13] reported no more than 15% reduction with CO or H_2 at 500°C. This difference in the reduction capacity explains why methane is easily oxidated by the extralattice oxygen in the case of CuMordenite, this oxygen being returned from the gas phase in a redox cycle, NO_x remaining unreacted. Instead, in the case of CoMordenite, the more moderate oxidating capacity of this solid could allow the formation of oxygenated intermediates for the subsequent NO_x reduction.

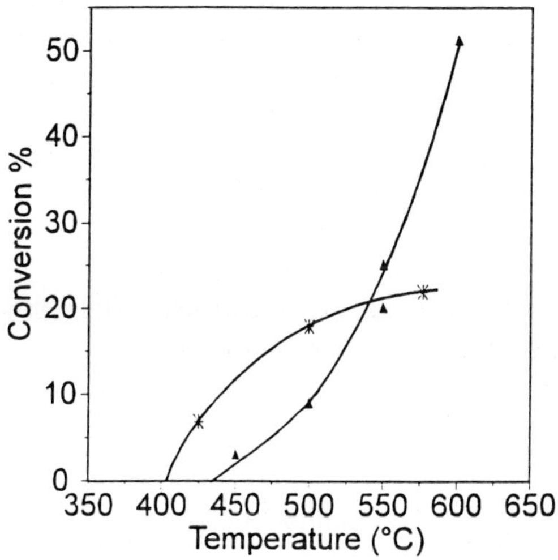

Figure 2: NO conversion in CH_4 + NO reaction over CuMordenite —▲— and CoMordenite —✳—

In the absence of oxygen (CH_4 + NO reaction) both catalysts, CuMordenite and CoMordenite present similar conversion values (Figure 2). Taking into account that these values have been obtained after approximately 30 minutes of reaction, it is possible that copper, due to the non-existence of oxygen in the gas phase, be partially reduced by methane. This reduction of copper could decrease its oxidating activity, thus allowing the reaction of methane with nitric oxide. Under the working conditions of Figure 2, the decomposition reaction of nitric oxide over CoM (without methane) practically does not occur. In the case of CuM, NO decomposition occurs, but at lower conversions than the observed for CH_4 + NO reaction.

3.2. Methanol as reducing agent

With the object of stressing the possibility of an oxygenated compound as intermediate of the selective NO reduction with methane, methanol was used as reducing agent. Figure 3.b shows the results of methanol and nitric oxide conversion at different temperatures for reactions A and B over CoMordenite. It can be observed that the conversion of nitric oxide to nitrogen begins to occur at temperatures similar to those for the selective reduction with methane. This fact clearly supports the existence of an oxygenated compound in the $CH_4 + O_2 + NO$ reaction over CoMordenite.

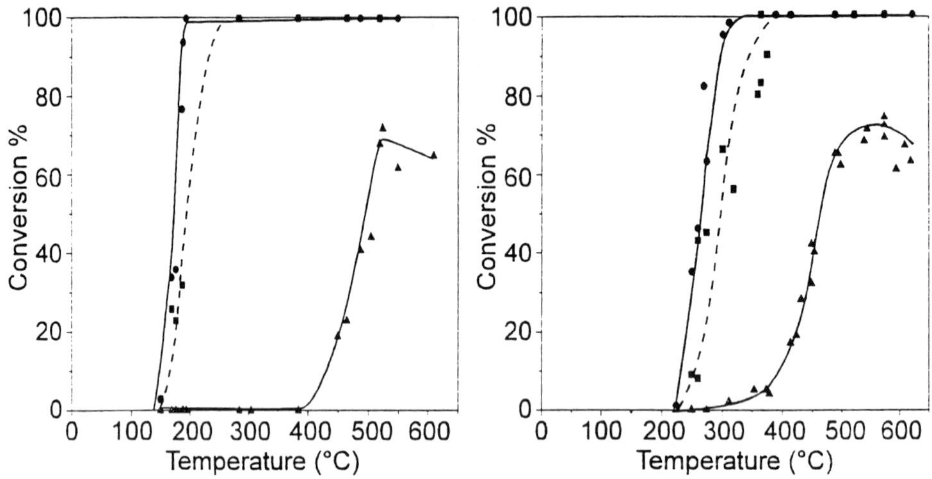

Figure 3.a: $CH_3OH + NO + O_2$ and $CH_3OH + O_2$ reactions over CuMordenite. Figure 3.b: $CH_3OH + NO + O_2$ and $CH_3OH + O_2$ reactions over CoMordenite. —●— Methanol conversion ($CH_3OH + O_2 + NO$), ---■--- Methanol conversion ($CH_3OH + O_2$), —▲— NO conversion

In the case of the CuMordenite catalyst, as stated before, no NO conversion is observed when CH_4 is used as reducing agent due to the high oxidating capacity of this solid. Concerning the role of a possible oxygenated intermediate in this catalyst, two hypotheses may be formulated: methane is partially oxidated but the intermediate is destroyed on the catalysts surface without reacting with the nitric oxide, or else methane oxidates to CO_x directly.

The first hypothesis can be sustained with the values of the activation energies of the $CH_3OH + O_2$ reaction, calculated for CoM and CuM. For the first catalyst E = 48.3 kcal/mol was obtained, and for CuM, E = 29.6 kcal/mol. These values indicate a greater stability of methanol over CoM. It seems reasonable then that in the case of the CoM catalyst, an oxygenated intermediate may acquire a mean life time enough to reduce nitric oxide, whereas in CuM this possibility is smaller.

However, the results shown in Figure 3.a support the second hypothesis (direct oxidation of CH_4 to CO_x), since it can be observed that methanol reduces nitric oxide in the presence of oxygen, indicating that if the oxygenated intermediate were formed during the $CH_4 + NO + O_2$ reaction, NO_x could favourably compete with oxygen to react with the said intermediate. The equilibrium between NO and NO_2, which might be an intermediate, may also play an important role in the reaction mechanism. Further experiments are in progress in order to dilucidate this phenomenon.

It is necessary here to draw attention on the way in which these experiments were conducted in what concerns to the oxygen feed. As already stated in the Experimental section, oxygen was incorporated to the gas phase very close to the catalyst bed. Experiences performed mixing oxygen at the reactor inlet, showed lower nitric oxide conversions. This is due to the fact that methanol oxidation takes place in the reactor void volume before reaching the catalyst bed, thus obtaining NO conversions similar to those of the non-catalytic gas phase reaction.

From the observation of Figures 3.a and 3.b, another interesting fact can be noticed. Conversion of NO to N_2 is produced at temperatures at which a 100% methanol conversion was already reached, that is, the conversion-temperature curves for the oxidation of methanol and for NO reduction are displaced approximately a DT of 100°C in both CoM and CuM catalysts. This differs from what is normally observed with methane. This delay in the NO reduction towards higher temperatures could be originated in the formation of a carbon deposit which would poison the sites necessary for NO adsorption. When an adequate temperature for the combustion of surface carbon with oxygen is reached (approximately 400°C), the catalyst becomes active for NO reduction. A similar effect is proposed in [14] using propylene and propane as reducing agents.

Hamada et al. [15] did not find nitric oxide conversion using methanol as reducing agent and CuZSM5 as catalyst, thus differing from our results. Probably this is due to the influence of the gas phase reactions mentioned above, or maybe these authors did not reach temperatures high enough to burn the carbon deposit.

Figure 4 shows transient experiments performed with CoM and CuM with the object of stressing the role of O_2 on the CH_3OH+O_2+NO reaction. In those experiments, once the steady state was reached, the oxygen flow was interrupted (marked as time = 0 in the figure), thus decreasing its concentration to negligible levels before the five minutes of the transient experiment. The curves shown indicate that initially, when the surface is clean due to the presence of oxygen, there is an appreciable conversion level (40% for CuM and 70% for CoM). When oxygen is eliminated, the rate of NO reduction slowly decreases until reaching a new steady-state value which coincides with the one observed for the non-catalytic reaction, without oxygen, in the gas phase (Figure 5). This result supports the role of oxygen in maintaining the surface clean of carbon deposits, since when O_2 is interrupted, the reducing activity is maintained for a time even though oxygen has disappeared from the gas phase. Similar results were reported by Petunchi and Hall (7) in their studies on the selective NO reduction with isobutane over CuZSM5. These authors found coke formation and the reduction of Cu^{2+} (disappearance of 100% of the Cu^{2+} signal by EPR).

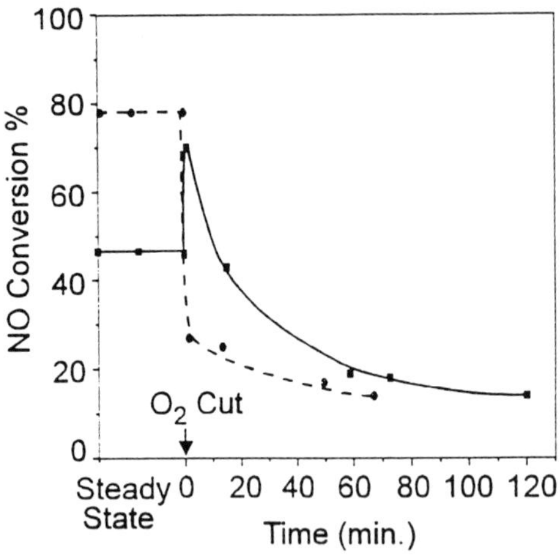

Figure 4: Transient experiments over CoMordenite ---•--- and Cu Mordenite —▪—. Oxygen is cut at time equal zero.

Figure 5 shows nitric oxide conversions at different temperatures for the NO + CH_3OH and NO + CH_3OH + O_2 reactions in the gas phase.

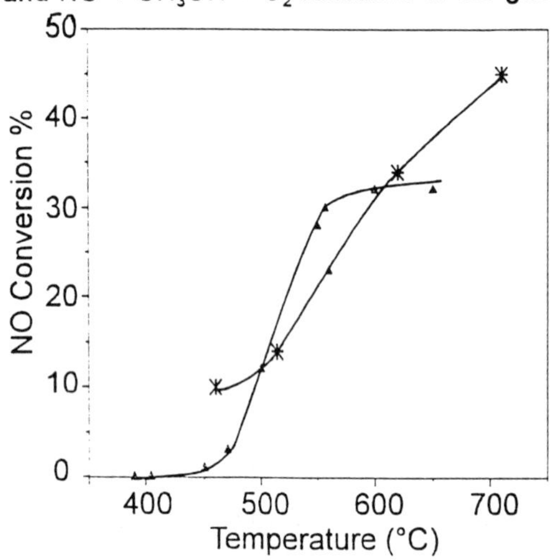

Figure 5: Gas phase reaction
—✳— CH_3OH + NO
—▲— CH_3OH + NO + O_2

In these experiments, the catalyst bed was replaced by a similar volume of quartz chips. It is interesting to observe that the NO conversions for both reactions are similar. Different from methanol, when methane or ethylene are used as reducing agents in the HC + NO + O_2 reaction in the gas phase, no nitric oxide conversion is observed [10]. These results seem to support the existence of a mechanism with oxygenated intermediates, the production of these intermediates via catalytic oxidation of HC with O_2 or with NO_2, being a fundamental role of the catalyst. Once the intermediate has been formed, its reaction with NO_x could take place on the catalyst surface or the gas phase as depicted in Figure 5. When the hydrocarbon is methanol, oxygen has already been incorporated into the molecule, and the partial oxidation stage is not necessary. For this reason, in the gas phase reaction, oxygen does not noticeably impinge on the NO to N_2 conversion. When the CH_3OH + NO + O_2 reaction is carried out through the catalytic way with CoM or CuM, the fundamental role of O_2 would be to maintain the surface free from carbon deposits.

4. CONCLUSION

The main conclusions from this study may be summarized as follows:

- NO, besides being reduced in the presence of O_2 with CH_4 or CH_3OH over CoM, acts as a catalyst in the oxidation of the said reactants.

- CH_3OH is a selective NO_x reducing agent in Cu and Co Mordenite.

- The greater redox activity of CuMordenite, compared to that of CoM, would justify why the first solid does not selectively reduce NO_x with CH_4.

- An oxygenated compound, of the CH_3OH type when the reducing agent is CH_4, would act as an intermediate in the case of CoM.

- Oxygen would perform the function of maintaining the catalyst surface free from carbon deposits.

ACKNOWLEDGMENTS

Finantial support was provided by CONICET and UNL (CAI+D 162). We are indebted to Prof. Elsa Grimaldi for her assistance in the edition of the manuscript.

REFERENCES

1 T.S. Truex, R.A. Searle and D.C. Sun, Plat. Met. Rev., 36(1) (1992).
2 W. Held and A. Koning, Ger. Offen, DE 3, 642018 (1989) assigned to Vokswagen.
3 S. Sato, Y. Yu-u, H. Yahiro, N. Mixuno and M. Iwamoto, Appl. Catal., 70 L1 (1991).
4 R. Burch and P.J. Millington, Appl. Catal. B: 2, 101 (1993).
5 E. Kukuchi, K. Yogo, S. Tanaka and M. Abe, Chem. Lett., 1063 (1991).
6 T. Inui, S. Iwamoto, S. Kojo and T. Yoshida, Catal. Lett., L3, 87 (1992).
7 J.O. Petunchi and W.K. Hall, Appl. Catal. B:2, L17, (1993).
8 M. Sasaki, H. Hamada, Y. Kintaichi and T. Ito, Catal. Lett., 15, 297(1992).
9 Y. Li, P. Battavio and J. Armor, J. Catal., 142, 561 (1993).
10 J. Vassallo, E. Miró and J.O. Petunchi, unpublished results.
11 B. McConkey and P. Wilkinson, I & E C Proc. Res. Des. and Dev., 6(4) 436 (1967).
12 E. E. Miró, E.A. Lombardo and J.O. Petunchi, J. Catal., 104, 176 (1987).
13 E.E. Miró and J.O. Petunchi, J. Chem. Soc. Faraday Trans., 88(8) 1219 (1992).
14 J. d'Itri and W. Sachtler, Appl. Catal. B:2, L7 (1993).
15 H. Hamada, Y. Kintaichi, M. Sasaki, T. Ito and T. Yoshinari, Appl. Catal. A: 88, L1 (1992).

A. Frennet and J.-M. Bastin (Eds.)
Catalysis and Automotive Pollution Control III
Studies in Surface Science and Catalysis, Vol. 96

SEPIOLITE BASED MONOLITHIC CATALYSTS FOR THE REDUCTION OF NITRIC OXIDE WITH PROPYLENE IN OXIDISING ATMOSPHERE

P. Avila, J. Blanco, J.M.R. Blas, O. Ruiz de los Paños and M.Yates.

Inst. Catálisis y Petroleoquímica, Campus UAM, Cantoblanco, 28049 Madrid.

ABSTRACT

This work deals with the analysis of the behaviour of new materials as catalysts for the selective reduction of nitrogen oxides in oxidising atmospheres. The catalysts were prepared using as the active phase oxides of copper and nickel, impregnated on different types of support whose principal component was sepiolite, a natural magnesium silicate. The catalysts were produced in the form of monoliths of 8 cells cm^{-2}, to obtain solids with high thermal stability and excellent mechanical properties, with which NO conversions of greater than 75% were obtained when operating at an Area Velocity of 1.6 mh^{-1}, with an O_2 concentration of 8% by volume. Significant textural differences have been found between the solids which could affect to a certain extent the activity and selectivity of the catalysts in the studied reaction.

1. INTRODUCTION

Since Iwamoto *et al.* [1] and Held *et al.* [2] in 1990, published their first studies on the elimination of nitrogen oxides in oxidising atmospheres using hydrocarbons as selective reducing agents, there has been a substancial growth in the number of studies which have appeared in the literature on this process, fundamentally due to the necessity of finding a solution to the control of emissions from diesel engines.

Most catalysts appearing in the literature since then are based on zeolites exchanged with transition metals, with those exchanged with copper the most active. Outstanding contributions have been made by Iwamoto *et al.* [3] and Hamada *et al.* [4] in this field generally with zeolite H-ZSM-5 exchanged with copper, using various reductors. Latterly, a range of studies have been published

*This work was supported by the project numbers AMB 92-0190 and AMB 93-0244 of the Spanish CICYT.

which likewise use copper or copper oxide exchanged or supported on various materials [5,6].

Although great advances in the development of catalysts for the elimination of nitrogen oxides in oxidising atmospheres have been made, serious drawbacks still exist before their adoption in industrial applications due to the instability of the catalysts and the necessity of operating at high hydrocarbon:nitrogen oxide ratios.

In this work the catalytic activity of a series of copper oxide catalysts supported on monolithic honeycomb supports in the reduction of nitrogen oxide with propylene in an oxidising atmosphere was studied. The monoliths were produced from acid washed sepiolite, sepiolite or a mixture of sepiolite and alumina in order to study the effect of the support on the activities and selectivities of the catalysts. The introduction of nickel oxide as a second active species on the overall activity was also determined. Finally the application of an alumina washcoat impregnated with the copper and nickel salts to increase the accessibility of the gases to be treated to the active phase was studied.

2. EXPERIMENTAL

The activity measurements were made in a continuous tubular reactor of 2.54 cm ID operating in an integral regime at a total pressure of 120 KPa. The NO and propylene inlet concentrations were 1000 ppm while the oxygen was varied between 1 and 8 vol.% with a nitrogen balance. The area velocity was varied between 1 and 5 mh^{-1} and the reaction temperature between 100 and 500°C. Analysis of the inlet and outlet gas concentrations were made using specific analyzers for each gas. Thus, the NO_x concentration was determined by chemiluminiscence with a Beckman analyser model 951A, the CO and CO_2, by nondispersive I.R. spectrometry in two Beckman model 880 analysers, the oxygen by paramagnetic analysis in a Beckman 755 analyser and the hydrocarbons by means of a Beckman flame ionization detector model 400A.

The textural characterization of the supports and catalysts: pore size distribution, pore volume, and surface areas were determined by use of mercury intrusion porosimetry using a Micromeritics Poresizer 9320 and nitrogen gas adsorption/desorption isotherms carried out on a Micromeretics ASAP 2000 respectively. For the porosimetry analysis a contact angle of 140° and surface tension of $480 mNm^{-1}$ for mercury were assumed.

The catalysts were prepared on monolithic structures of parallel channels of square section with a density of 8 cells/cm^2 and a wall thickness of 0.89 mm. Three types of monolithic support were used in this study: "ST" from sepiolite,

"SL" from sepiolite washed with a 2N HNO_3 aqueous solution and "SA" from sepiolite and alumina (60 and 40 wt% respectively). All of these monolithic supports were subsequently heat treated at 500°C for 4 hours in an air atmosphere.

The supported metal oxides which constitute the active phase of these catalysts were incorporated by impregnation on the support (SLCu, SLCuNi, STCuNi and SACuNi catalysts) or over a dispersion of alumina that was wash coated (5% by weight of the support) over the support (SA(RCuNi).

3. RESULTS AND DISCUSSION

The results of this study have been divided into three sections. In the first the influence of the incorporation of nickel oxide on the catalytic activity of a copper catalyst was studied, the second deals with the influence of the textural nature of the support while the third looks at the selectivity of the reduction of NO with propylene on changing the concentration of oxygen in the gas stream.

3.1 Effect of the incorporation of a second element in the active phase

Taking as a reference the activity of the catalyst of copper oxide supported on sepiolite washed with an acid solution, SLCu, the influence of the introduction of a second metal, Co or Ni, maintaining the Cu:M ratio at 9:1 wt% was studied.

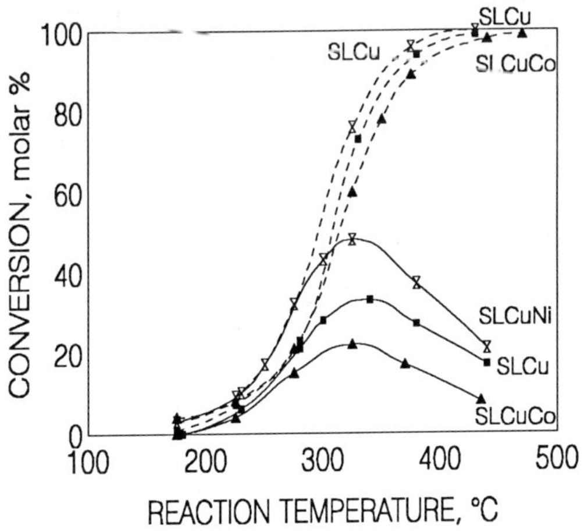

Fig.1 Influence of the nature of the active phase on the conversion of NO (____) and C_3H_6(-----)

In Figure 1 the conversion of nitric oxide (continuous lines) and propylene (dashed lines) *versus* the reaction temperature is presented. From this figure it may be observed that at temperatures below 280°C the propylene and nitric oxide react at a 1:1 molar ratio. At temperatures above this the oxidation of propylene begins to take place, due to the presence of oxygen (1% vol.) in the gas stream, with the subsequent separation of the two curves. Thus, the propylene conversion continues to increase with increasing temperature but the elimination of NO reaches a maximum at about 350°C and then progressively decreased.

With respect to the effect of the incorporation of a second component on the activity, it may be observed that the catalyst which included nickel, SLCuNi, was appreciably more active than the catalyst of copper only, SLCu, over the whole range of temperatures. However, the catalyst of copper and cobalt, SLCuCo, gave rise to NO conversion values inferior to the copper catalyst over the whole temperature range.

From these results it was clear that the incorporation of nickel was beneficial in both reactions. This increase was especially pronounced in the reduction of nitric oxide where conversion values of 50% could be achieved working at an area velocity of 4.7 mh^{-1}.

Similar results to those discussed above have been obtained by Blanco *et al.* using natural gas as a reducing agent [7], operating with catalysts supported on alumina. R. Hierl *et al.* [8], studying these catalysts, found that "the presence of Ni^{2+} in the catalyst leads to an enhanced surface segregation of Cu^{2+} accompanied by an increased tetrahedral site population by Cu^{2+}".

3.2 Effect of the physical nature of the support

Maintaining the concentration of the active phase constant (Cu and Ni), two further oxide based catalysts were prepared using as supports: sepiolite "STCuNi" and a mixture of sepiolite and alumina "SACuNi" respectively. These were compared with the catalyst described previously "SLCuNi" prepared on a support of sepiolite which had been washed with an acid solution.

In Figure 2 the conversion of NO and propylene *versus* the reaction temperature in the conditions stated previously are presented. Taking as a reference the behaviour of the catalyst SLCuNi, described previously, it may be observed that when sepiolite treated at 500°C was used as a support there was an appreciable decrease in the selectivity of the catalyst towards the conversion of NO. From Figure 2 it may be seen that the NO conversion separated from that observed for propylene at a temperature between 200° and 220°C. Under these conditions the maximum value of NO conversion achieved was only about 30%.

The catalyst prepared using a mixture of sepiolite and alumina as support "SACuNi", without doubt led to a pronounced increase in the activity compared

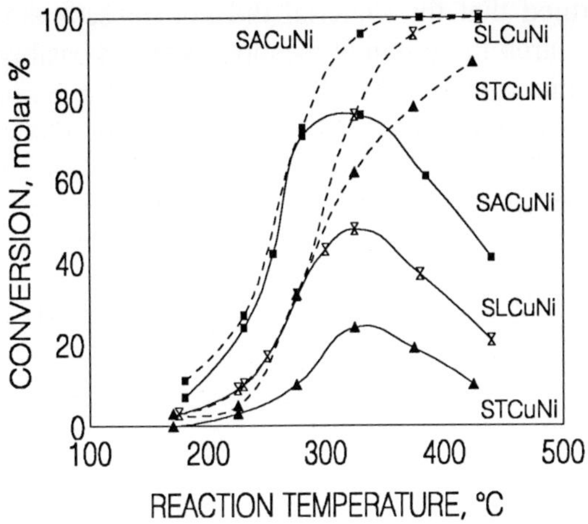

Fig. 2 Influence of the support on the activity of Cu-Ni catalysts.

to the other two catalysts tested. With this catalyst NO conversion values of the order of 80% were reached.

An attempt was made to explain the observed variations in the activities of the catalysts from their corresponding XRD and FTIR analyses, however, no marked differences in the spectra were observed. However, from the point of view of their textural characterization, the catalysts supported on mixtures of sepiolite and alumina possessed a porous structure which was markedly different from that of the catalysts supported on monoliths produced from sepiolite.

In Table 1 the results obtained from the textural characterization of the supports and catalysts by nitrogen adsorption and mercury intrusion porosimetry are presented. In the table the values of surface area obtained from the gas adsorption results, using the BET method for which the linear portion was usually located in the relative pressure range of 0.05 to 0.3 S_{BET} [9], and those from the intrusion curve of the porosimetry analysis, using a nonintersecting cylindrical pore model S_{Hg} [10], are shown. The pore volume V_p is that recorded at the highest intrusion pressure reached during the porosimetry analysis, and as such represents the pore volume of pores between *ca.* 30µm to 3nm pore radius. The pore radii were taken from the maxima of the curves of pore size distribution.

Taking as a reference the textural characteristics of the heat treated sepiolite support "ST", the data in Table 1 indicated that washing this natural silicate with an acid solution produced an appreciable increase in the BET surface area but had little effect on the surface area and pore volume measured by mercury porosimetry. Thus, this increase must have been in pores of between 1 and 3nm radius, since evaluation of the adsorption isotherm by use of the t-plot

[11], with a suitable standard, confirmed that the material did not possess any microporosity. The increase in surface area in the small mesopores was possibly due to the removal of impurities and the reduction in alkaline and alkaline earth cations, notably Mg and Ca, by the acid treatment.

Table 1 Textural Properties of Supports and Catalysts

Support	S_{BET} (m^2/g)	S_{Hg} (m^2/g)	V_p (cm^3/g)	r_{pore} (1) (nm)	r_{pore} (2) (nm)	r_{pore} (3) (nm)
ST	124	82	0.489	18.1	---	---
SL	199	86	0.517	16.1	---	---
SA	171	84	0.448	3.1	16.4	401.0
Catalysts						
SLCuNi	133	104	0.552	14.2	---	---
SACuNi	132	91	0.432	6.0	18.3	326.8
SA(RCuNi)	135	124	0.495	3.1	17.4	549.0

The incorporation of alumina into the support caused an apparent reduction in the total pore volume although the value of the BET surface area was located between that of the other two supports. This apparent fall in the pore volume was due to the fact that the alumina had a high volume of mesopores which were too small to be detected by the mercury porosimetry technique. Calculation of the mesopore volume of this material from the plateau in the desorption branch of the nitrogen isotherm at high relative pressure ($p/p^o = 0.95$) gave a mesopore volume of 0.484cm^3/g. The macropore volume of this material due to the interparticulate space between the alumina particles was 0.202cm^3/g, which gave a true total pore volume of 0.686cm^3/g. The most significant characteristic of the porosity in this material was the trimodal pore size distribution due to the characteristic porosities of the two base materials with maxima of pore diameters at 6, 18 and 400nm.

The solids prepared from heat treated sepiolite (ST, SL and the catalyst SLCuNi) all gave a Type II isotherm for the adsorption branch with a narrow desorption hysteresis, designated as Type IIb by Sing [10]. The narrow hysteresis was due to slit shaped mesopores which extend into the macropore range. The typical shapes of the isotherms obtained with these materials are shown in Figure 3 as that corresponding to the heat treated sepiolite support "ST".

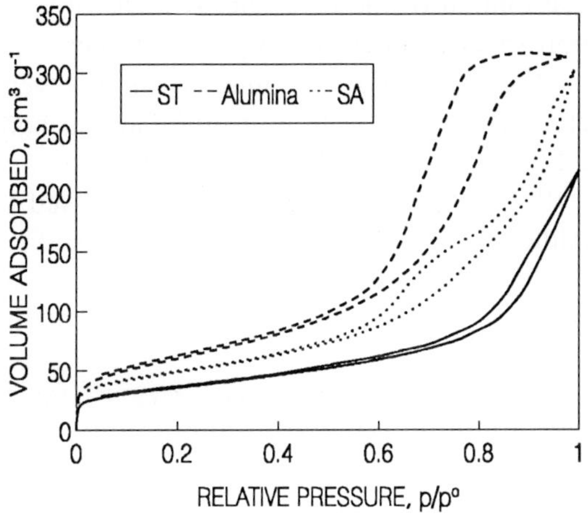

Fig. 3. Nitrogen isotherms of monoliths produced from heat treated sepiolite "ST", "Alumina" and a mixture of a two "SA".

In the same figure, for comparison, the isotherm obtained for the Alumina support after heat treatment at 500°C is shown. The isotherm was of Type IV, corresponding to a mesoporous material *i.e.* having pores between 1 and 25nm pore radii. The pore size distribution, calculated using the BJH [12] method on the desorption branch of the isotherm, indicated that the majority of the mesopores in this material had a radius of 3nm.

In the monoliths composed of a physical mixture of sepiolite-alumina (SA and SACuNi) the adsorption branch was of Type II but the desorption hysteresis was much more pronounced due to the presence of the mesopores of the alumina. The shape of the curve for these samples is shown as that corresponding to SA in Figure 3.

The catalysts prepared on these supports had similar properties to those corresponding to the support used. From the results presented in Table 1 the modifications in the properties which were caused by the introduction of the active phase on the supports may be seen. Thus, on the introduction of the active phases on the acid washed sepiolite, SLCuNi, the BET surface area was reduced, probably due to blocking of the narrower mesopores. Due to the deposition of the active phases on the pore walls the areas calculated from the intrusion curves of the porosimetry determinations, S_{Hg}, increased and also, surprisingly, the total pore volume, as measured by mercury porosimetry.

The most significant characteristic of the catalyst supported on sepiolite/alumina monolith which was thought to lead to the greater catalytic activity was the trimodal pore size distribution of the material. This distribution was due to the combination of the porosities of the two support materials. The resulting monolith not only had a high BET area due to the alumina but also very wide macropores (400nm) which increased the accessability to the internal pore network. With the catalysts supported on sepiolite based monoliths the absence of these wider pores in conjunction with the high area velocities probably led to the lower activities of the catalysts. Although the SA supports had higher surface areas which could have led to a better dispersion of the active phases, as mentioned previously this was not noted by differences in the X-ray diffraction patterns.

3.3 Effect of Oxygen

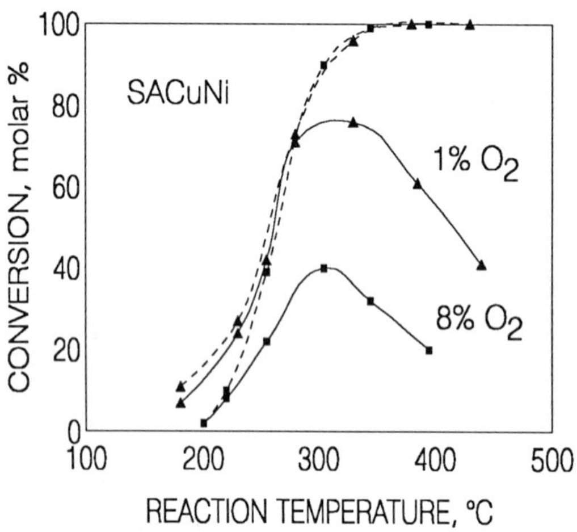

Fig.4. Influence of the concentration of O_2 in the gas stream

The principal object of this study was the development of catalysts which could achieve the selective reduction of NO in the presence of elevated concentrations of O_2. One of the most important parameters to be taken into account was the influence of the concentration of O_2 on the activity of the catalyst. Following the results described previously in this paper, it was found that the catalyst supported on a monolith based on sepiolite/alumina, SACuNi, demonstrated a higher activity and selectivity in the studied reaction when the O_2 concentration was of the order of 1% vol. Thus, further work at higher O_2 concentrations was based on this catalyst.

When the concentration of O_2 in the gas stream was increased there was a reduction in the selectivity of the process to NO conversion over the whole temperature range. In Figure 4 the activity results obtained with the SACuNi catalyst operating at two concentrations of O_2 in the gas stream (1 and 8%) are presented. From the figure it may be appreciated that when the reaction took place in the presence of 1% O_2 the conversion of propylene and NO followed a 1:1 molar ratio up to 280°C. At higher temperatures the curves separated due to the reaction of the propylene with the O_2 in the gas stream. The maximum NO conversion of 75% was reached at a temperature of 300°C.

When the O_2 concentration was increased to 8% the separation of the conversion curves for propylene and NO occured at a lower temperature (210°C), and the maximum NO conversion attained was only of the order of 40%. Also under these conditions the optimum operating temperature range was found to be narrower.

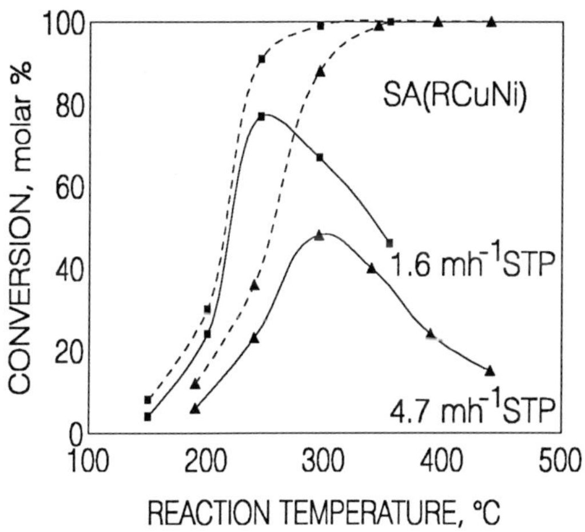

Fig.5. Influence of the Area Velocity.

The results obtained with SACuNi operating at 8 vol.% O_2 improved when the active phases were introduced onto the monolith in the form of a washcoat SA(RCuNi). In Figure 5 the activity results obtained with this catalyst operating at two area velocities are presented. Of note were the NO conversions of greater than 75% at relatively low reaction temperatures (200-300°C).

Although the operating temperature range was narrow and the area velocities low, this result may be considered as encouraging in offering alternatives to copper exchanged zeolites, nevertheless further work is required to improve the operating parameters.

4. CONCLUSIONS

From the results presented in this study it was found that the inclusion of nickel in the active phase caused a significant increase in the catalytic activity and selectivity of the catalyst in comparison to those with only copper.

Catalysts supported on monoliths based on a mixture of sepiolite and alumina had much better selectivities to NO conversion than the corresponding catalysts supported on monoliths prepared from sepiolite. This was thought to be due to the different textural properties of these mixed composition supports, where the presence of wide macropores and the higher surface areas improved the accessibility of the reaction gases to the internal pore network.

Increase in the concentration of O_2 in the feed stock produced a marked decrease in the selective reduction of NO due to the combustion of the propylene. Not only was the maximum NO conversion reduced but the useful operating temperature range also became narrower.

The incorporation of the active phases on to the monolithic support after first applying an alumina washcoat greatly increased the selectivity of the catalyst where NO conversions of 75% could be achieved. This was probably due to a reduction in any diffusional limitations.

REFERENCES

1 M. Iwamoto, Proc. of Meeting of Catalytic Technology for Removal of Nitrogen Monoxide, Tokyo, (1990) 17.
2 W. Held, A. König, T. Richter and L. Puppe, SAE Paper 900496 (1990).
3 M. Iwamoto, H. Yahiro, S. Shundo, Y. Yu-u and N. Mizuno, Applied Catalysis, 69,(1991) L15, L19.
4 H. Hamada, Y. Kintaichi and T. Yoshinari, Enviromental Industrial Catalysis, 1st European Workshop Meeting, (1992) 219.
5 M. Iwamoto, N. Mizuno and H. Yahiro, Catalysis Research Center, (1992) 213.
6 H. Hosose, H. Yahiro, N. Mizuno and M. Iwamoto, Chemistry Letters,(1991) 1859.
7 J. Blanco and A. Martínez, Química e Industria, 22, 8, (1976) 855.
8 R. Hierl, H. Knözinger and H-P. Urbach, J Catal., 69, (1981) 475.
9 S. Brunauer, P.H. Emmett and E. Teller, J. Am. Chem. Soc.,60 (1938) 309.
10 H.M. Rootare and C.F. Prenzlow, J. Phys. Chem., 71 (1967) 2733.
11 B.C. Lippens and J.H. de Boer, J. Catal., 4, (1965) 319.
12 K.S.W. Sing, Third Internacional Conference on Fundamentals of Adsorption, Engineering Foundation, New York, (1991) 67.
13 E.P. Barret, L.G. Joyner and P.H. Halenda, J. Amer. Chem. Soc., 73, (1951) 373.

Catalyst Aging and Poisoning

A. Frennet and J.-M. Bastin (Eds.)
Catalysis and Automotive Pollution Control III
Studies in Surface Science and Catalysis, Vol. 96

IMPACT OF SULFUR ON THREE-WAY CATALYSTS: COMPARISON OF COMMERCIALLY PRODUCED Pd AND Pt-Rh MONOLITHS

D. D. Beck and John W. Sommers

Physical Chemistry Department
General Motors NAO Research and Development Center
Warren, MI 48090

ABSTRACT

Commercially-prepared Pt-Rh and Pd monolith catalysts were thermally aged then characterized for catalytic performance using a laboratory reactor to evaluate the magnitude and reversibility of the impact of sulfur on three-way activity. The SO_2 concentration in the feedstream was varied from 0 ppm to 30 ppm, which was comparable to sulfur levels in gasoline ranging from 0 to 450 ppm. Tests were first conducted using propylene and repeated using propane to represent the hydrocarbon mixture in exhaust.

Of the two catalysts, Pd showed better propylene lightoff activity while Pt-Rh showed better propane lightoff activity, regardless of the sulfur content. For each catalyst, increasing the sulfur concentration from 0 ppm SO_2 to 30 ppm SO_2 of sulfur resulted in a lightoff temperature increase by 40 to 60°C. Under warmed-up conditions, the loss of activity for HC, CO and NOx due to the presence of sulfur was greater under slightly rich conditions than under lean conditions for both Pd and Pt-Rh, while the magnitude of the impact on NOx and particularly on HC activity under warmed-up stoichiometric conditions was significant and much greater for Pd than for Pt-Rh catalyst. Using propylene, the effect of SO_2 on the activity of the Pd catalyst was partly reversible, while the effect on Pt-Rh was completely reversible. Using propane, the effect of sulfur on the activity of both catalysts was larger than for propylene. The resulting decrease in activity due to the presence of sulfur was partly reversible on Pt-Rh, but the poisoning of the Pd catalyst was mostly irreversible. Part of the irreversible poisoning effect is attributed to a direct interaction or reaction between SO_2 and Pd, while the other part is attributed to the promotional effect of SO_2 in hydrocarbon coking of the catalyst when alkane hydrocarbons are present.

1. INTRODUCTION

Organo-sulfur compounds are commonly present in nearly all commercially produced gasoline blends which are intended for vehicle use. These compounds are involved in the combustion process in the engine and are converted mainly into sulfur dioxide before entering the catalytic converter or converters. This form of sulfur has been demonstrated to deactivate vehicle exhaust catalysts. Prior work performed in the 1970's focused on the impact of sulfur on Pt and/or Pd supported on alumina catalysts operating under net oxidizing, or "lean" conditions [1-3]. Currently used three-way catalyst systems are far more complicated in that their washcoat formulations contain in addition to Pt, Rh, and sometimes Pd noble metals, a number of base metal oxides of Ce, La, Ni, Ba, Fe, Nd, and others. The exhaust adds to the complexity by cycling rapidly between net reducing and net oxidizing conditions when the vehicle operates under closed loop control. The cycling between these conditions may not be balanced under certain operating conditions such as a quick acceleration or deceleration or when a load condition is imposed. Moreover, the vehicle calibration may be slightly biased toward net lean or net rich operation. All of these conditions influence how SO_2 in exhaust interacts with the support, base metal oxides, and noble metal components in the catalyst and, in turn, affects activity.

Laboratory studies of the effect of sulfur on three-way activity showed that activity was inhibited particularly under rich conditions and mostly for catalysts containing both Pt and Pd, but activity was not significantly affected when the feedstream was cycled about a net stoichiometric point [4-6]. Sulfur has been demonstrated to poison the water-gas shift reaction and steam reforming reactions, [6-7] which is more important during operation under net stoichiometric or rich conditions since the availability of oxygen under these conditions is limited (HC and CO are easily removed by simple oxidation under lean conditions). It is also known from surface science studies that SO_2 dissociates on Pt, Rh, and Pd, forming adsorbed oxygen and adsorbed sulfur, the latter being difficult to remove in rich conditions [8-13]. Small coverages of elemental sulfur can significantly poison the noble metal surface for the adsorption of adsorbates such as CO [13]. Vehicle studies indicated that sulfur deteriorates the performance of three-way catalysts [14], and the magnitude of the impact was found to be calibration-dependent [5,15].

When three-way catalyst formulations began incorporating larger amounts of Ce (up to ~30 wt.% Ce) to provide a number of beneficial effects, including the enhancement of the A/F "window" in which three-way activity occurs, H_2S emissions in the exhaust became noticeable. These events prompted a number of studies of the mechanism of H_2S formation in three-way catalysts to suggest methods of inhibiting the release of sulfur as H_2S in large "pulses". These studies showed that sulfur stored by the catalyst under net lean conditions was released under rich conditions [16]. Storage in lean conditions involves absorption and even reaction with the support and/or base metal oxide to form sulfates and sulfites [17,18]. Some of these species are strongly bound to the support and may be formed by the noble metal catalyzed oxidation of SO_2 to form SO_3^{-2} [9,10]. Previous studies have also shown that the uptake of SO_2 on CeO_2 is far greater than on Al_2O_3, further pointing to the role of CeO_2 in the storage of sulfur [16,19]. In reducing conditions, removal of sulfur occurs and involves reduction to H_2S, elemental sulfur, and reaction with CO to form COS. In Pt-Rh catalysts, the storage and removal of sulfur can be at least partially inhibited by a variety of methods, including the processing of the catalyst to reduce the amount of sulfur stored by the base metal oxide components, or adding a scavenging agent such as Ni or Cu [19-23].

The study of the impact of sulfur on three-way systems has recently gained interest again, as attention has turned to the reformulation of gasoline to reduce vehicle emissions as called for by the Clean Air Act amendments of 1990 and new regulations imposed by the state of California. One of the issues taken up by the Auto-Oil Industry Air Quality Study has been the effect of fuel sulfur on FTP converter efficiency in two vehicle fleets, one comprised of vehicles manufactured between 1979 and 1986, the other of vehicles manufactured in 1989. Both fleets showed a significant improvement in emissions when the sulfur content in fuel was reduced, leading to a recommendation that gasoline reformulation should include a reduction of sulfur level [15,24]. These and other vehicle studies also showed that the effects of sulfur are generally reversible when the fuel was changed from a relatively high sulfur content to a low sulfur content [25]. Ultimately, vehicle studies such as these are needed to better determine the effect of various sulfur content fuels and operating conditions on catalyst performance, but laboratory studies can provide a better understanding of sulfur storage, release and poisoning mechanisms and effects under well-controlled conditions to help in the design of these vehicle tests. Such studies have been carried out in our laboratory using newer technology three-way catalyst systems employing Pt and Rh [13, 26].

Recent attention has been placed on the use of Pd-based catalyst formulations because of their ability to catalyze the oxidation of hydrocarbons and CO at relatively low temperatures as a strategy to improve cold-start emissions and thus comply with low emission vehicle regulations, and because of the lower cost of Pd metal relative to Pt and especially Rh [27-32]. Three-way catalysts using Pd or Pd and Pt were evaluated in the late 1970's and early 1980's, but were found to be easily poisoned by sulfur [33,34]. Indeed, recent studies conducted in this laboratory have confirmed Pd-containing catalysts were particularly susceptible to irreversible sulfur poisoning under isothermal exhaust-like conditions when compared to similar catalysts consisting of Pt and Rh supported on alumina [26, 36].

Recent developments in Pd catalyst technology have been demonstrated to improve the durability of the low temperature lightoff properties of this catalyst after repeated exposure to relatively high temperature exhaust [30,32,35]. Thus, these new catalyst formulations show promise for application as a close-coupled catalyst which can reach HC lightoff rapidly after cold-start. These new formulations may also be more resistant to sulfur poisoning.

The effect of SO_2 on these Pd catalyst formulations, however, has not been widely discussed in the published literature; thus the present study was performed to gain a better understanding of how sulfur affects the performance of a current production Pd catalyst intended for use in high temperature (up to 1000°C) conditions such as a close-coupled converter, and how the impact of sulfur compares to a current technology Pt-Rh three-way converter having a similar noble metal loading. Changes in lightoff and isothermal performances were measured as a function of increasing SO_2 concentration in a simulated exhaust feedstream using a laboratory reactor. Tests were first conducted using propylene as a model of the hydrocarbon species in exhaust, and then repeated using propane in a similar function. Additionally, the effect of sulfur on the performance of the catalysts during an air/fuel ratio scan was also evaluated.

2. EXPERIMENTAL ASPECTS

2.1 Catalysts

Two catalyst formulations were used in this study. The first, a Pd catalyst having a loading of 50 g/ft.^{3} (0.29 wt.%), was obtained in monolith form (cell density 400 cells/in^3). This washcoat formulation is considered to represent a state-of-the-art Pd catalyst for use in a rapid HC lightoff close-coupled (within 10-15" of the exhaust manifold) converter, and has been processed with several additives which serve to promote the activity of the Pd and to thermally stabilize the support and supported metal. The Pt-Rh monolith catalyst was also obtained in monolith form, having a cell density of 400 cells/in^3 and a loading of 23.5 g/ft^3 Pt, 1.68 g/ft^3 Rh (0.13 wt.% Pt, 0.0093 wt.% Rh). This washcoat is designed for use in converters which will be exposed to exhaust temperatures consistent with an underfloor location (at least 25-30" from the exhaust manifold), which are significantly lower than those encountered in a close-coupled location.

2.2 Thermal Aging Treatment

All of the catalysts were evaluated following a thermal aging treatment. The thermal aging was performed by placing the sample in the center of a tube furnace and downstream from a heat exchange zone. In this apparatus, the catalyst is heated mainly by the treatment gases which pass through a heat exchanger. Constant treatment temperature was maintained at a constant value using a thermocouple contacting the catalyst bed to control the furnace. The aging treatment consisted of alternating the gas feed between 5% O_2/N_2 and 5% H_2/N_2 at a rate of 0.1 Hz while maintaining the temperature of the catalyst at 1000°C for 4 h.

2.3 Activity Measurements

Both Pd and Pt-Rh catalyst formulations were evaluated for lightoff and isothermal activity using a laboratory reactor [37]. The composition of the feedstream was chosen to simulate vehicle exhaust, and the composition was controlled using a computer which has been interfaced with a bank of mass flow controllers. In the first series of tests, propylene was chosen to represent the alkene hydrocarbon species in the emissions as it is one of the most abundant hydrocarbon species in exhaust [38] and it figures significantly in the determination of total NMOG emissions since its reactivity factor is relatively high. The tests were repeated using propane to represent the alkane hydrocarbon mass emissions as the alkanes are believed to be among the most difficult of all hydrocarbons in vehicle exhaust to oxidize catalytically. The reactor is configured with a computer-controlled switching valve which alternates between a net reducing feedstream composition and an oxidizing feedstream composition so that the dynamic characteristics of exhaust produced by a vehicle operating under closed-loop control are simulated [37].

For lightoff testing, a sample was first prepared by cutting a section of the monolith having a facial dimension of 10 x 10 cells and a length of 1", was then thermally aged, then loaded into the reactor and stabilized in a nitrogen flow for 1 h at 100°C, and finally exposed to the simulated exhaust at 9 l/min while the temperature was decreased from 600°C to 100°C at 20°C/min. Evaluations were performed using a feedstream which cycled about the stoichiometric point at A/F =14.1 with an amplitude of ±0.5 A/F and a cycling frequency of 0.5 Hz. The sulfur level was held at a constant level during each lightoff test, but was changed between tests to determine the effect of sulfur on lightoff. The gas composition used for the lightoff tests is listed in Table 1.

Isothermal tests were performed using the same samples, which were stabilized in a nitrogen flow for 1 h at 500°C prior to exposure to the simulated exhaust feed. During the isothermal test, the sulfur concentration was step-changed from 0 ppm to a predetermined level for 15 min, then step-changed to 0 ppm for an additional 30 min to determine the effect of sulfur on warmed-up activity and to determine the reversibility of the effect.

Table 1. Laboratory Reactor Simulated Exhaust, Cycled about Stoichiometry (Lightoff and Isothermal Tests)

Gas Component	Concentration
O_2	0.6 % (net)[1]
propylene (or propane)	300 ppm
CO	0.77 %
NO	500 ppm
H_2	0.2%
H_2O	10.0 %
CO_2	10.0 %
SO_2	0,10, 20, or 30 ppm
N_2	balance

1. The oxygen composition was switched between 0.2% O_2 and 1.0% O_2 every second.

Table 2. Isothermal Stoichiometry Scan Test

Gas Component	Concentration
O_2	0.20 % (net) to 1.0% (net)[1]
propylene (or propane)	300 ppm
CO	0.77 %
NO	500 ppm
H_2	0.2 % variable
H_2O	10.0 %
CO_2	10.0 %
SO_2	0,10, 20, or 30 ppm
N_2	balance

1. During this test, the oxygen composition was switched between a value lower than net and a value greater than net. During the scan, both values increase such that the net oxygen composition is increased from 0.2% to 1.0% in steps.

Additional tests were performed in which the net stoichiometry was scanned from a net reducing feedstream (A/F = 14.1) through stoichiometry (A/F = 14.6) to a net oxidizing feedstream (A/F = 15.1) at a given catalyst temperature. This test, sometimes called an A/F ratio sweep test, closely models the exhaust of a late model 3.8 L V-6 engine operating under closed-loop control (the gas composition used for this test is listed in Table 2). During this test, a cycling amplitude of ± 0.5 A/F and a cycling frequency of 0.5 Hz was used. These tests were performed to determine the impact of sulfur on catalyst activity as a function of feedstream stoichiometry. The exhaust gas concentrations and conditions used for all three of these tests are discussed in more detail elsewhere [35-37].

3. RESULTS AND DISCUSSION

3.1 Effect of Sulfur on Lightoff

The effect of sulfur dioxide on lightoff at stoichiometry of the thermally aged 0.29 % Pd catalyst is shown in Figures 1 and 2 using propylene and propane, respectively, to represent the hydrocarbon in the simulated exhaust feedstream. For the purpose of this work, lightoff activity is defined as the temperature at which 50% conversion efficiency occurs for a particular reactant. The lightoff activity was first obtained using a feedstream with no sulfur dioxide added. This experiment was repeated several times, using a different concentration of sulfur dioxide in the feedstream with each experiment. Lightoff activities were thus obtained using 10 ppm, 20 ppm, and 30 ppm sulfur dioxide. These particular sulfur concentrations were chosen as they represent realistic sulfur levels in current available fuels, approximately 150 ppm, 300 ppm, and 450 ppm, respectively [35,36]. As will be discussed later, each experiment was followed by a high temperature treatment in which the SO_2 was removed from the feedstream, and the sample temperature was increased to 700°C where it remained for 30 min, followed by cooling to 500°C. This treatment was found to restore the catalytic activity to the level originally measured prior to exposure to any SO_2.

The results obtained using either propylene or propane consistently show an increase in lightoff temperatures of HC, CO and NOx as the sulfur content in the feedstream is increased. The magnitude of the increase in the propylene and CO lightoff temperatures was on the order of 40°C as the SO_2 content was increased from 0 ppm to 30 ppm, while the increase in the propane and CO lightoff temperatures was on the order of 60°C for the same increase in SO_2 concentration. Regardless of the type of hydrocarbon used, the increase in the lightoff temperature of HC and CO appeared to be non-linear with increasing sulfur concentration: most of the increase in the lightoff temperature occurred when the SO_2 concentration was increased from 0 ppm to 10 ppm, while a smaller increase in the lightoff occurred with further increase in the SO_2 concentration from 10 ppm to 30 ppm. The NOx activity was markedly lower than HC or CO, and adding even small amounts of SO_2 resulted in the failure to reach the 50% conversion efficiency over the range tested (200 to 500°C). However, in comparing the activity profile over this temperature range, one can conclude that the NOx activity was also significantly affected by the presence of SO_2. We note that upon adding 10 ppm of SO_2 to the simulated exhaust, not only the beginning of the catalyst lightoff was increased (by roughly 40°C using propylene and 60°C using propane), but the warmed-up activity at 500°C was also decreased by roughly a factor of 2 using either hydrocarbon when the SO_2 concentration was increased from 0 ppm to 30 ppm. For NOx, most of the increase in the beginning of lightoff occurred upon increasing the SO_2 level from 0 ppm to 10 ppm, whereas further increases in SO_2 resulted in smaller increases in the temperature at which lightoff begins. Although not indicated in the figures, the magnitude of the effect of SO_2 on lightoff activity is **larger** than the effect of the thermal aging procedure on activity. The thermal aging procedure used in this study has been found to generally cause a 30-35°C increase in CO and HC lightoff temperature.

Similar lightoff tests were performed with the thermally aged 0.13% Pt, 0.0093% Rh production catalyst. In this case, the results obtained using either propylene (Figure 3) or propane (Figure 4) also consistently show an increase in lightoff temperatures of HC, CO and NOx as the sulfur content in the feedstream was increased. The magnitude of the increase in the propylene, CO, and NOx lightoff temperatures was on the order of 40°C as the SO_2 content was increased from 0 ppm to 30 ppm. When propane was used as the hydrocarbon, the increase in the propane lightoff temperature was on the order of 40°C while the increase in the CO and NO_x lightoff temperature was on the order of 60-70°C for the same increase in SO_2 concentration. Regardless of the type of hydrocarbon used, the increase in the lightoff temperature of HC and CO was also found to be non-linear with increasing sulfur concentration. For experiments conducted using propylene for the hydrocarbon, nearly all of the increase in the lightoff temperature occurred when the SO_2 concentration was increased from 0 ppm to 10 ppm: further increases in the SO_2 level up to 30 ppm resulted in very little additional increase in the lightoff temperature. For experiments in which propane was used, the increase in the lightoff temperature with but still very non-linear. Although not indicated on the figures, we note that the effect of sulfur on the lightoff performance was comparable to the effect due to the thermal aging treatment used in this work.

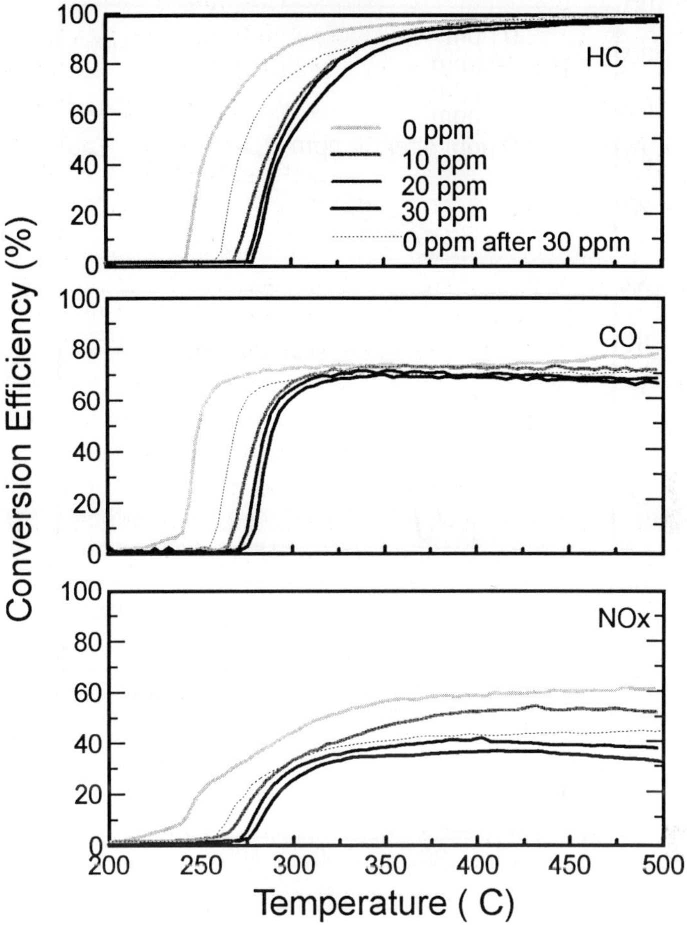

Figure 1. Lightoff curves for a thermally aged Pd catalyst sample using propylene as a hydrocarbon surrogate. Each curve was obtained using a different sulfur dioxide concentration in the synthetic exhaust feedstream. The dashed line was the result of a lightoff test using 30 ppm sulfur dioxide in the feedstream which was repeated using 0 ppm sulfur dioxide in the feedstream.

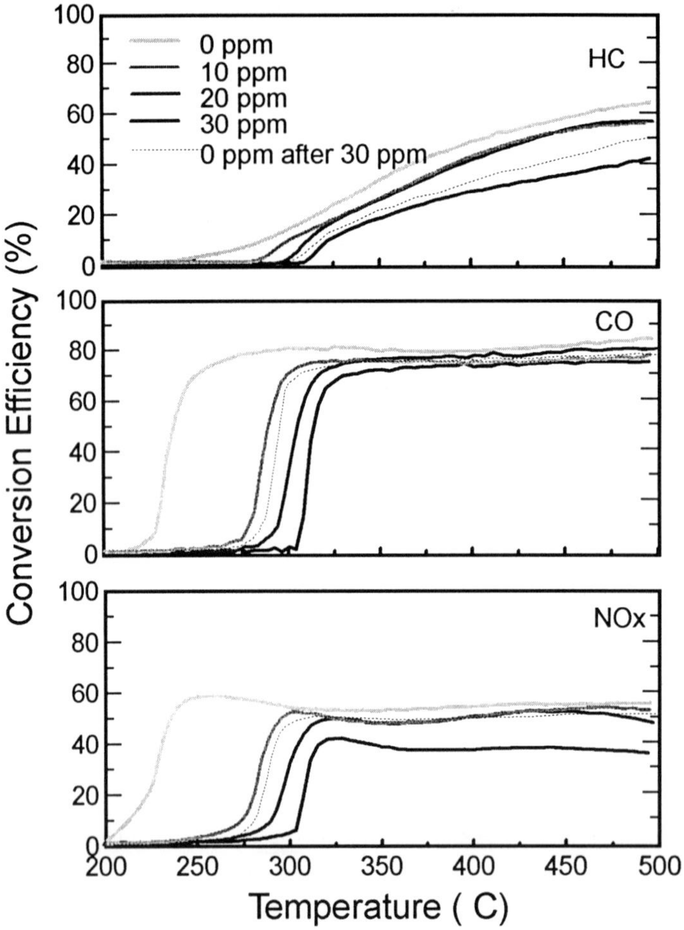

Figure 2. Lightoff curves for a thermally aged Pt-Rh catalyst sample using propylene as a hydrocarbon surrogate. Each curve was obtained using a different sulfur dioxide concentration in the synthetic exhaust feedstream. The dashed line was the result of a lightoff test using 30 ppm sulfur dioxide in the feedstream which was repeated using 0 ppm sulfur dioxide in the feedstream.

Because the Pt-Rh catalyst formulation was intended for positioning in an underfloor location in the vehicle, it is therefore not optimized for high temperature operation, and the high temperature aging treatment we used may have deteriorated the performance of the commercial Pt-Rh catalyst to a greater degree than the performance of the Pd catalyst, which has been formulated for operation at higher temperatures. Even considering this potential difference in the aged catalysts, a comparison of the lightoff performance of the thermally Pd catalysts vs. the Pt-Rh catalyst with no sulfur in the feedstream does point out a clear advantage of the Pd catalyst for reducing cold-start alkene hydrocarbon and CO emissions. Although the Pt-Rh catalyst shows better lightoff performance for alkane hydrocarbons in this comparison, the alkenes figure more prominently in the contribution to overall NMOG since they are more abundant and have higher reactivity factors; this is one of several reasons why Pd-based catalyst technology holds promise for use in close-coupled converter applications.

We note with interest that the non-linear relationship between the SO_2 concentration and the magnitude of the impact on lightoff performance for both Pd and Pt-Rh based catalysts is characteristic of the effect of sulfur on monolithic three-way catalysts observed in previously reported laboratory [26] and vehicle studies [24]. Both studies showed that as the sulfur content was increased, the degradation in the lightoff or warmed-up emissions increased at a high rate initially, but became more gradual at higher sulfur levels, suggesting that very small amounts of sulfur in the exhaust can have a significant effect on emissions. In a more recent laboratory study, the relationship between the concentration of SO_2 and the magnitude of the effect on the activity of model Pd catalysts (Pd supported on alumina alone or in the presence of one promoter such as ceria or lanthana) was found to be linear for CO and HC activity, although we note that the model catalysts used in that study were pelleted, not monolithic. Differences in mass transfer effects and the distribution of noble metals in pelleted vs. monolith catalysts may account for the linear effect of sulfur on pelleted catalysts vs. the non-linear effect of sulfur in monoliths. We note that the vehicle studies, in which the non-linear effect was also observed, were also conducted primarily with monolith catalysts.

For all of the aged catalysts, an additional lightoff experiment was performed with no SO_2 in the feedstream following the lightoff experiment using 30 ppm SO_2. The results, shown in dashed lines in Figures 1-4, shows a decrease in the lightoff temperatures from the 30 ppm SO_2 experiment for both Pd and Pt-Rh type catalysts, but only partial recovery of activity (relative to the catalyst prior to exposure to SO_2) has taken place with the Pd catalyst, whereas nearly complete recovery of the original activity has taken place with the Pt-Rh catalyst. This result, which was also observed in a prior study of the effect of SO_2 on model Pd catalysts [36], further emphasizes the non-reversibility of sulfur poisoning in Pd catalysts, even for a state-of-the-art formulation, compared to the complete reversibility of the effect of sulfur in Pt-Rh three-way catalysts. We have speculated that the non-reversible sulfur poisoning of Pd is related to a direct reaction between adsorbed S and the supported Pd metal, perhaps leading to the formation of a surface PdS compound [36], or possibly migration of S into the bulk of Pd. We have obtained evidence for the latter in studies of SO_2 adsorption on Pd foils [39], but further study is needed to better understand this phenomenon.

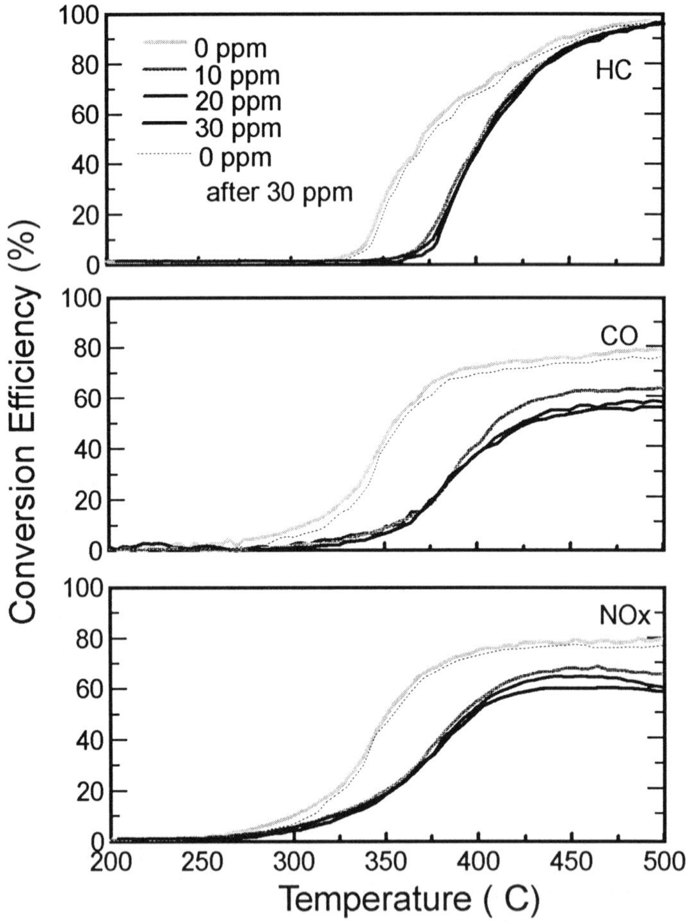

Figure 3. Lightoff curves for a thermally aged Pd catalyst sample using propane as a hydrocarbon surrogate. Each curve was obtained using a different sulfur dioxide concentration in the synthetic exhaust feedstream. The dashed line was the result of a lightoff test using 30 ppm sulfur dioxide in the feedstream which was repeated using 0 ppm sulfur dioxide in the feedstream.

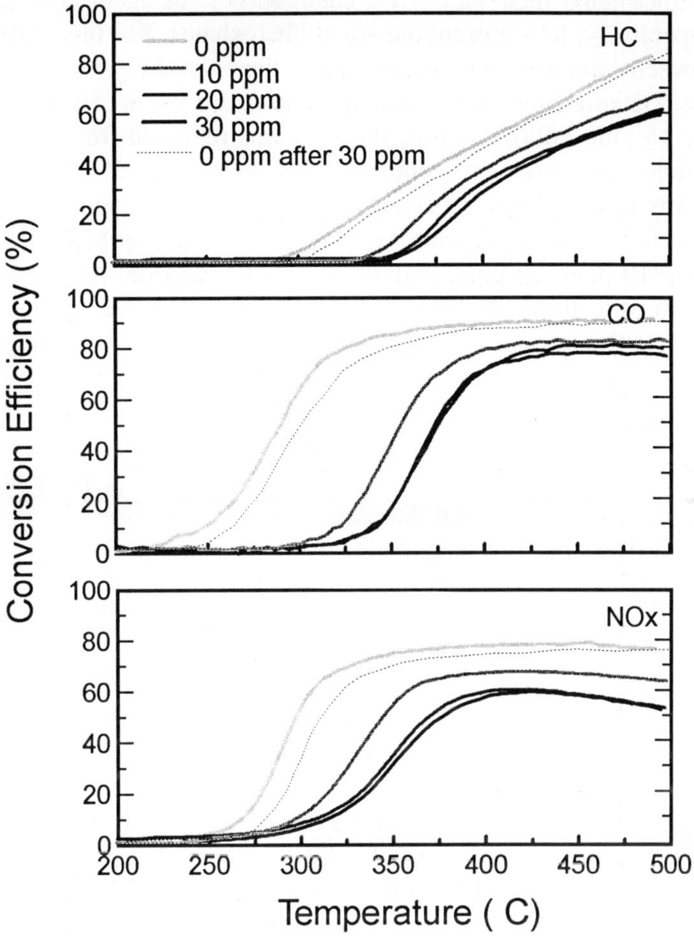

Figure 4. Lightoff curves for a thermally aged Pt-Rh catalyst sample using propane as a hydrocarbon surrogate. Each curve was obtained using a different sulfur dioxide concentration in the synthetic exhaust feedstream. The dashed line was the result of a lightoff test using 30 ppm sulfur dioxide in the feedstream which was repeated using 0 ppm sulfur dioxide in the feedstream.

3.2. Effect of Sulfur on Warmed-up Performance

The effect of sulfur dioxide on the isothermal (500°C), or "warmed-up" activity in which the exhaust was cycled about stoichiometry is shown in Figures

5 and 6 for thermally aged 0.29% Pd catalyst, first using propylene and then propane, respectively, to represent the HC in the exhaust. For this experiment, no SO_2 was present for approximately 5 min, then the SO_2 concentration was increased in a step-change to a desired level (10, 20 or 30 ppm) where it remained for 15 min. Following this, the SO_2 was removed from the feedstream and the activity measurement continued for an additional 25-35 min. This experiment was repeated several times, using a different concentration of sulfur dioxide in the feedstream with each experiment. Warmed-up activities were thus obtained using 10 ppm, 20 ppm, and 30 ppm sulfur dioxide. After each test, the SO_2 was removed from the feedstream, and the sample temperature was increased to 700°C where it remained for 30 min, followed by cooling to 500°C. This treatment was found to restore the catalytic activity to the level originally measured prior to exposure to any SO_2.

When propylene was used to represent the HC in this test (Figure 5) the results obtained for the 0.29% Pd catalyst indicate that when the sulfur level in the feed was increased, a gradual decrease in the activity was observed for HC, CO, and NOx. The absolute magnitude of the decrease was large for CO and NOx: an increase in the SO_2 concentration from 0 ppm to 30 ppm resulted in a decrease in CO conversion from 70% to 55% and in the NOx conversion from 65% to 30% after 15 min exposure to sulfur. Increasing the SO_2 concentration from 0 ppm to 30 ppm resulted in a decrease in the HC conversion from 98% to 95%, which in absolute magnitude does not appear to be as large an effect as with CO or NO_x, but we note that the 3% decrease in HC conversion translates into an increase in HC breakthrough of 150%. The rate of the decrease in conversion efficiency for HC, CO and NO_x slowed considerably after 15 min of exposure, but had not yet reached a stable level at this point.

When SO_2 was removed from the feedstream, the HC activity appeared to return to near its original level, but the recovery of CO and NO_x activity was incomplete, even following an exposure of 30 min in a "clean" feedstream. Although not shown, when the experiment was allowed to run for 3 h, very little additional recovery took place.

Similar experiments with the 0.29% Pd catalyst were performed using propane to represent the hydrocarbon (Figure 6). Again, a gradual decrease in the HC, CO, and NO_x activity was observed, although the rate of poisoning did not change during the 15 min exposure to sulfur in the feedstream, and therefore equilibrium was not achieved. Note also that the magnitude of the decrease in activity was significantly larger for propane than for propylene: for instance, exposure to 30 ppm SO_2 resulted in an 18% decrease in HC conversion, translating into a ~200% increase in HC breakthrough. Decreases in the CO (by 45%) and NOx (by 50%) were also more significant than that observed when

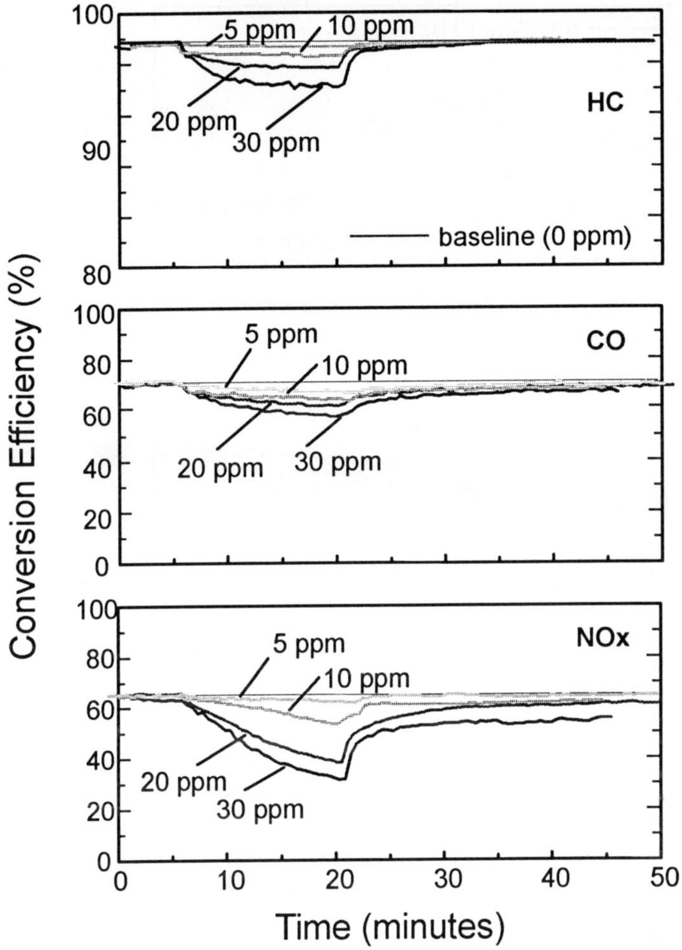

Figure 5. Isothermal (500 ℃) activity of the thermally aged Pd catalyst using propylene as a hydrocarbon surrogate. In these experiments, the sulfur dioxide concentration is maintained at 0 ppm for 5 min, then increased to either 5, 10, 20 or 30 ppm for 15 min, then decreased to 0 ppm for an additional 45 min.

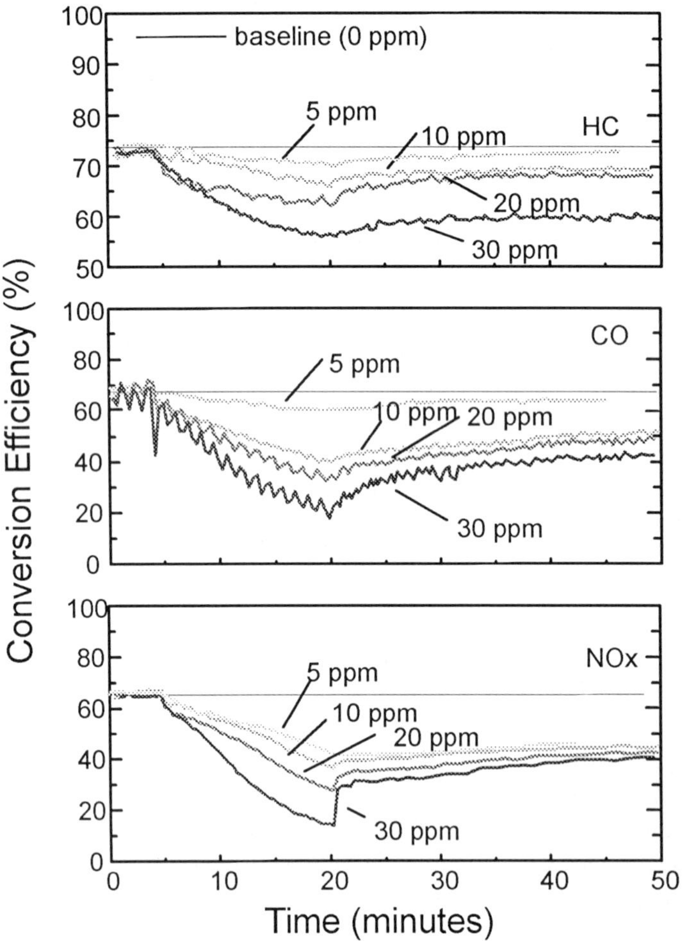

Figure 6. Isothermal (500°C) activity of the thermally aged Pd catalyst using propane as a hydrocarbon surrogate. In these experiments, the sulfur dioxide concentration is maintained at 0 ppm for 5 min, then increased to either 5, 10, 20 or 30 ppm for 15 min, then decreased to 0 ppm for an additional 45 min.

propylene was used to represent the hydrocarbon (15% and 40%, respectively). It is also important to point out that when sulfur was removed from the feedstream, very little recovery of activity took place, indicating that most of the poisoning is irreversible at this temperature. The reason for the absence of a change in the poisoning rate over the 15 min exposure period together with the dominance of the irreversible poisoning has yet to be identified, but at present we attribute part of this effect to hydrocarbon coking of the catalyst which is promoted by the presence of SO_2 in the feedstream.

In Figure 7, we show the effect of SO_2 on the warmed-up activity of the thermally aged 0.13% Pt, 0.0093% Rh production catalyst using propylene to represent the hydrocarbon. For this catalyst, the presence of SO_2 in the feedstream resulted in a more instantaneous decrease in the HC, CO, and NO_x conversion efficiencies, but the magnitude of the decrease was dramatically lower than observed with the 0.29% Pd catalyst exposed to the same level of sulfur. For example, exposure to 30 ppm SO_2 resulted in a decrease of roughly 1% in the HC conversion, translating to an increase of 30% in HC breakthrough. Decreases of 20% in the CO, and 30% in the NO_x conversions were also observed. Moreover, when SO_2 was removed from the feedstream, near complete recovery of the activity occurred within 15-30 min. The relative instantaneous behavior of poisoning as well as the complete reversibility were found to be consistent with those of a similar laboratory study in which the effect of SO_2 on a three-way Pt-Rh commercial monolith catalyst was compared with the effect on a commercial Pt-Pd-Rh pelleted catalyst [26].

Finally, in Figure 8 we show the effect of SO_2 on the warmed-up activity of the thermally aged 0.13% Pt, 0.0093% Rh production catalyst using propane to represent the hydrocarbon. For this case, the presence of sulfur at 30 ppm resulted in a larger decrease in the HC conversion efficiency, from 86% to 68%, translating into an increase in HC breakthrough of ~200%, while the impact on the conversion efficiencies of CO (97% to 80%) and NOx (92% to 75%) was smaller than when propylene was used as the hydrocarbon. Removal of sulfur from the feedstream did lead to recovery of most (~60%-70%) but not all of the original activity after exposure to a sulfur-free feedstream for 30 min. The incomplete recovery of activity observed with both the Pd and Pt-Rh commercial samples when propane was used as the hydrocarbon suggests coking of the catalyst is important. Further evidence for this rationalization was obtained by observing that treatment of the Pt-Rh catalyst in a net lean (1% O_2) feedstream at 700°C resulted in near complete recovery of activity. Similar treatment of the propane-exposed Pd catalyst resulted in partial recovery of activity, whereas subsequent treatment in a net rich feedstream at 700°C resulted in complete recovery of the original activity. Our findings that SO_2 enhances hydrocarbon coking (when propane is present in the feedstream) is consistent with earlier reports that SO_2 can increase the acidity of supports such as alumina, and thus enhance the role of this support as an acid catalyst in hydride abstraction from alkanes such as propane, which then leads to a number of different reactions including polymerization and further dehydrogenation. The resulting carbonaceous material is partially hydrogenated (so called "reversible" coking [40]) since it can be removed by oxidation. This coking phenomenon has been commonly observed on acid catalysts used for steam reforming [41,42].

3.3 Effect of Sulfur on Performance as a Function of Stoichiometry

Both thermally aged catalysts were evaluated using tests in which the net stoichiometry was scanned from a net reducing feedstream through stoichiometry to a net oxidizing feedstream at a catalyst temperature of 500°C. These tests were first performed without sulfur in the feedstream and then repeated with 30 ppm SO_2 in the feedstream to determine the impact of sulfur on catalyst activity as a function of stoichiometry.

Figure 9 shows the HC, CO and NO_x conversion efficiencies for the aged 0.29% Pd catalyst as a function of mean air/fuel ratio at 500°C using a test in which the net stoichiometry was scanned from a net reducing feedstream through stoichiometry to a net oxidizing feedstream. For this case, propylene was used to represent the hydrocarbon. These tests were performed first with no SO_2 and repeated with 30 ppm SO_2 in the feedstream. In the oxidative, or 'lean" exhaust environment between mean A/F ratios of 15.1 to 14.7, the presence of SO_2 results in a slight decrease in the conversion efficiency for HC, CO and NO_x. Below a value of 14.7, the presence of sulfur results in a slight decrease in the CO activity which continues until a rich extreme of 14.1 in A/F ratio. The effect of SO_2 on the conversion efficiencies of HC and NO_x, however, was significant below an A/F ratio of 14.7 and increases with decreasing mean A/F until the "rich" extreme of 14.1 was reached. Thus, the degree of the impact of sulfur on HC and NO_x conversion efficiency is greatest under stoichiometric and "rich" operating conditions. The maximum impact of 30 ppm SO_2 in terms of breakthrough in HC and NO_x occurred at a mean A/F ratio of 14.3 to 14.2. At this point, the HC breakthrough increased by 600-700% while the NO_xbreakthrough increased by 400-500%. This behavior was also observed in a similar study of the effect of SO_2 on model Pd three-way catalysts [36] and is consistent with suggestions that SO_2 directly poisons the noble metal surface in rich conditions [13].

Figure 10 shows the HC, CO and NO_x conversion efficiencies for the thermally aged 0.29% Pd catalyst as a function of mean air/fuel ratio at 500°C using propane to represent the hydrocarbon. For reasons stated earlier, the extent of the effect of sulfur on catalyst activity was difficult to determine due to the inability to reach equilibrium when sulfur was added to the feedstream.

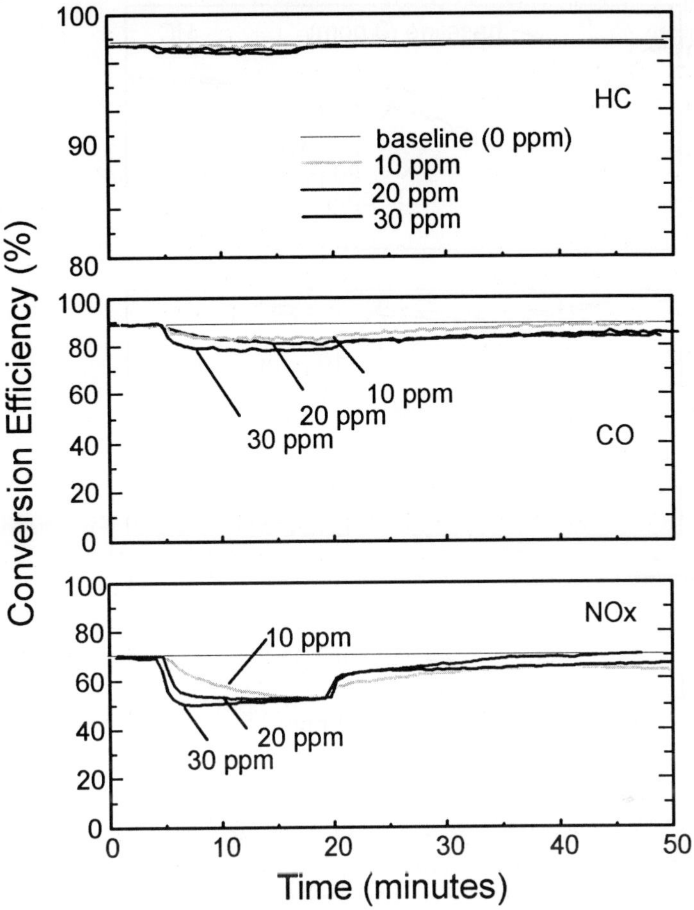

Figure 7. Isothermal (500℃) activity of the thermally aged Pt-Rh catalyst using propylene as a hydrocarbon surrogate. In these experiments, the sulfur dioxide concentration is maintained at 0 ppm for 5 min, then increased to either 5, 10, 20 or 30 ppm for 15 min, then decreased to 0 ppm for an additional 45 min.

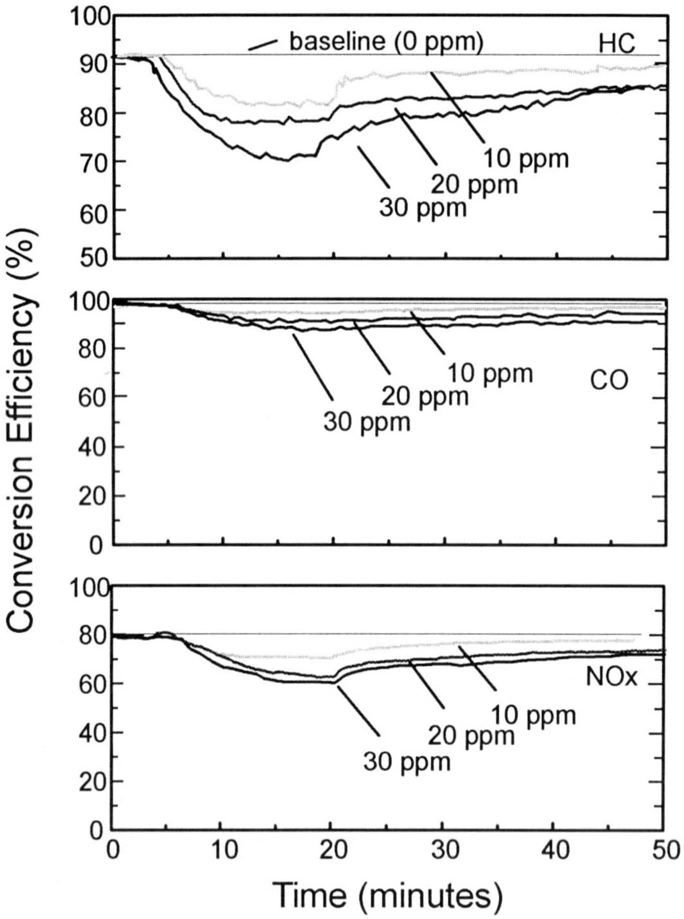

Figure 8. Isothermal (500℃) activity of the thermally aged Pt-Rh catalyst using propane as a hydrocarbon surrogate. In these experiments, the sulfur dioxide concentration is maintained at 0 ppm for 5 min, then increased to either 5, 10, 20 or 30 ppm for 15 min, then decreased to 0 ppm for an additional 45 min.

Consequently, when 30 ppm SO_2 was added to the feedstream, the A/F ratio scan experiment was repeated several times until the result was found to reproducible within 5%, but we note that the full impact of sulfur on activity is probably larger. As indicated in the figure, we also found that the first scan from a lean to rich A/F ratio produced a different result in the propane conversion than a first scan from a rich to lean A/F ratio when SO_2 was present. Repeated A/F ratio scans resulted in the lower curve showing relatively constant propane conversion of 40-50% over the entire A/F ratio range tested. Thus, operation in a rich exhaust containing propane and SO_2 results in an additional poisoning effect which is only evident during operation in a stoichiometric to lean environment, and which cannot be reversed at 500°C. We attribute this particular effect to SO_2-induced coking of the Pd catalyst in rich conditions: treatment in lean conditions with no sulfur in the feedstream results in significant recovery of activity in the lean A/F ratio region, but not in the rich A/F ratio region. We also found that the presence of SO_2 resulted in a slight decrease in the CO conversion and a significant decrease in the NO_x conversion in rich conditions, but the coking effect noticed in the HC activity did not significantly affect CO and NO_x activity. This observation suggests that the active site participating in the oxidation of propane differs from those sites catalyzing the oxidation of CO and reduction of NO

Turning now to the thermally aged 0.13% Pt, 0.0093% Rh production catalyst, we show the HC, CO and NO_x conversion efficiencies as a function of mean A/F ratio at 500°C using propylene to represent the hydrocarbon in Figure 11. It is interesting to note that for this catalyst, the effect of adding 30 ppm SO_2 is generally similar to the Pd catalyst: the HC and NO_x conversion efficiencies were significantly decreased in rich conditions, while the CO conversion efficiency was only moderately decreased, but again in rich conditions. With no sulfur present, it is interesting to note that in lean and stoichiometric conditions, the propylene conversion efficiency was slightly lower for the Pd catalyst in comparison to the Pt-Rh catalyst, while the Pd catalyst was significantly more active than the Pt-Rh catalyst for propylene oxidation in rich conditions, between 14.6 and 14.1 in mean A/F ratio units. This hydrocarbon activity advantage of the Pd over Pt-Rh was completely canceled when 30 ppm SO_2 is present in the feedstream, and became a disadvantage when sulfur was removed from the feedstream, as the Pt-Rh catalyst recovered all of its original activity (not shown in the figure), but the Pd catalyst did not. The rapid decrease in HC activity in extreme A/F ratio conditions (between 14.2 and 14.1) and that the presence of sulfur does not significantly affect the activity in this region. We also note that with 30 ppm SO_2 present, the decrease in the NO_x activity of the Pt-Rh catalyst was not as large as the decrease in the NO_x activity observed with the Pd catalyst. This result is consistent with the reported relative resistance of Rh to direct sulfur poisoning [13].

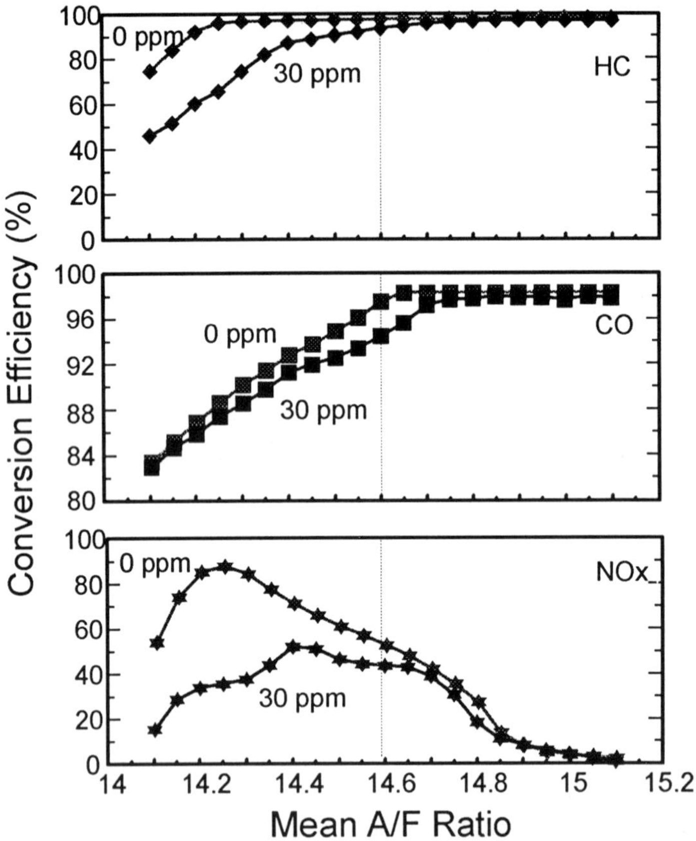

Figure 9. Effect of sulfur dioxide on the activity of thermally aged Pd in a cycled stoichiometric exhaust as a function of mean air/fuel ratio using propylene as a surrogate hydrocarbon.

The HC, CO and NO_x conversion efficiencies for the thermally aged 0.13% Pt, 0.0093% Rh catalyst as a function of mean air/fuel ratio at 500°C were again determined except that propane was used to represent the hydrocarbon. The result is shown in Figure 12 first with no SO_2, then with 30 ppm SO_2 in the feedstream. The effect of sulfur on catalyst activity followed a similar trend as with the previous case: the HC and NO_x activity was reduced in stoichiometric and rich conditions, although the HC activity was not significantly affected at extreme rich conditions of 14.2 to 14.1 in A/F ratio. The CO conversion efficiency was also similarly affected, being moderately reduced in rich conditions

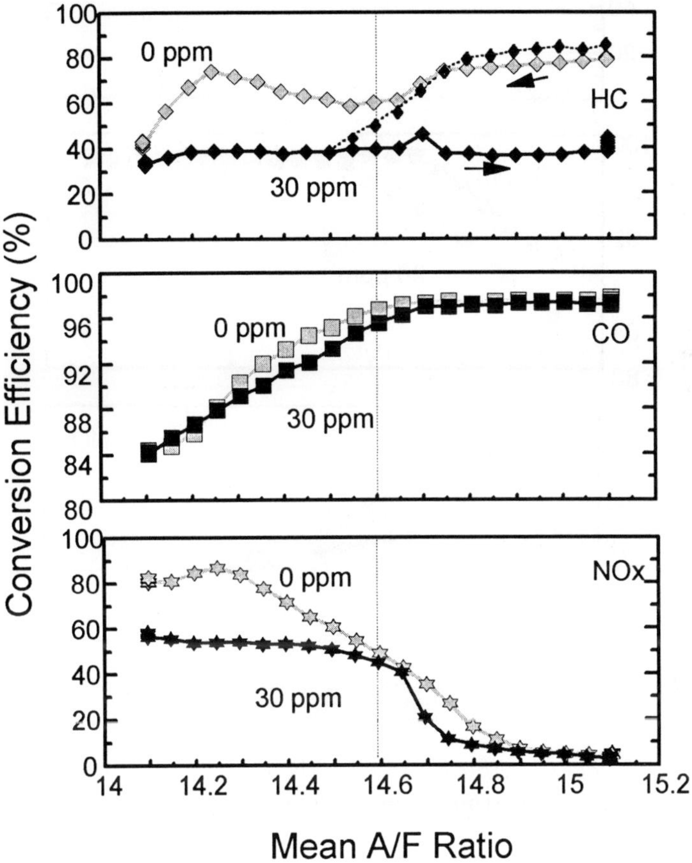

Figure 10. Effect of sulfur dioxide on the activity of thermally aged Pd in a cycled stoichiometric exhaust as a function of mean air/fuel ratio using propane as a surrogate hydrocarbon.

. Note that the magnitude of the effect of sulfur on the overall activity of the Pt-Rh catalyst was not as large as the effect on the Pd catalyst, particularly with regard to the HC conversion.

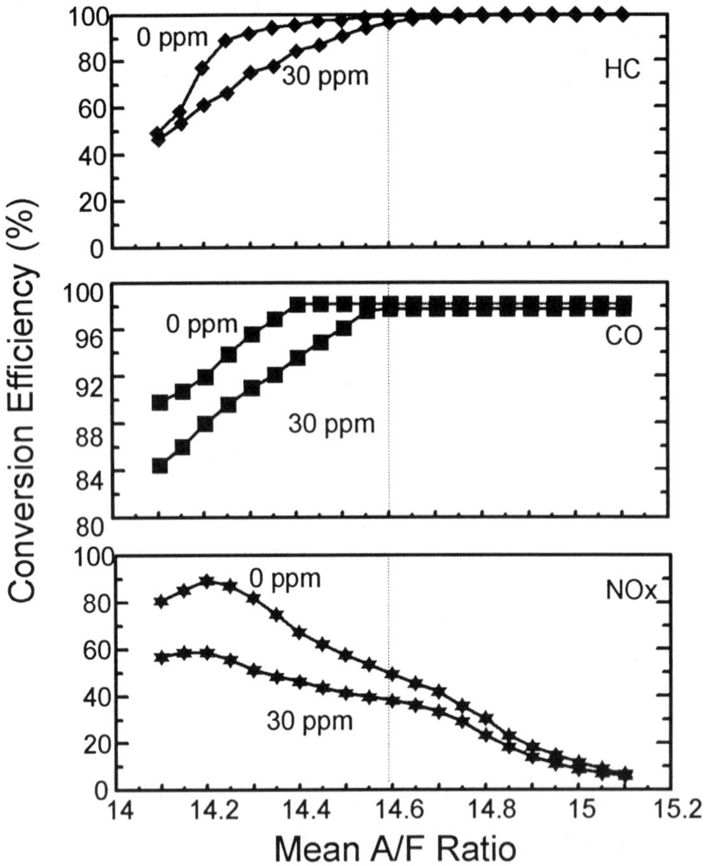

Figure 11. *Effect of sulfur dioxide on the activity of thermally aged Pt-Rh in a cycled stoichiometric exhaust as a function of mean air/fuel ratio using propylene as a surrogate hydrocarbon.*

There is one exception in these results using propane relative to those obtained when propylene was used to represent the hydrocarbon: in extremely lean conditions, the HC activity was enhanced by the presence of SO_2: this effect has been reported in previous laboratory studies of propane oxidation [26]. We suggested previously that SO_2 promotes acid catalysis of propane dehydrogenation, only in this case, the carbonaceous material may be more easily removed from Pt-Rh than from Pd under oxidizing conditions, thus complete oxidation of propane dominates over coking. Other factors, however, may also be responsible for these observations due to a number of differences in the properties of the two catalysts. For example, the dispersion of the Pt-Rh catalyst is generally significantly greater than the Pd catalyst, and it has been shown that highly dispersed noble metals are less likely to be poisoned by coke formation than similar catalysts with lower noble metal dispersion [43]. Certainly, further study is needed to gain a better understanding of this phenomenon.

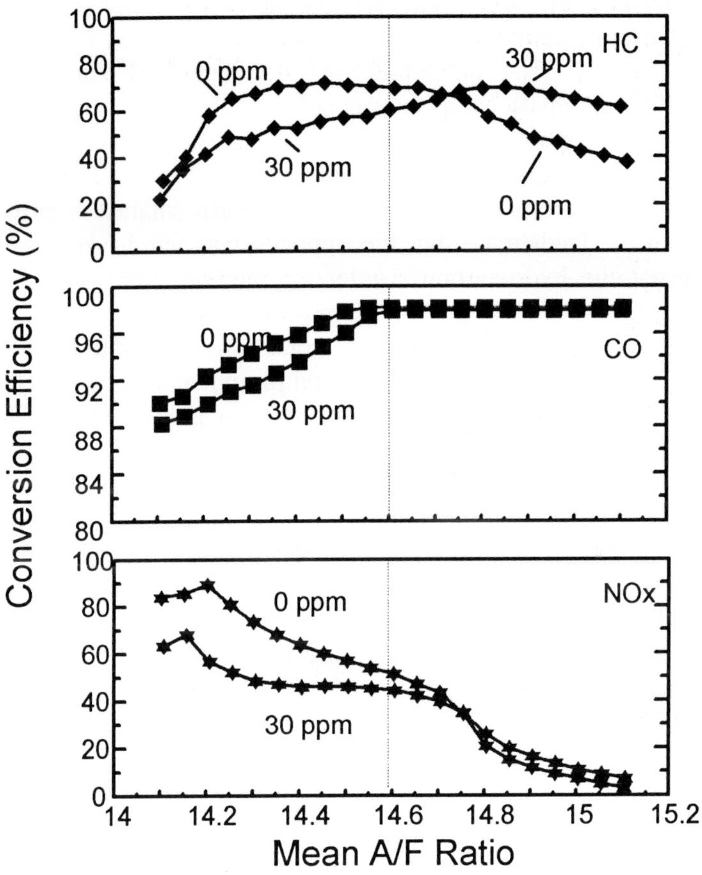

Figure 12. Effect of sulfur dioxide on the activity of thermaly aged Pt-Rh in a cycled stoichiometric exhaust as a function of mean air/fuel ratio using propane as a surrogate hydrocarbon.

4. SUMMARY

In this work, we have compared the effect of SO_2 on the three-way performance of a commercially prepared, thermally aged Pd three-way catalyst with that of a commercially prepared, thermally aged Pt-Rh three-way catalyst. The results obtained in the present study were generally consistent with previous laboratory studies of commercially prepared Pt-Rh catalysts [26] and model Pd three-way catalysts [36]. We found that an increase in the sulfur content in simulated cycled stoichiometric exhaust resulted in a loss of both lightoff and warmed-up activity. The magnitude of the impact of sulfur on lightoff activity of both Pd and Pt-Rh catalysts was comparable when propylene was used for the hydrocarbon contribution to the exhaust, and larger but still comparable when propane was used to represent the hydrocarbon in exhaust. The Pd catalyst showed generally better lightoff activity than the Pt-Rh catalyst regardless of the sulfur content in the feedstream, but the opposite was valid when propane was used to represent the hydrocarbon. Under warmed-up conditions, the loss of activity for HC, CO and NO_x due to the presence of sulfur was greater under slightly rich conditions than under lean conditions for both Pd and Pt-Rh catalysts, while the magnitude of the impact on propylene and NO_x activity under warmed-up stoichiometric conditions was significant greater for the Pd catalyst than for the Pt-Rh catalyst. When propane was used to represent the hydrocarbons in exhaust, the impact of sulfur on the HC conversion was comparable for both catalysts, although the impact on CO and NO_x was greater for Pd than for Pt-Rh. Finally, we found that when propylene was used for the hydrocarbon, the effect of SO_2 on the activity of the Pd catalyst was partly irreversible under the conditions used in this study, while the effect on Pt-Rh was completely reversible. When propane was used to represent the hydrocarbon, the effect on the activity of both catalysts was partly irreversible, but the magnitude of the irreversible poisoning was larger on Pd. Part of the irreversible poisoning effect is attributed to a direct interaction or reaction between SO_2 and Pd, while the other part is attributed to the promotional effect of SO_2 in hydrocarbon coking of the catalyst when alkane hydrocarbons are present.

5.ACKNOWLEDGMENT

The authors wish to acknowledge the assistance of Galen Fisher and David Monroe for helpful discussion and comments, and to Doug Ball of the AC Rochester Division for their discussion and assistance with this study.

REFERENCES

1 K. C. Taylor, Ind. Eng. Chem., Prod. Res. Dev. 15 (1976) 264.
2 J. C. Summer, Env. Sci. Tech. 13 (1979) 321.
3 H. C. Yao and J. Yu, J. Catal. 36 (1975) 266.
4 J. C. Summers and K. Baron, J. Catal. 53 (1979) 380.
5 W. B. Williamson, H. S. Gandhi, M. E. Heyde and G. A. Zawacki, SAE Paper No. 790942, 1979.
6 G. J. Joy, G. R. Lester and F. S. Molinaro, SAE Paper No. 790943, 1979.
7 J. C. Schlatter and P. J. Mitchell, Ind. Eng. Chem. , Prod. Res. Dev. 19 (1980) 288.
8 U. Kohler and H.W. Wassmuth, Surf. Sci. 126 (1983) 448, and references therein.
9 O. K. T. Wu and R. P. Burns, Surf. Int. Anal. 3 (1981) 29.
10 M. L. Burke and R. J. Madix, Surf. Sci. 194 (1988) 223.
11 R. C. Ku and P. Wynblatt, Appl. Surf. Sci. 8 (1981) 250.
12 St. Astegger and E. Bechtold, Surf. Sci. 122 (1982) 491.
13 D. D. Beck, M. H. Krueger and D. R. Monroe, SAE Paper No. 910844, 1991.
14 R. L. Furey and D. R. Monroe, SAE Paper No. 811228, 1981.
15 "Effects of Fuel Sulfur Levels on Mass Exhaust Emissions", Auto/Oil Air Quality Improvement Research Program, Technical Bulletin No. 2, February 1991.
16 A. F. Diwell, C. Hallett and J. R. Taylor, SAE Paper No. 872163, 1987.
17 A. V. Deo, I. G. Dalla Lana and H. W. Habgood, J. Catal. 21 (1971) 2710.
18 A. Datta, R. G. Cavell, R. M. Tower and Z. M. George, J. Phys. Chem. 89 (1985) 443.
19 H. G. Henke, J. J. White and G. W. Denison, SAE Paper No. 872134, 1987.
20 E. S. Lox, B. H. Engler and E. Koberstein, SAE Paper No. 890795, 1989.
21 J. C. Dettling, H. S. Hwang, S. Pudick and S. J. Tauster, SAE Paper No. 900506, 1990.
22 M. A. Harkonen, S. Salanne, T. -K. Rantakyla and V. J. Pohjola, SAE Paper No. 900498, 1990.
23 J. S. Rieck, W. Suarez and J. E. Kubsh, SAE Paper No. 892095, 1989.
24 "Effects of Fuel Sulfur on Mass Exhaust Emissions, Air Toxins, and Reactivity",Auto/Oil Air Quality Improvement Research Program, Technical Bulletin No. 8, 1992.
25 J. C. Summers, J. F. Skowron and W. B. Williamson, SAE Paper No. 920558, 1992.

26 D. R. Monroe, M. H. Krueger, D. D. Beck and M. J. D'Aniello, Jr., "the Effect of Sulfur on Three-Way Catalysts", in *Catalysis and Automotive Pollution Control II*, A. Crucq, ed., Amsterdam: Elsevier Science Publishers, 1991.
27 J. C. Summers, J. J. White and W. B. Williamson, SAE Paper No. 890794, 1989.
28 G. B. Fisher, M. G. Zammit and W. J. LaBarge, SAE Paper No. 920846, 1992.
29 J. C. Dettling and Y.-K. Lui, SAE Paper No. 920094, 1992.
30 H. Muraki, SAE Paper No. 910842, 1991.
31 T. Yamada, K. Kayano and M. Funabiki, SAE Paper No. 930253, 1993.
32 J. C. Summers, J. F. Skowron and M. J. Miller, SAE Paper No. 930386, 1993.
33 R. A. Giacomazzi and M. F. Homfeld, SAE Paper No. 730595, 1973.
34 G. L. Barnes and R. L. Klimisch, SAE Paper No. 730595, 1973.
35 D. D. Beck, J. W. Sommers and C. L. DiMaggio, Appl. Catal. B: Env., 3 (1994) 205.
36 S. Subramanian, R. J. Kudla, C. R. Peters and M. S. Chattha, Catal. Lett. 16 (1992) 323.
37 D. D. Beck, D. R. Monroe, C. L. DiMaggio and J. W. Sommers, SAE Paper No. 930084, 1993.
38 D. D. Beck, M. H. Krueger, D. R. Monroe, D. J. Upton, J. M. Lendway and D. R. Smith, SAE Paper No. 920099.
39 C. L. DiMaggio and D. D. Beck, manuscript in preparation.
40 S. M. Davis and G. A. Somorjai, Chem. Phys. Solid Surf. 217 (1982).
41 E.E. Wolf and F. Alfani, Catal. Rev. Sci. Eng. 24 (1982) 329.
42 J. Barbier and P. Marecot, J. Catal. 102 (1986) 21. G. A. Olah, G. K. S. Williams, J. D. Field, and D. Wade, "Hydrocarbon Chemistry", Wiley, New York, 1987.
43 H. Wise, J. McCarty, and J. Oudar, in "Deactivation and Poisoning of Catalysts", J. Oudar and H. Wise, eds., Marcel Dekker, New York, 1985.

A. Frennet and J.-M. Bastin (Eds.)
Catalysis and Automotive Pollution Control III
Studies in Surface Science and Catalysis, Vol. 96
1995 Elsevier Science B.V.

AN X-RAY ABSORPTION SPECTROSCOPIC INVESTIGATION OF AGED AUTOMOTIVE CATALYSTS

F. Maire[a], M. Capelle[b], G. Meunier[b], J.F. Beziau[b], D. Bazin[a], H. Dexpert[a], F. Garin[c], J.L. Schmitt[c] and G. Maire[c]

[a]*LURE, bât. 209D, Paris Sud, 91405 Orsay-cedex*

[b]*PSA, Centre Technique de Belchamp, 25420 Voujeaucourt et PSA Etudes et Recherches, 78140 Vélizy-Villacoublay*

[c]*LERCSI, URA1498, CNRS-ULP-EHICS, 4 rue Blaise Pascal, 67070 Strasbourg, France*

ABSTRACT

By XAS investigations of model and automotive catalysts (Pt, Rh / CeO_2 / Al_2O_3) associated with analytical electron microscopy, it is shown that for both model and aged catalysts treated in air at high temperatures, true alloyed phases were formed.
Despite sintering of the alloyed aggregates, the ECE Test Procedure indicated that the commercial aged catalysts were still active and gave emissions below the standard limits.

1. INTRODUCTION

A bimetallic catalyst system of industrial importance presently is the three-way catalyst (T.W.C.) used for the simultaneous conversion of carbon monoxide, hydrocarbons,and nitrogen oxides in automobile exhaust. Typical commercial T.W.C.'s contain platinum and rhodium associated with ceria and alumina. Platinum is an effective catalyst for the oxidation of CO and "HC". However, for the reduction of nitric oxide this metal is less effective [1,2]. Rhodium is the essential element in the T.W.C. for the conversion of NO into N_2. Ceria promotes platinum and rhodium by preventing sintering of the catalyst particles, increasing the dispersion of the catalyst [3], providing oxygen storage by shifting between

Ce_2O_3 and CeO_2 when passing from fuel-rich to fuel lean conditions [4], by enhancing the water-gas shift reaction [5,6], and stabilizing the alumina support [7].During ageing the T.W.C. displays a general loss of activity. The thermally induced sintering due to elevated temperatures, and various compositions of the exhaust gas, results in the support sintering and modifications (Al_2O_3, CeO_2), in the noble metal particles coalescence or migration of oxides on the surface or in the bulk in oxidative conditions [8,9,10]. Deactivation may also occur via chemically induced poisoning, which affects active metal sites due to strong deposition or chemisorption of contaminants on the catalyst such as P, S, Pb, Zn, Ca, Si, etc. [11]. Here too, the temperature and composition of the exhaust gas affect the degree of deactivation.

Over the last decade, several investigations of the deactivation processes were dedicated to the studies of noble metal particles present in "in-use" catalysts. Some studies indicate that the particles are clearly bimetallic and not separate clusters of Rh and Pt [12]. Reports of beneficial synergistic reactivity of Pt-Rh bimetallics toward catalyzing automotive exhaust reactions has motivated further research involving Pt_x-Rh_{1-x} alloys [13]. Studies on model surfaces of Pt-Rh alloys (single crystals) confirmed synergies for different reactions [14,15]. The major investigations published on aged automotive catalysts were obtained for metals deposited on Al_2O_3 in the abscence of CeO_2, or with simple laboratory feedstreams and accelerated catalyst ageing schedules [13,16]. Extrapolation of results obtained from laboratory reactor experiments to the actual exhaust environment should be done with caution. Commercial T.W.C.'s contain significant amounts of CeO_2. In addition to the Pt-Rh interactions, the interactions between base metal additives and the noble metals can play an important role in determining the performance and durability of T.W.C.'s.

To understand the general loss of activity during ageing of automotive exhaust catalysts we used X-Ray Absorption Spectroscopy (XANES and EXAFS) allowing "in-situ" measurements to characterize fresh and vehicle-aged commercial catalysts. Complementary techniques were necessary : XRD, TEM, STEM, XPS in association with catalytic tests performed in a laboratory reactor working under transient conditions [16], along with various cycles done at Belchamp (PSA) under more realistic conditions. We already have made preliminary investigations by X-Ray Absorption Spectroscopy on model catalysts with high loadings of Pt and Rh/Al_2O_3, treated under elevated temperatures and various compositions of gases, showing major changes in microstructure and microchemistry : increasing particle size, presence of bimetallics or alloys, interactions between Pt, Rh and CeO_2 or Al_2O_3, catalytic deactivation [17, 18]. We present in this paper results obtained for vehicle-aged, commercial automotive exhaust catalysts which have been studied using the aforementioned techniques.

2. EXPERIMENTAL

2.1. catalysts

Two sets of catalysts were selected for this study, "model" catalysts with high contents of noble metals supported on alumina (already discussed in reference 17) and the "commercial" catalysts furnished by PSA containing Pt, Rh and CeO_2 / Al_2O_3 (fresh and aged).

Model catalysts: Four supported catalysts were employed :

(a) 1%Rh/γ-Al_2O_3 calcined in air at 600°C, (b) 10%Pt/γ-Al_2O_3 reduced at 300°C in H_2, (c) 5%Pt-1%Rh/γ-Al_2O_3 calcined in air at 600°C, (d) 5%Pt-1%Rh/ γ-Al_2O_3 calcined in air at 1600°C.

The catalysts were prepared by co-impregnation of γ-alumina from Johnson Matthey with solutions of platinum sponge dissolved in HCl and of $RhCl_3$ in H_2O. Calcination at 600°C did not affect greatly the structure of γ-Al_2O_3 despite sintering, whereas at 1600°C in air the Al_2O_3 was exclusively of the alpha type as determined by XRD.

Commercial formulations: These catalysts contained Pt and Rh supported on alumina deposited in monoliths which also contained ceria.

The two vehicle-aged samples A_2 and B_2 were collected after 65,000 and 103,000 customer-operated kilometers respectively, both aged catalysts contained approximately equivalent amounts of precious metals and contaminants, as listed

Table 1: Characterization data (in wt. %) for fresh and aged catalysts

Catalyst	Specific area m^2/g BET	Pt	Rh	Ce	P	S	Pb	Zn
A1 Fresh	-	0.70	0.16	16.3	0.01	0.05	0.05	0.02
A2 Aged (103,000 km)	9	0.72	0.13	7.6	0.82	0.14	0.29	0.32
B1 Fresh	40	0.60	0.12	13.7	0.01	0.03	0.01	0.02
B2 Aged (65,000 km)	13	0.68	0.13	12.7	1.19	0.17	0.17	0.37

in Table 1, data for the fresh catalysts A_1 and B_1 are included for comparison. The powders, A_1, A_2, B_1, B_2, were extracted from monoliths using an "ad-hoc" tool to scratch the channels of the monoliths. The amount of catalyst needed for XAS analysis were 800 mg for the transmission mode and 2000 mg for the fluorescence mode.

The samples A_1, A_2 and B_1, B_2 were not obtained from the same monolith. The main difference concerned the aged A_2 sample having a much lower content of Ce due to erosion of the wash-coat at the entrance of the monolith.
The ECE Test Procedure indicated that both A_2 and B_2 aged catalysts were still active and under the standard limits.

2.2 X-Ray Absorption Spectroscopy (XANES and EXAFS)

X-Ray Absorption Spectroscopy is a technique in which the ejected photoelectron acts as a probe of the surrounding environment in a manner similar to electron scattering. Since the absorption edges of different elements are well separated in energy (which is the case for the L_{III} Pt edge (11,564 eV) and the K Rh edge (23,220 eV)) the technique is element specific and able to examine the surroundings of Rh or Pt in the presence of the support [19,20]. We used X-Ray Absorption Spectroscopy combining X-Ray absorption near edge structure (XANES) and extended X-Ray absorption fine structure (EXAFS) to extensively characterize our fresh and aged commercial catalysts. XANES is sensitive to the electronic data (oxidation state) and EXAFS data can provide information on bond distances, coordination numbers, disorder, and types of ligand for the first few coordination spheres [21].

We used the synchrotron radiation facilities of LURE (ORSAY) from the DCI storage ring running at 1.85 GeV with an average current of 250 mA. The XAS data (XANES and EXAFS) were collected on the EXAFS-4 station using a conventional step-by-step set up with a channel cut mono-chromator Si (111) for Pt and Si (311) for Rh and two ion chambers as detectors. In this work, all X-Ray absorption spectra were obtained at 25°C with the transmission detection configuration. Consequently all the catalysts listed above were studied at the L_{III} Pt edge but only the model catalysts were studied at the K Rh edge. The EXAFS spectra were analyzed in a conventional manner [22].

The amplitude and phase shift parameters associated with the backscattering process were extracted from reference compounds : Pt and Rh polycrystalline foils ; PtO_2 ; Rh_2O_3 ; $RhCl_3,nH_2O$; $H_2PtCl_6,6H_2O$. The structural parameters associated with each shell in the reference compounds were obtained from crystallographic data. For consistency, the phase and amplitude functions obtained from the reference compounds were compared with those generated by Mc Kale et al. [23] using curved wave models for scatterers of the same identity.

We measured the variation of σi namely the Debye-Waller factor, arising from both static and dynamic disorder, compared to its value in the standard. Finally, we used a two-shell least-squares fitting procedure to extract the values of N_j(coordination numbers), R_j(bond distances) and $\Delta\sigma i$. The photoelectron mean free path Γ has been fixed to its value in the standard. The XAS spectra recorded with the commercial formulations correspond to an averaging along the length of the monolith.

2.3. Analytical electron microscopy

Only the commercial formulations, fresh and aged, respectively A_1, B_1 and A_2, B_2 have been characterized. For the TEM measurements, ethanol was added to a few finely ground grains of the catalyst and the resultant suspension sonicated in an ultrasound bath before deposition of a few drops on a 200 mesh copper grid coated with a holey carbon film. Bright field micrographs were recorded at accelerating voltages of 200 kV using a JEOL 2010 instrument equipped with a spectrometer for energy dispersion Z-MAX from TRACOR (available at the I.F.P. in Rueil-Malmaison) to differentiate Pt, Rh, Pt/Rh aggregates from CeO_2, to determine the ratio Pt/Rh in the bimetallics and to localize other oxides (eg. Ba, Zr, La). Metallic particles are not visualized under 7 Å.

3. Results

The results from different techniques will be presented in the following sections and then discussed together

3.1 XAS ANALYSIS

Two kinds of information were extracted from the data. Qualitative information is gained from the change in the white line intensity. Quantitative results are extracted from the EXAFS analysis, and a common way to visualize the change of the metal environment when comparing different catalysts is to plot the magnitude of the Fourier transform (FFT) of the EXAFS oscillations. For the supported model catalysts studied it resulted that for all catalysts treated in air at 600 °C only well dispersed oxidic phases were observed on the L_{III} edge of Pt and the K edge of Rh as illustrated in Figure 1 for the 5%Pt/Al_2O_3, 1%Rh/Al_2O_3 and the 5%Pt-1%Rh/Al_2O_3 catalysts treated in air at 600 °C [17,18]. The Fourier transform curves are compared with the references PtO_2 and Rh_2O_3. It appears that the contributions of the second and third shells are very weak compared to bulk PtO_2 or Rh_2O_3 in agreement with the recent work of Beck et al. [24] for Rh/Al_2O_3 catalysts after treatment in high temperature oxidizing environments. The 5%Pt-1%Rh/γ-Al_2O_3 catalyst treated in air at 1600 °C reveals the formation of an alloy $Pt_{51}Rh_{49}$ with a mean particle size ≤10 Å as deduced from EXAFS data in agreement with XRD measurements ($Pt_{52}Rh_{48}$).The composition of the alloy determined by EXAFS and XRD seems to disagree with the microanalysis of the 5%Pt-1%Rh/Al_2O_3 catalyst treated in air at 1600°C for which a composition of $Pt_{73}Rh_{27}$ was expected. To account for the coincidence of EXAFS, XRD with microanalysis it is necessary to consider monometalllic phases almost atomically dispersed. In Table 2 we report quantitative EXAFS analysis data for the model catalysts indicating Pt-O and Rh-O distances higher than for the reference, materials, PtO_2 and Rh_2O_3, in the case of the catalyst treated at 1600°C in air and ascribed to metal-support interface interactions for the highly dispersed particles.

Table 2: Quantitative EXAFS analysis data for the model catalysts.

Catalysts	Edge	Neighbours	Coordination number	Distance Å
5%Pt/Al_2O_3 air 600 °C	Pt L_{III}	O	6.0	2.04
1%Rh/Al_2O_3 air 600 °C	Rh K	O	6.0	2.00
5%Pt-1%Rh/Al_2O_3 air 600 °C	Pt L_{III}	O	5.4	1.98
	Rh K	O	6.4	2.00
5%Pt-1%Rh/Al_2O_3 air 1600 °C	Pt L_{III}	O	1.2	2.10
		Pt	2.2	2.73
		Rh	2.9	2.73
	Rh K	O	1.7	2.15
		Rh	2.0	2.73
		Pt	3.0	2.73
Pt foil	Pt L_{III}	Pt	12.0	2.77
PtO_2	Pt L_{III}	O	6.0	2.04
Rh foil	Rh K	Rh	12.0	2.69
Rh_2O_3	Rh K	O	6.0	2.05

Furthermore the 5%Pt-1%Rh/Al_2O_3 catalyst treated in air at 1600 °C reveals a very strong metallic character as deduced from quantitative EXAFS data and qualitative XANES spectra on both Pt L_{III} and K Rh edges. The Figure 2 shows that on the Rh K edge, the white line changes after normalization for Rh

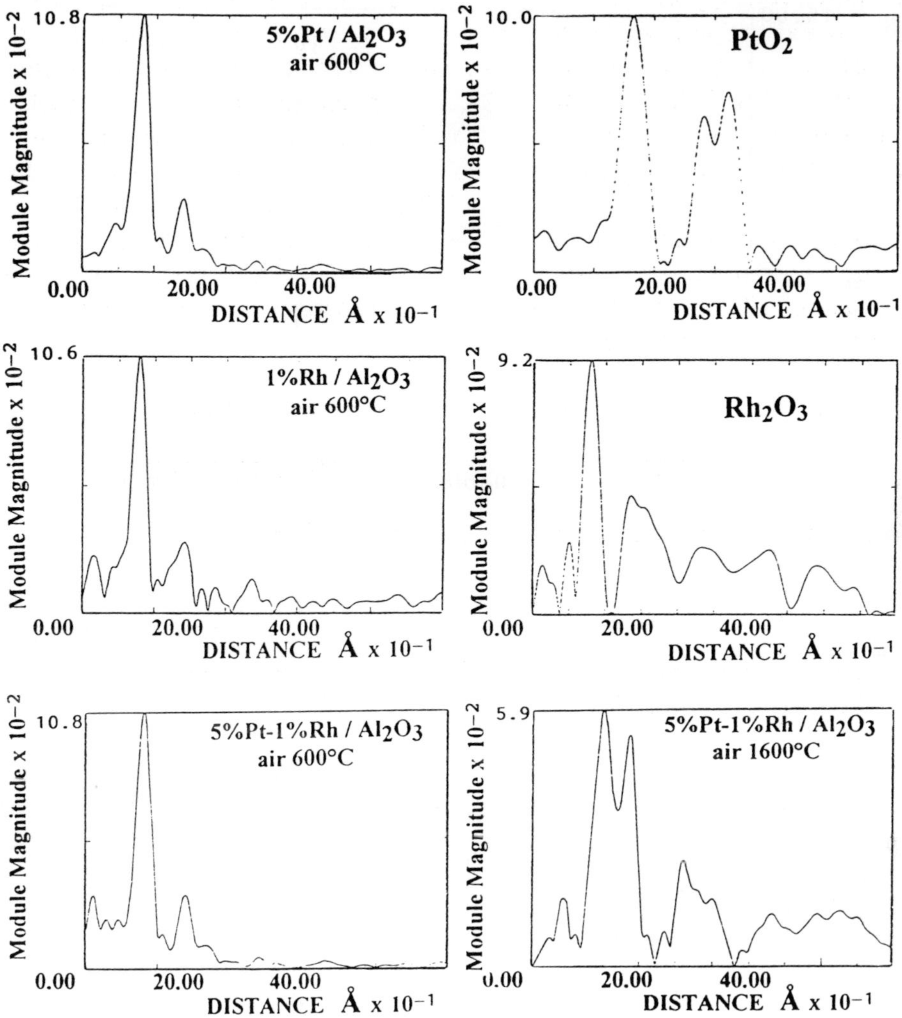

Figure 1 Model catalysts : different FFT moduli are shown.
Curve (1) corresponds to the 5%Pt/Al_2O_3 treated in air at 600 °C (Pt L_{III} edge). Curve (2) to bulk PtO_2. Curve (3) to the 1%Rh/Al_2O_3 treated in air at 600 °C (Rh K edge). Curve (4) to Rh_2O_3. Curve (5) to the 5%Pt-1%Rh/Al_2O_3 treated in air at 1600 °C (Pt L $_{III}$edge).

metal, Rh_2O_3, 1%Rh/γ-Al_2O_3 and two model catalysts. The 5%Pt-1%Rh/γ-Al_2O_3 catalyst treated in air at 600°C is constituted of PtO_x and RhO_x entities only.

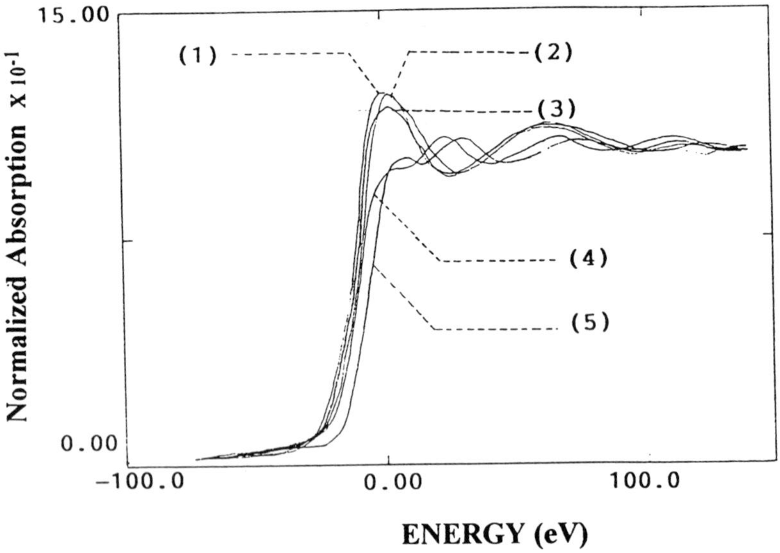

Figure 2 Model catalysts. Changes in the XANES (Rh edge K) as a function of the various samples : Curve (1) Rh_2O_3 reference. Curve (2) 1%Rh/Al_2O_3 air 600 °C. Curve (3) 5%Pt-1%Rh/Al_2O_3 air 600 °C and curve (4) 5%Pt-1%Rh/Al_2O_3 air 1600 °C. Curve (5) Rh metal reference

For the commercial formulations both the fresh and aged automotive exhaust catalysts were studied by XAS. Important differences were observed between the fresh and the aged catalysts as shown in Figure 3 where the different Fourier transform moduli are represented. Here too the A_1 and B_1 fresh catalysts are well dispersed in oxidic forms. The aged catalysts A_2 and B_2 reveal great similarities, firstly an increase in the average particle size (d=20-25 Å) and secondly the presence of the same alloy $Pt_{80}Rh_{20}$ probably inhomogeneous and different in composition from the previous model $Pt_{51}Rh_{49}$ catalyst. A_2 and B_2 are constituted of bimetallic (alloy) clusters mainly in the metallic state as deduced from EXAFS and XANES. Neither a contribution from oxygen, nor cerium, atoms was necessary to improve the EXAFS fit. Complementary EXAFS data, obtained for the A_2 aged catalyst by "in situ" measurements under hydrogen at atmospheric pressure at 430°C for 4 hours, are similar to the A_2 catalyst. The quantitative EXAFS data obtained on the Pt L_{III} edge are indicated in Table 3.

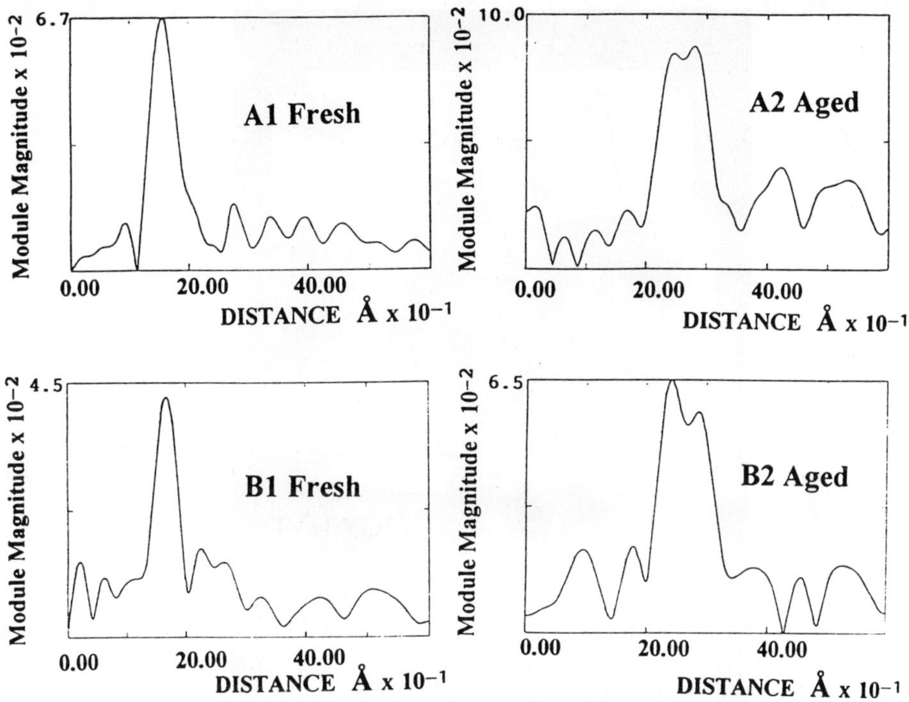

Figure 3 Commercial vehicle aged automotive exhaust catalysts (Figure 3). FFT moduli : Curve (1) corresponds to the A1 fresh catalyst. Curve (2) to the A2 aged 103,000 km. Curve (3) to the B1 fresh catalyst and Curve (4) to the B2 aged 65,000 km.

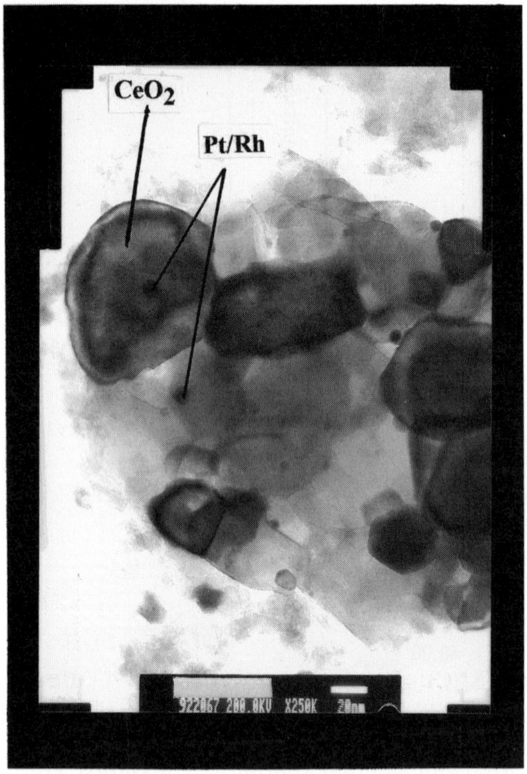

Figure 4 B2 aged 65,000 km Particles of noble metals associated with CeO_2 G = 250k

Table 3: EXAFS data for commercial formulations

Catalysts	Neighbours	Coordination number	Distance Å
A_1 fresh	O	2.5	2.06
A_2 aged	Pt	7.0	2.76
	Rh	2.0	2.76
B_1 fresh	O	4.2	2.02
B_2 aged	Pt	6.0	2.75
	Rh	2.0	2.76

3.2 Electron microscopy analysis

The alumina supports of the fresh commercial catalysts were originally γ-Al_2O_3 which became more crystallized during the ageing process : (δ and θ)-

Al_2O_3. From complementary XRD and XPS measurements $CePO_4$ and $CeAlO_3$ phases were observed for the aged catalysts. The sintering of the ceria particles observed for the aged catallysts A2 and B2 were similar : mean particle sizes of ceria (Å) : A_1=130 ; A_2=320 ; B_1=60 ; B_2=350. Both vehicle aged catalysts display average metal particle sizes (d(Å)) which are larger than those of the fresh catalysts, but still moderate in agreement with their catalytic activity.

d(Å) A_1=20 ; B_1=10 (after reduction in H_2 at 300°C befor TEM measurements); A_2=80 ; B_2=70. For sample B_2, bimetallic particles of Pt and Rh (alloy) between 30 and 300 Å, have been shown by EDS to always be associated with the presence of Ce. The atomic ratio Pt/Rh increases with the size of the particles. The average being around 1.6 for a mean particle size of 70 Å. Here again as for the EXAFS data, to explain the divergence with the expected stoichiometry, one needs to assume the presence of bimodal distributions of particles containing very small particles and bigger ones (12).

4. Discussion and Conclusion

From the results presented in this publication three points have to be underlined :

1. In both the case of the model catalysts treated in air at high temperature, and of the commercial aged catalysts, the formation of alloyed phases are shown.
2. Some discrepancies seem to exist between XAS data (EXAFS) and TEM-EDS data concerning essentially with the size of the noble metal particles.
3. Is it necessary to establish a special relationship between the mono or bimetallic phases (Pt, Rh, Pt-Rh) and the CeO_x phases in order to improve the resistance to the general loss of activity during ageing which results from the thermally induced sintering due to elevated temperatures, and under various compositions of the exhaust gas ?

Considering first the formation of alloys during the ageing of the T.W.C. it is clear, in our case, that these alloy phases observed in the aged commercial catalysts A_2 and B_2 represent the majority of the noble metal particles even if some monometallic Pt or Rh particles are present, meaning that the remaining catalytic activity, must be attributed to the alloy phases. Thus implying that alloying is benefical in practical catalysts. But if so, the questions remain, when are the effects most noticeable?, what are the conditions? and what are the metal ratios in the alloy, as already mentioned by Joyner (25)? The answer can only come from extended work on model Pt-Rh alloys. Studies on model surfaces of Pt_x Rh_{1-x} alloy single crystals have already confirmed synergies for different reactions (14,15). Furthermore, it is worthwhile to underline the relative low sintering of these alloy particles around 30 Å (EXAFS data) - 80 Å (TEM data).

Concerning the estimates of the mean particle size of noble metals determined from XRD, EXAFS or TEM it is necessary to consider that the various techniques used have intrinsic strengths and weaknesses as pointed out recently (26). XRD gives no indication of the range of particles sizes present in the catalyst. Very small particles are likely to produce diffraction peaks which are too broad to be observed. EXAFS is not phase-specific, but is element specific. As with XRD, the EXAFS method is insensitive to the range of particle sizes. However, XAS is well adapted for "in-situ" measurements, and provides physical and chemical information on local environments even if averaged. Electron microscopy is neither element nor phase specific, but provides a much more direct measurement of particle sizes. But it only involves the study of a relatively small proportion of the whole catalyst particle (and the limitation in the size determination is 10 Å for TEM and 100 Å for STEM). We consider in our case that XAS and analytical electron microscopy are complementary and that the tendancies observed with both techniques are in good agreement.

Concerning any relationship between the mono-or bimetallic phases, and the CeO_x phases present in the T.W.C., it is clear from the literature that ceria improves the resistance to the coalescence of the noble metal particles [28,29]. From our results, it appears for the commercial T.W.C. A_2 and B_2, that the noble metal particles (alloys) are always associated with CeO_x as deduced from TEM-EDS data (Figure 4). On the other hand it is impossible, by EXAFS, to correlate the alloy particles with the wash coat via an interface constituted of Pt-Ce or Rh-Ce bonds which does not exclude an interface between the alloy particle and ceria via Pt-O and Rh-O bonds with oxygen from the CeO_x support. In this case, due to the size of the alloy particle, EXAFS being more an average volume technique, is unable to provide evidence for the interaction with the substrate CeO_x. Some results obtained recently in our laboratory by "in-situ" EXAFS measurements on fresh Pt, $Rh/CeO_2/Al_2O_3$ catalysts show the epitaxy of the noble metal particles on CeO_x entities in agreement with recent work of Bernal et al. on ceria-supported Rh catalysts [27]. The structural nature of such epitaxial relationship might well be interpreted as a kind of SMSI [18].

ACKNOWLEDGEMENTS

The authors (M.C.[b]) would like to thank the staff members who contributed to the use of the TEM-EDS facilities at the Institut Français du Pétrole, Rueil-Malmaison.

REFERENCES

1 K.C. Taylor, Catal. Sci. Technol. 5 (1984) 119
2 K.C. Taylor, in Catalysis and Automotive Pollution Control, vol 30 of Stud. Surf. Sci. Catal., ed. A. Crucq and A. Frennet (Elsevier, Amsterdam, (1987) p.97
3 J.C. Summers and S.A. Ausen, J. Catal. 58 (1979) 131
4 R.K. Herz, Ind. Eng. Chem. Prod. Res. Dev. 20 (1981) 451
5 G. Kim, Ind. Eng. Chem. Prod. Res. Dev. 21 (1982) 267
6 J.C. Schlatter and P.J. Mitchell, Ind. Eng. Chem. Prod. Res. Dev. 19 (1980) 288
7 B. Harrison, A.F. Diwell and G. Hallett, Platinum Met. Rev. 32 (1988) 73
8 H.C. Yao, S. Japar and M. Shelef, J. Catal. 50 (1977) 407
9 C. Wong and R.W. McCabe, J. Catal. 119 (1989), 47
10 H.C. Yao, M. Sieg and H.K. Plummer, Jr., J. Catal. 59 (1979) 365
11 M. Shelef, K. Otto and N.C. Otto, Adv. Catal. 27 (1987) 311
12 S. Kim and M.J. D'Aniello, Jr, Appl. Catal. 56 (1989) 23
13 S.H. Oh and J.E. Carpenter, J. Catal. 98 (1986) 178
14 R.M. Wolf, J. Siera, F.C. van Delft and B. Nieuwenhuys, Farad. Disc.Chem. Soc. 87 (1989) 275
15 G.B. Fisher, C.L. Di Maggio and D.D. Beck, Proc.10th Internat. Congr. on Catalysis, Budapest, p.383 (1993). Elsevier Science Publishers. Guczi et al. (Editors), New Frontiers in Catalysis.
16 M. Weibel, Ph.D thesis, Univ. L. Pasteur, Strasbourg, (1991)
17 F. Maire, H. Dexpert, G. Meunier, M. Capelle, F. Garin and G. Maire, submitted to Catal. Lett., apr. (1994)
18 F. Maire, Ph.D thesis, LURE (Orsay) - Univ. L. Pasteur (Strasbourg), (1994)
19 G.H. Via, K.F. Drake, G. Meitzner, F.W. Lytle and J.H. Sinfelt, Catal. Lett. 5 (1990) 25
20 D. Bazin, H. Dexpert, J.P. Bournonville and J. Lynch, J. Catal. 123 (1990) 86
21 F.W. Lytle, R.B.Greegor, E.C. Marques, D.R. Sandstrom, G.H. Via and J.H. Sinfelt, J. Catal. 95 (1985) 546
22 P. Lagarde, F. Murata, G. Vlaic, E. Freund, H. Dexpert and J.P. Bournonville, J. Catal. 84 (1983) 333
23 A.G. McKale, B.W. Veal, A.P. Paulikas, S.K. Chan and G.S. Knapp, J. Am. Chem. Soc. 110 (1988) 3763
24 D.D. Beck, T.W. Capehart, C. Wong and D.N. Belton, J. Catal. 144 (1993) 311

25 R. Joyner, Proc. 10th Internat. Congr. on Catalysis, Budapest, (1993), Discussion Paper of G.B. Fischer, C.L. Di Maggio and D.D. Beck p.383 Elsevier Science Publishers. Guczi et al. (Editors), New Frontiers in Catalysis.
26 A.T. Ashcroft, A.K. Cheetham, P.J.F. Harris, R.H. Jones, S. Natarajan, G. Sankar, N.J. Stedman and J.M. Thomas, Catal. Lett. 24 (1994) 47
27 S. Bernal, F.J. Botana, J.J. Calvino, M.A. Cauqui, G.A. Cifredo, A. Jobacho, J.M. Pintado and J.M. Rodriguez-Izquierdo, J. Phys. Chem.,97 (1993) 4118
28 C. Serre, Ph.D thesis, Univ. L. Pasteur, Strasbourg, (1991)
29 C. Serre, F. Garin, G. Belot and G. Maire, J. Catal. 141 (1993) 1 and 141 (1993) 9

A. Frennet and J.-M. Bastin (Eds.)
Catalysis and Automotive Pollution Control III
Studies in Surface Science and Catalysis, Vol. 96

SULFUR ADSORPTION AND DESORPTION ON FRESH AND AGED Ce CONTAINING CATALYSTS

S. Lundgren*, G. Spiess*, O. Hjortsberg*, E. Jobson*, I. Gottberg**, and G. Smedler***

**AB Volvo, Technological Development, Applied Physics Department, S-412 88 Göterborg*
*** Volvo Car Cooperation S- 405 08 Göteborg*
**** Current adress: Johnson Matthey/Svenska Emissionsteknik AB, S-42131 Göteborg.*

ABSTRACT

The effect of field ageing on the transient H_2S emissions from Ce-containing three-way catalysts was studied with TPR experiments and XPS analysis. A set of differently aged catalyst samples was analyzed by SO_2 exposure followed by TPR treatment with H_2, in combination with successive XPS analysis of the oxidation state of Ce. TPR studies were also made using H_2, CH_4, C_2H_2, C_3H_6 and C_3H_8 as reducing agents. It was found that the H_2S desorption declines rapidly as a function of field ageing. Simultaneously, the TPR (with H_2) desorption peak temperature of H_2S increases and the SO_2 peak temperature decreases. For the catalyst with high activity the Ce3d XPS spectra indicated a shift in the oxidation state from (III) to (IV) during sulfur adsorption. The severely aged catalyst stayed in the trivalent state, during sulfur storage and TPR. The temperature for maximum H_2S formation strongly depend on the type of reducing agent. A fullscale engine bench test was performed, by use of chemical ionization mass spectrometry. The results are dicussed in the light of the flow reactor results. Possible explanations are discussed, as well as a possible mechanism for changes in the H_2S formation, due to deactivation.

1. INTRODUCTION

One of the most extensively used metals for oxygen storage in TWC systems is Ce, which however also has an ability to adsorb, store and release sulfur (1)

This sulfur, originating from the gasoline as well as from the oil additives, reaches the catalyst mainly as SO_2 under normal operation conditions, e.g.(2).

During oxidizing (lean) conditions sulfur , adsorbed on the PM particles is oxidized to SO_3 (3). Sulfur can then react with the metal oxides of the washcoat, e.g. with Al_2O_3 or CeO_3 etc., and form metal sulfates (4-6). Both Al and Ce sulfates are formed but sulfur is known to react preferentially with CeO_x (4), (7) through one of the following reactions:

$$2\ CeO_2 + 3\ SO_2 + O_2 \longrightarrow Ce_2(SO_4)_3 \quad (1)$$
$$6\ CeO_2 + 3\ SO_2 \longrightarrow Ce_2(SO_4)_3 + 2\ Ce_2O_3 \quad (2)$$
$$2\ SO_2 + CeO_2 \longrightarrow Ce(SO_4)_2. \quad (3)$$

During reducing conditions, cerium sulfate decomposes around 550-750 K (1) to SO_2, which then easily reacts with adsorbed hydrogen atoms on the well dispersed PM particles to form H_2S, according to:

$$SO_2 + 3H_2 \longrightarrow H_2S + 2H_2O \quad (4)$$

The produced H_2S then desorbes from the catalyst. Stored sulfur can in this way rapidly be released during retardation or acceleration, induced by changes in the air-fuel ratio in the exhaust gas. The formation of H_2S can during short periods reach concentrations well above the human smell threshold of 0.02-0.03 ppm (5). This study was designed to investigate the sulfur uptake and release for differently deactivated catalysts and the influence of different reducing agents.

2. Experimental

2.1 Sample preparation

Both fresh and aged Ce-containing three-way catalyst samples were used in this study, see Table I. All TWC samples were of Pt/Rh (5/1) impregnated Ce-stabilized g-alumina washcoat with a PM loading of 0.2 at%. The model catalyst was impregnated with a pure Ce washcoat with a Pt loading of 1 at%. All catalysts had a cell density of 400 cpsi (Corning cordierite monolith). The fresh samples had an washcoat surface area of approximately 100 m^2/g. Samples with a weight of 5 g were drilled out 5 mm from the front center of the 4.66" catalyst monoliths. The aged catalysts were collected from the swedish market (years 1989-1990).

#	formulatio n	ageing (km)	methods
1	TWC	fresh	1, 2, 3
2	Model	fresh	1, 2
3	TWC	4800	1, 2
4	TWC	10 000	1
5	TWC	34 600	1, 2
6	TWC	71 800	1, 2
7	TWC	121 000	1, 2
8	TWC	133 000	1, 2
9	TWC	200 000	1, 2, 3
10	TWC	serverly. aged	1, 2

Table I: Characterization methods: 1) TPR, 2) XPS, and 3) CIMS

2.2 TPR

<u>TPR with H_2</u>

The TPR experiments were performed in a quartz flow reactor system, which was connected to an XPS system (8).The reactor was also connected to an MS for continous gas analysis by a quartz capillary inlet probe system (9). The probe was positioned 10 mm from the sample outlet. The inlet gas and the reactor were heated by a heating coil outside the reactor. The temperature of the samples were measured at the sample outlet by a chromel-alumel thermocouple.

The sulfur storage was made by an exposure of 1% SO_2 in Ar for 60 min at 673K in a gas flow of 100 ml/min (or a space velocity of 7 000 h^{-1}). This period was followed by 13 min exposure of pure Ar at 373 K. The first 10 minutes in the samples were exposed to a gas flow of 1000 ml/min and the last three minutes they were exposed to a gas flow of 100 ml/min. The low exposure during the 3 min period were performed in order to reach the background level of SO_2 concentration and to stabilize the temperature at the TPR start temperature. The TPR-sequence was then conducted in a gas flow of 4% H_2 in Ar at 100 ml/min with a linear temperature increase of 0.2 K/second from 373K to 973 K. The low gas velocity was chosen in order to enhance the detection of desorbed sulfur species. SO_2, H_2S, O_2 and H_2O were continously recorded during the TPR sequence.

TPR with different reducing agents .

TPR experiments were also made on samples from a fresh catalyst with H_2, CH_4, C_2H_2, C_3H_6 and C_3H_8 as reducing agents. The reducing agents were used in separate runs with the following concentrations in Ar: H_2:4 %, CH_4 :1 %, C_2H_2 : 0.8 %, C_3H_6: 0.44% and C_3H_8 : 0.4 % to keep the reducing capacity constant. The sulfur storage and the TPR sequence were performed as described above.

TPR details

All gases used were of high purity (99.9997%). The relative sensitivity of the MS for the different gases used was calibrated by the use of known gas mixtures. Blank experiments, without a sample in the reactor, were run in order to check the inertness and the background levels of the reactor system.

2.3 XPS analysis

Four different samples were analyzed by XPS: samples # 1, 2, 9 and 10, see Table I. The surface analysis was performed in an XPS instrument (VG ESCA-lab II) with an energy resolution of 0.7 eV. Spectra were recorded on an analyzed area of 2 * 5 mm, using Mg Ka radiation. The XPS base pressure during the analysis was 10^{-9} mbar. The XPS study was focused on the Ce3d spectrum. The XPS spectra were recorded at all three stages of treatment, described in section 2.1: before the sulfur exposure, after the sulfur exposure and after the TPR treatment. The samples were transferred from the flow reactor to the XPS analyser chamber under vacuum conditions (10^{-7} mbar), after cooling in a pure Ar flow. The spectra were recorded by accumulation of scans, followed by a smoothing procedure. Possible drift in the binding energy was checked after the first and the last scan, to ensure that the charge shift remained stable during the period of signal accumulation. The binding energy scale of the recorded spectra was adjusted to literature data value of Al2p=74.2 eV (10) in a g-Al_2O_3 spectrum. Sample #2, with pure Ce washcoat was also referenced in this way, since the analysed surface of the sample includes areas of broken cordierite walls.

2.3 Full scale CIMS measurement

The formation of SO_2, COS and H_2S was also studied during exhaust conditions. Chemical Ionization Mass Spectrometry (CIMS) was used for on-line exhaust gas analysis (11). In the CIMS instrument (CIMS 500, V & F) the exhaust gas is soft ionized by a pre-ionized Xe gas. SO_2, H_2S and COS in the exhaust gas was continously analyzed. The exhaust gas was generated by a Volvo 234 FT engine, with a fuel containing 629 ppm sulfur. The engine speed was 2700 rpm with a load of about 15 Nm and the outlet temperature was 873 K. The engine was run with lambda control. The exhaust stream was diluted with about 0.6% O_2 during the sulfur storage phases of 5 min and with 0.6% propane C_3H_8 during the reducing phases.

3. RESULTS

3.1 Ageing effects

3.1.1 TPR and CIMS measurements

Fig. 1 a and b, show H_2S and SO_2 desorption at two TPR runs with H_2 for the samples #1 and #9 in Table I. One H_2S peak and two separate SO_2 peaks were identified and denoted SO_2 I and SO_2 II in fig. 1 b. No further sulfur desorption was detected when the TPR sequence was repeated. Fig. 1 c and d show the corresponding formation of H_2S, COS and SO_2 as a function of time from the full scale exhaust gas measurements. The exhaust flow conditions were manipulated from lean to rich (l=1.05 to l=0.86), according to the sequence described in section 2.1, for a fresh catalyst (similar to sample #1) and for an aged catalyst.

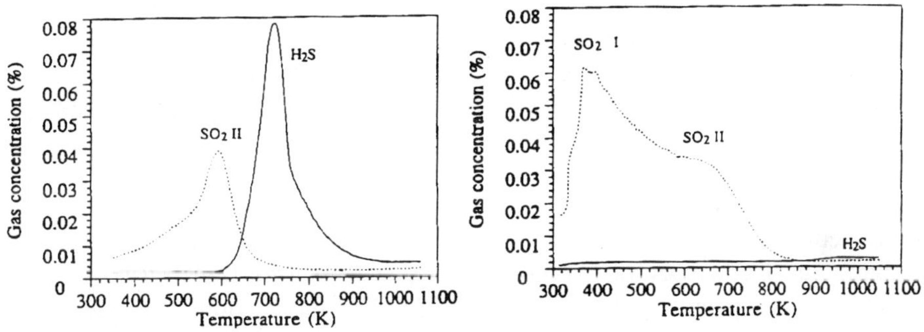

Fig. 1 (a)-(b): H_2S and SO_2 desorption during the TPR sequence (with H_2) as a function of temperature for a fresh (a) and an aged catalyst (b).

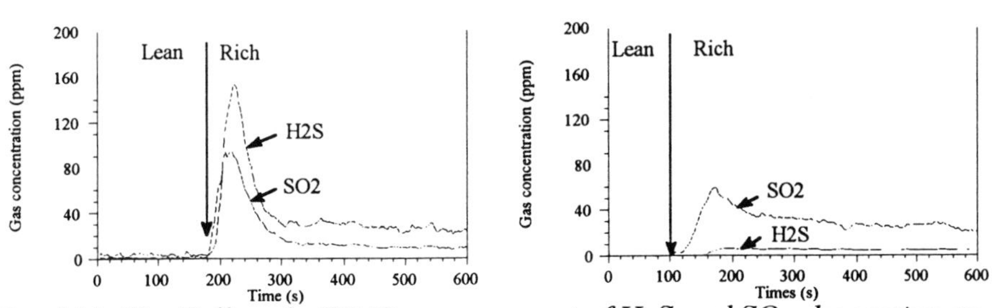

Fig. 1(c)-(d) : Full scale CIMS measurement of H_2S and SO_2 desorption as a function of time for a fresh (c) and an aged catalyst (d).

Fig. 2 shows the integrated H_2S and SO_2 adsorption/desorption and formation peak temperature, during the TPR in H_2 sequence, as a function of field ageing. The adsorbed and desorbed amounts are shown as a fraction of the adsorbed amount of SO_2 on the fresh catalyst, during the preceding storage phase. The total sulfur adsorption was about 0.6 mmole SO_2 on the fresh catalyst.

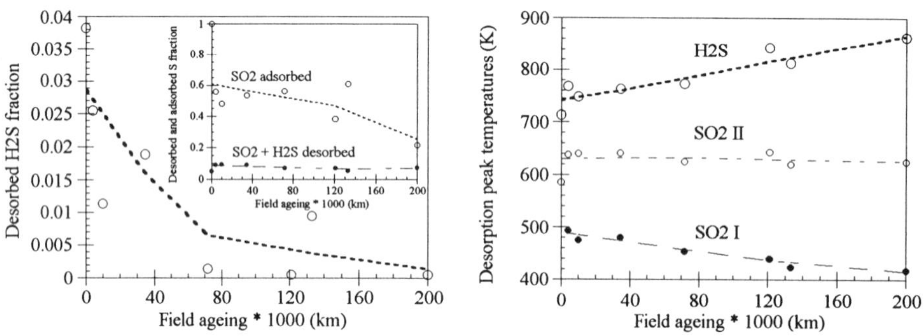

Fig. 2 (a): H_2S desorption, SO_2 adsorption and total S desorption as a function of field ageing. Fig. 2 (b): H_2S and SO_2 I and II desorption peak maximum as a function of field ageing.

3.1.2 XPS

Fig. 3 (a) shows the Ce3d spectrum for the pure Ce washcoat catalyst in its fresh state. The Ce3d spectra is complicated due to charge transferring processes between Ce3d, O2p and Ce4f orbitals (12,13). The most common assignment in literature for the six peaks associated to the Ce3d spectra are v, v,' v" and v'" for Ce3d5/2 and u, u', u" and u'"(marked with an arrow) for Ce3d3/2 transitions (14). There is, however, some uncertainty in the peak assignment (12).The spectra were recorded before sulfur exposure (a), after sulfur exposure (b) and after the TPR treatment (c) as described above. The sulfur concentration calculated from the S 2p spectra recorded before and after the TPR sequence decrease from 11 at% to 0.9 at% between the two states. The binding energy for the S 2p transition decreased from 169.5 eV after the SO2 exposure to 168.5 eV after the TPR sequence. The corresponding spectra for the fresh fully formulated three-way catalyst, for the field aged 200 000 km catalyst and from the severely aged catalyst (samples # 1, 9 and 10 in table I), are shown in fig. 4 , 5 and 6 respectively.

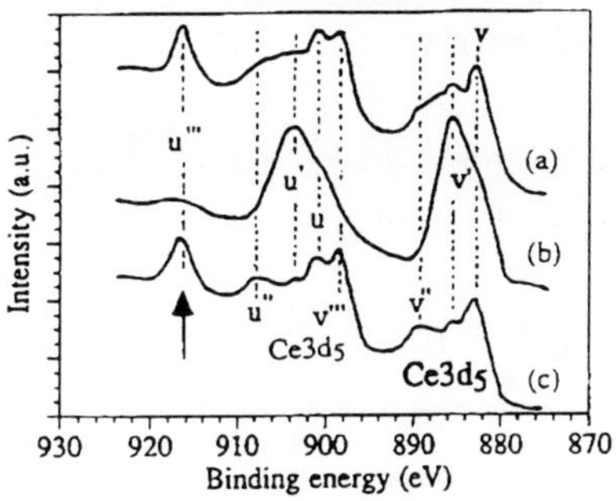

Fig. 3: Ce model sample, #2

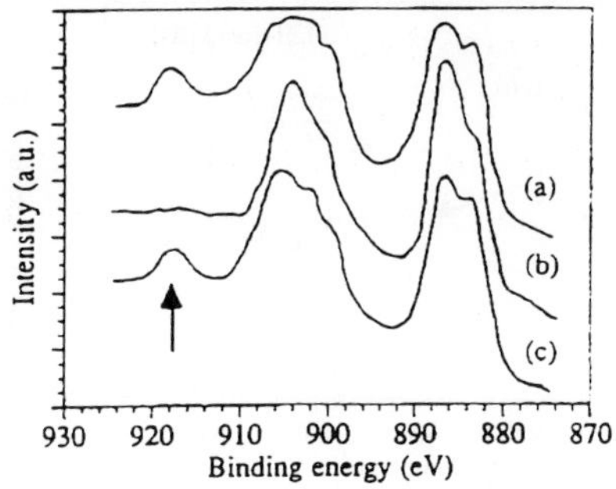

Fig. 4: Fresh TWC sample, #1

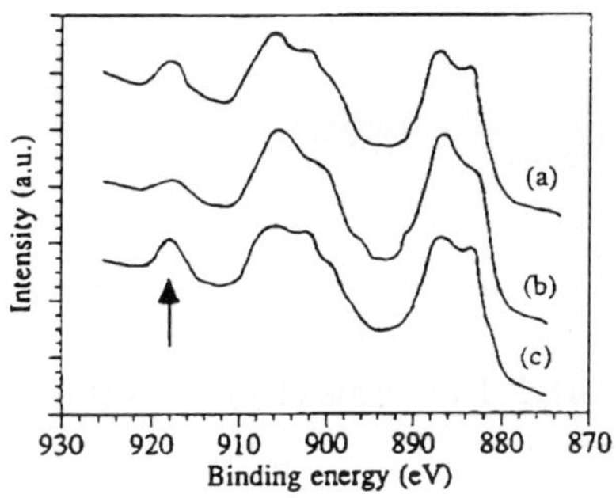

Fig. 5: Aged TWC sample, #9

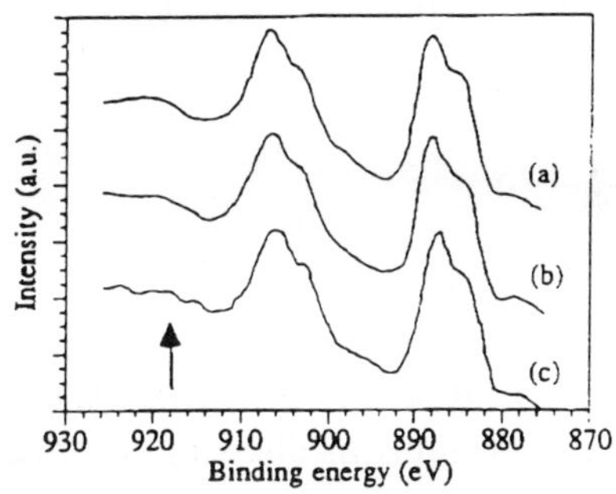

Fig. 6: Severly aged TWC sample, #10

Fig. 3-6 The Ce3d XPS spectra of samples # 1, 2, 9 and 10 after: (a) initial oxidation, (b) sulfur exposure and (c) TPR treatment.

3.2 Effects of different reducing agents during TPR

Fig. 7 show the H_2S and SO_2 desorption for the five different reducing agents on a fresh catalyst. The following peak temperatures were identified: H_2: 620 K, CH_4 : 860 K, C_2H_2 : 640 K, C_3H_6: 560 K and C_3H_8 : 750 K. Note, that COS desorption (not quantified here) can contribute to the total sulfur desorption when carbon containing reducing agents are used.

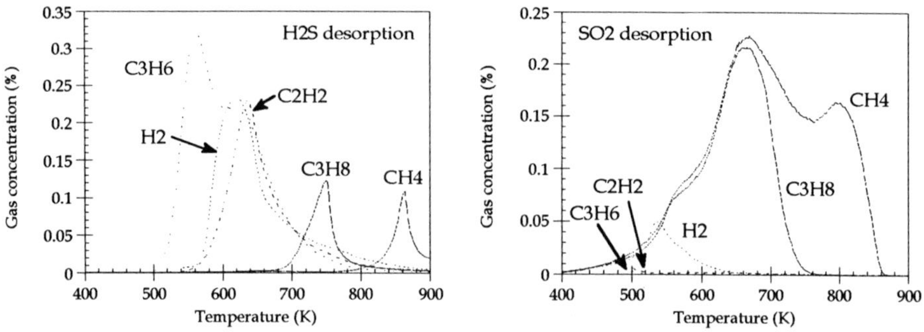

Fig. 7 (a) and (b) show the H_2S and SO_2 desorption during TPR with different reducing agents on a fresh catalyst.

4. DISCUSSION

As seen in fig. 1 and 2 the sulfur desorption pattern is changed dramatically when the catalyst is aged. The formation of H_2S has almost totally vanished after 200 000 km of field ageing. However, even if the H_2S formation decreases the total release of sulfur remains almost unchanged according to fig 2. The full scale engine bench exhaust gas measurements show in principal the same behaviour. The rapid change in stoichiometry at 873 K gave a direct reduction of desorbed SO_2 forming an H_2S peak, which rapidly declines to the steady state level of H_2S formation through direct reduction of SO_2 to H_2S by exhaust HC and H_2. On the aged catalyst almost no H_2S formation is observed. SO_2 originating from decomposed sulfate is however still released rapidly. The SO_2 storage/formation mechanism is better understood by the use of a mechanistic model. We are assuming the following reaction scheme in the discussion below:

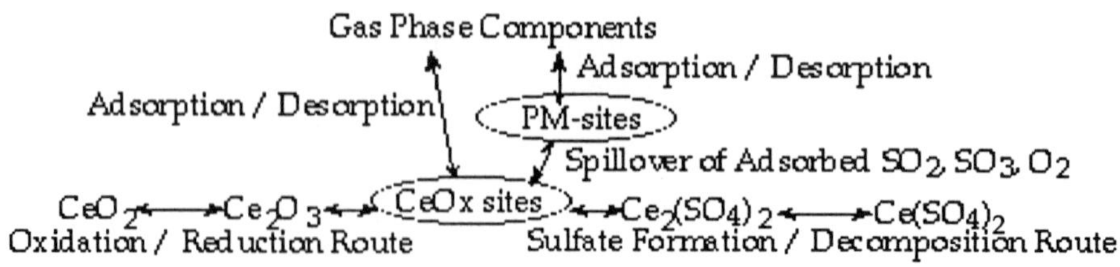

The scheme covers the adsorption and desorption of the components relevant for the interaction between the surface bounded components and the Ce-compounds. With this reaction scheme the H_2S release is fully dependent on the amount of free precious metal (PM) sites at the surface for H_2 adsorption and H_2S reaction. The CeO_2 will not convert to $Ce_2(SO_4)_3$ unless SO_2 can be found at the surface as adsorbed SO_2. At reducing conditions, free PM-sites are needed for the adsorption of H_2 and SO_2. When the amount of free PM-sites decrease due to field ageing adsorption of both components decreases, leading to a decrease in H_2S formation.

From the TPR experiments, it was concluded that the temperature for maximum rate in the H_2S formation increase with increasing age of the catalyst. The SO_2 (I) desorption peak temperature (with sulfate origin) decreased with increasing field ageing.

The increase in the desorption peak temperature for H_2S can be explained by the change of the active sites. The aged sites will partly consist of blocked PM-material and PM-material embedded in inorganic compounds, and partly of a reduced number of the original PM-sites. The aged surface is less active than the fresh, leading to slower kinetics for the adsorption of hydrogen and the reaction between hydrogen and SO_2, and thus a lower desorption rate for H_2S.
The decrease in the SO_2 (I) desorption peak temperature is more difficult to explain with this model.

The importance of the PM site reaction kinetics is also clearly demonstrated by the large influence of different reducing agents, as shown in fig. 7. Except H_2, the unsaturated C_2H_2 and C_3H_6 molecules gave a low temperature desorption while the saturated C_3H_8 and CH_4 molecules gave a high temperature desorption of H_2S. The degree of unsaturated hydrocarbons in the raw emission (such as e.g. aromatics) could thus have an important influence on the H_2S formation.

The amount of sulfur adsorbed on the fresh catalyst in fig 1 (a) corresponds to about $3*10^{20}$ SO_2 molecules. If all adsorbed sulfur is preferentially (4) converted to $Ce_2(SO_4)_3$ (neglecting Al_2O_3-interaction), $6*10^{20}$ washcoat CeO_2 molecules would be involved, according to reaction scheme (2). This amount corresponds to an involved bulk depth of 6 nm of CeO_2 (20 monolayer) for the fresh 5 g* sample. The total CeO_2 "layer" bulk depth is however of about 5 nm** which then implies that the Ce containing part of the washcoat was totally sulfated.

The amount of desorbed H_2S and SO_2 molecules during the TPR treatment corresponds to about 1.5 monolayer of recovered CeO_2 sites. From these rough calculations we can thus conclude that only a part of the available sulfate source was released as H_2S/SO_2. This is supported by the findings of Diwell (4) and Beck (7),who state that the reaction only involves the surface region of the washcoat. The XPS results verifies the role of the surface. It was found that a residual sulfur concentration of 0.9 % remained after the TPR sequence, thus deriving from bulk sulfate species, simultaneously as the oxidation state of Ce was strongly influenced, as shown in fig. 4.

The spectrum 3 (a) is consistent with literature data for the $Ce^{(IV)}$ state and reference spectra (3) for pure CeO_2. The subsequent sulfur exposure at 673K altered the oxidation state to Ce $^{(III)}$ due to formation of $Ce_2(SO_4)_3$ (13) and Ce_2O_3, in 3 (b). The final TPR sequence then decomposed the surface sulfate species and brought the Ce state back to the (IV)- state, in 3 (c), by re-oxidation of released oxygen from decomposed and reduced SO_2 species. The Ce 3d spectra has becomes slightly changed for the mixed Al_2O_3/CeO_2 washcoat sample, but the changes in oxidation state is still clearly seen in fig 4.

For the two field aged catalysts the change in oxidation state decreases with an increasing contamination of the catalyst. The spectra for the field aged 200 000 km sample (in fig. 5) , however, still show a small changed in the weak u''' peak in spectrum 5 (b) after sulfur exposure. The absence of change is even more pronounced in the spectra from the severely aged catalyst. The oxidation states in spectra 6 (a), (b) and (c) are all very close to the $Ce^{(III)}$ state, thus the u''' is totally absent.

---

* The value is based on $3*10^{19}$ surface sites of CeO_2, calculated from a surface area of 20 m^2/g monolith, a sample weight of 5 g, a surface concentration of 9 at% and a surface site density of $3*10^{14}$ cm^{-2}.

** Calculated using a CeO_2 density of 7.13 g/cm^3, a wachcoat area of 100 m^2/g washcoat, a sample weight of 5 g and a atom layer distance of 314 pm.

Neither the sulfur exposure nor the TPR treatment change the shape of the spectrum significantly. This rigidity of the oxidation state is probably influenced by the degree of oil and fuel additive poisons or by thermal effects (16) or a combination of both these effects. One suggestion is that cerium phosphate was formed, since P was the largest contaminant with a surface concentration of 14 at %. It should be noted that the oxygen storage capacity was strongly reduced, as discussed in detail elsewhere (16). The relative constant desorption of sulfur during the TPR sequence, no matter the degree of field ageing, is thus an evidence for that Ce-sulfate species embedded, or situated deeper down in the bulk, still are available for sulfur or oxygen storage.

5. CONCLUSIONS

We have shown that:
i) The released amount of H_2S, drops during TPR strongly with increasing catalyst age, for the aged catalysts used in this work in similarity to our earlier findings [16]. The total amount of sulfur desorbed remains relatively constant.
ii) The H_2S and SO_2 I peaks increase and decrease respectively with increasing field ageing.
iii) A simplified model based on a loss of PM surface area can explain the H_2S decrease.
iv) Only a small part of the Ce-bulk is involved in the decomposition and release of sulfur on both fresh and aged catalysts.
v) The surface region is sulfated at 673 K, during oxidizing conditions.
vi) The XPS Ce 3d spectrum (from the surface region) is locked in a Ce $^{(III)}$ state for strongly aged catalysts. Sulfur is however still released as SO_2.
vii) The type of reducing agent has a large effect on the formation of H_2S, with the peak temperature of formation increasing in the order
$C_3H_6 < H_2 < C_2H_2 << C_3H_8 << CH_4$.

ACKNOWLEDGEMENTS

The authors kindly acknowledges: Göran Wirmark, Volvo Technological Development and Prof. Bengt Kasemo, Chalmers University of Technology, for contributing effort and valuable discussion. The financial support from the National Swedish Board for Technical Development (contract no. 89-01176).

REFERENCES

1 M. G. Henke, J. J. White, G. W. Denison, *SAE 872134,* (1987).
2 M. A. Härkönen, T.-K. Rantakylä, V. J. Pohjola, S. Salanne, *SAE 900498*, (1990).
3 C. C. Chang, *J. Catal.* **53**, 374 (1978).
4 A. F. Diwell, C. Hallet, J. R. Taylor, *SAE 872163*, (1987).
5 I. Gottberg, E. Högberg, K. Weber, *SAE 890491*, (1989).
6 T. J. Trux, H. Windawi, P. C. Ellgren, *SAE 872162*, (1987).
7 D. D. Beck, M. H. Krueger, D. R. Monroe, *SAE 910844,* (1991).
8 S. Lundgren, K.-E. Keck, B. Kasemo, (To be published).
9 B. Kasemo, *Rev. Sci. Instrument* **50**, 12 (1979).
10 D. M. Bickerman, C. D. Wagner, . (U.S Department of commerse, 1989).
11 F. J. von Carlowitz, M. G. Henk, P. H. Gagneret, *SAE 900272* , (1990).
12 P. Lööf, B. Kasemo, L. Björnqvist, S. Andersson, A. Frestad, Catalysis and automotive pollution control II, A. Crucqs, Eds., 1991),
13 F. L. Normand, L. Hilaire, G. Krill, G. Marie, *J. Phys. Chem* **92**, 2561(1988).
14 D. Briggs, J. C. Riviere, *Practical Surface Analysis*. D. Briggs, M. P. Seah, (1983).
15 E. S. Lox, B. H. Engler, E. Koberstein, *SAE 890795* , (1989).
16 G. Smedler, S. Eriksson, M. Lindblad, H. Bernler, S. Lundgren, E. Jobson, *SAE 930944* (1993).

A. Frennet and J.-M. Bastin (Eds.)
Catalysis and Automotive Pollution Control III
Studies in Surface Science and Catalysis, Vol. 96
1995 Elsevier Science B.V.

INHIBITION OF POST-COMBUSTION CATALYSTS BY ALKYNES: A CLUE FOR UNDERSTANDING THEIR BEHAVIOUR UNDER REAL EXHAUST CONDITIONS

G. Mabilon, D.Durand and Ph. Courty

Institut Français du Pétrole, BP311, 92506 Rueil Malmaison Cedex, France

ABSTRACT

The exhaust gas of spark ignited engines contains a hydrocarbon mixture of complex composition ranging from C_1 to C_9 molecules including alkanes, olefins, diolefins, alkynes and aromatics.

It has been evidenced that among the various unsaturated hydrocarbons, alkynes show the largest inhibiting effect on all oxido-reduction reactions occuring on a PtRh catalyst, i.e. non-alkyne hydrocarbons, CO and NO are not converted until alkynes are eliminated.

A scheme based on competitive adsorption has been proposed which accounts for the effect of the nature of the hydrocarbons on their own oxidation and on CO oxidation.

It has also been shown that the inhibition effect exists even for very low C_2H_2 concentrations, representative of real exhausts gases. Furthermore, the addition of 15 ppm SO_2 slows down C_2H_2 oxidation, which increases the C_2H_2 inhibiting effect.

From these results a laboratory test has been defined enabling the simulation of engine bench tests.

1. INTRODUCTION

The combustion of gasoline air mixtures in the combustion chamber of spark ignited engines leads essentially to the formation of total oxidation products, but also to CO, H_2, NO, a hydrocarbon (HC) mixture and SO_2. Several HC emissions formation mechanisms are possible to explain the origin of the hydrocarbon mixture [1,2], such as flame quenching at the cylinder walls or at crevice entrance, adsorption-desorption in the oil film and incomplete combustion (partial or complete misfire) particularly during transient operations. The HC that are not combusted (about 1 % of the gasoline) are either exhausted unmodified or undergo pyrolitic transformations. Unmodified HC are essentially C_4 to C_8 alkanes (including iso-alkanes) and C_6 to C_9 aromatics. The most abundant HC coming from pyrolitic processes are methane, ethane, ethylene, acetylene, propene, propyne, butenes and butadiene. These hydrocarbons represent about one third of the total HC emissions (about 0.15-0.2 % C) from an engine operated at 2300 rev/min and at half load [3].

The catalytic post-treatment of automotive exhaust gases aims at eliminating CO, NO_x (NO plus NO_2) and HC. Precious metals are recognised as the best catalysts for the simultaneous elimination of these pollutants [4]. Their performances are determined on engine benches and vehicles but preliminary screening tests are performed at the laboratory scale. These tests aim at determining the half conversion temperatures for CO, NO and HC in mixtures containing O_2, CO, NO, one or two hydrocarbons, CO_2, H_2O, N_2, and eventually H_2 and/or SO_2. The most commonly used HC is propane although it is nearly absent from real exhaust gas. It is eventually replaced or mixed with propene [5]. Propane and propene are easy to handle and are supposed to correctly model the behaviour of the different classes of hydrocarbons that are present in exhaust gas.

Nevertheless, the light-off temperatures determined on laboratory tests are inferior by about 100°C to those determined on engine bench tests or on vehicle under similar conditions [6]. This difference could only be the consequence of major phenomena that have not yet been identified. It was thought that, under real conditions, there could be mixture effects due to interactions between the catalyst active sites and hydrocarbons belonging to different families, in the presence of CO, NO, O_2, CO_2, H_2O and SO_2. Thus, a study was undertaken to determine the influence of the nature of various HC on the oxidation reactions of CO and HC by O_2 and NO. Special emphasis was directed toward HC able to strongly coordinate to the catalyst surfaces.

2. EXPERIMENTAL

Catalyst preparation

The catalyst was prepared on a cordierite monolith having 62 cells per cm^2. The support was coated with a promoted alumina-ceria (6 % CeO_2) wash-coat and impregnated with 1.06 g/l of platinum plus rhodium, with a Pt/Rh mass ratio of 5. After impregnation, the catalyst was calcined for 2 hours at 500°C.

Activity measurement

The catalytic performances were determined for cylindrical samples (diameter 30 mm, height 76 mm) on a laboratory test working at stoichiometry and under programmed temperature increase at 5°C/min between 120°C and 450°C unless otherwise specified. Gas flowmeters and a syringe pump for liquid hydrocarbons were used to prepare mixtures generally containing (vol % in N2): CO 0.9, NO 0.2, HC 0.15 (as CH_4), CO_2 10, H_2O 7, 0.62 O_2. Oxygen flow was adjusted to maintain lambda to 1 with control by a lambda sensor. Total gas flow was 2650 l/h (GHSV 50 000 h^{-1}). On-line analyzers allowed conversions of CO (IR detection), HC (FID) and NO (chemiluminescence detection) to be followed as a function of temperature. Before testing, the catalyst sample was activated under a full component test mixture (HC being a mixture of CH_4, C_2H_2 and C_2H_4) at 600°C for 2 hours. Between two tests, the catalyst was cooled with a full component mixture in the absence of HC to prevent HC adsorption.

The catalyst stability was periodically checked with a full mixture containing propane as an HC.

The light-off temperature of a pollutant has been defined as the reactor inlet temperature at which 50% conversion of the pollutant occurs.

3. RESULTS AND DISCUSSION

3.1 Oxidation of CO by NO + O_2 in the absence of hydrocarbon

In the absence of hydrocarbon, but with O_2 and NO concentrations equivalent to that used at stoichiometry with propane, the CO light-off temperature is 155°C. At 170°C, CO is totally converted while NO conversion only reaches 35 %.

CO is known to strongly adsorb on metallic surfaces [7]. It prevents O_2 adsorption at low temperature. Partial thermal desorption of CO allows the adsorption of O_2 and the oxidation of remaining adsorbed CO [7,8].

3.2 Oxidation of CO and HC by NO + O_2

Influence of the nature of the hydrocarbon (Fig. 1 a,b)

Each test was performed with a mixture containing all reactants including 0.15 % C of one of the selected hydrocarbons found in real exhaust gases: n-, iso- and cyclo-alkanes , n- and iso-olefins, diolefins, alkynes and aromatics.

Alkanes

The light-off temperature for n-alkanes decreases very rapidly as the chain length increases from C_1 to C_6 (450 to 226°C). The branched alkanes 2-methyl propane and 2,2,4-trimethylpentane (isooctane) and cyclohexane have similar light-off temperatures to n-butane. The oxidation of CO occurs at about the same temperature in the presence or the absence of alkanes.

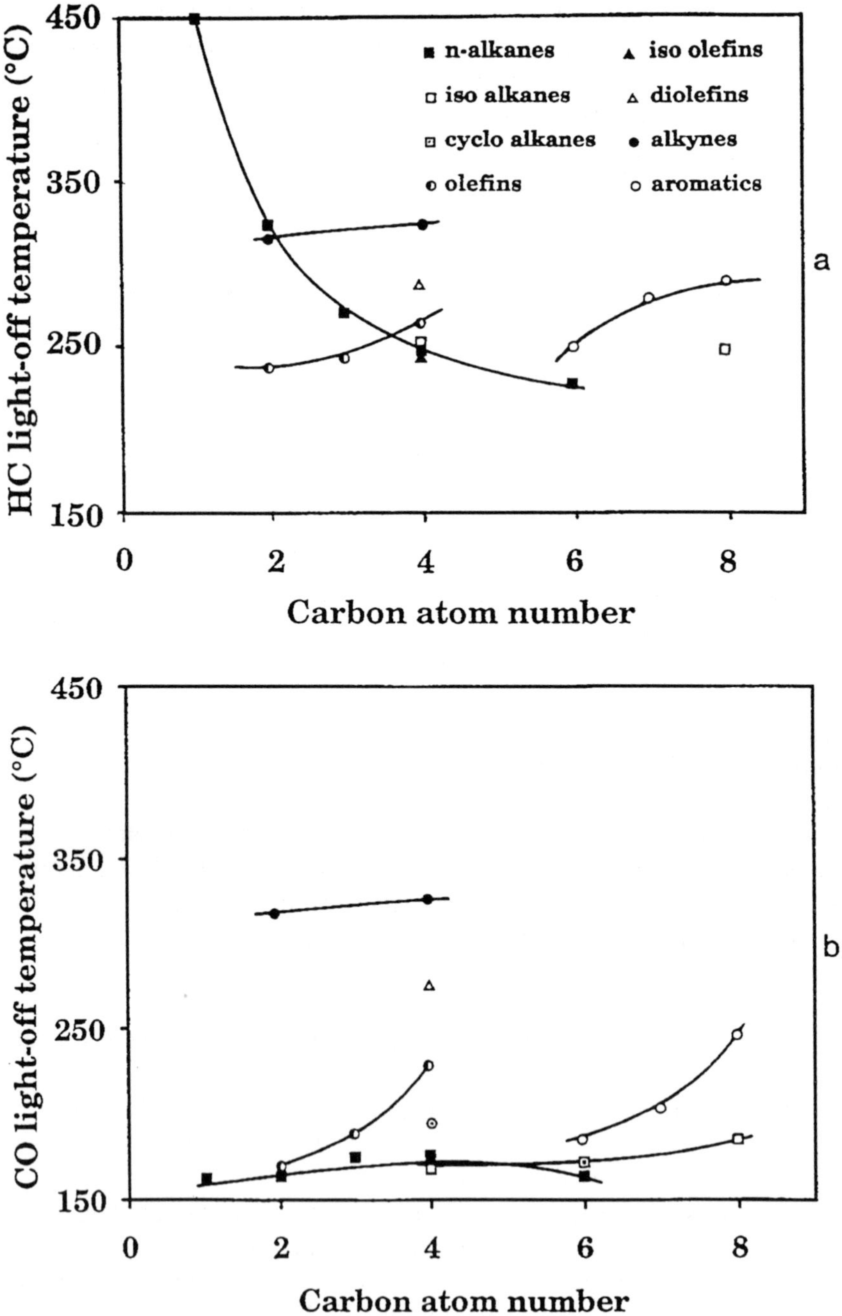

Fig.1 Effect of the HC carbon number on the HC (a) and CO (b) light-off temperature

Oxidation of alkanes requires the abstraction of a hydrogen atom. This is favored by longer distances between C-H bond to be broken and methyl groups [9]. This explains why long chain alkanes are more easily oxidised, and why alkanes containing tert-butyl groups, such as 2,2,4-trimethylpentane, are more difficult to oxidise than n-alkanes of same carbon number. Although not evaluated, n-octane should be oxidised below 226°C, whereas 2,2,4-trimethylpentane is oxidised at 252°C.

Alkanes adsorb very poorly on metallic surfaces. Thus they do not modify the adsorption of CO, O_2 and NO. As would be expected CO oxidation is practically insensitive to the presence of alkanes.

Olefins

The light-off temperature for 1-olefins and for CO increases with increasing chain length in the series ethylene to 1-butene. The oxidation of 2-methyl propene begins slightly before that of 1-butene.

The electron donor behaviour of olefins is moderate but increases with chain length due to the inductive effect of the alkyl group linked to the double bond. Olefins can compete with the other reactants, which explains the inhibiting effect for CO oxidation and the negative partial order for their oxidation [10].

Diolefins

1,3-butadiene has a rather high light-off temperature: 287°C. In the presence of butadiene the CO light-off temperature is only 9°C below that of the hydrocarbon.

Diolefins are strong electron donor molecules and can adsorb very easily on metallic surfaces thereby preventing the adsorption of other reactants as already evidenced in the selective hydrogenation of diolefins [11]. Thus, they are difficult to oxidise and they strongly inhibit CO oxidation.

Alkynes

Acetylene, propyne and traces of butyne are found in exhaust gases (3).

Acetylene is so difficult to oxidise that its light-off temperature is nearly the same as that of ethane (315 against 323°C). In the presence of acetylene, CO oxidation occurs slightly after that of the hydrocarbon. These effects are even more pronounced in the presence of butyne.

Alkynes have an even stronger electron donor behaviour than diolefins [11]. The acetylene sticking coefficient on palladium is close to 1, even on an oxygen precovered surface [12]. Acetylene prevents the adsorption of other molecules and displaces pre-adsorbed molecules [13].

At low temperature, in the presence of alkynes, the number of free sites must be extremely low. The number of free sites increases with temperature so that oxidizing species can adsorb and allows alkyne oxidation above 250°C. Above 300°C, when alkyne oxidation is rapid, the number of free sites becomes sufficiently high to enable CO oxidation.

Aromatics

C_6-C_8 aromatics show a similar behaviour to C_2-C_4 olefins. Benzene is easily oxidised and has a limited effect on CO oxidation, while toluene and m-xylene are more difficult to oxidise and more strongly affect CO oxidation. Inhibition of CO oxidation by m-xylene is intermediate between that of 1-butene and 1,3-butadiene.

The inductive effect of methyl groups on aromatic rings enhances the electron donor behaviour of the molecule [14] and strengthens their adsorption on metallic sites. This leads to inhibition of the oxidation of the aromatic molecule and of CO oxidation.

Comparison of hydrocarbons of increasing degree of unsaturation

The HC, CO and NO conversions as a function of temperature has been plotted for C_4 hydrocarbons with increasing degree of unsaturation, Fig. 2.

In this series, the HC oxidation and CO oxidation become more and more difficult with increasing unsaturation degree of the HC. The inhibiting effect of butyne is far more important than that of 1,3-butadiene. With butadiene, CO conversion starts at about 150°C, and reaches 50 % at 278°C, whereas with butyne it only starts very slowly at 250°C, and increases abruptly above 310°C.

In the presence of butane, the NO reduction profile can be decomposed into two curves corresponding to the oxidation respectively of CO and of butane. With butene, butadiene and butyne it is not possible to separate the contribution of CO and HC in the reduction of NO.

Proposed scheme for interpreting inhibition effects

A tentative classification of the effect of the nature of the hydrocarbons on the catalytic activity for their oxidation and for CO oxidation is proposed in figure 3. As catalytic activity decreases when the light-off temperature increases, the catalytic activity has been expressed by the light-off temperature in a decreasing scale. It is plotted as a function of the hydrocarbon adsorption strength (arbitrary units).

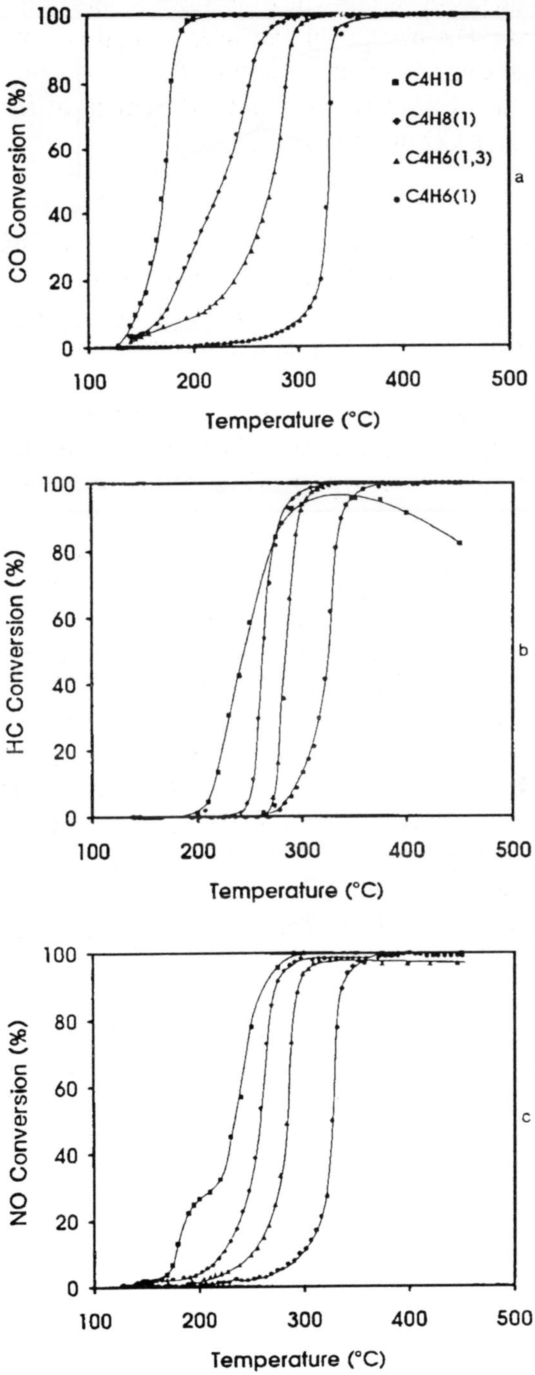

Fig.2 effect of the HC unsaturation degree on the CO (a), HC (b) and NO (c) conversion as a function of temperature

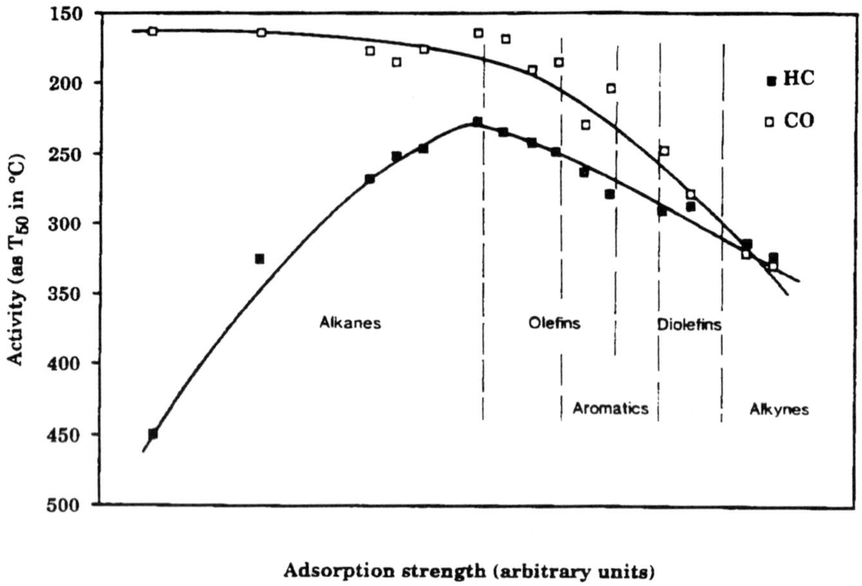

Figure 3. Activity (as decreasing light-off temperature) for HC and CO oxidation as a function of HC adsorption strength

In the absence of hydrocarbon or in the presence of alkanes, CO oxidation activity is high. At low temperatures, CO coverage on metallic sites is probably very high [7]. It decreases when temperature increases, which allows the adsorption of oxidizing species and the subsequent CO oxidation. Alkanes are poorly adsorbed on metallic sites. They can not compete with CO and do not modify its oxidation rate. They should adsorb only when CO is eliminated. Even under these conditions they have to compete with oxygen adsorption, in the case of platinum catalysts [15]. Their adsorption is favored by an increase of their chain length.

In the presence of hydrocarbons of increasing adsorption strength (increasing unsaturation degree and presence of alkyl groups on double bonds or aromatic nucleus), CO oxidation activity decreases so as HC oxidation activity. The HC coverage on the metallic sites should increase with their adsorption strength, at the expense of the coverage of species that have lower adsorption strength, particularly oxygen.

As a result of the competitive adsorption between HC and oxidizing species, the HC oxidation activity versus adsorption strength is a typical volcano curve. It shows a maximum with long chain alkanes and short chain olefins, and minima with short chain alkanes and alkynes. The decrease of the CO oxidation activity curve should result from the competitive adsorption between unsaturated hydrocarbons, CO and oxidizing species.

3.3 Insight on the inhibition by acetylene

Influence of the C_2H_2 concentration

As previously described, the tests with acetylene were performed with 1500 ppmC. This is much higher than exhaust gas concentrations which are of the order of 100 to 150 ppmC. The effect of C_2H_2 concentration was studied over a large range from 5 to 1500 ppmC C_2H_2. Figure 4 shows that the CO, NO and HC light-off temperatures increase linearly with the logarithm of the C_2H_2 concentration. In all the studied range, the C_2H_2 light-off occurs before that of CO or NO. The HC light-off temperature is about 260°C in the presence of 100 ppmC C_2H_2.

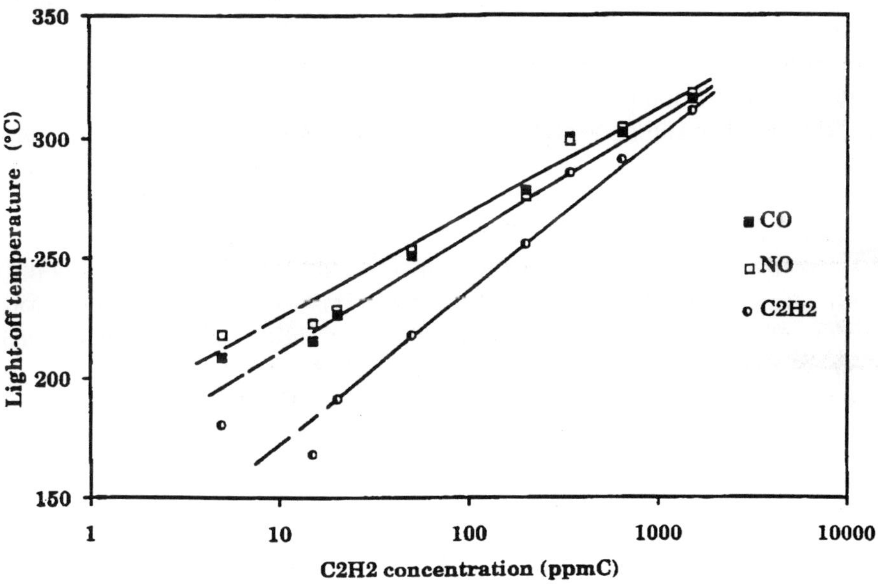

Figure 4. Effect of C_2H_2 concentration on the HC, CO and NO light-off temperature

Toxicity

The toxicity of a catalyst poison is defined as the number of active sites which are blocked by one poison molecule. Toxicity of acetylene was determined using the Maxted method described by Barbier [16]. A mixture containing all reactants except acetylene was passed over the catalyst at 152°C, so that CO conversion was 21 %. A rapid step change in acetylene concentration from 0 to 7 ppmC was done. From the slope of the conversion decrease versus time, the molar flow of acetylene, and the metal dispersion of the catalyst (40 % from CO chemisorption) C_2H_2 toxicity was determined to be 3.6. This clearly confirms that C_2H_2 is a very strong poison, since toxicity values of 3 are only reported for a few sulfur and lead compounds [16].

Influence of acetylene on the oxidation of a hydrocarbon mixture

Acetylene can retard CO oxidation and is so strongly adsorbed that it inhibits its own oxidation. A remaining question is the effect of acetylene on the oxidation of other hydrocarbons. To examine this effect 1500 ppm C of a mixture containing CH_4, C_2H_2, C_2H_4 and C_3H_8 (respectively 94, 100, 586 and 720 ppmC) were used instead of pure C_2H_2. Figure 5 shows that in contrast to what was obtained with ethylene alone (T50 = 234°C), the oxidation of the hydrocarbon mixture does not begin before 250°C. This is nearly the same temperature as that obtained in the presence of 100 ppmC C_2H_2.

This confirms that acetylene inhibits all the oxidation reactions on the catalyst surface.

Influence of the addition of SO_2

Sulfur dioxide is present at very low concentrations (about 5-30 ppmv) in exhaust gases. It has been shown that it poisons post-combustion catalysts for the oxidation of olefins, but promotes the oxidation of short chained alkanes such as propane [5,17].

A small amount of SO_2 (15 ppmv) was introduced into the reaction mixtures containing various amounts of acetylene. Figure 6 shows that the addition of SO_2 causes a small increase (10-20°C) in the CO light-off temperature when low C_2H_2 concentrations were used. However, a surprisingly large increase was observed when using 1500 ppm C_2H_2. The resulting light-off temperature was over 350°C. This increase is caused by SO_2 poisoning effect on the oxidation of C_2H_2.

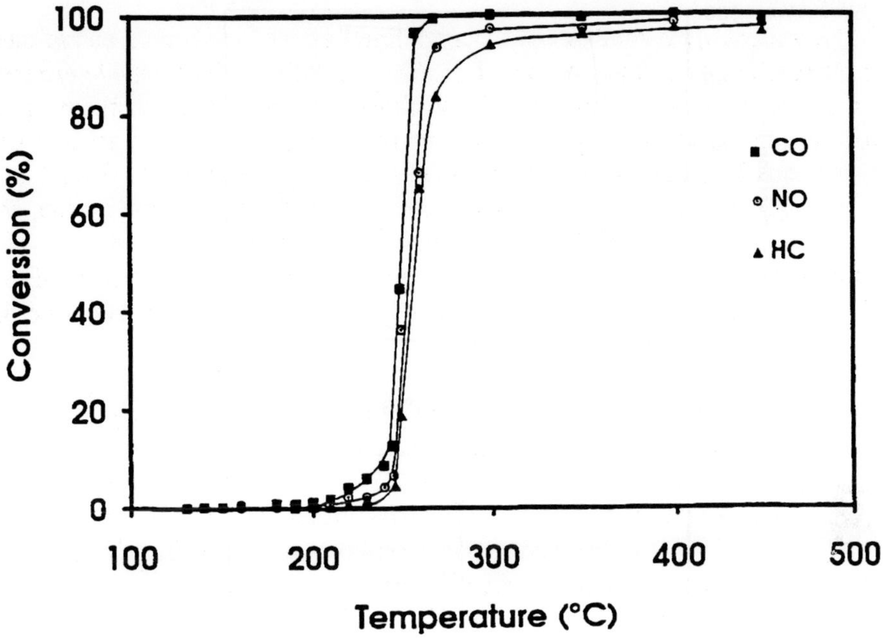

Figure 5. HC, CO and NO conversion as a function of temperature when the HC mixture contains CH_4, C_2H_2, C_2H_4 and C_3H_8

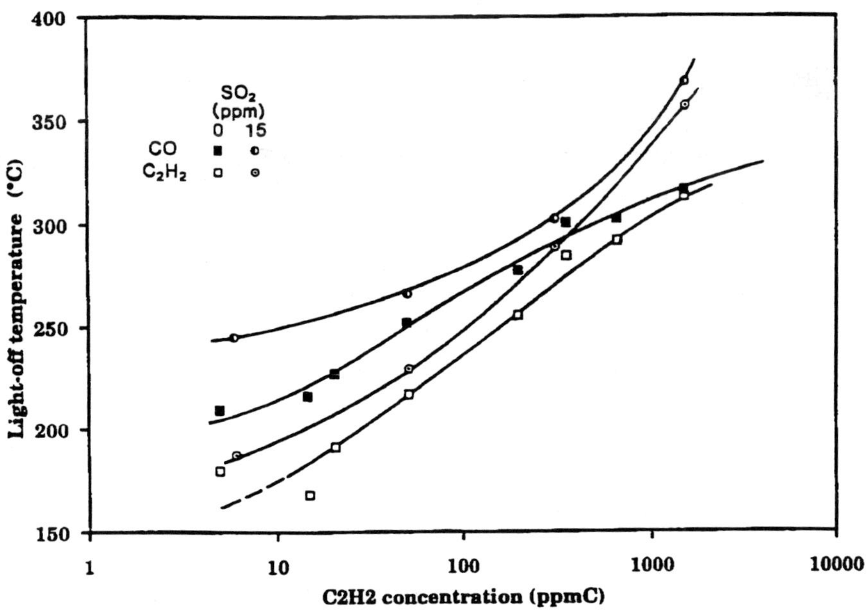

Figure 6. Effect of SO_2 on C_2H_2 and CO oxidation

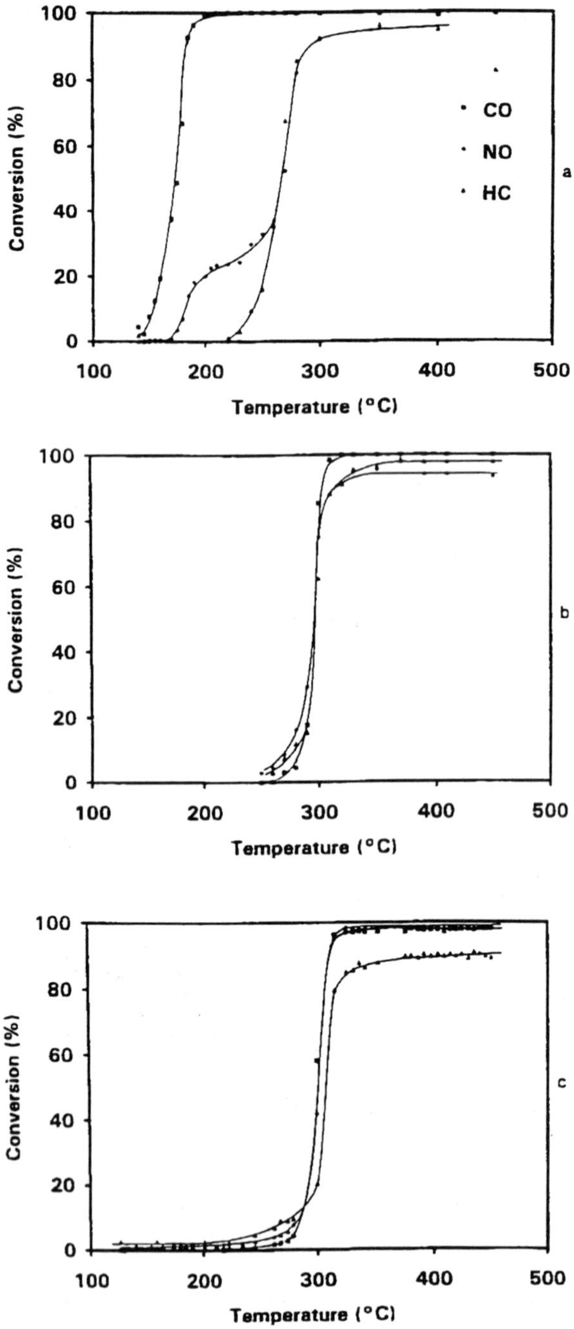

FIG.7 HC, CO and NO conversion as a function of temperature. a) HC = propane; b) real exhaust gas; c) HC = simulated exhaust gas

3.4 Consequences of the predictivity of laboratory tests

As indicated previously, propane is frequently used to model the behaviour of the hydrocarbon mixture of exhaust gas. With this hydrocarbon, the oxidation of CO occurs at about the same temperature (176°C, Fig. 7-a) as without hydrocarbon. Propane is oxidised at 268°C, and NO conversion curve shows two waves corresponding to CO and C_3H_8 oxidation. Using propene instead of propane would slightly increase CO light-off (190°C) and slightly decrease HC light-off (241°C).

On an engine bench test performed with a catalyst of the same composition, under the same space velocity, the light-off temperatures for CO, NO and HC are identical and very high: 305°C (Fig. 7-b). This confirms that neither propane nor propene is adequate to model the properties of the real hydrocarbon mixture.

Then, the catalyst was tested with a hydrocarbon mixture containing CH_4, C_2H_2, C_2H_4, C_6H_8, C_7H_8, C_8H_{10} and 15 ppmv SO_2 (Fig. 7-c). In that case, as was observed on the engine bench test, all pollutants are converted at about 300°C. A chromatographic analysis confirmed that acetylene is the first hydrocarbon to be converted. The oxidation of the other hydrocarbons, except methane, begins out after the near total elimination of C_2H_2.

4. Conclusion

Exhaust gases from spark ignited engines contain alkynes, essentially acetylene. Alkynes strongly coordinate on precious metal catalysts and inhibit all post-combustion reactions. This occurs even at the low alkyne concentrations characteristic of real exhaust gases. Three major consequences have been evidenced:

- the oxidation, by O_2 and NO, of non alkyne hydrocarbons and of CO, is inhibited as long as alkynes are not converted,
- around 300°C, the oxidation of alkynes causes a rapid light-off for all pollutants except methane,
- as methane oxidation occurs above the light-off temperature of alkynes it is practically not affected.

These results have been used to define laboratory testing conditions enabling the correct simulation of engine bench results. In addition they open the way to new kinetic and surface studies dedicated to a better description of active site inhibition by alkynes.

Acknowledgements

The authors acknowledge G. Duchesne and Ph. Villeret for their technical support.

References

1 J. B. Heywood, Internal combustion engine fundamentals, Mc Graw Hill Book Co, New York, 1988
2 P. Degobert, Automobile et Pollution, Technip, Paris, 1992
3 N. Pelz, N.M. Dempster, G.E. Hundleby and P.R. Shore, SAE paper n°902074
4 K. Taylor, Catalysis Science and Technology, vol 5, p119, Springer Verlag, Berlin, 1984
5 D.R. Monroe, M.H. Krueger, D.D. Beck and M. J. D'Aniello Jr, Stud. Surf. Sci. Catal., vol 71, p 593, Elsevier, Amsterdam, 1991
6 M. Prigent, G. Mabilon, R. Dozière and D. Durand, Exposé SIA 89077, p 277
7 T. Engel and G. Ertl, Adv. Catal. vol 28, 1979, 1
8 R.G. Compton, Kinetic Models of Catalytic Reactions, Comprehensive Chemical Kinetics, vol 32, p 311, Elsevier, 1991,
9 G.I. Golodets, Heterogeneous Catalytic Reactions Involving Molecular Oxygen, Stud. Surf. Sci. Catal., vol 15, Elsevier, Amsterdam, 1983
10 Y. F. Yu Yao, J. Catal., 87 (1984) 152
11 J. P. Boitiaux, J. Cosyns and S. Vasudevan, Appl. Catal., 6 (1983) 41
12 H. Dannetun, I. Lundström and L.-G. Petersson, Surf. Sci., 193 (1988) 109
13 L. Cider and N.-H. Schöön, Appl. Catal., 68 (1991) 191
14 J. Massardier, J. C. Bertolini, T. M. Tri, P. Gallezot and B. Imelik, Bull. Soc. Chim. Fr., 3 (1985) 333
15 Y.F. Yu Yao, Ind. Eng. Chem. Prod. Res. Dev. 19(1980)293
16 J. Barbier, Catalyse par les métaux, Editions du CNRS, Paris, 1984, p 305
17 H. C. Yao, H. K. Stepien and H. S. Gandhi, J. Catal., 67 (1981) 231

A. Frennet and J.-M. Bastin (Eds.)
Catalysis and Automotive Pollution Control III
Studies in Surface Science and Catalysis, Vol. 96
1995 Elsevier Science B.V.

METAL SURFACE AREA MEASUREMENT OF FRESH AND AGED AUTOMOTIVE CATALYSTS BY CO METHANATION

R. K. Usmen, R. W. McCabe, And M. Shelef

Ford Research Laboratory
Ford Motor Co.
Dearborn, Michigan, U.S.A.

ABSTRACT

The CO chemisorption method of Komai et al. [J. Catal. 120 (1989) 370], based on methanation of adsorbed CO and chromatographic quantification of the product methane with a flame ionization detector, was used to determine noble metal particle sizes of two 80,000 km vehicle-aged automotive catalysts, a Pt/Rh formulation and a Pd/Rh formulation. The aged Pt/Rh catalyst gave a spherical particle diameter of 72±6 Å while the aged Pd/Rh catalyst gave a spherical particle diameter of 366±44 Å. These values compare closely with particle sizes of 110±20 and 470± 120 Å obtained for the Pt/Rh and Pd/Rh catalysts, respectively, from line-broadening in x-ray diffraction. In contrast, conventional CO chemisorption could not be successfully applied to either of the commercial catalysts because of interference from support components. Overall, the results of this study indicate that the chemisorption technique based on CO methanation is particularly useful for aged automotive catalysts; it is fast, highly sensitive, and free of complications associated with adsorption on support components.

1. INTRODUCTION

Determination of the metal surface area, or the dispersion, is a standard characterization practice for supported metal catalysts [1-4]. A variety of methods has been employed to measure metal surface areas, each with advantages and disadvantages. Conventional x-ray powder diffraction (XRD) gives a measure of the size of supportedcrystallites, but with a number of inherent and instrumental limitations [5] which effectivelylimit the detectable particle size to about 40 Å with ordinary x-ray sources. High-resolution transmission electron microscopy (HRTEM) is, in principle, the most definitivetechnique for obtaining a complete particle size distribution. However, a number of practical limitations prevent the routine use of HRTEM for particle size analysis. These include the problems of clearly imaging small particles on high contrast supports such as alumina, the tendency of all but the highest resolution instruments to undercount the smallest particles, and the large number of images needed to obtain a representative particle histogram for a given sample. The selective chemisorption methods, such as static volumetric adsorption [6] and pulse techniques [7], have been used extensively, and, unlike XRD and HRTEM, do not have an inherent bias toward larger particles. However, they have their own shortcomings, especially uncertainty over adsorption stoichiometries (i.e. adsorbate to metal atom ratios), adsorption of the probe molecule on sites other than the metal particles (a particular concern in commercial automotive catalysts which often contain 4 or 5 metal oxide components in addition to the noble metals), and complications associated with specific adsorbate/substrate interactions (such as hydrogen dissolution and hydride formation in H2 chemisorption on Pd- containing catalysts). In addition, the chemisorption techniques are often difficult to employ with low loaded catalysts or catalysts with low dispersion. In those cases, either large catalyst volumes, small gas volumes, or low gas pressures are required to ensure measurable pressure changes during chemisorption.

The automotive catalysts employed in this study contain only 0.2 wt% (Pt/Rh 5:1 ratio catalyst) to 0.45 wt% (Pd/Rh 9:1 ratio catalyst) total noble metal. After vehicle aging, metal dispersions of 10% or lower are common [8-10]. The low initial loadings together with the high degree of thermal sintering experienced by many automotive catalysts produce extremely low concentrations of exposed surface atoms after vehicle aging. Komai et. al. [4] have developed a method for the determination of the dispersion of supported catalysts with very low metal surface areas, and we report here the first application of this technique to automotive catalysts. The method gives the amount of adsorbed CO by detecting CH_4 formed during the titration of preadsorbed CO with H_2 [4]. The method has high accuracy and high sensitivity for three basic reasons: 1) unlike conventional chemisorption techniques it is based on the detection of a new species (methane) not present prior to the chemisorption measurement, 2) the methane is quantified by chromatographic separation and flame ionization detection, thus benefiting from the high sensitivity of those techniques, and 3) for automotive catalysts, the method appears to be less subject to complications due to CO chemisorption on sites other than noble metal sites than is the conventional volumetric chemisorption method.

The CO-methanation technique, like conventional chemisorption techniques, is subject to uncertainty over CO/metal-atom stoichiometries. However, for our objective of characterizing aging effects in large numbers of automotive catalysts, we are interested in measuring relative changes in dispersion/surface area rather than obtaining an absolute dispersion/surface area measurement for any given catalyst. Thus, the choice of a stoichiometric factor is not critical.

In the present study, the method developed by Komai et. al. has been modified slightly and used to characterize the noble metal particle sizes of commercial Pt/Rh and Pd/Rh three-way automotive catalysts (TWC). Good agreement was obtained with x-ray diffraction particle size analysis of the same catalysts, indicating general utility of the CO methanation technique for rapidly determining the particle sizes of both fresh and aged automotive catalysts.

2. EXPERIMENTAL

2.1 Catalysts:

Details of the catalyst histories, as well as results of other characterization studies conducted on these particular catalysts have been reported elsewhere [8]. The automotive catalysts used in this study are production three-way catalysts consisting of a cordierite honeycomb substrate covered with an alumina washcoat containing various stabilizers and promoters as well as the active noble metals. Ceria is the principal promoter employed in these catalysts, typically at a concentration 15-25 times greater than the noble metal concentration. Both a Pt/Rh catalyst formulation and a Pd/Rh formulation were examined, and both were analyzed in their fresh state and after vehicle aging to 80,450 km (50,000 miles). For the Pt/Rh vehicle aged catalysts, samples were taken from the outlet section of the catalyst brick located closest to the engine. The outlet section was chosen to ensure that concentrations of contaminant species such as phosphorus and zinc (from engine oil additives) and lead (from residual tetraethyl lead contamination of unleaded fuel) were negligibly low (thus allowing us to focus on thermal deactivation of the catalysts).

Both of the vehicle-aged catalysts were driven a total of 80,450 km on the AMA 223 durability driving cycle. Note, however, that the vehicles were different makes - the Pt/Rh catalyst came from a 1990 5.0L Crown Victoria, while the Pd/Rh catalyst came from a 1989 3.8L Thunderbird. Despite the same durability schedule, the thermal histories of the two catalysts may have been quite different due to differences in engine load factors, catalyst mounting locations, air-fuel control strategy, etc. Consequently, no conclusions can be drawn regarding the relative sintering rates of Pt/Rh and Pd/Rh TWCs.

Blank runs were also carried out on fully washcoated cordierite containing various stabilizers and promoters (without any noble metals) to determine possible contributions due to CO adsorption and subsequent methanation on sites other than the noble metals. As a result of these runs, it was confirmed that only negligible amounts of CH4 were formed on the support.

In addition to the automotive catalysts, chemisorption measurements were also carried out on a sample of EuroPt-1 (6.3%Pt/silica) catalyst. The measurements on the EuroPt-1 catalyst were undertaken to ensure that our CO methanation technique yielded values consistent with particle size measurements obtained by other groups using different techniques.

2.2 Adsorption Measurements:

Figure 1 shows a schematic diagram of the experimental apparatus. A small amount of powdered catalyst (typically 60 to 120 mg, depending on the catalyst loading and dispersion) is packed in a quartz glass u-tube (6 mm o.d.; 5.7 cm^3 volume) which is heated using a single-zone 5.5 cm diameter furnace equipped with a programmable temperature controller (Omega, model 2010). No special catalyst preparation is required. Monolith samples are simply crushed and loaded in the quartz u-tube reactor.

The standard procedure involves: 1) oxidizing the sample in a flow of dry O_2 (40 cm^3/min) for 30 min at 400°C, 2) purging in flowing He for a few minutes, 3) reducing in a flow of H2 (40 cm^3/min) for 30 min at 400°C, and 4) cooling to room temperature in flowing H_2. The sample is then exposed to 2 pulses of CO (2 cm^3 each) through H_2 carrier gas. After purging the excess CO with flowing H_2 for 10 min, the reactor is sealed with three-way ball valves to trap the adsorbed CO along with the gaseous H_2. The catalyst is then heated to 400°C to hydrogenate the preadsorbed CO. After 30 minutes of reaction time, the entire contents of the reactor are flushed directly into a Varian 3400 GC and the concentration of CH_4 is analyzed with a flame ionization detector. No higher hydrocarbons or oxygenated hydrocarbons were detected. The amount of methane was quantified by comparing the integrated area of the methane in the reactor gas to the integrated areas of methane samples admitted from a calibrated volume sample loop.

Special precautions were taken to avoid sample contamination. High-purity CO(>99.99%) was passed through a carbonyl trap. The helium feed to the reactor was purified by passing it through an oxygen trap and also through a silica-gel trap at liquid nitrogen temperature. H_2 (>99.999%) was further purified with a hydrocarbon filter.

The apparatus shown in Fig. 1 and procedure listed above represent only a slight departure from the method first described by Komai et al. [4]. The main differences are:

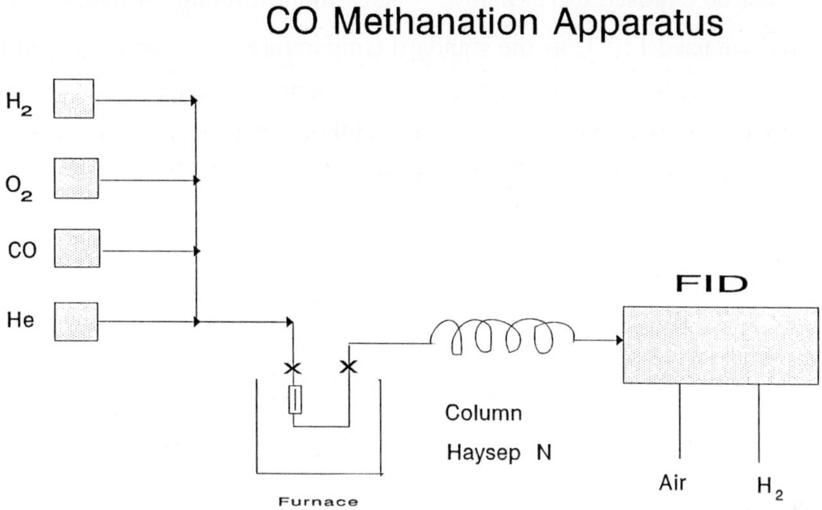

Figure 1. Schematic diagram of the apparatus.

1) the samples are oxidized in a flow of O_2 for 30 minutes instead of 1 hour at 400°C,

2) the samples are reduced in a flow of H_2 for 30 minutes instead of 1 hour at 400°C, and

3) 2 pulses of CO (2 cm^3 each) are used instead of a single large dose. Whereas Komai et al. used the technique only for Pt on alumina catalysts, we have extended it to Pd and Rh containing catalysts. Preliminary experiments, carried out with single-component Pd/Al_2O_3 and Rh/ Al_2O_3 confirmed that reaction under the conditions outlined above is sufficient to convert all of the adsorbed CO to CH_4. Moreover, we have found that Pd- containing catalysts require a modified procedure from that used for the Pt based catalysts. In other studies carried out in parallel with this study, we have found that even relatively mild reducing conditions (ca. 300°C or higher) cause significant sintering of Pd particles, particularly in fresh catalysts with relatively high initial dispersions. Others have reported similar observations [11]. In addition, Pd-containing

catalysts cannot be exposed to H_2 below 70°C without forming Pd hydride [12]. Consequently, we used 150°C as the standard temperature for reducing the Pd/Rh catalyst. Thus, after 400°C oxidation for 30 min, the palladium catalysts were cooled to 150°C in flowing O_2 (40 cm^3/min) and then reduced at 150°C in H_2 for 30 min followed by a He purge at 150°C for 30 min.

2.3 X-ray diffraction (XRD):

XRD patterns were recorded on a Philips X-ray generator with a Debye-Scherrer camera. CuKα X-rays (λ = 1.5418 Å) were used as the X-ray source. XRD measurements were performed on powdered washcoat scraped from the corner region of exposed channels on the outer surface of the cores. The XRD apparatus used to analyze the samples have been described in detail previously [8]. Particle sizes were estimated using the Scherrer equation with correction for instrumental line broadening.

3. Results

Table 1 compares our results obtained for the EuroPt-1 catalyst using the CO methanation technique with results reported by other groups [13-17] using x-ray diffraction, conventional chemisorption, and TEM. A dispersion of 45% was calculated using the spherical particle approximation and an assumed adsorption stoichiometry of 1 CO molecule per exposed Pt atom. If the adsorption stoichiometry is instead assumed to be 0.7, in keeping with previous studies of CO chemisorption out on Pt catalysts [17,18], then the dispersion becomes 65% in close agreement with the studies shown in Table 1 employing other techniques.

The results obtained for the fresh and aged commercial Pt/Rh and Pd/Rh TWCs are shown in Table 2. The first column contains the dispersions and calculated spherical particle sizes (in parentheses) derived from the CO methanation technique based on an assumed adsorption stoichiometry of 1 CO per exposed noble metal atom. The arbitrary choice of a stoichiometric factor of 1, rather than the value of 0.7 suggested by the EuroPt-1 catalyst, was made on the basis of several factors. The main reason is that the presence of Rh in these catalysts (16% and 10% of the noble metal weight in the Pt/Rh and Pd/Rh catalysts, respectively) is likely to increase the average stoichiometric factor above 0.7 due to the presence of gem-dicarbonyl species on Rh. Bimetallic Pt/Rh particles have been found in automotive catalysts, sometimes with surface enrichment by Rh [20,21] or even bulk enrichment of selected particles as

evidenced by TEM [9]. Thus the Rh in automotive TWCs may affect the adsorption stoichiometry much more strongly than the overall loading ratio would suggest. With the assumed stoichiometric factor of 1, the dispersion of the Pt/Rh TWC went from about 57% in the fresh state to about 13% after vehicle aging, corresponding to a growth in particle size from 16±1 Å to 72±6 Å.
The Pd/Rh catalyst started out at about 42% dispersion and decreased to 2.4% dispersion after vehicle aging. This corresponds to a growth in particle size from 21±2 Å to 366±46 Å assuming spherical particles.

Table 1.: Chemisorption, XRD, and TEM results of EuroPt-1.

Method	Dispersion (%)	Ref.
XRD	65	[13]
TEM	56-66	[14]
O_2 adsorption	50-62	[15]
H_2 adsorption	65	[16]
CO-H_2 titration	45 (65)[a]	This study

a Assuming CO:Pt=0.7

Column 2 of Table 2 shows the dispersion values which we measured previously using conventional volumetric adsorption techniques. For the Pt/Rh catalyst, where it was necessary to use H_2 as the probe molecule due to a strong and irreversible interaction of CO with one of the washcoat components (not present in the Pd/Rh catalyst), the volumetric chemisorption data agree fairly well with the CO-H_2 methanation technique. However, for the Pd/Rh catalyst, the volumetric chemisorption method employing CO gave a nonsensical dispersion of 174% for the fresh catalyst and an implausible dispersion of 96% for the vehicle-aged catalyst.

Column 3 of Table 2 lists the apparent particle sizes obtained from x-ray diffraction. For both the Pt/Rh and Pd/Rh catalysts, particles were too small to detect with our x-ray system. The vehicle aged Pt/Rh catalyst gave a particle size of 110±20 Å, in good agreement with both the CO-H_2 methanation and volumetric chemisorption results, particularly considering that only the particles greater than about 40 Å contribute to the x-ray data. Similarly, the XRD particle size of the aged Pd/Rh catalyst (470±120 Å) is consistent with the particle size of 366±44 Å obtained with the $CO-H_2$ titration technique.

Table 2.: Chemisorption and XRD results of commercial TWC catalysts.

Sample	Dispersion, % (Particle size, Å)[a] $CO-H_2$ Methanation[b]	Volumetric chemisorption[c]	XRD (Å)
Pt-Rh fresh	56.6 (16±1)	73 (12)	n.d.[d]
Pt-Rh vehicle aged	12.7 (72±6)	8 (114)	110±20
Pd-Rh fresh	41.9 (21±2)	174 (--)	n.d.[d]
Pd-Rh vehicle aged	2.4 (366±44)	96 (--)	470±120

a Particle size estimated assuming spherical particles.
b Assuming 1 CO per noble metal surface atom.
c From Ref.[8], assuming 1 CO or 1 H per noble metal surface atom.
d Not detected; below detectable limit (< 40 þ).

Note: Error band in $CO-H_2$ methanation particle sizes based on range of values obtained in repeat experiments. Typically 5-6 runs were made on each sample. Error bands in the XRD particle sizes result from particle size estimates made from different x-ray lines.

4. Discussion

Table 2 clearly shows that of the three techniques compared (CO methanation, volumetric chemisorption, and XRD) only the CO methanation technique yielded a complete set of realistic dispersion/particle-size data for all four of the fresh and aged commercial catalyst samples analyzed. Volumetric chemisorption was very difficult to perform on these samples because of interferences from the metal oxide components of the catalyst, while XRD was limited by its inability to detect small particles with our conventional source and detector. The CO methanation technique is relatively fast (less than two hours for a complete test) and requires only standard equipment found in most catalysis laboratories. (Even though we used a GC, the analysis can be carried out using a standard total hydrocarbon analyzer equipped with a flame ionization detector.) The high sensitivity of the technique is clearly evident in its ability to easily measure CO uptakes (i.e. CH4 yields) on the 0.46 wt% Pd/Rh catalyst which was only slightly more than 2% dispersed. In contrast, the 65% dispersed 6.3 wt% EuroPt-1 catalyst presented the greatest challenge with the CO methanation technique; the methane yield was so high that the catalyst charge had to be decreased to 3 mg to avoid overloading the GC column. Another indication of the sensitivity of the CO methanation technique is that even highly sintered catalysts only required about 100 mg of sample versus sample sizes near 10 g required in pulse CO chemisorption studies of automotive catalysts by Dalla Betta et al. [10].

Aside from the high sensitivity of the CO methanation technique, its most attractive feature for the analysis of complex automotive catalysts is the apparent absence of complications due to adsorption of CO on sites other than noble metal sites. One possible explanation raised by Bozon-Verduraz et al. [22] is that any CO which interacts strongly with basic washcoat components such as ceria, lanthana and baria to form stable compounds such as carbonates, does not get counted because it does not react with hydrogen to form methane. On the other hand, any CO which adsorbs weakly on the support will get counted, because it will desorb during heating, re-adsorb on the noble metal particles, and react with H2 to form methane. Apparently, the techniques employed in the CO-methanation experiment, namely quick dosing of CO from flowing pulses and purging of the reactor cell with flowing H_2 after CO dosing, prevent significant accumulation of weakly adsorbed CO on non-noble metal washcoat components. In this respect, the CO methanation technique closely resembles pulse adsorption methods, as noted in a follow-up study to the original work by Komai et al. employing the same set of Pt/Al_2O_3 catalysts [23]. IR spectroscopy was used in the follow-up study to confirm the absence of CO on the support following the H_2 purge step. In the early days of automotive catalysis, the oxidation catalysts had washcoats consisting essentially only of γ Al_2O_3. Thus the catalyst formulations were simple enough that the measurement of metal surface areas, although laborious [10], could be performed. With the advent of complex washcoats in TWCs, post mortem analyses of field-aged automotive catalysts have relied largely on BET surface area measurements as a probe of the extent of thermal damage. As noted in our previous study, however, BET measurements are useful only for identifying catalyst which have suffered gross thermal abuses, usually from engine misfires. In the case of the Pt/Rh and Pd/Rh catalysts examined here, BET surface areas were very close to fresh catalyst levels, indicating the absence of severe thermal damage. Nevertheless, it is clear from the CO methanation data that thermal damage far short of that required to cause significant phase change in the alumina support can lead to dramatic decreases in noble metal surface areas (and dispersions).

Although XRD and TEM have been employed for in-depth analysis of selected catalysts [9,21], very little chemisorption data have been published for field-aged automotive catalysts [10]. The chemisorption technique based on CO methanation opens the door to in-depth noble metal particle size characterization of large numbers of field aged catalysts. We are currently in the process of doing this on a series of catalysts with well documented thermal histories, with the eventual goal of establishing a quantitative link between noble metal particle size and catalytic activity. We are also continuing our characterization studies of the two commercial catalysts employed in the present study, using HRTEM to obtain a complete particle size distribution. Preliminary indications are that the particle size distribution is broad in field-aged catalysts; consequently the spherical particle assumption normally used to estimate particle sizes from chemisorption data is not particularly meaningful. Rather than estimate a particle size from chemisorption data and compare it with a particle size histogram from HRTEM, our preliminary results suggest that a more meaningful correlation can be obtained by measuring the total noble metal surface area in chemisorption and comparing it to the total noble metal surface area obtained by calculating and summing the surface areas of individual particles in the HRTEM particle size histogram.

5. SUMMARY

A chemisorption technique developed by Komai et al., based on CO methanation, was successfully used to analyze noble metal dispersions of both fresh and vehicle-aged Pt/Rh and Pd/Rh commercial automotive three-way catalysts. The technique is relatively rapid ($<$ 2 hours), extremely sensitive, and largely free from complications due to adsorption of CO on non-noble metal components of the washcoat (support, promoters, stabilizers, etc.). Particle sizes of the vehicle-aged catalysts, calculated by applying the spherical particle assumption to the dispersions measured by the CO methanation method, agreed well with particle sizes calculated from x-ray diffraction line-broadening data. These results indicate that the CO methanation technique can be applied routinely to obtain fast and accurate measurements of noble metal surface areas in automotive catalysts retrieved from the field, even those with metal dispersions ca. 2% or less.

ACKNOWLEDGMENT

We gratefully acknowledge Prof. A. Frennet for his help in obtaining the sample of EuroPt-1. The authors also acknowledge the assistance of the late C. R. Peters in providing the x-ray diffraction data.

REFERENCES

1 J.R. Anderson, "Structure of Metallic Catalysts," Academic Press, New York, NY, 1975.
2 T.R. Hughes, R.J. Houston, and R.P. Sieg, Ind. Eng. Chem., Proc. Design and Dev.,1 (1962) 336.
3 J.E. Benson and M. Boudart, J. Catal., 4 (1965) 704.
4 S. Komai, T. Hattori, and Y. Murakami, J. Catal. 120 (1989), 370.
5 J.B. Cohen, Ultramicroscopy, 34 (1990) 41.
6 L. Spenadel and M. Boudart, J. Phys. Chem., 64 (1952) 204.
7 J. Prasad, K.R. Murthy, and P.G. Menon, J. Catal., 52 (1978) 515.
8 R.K. Usmen, R.W. McCabe, G.W. Graham, W.H. Weber, C.R. Peters, and H.S. Gandhi, SAE paper 922336 (1992).
9 J.-O. Malm and J.-O. Bovin, Microscopy, Microanalysis, Microstructures, 1 (1990) 387.
10 R.D. Dalla Betta, R.C. McCune, and J.W. Sprys, Ind. Eng. Chem., Prod. Res. Dev.,15 (1976) 169.
11 J.J.F. Scholten and A. Van Montfoort, J. Catal., 1 (1962) 85.
12 P.C. Aben, J. Catal., 10 (1968) 224.
13 V. Gnutzmann and W. Vogel, J. Phys. Chem., 94 (1990) 4991.
14 J.W. Geus and P.B. Wells, Appl. Catal., 18 (1985) 231.
15 P.B. Wells, Appl. Catal., 18 (1985) 259.
16 A. Frennet and P.B. Wells, Appl. Catal., 18 (1985) 243.
17 K. Kunimori, T. Uchijima, M. Yamada, H. Matsumoto, T. Hattori, and Y. Murakami, Appl. Catal., 4 (1982) 67.
18 J.E. Benson, H.S. Hwang, and M. Boudart, J. Catal., 30 (1973) 146.
19 W.B. Williamson, H.S. Gandhi, P. Wynblatt, T.J. Truex, and R.C. Ku, AIChE Symposium Series, 76 (1980) 212.
20 F. Bozon-Verduraz, D. Tessier, and A. Rakai, J. Catal., 127 (1991) 457.
21 B.R. Powell and Y.L. Chen, Appl. Catal., 53 (1989) 233.
22 T. Hattori, S. Komai, E. Nagata, and Y. Murakami, J. Catal. 127 (1991), 460.

A. Frennet and J.-M. Bastin (Eds.)
Catalysis and Automotive Pollution Control III
Studies in Surface Science and Catalysis, Vol. 96
1995 Elsevier Science B.V.

EFFECTS OF SINTERING AND OF ADDITIVES ON THE OXYGEN STORAGE CAPACITY OF PtRh CATALYSTS.

D. Martin, R.Taha and D. Duprez

Laboratoire de Catalyse en Chimie Organique, URA 350 CNRS
40 Av. du Recteur Pineau, 86022 Poitiers Cedex, France

ABSTRACT :

PtRh/Al_2O_3 and PtRh/CeO_2-Al_2O_3 bimetallic catalysts (Pt+Rh ≈ 60 $\mu molg^{-1}$) were prepared via chlorine-free precursors. Oxygen storage capacities (OSC) were measured on the fresh (calc. 723K) and on the sintered catalysts (1%O_2 + 10%H_2O, 2h, 973K and 1173K).

On alumina catalysts, only Rh can promote OSC which is extremely sensitive to sintering. OSC values are higher on alumina-ceria catalysts, but do not depend on the composition of the bimetallics. Moreover ceria renders the catalysts resistant to sintering.

PtRh/Al_2O_3 and PtRh/CeO_2-Al_2O_3 were modified by Cl, SO_4^{2-} and K. On Al_2O_3, OSC variations due to the additives follow the same trend as the variations of oxygen mobility (deduced from $^{18}O/^{16}O$ isotopic exchange). Chlorine and sulfur are inhibitors of OSC while K, at low content, is a promotor.

1. INTRODUCTION

Rare earth oxides, especially cerium oxide, are used to improve the oxygen storage capacity (OSC) of three-way catalysts. Noble metals, in particular rhodium, play an active role in promoting the OSC of the support [1-5]. The mechanism of OSC can be described in terms of three principal steps :

- dissociative adsorption and desorption of molecular oxygen on the metals
- transfer of active oxygen species from the metal to the support and surface migration of these species on the support
- storage of oxygen by cerium oxide.

Duprez and al. investigated the first and the second step by means of $^{16}O/^{18}O$ isotopic exchange [6-8]. The rate of step 1 can be deduced from the rate of the $^{16}O_2/^{18}O_2$ equilibration reaction that occurs on the metal :

$$^{18}O_{2(gas)} + {}^{16}O_{2(gas)} \rightleftharpoons 2\ {}^{18}O^{16}O_{(gas)} \qquad (1)$$

while the rate of step 2 is in direct relation with the rate of isotopic exchange of $^{18}O_2$ with the ^{16}O of the support :

$$^{18}O_{2(gas)} + {}^{16}O_{(sup.)} \rightarrow {}^{18}O^{16}O_{(gas)} + {}^{18}O_{(sup.)} \qquad (2)$$

The surface of ceria being readily reduced at low temperature (< 473K) in the presence of noble metals [1,4], step 3 is expected to be very rapid under the usual conditions of exhaust gas catalysis (T=673-773K).

Duprez and Kacimi [7] showed that rhodium was the metal which has the highest intrinsic rate in reaction (1). Accordingly the rate determining step (rds) of the mechanism of OSC on rhodium should be the transfer and the migration of oxygen at the surface of the support.

Our aim, in the first part of this paper, is to study the effect of the sintering of $PtRh/Al_2O_3$ and $PtRh/CeO_2$-Al_2O_3 catalysts on the OSC and to correlate the OSC with the surface migration of oxygen measured by $^{18}O_{2(gas)}/^{16}O(support)$ isotopic exchange [6,8]. In the second part of this paper, we study the effects of certain additives (SO_4^{2-}, Cl) on the OSC values and on the oxygen mobility in the catalysts. As we have shown recently that the surface mobility of oxygen was linked to the basicity of oxides used as supports [8], we have also investigated the effect of potassium.

2. Experimental

The support was a γ-Al_2O_3 (100 m^2g^{-1}) supplied by IFP. A 12wt.% CeO_2-Al_2O_3 support was prepared by impregnating the γ-Al_2O_3 with aqueous solutions of ceric ammonium nitrate. The catalysts used in the sintering studies were prepared by successive impregnations with aqueous solutions of dinitrodiammine platinum(II) and of rhodium(III) nitrate. After impregnation, the catalysts were dried and calcined at 723K (fresh catalysts). They are referred to here as PtRhXA and PtRhXCA, where A designates alumina, CA, ceria-alumina and X is the atomic percentage of rhodium : % (Rh / Pt+Rh).

Aliquot samples of each catalyst were sintered for 2 h at 973K and at 1173K under a continuous flow of 1%O_2 +10%H_2O (vol%) in nitrogen.

In the second phase of the experiments (effect of additives), Rh catalysts prepared on two different supports : γ-Al_2O_3 (100 m^2g^{-1}, IFP), CeO_2/γ-Al_2O_3 (93 m^2g^{-1}, on IFP alumina) were used. All the modified catalysts were prepared by impregnating a fresh catalyst with aqueous solutions of diammonium sulfate, hydrochloric acid or potassium nitrate. After modification, the catalysts were dried and calcined at 723K. They are referred to here as RhSyP , where y is the weight percentage of the additive P (P=SO_4^{2-}, Cl or K) and S is for the type of support (S=A for alumina and CA for ceria-alumina). The weight percentages of rhodium are 0.51 for catalysts supported on alumina and 0.53 on ceria-alumina.

Dispersion measurements were carried out on fresh catalysts in a pulse chromatographic system previously described [9,10]. The catalysts were reduced at 723K in H_2 degassed in Ar (impurities < 1ppm) and cooled down to room temperature. Hydrogen chemisorption (H_C), oxygen (O_T) and hydrogen (H_T) titrations were successively recorded. Metal dispersion was deduced from oxygen titration of the chemisorbed hydrogen using the following stoichiometries : H/Rh_S =1, O/Rh_S = 1.5 and H/Pt_S = O/Pt_S = 1.

OSC and OSCC (Oxygen Storage Capacity Complete) measurements were carried out in a pulse chromatographic system. The catalyst sample (10 to 50 mg) was inserted in a U-quartz reactor, heated to 723K (4 $Kmin^{-1}$) in a helium flow (30cm^3min^{-1} ; less than 1 ppm impurities) and then oxidized to saturation by O_2 pulses (0.268 cm^3) injected every other minute ("oxygen storage"). The CO pulses were then injected, at T_{OSC} (423K< T_{OSC} <723K), every other minute. OSC values were determined from the amount of CO_2 produced at the first pulse of CO (titration of the "fast" oxygen) and OSCC, from the amount of CO_2 produced at the five first pulses of CO (titration of the "complete" oxygen storage).

Isotopic exchange experiments were carried out in a recycle reactor coupled to a mass-spectrometer [11,12]. “In situ” pretreaments were : $^{16}O_2$ at 723K for 0.25 h, H_2 at 723K for 0.25 h, outgassing at 723K for 0.5 h and cooling down to the temperature of exchange before admission of $^{18}O_2$. Mass spectra ($^{18}O_2$, $^{18}O^{16}O$ and $^{16}O_2$, plus mass 18, 28 and 44 to detect possible leaks) were recorded every nine seconds, which allowed us to determine the coefficient D_S of surface diffusion [8,12] from equation (3) :

$$Ne = 4\,N\,a\ {}^{18}O\ \sqrt{\pi^{-1} D_S\, t}\ \ C \qquad (3)$$

where Ne is the number of diffusing species, a is the radius and $C\ {}^{18}O$ the concentration of ^{18}O of the N circular metal particles.

3. RESULTS AND DISCUSSION

3.1 Catalyst characteristics :

Table 1 gives the characteristics of the alumina-supported catalysts. Fresh catalysts and particularly Rh/Al_2O_3 are well-dispersed. During treatments at 973 - 1173K in an oxidizing medium, two phenomena can occur with rhodium catalysts : (i) surface sintering by coalescence of Rh_2O_3 particles, (ii) formation of a non-reducible oxide phase in strong interaction with alumina (diffusion of Rh^{3+} ions in the matrix [13] or refractory rhodium oxide [14]). On platinum only the mode of deactivation (i) can be usually observed. However Pt is significantly more sensitive than Rh to surface sintering in an oxidizing medium, which explains why PtA deactivates more rapidly at 973K. At higher temperatures, a second mode of deactivation (non-reducible oxide phase) predominates on Rh/Al_2O_3. However one may note the synergy effect between Pt and Rh which renders the bimetallic catalysts more resistant to sintering than the monometallics.

Table 1 $PtRh/Al_2O_3$ fresh and sintered catalysts characteristics

Catalysts	%Pt	%Rh	Fresh catalysts		Relative dispersion D/Do of sintered catalysts at:	
	wt. %	wt. %	Am m^2g^{-1}	Do %	973K	1173K
PtA	1.17	0	1.63	57	0.12	0.067
PtRh09A	1.01	0.05	1.63	60	0.29	0.055
PtRh22A	0.75	0.11	1.55	67	0.22	0.103
PtRh35A	0.70	0.21	2.07	80	0.25	0.101
PtRh60A	0.45	0.36	1.74	65	0.30	0.103
RhA	0	0.51	1.95	87	0.29	0.089

The characteristics of the fresh CeO_2-Al_2O_3 supported catalysts are reported in Table 2. The stoichiometries for oxygen (O_T) and hydrogen (H_T) titrations were abnormally high, which could be explained by the fact that the ceria surface initially reduced in H_2 at 723K, can be partially re-oxidized during O_T. On the bare support, there is no hydrogen uptake on chemisorption (H_C) or titration (H_T). This is not the case of the metal catalysts on which the oxygen (O_T) taken both by the metal and by the support is titrable by H_2. Accordingly, on ceria-alumina catalysts, only H_C values can be used for dispersion measurements. Moreover the stoichiometries H/Pt_S and H/Rh_S having not been verified by other techniques, chemisorption was not applied to sintered catalysts.

Table 2 PtRh/CeO_2-Al_2O_3 fresh catalysts characteristics

Catalysts	%Pt	%Rh	chemis. and titrat. (µmol at. H or O g^{-1})			H_C/M	Am (based on H_C/M=1)
	wt. %	wt. %	H_C	O_T	H_T		m^2g^{-1}
CA (sup.)	/	/	0	90	0	/	/
PtCA	1.00	0	43	150	245	0.8	1.9
PtRh10CA	0.90	0.05	46	113	234	0.9	2.3
PtRh25CA	0.75	0.13	49	152	257	1.0	2.4
PtRh40CA	0.60	0.21	50	160	311	1.0	2.3
PtRh70CA	0.30	0.37	43	146	282	0.8	1.8
RhCA	0	0.53	43	161	305	0.8	1.9

3.2 Effect of the temperature T_{OSC} on the OSC metals' activities :

In these experiments, carried out on the monometallics, the temperature of oxidation (T_{ox} = 723K) was kept constant while the temperature of OSC measurements (T_{OSC}) was increased by step of 50K from 423K to 723K.

The results shown in Fig.1 confirmed ceria as a good promotor of the oxygen storage [1-5,15].

Two temperature ranges can be distinguished for the two series of catalysts. For alumina catalysts, at T_{OSC} < 523K platinum is more active than rhodium and at T_{OSC} > 523K the opposite order is observed. For ceria-alumina catalysts, the same phenomenon is observed but the inversion temperature occurs for T_{OSC} = 623K.

After reduction and re-oxidation at 723K, it was shown by temperature programmed reduction in H_2 that surface PtO_x ($1<x<2$) and bulk Rh_2O_3 were

formed in the alumina supported catalysts [16]. Because of the reduction of CeO_2 together with the metals, no O/M stoichiometry could be proposed for the metals on ceria-alumina [16]. At low OSC temperature, on Al_2O_3, the platinum surface is readily reduced by CO and practically no oxygen uptake by the support can be observed whatever the temperature (PtA on Fig.1). Reduction of Rh_2O_3 yields 75 µmol CO_2 g^{-1} and is achieved at about 600K (RhA, Fig.1). However in RhA, alumina can store a significant amount of active oxygen at high temperature.
On ceria-alumina, the metals (Pt and Rh) are readily reduced by CO at low temperature (^{2}470K) and the support soon begins to store active oxygen.

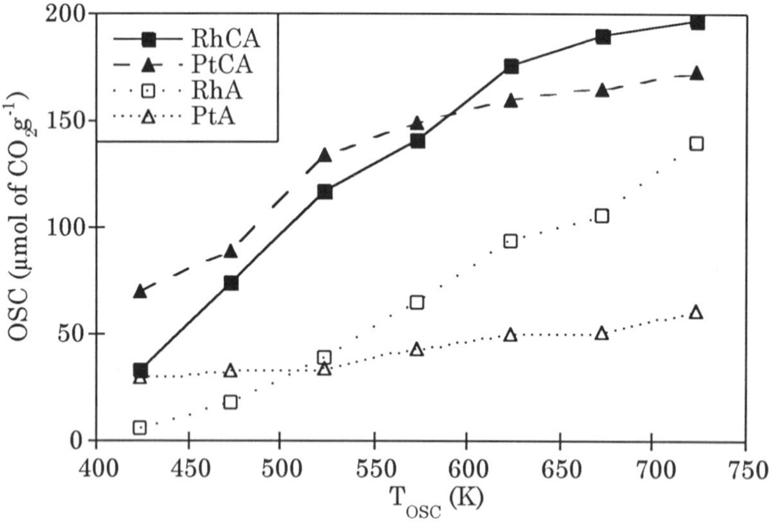

Figure 1. Effect of T_{OSC} on the OSC of Rh or Pt catalysts supported on ceria-alumina and alumina

3.3 Effect of the sintering on the OSC values of PtRhA and PtRhCA :

In what follows, the temperature of oxidation (T_{ox}) and of OSC measurements were kept constant (723K). The results obtained on $PtRh/Al_2O_3$ catalysts are shown in Fig.2.

On fresh samples, there is a quasi-linear increase of OSC values from pure Pt to pure Rh, which seems to indicate that there is no significant change in the surface composition of the bimetallics. Sintering at 973K induces a decrease of the OSC values more marked on Pt-rich samples than on those rich in Rh. Apparently, there is a profound change in the surface composition, bimetallics with X < 25 at.%Rh behaving like pure Pt. Sintering at 1123K causes a dramatic decrease in the OSC values both on Pt and on Rh. However the bimetallics and particularly those in the 20 at.%Rh's region are more resistant to sintering than the other samples.

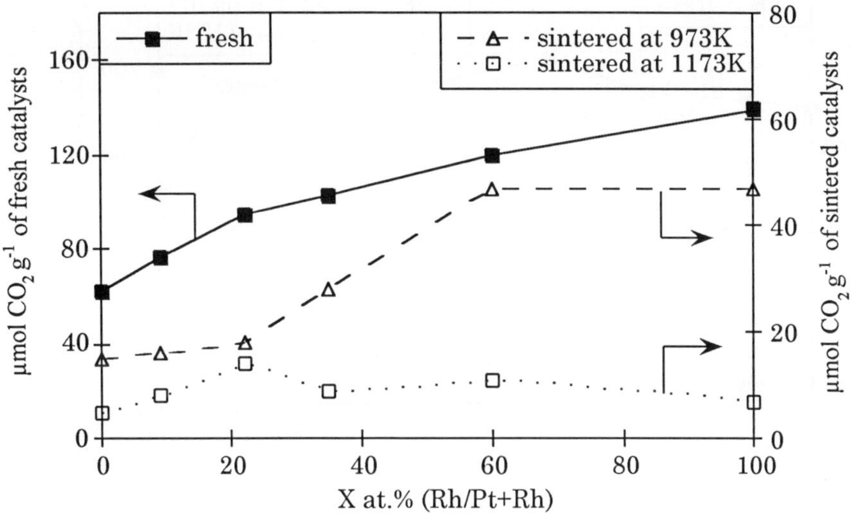

Figure 2. Effect of sintering on the OSC of PtRh catalysts supported on alumina

The effect of sintering on PtRh/CeO_2-Al_2O_3 catalysts is shown in Fig. 3. Contrarily to what was observed on Al_2O_3-supported samples, ceria catalysts resist well to sintering (Table 3).

Table 3 Relative variations of OSC values with sintering.

Catalysts sintered at :	OSC sint. / OSC fresh (%)			
	PtA	RhA	PtCA	RhCA
973K	24%	32%	68%	64%
1173K	8%	5%	36%	39%

The decrease in the OSC values of ceria-catalysts is the result of the sintering of both metal and ceria. These two components participating in the oxygen storage are relatively little affected by sintering : ceria stabilizes the metals but is in turn stabilized by alumina.

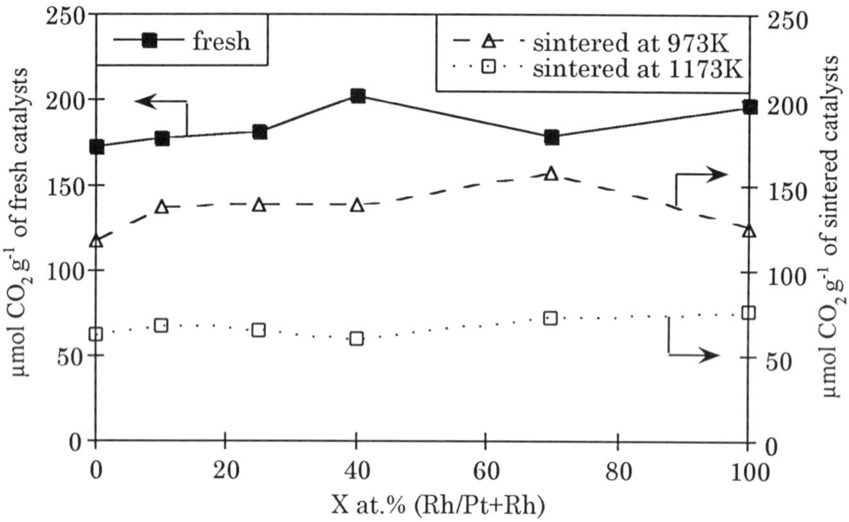

Figure 3. Effect of sintering on the OSC of PtRh catalysts supported on ceria-alumina

3.4 Effect of additives on the OSC and the oxygen mobility of Rh/Al_2O_3, Rh/CeO_2-Al_2O_3 :

Five catalysts were prepared by addition of chlorine (0.1 and 0.5 wt.% Cl), sulfur (4.5 wt.% SO_4^{2-}) or potassium (0.2 and 1.5 wt.% K^+) on the RhA parent catalyst. OSC values and coefficient Ds of surface diffusion (deduced from the initial rates of exchange of $^{18}O_2$ with the ^{16}O of the support) were measured on these samples. Fig. 4 gives the relative variations (Æ%) in the values of OSC and of Ds of the modified catalysts compared to the parent catalyst.

These variations of OSC and Ds follow the same tendency, which justifies the idea that oxygen storage and $^{18}O/^{16}O$ are both controlled by the same rate determining step (transfer metal to support and/or surface diffusion in the case of RhA).

Potassium (at low content) is a promotor of OSC and of oxygen mobility while chlorine, sulfur and potassium at high content have inhibiting effects. The location of the additive at the surface has a determining effect. It was shown that chlorine tends to concentrate, even at low content, near the metal particles. This concentration amplifies inhibiting effect of Cl by blocking the transfer step.

Potassium at low content, would be essentially located on the support ; it increases the basicity and thus the surface mobility of oxygen [8]. At higher content K probably interacts strongly with the metal and decreases the catalyst activity.

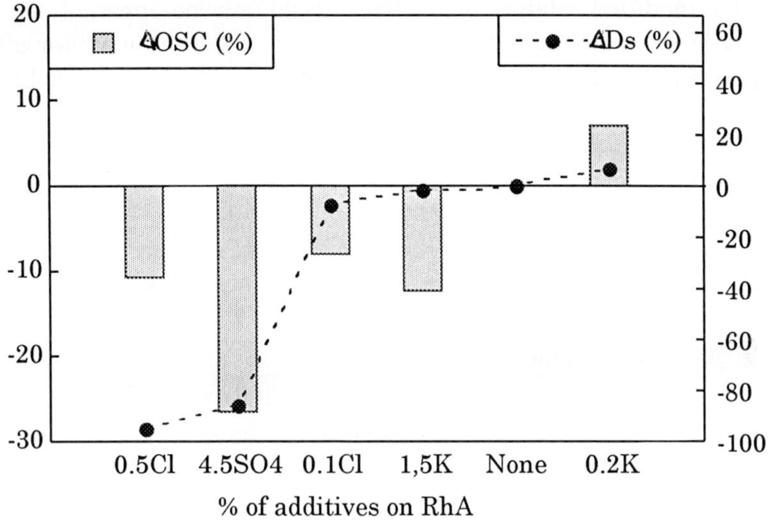

Figure 4. Effect of additives on the OSC and the mobilities of oxygen on RhA catalyst.

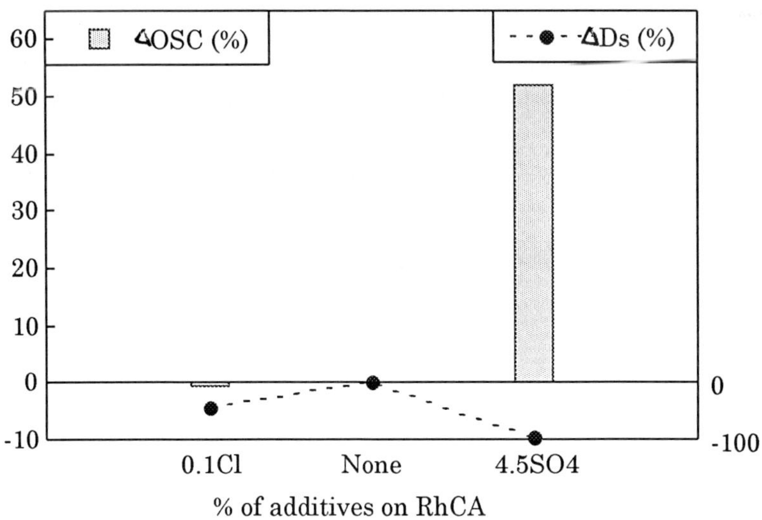

Figure 5. Effect of additives on the OSC and the mobilities of oxygen on RhCA catalysts.

The behavior of modified Rh/CeO_2-Al_2O_3 catalyst is shown in Fig.5. Contrarily to what was found on Al_2O_3, for modified catalysts supported on ceria-alumina, chlorine has only a slight effect on the OSC values. This difference would be result of the presence of ceria that could promote a better distribution of chlorine at the carrier surface. Sulfates would increase the oxygen storage capacity of the modified catalyst. This effect would only be apparent and would be due partly to the sulfur reduction by CO which masks the inhibitor effect of sulfur on OSC. This can be observed by comparison with the oxygen mobilities.

4. CONCLUSION

Noble metals play a definite role in storing oxygen on exhaust gas catalysts. Pt and Rh are able to store oxygen on CeO_2-Al_2O_3 but only Rh is active on Al_2O_3 catalysts.

On Al_2O_3, OSC values depend on the metal used (Rh > Pt) and are extremely sensitive to sintering (1%O_2 + 10%H_2O, at 973 and 1173K).

On CeO_2-Al_2O_3, OSC values depend little on the metal used (Rh ≈ Pt) and the catalysts resist well to sintering.

On Rh/Al_2O_3, both OSC and oxygen mobility (deduced from $^{18}O/^{16}O$ isotopic experiments) are inhibited by chlorine and sulfate and conversely promoted by potassium. Similar results are obtained with Rh/CeO_2-Al_2O_3 except the apparent promotion of the OSC by sulfates.

ACKNOWLEDGMENT:

This work was carried out within the Groupement Scientifique Catalyseurs de Postcombustion funded by the Centre National de la Recherche Scientifique, the Institut Français du Pétrole and the Agence de l'Environnement et de la Maîtrise de l'Energie.

REFERENCES :

1 H.C.Yao and Y.F.Yu Yao, J. Catal., 86 (1984) 256.
2 E.C. Su, C.N.Montreuil and W.G.Rothschild, Appl. Catal., 17 (1985) 75.
3 E.C. Su, and W.G.Rothschild, J. Catal., 99 (1986) 506.
4 B.Harrison, A.F.Diwell and C.Hallett, Platinum metals Rev., 32 (1988) 73.
5 B.Engler, E.Koberstein and P.Schubert, Appl. Catal., 48 (1989) 71.
6 D. Duprez, H. Abderrahim, S. Kacimi and J. Rivière in K.H. Steinberg, Editor, Proc. 2nd Int. Conf. Spillover, Leipzig, June 12-16, 1989, Karl-Marx-Universitat, Leipzig, 1989, p. 127.
7 S. Kacimi and D. Duprez in A. Crucq, Editor, Catalysis and Automotive Pollution Control II, Proc. 2nd Int. Symp. CaPoC 2, Brussels, Sept. 10-13, 1990, Stud. Surf. Sci. Catal., Vol.71, Elsevier, Amsterdam, 1991, p. 641.
8 D.Martin and D.Duprez, in T. Inui, K. Fujimoto, T. Uchijima and M. Masai, Editors, New Aspects of Spillover Effects in Catalysis, Proc. 3rd Int. Conf. Spillover, Kyoto, Aug. 17-20, 1993, Stud. Surf. Sci. Catal., Vol. 77, Elsevier, Amsterdam, 1993, p. 201.
9 D. Duprez, J. Chim. Phys., 80 (1983) 487.
10 R.Taha and D.Duprez, Catal. Letters, 14 (1992) 51.
11 H. Abderrahim and D. Duprez, in A. Crucq and A. Frennet, Editors, Catalysis and Automotive Pollution Control, Proc. 1st Int. Symp. CaPoC 1, Brussels, Sept. 8-11, 1986, Stud. Surf. Sci. Catal., Vol.30, Elsevier, Amsterdam, 19871, p. 359.
12 D. Martin and D. Duprez, in S. Goldstein, E. Soulie and P. Louvet, Editors, Proc. Symp. "Les Isotopes Stables Applicatoins-Production", Saclay, Nov. 24-25, 1993, Commissariat à l'Energie Atomique, in press.
13 H.C. Yao, S. Japar and M. Shelef, J. Catal., 50 (1970) 407.
14 D.D. Beck, T.W. Capehart, C. Wong and D.N. Belton, J. Catal., 144 (1993) 311.
15 S. Kacimi, J. Barbier Jr., R. Taha and D. Duprez, Catal. Lett., 22 (1993) 343.
16 R. Taha, Thesis, Poitiers, 1994.

A. Frennet and J.-M. Bastin (Eds.)
Catalysis and Automotive Pollution Control III
Studies in Surface Science and Catalysis, Vol. 96
1995 Elsevier Science B.V.

STUDY OF HYDROCARBONS REMOVAL WITH A THREE-WAY AUTOMOTIVE CATALYST AFTER SEVERE THERMAL AGING

J.M. Bart[(a)*], M. Prigent[(a)] and A. Pentenero[(b)]

[(a)] *Institut Français du Pétrole,*
BP 311, 92506 Rueil-Malmaison, France
[(b)] *Université de Nancy I, Vandoeuvre les Nancy, France*

ABSTRACT

Frequently exposed to very high temperatures, the catalysts used for depollution of motor vehicle exhaust gases are deactivated mainly by structural and textural evolution processes. This paper describes how the catalytic activity of a typical three-way catalyst (platinum-rhodium on a wash-coated cordierite monolith) was determined for the removal of hydrocarbons of various types before and after high-temperature treatments. Electron microscopy (CTEM and STEM) was used to determine metal particle size in fresh and aged catalysts.

1. INTRODUCTION

Exhaust gases from spark ignition engines contain several hundred hydrocarbons and various oxygen-containing derivatives [1-4]. The nature of these compounds depends partly on the formulation of the fuel [5-7], which contains a wide variety of paraffins, olefins and aromatics and sometimes various oxygenated compounds, which are added to gasoline to improve its octane value. Other hydrocarbons that are not present initially in the fuel (methane, ethane, ethylene, acetylene, ...), are also found in engine exhaust.

Under actual use conditions, automotive catalysts of the three-way type make it possible to remove a high proportion of these unburnt hydrocarbons (but with the exception of methane). Being subjected to very severe thermal stresses, since

* present adress: PSA Peugeot-Citroën, La Garenne Colombes, France

their working temperature can go up to 1000°C or more, these catalysts undergo a gradual loss of activity. This loss of activity, generally attributed to a decrease in the accessible area of precious metal due to sintering, constitutes a major problem in motor vehicle exhaust aftertreatment [8].

The present study was performed with the aim of improving the understanding of, and of quantifying, the thermal aging of three-way catalysts based on Pt and Rh deposited on an Al_2O_3-CeO_2 support and of examining the consequences of this aging for the removal of unburnt hydrocarbons of various types.

2. EXPERIMENTAL PART

The device used for measuring catalyst activities has been described previously [9,10]. The composition of the gas mixture used for these tests, which were carried out at and near stoichiometry, is shown in Table 1.

Table 1 Feed gas composition used in the test (% vol)

CO	0.61	SO_2	0 or 0.002
HC (C_1)	0.15	CO_2	10
H_2	0.2	H_2O	10
NO	0.048	N_2	balance
O_2	~ 0.63		

The reactor consisted of a quartz tube ($\varnothing_i$ = 10 mm) packed, upstream from the catalyst, with quartz scrap to help with inlet gas preheating. The gas flow was 400 standard litres per hour, giving a gas space velocity in the catalyst of 50 000 h^{-1}. A thermocouple placed 5 mm in front of the monolith inlet side was used for temperature measurement.

The catalyst used was a 46 mm long cylindrical section of a cordierite monolith (400 cells/in^2) with an alumina-based wash-coat (100 g/l) containing 4.5% of CeO_2, an iron promoter and 1.41 g/l (40 g/CF) of Pt + Rh. The Pt and Rh weight ratio was 5:1.

CO and NO conversions were measured using conventional nondispersive infrared and chemiluminescence analyzers respectively. Hydrocarbon conversions were measured with a heated flame ionization detector analyzer.

The measurements were made in the steady state; the conversion is defined as the fractional reduction of the inlet concentration, i.e.: $X_i = (C_i^\circ - C_i / C_i)$, where C_i and C_i° denote outlet and inlet concentrations of component i respectively.

The outlet concentration refers only to the signal value given by the FID analyzer, regardless of the nature of the compound present in the gas. The catalyst light-off temperature is defined as the minimum temperature required to obtain 50% conversion of the compound considered.

The results reported here were obtained with the catalyst in fresh condition (activated for 2 h at 500°C in a stoichiometric mixture), with thermally aged catalyst samples. These aging operations were performed for 16 hours at 900°C or 980°C under N_2 + 10% H_2O.

H_2/O_2 titration is frequently used for measuring the dispersion of precious metals. Unfortunately, the presence of CeO_2 on the support makes this method unusable because CeO_2 is reduced by H_2 [11]. Since determination using CO adsorption did not give a satisfactory result either, electron microscopy was finally used to obtain the metal dispersion [8,12,13].

3. Characterisation of Catalysts by Electron Microscopy

Wash-coat separated from the cordierite monolith by scraping was examined directly by two electron microscopy techniques: CTEM (Conventional Transmission Electron Microscopy) and STEM (Scanning Transmission Electron Microscopy).

CTEM enables particle size to be measured but without making it possible to distinguish between the precious metal particles and the particles of Ce and/or Fe oxides (resolution of the apparatus used = 0.7 nm).

STEM, on the other hand, provides access both to the size of the particles and to their chemical composition (particles > 5 nm in size). However, bearing in mind the greater complexity of STEM when compared with CTEM, the number of crystallites examined by this technique remained fewer (1569 particles measured using CTEM, against 100 particles measured and analyzed using STEM).

STEM analysis shows that three particle populations exist: Pt, Pt-Rh and CeO_2-Fe_2O_3 (Table 2). It also shows that aging results in an enrichment in Rh of the mixed Pt-Rh particles at the expense of the particles of Rh, which exist in a hyperdispersed manner in the new catalyst. Particles of Pt by itself are no longer detected in the catalyst aged at 980°C.

Table 2 Number, size and composition of the particles observed using STEM in the wash-coat of new catalyst and of catalysts aged at 900°C or at 980°C

State	Characteristics	Nature of particles			
		Pt	Rh	Pt+Rh	CeO_2 + Fe_2O_3
New	number/30	8	0	16	6
	size (nm)	6-13	HD (2)	6.5-11	3-20
	Pt/Rh (1)	-	-	3.6-33.3	-
Aged 16 h at 900°C	number 32	5	0	19	8
	size (nm)	2-20	-	2-40	3-50
	Pt/Rh (1)	-	-	1.4-10	-
Aged 16 h at 980°C	number/38	0	0	24	14
	size (nm)	-	-	14-80	10-40
	Pt/Rh (1)	-	-	1.45-15.	-

(1) Atomic ratio
(2) HD = hyperdispersed

Three CTEM photographs illustrating the particles present in the wash-coat of new catalyst and of catalysts aged at 900°C or at 980°C are shown in Figure 1.

The shape of the particles (dark spots) has been taken to be sphere-like and, from the measurement of their diameters, histograms have been plotted representing the percentage of particles of a given size relative to the total number of particles. Figure 1 shows that the particle diameter varies between 0.7 and 16 nm in new catalyst. In catalysts aged at 900 or 980°C the distribution is more heterogeneous, with the sizes varying between 1 and 65 nm or between 1 and 81 nm respectively. It can be seen that in the new catalyst 70% of the particles observed are smaller than 3 nm in size, whereas in catalysts aged at 900°C or at 980°C particles less than 3 nm in size now represent only 30 and 10% of the total, respectively.

It is found, furthermore, that the longer the catalyst is aged at high temperature the larger the particle size, but that small-sized particles whose diameter is between 1 and 5 nm still exist in these aged catalysts. This finding is in good agreement with the sintering mechanism known as the "Ostwald ripening

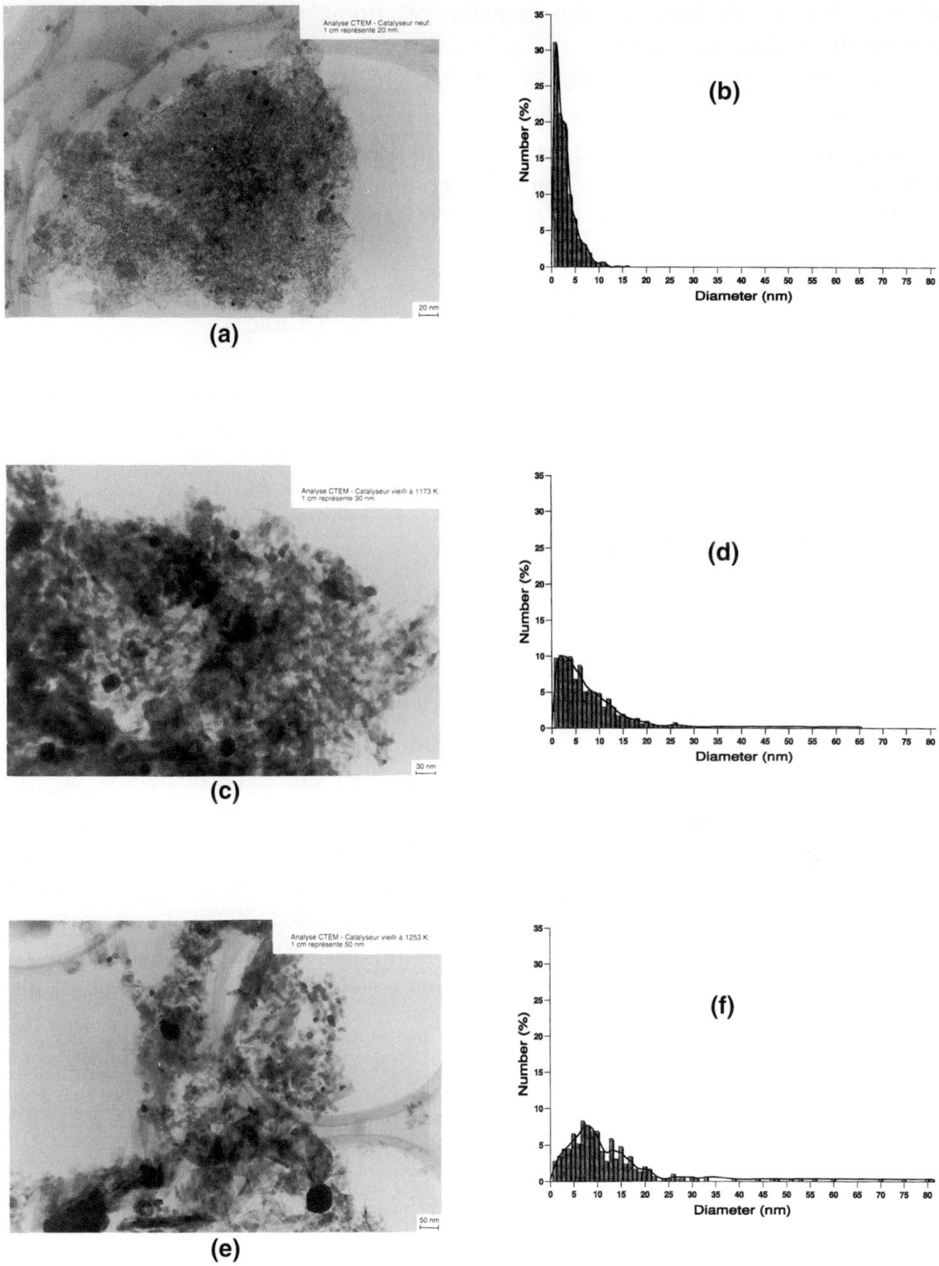

Figure 1. CTEM photography and diameter histograms for particles present in washcoat a,b: fresh catalyst;c,d: catalyse aged at 900°C;e,f: catalyst aged at 980°C.

mechanism", which involves the transfer of atoms from the small particles towards the larger ones.

Taking into account the size of all the particles observed by CTEM, a calculation was made to estimate the arithmetic mean diameter of these particles, the equivalent diameter d_{eq}, comparable to the mean diameter obtained by chemisorption methods and calculated by comparing surface and volume of the particles, the specific surface area, and the dispersion, which represents the number of accessible metal atoms [14-16].

The results of these various calculations are given in Table 3.

Table 3 Number and characteristics of the particles observed by CTEM in the wash-coat of new catalysts and of catalysts aged at 900°C or at 980°C

catalyst	n_i	d (nm)	d_{eq} (nm)	D (%)	S_{aged}/S_{new}
new	768	2.9	6.2	18	
aged 900°C	513	7.5	20.6	6	0.301
aged 980°C	306	12.5	42.1	3	0.147

The equivalent diameter of 6.2 nm observed in the new catalyst is probably an overestimate, due to the existence of particles that are hyperdispersed on the support and that have not been taken into account in these calculations.

Furthermore, all these values constitute only an approximation, bearing in mind the fact that it has not been possible to distinguish between precious metal particles and particles of Ce and Fe oxides. Nevertheless, these oxide particles were a minority in number, although not insignificant (20 to 37% of the total of the particles analyzed by STEM), and their sizes were comparable to those of the metal particles.

4. OXIDATION OF HYDROCARBONS AT STOICHIOMTRY

The degrees of conversion of various C_2, C_3 and C_6 hydrocarbons are given simultaneously for saturated hydrocarbon and for unsaturated hydrocarbon, in Figures 2a, b and c, as a function of temperature. These determinations were carried out with gaseous mixtures of stoichiometric redox composition and with new catalysts and catalysts aged at 900°C or at 980°C.

These light-off temperatures (T 50 = temperature at which 50% of the hydrocarbon is oxidized) and the mean conversions C_M in the 100-600°C range, calculated using the equation below, are given in Table 4.

Mean conversion: $$C_M = \frac{\int_{100}^{600} \frac{\overset{\circ}{c}_{HC} - c_{HC}}{\overset{\circ}{c}_{HC}} dT}{\int_{100}^{600} dT}$$

When the light-off temperatures of these hydrocarbons are considered, the following ranking is obtained with new catalyst:

$C_3H_6 > C_6H_{14} > C_2H_4 \sim C_6H_6 >>> C_3H_8 >>>> C_2H_6$

With catalyst aged at 900°C, this ranking becomes:

$C_6H_6 \sim C_6H_{14} > C_3H_6 > C_2H_4 >>>> C_3H_8 >> C_2H_6$

and with catalyst aged at 980°C:

$C_2H_4 \sim C_6H_6 > C_3H_6 >>>>> C_3H_8 >> C_2H_6$

It can be observed that the difference in reactivity for easy to oxidize compounds decreases when the catalyst ages (the mean activities are of the order of 74% for the catalyst aged at 900°C and of the order of 68% for the catalyst aged at 980°C). It is found that the differences in conversion between the easy to oxidize compounds are smaller with aged catalyst than with new catalyst; on the other hand, the difference between the conversions with easy to oxidize compounds and those that are difficult to oxidize increases appreciably.

The drop observed in overall activity can be attributed to a decrease in the number of active sites or may result from a decrease in the individual activity of each site. In the former case the reaction is not "structure-sensitive"; in the latter case, on the other hand, it is.

The specific rate ($mol.g.m^{-2}.h^{-1}$), which varies like the Turn Over Number (TON), can be calculated from the following equation: R = F/S.X, F being the initial molar flow rate of hydrocarbon ($mol.h^{-1}$), S the specific surface area ($m^2.g^{-1}$) and X the conversion.

By arbitrarily choosing the specific surface area of new catalyst as being equal to 1, it is possible to evaluate a relative specific rate for aged catalysts (Figure 3). It is found that, for propane and ethane, R increases with catalyst aging, whereas R decreases for the other hydrocarbons

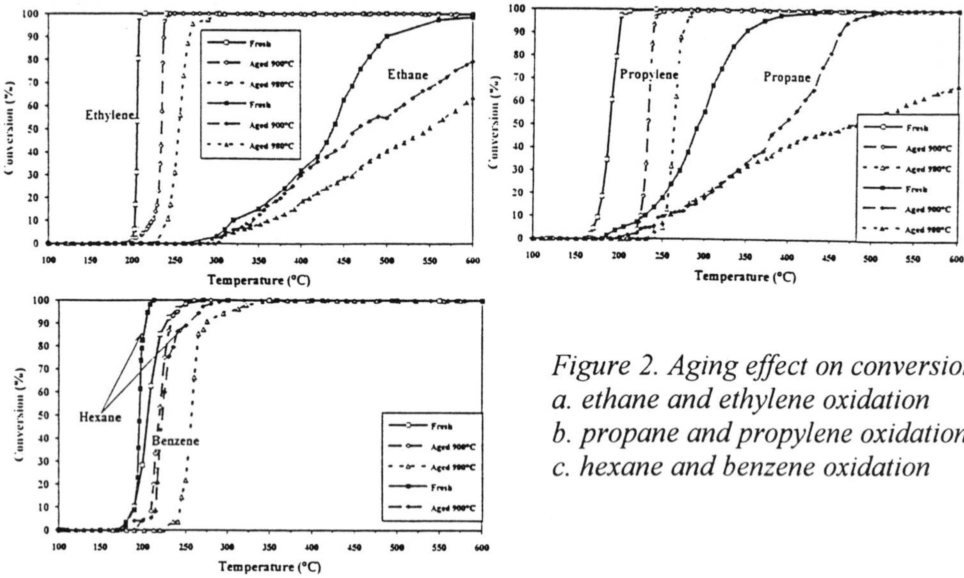

Figure 2. Aging effect on conversion
a. ethane and ethylene oxidation
b. propane and propylene oxidation
c. hexane and benzene oxidation

Table 4 Light-off temperatures (T 50) and mean activities C_M observed with various hydrocarbons for new and aged catalysts

HC	T50 (°C)			Mean activity (%)		
	new	aged at 900°C	aged at 980°C	new	aged at 900°C	aged at 980°C
ethane	437	467	547	35	25	18
propane	292	392	471	62	45	29
hexane	197	222		81	74	
ethylene	207	237	257	79	74	69
propylen	187	232	262	83	74	67
benzene	207	219	257	79	76	68

From this it is consequently deduced that the specific rate increases with the particle size for C_3H_8 and C_2H_6 whereas it decreases for the other hydrocarbons (hexane, ethylene, propylene, benzene). In contrast to CO, for which the specific rate is independent of the particle size [17], the hydrocarbon oxidation reaction is therefore structure-sensitive.

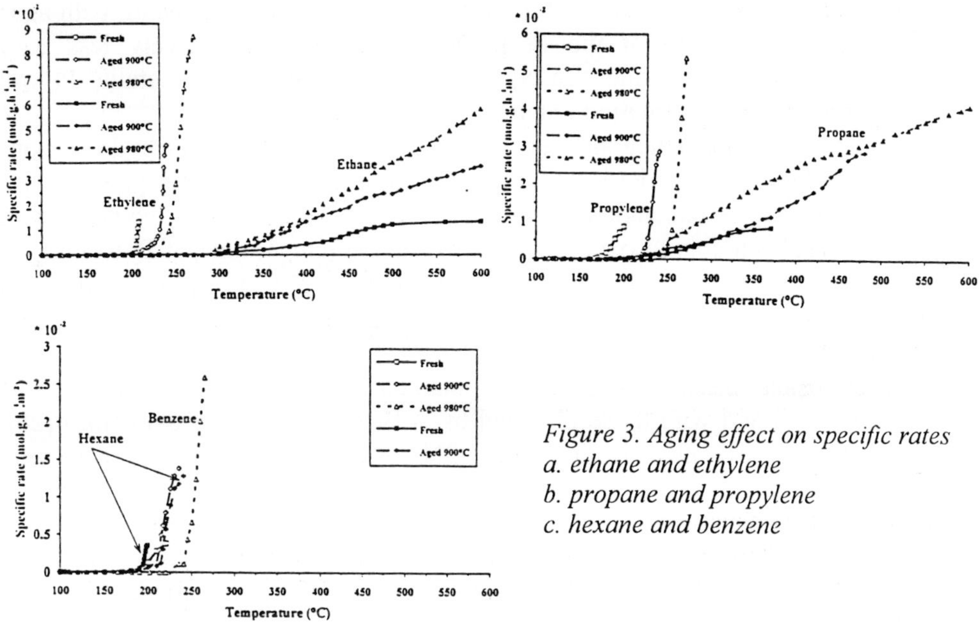

Figure 3. Aging effect on specific rates
a. ethane and ethylene
b. propane and propylene
c. hexane and benzene

It is known that the oxidation of short-chain alkanes on Pt/Al_2O_3 and on Pd/Al_2O_3 is structure-sensitive [18-20]: the Turn Over Number is higher for large particles than for smaller ones. For light alkanes, the increase in the specific rate with particle size which has been observed can be explained by considering the reactivity of the oxygen adsorbed on Pt. The interaction of oxygen with Pt-based supported catalysts has been studied in various laboratories [21-25]. When the Pt is completely dispersed, it is already oxidized to PtO_2 at 300°C [21,22]. On the other hand, when it is in the form of larger crystallites, only the surface of these crystallites is oxidized, and this takes place at a higher temperature (500°C) [21]. Dispersed PtO_2 is less active in the hydrocarbon oxidation reaction than Pt covered with oxygen at the surface [21,26]. It is found that, at an identical temperature, oxygen covering large crystallites is more reactive than oxygen bound on hyperdispersed Pt. This result leads to believe that the surface of hyperdispersed Pt is then in the form of PtO_2; on the other hand, for the large particles the oxygen is merely chemisorbed onto the catalyst. This conclusion is similar to that of R.F. Hicks [18].

In the temperature range in which the oxidation of propane and of ethane is initiated on new catalyst (170-377°C and 287-527°C respectively), it may be thought that Pt is oxidized before the hydrocarbons. On the other hand, with aged catalysts, the oxidation of the Pt particles will be progressively less easy (particles of larger size) and consequently proportionally more metallic Pt will remain to activate the oxidation reaction of these hydrocarbons.

In contrast to short-chain alkanes, the specific rate for the other hydrocarbons (hexane, ethylene, propylene and benzene) decreases when the particle size increases. It may be concluded that these hydrocarbons react with oxygen before the latter has been able to oxidize the Pt (oxidation temperature of Pt to PtO_2 of the order of 300°C for a hyperdispersed catalyst [21,22] and a temperature range where the oxidation is initiated with new catalyst for these compounds are between 157 and 227°C).

These results disagree with those obtained by L.M. Carballo (propylene oxidation) [27] and with those obtained by A.B. Kooh (heptane oxidation) [28]. In fact, in these two investigations the TONs or the specific rates increase with the particle size. This increase is attributed to a decrease in the fraction of Pt atoms on the ridges and to an increase in these atoms on the terraces [27]. It should be noted, however, that these two investigations, though carried out with catalysts based on Pt or Pd supported on Al_2O_3 or ZrO_2, were conducted in the presence of an excess of oxygen and with simple gas mixtures (helium, oxygen and propylene or heptane).

5. INHIBITION EFFECTS

Figure 4a shows the variation in the conversion of propane on new catalyst and on catalyst aged at 900°C as a function of the O_2 concentration. It is found that inhibition by O_2 in an overall oxidizing gas mixture exists both for new catalyst and for aged catalyst. The conversion maximum is situated in a slightly reducing gas mixture.

For hexane and benzene, no inhibition appeared in an oxidizing gas mixture for new catalyst [9,10]. Their removal was already complete at stoichiometry. For aged catalyst, on the other hand, a small decrease in conversion is observed; this is of the order of 5% for hexane and 15% for benzene (Figure 4b).

Since inhibition due to oxygen is observed only with the compounds that are difficult to oxidize, it may be attributed to a competition between the chemisorptions of the different compounds. O_2 would appear to be adsorbed more rapidly than light alkanes and, as it then covers the whole surface, no more active sites permitting hydrocarbon chemisorption would appear to be left. In other words, this is equivalent to stating that since the precious metal crystallites are oxidized by O_2, propane and ethane would appear to be no longer sufficiently reactive to reduce the crystallites, while the other hydrocarbons are. It may therefore be concluded that short-chain alkanes are oxidized by a mechanism of the Langmuir-Hinshelwood type (the hydrocarbon and oxygen react once they are adsorbed) and that the other compounds can react by a mechanism of the Eley-Rideal type (unadsorbed hydrocarbon reacts with adsorbed oxygen).

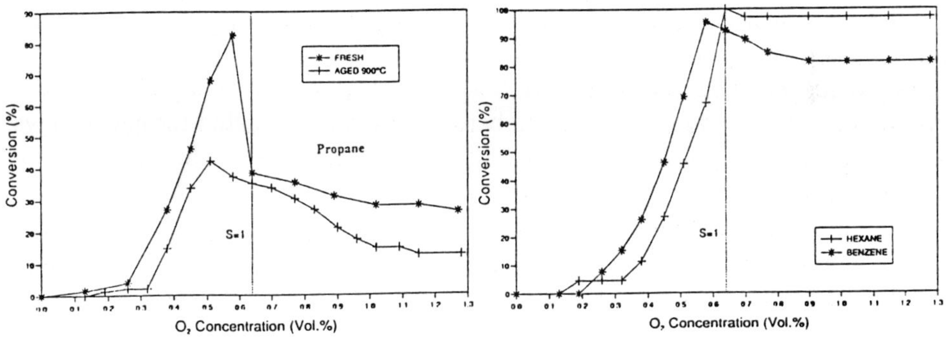

Figure 4. Conversion at 260°C of propane, hexane and benzene as a function of O_2 concentration on new catalyst or on catalyst aged at 900°C.

It has also been observed that the inhibiting effect of oxygen in an oxidizing gas mixture is slightly increased by the addition of 1000 ppm of NO. To verify whether these inhibitions by O_2 and NO are rapidly reversible, the following test was performed. The catalyst was first of all placed in contact with a gas mixture with the following composition:

O_2	NO	CO	H_2	C_3H_8	CO_2	H_2O	N_2
0.77%	480 ppm	0.6%	0.2%	1500 ppmC	10%	10%	78.3%

After this mixture had been passed over the catalyst for 5 minutes, the NO feed was cut off and the conversion of propane was allowed to stabilize for 10 minutes. When it was judged to be stable, the O_2 concentration was lowered from

0.77% to 0.45% to change over to reducing conditions. Figure 5 shows the result obtained.

Qualitatively the same effect is observed for a catalyst aged at 900°C and for the new catalyst. However, the removal of NO increases the propane conversion by only 7%, whereas it increased by approximately 20% with new catalyst. The NO desorption time itself is also shorter (3 minutes instead of 5). From these two points it can be assumed that there is less NO adsorbed on the surface and that consequently the inhibition by NO is less marked than for new catalyst. When the O_2 concentration decreases, propane conversion increases by 50% and then drops instantaneously for new catalyst. The same phenomenon is encountered with aged catalyst, except that the increase in conversion is only 30%. As in the case of NO, it may be concluded from this that the inhibition by O_2 is less marked for new catalyst than for aged catalyst. The drop in activity of the aged catalyst in oxidizing conditions would not therefore appear to be due to a stronger inhibition by NO and by O_2, but indeed to be the consequence of sintering.

Likewise, the inhibiting effect of SO_2 on the oxidation of hydrocarbons of different types (propane, propene and benzene) was investigated for new catalyst and for catalysts aged at 900°C or 980°C. The activities in the presence and absence of SO_2 of fresh and aged catalysts are compared in Table 5.

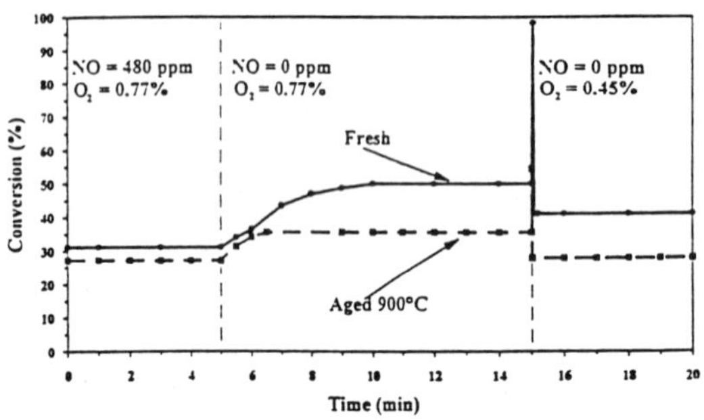

Figure 5. Transient conversion of propane, as a function of time, on a catalyst aged at 900°C.

The differences in the light-off temperatures with and without SO_2 are of the order of 10 to 20°C. For propylene and benzene, the inhibiting effect observed for this aged catalyst is less marked than for new catalyst [9,10].

The inhibiting effect of SO_2 on alkenes has also been observed [29-31]. The inhibiting effect of SO_2 on the oxidation of propylene and benzene can be explained by the competition between these compounds for access to the active sites. In fact, SO_2 can already be adsorbed on precious metals at room temperature, (SO_2 or SO_3^{2-}) [32], like hydrocarbons (alkenes and aromatics).

Table 5 : Inhibiting effect of SO_2 on the oxidation of C_3H_8, C_3H_6 and C_6H_6 for new Pt+Rh catalyst or catalyst aged for 16 h at 900°C

HC	SO_2 ppm	T50 (°C)		Mean activity from 100 à 600°C (%)	
		new	aged	new	aged
C_3H_8	0	292	299	62	53
	20	267	309	67	53
C_3H_6	0	188	222	83	76
	20	224	237	75	73
C_6H_6	0	196	219	81	76
	20	209	237	78	72

On the other hand, the accelerating effect of SO_2, seen during the oxidation of propane on new catalyst [9,10], disappears for aged catalyst. A very slight inhibition is found at low conversion and, at high temperature, a slight acceleration of the oxidation of propane by SO_2. Overall, it can be said that there is no inhibition or acceleration of the reaction for aged catalyst.

For new catalyst the acceleration of the oxidation reaction of short-chain alkanes is due to the formation of new adsorption sites that did not exist in the absence of SO_2 [29]. Studies using infrared [30] have shown the presence of sulphates on the alumina support, which are formed by adsorption of SO_2 that is already oxidized at 200°C. These sulphates could be responsible for the formation of these new sites promoting the dissociative adsorption of short-chain alkanes, resulting in a higher rate of oxidation. These new adsorption sites would appear to be located at the junction between the particles of precious metals and the sulphated support [29]. With aged catalyst these junctions are rarer because of the sintering of the particles. Fewer new sites are therefore formed and thus the acceleration effect is less marked. It can also be assumed that for aged catalyst the oxidation rate of SO_2 to SO_3 is decreased. Less sulphate would therefore appear to be formed and consequently the adsorption of alkanes would be less easy. This second hypothesis also makes it possible to explain why the inhibition of propylene and benzene is lower with aged catalyst. Since SO_2 is adsorbed less rapidly on aged catalyst, the competition between this compound and the hydrocarbons is consequently reduced and thus the inhibition is less marked.

6. CONCLUSION

Thermal aging of 3-way catalysts based on Pt+Rh on alumina promoted by Ce and Fe oxides has been followed by examinations using electron microscopy (CTEM and STEM). Three particle populations exist in the wash-coat of catalysts of this type: Pt, Pt/Rh and Ce/Fe. The rhodium that is not associated with the Pt is also detected for new catalyst, but in a form that is hyperdispersed in the atomic state or in the form of very small particles that are not visible by the CTEM technique. The heat treatments performed (16 h in wet nitrogen at 900 or 980°C) result in an increase in the size of the particles, whose mean diameter changes from 3 to 7.5 and then to 12.5 nm. The mixed Pt/Rh particles are also enriched in Rh at the expense of the very small Rh particles, which eventually disappear.

The activity of this type of catalyst in the oxidation of different hydrocarbons in gas mixtures at stoichiometry was next determined in the new or aged state. Irrespective of the nature of the hydrocarbons, their conversion gradually decreases in keeping with the decrease in metal dispersion. This decrease in conversion can be explained by the loss of active sites when sintering takes place.

It has been shown, however, that the oxidation of hydrocarbons is structure-sensitive, *i.e.* that the specific rate relative to the area of the particles is dependent on the size of the latter. These observations are explained by considering the reactivity of oxygen towards Pt crystallites. The larger the particles, the more difficult they are to oxidize. Since metallic Pt is more active than its oxide in catalyzing hydrocarbon oxidation, it is consequently easier to understand that the specific rate increases with particle size.

For aged catalyst the inhibition of propane oxidation by nitrogen monoxide and by oxygen is, furthermore, less pronounced in oxidizing conditions. This can be explained by the decrease in the stability of adsorbed oxygen (or of Pt oxide) as a function of the increase in the size of the metal particles. An inhibiting effect of oxygen on the conversion of benzene and hexane in oxidizing conditions was also noted, whereas no such effect existed for new catalyst.

Lastly, the effect of SO_2 is less marked with catalyst aged at 900°C than with new catalyst. The promoting effect of SO_2 on the oxidation of propane no longer exists and the inhibiting effect on the oxidation of propylene and of benzene is less marked.

ACKNOWLEDGMENTS

The financial support of this research by Octel SA, PSA Peugeot-Citroën, and Renault is gratefully aknowledged.

REFERENCES

1 P. Degobert: "Automobile et Pollution", Ed. Technip, Paris 1992.

2 J.K. Walker, C.L. O'Hara: "Analysis of Automobile Exhaust Gases by Mass Spectrometry", Anal. Chem. 27 (5), 825-828, 1955.

3 E.S. Jacobs: "Rapid Gas Chromatographic Determination of C1 to C10 Hydrocarbons in Automotive Exhaust Gas", Anal. Chem. 38 (1), 43-47, 1966.

4 M.W. Jackson: "Effect of Catalytic Emission Control on Exhaust Hydrocarbon Composition and Reactivity", SAE Technical Paper 780624, 1978.

5 D.E. Seizinger, W.F. Marshall, F.W. Cox, M.W. Boyd: "Vehicle Evaporative and Exhaust Emission as Influenced by Benzene Content of Gasoline", SAE Technical Paper 860531, 1986.

6 W.F. Marshall, M.D. Gurney: "Effect of Gasoline Composition on Emission of Aromatic Hydrocarbons", SAE Technical Paper 892076.

7 A. Gorse, J.D. Benson, L.J. Painter, V.R. Burns, R.M. Reuter, A.M. Hochlauser, B.H. Ripon: "Toxic Air Pollutant Vehicle Exhaust Emissions with Reformulated Gasolines", SAE Technical Paper 912324.

8 B.R. Powell: "Analytical Electron Microscopy of a Vehicle Aged Automotive Catalyst", Applied Catal., 53, 233-250, 1989.

9 J.M. Bart: "Oxydation sur catalyseur trois-voies des différents hydrocarbures et dérivés oxygénés présents dans un gaz d'échappement issu d'un moteur à allumage commandé", thèse Nancy, 1992.

10 J.M. Bart, A. Pentenero, M.F. Prigent: "Experimental Comparison Among Hydrocarbons and oxygenated Compounds for their Elimination by Three-Way Automotive Catalysts", Catalytic Control of Air Pollution-Mobile and Stationary Sources, ACS Symposium Series 495, pp. 42-60, Washington, 1992.

11 R.J. Matyi, H. Schwartz, J.B. Butt: "Particle Size, Particle Size Distribution and Related Measurements of Supported Metal Catalyst". Catal. Rev. - Sci. Eng., 1987, 29 (1), 44-91.

12 H. Dexpert, P. Gallezot, C. Leclercq: "Caractérisation des Catalyseurs par Microscopie Electronique Conventionnelle et Analytique à Haute Résolution", les Techniques Physiques d'Etudes des Catalyseurs, Ed. Technip, Paris, 1988, 655-721.

13 A. Rochefort, F. le Peltier: "Les particules Métalliques Supportées", Revue IFP, 46 (2) 1991.

14 P. Duneau : "Effet de support sur la stabilisation du platine en conditions oxydantes", thèse Paris, 1989.

15 G.R. Anderson: "Structure of Metallic Catalyst", Academic Press 1975.

16 P. Stonehart: "Electrocatalyst Advances for Hydrocarbon Oxidation in Phosphoric Acid Fuel Cells", J. Hydrogen Energy, 9 (11), 921-928, 1984.
17 J.T. Kummer: "Use of Noble Metals in Automobile Exhaust Catalysts", J. Phys. Chem. 90 (20), 1986.
18 R.F. Hicks, H. Qi, M.L. Young, R.G. Lee: "Structure Sensivity of Methane Oxidation over Platinum and Palladium", J. Catal. 122, 280-294, 1990.
19 R.F. Hicks, H. Qi, M.L. Young, R.G. Lee: "Effect of Catalyst Structure on Methane Oxidation over Platinum and Palladium", J. Catal. 122, 295-306, 1990.
20 K. Otto, J.M. Andino, C.L. Parks: "The Influence of Platinum Concentration and Particle Size on the Kinetics of Propane Oxidation over Pt/α Alumina", J. Catal. 131, 243-251, 1991.
21 R.W. Mc Cabe, C. Wong, H.S. Woo: "The passivating Oxidation of Platinum", J. Catal. 114, 354-367, 1988.
22 H. Lieske, G. Lietz, H. Spindler, J. Volter: "Reactions of Platinum in Oxygen and Hydrogen Treated Pt/Al_2O_3 Catalysts", J. Catal. 81, 8-16, 1983.
23 C.F. Cullis, B.M. Willat: "Oxidation of Methane over Supported Precious Catalysts", J. Catal. 83, 267, 1983.
24 N. Niwa, K. Awano, Y. Murakami: "Activity of supported Catalysts for Methane Oxidation", J. Catal. 7, 317, 1983.
25 R.K. Nandi, F. Molinaro, C. Tang, J.B. Cohen, J.B. Butt, RL.L. Burwell: "The effects of Pretreatment on Structures", J. Catal. 78, 289, 1982.
26 H.C. Yao, M. Sieg, H.K. Plummer: "Surface Interactions in the Pt/α Al_2O_3", J. Catal. 59, 365, 1979.
27 L.M. Carballo, E.E. Wolf: "Crystallite Size Effect during the Catalytic Oxidation of Propylene on Pt/α Al_2O_3", J. Catal., 53, 366-373, 1978.
28 A.B. Kooh, W.J. Han, R.G. Lee, R.F. Hicks: "Effect of Catalyst Structure and Carbon Deposition on Heptane Oxidation over Supported Platinum and Palladium", J. Catal. 130, 374-391, 1991.
29 H.S. Gandhi, M. Shelef: "Effect of Sulphur on Noble Metal Automotive Catalyst", Appl. Catal., 77, 175-186, 1991.
30 H.C. Yao, H.K. Stepien, H.S. Gandhi: "The Effects of SO_2 on the Oxidation of Hydrocarbons and Carbon Monoxide over Pt/Al_2O_3 Catalyst", J. Catal. 67, 231-236, 1981.
31 V.I. Panchishnyi, N.K. Bondareva, A.V. Sklyarov, V.V. Rozanov and G.P. Ghadina: "Oxidation of Carbon Monoxide and Hydrocarbons on Platinum and Palladium Catalyst in the Presence of Sulfur Dioxide", Zhurnal Prikladdnoi Khimii, 61 (5), 1093-1098, 1988.
32 C.C. Chang: "Infrared Studies on γ Alumina", J. Catal. 53, 374-385, 1978.

A. Frennet and J.-M. Bastin (Eds.)
Catalysis and Automotive Pollution Control III
Studies in Surface Science and Catalysis, Vol. 96

SIMULTANEOUS ATMOSPHERE AND TEMPERATURE CYCLING OF THREE-WAY AUTOMOTIVE EXHAUST CATALYSTS

S. Humbert[a], A. Colin[a], L. Monceaux[a], F. Oudet[b] and P. Courtine[a]

[a]*Département de Génie Chimique*

[b]*Service d'Analyse Physico-Chimique, Université de Technologie de Compiègne, BP 649, 60206 Compiègne cedex, France*

ABSTRACT

As an attempt to simulate real operating conditions of automotive converters, a laboratory bench has been designed and ageing procedures determined to reproduce simultaneous chemical and thermal modifications encountered by catalysts in the exhaust line. Characterization of commercial samples after ageing according to different temperature cycles evidences formation of both platinum/rhodium alloys and cubic perovskite-type compound, $CeAlO_3$. Simultaneously with the formation of cerium aluminate, a thermal stabilization of catalysts is observed, in terms of mean noble metal particles size and concentration of rhodium in alloyed phases. An interpretation based on the crystallographic adaptation of alumina, cerium aluminate and ceria is proposed.

1. INTRODUCTION

From the automotive industry point of view, study of exhaust catalysts and more particularly their behaviour in varying environments are of primary importance [1]. In this field, laboratory procedures able to simulate operating conditions

constitute a practical and economical alternative to motor benches and real driving testing [2-4]. To keep close to realistic condition, i.e., simultaneous thermal and chemical cyclings submitted to converters in the exhaust line, an automated laboratory bench has been specifically designed which can simulate repeated fuel cut-off cycles.
In the following are reported characterization results of commercial three-way catalysts after different treatments in this apparatus by means of X-ray diffraction and transmission electron microscopy.

2. EXPERIMENTS

2.1. Samples

Samples under study are commercial cordierite honeycomb catalysts. The catalytic washcoat is conventionally composed of ceria-promoted transition alumina-supported 5Pt/1Rh (weight %). Samples are cut from original converters in the form of cylinders of 1 inch (diameter) X 3 inches (length) to meet the geometrical requirements of the experimental set-up (see below).

2.2. Procedures and apparatus

Catalysts are systematically treated for four hours under the fifty times repeated identical following cycle:

step 1: warming between low and high temperatures of the cycle under reducing atmosphere

step 2: steady state at high temperature; atm.: N_2, 10% H_2O

step 3: cooling to low temperature under oxidizing atmosphere

step 4: steady state at low temperature; atm.: N_2, 10% H_2O.

To realize temperature variations, samples are placed in a quartz tube (3 inches internal diameter) surrounded by two adjacent tubular furnaces. One furnace is maintained at the low temperature of the cycle and contains the catalyst under study. A second one is held at 1200°C and can be rapidly moved in contact with the first one by a pneumatic jack.

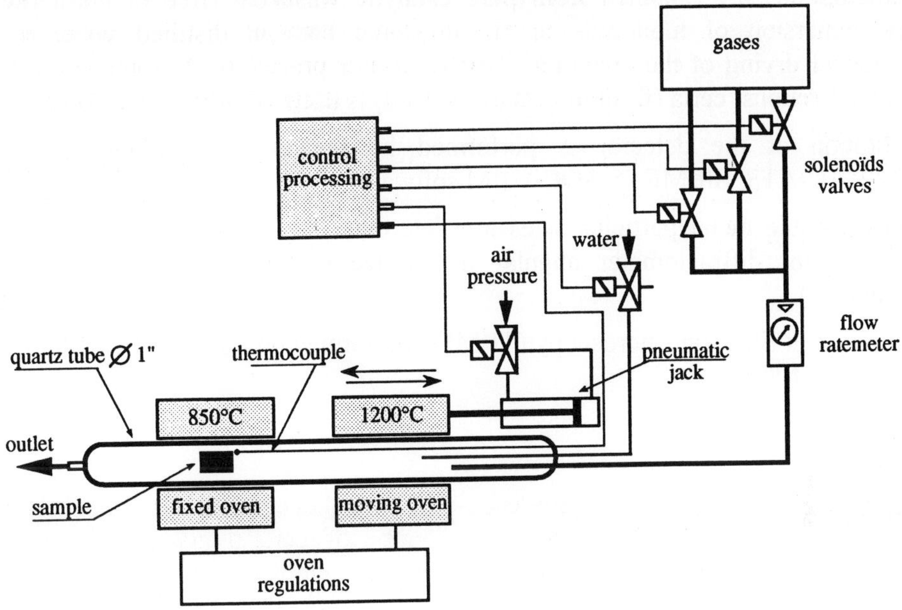

Figure 1. Schematic representation of the laboratory bench. Samples are placed in a quartz tube in a fixed oven. Moving a second oven determines temperature modification. Simultaneous temperature and atmosphere cycles are automated.

The position of this second oven relatively to the first one determines the high temperature of the cycle. Motion of the second furnace and simultaneous atmosphere cyclings are completely automated (TSX-T407 TELEMECANIQUE, GRAFCET programming). The space velocity of gases is 20,000h^{-1}.
Figure 1 shows a schematic representation of this apparatus.
In the present experiments, the low temperature of the cycle is maintained at 850°C, and three high temperature differing cycles were tested: 850°C (no «thermal shock», cycle 1), 950°C (cycle 2) and 1050°C (cycle 3). These reference temperatures are measured at the immediate inlet of samples.

2.3.Characterization

X-ray diffraction experiments (XRD) are performed on a curve position sensitive detector INEL CPS 120 which allows a simultaneous collection of diffracted beams in the range $5° < 2\theta < 125°$, with the monochromatized CuKα1 radiation (λ=1,5406Å, CGR focusing quartz single crystal).

Diffractograms are obtained from pure catalytic washcoat (free of cordierite) after immersion of monoliths in an ultrasonic bath in distilled water and subsequent drying of the resulting suspension. For precise measurements of X-ray line positions, ceria (from the catalysts itself) is used as internal standard.

Calibration of the detector is performed by means of a spline function (DIFFRAC-AT3, SIEMENS/SOCABIM software).

Average noble metal particles sizes are determined by X-ray line broadening, corrected for diffractometer angular divergence (Warren-corrected Sherrer's formula (5)).

Transmission Electron Microscopy (TEM) is performed on a JEOL JEM 1200EX operated at 120 kV.

3. RESULTS

Figure 2 presents a part of typical X-ray diffraction spectra around the (111) reflexion of noble metals for catalysts aged according to cycles 1-3 procedures. Together with these lines are indicated the theoretical positions of the (111) reflexion of pure platinum and rhodium (both fcc structures, a_{Pt} = 3.924 Å and a_{Rh} = 3.804 Å).

Measurement of this line position for treated catalysts, averaged on five samples for each treatment leads to lattice parameters of 3.914 Å (cycle 1,

T_{max} = 850°C), 3.903 Å (cycle 2, T_{max} = 950°C) and 3.901 Å (cycle 3,

T_{max} = 1050°C).

These line shifts suggest the presence of Pt/Rh alloyed particles in the samples. According to Darling [6], application of Vegard's law [7] allows to evaluate the atomic rhodium concentration in alloyed phases. For cycle 1, cycle 2 and cycle 3-based procedures, these concentrations are 8%, 17% and 19% respectively. Since the nominal Rh/Pt+Rh atomic ratio of samples is 27%, these results indicate that whereas these two noble metals are completely miscible [6], some rhodium remains unalloyed in catalysts after ageing in the present experimental conditions. Moreover, no significant modification of the rhodium concentration in alloyed phases is detected after ageing under cycle 2 and cycle 3 conditions (17% and 19% respectively).

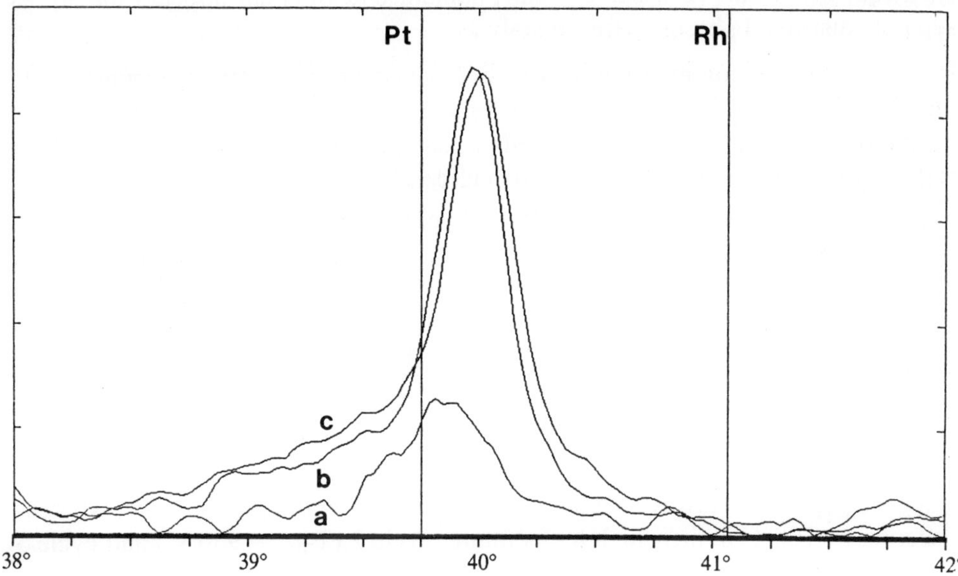

Figure 2. X-ray diffraction spectra around the (111) reflexion of noble metals for catalysts aged according to cycles 1 (a), 2 (b) and 3(c) procedures. Lines shifts suggest Pt/Rh alloying.

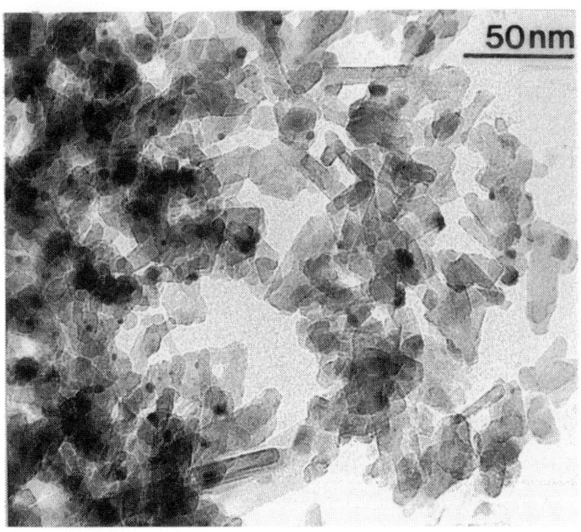

Figure 3. TEM micrograph of fresh catalyst.

As determined by TEM (figure 3), fresh catalysts present very small noble metals particles size, and an important X-ray line broadening which precludes precise measurement of line positions. It thus appears delicate to determine whether alloying pre-exists or is directly related to the present treatments.

To examinate more precisely the conditions of formation of Pt/Rh alloys, the atmosphere cycle was modified, while temperature profile remained unchanged. These experiments reveal that in the absence of a reducing atmosphere step (step 1) in the cycle, no X-ray line shift of the Pt (111) reflection can be detected, neither in the case of cycle 1 nor cycle 2 or cycle 3 high temperatures conditions. Formation of Pt/Rh alloys is thus most probably due to the present experimental conditions.

Consequently it can be assumed that platinum and rhodium appears mainly as separate dispersed phases in fresh catalysts.

X-ray line broadening measurement is a convenient way to derive average noble metal particles size in specific (hkl) crystallographic directions. Because of the very low concentration of noble metals in samples, only the most intense (111) reflexion is detected in diffractograms and can be analyzed in the present work. Table 1 presents these determinations, t_{111} expressed in nanometers, averaged on five samples, for catalysts tested according to the three different procedures. Together with these data are given the corresponding rhodium atomic concentration in alloyed phases.

Table 1 confirms thermal sintering of supported noble metals after exposure to high temperatures. From 5 nm before ageing, catalysts present average particles sizes of 28 nm after treatment according to cycle 3 conditions (T_{max} = 1050°C). This thermal evolution of supported metal catalysts has been largely reported and analyzed (4, 8-11). However, it should be noted that in the present case, whereas differing by 100°C in their high temperature, cycle 2 (T_{max} = 950°C) and cycle 3 (T_{max} = 1050°C) ageing procedures lead to identical mean noble metal particles size (28 nm).

Table 1 Average noble metal particles size (nm) and corresponding rhodium atomic concentration in alloys for fresh and aged catalysts as determined by XRD.

Treatment	Fresh	Cycle 1	Cycle 2	Cycle 3
t_{111}	< 5*	20	28	28
%Rh	-	8	17	19

*determined by TEM

Such an inhibition of high temperature noble metal particles sintering and stabilization of the rhodium concentration in alloyed phases suggest that thermal and chemical conditions of cycle 2 and cycle 3 strongly modify some physical characteristics of catalysts.

Figure 4 presents typical X-ray diffraction spectra in the range 20° < 2θ < 35°

for aged catalysts according to cycle 1, cycle 2 and cycle 3 procedures. Formation of a cubic perovskite-type oxide $CeAlO_3$ is clearly evidenced after ageing under cycle 2 and cycle 3 conditions.

Similarly to Pt/Rh alloying, modifications of step 1 duration show that cerium aluminate is formed only when reducing conditions are present in the cycle, most probably to ensure reduction of Ce^{4+} (CeO_2) to Ce^{3+} ($CeAlO_3$). Moreover, as deduced from X-ray intensities measurements, the $CeAlO_3/CeO_2$ ratio increases with duration of the reducing step (for these experiments, the number of cycles was kept constant to 50 in four hours by modifying neutral phases durations, steps 2 and 4).

Numerous characterizations of catalysts after real driving conditions and various mileages on different vehicles, show, among other features, Pt/Rh alloying and cerium aluminate formation [12]. The similarity of these results with our own observations on commercial samples suggest that the laboratory procedures described herein allow to reproduce significant behaviour of catalysts under real driving conditions.

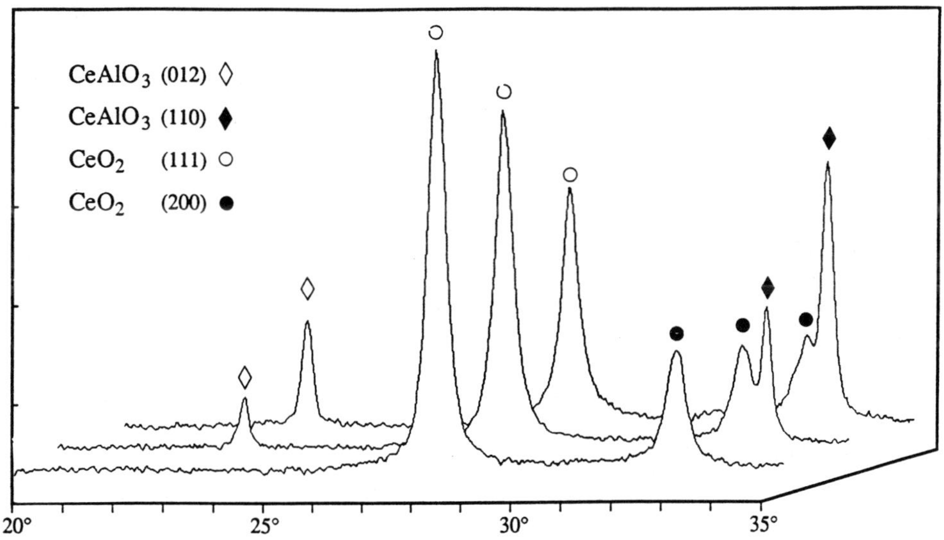

Figure 4. X-ray diffraction spectra in the range 20° < 2θ < 35° for aged catalysts in the conditions of cycle 1 (a), cycle 2 (b) and cycle 3 (c). Formation of CeAlO3 is observed upon treatments 2 and 3.

4. DISCUSSION

Local non-stoichiometry of cerium oxide particles in varying oxidizing and reducing atmospheres is well known in the field of automotive exhaust purification [13-16] and must be taken into account in the present context. Since it involves both trivalent cerium and aluminum ions, $CeAlO_3$ nucleation certainly takes place at the Al_2O_3/CeO_2 interface, most probably by diffusion of relatively (to Ce^{3+}) small Al^{3+} ions ($r_{Al^{3+}}$ = 0.51Å, $r_{Ce^{3+}}$ = 1.07Å) into partially reduced CeO_2 resulting from the exposition to conditions of step 1 of cycle 2

(T_{max} = 950°C) or cycle 3 (T_{max} = 1050°C).

According to Harrison et al. [13] and Diwell et al. [17], formation of local non-stoichiometric ceria (unsatured surface cerium ions) can result from interactions of rhodium and platinum with ceria under reducing atmosphere (CO). To examinate the role of noble metals in the formation of cerium aluminate in the conditions of the present study, we prepared specifically (by conventional impregnation techniques) transition alumina-supported ceria. Ageing such powders according to the three cycles defined above leads to the same observations as in the case of noble metals commercial catalysts, i.e., formation of cerium aluminate in the conditions of cycle 2 and cycle 3. Accordingly, whereas eventually catalyzed by noble metals, formation of $CeAlO_3$ mainly results from ceria/transition alumina solid state reactions.

The above results indicate that formation of cerium aluminate and absence of modification of catalysts, in terms of noble metal particles sizes and amount of rhodium in alloyed phases, are simultaneous. Consequently, it can be assumed that this stabilization of catalysts in the present conditions must be related to the presence of the complex oxide $CeAlO_3$.

We previously proposed an interpretation of thermal stabilization of supported catalysts by rare-earth elements [10, 11, 18] by means of interfacial structural coherence between alumina and rare-earth aluminates. This model, developped for lanthanum and neodymium, can be easily extended to cerium since $LaAlO_3$, $NdAlO_3$ and $CeAlO_3$ are iso-structural mixed oxides with identical lattice parameters (a = 7.6 Å).

In the present case, the solid state reaction between ceria and alumina is not complete and the model takes into account remaining CeO_2 interfaced with $CeAlO_3$.

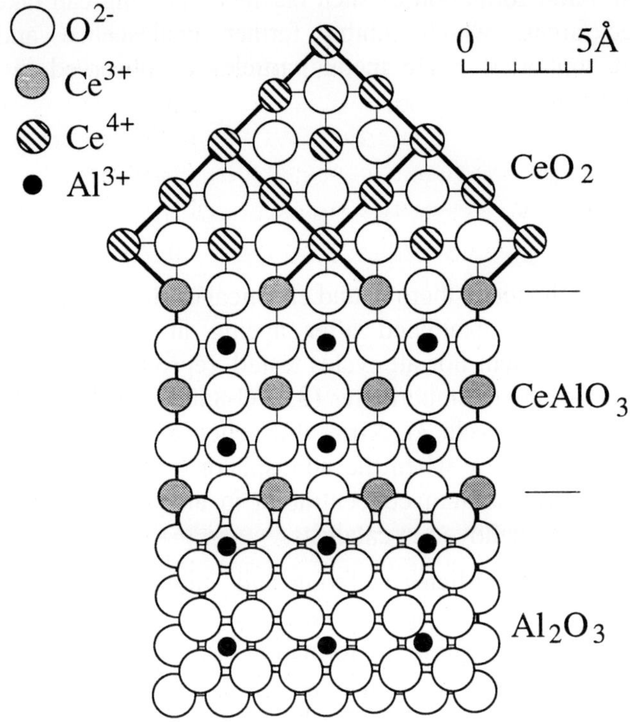

Figure 5. Modelization of the $Al_2O_3/CeAlO_3/CeO_2$ structural relationship along the c-axis of transition alumina [18]. Coherence of the interface results from similarity of oxygen packings and continuity of cations coordination(Al^{3+}:$Al_2O_3/CeAlO_3$ and Ce^{3+}/Ce^{4+}: $CeAlO_3/CeO_2$).

Figure 5 presents the modelization of the $Al_2O_3/CeAlO_3/CeO_2$ structural relationship. The continuity of the three anionic frameworks is a result of the common nature of oxygen geometrical arrangement in all structures. Similarly, cationic environments are respected for octahedral Al^{3+} ions ($Al_2O_3/CeAlO_3$) and cubic Ce^{3+}/Ce^{4+} ions ($CeAlO_3/CeO_2$).

In our sense, as previously proposed [10, 11, 18], these remarkable crystallographic adaptations provide strong anchoring areas for composite $CeAlO_3/CeO_2$ microdomains, considered as physically impassable barriers for surface diffusion. After formation of such barriers, sintering can take place only within delimited areas, which inhibit further coalescence and important modifications of supported noble metal particles as observed in the present experimental conditions.

5.CONCLUSION

The influence of simultaneous thermal and chemical cycling on commercial three-way catalysts has been examined after ageing in a specifically designed automated laboratory bench. For all cycles tested, reproducing repeated fuel cut-off procedures between two temperatures (850°C-850°C, cycle 1 ; 850°C-950°C, cycle 2 ; 850°C-1050°C, cycle 3), X-ray diffraction evidences the formation of platinum/rhodium alloys only when the atmosphere cycle comprises a reducing step. Evaluation of the rhodium concentration in alloyed phases suggests that some rhodium remains unalloyed in catalysts.

For both noble metal particles sintering and amount of allied rhodium, no differences are observed after ageing under cycle 2 and cycle 3 conditions. Moreover, both procedures lead to the formation of cerium aluminate, $CeAlO_3$.

The stabilizing effect of the latter is interpreted as a consequence of structural coherences between transition Al_2O_3, $CeAlO_3$ and CeO_2.

Further experiments, based on analytical scanning transmission electron microscopy and high resolution electron microscopy will be devoted to complete these observations.

On a practical point of view, the similarity of these results (Pt/Rh alloying and formation of cerium aluminate) with those obtained on catalysts after real driving periods, suggest that the laboratory apparatus and related procedures described above constitute promising tools for the study of automotive converters.

Running experiments concern mechanical and morphological modifications of catalytic washcoat resulting from simultaneous temperature and atmosphere cyclings.

ACKNOWLEDGEMENTS

The authors are grateful to PEUGEOT SA for financial support and helpful discussions and information.

REFERENCES

1 G. Belot, Peugeot SA, private communication.
2 G. Mabilon, D. Durand and M. Prigent, The Science of the Total Environment, 93 (1990) 223.
3 G. Mabilon, D. Durand and M. Prigent, Catalysis and Automotive Pollution Control II, Studies in Surface Science and Catalysis, vol. 71, p.569, A. Crucq (ed.), Elsevier, Amsterdam, 1991.
4 H. Shinjoh, A. Muraki and Y. Fujitani, Catalysis and Automotive Pollution Control II, Studies in Surface Science and Catalysis, vol. 71, p.617, A. Crucq (ed.), Elsevier, Amsterdam, 1991.
5 P. Gallezot, Catalysis, 5 (1984) 221.
6 A.S. Darling, Platinum Met. Rev., 5 (1961) 58.
7 B.D. Cullity, Elements of X-ray Diffraction, Addison-Wesley, Reading, 1978.
8 Y.F. Chu and E. Ruckenstein, J. Catal., 55 (1978) 281.
9 B.J. Cooper, W.D.J. Evans and B. Harrison, Catalysis and Automotive Pollution Control, Studies in Surface Science and Catalysis, vol. 30, p. 117, A. Crucq and A. Frennet (eds.), Elsevier, Amsterdam, 1987.
10 F. Oudet, A. Véjux and P. Courtine, Applied Catal., 50 (1989) 79.
11 F. Oudet, E. Bordes, P. Courtine, G. Maxant, C. Lambert and J.P. Guerlet, Catalysis and Automotive Pollution Control, Studies in Surface Science and Catalysis, vol. 30, p. 313, A. Crucq and A. Frennet (eds.), Elsevier, Amsterdam, 1987.
12 G. Belot, J.F. Beziau and G. Meunier, Peugeot S.A., unpublished results.
13 B. Harrison, A.F. Diwell and G. Hallet, Platinum Met. Rev., 32 (1988) 73.
14 P. Lööf, B. Kasemo and K.E. Keck, J. Catal.,118 (1989) 339.
15 P.K. Herz, J.B. Kiela and J.A. Sell, Ind. Eng. Chem. Prod. Res. Dev., 22 (1983) 387.
16 E.C. Su, C.N. Montreuil and W.G. Rothschild, Appl. Catal., 17 (1985) 75.
17 A.F. Diwell, R.R. Rajaram, H.A. Shaw and T.J. Truex, Catalysis and Automotive Pollution Control II, Studies in Surface Sciences and Catalysis, vol. 71, p. 139, A. Crucq (ed.), Elsevier, Amsterdam, 1991.
18 F. Oudet, P. Courtine and A. Véjux, J. Catal., 114 (1988) 112.

A. Frennet and J.-M. Bastin (Eds.)
Catalysis and Automotive Pollution Control III
Studies in Surface Science and Catalysis, Vol. 96
1995 Elsevier Science B.V.

THE EFFECT OF "SPARK RETARD" AS A METHOD OF RAISING EXHAUST GAS TEMPERATURE FOR AUTOMOTIVE CATALYST DYNO AGEING

J Kisenyi[a], N Pallin[a], C Tooby[a], R G Hurley[a], P Atherton[b], D Webb[b]
[a]*Ford Motor Company*
[b]*Tickford Limited*

ABSTRACT

Retarding spark, results in a greater likelihood of incomplete combustion in the engine. At sufficiently high temperatures, the combustion process continues as the exhaust gases travel between the manifold and the catalyst face. This results in a change in the feedgas composition. The concentrations of oxygen and the three legislated exhaust emission gases HC (hydrocarbons), CO (carbon monoxide), and NOx (Oxides of nitrogen) which get to the catalyst face, are reduced.
On a proprietary, lean spike, rapid ageing cycle for automotive catalysts, at sufficiently high temperatures, retarding spark affects the temperature profiles both feedgas and midbed, and may ultimately result in more severe catalyst deactivation.

1 INTRODUCTION

There is considerable interest in high temperature stable catalyst technology [1,3], This is expected to cater for a calibration strategy which allows increased catalyst gas-inlet temperatures. The catalysts must be evaluated using a vehicle-representative high temperature ageing cycle. Retarding Spark doesn't involve changing the overall calibration strategy of the engine. It is therefore a convenient method of raising the catalyst gas-in temperature of exhaust emissions for rapid dyno catalyst ageing.

In the combustion chamber, a retarded Spark results in delayed heat-release, allowing only limited heat exchange and thus higher exhaust gas temperatures. However, there is a greater likelihood of incomplete combustion products being available when the exhaust valve opens.
At sufficiently high temperatures (>=750 °C, 25 cm from the catalyst substrate face),

the combustion process continues as the exhaust gases travel between the exhaust valve and the catalyst substrate face. This results in a change in the feedgas composition. There should be a conceivable feedgas difference between close-coupled and underfloor applications. In the close coupled application being discussed below, the concentrations of Oxygen and the legislated emission gases : Hydrocarbons(HC), carbon monoxide(CO), and oxides of nitrogen(NOx), are all reduced. The resulting proportions may give greater catalyst deactivation than would be expected if the higher exhaust gas temperatures had been achieved by better management of Speed, Load and Spark advance.

If the exhaust gas temperatures are relatively low, (<700 °C, 25 cm from the catalyst substrate face), there is minimal combustion in the downpipe, the catalyst has to do more to meet the legislated levels.

On a lean spike cycle for rapid catalyst ageing, the air fuel ratio may be described by the following four events.

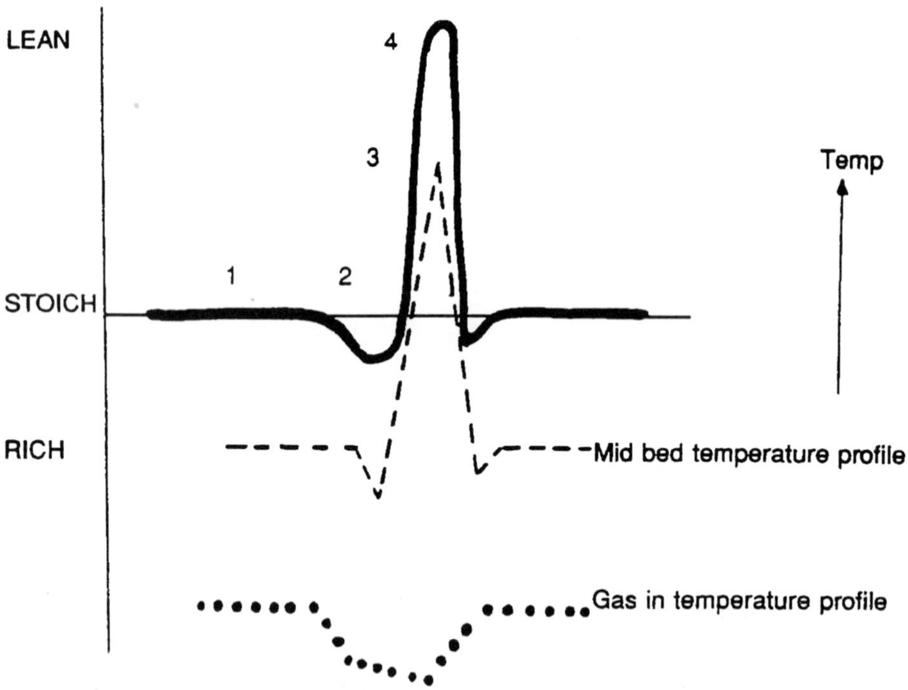

Event 1 - Stoich
Event 2 - Rich Spike - Incomplete combustion from wall wetting, crevices, etc.
Event 3 - Point when rich spike products burn on the catalyst.
Event 4 - Lean

The spikes are repeated over a number of hours to simulate vehicle ageing The expected temperature profiles are shown. At temperatures less than 700 °C, the gas-in temperature profile is as shown, while at higher temperatures, it is similar to the mid bed profile but 60 - 80 °C lower.

This paper describes the effect of retarding Spark in order to raise the temperature to 900 °c and 950 °C on

- Gas in and mid bed temperature profiles.
- Catalyst ageing

in a proprietary catalyst ageing cycle which simulates the lean spike condition and allows measurement of temperatures.

2 EXPERIMENTAL

In the cycle under consideration, the events take place every minute.

1 Catalysts were aged for 75 hours, at gas-in temperatures of 900 and 950 °C achieved with and without spark retard.

1 They were then evaluated by perturbed scan :

Range	:	0.950 - 1.030	Frequency	:	1 Hz
Amplitude:	:	± 1 A/F ratio	Temperature °C	:	450

1 Light off tests were done under steady state Lambda conditions :
Lean : 1.020 Rich : 0.984

Engine : Ford , 2.0 L DOHC, 4 Cylinder, 8 Valve, 115 PSI
Speed : 5,500 rpm
Close Coupled Catalyst

3 RESULTS AND DISCUSSION

As Spark retard is employed to raise the temperature, the concentrations of oxygen and the three legislated exhaust emission gases HC (hydrocarbons), CO (carbon monoxide), and NOx (Oxides of nitrogen) which get to the catalyst face at stoich, are reduced. The HC concentration during events 2 and 3 of the cycle is also reduced.

TABLE1: Factors that change when the gas in temperature is raised by retarding Spark during Event 1.

TEMPERATURE (°C)	850	900	950
Spark Advance °	27.0	21.2	15.0
CO%	0.74	0.58	0.53
O_2%	0.81	0.63	0.43
HC ppm	1896	718	573
NOx ppm	2942	2010	974
Lambda	1.001	1.006	1.000
HC : event 2 (ppm)	2657	2323	2027
HC : event 3 (ppm)	2387	2120	1822

The important aspects of "Lean Spike" catalyst ageing are :

- Temperature
- Air fuel ratio

The most deactivating conditions for a Platinum/Rhodium catalyst are high temperature lean(1).

Light-off temperature and Perturbed Scan data is shown in figures 1 - 18. Perturbed scan data shows that ageing at 900 °C achieved by spark retard results in more deactivation than ageing without spark retard, for rich HC (Fig. 2 and less deactivation for CO and NOx (Fig. 1 and 3). Ageing at 950 °C achieved by spark retard results in more deactivation than ageing without spark retard for HC, CO and NOx.

Light off data shows that the catalyst is less deactivated after ageing with Spark retard at 900 °C, but more severely deactivated after ageing with spark retard at 950. This may suggest that at 900 °C, retarding spark does not result in feedgas with such a lean air fuel ratio for Events 3 and 4 so the dominant ageing effect is that of having reduced the quantity of reactants that reach the catalyst's face. At 950 °C retarding Spark results in more severe deactivation because not only is the temperature higher but the resulting air fuel ratio is considerably leaner.

EFFECT OF GAS-IN TEMPERATURES ON THE TEMPERATURE PROFILES

Gas in Temp	Event	Precat Temp.	Midbed Temp.	Comments
750	1	750	780	Air injection after HEGO
	2	680	765	
	3	730	860	
	4	735	780	
850	1	850	900	Air injection after HEGO
	2	815	885	
	3	770	990	
	4	820	890	
850	1	850	910	Air Injection before HEGO
	2	805	875	
	3	845	985	
	4	810	905	
900	1	900	970	with spark retard
	2	870	940	
	3	880	965	10 C gas-in exotherm,
	4	840	920	25 C mid-bed exotherm
900	1	895	960	without spark retard
	2	870	920	
	3	960	1040	90 C gas-in exotherm,
	4	880	950	120 C mid-bed exotherm
900	1	900	965	without spark retard
	2	875	930	
	3	955	1025	80 C gas-in exotherm
	4	985	960	95 C mid-bed exotherm
950	1	950	1000	with spark retard
	2	910	930	
	3	920	930	10 C gas-in exotherm
	4	870	890	no mid-bed exotherm
950	1	945	990	without spark retard
	2	910	945	
	3	1005	1065	95 C gas-in exotherm
	4	920	970	120 C mid-bed exotherm

WITH vs WITHOUT SPARK RETARD

FIG 1

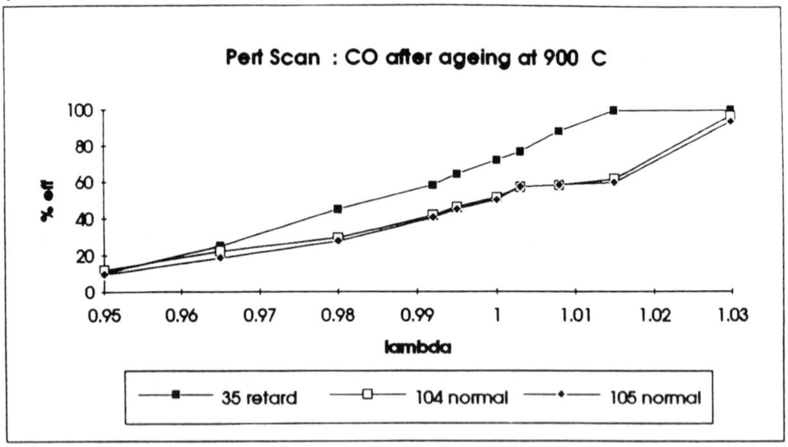

FIG 2

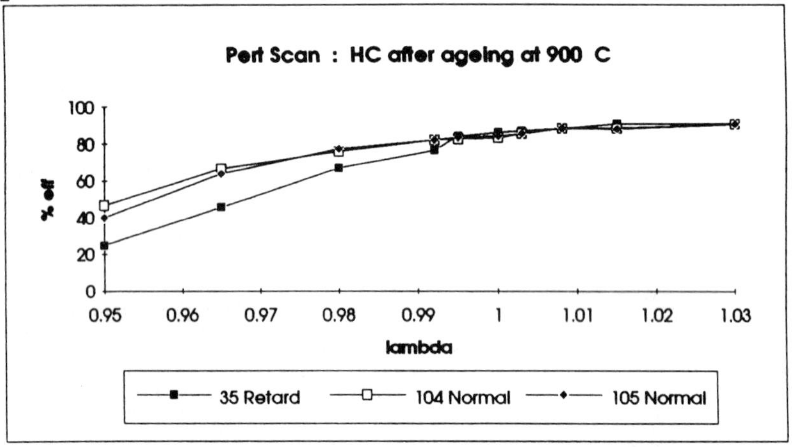

FIG 3

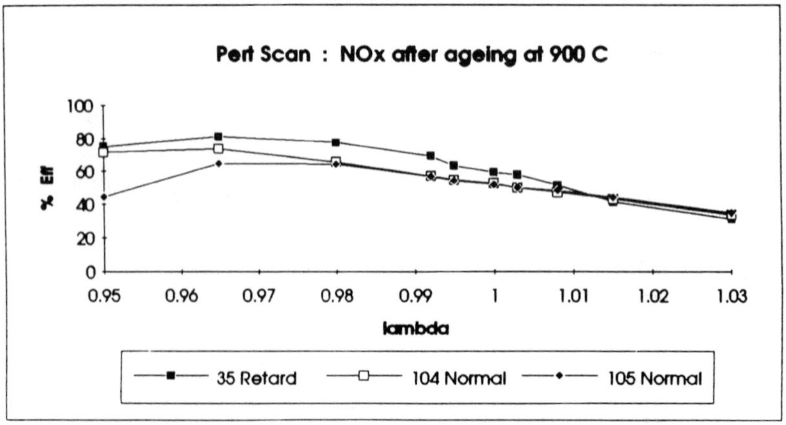

WITH vs WITHOUT SPARK RETARD

FIG 4

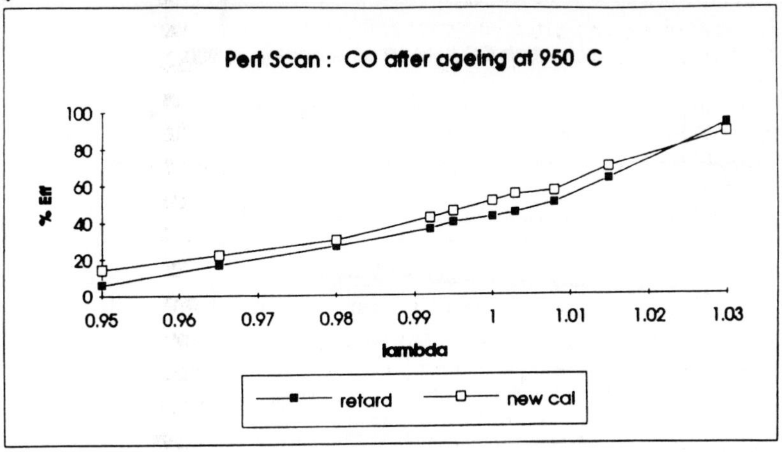

FIG 5

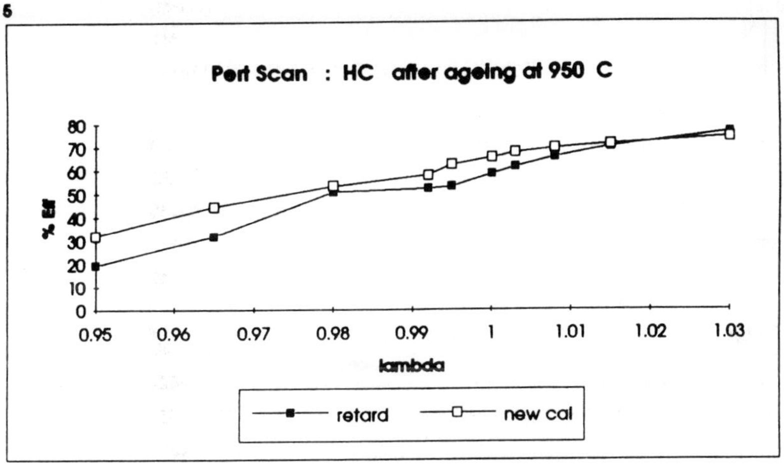

FIG 6

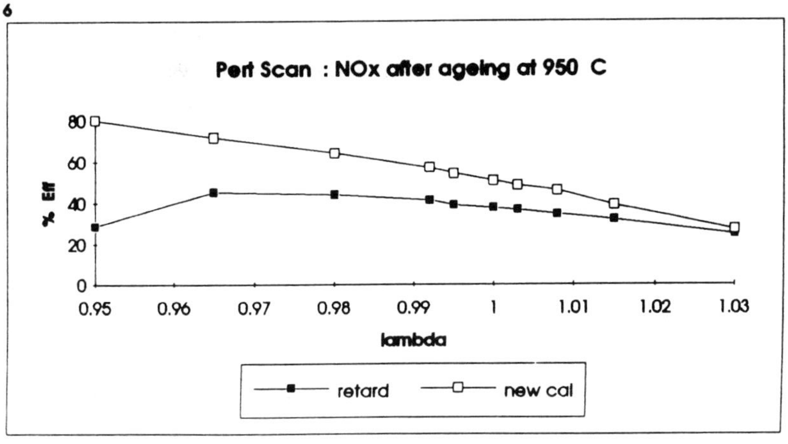

WITH vs WITHOUT SPARK RETARD

FIG 7

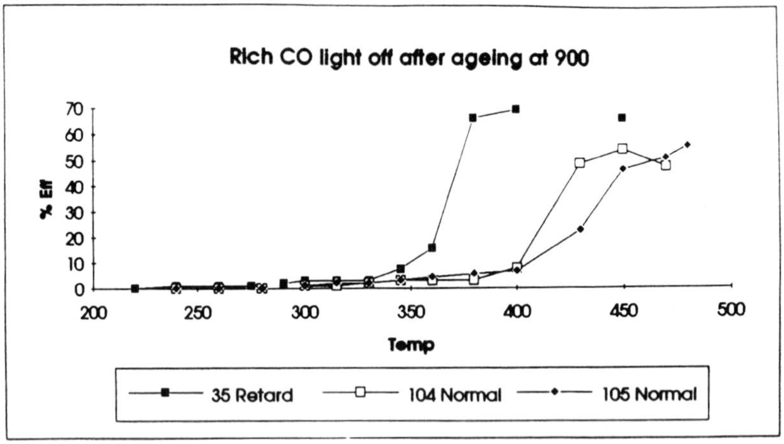

FIG 8

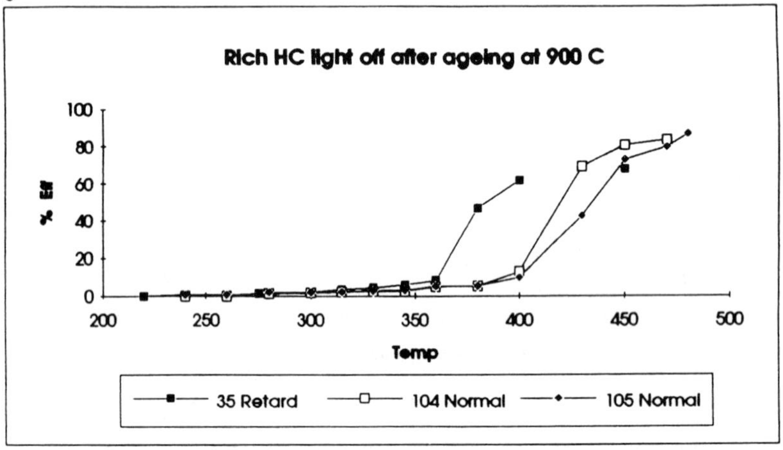

FIG 9

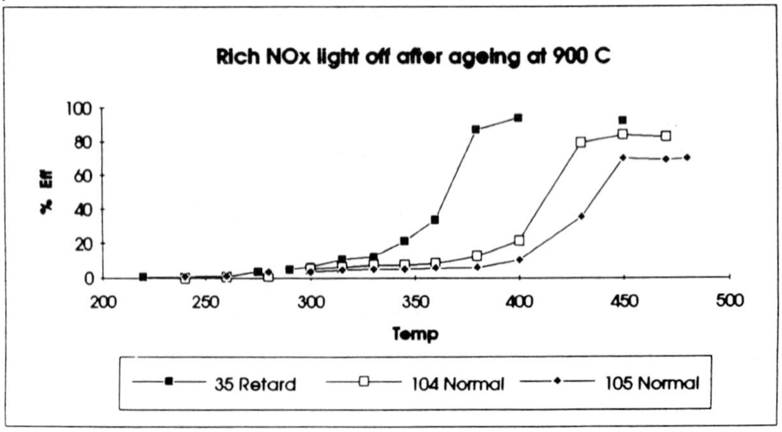

WITH vs WITHOUT SPARK RETARD

FIG 10

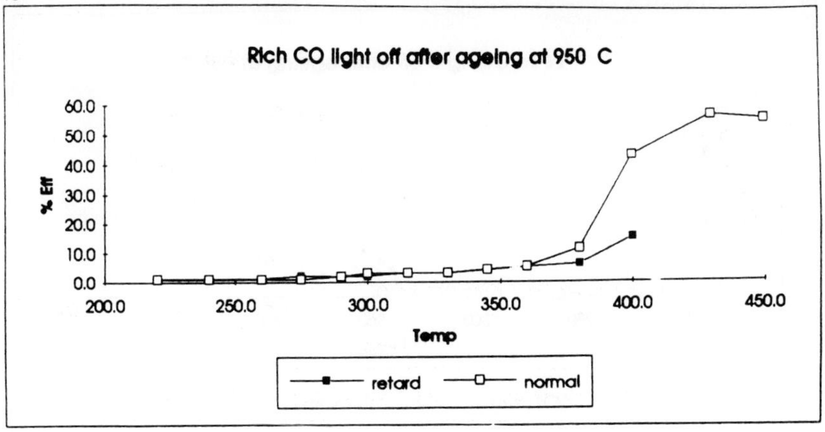

FIG 11

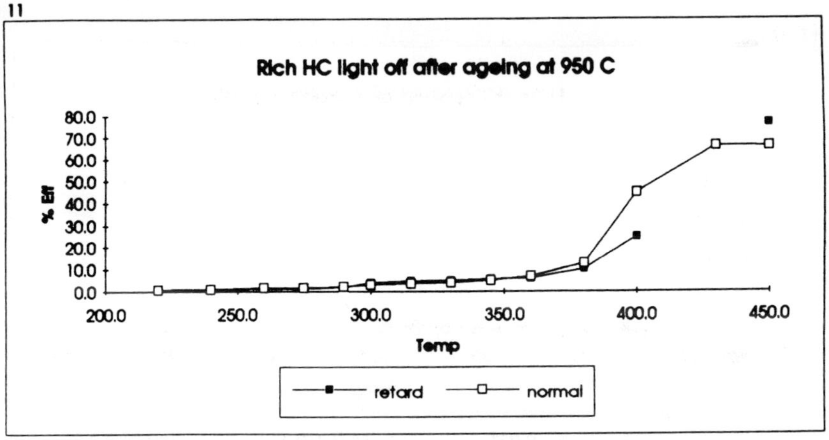

FIG 12

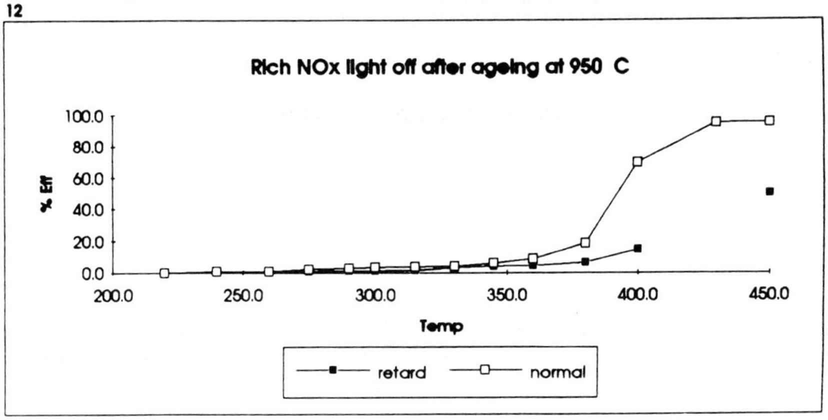

WITH vs WITHOUT SPARK RETARD

FIG 13

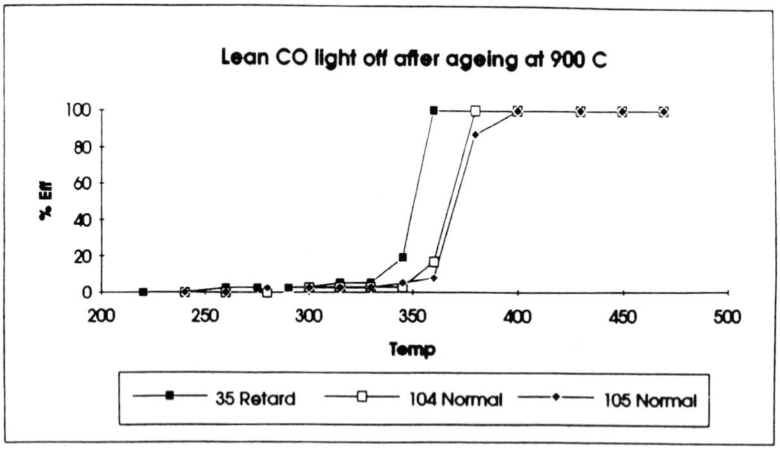

FIG 14

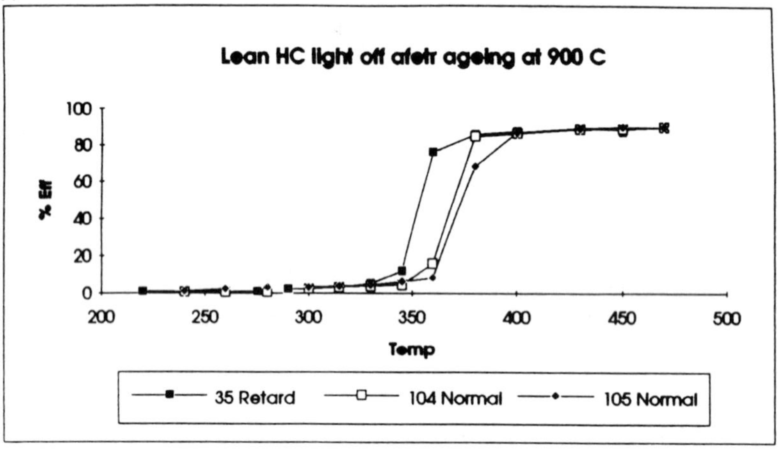

FIG 15

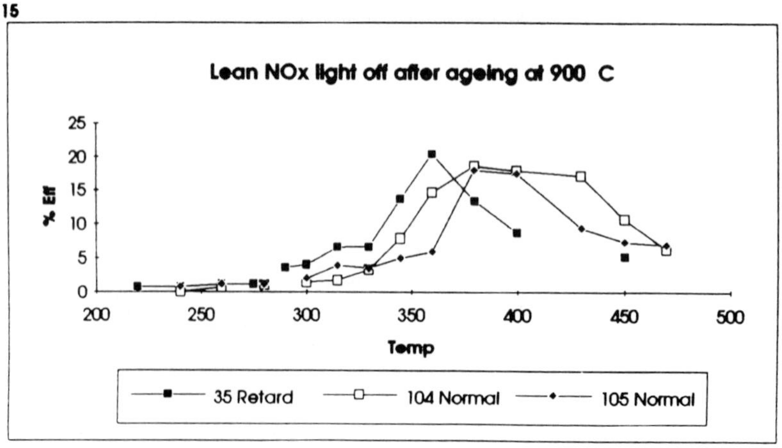

WITH vs WITHOUT SPARK RETARD

FIG 16

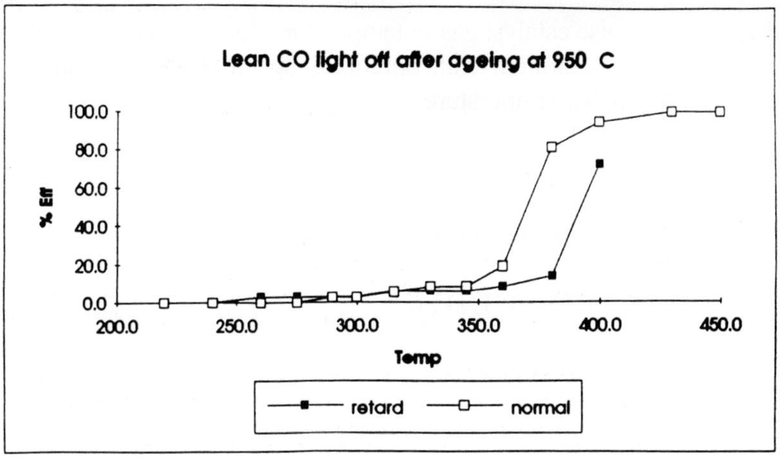

FIG 17

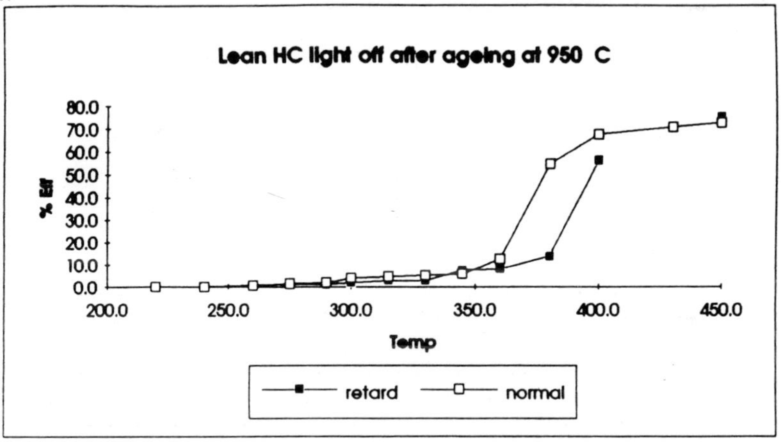

FIG 18

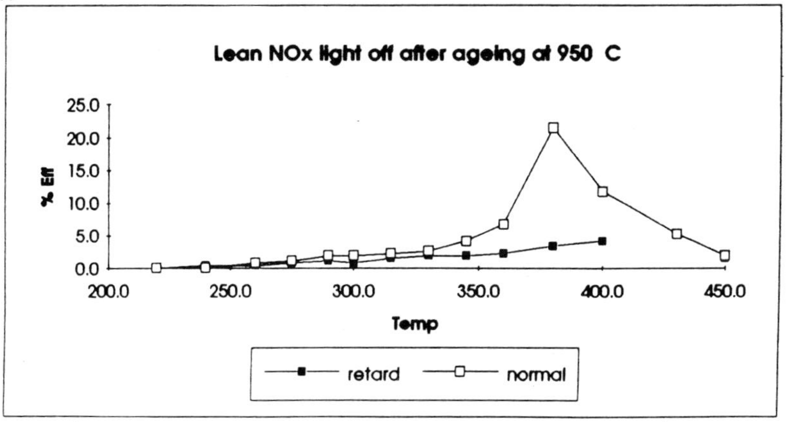

4 Conclusion

Retarding spark to raise catalyst gas-in temperature does not deactivate the catalyst in the same way as a calibration with optimised Speed, Load, and Spark advance parameters for the higher temperature.

References

1 R J Brisley, R D O'Sullivan, and A J J Wilkins
SAE 910175.

2 M L Church, J E Thoss, L D Fizz
SAE 910845.

3 J F Skowron, W B Williamson, J C Summers
SAE 892093

Alternative Fuels and Appropriate Catalysts

A. Frennet and J.-M. Bastin (Eds.)
Catalysis and Automotive Pollution Control III
Studies in Surface Science and Catalysis, Vol. 96

CONTROL OF UNREGULATED EMMISSIONS FROM ETHANOL-FLUELLED DIESEL ENGINES - A STUDY OF THE EFFECT OF CATALYST SUPPORT ON THE LOW TEMPERATURE OXIDATION OF ETHANOL AND ACETALDEHYDE USING PRECIOUS METALS

L.J. Pettersson[a], S.G. Järås[a], S. Andersson[b] and P. Marsh[b]

[a]*Royal Institute of Technology, Department of Chemical Engineering and Technology, Chemical Technology, S-100 44 Stockholm, Sweden*

[b]*Svenska Emissionsteknik AB, Victor Hasselblads gata 8, S-421 31 V:a Frölunda, Sweden*

ABSTRACT

Catalytic oxidation of ethanol and acetaldehyde over precious metal catalysts applied on monolithic cordierite substrates was studied. Platinum or palladium were applied onto a support consisting of either aluminum oxide, cerium dioxide, silicon dioxide or titanium dioxide. The catalysts were tested in a quartz reactor in synthetic exhaust from ethanol-fuelled diesel engines using mixtures of bottled gases. The catalysts were characterized using techniques such as BET surface area measurement, gas porosimetry, hydrogen-oxygen titration, pulse chemisorption of carbon monoxide, and X-ray diffraction. The effect of support material on the low temperature oxidation of ethanol and acetaldehyde in the presence of nitric oxide and carbon monoxide was investigated.

The results indicate that the choice of support material influences the activity and the product distribution in catalytic oxidation of ethanol. Both platinum and palladium are active materials for ethanol oxidation, but the selectivity for acetaldehyde production is high at low temperatures. The selectivity for acetic acid formation is low for the palladium catalysts. The yield of acetic acid in the platinum catalyst experiments is of the same magnitude as the odour threshold. Acetic acid is not formed via oxidation of acetaldehyde.

1. INTRODUCTION

Alcohol fuels have a potential to improve air quality, especially in densely populated areas (Gushee, 1992). A transition to a new fuel will affect the composition of the engine-out emissions. Primarily, the emissions of reactive hydrocarbons, aromatic compounds, particulates and nitrogen oxides will decrease. The emission standards are going to be more stringent for diesel vehicles in the future (Porter *et al.*, 1991). A special problem is meeting the future standards for particulates. This makes the use of low-molecular-weight alcohols, such as methanol or ethanol interesting, since they produce small amounts of particulates. Decreasing the carbon dioxide emissions from the road traffic sector is also regarded as important, which forwards fuels produced from renewable feedstocks, such as alcohols. On the other hand, alcohol fuels produce higher emissions of unburned alcohols and aldehydes compared with petroleum-derived fuels, such as diesel oil.

An important vehicle for public transportation is the city bus, usually fuelled by diesel oil. Combustion of diesel oil causes environmental problems, such as emissions of soot particles, polyaromatic compounds, hydrocarbons, nitrogen oxides and sulfur oxides. To address air pollution problems in the Swedish big city areas, local transport companies have set up fleet tests with alternative-fuelled vehicles. One example is the gas buses in Gothenburg fuelled by compressed natural gas. In Stockholm, Stockholm Transport has a fleet test with 32 ethanol buses (Rydén and Berg, 1991). This is the largest fleet test of its kind in the world.

The typical stop-and-go driving pattern of a city bus gives rise to technical problems. Due to traffic jams, waiting at red lights in street crossings and halting at bus stops, the engine is running idle for a substantial part of its operating time. This means that the exhaust temperature can decrease to below 200°C, with low catalyst activity as a result. Experience from the Stockholm fleet project and other ethanol engine development projects show that substantial amount of by-products can be formed at certain points of operation (Egebäck, 1992). Some of the by-products, such as acetic acid, have a characteristic smell which differs from the smell of the exhaust gas from diesel-fuelled buses. In order to gain acceptance from the public for equipping ethanol-fuelled buses with oxidation catalysts these problems must be solved.

The exhaust gas from ethanol-fuelled vehicles contains unburned alcohols and aldehydes (Chui, 1979; Goodrich, 1982). Aldehydes are formed in the combustion chamber, in the exhaust pipe or over the oxidation catalyst by partial oxidation of unburned alcohol (Windawi, 1992). In California, formaldehyde already belongs to the regulated emissions together with hydrocarbons, carbon monoxide, nitrogen oxides and particulates. Acetaldehyde is one of the five candidates on the list for future regulated emissions proposed by the US Environmental Protection Agency (Knapp, 1992). Acetic acid, which with its characteristic smell can be experienced as an irritant far below directly hazardous concentrations, can also appear in the exhaust from ethanol-fuelled vehicles. Consequently, it is important to minimize these emissions.

Lox *et al.* (1991) have given an overview on emission control for diesel-fuelled vehicles. McCabe and Mitchell (1983 and 1984) have in two articles presented studies on the oxidation of ethanol and acetaldehyde over precious metal and metal oxide catalysts. The most active catalysts were commercial Hopcalite ($CuO-MnO_2$) and 0.1 wt% Pt/Al_2O_3. Yu-Yao (1984), performed a study of catalytic oxidation of ethanol at 100-450°C over various precious metal and base metal oxide catalysts. Gonzalez and Nagai (1985) studied ethanol oxidation over a series of silica-supported noble metal catalysts. In a recent paper Rajesh and Ozkan (1993) have investigated total oxidation of ethanol, acetaldehyde and a methanol-ethanol mixture over supported copper oxide, chromium oxide, and copper oxide/chromium oxide in a gradientless external recycle reactor. Research on formaldehyde oxidation has been reported by McCabe and Mitchell (1988), and McCabe *et al.* (1990). Many of the investigations related above have used packed bed configurations of catalyst particles, which makes a comparison with monolithic flow-through type catalysts difficult. Most of these studies have not included nitric oxide and carbon monoxide in the reactor feed. For a review of exhaust gas catalysts for alcohol vehicles we refer to Pettersson (1991).

The work presented in this paper is the first part of a project aiming at the development of tailor-made oxidation catalysts for diesel engines fuelled by alcohol fuels, ethanol or methanol. The investigation is focused on the influence of support material on the low temperature oxidation of ethanol and acetaldehyde. The study presents results from an experimental investigation with precious metal catalysts applied on monolithic cordierite substrates. Platinum or palladium were applied onto a support consisting of either aluminum oxide, cerium dioxide, silicon dioxide or titanium dioxide.

2. EXPERIMENTAL

2.1. Catalyst preparation

The substrate consisted of ceramic monolith made of cordierite (2 MgO*5 SiO_2*2 Al_2O_3) manufactured by Corning with a cell density of 62 cells/cm^2. The catalyst volume was ca 12 cm^3 and the wall thickness was ca 180 μm as determined by scanning electron microscopy.

The catalysts were prepared by using the following support materials: aluminum oxide (Condea PX 140), cerium dioxide (Molycorp HSA 5315), silicon dioxide (EKA Nobel Bindzil 50/80), and titanium dioxide (Thann et Mulhouse OT 51). According to the manufacturers′ specifications, the BET surface areas were 140, 145, 80 and 80-100 m^2/g, respectively. The monolithic catalyst samples were prepared according to standard techniques at Svenska Emissionsteknik AB. The preparation of the platinum and palladium catalysts was made on an equimolar basis and they contained 9.5 $mmol/dm^3$ monolith, which corresponds to 1.8 g Pt/dm^3 and 1.0 g Pd/dm^3.

2.2 Test gas composition

The catalysts were tested in a gas stream simulating ethanol combustion at 100% excess air (λ=2). The composition was: 10 vol.% oxygen, 10% steam and 6.5% carbon dioxide with nitrogen as the remainder (73.4%). The concentrations of the reactive gases were: 300 ppm carbon monoxide, 600 ppm nitric oxide and either 200 ppm ethanol or 100 ppm acetaldehyde. The experimental conditions were a space velocity of 100,000 h^{-1} and temperatures between 75 and 500°C.

The liquid ethanol used was of spectrographic grade, minimum 99.5 vol.% from Kemetyl. Acetaldehyde was added by means of a gas mixture consisting of 1.00 ± 0.03 vol.% CH_3CHO in nitrogen (99.9995%). The acetaldehyde-nitrogen gas mixture, as well as other gases used, were supplied by AGA Gas Co. and they were of high purity grade (>99.995%).

Prior to activity evaluation all catalysts were subjected to a standardized pretreatment procedure in situ in the reactor in three steps. The samples were heated to 500°C under operating conditions with a heating rate of approximately 10°C/min. The samples were then kept at 500°C for 1 h followed by cooling to reaction temperature in a gas mixture consisting of 10% O_2, 10% H_2O and 80% N_2 with a space velocity of 25,000 h^{-1}.

2.3 Equipment

A laboratory-scale equipment was constructed in order to perform catalyst testing in a gas mixture simulating exhaust gas from ethanol-fuelled diesel engines (see Figure 1). The catalysts were tested in a tubular quartz reactor with an inner diameter of 24.8 mm placed in a furnace. All feed gas flows are regulated by mass flow controllers. Ethanol is evaporated by bubbling nitrogen through two specially constructed consecutive glass cylinders equipped with porous glass filters kept at constant temperature. For details on equipment and analysis we refer to Pettersson and Järås (1993).

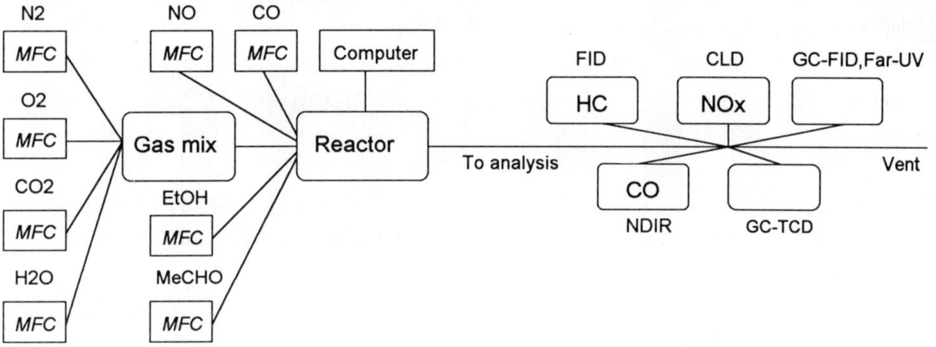

Figure 1. Experimental set-up. MFC: mass flow controller, FID: flame ionization detector, CLD: chemiluminescence detector, GC: gas chromatograph, Far-UV: far ultra-violet detector, NDIR: non-dispersive infra-red instrument, TCD: thermal conductivity detector.

The concentrations of total hydrocarbons, nitrogen oxides and carbon monoxide in the effluent were monitored on-line using conventional continuous analysis instruments. Detailed analysis of the concentrations of oxygen, nitrogen, carbon dioxide, oxygenated compounds and hydrocarbons were made on-line in a semi-continuous manner by gas chromatographs equipped with thermal conductivity, flame ionization and far ultra-violet detectors.

3. Results

3.1 Catalytic oxidation of ethanol

In these experiments oxidation of ethanol with a gas phase concentration of 200 ppm was studied. Figure 2 shows the light-off temperatures (temperature at which 50% conversion of the compound in question occurs) for ethanol oxidation over eight supported platinum and palladium catalysts as measured with a continuous flame ionization detector. Depending on by-product formation the signal from the instrument changes. The sensitivity for ethanol is for example about twice as high as for acetaldehyde. A marked difference can be observed for the activities of the different supported Pt catalysts.

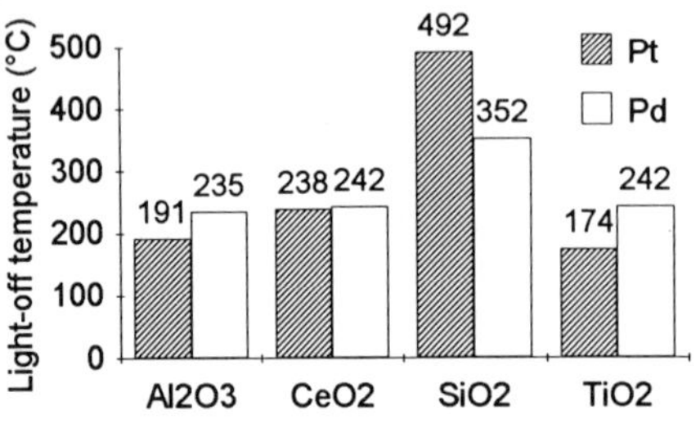

Figure 2. Light-off temperatures for ethanol oxidation using Pt and Pd catalysts as measured by a continuous flame ionization detector

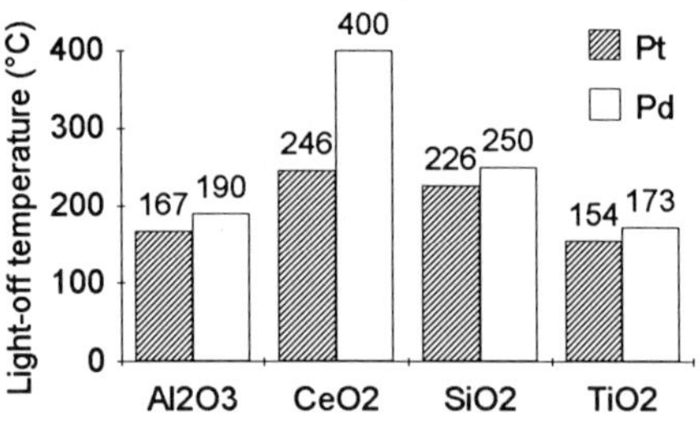

Figure 3. Light-off temperatures for CO oxidation using Pt and Pd catalysts as measured by a continuous NDIR instrument.

In Figure 3 the light-off temperatures for carbon monoxide oxidation for the Pt and Pd catalysts in ethanol oxidation are presented. The curves for CO conversion follow the same pattern as the conversion curves for ethanol oxidation, except for Pt/SiO_2 and the ceria catalysts (cf Figure 4). The light-off temperatures for the cerium dioxide catalysts are rather high. This seems reasonable, since it can be expected that the activity of the palladium catalyst can be strongly affected by ceria at high oxygen concentrations (Shyu *et al.*, 1988).

Figure 4 shows the total conversion of ethanol as a function of temperature as measured by gas chromatography. Except for the silica catalysts, the platinum catalysts exhibit equal or lower light-off temperatures than the supported catalysts with palladium as active material (compare with Figure 7). The platinum on alumina and platinum on titania catalysts are more active than the other catalyst combinations. The conversion curves for the Pd and Pt on ceria catalysts practically coincide, which implies that ceria would be a more suitable support material for a palladium catalyst than for a platinum catalyst. The activities of the silica catalysts are low. This observation is consistent with recent results in another research project using the same type of silica sol (Zwinkels *et al.*, 1994). According to these experiments, it is crucial to reduce the alkali content to a very low level in the support, since sodium increases the mobility of silica, which poisons the active platinum and palladium sites. Platinum is apparently more sensitive to this phenomenon than palladium.

In Figure 5 the yields of acetaldehyde over the Pt catalysts are plotted. Both the alumina and titania catalysts are active at low temperatures, but the titania catalyst produces less aldehydes. The platinum on alumina catalyst produces a significant amount of acetaldehyde below 250°C. Pt/CeO_2 produces less acetaldehyde than Pt supported on alumina over the temperature interval studied but more than Pt supported on titania.

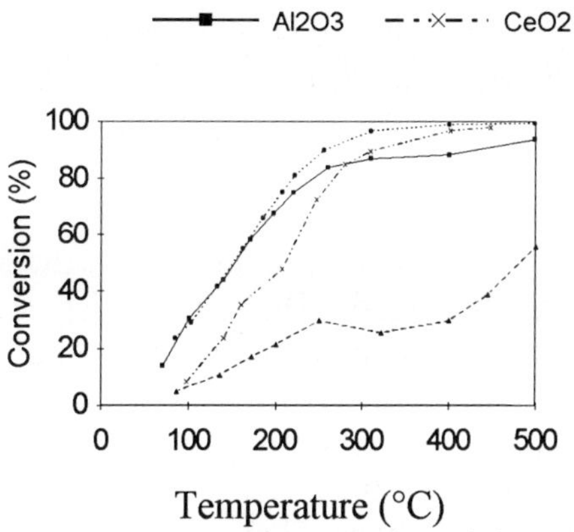

Figure 4. Conversion of ethanol for platinum catalysts analyzed by GC

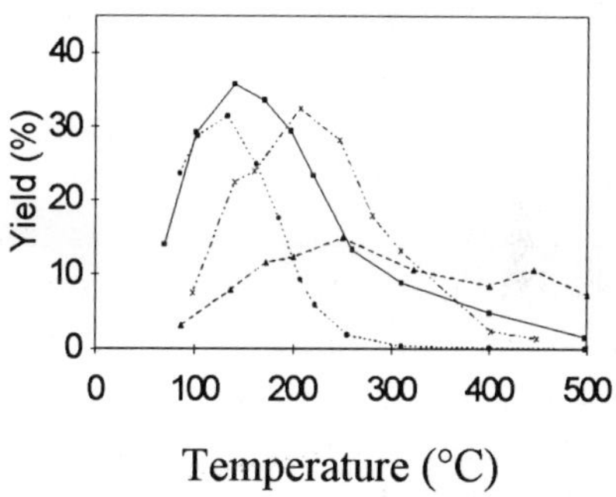

Figure 5. Acetaldehyde yield over platinum catalysts in ethanol oxidation.

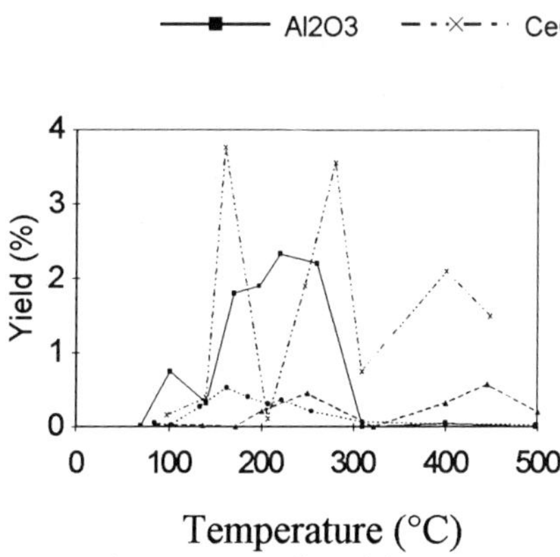

Figure 6 acetic acid yield over platinium catalysts in ethanol oxidation

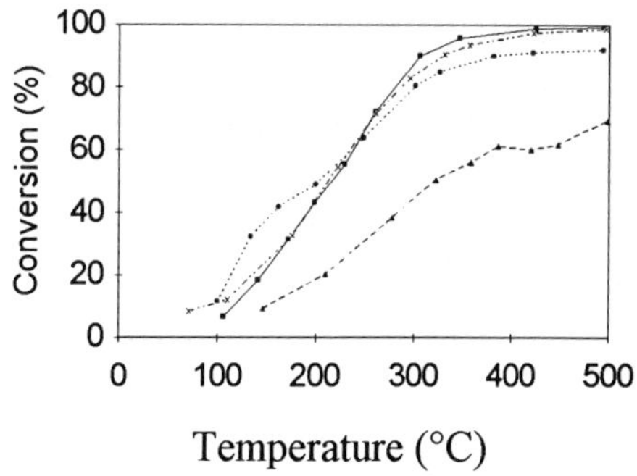

Figure 7 Conversion of ethanol over palladium catalysts analysed by GC

Figure 6 shows the yield of acetic acid over platinum catalysts. Pt supported on alumina and ceria gives rise to the highest production of this compound. Here it must be pointed out that the scattering of data is high especially for the CeO_2 catalyst. The level of acetic acid emissions is rather low for these catalysts, but the highest values are nevertheless of the same magnitude as the odour threshold.

The palladium catalysts do not exhibit the pronounced difference in ethanol oxidation activity between the different support materials as is observed for the platinum catalysts (see Figure 7). Palladium on titania is the most active catalyst below 200°C. As in the case of the platinum catalysts, the activity of the silica catalyst differs from the other three. The activity of Pd/SiO_2 levels off at a higher conversion, though.

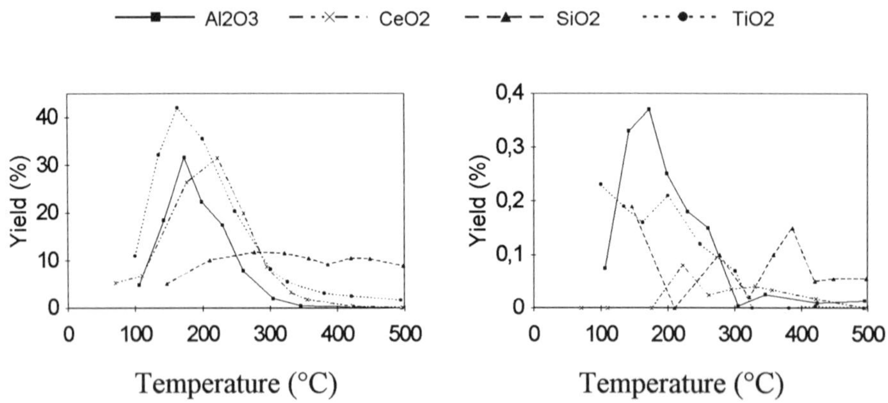

Figure 8. Yield of acetaldehyde for palladium for palladium catalysts in ethanol oxidation.

Figure 9. Yield of acetic acid catalysts in ethanol oxidation.

The yield of acetaldehyde over the palladium catalysts (see Figure 8) is qualitatively the same as for the platinum catalysts shown in Figure 5, though the yield of acetaldehyde is higher for palladium on titania than with platinum as active material.

In Figure 9, it can be observed that the yield of acetic acid for the palladium catalysts is an order of magnitude lower compared with the yield of acetic acid for the platinum catalysts in this study. Pd/Al_2O_3 produces the highest amounts of acetic acid of the Pd catalysts evaluated.

3.2 Catalytic oxidation of acetaldehyde

In these experiments, oxidation of acetaldehyde with a gas phase concentration of 100 ppm was studied. Using the catalysts in this study, acetaldehyde is oxidized without any side reactions of importance. Only the HC-FID curves are shown, since the conversion curves for the continuous FID instrument and the gas chromatograph coincide. The light-off temperatures for acetaldehyde oxidation are higher than for ethanol oxidation. This means that if acetaldehyde is formed it is more difficult to eliminate than unburned ethanol.

The Pt catalysts are more active for acetaldehyde oxidation than the Pd catalysts (see Figures 10 and 11), especially when titania is used as support material. Below 300°C, Pt supported on TiO_2 and Al_2O_3 has a higher activity than Pt supported on CeO_2. Pd on alumina is more active than the other Pd catalysts. The silica catalysts exhibit the same plateau-shaped curves as for the ethanol oxidation experiments (see section 3.1). Compared with the ethanol oxidation experiments the difference is here that the Pt catalyst is slightly more active than Pd on silica.

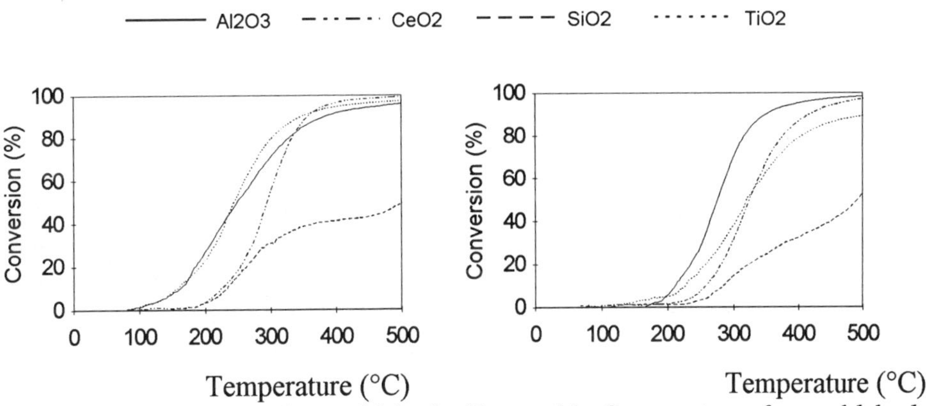

Figure 10. Conversion of acetaldehyde over platinum catalysts

Figure 11. Conversion of acetaldehyde over palladium catalysts

4. Catalyst Characterization

The catalysts were characterized using the following techniques: BET surface area measurement, gas porosimetry, hydrogen-oxygen titration, carbon monoxide chemisorption, scanning electron microscopy, and X-ray diffraction. Prior to characterization, the entire monoliths were crushed and sieved to a particle size less than 250 μm (60 mesh).

4.1. BET surface area measurements and X-ray diffraction analysis

The BET surface area measurements and pore volume measurements were performed on a Micromeritics ASAP 2000 by nitrogen adsorption using a volumetric method. The BET surface areas for the fresh alumina, ceria, silica and titania catalysts were 124, 91, 64 and 87 m^2/g washcoat, respectively. Spent catalysts from the ethanol oxidation experiments have been characterized. The results from these measurements show that the platinum on ceria catalyst has lost more than half of its original surface area, while Pd/CeO_2 exhibits stability towards sintering (see Table 1). The titania catalysts have lost ca 30% of their surface areas, while the surface areas of the alumina and silica catalysts have remained practically unchanged.

X-ray diffraction has been performed to clarify the structure of the titanium dioxide after the activity test. The measurements were performed on a Siemens D5000 X-ray diffractometer. The crystallographic structure shows that the sample contained only anatase, which means that no structural changes had occurred during catalyst preparation and activity evaluation.

4.2. Metal dispersion measurements

The Micromeritics TPD/TPR 2900, temperature-programmed chemisorption instrument, was used for metal dispersion measurements. Oxygen traps (Chrompack 7970) were used for the helium and argon carrier gases. Pulse chemisorption by carbon monoxide was utilized for the palladium catalysts as well as for the platinum on ceria catalyst and hydrogen-oxygen titration was utilized for the rest of the platinum catalysts. The catalysts were subjected to a pretreatment procedure where the catalysts were reduced in 45 cm^3 H_2/min in three steps; 30 min at 120°C, 30 min at 220°C and 2 h at 350°C. The heating rate was 10°C/min. Afterwards, the sample was flushed with 45 cm^3 helium/min at 350°C for 90 min, after which the temperature was subsequently decreased to 35°C in flowing helium. CO chemisorption measurements were performed using the dynamic pulse method (Sarkany and Gonzalez, 1982). Pulses were given at 35°C. The volume of the injected pulses in these measurements was 50 μl. CO/Pd stoichiometry was assumed to be 1. The hydrogen-oxygen titration measurements were carried out according to principles developed by Benson and Boudart (1965) and Prasad *et al.* (1978).

The results are summarized in Table 1. The metal dispersion is defined as the ratio of the number of surface atoms exposed to the gas phase and the total number of atoms present in the catalyst. The low metal dispersion indicates that some sintering of the precious metal particles has occurred during the experiments (cf Lööf *et al.*, 1993).

Table1. Catalyst characterization of spent catalysts in the ethanol oxidation experiments

	BET Surface area (m^2/g washcoat)				Metal dispersion (%)			
	Al_2O_3	CeO_2	SiO_2	TiO_2	Al_2O_3	CeO_2	SiO_2	TiO_2
Pt	114	41	63	58	2	12[1]	0.5	4
Pd	123	91	64	62	22	74	2	3

[1]Pulse chemisorption of CO, CO/Pt=1. H_2-O_2 titration was not applicable.

5. DISCUSSION

Combustion of fuels in internal combustion engines, causes emissions of aldehydes. Aldehydes are either formed in the cold zones of the combustion chamber, in the exhaust pipe or over the oxidation catalyst by partial oxidation of unburned alcohol. Acetaldehyde is the principal aldehyde formed in ethanol combustion and it is one of the compounds, which the US Environmental Protection Agency proposes as a future regulated pollutant. Acetic acid, which has a characteristic smell and can be experienced as an irritant far below directly hazardous concentrations, can also appear in the exhaust from ethanol-fuelled vehicles. Consequently, it is important to minimize these emissions.

The oxidation catalyst lowers the activation energy for the total oxidation reaction and efficiently converts ethanol to carbon dioxide and water at low temperatures (reaction 1). However, if the selectivity for this reaction is low, by-products will be formed. One example is acetaldehyde, which can be formed by partial oxidation of unburned ethanol or by decomposition of unburned ethanol (reactions 2 and 3). Considering that there is a 100% excess of air in our experiments, the decomposition of ethanol is not so likely to occur. The oxidative dehydration is a more plausible reaction in this case.

Experience from fleet tests indicates that at certain points of operation the concentrations of acetic acid can cause odour problems with the type of oxidation catalysts that has been used so far. If we study the possible reaction pathways, we see that acetic acid can be formed directly from ethanol or via acetaldehyde (reactions 4 and 5). The results from the acetaldehyde oxidation experiments give, however, clear evidence that acetic acid is not being formed via the acetaldehyde route (see reaction 5) over these catalysts.

$$C_2H_5OH + 3\,O_2 \rightarrow 2\,CO_2 + 3\,H_2O \quad (1)$$
$$C_2H_5OH + 0.5\,O_2 \rightarrow CH_3CHO + H_2O \quad (2)$$
$$C_2H_5OH \rightarrow CH_3CHO + H_2 \quad (3)$$
$$C_2H_5OH + O_2 \rightarrow CH_3COOH + H_2O \quad (4)$$
$$CH_3CHO + 0.5\,O_2 \rightarrow CH_3COOH \quad (5)$$

The differences in acetic acid yields appear to be small between the Pt and Pd catalysts, which are investigated in this study (see section 3.1). We have, however, indications from full-scale tests which show that these differences are quite significant for the possible elimination of odour problems (Pettersson *et al.*, 1994).

The criteria for selection of a suitable oxidation catalyst for heavy-duty ethanol-fuelled vehicles in city traffic involve aspects such as regulated emissions (CO, NO_x, hydrocarbons and particulates), emissions of unburned ethanol, formation of acetaldehyde, acetic acid, and nitrogen dioxide. The oxidation of nitric oxide to nitrogen dioxide can be a problem when using highly active precious metal catalysts. This is especially the case when using platinum as active material (Pettersson *et al.*, 1994). NO_2 is more toxic than NO and should be minimized at street levels. The catalyst with the highest activity for ethanol conversion is not necessarily the best choice if the minimum environmental impact is the objective.

The most promising catalysts in this study are currently being evaluated in full scale in bench tests with an ethanol-fuelled diesel engine at the Chalmers University of Technology in Gothenburg. At the Department of Thermo and Fluid Dynamics a 7 dm^3 Volvo truck engine, which is used for the tests, has been developed for ethanol operation (Gjirja and Olsson, 1993).

6. CONCLUSIONS

This study shows that the choice of support affects product distribution at low temperature catalytic oxidation of ethanol. Especially Pt and Pd supported on titania exhibit high activities at temperatures in the interval 75-150°C compared with Al_2O_3, CeO_2 and SiO_2 as support materials. When using ceria the activity for ethanol oxidation was nearly identical for Pd and Pt as active materials. Both platinum and palladium are active for ethanol oxidation, but the selectivity for total oxidation is low at temperatures below 250°C and, consequently, a lot of by-products are formed at these temperatures. The most important by-product is acetaldehyde, which is formed by oxidative dehydration of ethanol. There is clear evidence that acetic acid is not formed via acetaldehyde at the experimental conditions used. The yields of acetic acid are rather low, though the highest values are of the same magnitude as the odour threshold.

The selection of the optimal catalyst for heavy-duty ethanol vehicles in city traffic from the environmental point-of-view is a question of choosing the oxidation catalyst which produces the least hazardous combination of reaction products. It is not sufficient to consider only the activity for eliminating the regulated emissions. Considering formation of acetaldehyde, acetic acid and nitrogen dioxide, as well as emissions of unburned ethanol and carbon monoxide, there are some interesting catalyst combinations, which will be investigated further both in laboratory and full-scale experiments.

ACKNOWLEDGEMENTS

Financial support to this work given by the Swedish Transport and Communications Research Board (KFB) and the Swedish National Board for Industrial and Technical Development (NUTEK) is gratefully acknowledged. The authors are indebted to Professor Govind Menon for valuable discussions and suggestions.

REFERENCES

1 Benson, J.E. and Boudart, M. (1965). Hydrogen-oxygen titration method for the measurement of supported platinum surface areas. J. Catal. 4, 704-710.

2 Chui, G.K., Anderson, R.D., and Baker, R.E. (1979). Brazilian vehicle calibration for ethanol fuels. Proc. III Int. Symp. Alcohol Fuels Technol., Asilomar, California, May 28-31, Vol. II.

3 Egebäck, K.-E. (1992). Fleet Test with 32 Ethanol Buses at Stockholm Transport. Emissions from Ethanol-Fuelled Buses. AB Svensk Bilprovning, Haninge, Sweden, Report No MTC 9056 S. In Swedish.

4 Gjirja, S. and Olsson, E. (1993). Development of An Ethanol Fuelled Version of the Volvo TD73 Diesel Engine. Chalmers University of Technology, Dept. of Thermo and Fluid Dynamics.

5 Gonzalez, R.D. and Nagai, M. (1985). Oxidation of ethanol on silica supported noble metal and bimetallic catalysts. Appl. Catal. 18, 57-70.

6 Goodrich, R.S. (1982). Brazil´s alcohol motor fuel program. Chem. Eng. Prog. 78 (1), 29-34.

7 Gushee, D.E. (1992). Alternative fuels for cars: Are they cleaner than gasoline? Part 2. CHEMTECH 22, 470-476.

8 Knapp, K. (1992). Sampling and analysis of aldehydes. Proc. Seminar on Sampling and Analysis of Unregulated Automotive Emissions, Dalarö, Sweden, April 29-30.

9 Lox, E.S., Engler, B.H., and Koberstein, E. (1991). Diesel emission control, in Crucq, A. (Ed.), Catalysis and Automotive Pollution Control II, Elsevier Science Publishers, Amsterdam.

10 Lööf, P., Stenbom, B., Nordén, H., and Kasemo, B. (1993). Rapid sintering in NO of nanometre-sized Pt particles on γ-Al_2O_3 observed by CO temperature-programmed desorption and transmission electron microscopy. J. Catal. 144, 60-76.

11 McCabe, R.W. and Mitchell, P.J. (1983). Oxidation of ethanol and acetaldehyde over alumina-supported catalysts. Ind. Eng. Chem. Prod. Res. Dev. 22, 212-217.

12 McCabe, R.W. and Mitchell, P.J. (1984). Reactions of ethanol and acetaldehyde over noble metal and metal oxide catalysts. Ind. Eng. Chem. Prod. Res. Dev. 23, 196-202.

13 McCabe, R.W. and Mitchell, P.J. (1988). Exhaust-catalyst development for methanol-fueled vehicles. III. Formaldehyde oxidation. Appl. Catal. 44, 73-93.

14 McCabe, R.W., King, E.T., Watkins, W.L.H., and Gandhi, H.S. (1990). Laboratory and vehicle studies of aldehyde emissions from alcohol fuels. SAE Paper 900708.

15 Pettersson, L.J. (1991). Catalytic treatment of emissions. State of the art for alcohol and natural gas-fueled vehicles. Royal Institute of Technology, Dept. of Chemical Technology, Stockholm (Under contract by the IEA), ISRN KTH/KT/FR--91/9--SE.

16 Pettersson, L.J. and Järås, S.G. (1993). Catalytic abatement of emissions from alcohol-fueled diesel engines. Final Report, Phase 1. Swedish National Board for Industrial and Technical Development, Stockholm, NUTEK Report No U823-91-02088. In Swedish.

17 Pettersson, L.J., Järås, S.G., Andersson, S., Marsh, P., Gjirja, S., and Olsson, E. (1994). Development of exhaust gas catalysts for ethanol-fueled heavy-duty vehicles in city traffic. Proc. 6th Nordic Symposium on Catalysis, Hornbaek, Denmark, June 1-3.

18 Porter, B.C., Doyle, D.M., Faulkner, S.A., Lambert, P., Needham, J.R., Andersson, S.E., Fredholm, S., and Frestad, A. (1991). Engine and catalyst strategies for 1994. SAE Paper 910604.

19 Prasad, J., Murthy, K.R., and Menon, P.G. (1978). The stoichiometry of hydrogen-oxygen titrations on supported platinum catalysts. J. Catal. 52, 515-520

20 Rajesh, H. and Ozkan, U.S. (1993). Complete oxidation of ethanol, acetaldehyde, and ethanol/methanol mixtures over copper oxide and copper-chromium oxide catalysts. Ind. Eng. Chem. Res. 32, 1622-1630.

21 Rydén, C. and Berg, R. (1991). Extensive tests on ethanol-operated buses fitted with diesel engines. Proc. IX Int. Symp. on Alcohol Fuels, Firenze, Nov. 12-15, pp. 792-796.

22 Sarkany, J. and Gonzalez, R. (1982). On the use of the dynamic pulse method to measure metal surface areas. J. Catal. 76, 75-83.

23 H.S. (1988). Characterization of Pd/γ-alumina catalysts containing ceria. J. Catal. 114, 23-33.

24 Windawi, H. (1992). Controlling the exhaust emissions from alternative fuel vehicles. Platinum Metals Rev. 36, 185-195.

25 Yu-Yao, Y.F. (1984). Catalytic oxidation of ethanol at low concentrations. Ind. Eng. Chem. Process Des. Dev. 23, 60-67

26 Zwinkels, M.F.M., Järås, S.G., and Menon, P.G. (1994) Preparation of combustion catalysts by washcoating alumina-whiskers-covered metal monoliths using a sol-gel method. Proc. 6th Int. Symp. on Scientific Bases for the Preparation of Heterogeneous Catalysts, Louvain-la-Neuve, Belgium, Sep. 5-9, Vol. I, pp. 85-94.

A. Frennet and J.-M. Bastin (Eds.)
Catalysis and Automotive Pollution Control III
Studies in Surface Science and Catalysis, Vol. 96

CATALYSTS FOR NATURAL GAS EMISSION CONTROL APPLICATIONS

R. G. Silver and J. C. Summers+

AlliedSignal Environmental Catalysts, P.O. Box 580970, Tulsa, Oklahoma 74158-0970

+*Present Address: Rhone Poulenc, Cranbury Research Center, CN 7500, Cranbury, New Jersey 08512-7500*

ABSTRACT

Several parameters that affect catalytic methane oxidation on a natural gas vehicle were investigated with laboratory aged noble metal catalysts and a simulated vehicle exhaust. These include the air/fuel control strategy, the noble metal loading, the role of base metals and the exhaust hydrocarbon composition. The catalytic performance of several formulations was compared both slightly fuel-rich of the stoichiometric point and under extreme lean conditions.

Catalyst formulations different from those used near the stoichiometric point appear to be required if the vehicle is run under extreme lean conditions, such as those found with diesel engines. Under diesel conditions low exhaust temperatures may pose a challenge, since methane requires relatively high temperatures for catalytic oxidation. Spark ignition lean burn engines might offer a higher temperature exhaust, but NOx conversion becomes more problematic in this environment.

Increasing the Pd load improves methane and NOx conversions near the stoichiometric point and methane activity under lean conditions. Ce addition was found to be beneficial for aged catalyst performance and increasing the Ce loading resulted in improved methane activity over Pd near the stoichiometric point.

For natural gas vehicles run under closed loop control near the stoichiometric point, the composition of the controlled exhaust will modulate around the set point. The effect of the amplitude and frequency of these modulations on methane oxidation was explored. Increasing the frequency or decreasing the amplitude of exhaust modulations results in improved methane conversions, especially near the maximum methane conversion point.

The effect of natural gas fuel composition was investigated with feedstreams containing various hydrocarbon mixtures. Small amounts of propane in the feedstream led to increased hydrocarbon conversions lean of the stoichiometric point. This improvement may be due to propane reacting with and removing surface oxygen species, which otherwise block methane adsorption on the catalyst surface. Larger amounts led to improvements in hydrocarbon conversion rich of the stoichiometric point as well, due to the easier oxidation of propane.

The information gained from the above studies was used to design a catalyst for a natural gas vehicle tested by the U. S. EPA. With this catalyst, the vehicle achieved California Ultra Low Emmission Vehicule standards.

1. INTRODUCTION

Natural gas (NG) has recently become a prominent alternative fuel for light-duty vehicles, because it offers a combination of economic, technical and environmental advantages. NG is less expensive than gasoline on an energy equivalent basis and is readily available from many sources [1,2] It also has better knock resistance and can be used over a wider range of air-fuel (A/F) ratios than gasoline, allowing engines to operate at high compression ratios with fuel-lean mixtures. This results in higher fuel efficiencies than gasoline engines [2]. Consisting primarily of methane (85-95%), NG also provides environmental benefits such as low photochemical reactivity [3] and zero evaporative emissions. Cold start CO emissions are lower than from comparable gasoline vehicles [4]. Methane's high H/C ratio results in lower CO_2 emissions (the most common "greenhouse gas"), but methane itself is a more powerful greenhouse gas than CO_2 [5,6].

An NG vehicle can run either near the stoichiometric point, or under fuel-lean conditions, due to its wide ignition range. If the vehicle is run near the stoichiometric point, optimum conversion of methane is obtained under slightly rich conditions. Running the engine under fuel-lean conditions provides better fuel economy and lower CO and NO_x emissions [7]. Adequate NO_x reduction under fuel lean conditions is not yet practical, so tailpipe NO_x emissions are likely to be lower with operation near the stoichiometric point. This study primarily focused on controlling emissions from vehicles running near the stoichiometric point, which is where conventional three-way catalysts will be the most useful.

The exact composition of NG is highly variable, but it generally contains significant amounts of C_2H_6 and C_3H_8 in addition to the primary constituent, methane [8,9]. It is also very low in sulfur, with an industry-wide average content of 10 ppm (mass basis). Sulfur content ranges from a minimum of 3 ppm to a maximum of 30 ppm. In addition, over 67% of this sulfur is in the form of odorants added to the gas for leak detection [10]. Therefore, the impact of SO_2 is expected to be small since the sulfur content of gas is so low [11].

Methane is one of the most difficult hydrocarbons to oxidize [7,12]. A recent paper indicated that activity for methane oxidation followed the trend of Pd > Rh > Pt in the absence of CeO_2 for both oxidizing and reducing conditions [11]. The activity ranking for methane conversion in reducing feedstreams was not affected by the presence of CeO_2. Due to their superior activity for methane oxidation, Pd catalysts were the main focus of this study.

Methane oxidation over supported Pd catalysts has been investigated previously. It is thought to be a structure sensitive reaction [13]. Cullis and Willat reported that reduced palladium on alumina is more active than the oxidized form [14]. Hicks, *et al.* showed increasing Pd dispersion led to increased formation of PdO and a simultaneous decrease in the reaction rate [15]. A more recent paper claims the optimum state of palladium for methane oxidation is bulk reduced Pd with surface PdO [16]. Methane activity has been observed to increase with time on stream [17]. An increase in Pd/PdO particle size is observed concurrently with the increase in activity [17,18]. Briot, *et al.* [17] suggest that the reactivity of adsorbed oxygen is greater for oxygen adsorbed on large Pd/PdO particles than on small particles. They note that this may be supported by the work of Chou and Vannice [19] who found that the heat of oxygen adsorption on Pd decreases as the metal particle size increases. In contrast, Hicks' results [15] seem to imply that bulk PdO is less likely to form with larger Pd particles and this inhibition also leads to greater activity since the Pd is maintained in a more active state. Whatever the explanation, there appears to be general agreement among investigators that turnover number increases with decreasing Pd dispersion.

Under oxidizing conditions, ceria may promote PdO formation [20]. Summers and Ausen found ceria addition led to decreased Pd dispersion for samples aged in air at 900°C for 6 hours [21]. Ceria appears to inhibit the formation of CO and retards the oxidation of CH_4.

Most of the studies of methane oxidation in the literature utilize simplified feedstreams and fresh catalyst samples, generally in the form of powders or pellets. A recent paper [22] used a laboratory simulated NG vehicle exhaust to study the removal of methane, NO and CO using a Pd-only monolith catalyst. They found that optimum conversion of all three constituents occurred slightly rich of stoichiometry. These results appear to have been obtained over fresh catalyst samples. The present work utilizes monolith catalysts and laboratory simulated NG vehicle exhaust to study the effect of catalyst loading and space velocity, CeO_2 addition and variations in hydrocarbon composition. The effect of modulation amplitude and frequency around the stoichiometric point was also studied. Unlike previous studies, most of the work reported here utilized laboratory aged samples.

2. Experimental Procedure

Catalysts were prepared by impregnating the noble metal chloride onto either an alumina washcoat or a proprietary washcoat containing alumina, ceria and other base metals. The catalyst was supported on a monolithic cordierite substrate with 64 square cells/cm^2. Cylindrical cores used for laboratory evaluations were 2.5 cm in diameter and, unless otherwise noted, 5 cm in length. The length of each core was composed of smaller segments taken from various locations down the monolith bed in order to minimize sampling biases.

Catalysts were aged either in the lab or on an engine. In the lab, sample cores were hydrothermally aged by heating at 1000°C for 4h in a gas stream of 10% water in air in a Lindberg 3-zone tube furnace. Engine aging was accomplished using an engine operating at the stoichiometric point with an inlet temperature of 850°C and with fuel cuts resulting in average bed temperatures of 910°C. The engine used gasoline containing 4 mg Pb/L, 0.5 mg P/L and 90 ppm sulfur and the catalysts were aged for 75h.

Evaluations were done at various air/fuel ratios (or lambdas), using an HORIBA automated exhaust system analyzer. Lambda is defined as the air/fuel ratio of the exhaust divided by the air/fuel ratio of the exhaust at the stoichiometric point. Thus lambda less than 1 corresponds to a fuel-rich exhaust, while lambda greater than 1 corresponds to an oxygen-rich exhaust.

Lightoff tests were run from 20 to 600°C over typical diesel and NG catalysts using simulated diesel exhaust, then repeated by replacing the HC in the diesel test with methane and removing the sulfur. This substitution was to approximate the change in exhaust composition that might occur in switching a diesel engine to run on natural gas. Line-out tests were run at 500°C using simulated exhaust from a lean burn spark ignition engine, and substituting CH4 for the more usual hydrocarbons.

Air/fuel sweep tests were run at 500°C from lambda = 1.03 to lambda = 0.97 using simulated exhaust from a **CNG** vehicle running near the stoichiometric point. The exhaust at the leanest set point consisted of .06% NO_x, .038% HC (usually CH_4), .047% O_2, .05% CO, 6% CO_2 and 5% water. Sulfur was not included for reasons noted in the introduction. Consecutive set points were obtained by lowering the O_2 concentration and raising the CO concentration, while keeping the rest of the exhaust composition constant. Some catalysts were evaluated using line-out tests at a lambda of 0.995. Exhaust composition was the same as for the sweep tests, with 1000 ppm O_2 and 1250 ppm CO.

During stoichiometric testing, modulations in the exhaust gas composition, due to closed loop control on a vehicle, were simulated by pulsing additional O_2 to simulate lean and CO to simulate rich perturbations into the reactor. For most tests the amplitude of these modulations was +/- 0.5 air/fuel ratio units and the frequency was 0.5 Hz. Space velocity was $46000h^{-1}$

3. Results and Discussion

3.1. Effect of Exhaust Gas Stoichiometry

Two fresh 2.6 g/L Pd catalysts were tested at 450°C slightly rich of the stoichiometric point (lambda = 0.995) and also under extremely lean (oxygen rich) conditions (lambda = 1.25). One catalyst utilized an alumina washcoat (Pd/Al) while the other utilized a proprietary washcoat containing base metals (Pd/BM).

Table 1 Effect of Stoichiometric and Rich Exhaust on Pd-only conversion

Formulation	Exhaust Lambda	CO Converted	NO_x Converted	CH_4 Converted
Pd/Al	0.995	92%	70%	35%
Pd/ BM	0.995	94%	84%	56%
Pd/Al	1.25	96%	7%	23%
Pd/ BM	1.25	96%	0%	8%

These data reveal uniformly high CO conversions, but NO_x conversions are high only for lambdas near the stoichiometric point. Near the stoichiometric point the Pd/ BM catalyst outperforms the Pd/ Al catalyst. More HC and CO oxidation occurred than could be accounted for by O_2 and NO_x, so it seems likely that water-gas shift conversion took place over both samples. Water-gas shift conversion is known to occur on fresh Pd catalyst. Under lean conditions the Pd/ Al is preferred to Pd/ BM. Therefore, for the remainder of this study Pd/Al catalysts were used under lean conditions and Pd/ BM samples were used with exhaust near the stoichiometric point.

3.2. Catalysts for Lean CNG

Vehicles can be converted to run on NG under lean conditions using either a diesel engine or a spark ignition engine. The first option can take advantage of the generally lower emissions from the diesel engine, but diesel exhaust temperatures are well below those required for good CH_4 conversion. From a fuel distribution and marketing standpoint, a dual fuel vehicle capable of running either diesel or natural gas would be most attractive. Figure 1 shows that for a diesel/ NG dual fuel engine, it is not possible to obtain good HC lightoff for both exhausts with just one catalyst. In addition, the temperature for 50% conversion of CH_4 is nearly 450°C, which is higher than the normal exhaust temperature when run on diesel fuel. It may be possible to raise the exhaust temperature enough to achieve significant methane conversions when burning natural gas, but this is yet to be determined.

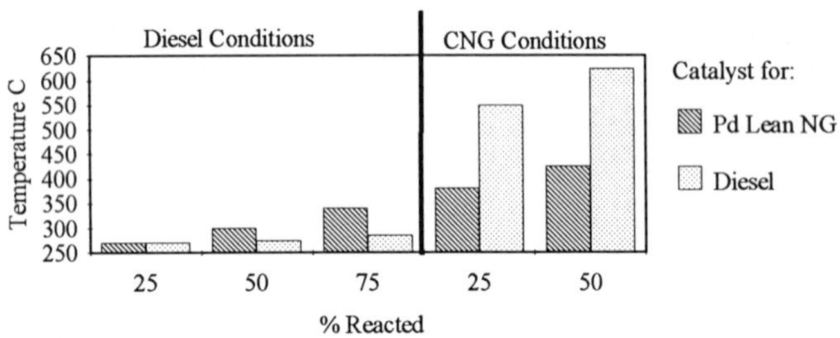

Figure 1. Effect of Aged Catalyst Type and Lean Exhaust Conditions on Hydrocarbon Lightoff

The second option for converting vehicles to run on NG under lean conditions is to convert a spark ignition engine. This engine produces higher emissions, but also runs at a higher temperature. Table 2 illustrates the potential performance of a Pd/Al catalyst.

Table 2 Activity over lab aged proprietary catalyst optimized for lean CNG

Emission	Lab Conversion at 500 C	Potential Vehicle Emission* (g/km)	Current Standards (g/km)	1996 Phase 1 Standards (g/km)
HC	92%	0.11	0.97 HC+NO_x	0.5 HC + NO_x
NO_x	2%	0.54	-	-
CO	99.9%	0.01	2.72	2.2

* Assuming Lab exhaust composition and temperature is representative of an actual natural gas vehicle and that the engine out volumetric flowrate is 15 l/sec. HC is 90% methane.

It is evident from this table that although a natural gas vehicle could meet current standards, Phase 1 standards pose a challenge for this lean CNG catalyst due to the lack of NO_X control. Lean NO_X catalysts, which generally use HC as the reductant for the NO_X [23], may provide a solution in this regard as long as enough natural gas HC will participate in the HC-NOx reaction under automotive conditions.

3.3. Effect of Pd Loading

A series of Pd catalysts was engine aged and then evaluated in the lab under rich and lean conditions to explore the effect of Pd loading on conversions. Figure 2 shows the results for CH_4 and NO_X. On the rich side, NO_X reduction increases with increasing Pd load up to 3.5 g/L, but then appears to level off. Lean NO_X conversions are insignificant. Methane conversion under rich and lean conditions increases continuously as the Pd load is increased. One hypothesis for the increased methane activity is that the lower loadings of Pd would tend to more readily form PdO, which is less active for methane oxidation [15]. As the Pd content is increased, the tendency to form PdO might go down and thus the catalyst activity would appear to increase. This idea appears to be refuted by the results shown in the figure for lean conversions, since the Pd under the lean conditions is likely to be mainly PdO. Another idea is that increasing the Pd load leads to increasing Pd crystallite size, which results in increased methane conversion.

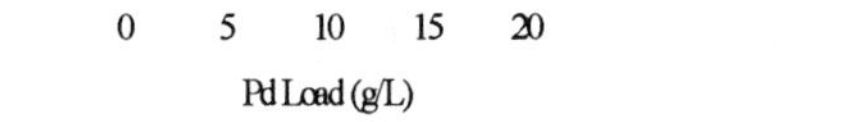

Figure 2. Increasing Pd Load improves Methane conversions at 500 C. Rich evaluations were done at a lambda of 0.995 and used a Pd/BM catalyst; lean evaluations were done at a lambda of 1.30 and used a Pd/Al catalyst.

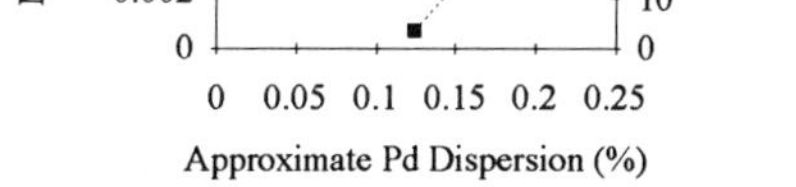

Figure 3. Increasing Pd dispersion improves CH_4 reaction rate per mole of noble metal, but decreases overall conversion. (n) is moles CH_4 reacted/ mole Pd; (D) is overall conversion.

A qualitative test of the second hypothesis utilized CO uptake studies, over the catalysts used under rich conditions, to obtain approximate average noble metal particle sizes. The results are shown in Figure 3 which relates the CH_4 reaction rate per mole of noble metal and overall CH_4 conversions to the Pd dispersion, using the same catalyst as shown in Figure 2. Increasing the Pd load resulted in decreased NM dispersion, which in turn decreases the amount of CH_4 reacted per mole of Pd. At the same time, the conversion per monolith part increases with increased noble metal load. This disagrees with reports in the literature over fresh catalysts that increasing Pd crystallite size increases methane conversion [17,18]. In this case the observed increase in activity may be due to greater numbers of larger crystallites and these greater numbers compensate for the reduction in specific rate in going from high to low dispersion. It is also possible that after engine aging the Pd in all of these samples is in what would be considered large crystallites under fresh conditions, and that these results merely show that larger numbers of big crystallites convert more methane than smaller numbers.

The remainder of the discussion will be limited to catalysts which are designed for near stoichiometric operation, since the best overall conversions can currently be obtained there.

3.4. Effect of Cerium Loading

One of the most common base metals used in commercial automotive catalysts is cerium as CeO_2. The effect of CeO_2 loading on NG conversions near the stoichiometric point was explored using aged 2.6 g/L Pd-only catalysts supported on alumina. Figure 4 shows the amount of methane converted for the given CeO_2 loading.

Increasing the CeO_2 loading initially improves the methane conversion of the Pd. As the CeO_2 loading is increased, the Pd performance appears to level out. It is speculated that ceria promotes the oxidation of CH_4 by making oxygen more readily available for reaction, so initial conversions increase. Previous work has shown that high levels of CeO_2 decrease Pd dispersion for catalysts aged similarly to those in this study [21]. As the CeO_2 load is increased, Figure 3 suggests that a decrease in dispersion at constant noble metal loading should result in a decrease in overall methane conversion. Increasing the oxygen availability for the reaction at the same time as the Pd dispersion is decreasing could result in the observed flattening of the methane oxidation curve with increasing CeO_2 load.

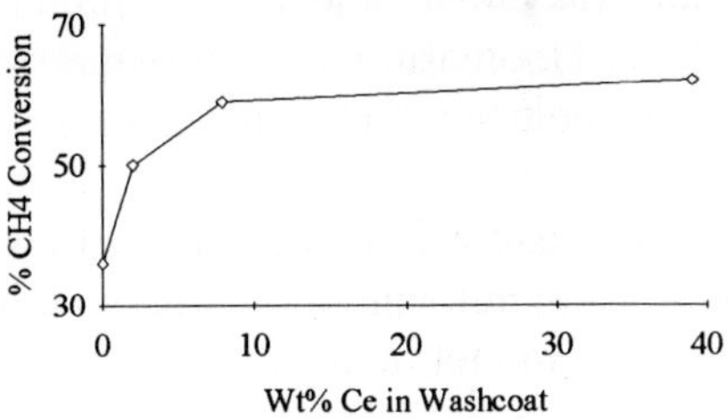

Figure 4. Increasing CeO_2 improves CH_4 conversions over Pd catalyst at the stoichiometric point.

3.5. Effect of Feedstream Modulation

The effect of varying feedstream modulation amplitude and frequency on NG conversions was tested using aged 3.5g/L Pd on the proprietary base-metal washcoat, using two sets of A/F sweep tests at 500°C. During the first set of A/F sweeps, the frequency was held constant at 0.5 Hz, while the amplitude was varied from 0.25 to 1 A/F units. A second set of A/F sweep tests was run where the frequency of modulation of the feedstream was varied from 0.25 to 1 Hz at a constant amplitude of +/- 0.5 A/F units.

Figure 5a illustrates the effect of changing amplitude at constant frequency on conversions over Pd/ base metal catalyst. Near the stoichiometric point (lambda = 0.995), decreasing the modulation amplitude from +/- 1 A/F ratio to 0.25 A/F ratio increases CH_4 activity by a factor of three. CO and NO_x conversions are also improved by the decreasing amplitude.

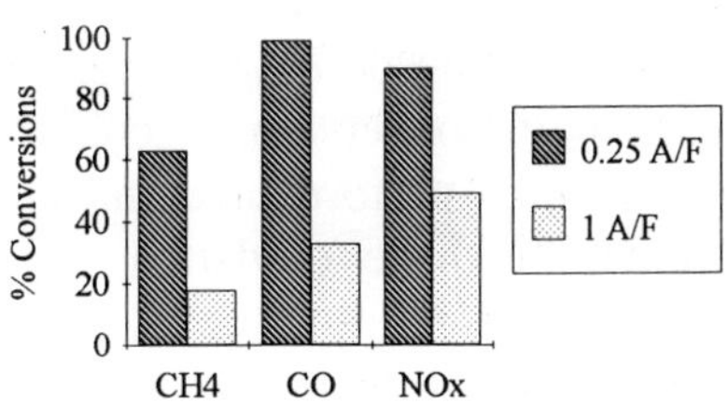

Figure 5a. Increasing feedstream modulation amplitude decreases conversions at the lambda for optimum CH_4 conversion (0.995).

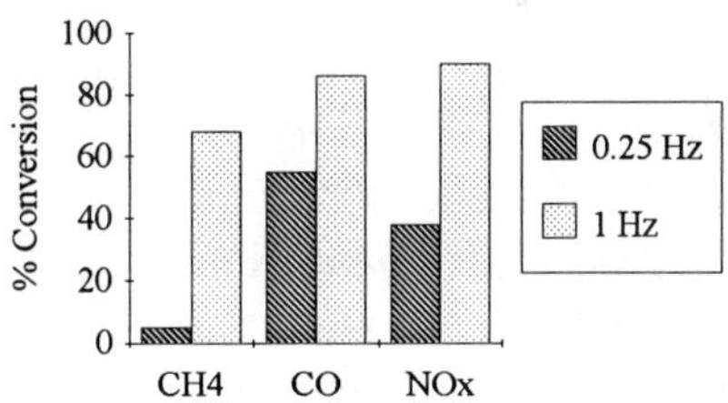

Figure 5b. Increasing feedstream modulation frequency increases conversions at the lambda for optimum CH_4 conversion (0.995).

Figure 5b illustrates the effect of increasing frequency with constant amplitude on Pd/BM catalyst. Near the stoichiometric point, increasing the modulation frequency from 0.25 to 1 Hz improved CH_4, CO and NO_x conversions. CH_4 conversions saw the largest improvements.

Methane conversions tend to drop off rapidly on either side of an optimum methane conversion point near lambda = 0.995. Decreasing the amplitude of modulation would tend to increase the time spent near this optimum conversion point.

At lower modulation frequencies, the catalyst will spend longer periods of time under extreme lean or rich conditions. It is speculated that a relatively long exposure to these extreme conditions may allow reversible changes to occur on the catalyst surface which are detrimental to activity. Some of the time spent near the stoichiometric point may then be necessary to recover catalyst activity after these exposures, thus decreasing overall conversions. Thus increasing modulation frequencies should improve performance.

The results of this feedstream modulation study suggest emissions from a natural gas vehicle will be minimized if the engine exhaust stoichiometry is tightly controlled near the optimum methane conversion point.

3.6. Effect of Fuel Hydrocarbon Composition

Natural gas contains significant amounts of ethane and propane in addition to methane. For this reason, methane oxidation studies alone may not be accurate predictors of the catalytic hydrocarbon oxidation required for NG vehicles. The effect of propane addition was studied with aged 2.6 g/L Pd catalysts supported on alumina and the proprietary base metal washcoat. Using the simulated natural gas exhaust, three different hydrocarbon compositions were evaluated using A/F sweep tests at 500° C. The composition of the hydrocarbon was varied from pure methane to a 9/1 ratio of methane to propane, to a 7/3 ratio of methane to propane. The total hydrocarbon concentration was held constant for all tests.

Figure 6a shows the results for a Pd/Al catalyst. Addition of 10% propane to the hydrocarbon in the exhaust improved lean-side conversion. Increasing propane's contribution to 30% further increased lean-side performance and also improved hydrocarbon conversions at the optimum methane oxidation point. Unlike methane, propane is not expected to be oxygen poisoned on the lean-side. It is also expected to be easier to oxidize than methane [12]. Increased lean-side HC conversions are believed to be due to propane reacting with and removing surface oxygen species, which otherwise block methane adsorption on the catalyst surface. Improved rich performance is likely due to easier oxidation of propane.

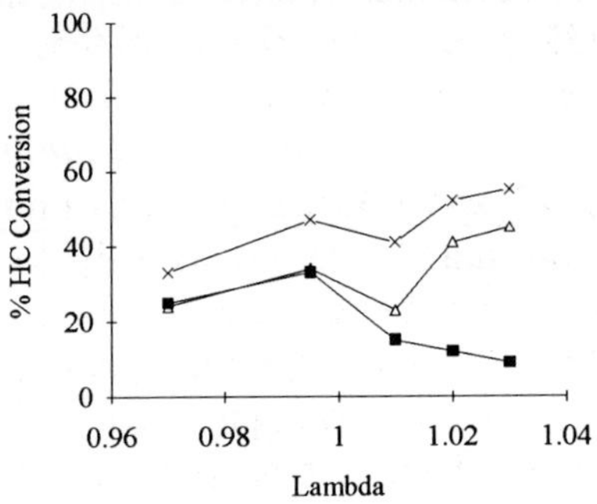

Figure 6a. Increasing non-methane component of natural gas HC improves conversions over Pd on alumina. (n) is 100% CH_4; (D) is 10% non-methane C_3H_8; (5) is 30% non-methane C_3H_8.

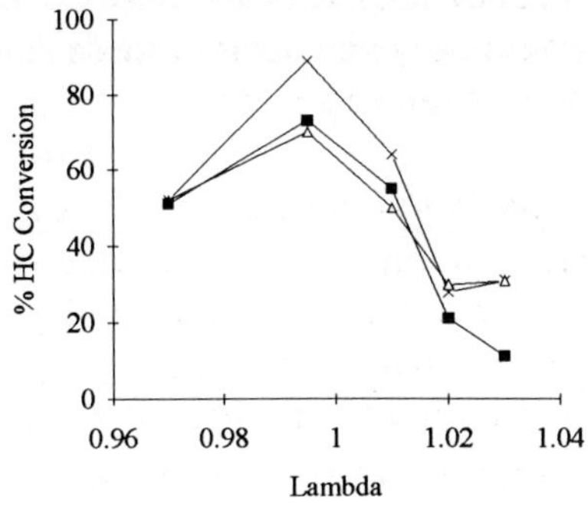

Figure 6b. Increasing non-methane component of natural gas HC slightly improves lean-side conversions over Pd on base metals. (n) is 100% CH_4; (D) is 10% non-methane C_3H_8; (5) is 30% non-methane C_3H_8.

Results for the Pd base-metal catalyst are shown in Figure 6b. Some improvement in hydrocarbon conversion on the lean side is seen with 10% propane, but the improvement is not as great as seen with Pd/Al. Increasing the propane content to 30% results in improved hydrocarbon conversion at the maximum methane oxidation point, but does not show improvement on the lean side relative to the 10% mixture. Adding propane to the hydrocarbon does not appear to aid the hydrocarbon conversion of the Pd/BM catalyst as much as it did the Pd/Al catalyst. This may be because CeO_2 in the Pd/BM catalyst is more efficient at replacing surface oxygen species which had been removed by reaction with propane. The oxygen species then block methane adsorption on the surface. However, the Pd/BM catalyst still had better stoichiometric hydrocarbon conversion than the Pd/Al, even in the presence of 30% propane, due to its higher baseline activity for methane oxidation.

3.7. Vehicle Results from U. S. EPA

The lessons learned from the preceding experiments were utilized in the design of a catalyst for a CNG vehicle for testing by the U. S. EPA [24]. The vehicle was a Dodge Dakota pickup truck with a 5.2 L engine. Table 3 indicates that a fresh catalyst, in combination with a carefully designed exhaust system, can achieve California ULEV standards. ULEV standards do not count methane emissions. In contrast, European Phase 2 will require methane conversion, but the optimized system can even achieve these standards.

Table 3EPA Test Results obtained over an Optimized CNG Catalyst and Future Emission Standards

	Engine Out Emissions	Tailpipe Emissions	California ULEV Standards	Europe Phase 2 Standards (Gasoline)
HC (g/km)	0.92	0.17		0.5 ($HC + NO_x$)
NO_x (g/km)	1.43	0.120	0.125	
CO (g/km)	8.1	0.5	1.056	2.2
CH_4 (g/km)	0.86	0.17		
NMHC (g/km)	0.06	0.01	.025	

4.SUMMARY AND CONCLUSIONS

Some of the parameters that can affect catalyst performance on a CNG vehicle were investigated with Pd-only three-way catalysts and a simulated vehicle exhaust. Formulations were primarily tested near the stoichiometric point under spark ignition engine conditions, in order to retain NO_x control with conventional catalysts and ensure that a high enough temperature was attained. Improved conversions can be obtained by tightly controlling the air/fuel mixture near the optimum conversion point for methane (lambda = 0.995). Increased Pd loading improved methane conversions under both stoichiometric and lean conditions, but decreased dispersion does not appear to correlate with improved activity per mole of Pd for aged samples. The addition of CeO_2 was found to be beneficial for engine aged catalyst performance near stoichiometric conditions and increasing the CeO_2 load resulted in improved methane activity over Pd-only formulations. Including non-methane hydrocarbon in the natural gas exhaust initially led to improved hydrocarbon conversions on the lean (oxygen rich) side, since the higher HCs remove surface oxygen which block methane adsorption sites. Increasing the ratio of non-methane to methane hydrocarbon led to improvements in HC conversion near the stoichiometric point as well.

In conclusion, it is possible to meet stringent emission standards, including a total hydrocarbon standard, with an alternative fuel vehicle. These standards can be met by careful design of a three-way catalyst and its environment in a CNG vehicle, as illustrated by the EPA test results.

ACKNOWLEDGMENTS

D. M. Thomason for catalyst preparation, E. O. Oates and E. Hofstetter for lab tests, K.J. Foley for engine tests, H. J. Robota for Pd dispersion results and helpful discussion, J. E. Sawyer, K.C.C. Kharas and W. B. Williamson for helpful discussions.

REFERENCES

1 W. F. Martin and S. L Campbell, " Natural Gas: A Strategic Resource for the Future", Washington Policy and Analysis, Washington, D. C., 1988.
2 C. S. Weaver, SAE Paper No. 892133 (1989).
3 D. Golomb and J. A. Fay, "The Role of Methane in Tropospheric Chemistry", Energy Laboratory, Cambridge, MA, 1989.
4 T. Sakai, B.-C. Choi, R. Osuga, Y. Ko and E. Kim, SAE Paper No. 920556 (1992).
5 M. A. DeLuchi, R. A. Johnston and D. Sperling, SAE Paper No. 881656 (1988).
6 B. Hillemann, Chem & Eng. News, 67 (1989) p. 25.
7 J.C. Summers, A.C. Frost, W. B. Williamson and I. M. Freidel, "Control of NOx/CO/HC Emissions from Natural Gas Fueled Stationary Engines with Three-Way Catalysts", presented at the 84th Annual Meeting of the Air & Waste Management Assoc. (1991).
8 S.R. King, SAE Paper No. 920593 (1992).
9 "Variability of Natural Gas Composition in Select Major Metropolitan Areas of the United States", Interim Report of the Gas Research Institute, August 1990 - March 1991.
10 W. E. Liss and W. H. Thrasher, SAE Paper No. 912364 (1991).
11 S. H. Oh, P.J. Mitchell and R. M. Siewert, J. Catal. 132, (1991) 287.
12 Y. -F. Yao, Ind. Eng. Chem. Prod. Res. Dev. 19, (1980) 293.
13 R. F. Hicks, H. Qi, M. L. Young and R.G. Lee, J. Catal. 122, (1990) 280.
14 C. F. Cullis and B.M. Willatt, J. Catal. 83, (1983) 267.
15 R. F. Hicks, H. Qi, M. L. Young and R.G. Lee, J. Catal. 122, (1990) 295.
16 R. J. Farrauto, M. C. Hobson, T. Kennelly and E. M. Waterman, Applied Catalysis A: General 81, (1992) 227.
17 P. Briot and M. Primet, Applied Catalysis 68, (1991) 301.
18 T. Baldwin and R. Burch, Applied Catalysis 66, (1990) 337.
19 P. Chou and M. Vannice, J. Catal. 105, (1987) 342.

20 R. F. Hicks, C. Rigano and B. Pang, Catalysis Letters 6, (1990) 271.
21 J. C. Summers and S. A. Ausen, J. Catal. 58 (1979) 131.
22 S. Subramanian, R. J. Kudla and M. S. Chattha, SAE Paper No. 930223 (1993).
23 M. Iwamoto and H. Hamada, Catalysis Today 10, (1991) 57.
24 K. Hellman, G. Piotrowski and R. Schaefer, SAE Paper No. 940473 (1994).

Miscellaneous

A. Frennet and J.-M. Bastin (Eds.)
Catalysis and Automotive Pollution Control III
Studies in Surface Science and Catalysis, Vol. 96
1995 Elsevier Science B.V.

MODELLING THREE-WAY MONOLITHIC CATALYTIC CONVERTER: COMPARISON BETWEEN SIMULATION AND EXPERIMENTAL DATA

S.Siemund, D.Schweich, J.P.Leclerc and J.Villermaux

C.N.R.S., L.S.G.C., E.N.S.I.C
1, rue Grandville, B.P. 451
54001 Nancy Cedex

ABSTRACT

A two-dimensional model for the three-way catalytic converter is presented. The proposed model with the necessary simplifying assumptions are represented. Then we compare the simulation results and experimental data, which are obtained by engine test bench. By adjusting the kinetic parameters we obtain a good agreement for the temperature evolution less for the conversion. Finally we briefly discuss the necessary improvements.

1. INTRODUCTION

The basic goal of the three-way catalytic converter system, is the simultaneous conversion of carbon monoxide, unburned hydrocarbons and nitrogen oxides, that are the origin of many health and environmental problems.
All the simulation programs up right now suffers from a lack of reliable kinetic data, especially concerning the reduction reactions. To our knowledge simultaneous oxidation and reduction reactions have rarely been considered when modelling three way catalytic converters. Furthermore, most authors simulate oxidizing conditions, which are in conflict with the usual operating conditions of the three-way catalytic converter.

2. THE PROPOSED MODEL

Based on previous published results[1-17], we present a model relying on the following assumptions:

1. The wash-coat thickness, w, is much smaller than the hydraulic diameter of the channel D_h.
2. The thermal conductivity and heat capacity of the was-coat and of the monolith are identical.
3. Pressure is uniform and constant.
4. Thermal conduction in the gas phase is neglected.
5. The thermal conductivity of the monolith is independent of temperature.
6. The gas phase is ideal
7. The monolith is cylindrical and it has uniform properties in a cross-section.
8. Heat transfer by radiation is neglected.
9. Transverse diffusion and laminar flow in a channel are accounted for by the film model.
10. The catalyst does not deactivate.
11. The reacting species are much diluted by an inert gas, thus the chemical expansion flux is negligible.
12. The accumulation rates in the gas phase are negligible. This assumptions forbids to simulate engine stop.

2.1. The chemical reactions

Although crucial for modeling purpose, reliable kinetic rate expression are scarce. For our model we have chosen those proposed by Chen and Coll.[5] for the oxidation reactions and Subramanian and Coll.[17] for the reduction reactions. The following oxidation reactions are accounted for

$$CO + \frac{1}{2} O_2 \longrightarrow CO_2 \qquad \Delta H_1(T^\theta) = -2.832\ 10^5\ J\ mol^{-1} \tag{2.1}$$

$$CH_x + \left(1 + \frac{x}{4}\right) O_2 \longrightarrow CO_2 + \frac{x}{2} H_2O \qquad \Delta H_2(T^\theta) = -6.424\ 10^5\ J\ mol^{-1} \tag{2.2}$$

$$H_2 + \frac{1}{2} O_2 \longrightarrow H_2O \qquad \Delta H_3(T^\theta) = -2.42\ 10^5\ J\ mol^{-1} \tag{2.3}$$

The rate laws for the reactions are Langmuir type expressions depending on mole fraction of reactants:

$$r_{CO} = \frac{k_{CO}\, x_{CO}\, x_{O_2}}{D(T,x)} \tag{2.4}$$

$$r_{CH_x} = \frac{k_{CH_x}\, x_{CH_x}\, x_{O_2}}{D(T,x)} \tag{2.5}$$

$$r_{H_2} = \frac{k_{H_2}\, x_{H_2}\, x_{O_2}}{D(T,x)} \tag{2.6}$$

$$D(T,x) = T_s\left(1 + k_{a1}\, x_{CO} + k_{a2}\, x_{CH_x}\right)^2 \left(1 + k_{a3}\, x_{CO}^2\, x_{CH_x}^2\right)\left(1 + k_{a4}\, x_{NO}^{0.7}\right) \tag{2.7}$$

And for the reduction reaction the following expression is considered:

$$\mathrm{NO} + \mathrm{CO} \rightarrow \mathrm{CO2} + \frac{1}{2} N_2 \qquad \Delta H_1\left(T^{\theta}\right) = -3.736\,10^5\ J\ mol^{-1} \tag{2.8}$$

The rate law for the reaction is also Langmuir type expression depending on mole fraction of reactants:

$$r_{NO1} = \frac{k_{NO1}\, x_{CO}^{1.4}\, x_{O_2}^{0.3}\, x_{NO}^{0.13}}{D_{NO1}(T,x)} \tag{2.9}$$

$$D_{NO1}(T,x) = T_s^{-0.17}\left(T + k_{a5}\, x_{CO}\right)^2 \tag{2.10}$$

2.2. The film model

In the film model, radial temperature and concentration profiles are assumed to be uniform in the bulk fluid whereas heat and mass transfer resistances are located close to the solid surface. Surface and bulk fluid properties being different, the specific radial mass and heat fluxes are given by:

$$\Phi_{Dj} = k_{Dj} c_{Tg}\left(x_{jg} - x_{js}\right) \tag{2.11}$$

$$q_D = h\left(T_g - T_s\right) \tag{2.12}$$

where c_{Tg}, x_{jg}, x_{js}, T_g and T_S are the total gas phase concentration, the mole fraction of species j in the bulk fluid and in the gas phase in contact with the channel wall, the gas temperature, and the solid temperature respectively. k_{Dj} and h are the mass and heat transfer coefficients which are obtained from the non dimensional Sherwood and Nusselt numbers respectively.

2.3. The equations

Figure 0 illustrates the control volume over which the balance equations are written. A detailed description of the basic equations have been already presented in Leclerc and Schweich[8]. Thus the equations will be written only in their final form.

These are:

- the mass balance equation for species j

$$u\, c_{Tg} \frac{\partial x_j}{\partial z} + \frac{4}{D_h} k_{Dj}\, c_{Tg} \left(x_{jg} - x_{js}\right) = 0 \tag{2.13}$$

- the mass balance equation in the wash-coat

$$k_{Dj}\, c_{Tg} \left(x_{jg} - x_{js}\right) + \beta\, \rho_{wc} w \sum_{i=1}^{R} \nu_{ij}\, r_{is} = 0 \tag{2.14}$$

- the heat balance equation in the gas phase

$$u\, \rho_g\, c_{pg} \frac{\partial T_g}{\partial z} + \frac{4}{D_h} h \left(T_g - T_s\right) = 0 \tag{2.15}$$

- the heat balance equation in the solid phase

$$\frac{4\varepsilon}{D_h} \beta\, \rho_{wc}\, w \sum_{i=1}^{R} \left[r_{is} \left(-\Delta H_i\right) \right] = \\ - \lambda_z \frac{\partial^2 T_s}{\partial z^2} - \lambda_r \left(\frac{\partial^2 T_s}{\partial r^2} + \frac{1}{r} \frac{\partial T_s}{\partial r} \right) + \frac{4\varepsilon}{D_h} h \left(T_s - T_g\right) + \left(1 - \varepsilon\right) \rho_s\, c_{ps} \frac{\partial T_s}{\partial t} \tag{2.16}$$

where u is the fluid velocity in a channel. The total gas phase concentration, c_{Tg}, is calculated from the gas temperature by the ideal gas law, b is the specific area of noble metal per unit mass of wash-coat, r_{wc} is the wash-coat density, w the thickness of the wash-coat, n_{ij} the stoichiometric coefficient of species j in reaction i, and r_{is} the i-th apparent chemical reaction rate per unit surface area of noble metal. c_{pg} is the heat capacity of the gas phase, e is the monolith void fraction, DH_i the i-th reaction enthalpy, c_{ps} the heat capacity of the solid per unit mass, r_s the real density of the matrix (channel excluded). land l_z are effective radial and axial heat conductivities referred to monolith area. A structured finite volume method is used for the discretization of the equations.

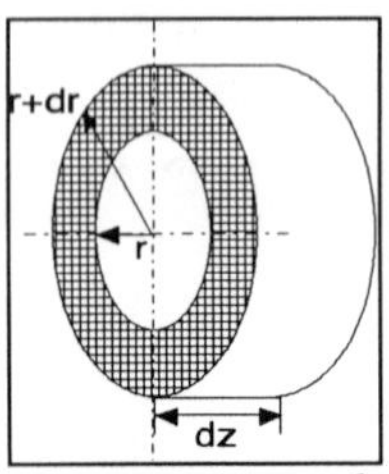

Figure1:Typical control-volume

3. COMPARISON EXPERIMENT-SIMULATION

The experimental tests took place on a test bench at I.F.P. Energy Application Techniques Department. The experimental set-up and testing method are the same as used by Germidis and Castagna[18], except that we have used ceramic monolith instead of a metallic monolith. Additionally the test bench was equiped with standard gas analyzers namely for HC, NO_X, CO.

3.1. The step change inlet temperature rise

Figure 2 shows the time evolution of experimental and the simulated temperatures at the outlet of the monolith for the step change experiment. For better understanding the feed temperature has been added. With respect to the published kinetic data[5,17], the simulation curve is obtained by reducing the activation energies of the CO-, HC- and NO-reaction slightly and increasing the frequency factor of the last mentioned reaction. Thus we obtain a nearly perfect agreement with the experimental curve. The lightoff behavior is calculated within an accuracy of 10% and the steadystate simulated and experimental temperatures are nearly identical.

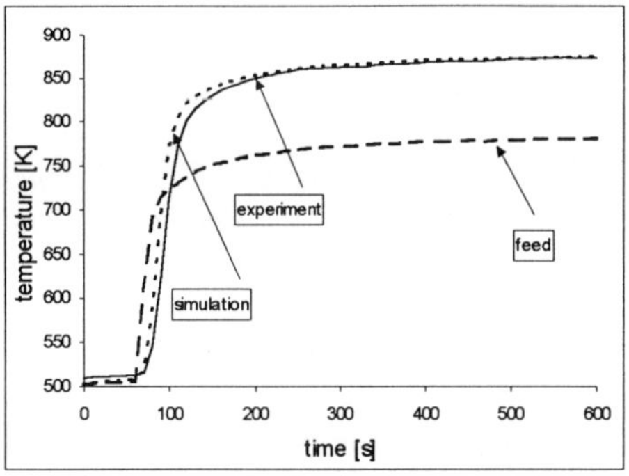

fig. 2: outlet temperature of experiment (continues line) and simulation (dotted line), feed temperature (dashed line)

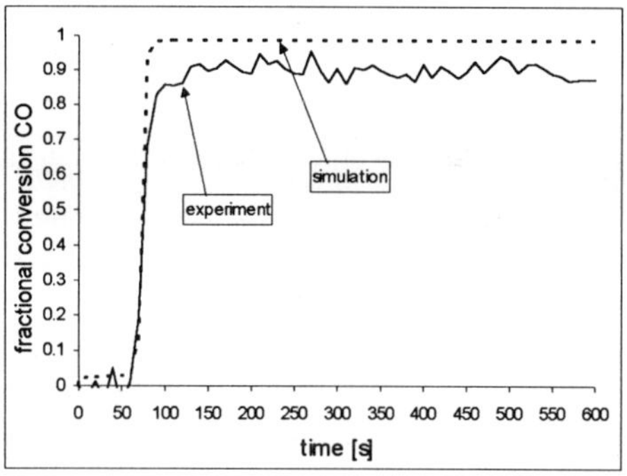

fig. 3: CO conversion, experiment continues line, simulation dotted line

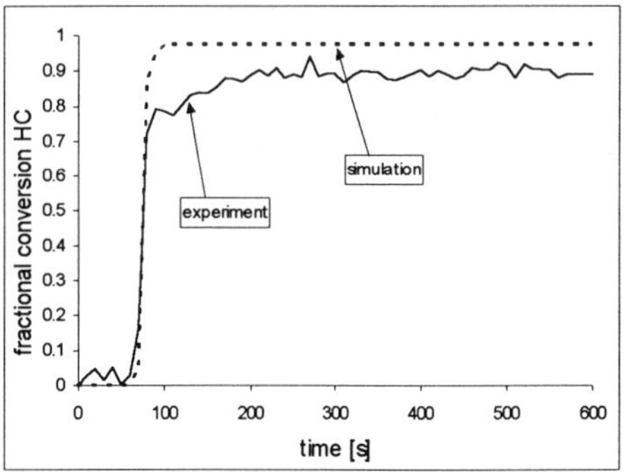

fig. 4: HC conversion, experiment continues line, simulation dotted line

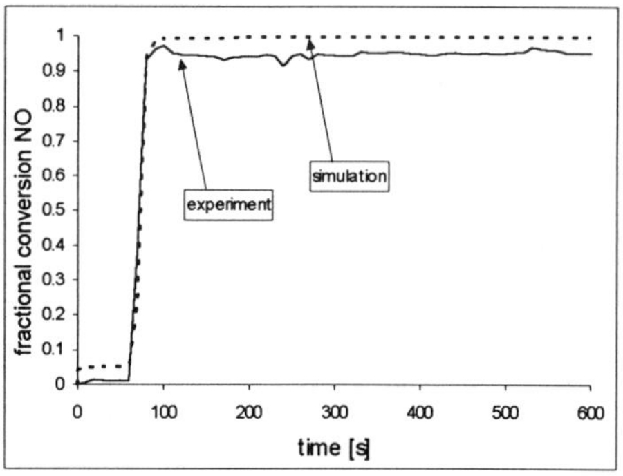

fig. 5: NO conversion, experiment continues line, simulation dotted line

Figures 3 to 5 show the conversion of CO, HC and NO respectively. For all three curves we observe, like for the temperature, a nearly perfect agreement concerning lightoff with the experimental curve. However, the model overestimates steadystate conversions.

3.2. The quasi-stationary inlet temperature rise

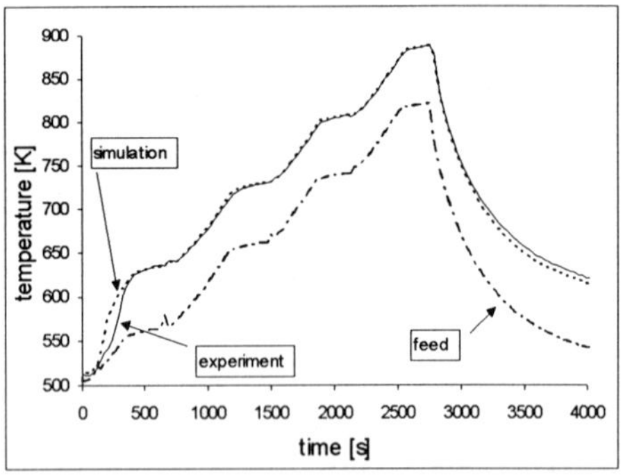

fig. 6: Outlet temperature: experiment (con-tinues line), simulation (dotted line), feed temperature (dashed line)

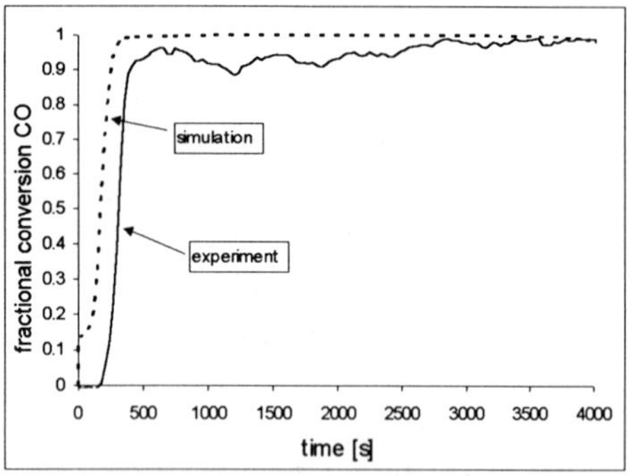

fig. 7: CO conversion, experiment continues line, simulation dotted line

Figure 6 shows the time evolution of the quasi-stationary inlet temperature rise for the experiment and the simulation at the outlet of the monolith. For better understanding the feed temperature has been added. The simulation curve is obtained with the same data modification as mentioned under section 3.1.

We obtain a nearly perfect agreement with the experimental curve. Figures 7 to 9 show the CO, HC and NO conversions respectively. For all three curves we observe an immediate rise of conversion, more developed for CO and NO, less

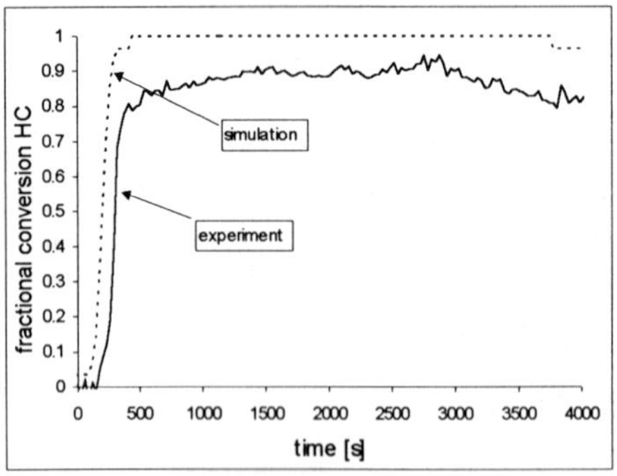

fig. 8: HC conversion, experiment conti-nues line, simulation dotted line

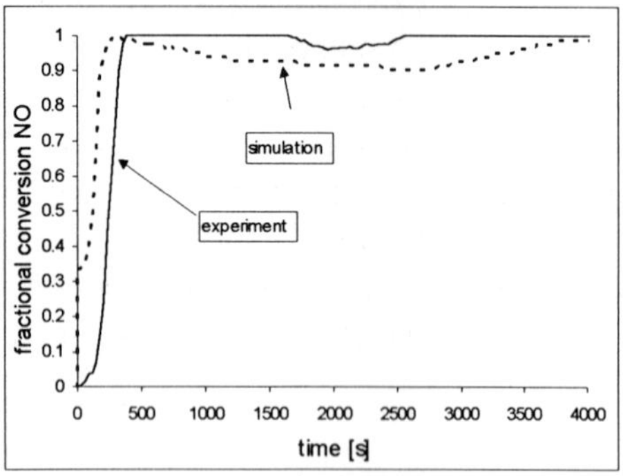

fig. 9: NO conversion, experiment conti-nues line, simulation dotted line

for HC. The model predicts lightoff earlier than the experimental lightoff, and the steady state conversions are again overestimated. However, remark that the model predicts the drop of conversion for HC at the end of experiment, and the undershoot of NO conversion in the middle of the experiment.

4. CONCLUSION

The good agreement between experimental and simulation curves confirms that working with fixed values for Nusselt and Sherwood (Nu=Sh=2.89) is sufficient. Furthermore the choice of the film model instead of the more complicated Greatz-Nusselt model seams to be justified and gives a good representation at monolith scale. The slightly earlier light-off observed for the temperature can be explained by the high sensitivity of light-off to the temperature of the feed flow and to the activation energy. So we have observed for the quasi-stationary inlet temperature rise, that a variation of activation energy of 5% shifts the light-off to the second temperature step change. Using rate expressions coming from different authors and obtained with different catalysts is debatable. However, our simulations proves that this method is acceptable provided that the activation energies and frequency factors are suitably adapted. In general, we have a good representation of the conversion evolution. The higher steady-state conversion of HC can be explained by the presence of fast and slow oxidizing hydrocarbons. Here, an additional reaction seems necessary for a better representation of the hydrocarbons. The simplest method to account for this couple of hydrocarbons families would be to employ the very same kinetic expressions for the two reactions using, however, different frequency factors.

All the models published until now for the three-way catalyst suffers from a lack of reliable kinetic data. But the difficulty is not only to deduce the oxidation and reduction reaction kinetics. There are also other influencing processes like poisoning, deactivation of the noble metal surface or the memory effect of the catalytic surface due to the O_2-storage induced by CeO_2. Long term deactivation processes could be accounted for by a suitable adjustment of the kinetic constants. Conversely, short term reversible poisoning and O_2-storage should be accounted for by appropriate generalized kinetic expressions. For the memory-effect of O_2 in CeO_2 we propose an additional balance

$$n_0 \frac{\partial \Theta_{O_2}}{\partial t} = r(O_2, HC, NO, \ldots)$$

where Q_{O2} is a normalized concentration for adsorbed and reactive O_2. The rates of the main reactions are then function of Q_{O2} and of the gas composition. Reaction rates of this type will certainly bring catalyst models closer to real catalyst behavior when they are controlled by chemical reaction.

Acknowledgments

This work was carried out within the 'Groupement Scientifique Pots Catalytiques' funded by the 'Centre National de la Recherche Scientifique', the 'Institut Français du Pétrole' and the AFME (Agence Française pour la Maîtrise de l'Energie). The 'JRC' (Volvo, Fiat, Rover, Renault, Peugeot PSA) is gratefully acknowledged for their support.

REFERENCES

1 S.H Oh, J.Cavendish 'Transients of monolithic catalytic converters: response to step changes in feedstream temperatures as related to controlling automobile emissions', Ind.Eng.Chem.Proc. Des.Devel.,21,29-37 (1981).
2 R.H. Heck, J.Wei, J.R. Katzer 'Mathematical modeling of monolithic catalyst', AIChE J., 22, 3, 477-484 (1976).
3 L.C. Young, B.A. Finlayson 'Mathematical models of the monolith catalytic converter, Part1, development of model and application of the orthogonal collocation', AIChE J., 22, 2, 331-343 (1976).
4 L.C. Young, B.A. Finlayson 'Mathematical models of the monolith catalytic converter, Part2, Application to automobile exhaust', AIChE J., 22, 2, 343-353 (1976)
5 D.K.S. Chen, S.H.Oh, E.J. Bisset, D.L. van Ostrom ' A three dimensional model for the analysis of transient thermal and conversion characteristics of monolithic catalytic converters', SAE Paper 880282 (1988).
6 D.K.S. Chen, C.E. Cole, 'Numerical simulation and experimental verification of conversion and thermal response for a Pt/Rh metal monolithic converter', SAE Paper 890798 (1989).
7 K. Zygourakis ' Transient operation of monolith catalytic converters: a Two-dimensional reactor model and the effect of radial non-uniform flow distribution', Chem. Eng. Sci. 44, 9, 2075-2086 (1989).
8 J.P. Leclerc, D.Schweich, 'Modeling Catalytic Monoliths For Automobile Emission Control', H.I. de Lasa et al. (eds.) Chemical Reactor for Environmentally Safe Reactors and Products, 547-576, (1993) Kluwer Academic Publishers.
9 R.K. Shah, T.C. London (1978), ' Flow forced convection in ducts. Advances in heat transfer - Laminar. ', Academic Press, New-York
10 R.H. Heck, J. Wei, J.R. Katzer, ' The transient response of a monolithic catalyst support ', Chem.React.Eng.Adv in Chem. Series 133,34-45,(1974) Amer. Chem. Soc.

12 J. Vortuba, O. Mikus, K. Nguen, V. Hlavacek,J. Skrivanek, ' Heat and mass transfer honeycomb catalyst - II ', Chem.Eng.Sci.,30, 201-206 (1975).
13 S.P. Waldram, U. Ullah, C.J. Bennet, T. Truex, 'Monolithic Reactors: Mass Transfer Measurements Under Reacting Conditions', Chem.Eng.Sci.,47,No9-11, 2431-2418, (1992).
14 S.T. Lee, R. Aris, ' On the effect of the radiative heat transfer in monoliths' Chem.Eng.Sci.,32,827-837. (1977).
15 S.E Voltz, C.R. Morgan, D. Liederman, S.M. Jacob, ' Kinetic study of carbon monoxide and propylene oxidation on platinum catalyst ', Ind.Eng.Chem. Proc. Des. Devel., 12, 4, 294-301 (1973).
16S.T. Lee, R. Aris, ' Poisoning in monolithic catalysts ', ACS Symp.Series,65, 110-121 (1978).
16 S.H. Oh, J.C. Cavendish,' Design aspects of poison-resistant automobile monolithic catalysts ', Ind.Eng.Chem.Prod.Res.Dev.,509-518 (1983).
17 B. Subramanian, A. Varma, 'Reactions of CO, NO, O_2, and H_2O on three-way and Pt/Al_2O_3 catalysts 'Frontiers inchemical engineering, Proceedings of the International Chemical Engineering Conference, 1, 231-240 (1984).
18 A. Germidis, F. Castagna, J. Banaigs, ' Thermal measurement inside a three-way catalytic converter on engine bench.', SAE paper 930624, 'New Engine Designe and Engine Technology' 67-78 (1993).

A. Frennet and J.-M. Bastin (Eds.)
Catalysis and Automotive Pollution Control III
Studies in Surface Science and Catalysis, Vol. 96

Behaviour of Three-Way Catalysts in a Hybrid Drive System: Dynamic Measurements and Kinetic Modelling

S. Tagliaferri, L. Padeste and A. Baiker

Department of Chemical Engineering and Industrial Chemistry, Swiss Federal Institute of Technology (ETH), CH - 8092 Zürich, Switzerland

The dynamic behaviour of a Pt-Rh/CeO_2-Al_2O_3 three-way catalyst (TWC) under forced λ-cycling and pulsed-flow (intermittent) operation, as occurring in a hybrid drive system, has been investigated. The experiments were carried out using a simulated exhaust gas containing CO, H_2, O_2, NO, C_3H_6, C_3H_8, CO_2, H_2O and N_2. The benefits of λ-cycling were lower C_3H_8 and NO_x light-off temperatures and lower N_2O production at temperatures below 350°C. The air plug preceding the exhaust gas pulse under pulsed-flow operation (intermittent operation of engine in hybrid system) had a negative influence on the performance of the TWC. Asymmetric cycling with longer fuel rich half cycles suppressed the negative effect of the air plug.

A simple semi-empirical model derived from a Langmuir-Hinshelwood model for CO oxidation, which was extended by introducing CO and O_2 equivalents to mimic the complex exhaust gas, was used to describe the dynamic behaviour of the catalytic converter. The model proved to be useful for finding the optimal operating conditions for eliminating the detrimental effect of the air pulses passing through the catalyst during the intermittent operation of the Otto-cycle engine.

1. INTRODUCTION

Hybrid drive systems are becoming increasingly important as an alternative to the conventional propulsion systems of cars [1]. The hybrid system results in a synergic combination of a combustion engine with an electric machine. Basically, hybrid concepts can be divided into parallel and serial configurations. Serial concepts allow the combustion engine to operate at its best efficiency. This minimises exhaust emissions, but the overall efficiency is reduced by the losses in the generator-battery-motor energy chain. In contrast, parallel concepts achieve a higher overall efficiency, but at higher emission levels.

As a part of an interdisciplinary study [2] we are working on emission control catalysis for an extended hybrid concept (fig. 1). The heart of this hybrid system is a flywheel, equipped with a continuous variable transmission. This permits the flywheel to be accelerated with the intermittently operated combustion engine or the electrical engine independently of the speed of the vehicle. The flywheel is then used to accelerate the car and to store braking energy. The flywheel concept allows the recuperation of braking energy and operation of the combustion engine under optimal efficiency conditions.

The intermittent operation mode of the Otto-cycle engine leads to a pulsed-flow operation of the three-way catalyst (TWC). Because air always enters into the cylinders when the engine is shut off and restarted, an air plug (cylinders filled with air) will precede the exhaust gas pulse (fig. 2). The objective of our study was to gain information about the dynamic behaviour of TWC's under pulsed-flow operation and to find a suitable model, that enables a prediction of the optimal operation mode of the catalytic converter.

The behaviour of TWC's under continuous operation has been extensively studied. Due to the step-like response of the oxygen sensors the gas composition oscillates with a frequency of about 1 Hz around the stoichiometric set point. Therefore, most studies focus on the behaviour of catalytic converters under oscillating exhaust gas composition. In particular, the contribution of ceria to the dynamic behaviour of automotive catalysts under transient air/fuel conditions [3, 4, 5, 6, 7, 8] has been investigated. Binary gas mixtures have been applied to clarify the mechanisms of the periodic operation effects over different noble metal catalysts [9, 10, 11, 12]. Muraki et al. used simulated exhaust gas to examine the performance of noble metals on α-Al_2O_3 [9].

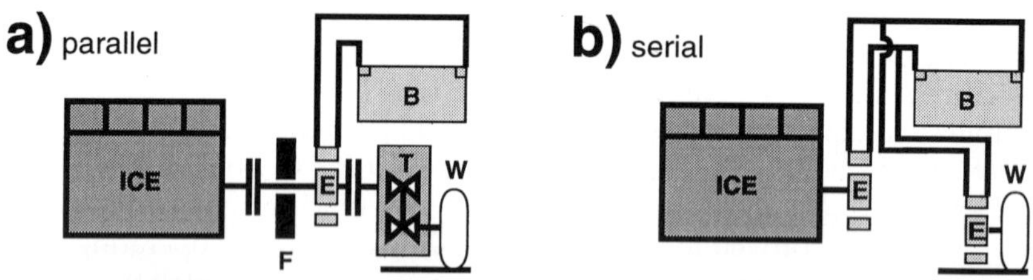

Figure 1. Extended parallel (**a**) and serial hybrid (**b**) concepts.
B: battery, E: electric motor / generator, F: flywheel, ICE: internal combustion engine, T: transmission, W: wheel

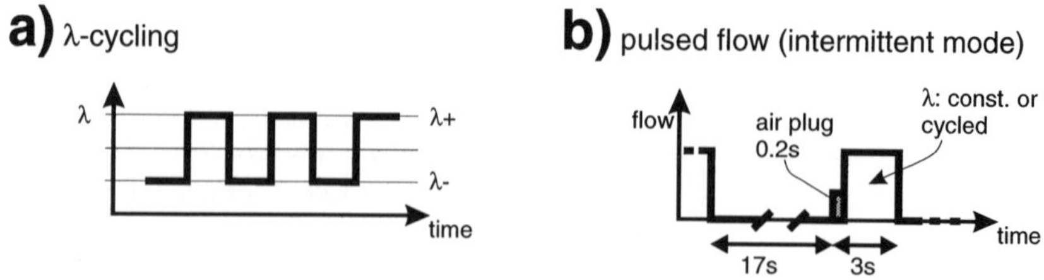

Figure 2. Time dependent λ-value and flow pattern during: (**a**) λ-cycling between λ+ and λ- with constant or variable length of the half cycles and constant gas flow; (**b**) Intermittent mode, 17 s at flow = 0 (engine shut off), 0.2 s pure air (air plug at engine start), 3 s exhaust gas (engine operation)

Mathematical models to describe the dynamic behaviour of the CO oxidation over noble metals have been developed by various authors [13, 14, 15]. All these studies were performed over Pt/Al_2O_3 catalysts in differential reactors below 200°C. In contrast, this work was carried out over a Pt-Rh/CeO_2-Al_2O_3 catalyst in an integral reactor at temperatures ranging from 200°C to 600°C, using not only a binary CO/O_2 mixture but also simulated exhaust gas.

2. EXPERIMENTAL

Apparatus: All experiments were carried out in a fully computer controlled apparatus, which is described in detail in ref. [16]. The simulated exhaust gas contained a mixture of CO and H_2 (3:1), O_2, NO (2000 ppm), C_3H_6 (500 ppm), C_3H_8 (500 ppm), CO_2 (12%), H_2O (10%) and N_2 (balance). The λ-value of the gas mixture, which is defined as the normalised molar air/fuel ratio, in this case:

$$\lambda = \frac{2c_{O_2} + 2c_{CO_2} + c_{H_2O} + c_{CO} + c_{NO}}{2c_{CO} + c_{H_2} + 10c_{C_3H_8} + 9c_{C_3H_6} + 2c_{CO_2} + c_{H_2O}} \quad (1)$$

could be altered by changing the flows of O_2 and the CO/H_2 mixture [ref. 16, Figure 2].

Analysis: The product stream was analysed using a system which consisted of an FT-IR spectrometer (Bruker IFS 66) and a quadrupole mass spectrometer (Balzers GAM 400). NO, NO_2, N_2O, NH_3, CH_4, C_3H_6, C_3H_8, CO, CO_2 and H_2O were analysed by FT-IR spectroscopy, O_2 and H_2 by mass spectroscopy with a time resolution of 0.2 s and 0.25 s, respectively [ref. 16, fig. 4].

Catalyst: The study was performed using Pt-Rh/CeO_2-Al_2O_3 coated on a monolithic cordierite honeycomb carrier with 400 cells per square inch and a washcoat loading of 110 g l^{-1}. The washcoat had the following composition (wt%): 1% Pt, 0.2% Rh, 12% CeO_2, 86.8% Al_2O_3. The honeycomb had a length of 15 cm and a diameter of 2.5 cm. To reduce the volume to 12.75 cm^3, the outermost channels were sealed with a ceramic paste. The fresh catalyst was conditioned at 600°C and λ = 1 for 10 h before use.

Experimental procedure: To characterise the dynamic behaviour of the catalyst, λ-cycling (fig. 2a) and pulsed-flow (fig. 2b) experiments were carried out. λ-Cycling was achieved by periodically changing the stoichiometry of the feed composition. Temperature (200-600°C), frequency (2 - 0.1 Hz) and amplitude (λ = 1 ±0.05, ±0.035, ±0.02) were varied, whereas the pressure was kept constant at 1 bar. The flow rate of the simulated exhaust gas was 10.625 l min^{-1} (NTP) corresponding to a gas hourly space velocity of 50'000 h^{-1}.

Experiments with pulsed flow (fig. 2b) were carried out at 1.7 bar and 310°C or 400°C. An air pulse of 0.2 s and 3.187 l min^{-1} preceded a simulated exhaust pulse of 3 s with different variation of the λ-value with time. A period of 17 s followed, with no gas flow through the converter.

Measurements at steady-state conditions were used to determine the axial concentration profiles, which were established when a stoichiometric CO/O_2 mixture (2 vol% CO, 1 vol% O_2) or a simulated exhaust gas with $\lambda = 1$ were fed into the converter. Stepwise reduction of the honeycomb length resulted in 12 axial data points, which were used to estimate the model parameters.

A more detailed discussion of the experimental conditions is given in ref. [16].

3. KINETIC MODELLING

The aim was to find a suitable model which aids in the search for the optimal operation mode of the catalytic converter. In addition the model should also be applicable in a closed-loop λ-control. With the latter in mind, we looked for a simple model which provides a satisfactory description of the dynamic behaviour.

The semi-empirical model used to describe the experimental results, measured with CO/O_2 feed gas and simulated exhaust gas, was based on Langmuir-Hinshelwood kinetics for the CO - O_2 reaction. The simplifying assumptions imposed on the model were:

(i) Reducing components (CO, H_2, CH_4, C_3H_6, C_3H_8, NH_3) exhibit the same behaviour as an equivalent of CO, and correspondingly, oxidising components (O_2, NO, N_2O, NO_2) the same as an equivalent of O_2.
(ii) CO is reversibly and molecularly adsorbed (CO_{ads}) on active surface sites, whereas O_2 is irreversibly and dissociatively (O_{ads}) adsorbed.
(iii) CO and O_2 adsorption are competitive.
(iv) The reaction $CO_{ads} + O_{ads} \rightarrow CO_2$ is irreversible, CO_2 is not adsorbed on active surface sites.
(v) No volume change due to reaction is considered (diluted gases).

CO and O_2 equivalents can be defined (assumption i.), which are not distinguished from CO and O_2 in all subsequent considerations:

$$\left[c_{CO}\right]_{eq.} = c_{CO} + c_{H_2} + 4c_{CH_4} + 9c_{C_3H_6} + 10c_{C_3H_8} + 1.5c_{NH_3} \tag{2}$$

$$\left[c_{O_2}\right]_{eq.} = c_{O_2} + 0.5c_{NO} + 0.5c_{N_2O} + c_{NO_2} \tag{3}$$

The reaction pathway is considered to involve the following steps (assumptions ii.-iv.):

$$CO \rightleftarrows CO_{ads} \tag{4}$$

$$O_2 \rightarrow 2O_{ads} \tag{5}$$

$$CO_{ads} + O_{ads} \rightarrow CO_2 \tag{6}$$

and the reaction rates r_i are described by the equations:

$$r_{CO,ads} = \sqrt{\frac{RT}{2\pi M_{CO}}} S\alpha_{CO}\Theta_V c_{CO} = k_{CO,ads}\Theta_V c_{CO} \tag{7}$$

$$r_{CO,des} = k_{CO,des_0} \exp\left(\frac{-E_{aCO,des}}{RT}\right)\Theta_{CO} = k_{CO_D}\Theta_{CO} \tag{8}$$

$$r_{O,ads} = 2\sqrt{\frac{RT}{2\pi M_{O_2}}} S\alpha_{O_2}\Theta_V^2 c_{O_2} = k_{O_2,ads}\Theta_V^2 c_{O_2} \tag{9}$$

$$r_{LH} = k_{LH_0} \exp\left(\frac{-E_{aLH}}{RT}\right)\Theta_{CO}\Theta_O = k_R\Theta_{CO}\Theta_O \tag{10}$$

where the root expression is the average velocity of gas molecule *i* in direction of the adsorbing surface. S is the area of 1 mol of active surface sites and α_i is the sticking probability of component i. $E_{aCO,des}$ and $k_{CO,des}$ are the activation energy and the pre-exponential factor for CO desorption, whereas $E_{a,LH}$ and k_{LH} represent the corresponding parameters for the surface reaction. Θ_i is the fractional occupancy of the active surface sites by component i, and the fraction of free active surface sites Θ_V is expressed by:

$$\Theta_V = (1 - \Theta_O - \Theta_{CO}) \tag{11}$$

Balancing of all space- and time-dependent variables over an infinitesimal element of a plug-flow reactor leads to the subsequent system of partial differential equations:

$$\frac{\partial}{\partial t}c_{CO} = -\frac{\dot{V}}{A}\frac{\partial}{\partial x}c_{CO} + c_S k_{CO_D}\Theta_{CO} - c_S k_{CO,ads}\Theta_V c_{CO} \tag{12}$$

$$\frac{\partial}{\partial t}c_{O_2} = -\frac{\dot{V}}{A}\frac{\partial}{\partial x}c_{O_2} - c_S k_{O_2,ads}\Theta_V^2 c_{O_2} \tag{13}$$

$$\frac{\partial}{\partial t}c_{CO_2} = -\frac{\dot{V}}{A}\frac{\partial}{\partial x}c_{CO_2} + c_S k_R\Theta_O\Theta_{CO} \tag{14}$$

$$\frac{\partial}{\partial t}\Theta_{CO} = k_{CO,ads}\Theta_V c_{CO} - k_{CO_D}\Theta_{CO} - k_R\Theta_O\Theta_{CO} \tag{15}$$

$$\frac{\partial}{\partial t}\Theta_O = 2k_{O_2,ads}\Theta_V^2 c_{O_2} - k_R\Theta_O\Theta_{CO} \tag{16}$$

where A is the free cross section of the catalyst and $\dot{V}$ the gas flow rate.

The apparent activation energy of the surface reaction E_{aLH} was calculated from the slope of Arrhenius plots at low temperatures and conversions, where the surface reaction was assumed to be the rate limiting step [17].

The model parameters $k_{CO,ads}$, $k_{O_2,ads}$, k_{CO,des_0}, k_{LH_0} and $E_{aCO,des}$ were estimated from the concentration profile measurements at different temperatures by non-linear regression analysis using equations (16-20) for steady-state conditions ($\partial c_i / \partial t = 0$, $\partial\Theta_i / \partial t = 0$), which were solved using the simulation software 'Simusolv' (V 2.2, Dow Chemical Company). The procedure of 'generalised reduced gradients' was used to minimise the objective function, i.e. the residual sum of squares of the O_2 equivalent concentration (c_{O_2}):

$$RSS = \sum_{i=1}^{n} (c_{O_2,predicted_i} - c_{O_2,experimental_i})^2 \tag{17}$$

The mass balance was integrated using Gear's method, taking into account the temperature profile by linear interpolation between the experimental points. The significance of the model was judged by variance analysis [18]. The F-test revealed that the model cannot be rejected at a significance level of $\alpha = 0.05$ for both the CO/O_2 mixture and the simulated exhaust gas (fig. 3).

The concentration of the active surface sites c_S, which is strongly correlated with k_{CO_D}, $k_{CO,ads}$, $k_{O_2,ads}$ and k_R (eq. 12 - 16), was separately estimated. As the storage capacity of the catalyst is mainly determined by c_S, data from λ-cycling experiments were used for its estimation.

The estimated model parameters are given in table 1. Note that the estimated model parameters cannot be considered to represent intrinsic kinetic constants. They represent lumped parameters which can be disguised by possible heat and mass transfer effects which are not accounted for in the model.

Table 1
Estimated model parameters for CO/O_2 mixture and for simulated exhaust gas

	Parameter	Value	95% Confidence Limits	Units
CO/O_2 mixture	$k_{CO,ads}$ / $T^{½}$	5.2	[4.7 - 5.8]	$m^3\,mol^{-1}s^{-1}K^{-½}$
	$k_{O_2,ads}$ / $T^{½}$	4.1	[3.6 - 4.4]	$m^3\,mol^{-1}s^{-1}K^{-½}$
	k_{CO,des_0}	$3.8\ \ 10^7$	[$3.2\ \ 10^7$ - $4.4\ \ 10^7$]	s^{-1}
	k_{LH_0}	$3.7\ \ 10^9$	[$2.9\ \ 10^9$ - $4.5\ \ 10^9$]	s^{-1}
	$E_{aCO,des}$	60	[59 - 61]	kJ mol^{-1}
	E_{aLH}	75	[73 - 78]	kJ mol^{-1}
Simulated exhaust gas	$k_{CO,ads}$ / $T^{½}$	5.8	[4.9 - 6.6]	$m^3\,mol^{-1}s^{-1}K^{-½}$
	$k_{O_2,ads}$ / $T^{½}$	8.7	[6.9 - 10.5]	$m^3\,mol^{-1}s^{-1}K^{-½}$
	k_{CO,des_0}	$2.2\ \ 10^4$	[$1.5\ \ 10^4$ - $2.8\ \ 10^4$]	s^{-1}
	k_{LH_0}	$1.9\ \ 10^7$	[$0.7\ \ 10^7$ - $3.2\ \ 10^7$]	s^{-1}
	$E_{aCO,des}$	23	[20 - 26]	kJ mol^{-1}
	E_{aLH}	58	[56 - 60]	kJ mol^{-1}
	c_S	15.6		mol m^{-3}

4. RESULTS AND DISCUSSION

Stationary experiments: Axial concentration profiles, measured and calculated at different temperatures, for reactions with CO/O_2 feed (A) and simulated exhaust gas (B) are presented in fig. 3. The model describes well the O_2 concentration profiles for the binary CO/O_2 mixture (fig. 3 A) at all temperatures. In contrast, the corresponding calculated oxygen equivalent profiles of the exhaust gas mixture (fig. 3 B) differ from the experimental results in the last section of the catalyst at lower temperatures. This is due to the poor conversion of C_3H_8 at low temperatures and the rigorous simplification in our model which does not differentiate between the reduction behaviours of the differ-

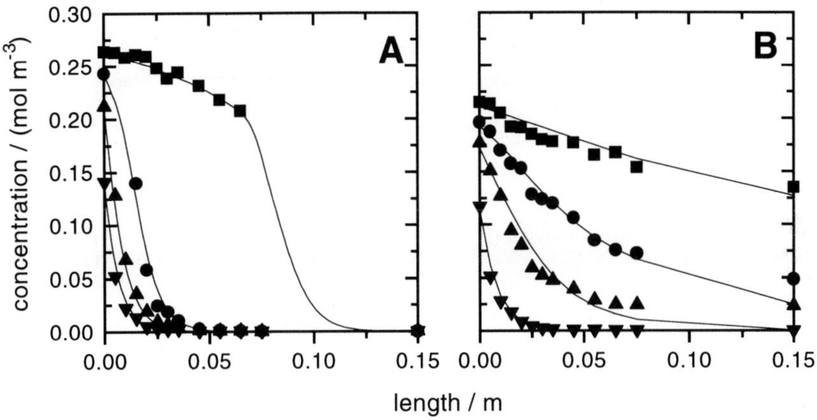

Figure 3. Axial O_2 (A) and O_2 equivalent (B) concentration profiles measured and calculated for different temperatures. **A**: CO/O_2 mixture (2 vol% CO, 1 vol% O_2), **B**: simulated exhaust gas ($\lambda = 1$)
Symbols: Experimental results at ■ 200°C, ● 250°C, ▲ 300°C, ▼ 600°C; — Calculated

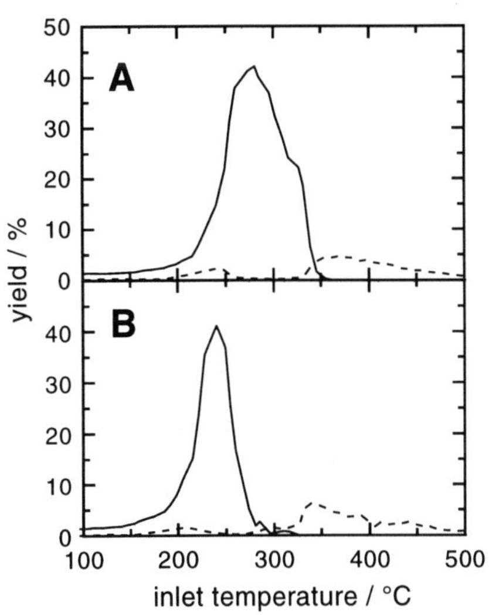

← Figure 4. Influence of temperature on N_2O and NH_3 formation observed with Pt-Rh/CeO_2-Al_2O_3. Yields of N_2O and NH_3, defined as

$\text{Yield}_{NH_3} = (c_{NH_3} / c_{NO_0}) \cdot 100$

$\text{Yield}_{N_2O} = (2\, c_{N_2O} / c_{NO_0}) \cdot 100$

are shown as a function of temperature for steady (**A**) and λ-cycled (**B**: $\lambda = 1 \pm 0.02$, 0.67 Hz) feed with the same time-averaged composition.
Symbols: —— N_2O, ---- NH_3

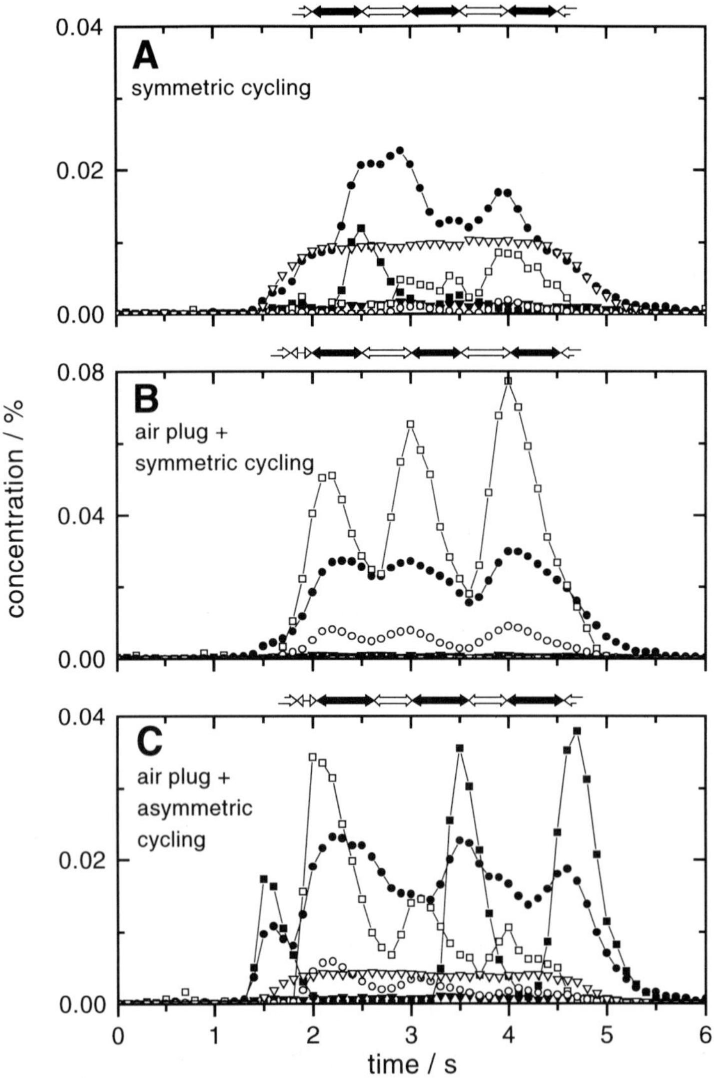

Figure 5. Concentrations of reaction products during an exhaust pulse of 3 s over Pt-Rh/ CeO_2-Al_2O_3 at 310°C with λ-cycling during pulse. Influence of air plug (⇦⇨) and its compensation. **A**: exhaust pulse with λ = 1 ± 0.05, 1 Hz; **B**: air plug followed by exhaust pulse with λ = 1 ± 0.05, 1 Hz; **C**: air plug followed by exhaust pulse with λ = 1 ± 0.05, 1 Hz (asymmetric cycling: 0.6 rich / 0.4 lean). The black arrows on top indicate the rich, the white arrows the lean half-cycles of which the exhaust pulse is composed. ⇨ represents the fraction of the previous exhaust pulse that remained in the catalytic converter during the period of 17 s with no gas flow. ⇦⇨ represents the air plug.
Symbols: ■ CO, ● C_3H_8, ▼ CH_4, □ NO, ○ N_2O, ▽ NH_3

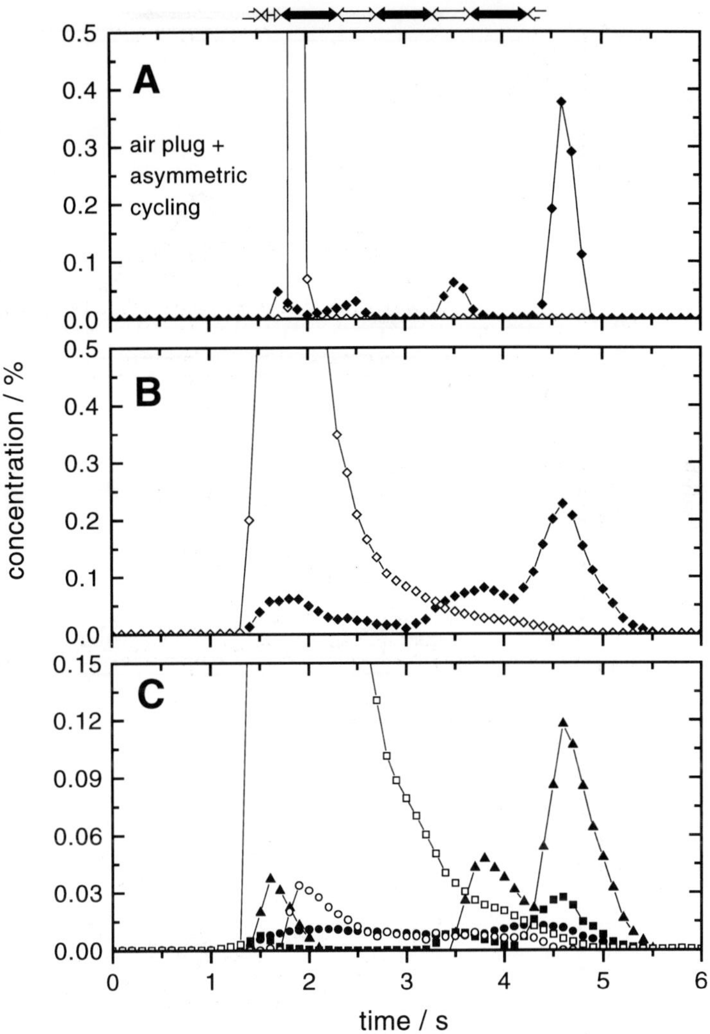

Figure 6. Concentrations of reaction products during an exhaust pulse of 3 s at 400°C with asymmetric λ-cycling (1 Hz, asymmetric cycling: 0.6 rich / 0.4 lean). **A**: *Simulated* O_2 and CO equivalent concentrations, **B**: *Measured* O_2 and CO equivalent concentrations, **C**: *Measured* concentrations of reaction products. The black arrows indicate the rich (λ = 0.95), the white arrows the lean half-cycles (λ = 1.05) of which the exhaust pulse of 3 s is composed. ⇒ represents the last fraction of the previous exhaust pulse that stayed in the catalytic converter during the period of 17 s, when no gas flowed through the converter. ⇔ is the air plug. Oxygen equivalent peak heights: A 21%, B 1.75%.

Symbols: ■ CO, ▲ H_2, ● NH_3, □ O_2, ○ NO, ◆ CO equivalent (eq. 2), ◇ O_2 equivalent (eq. 3)

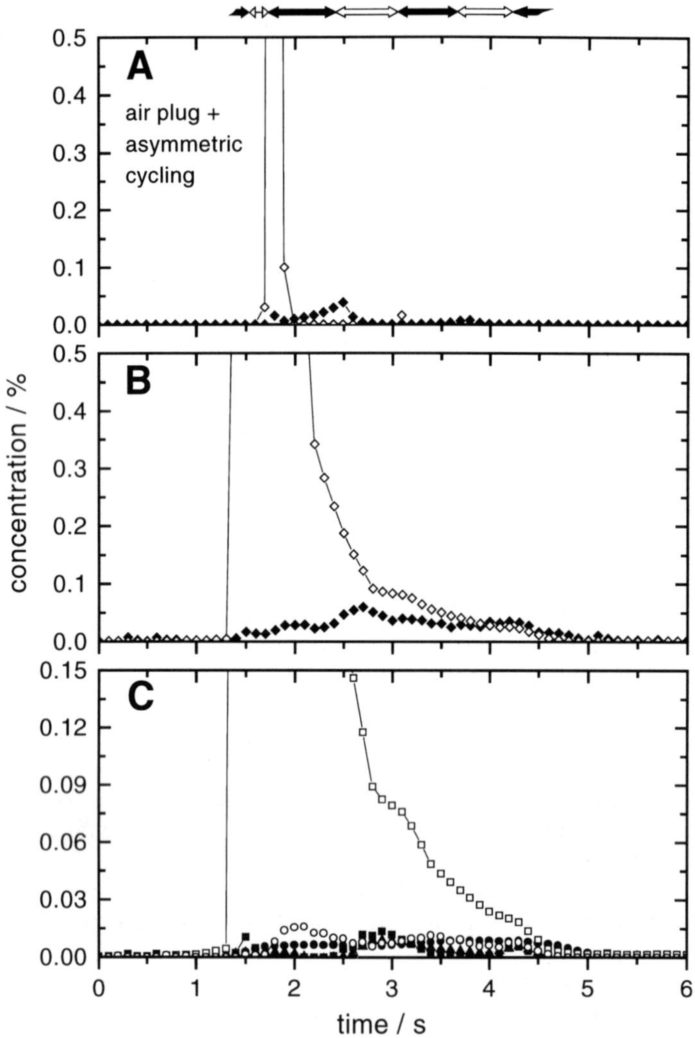

Figure 7. Concentrations of reaction products during an exhaust pulse of 3 s at 400°C with asymmetric λ-cycling (0.68s rich / 0.64s lean / 0.60s rich / 0.56s lean / 0.52s rich) during pulse. **A**: *Simulated* O_2 and CO equivalents, **B**: *Measured* O_2 and CO equivalent concentrations, **C**: *Measured* concentrations of reaction products. The explanation of the meaning of the different arrows is given in caption of fig. 6. Oxygen equivalent peak heights: A 21%, B 1.75%.
Symbols: ■ CO, ▲ H_2, ● NH_3, □ O_2, ○ NO, ◆ CO equivalent, ◇ O_2 equivalent

ent reducing species.

In the experiments at steady-state conditions, the reaction zone was found to be stationary, leading to the development of a distinct temperature profile. In contrast, the reaction zone moved during dynamic operation and the concentration and temperature profiles were much less pronounced. Therefore the temperature was set to the outlet temperature and held constant during the dynamic simulations.

Experiments with λ-cycling: With λ-cycling, the light-off temperatures for NO_x and C_3H_8 were drastically reduced, whereas CO and C_3H_6 conversion were only slightly influenced [16]. The temperatures necessary for 50% conversion decreased from 315 to 260°C for C_3H_8 and from 330 to 255°C for NO_x. In addition, λ-cycling lowered the undesired formation of N_2O at temperatures between 200°C and 350°C as illustrated in fig. 4.

Experiments with pulsed flow: As illustrated in fig. 5, the air plug had a dramatic influence on conversions, especially on those of NO_x and C_3H_8 (fig. 5 B), and the N_2O production increased. Using asymmetric cycling with longer rich half cycles reduced the negative influence of the air plug (fig. 5 C), but it was not possible to reach the same conversions as measured without the air plug (fig. 5 A).

The concentration peaks in fig. 5 result from depletion of oxidizing or reducing species stored on the catalyst surface. A detailed discussion of the influence of λ-cycling, air plug, ceria-promotion and other factors is given in ref. [16].

To exclude total poisoning of the active surface sites by oxygen, an additional assumption was imposed on the model: it was assumed that 5% of the active surface sites occupied by O_{ads} are accessible to CO. This assumption seems to be reasonable in the light of the experimental finding that CO still can adsorb on Pt surfaces saturated by oxygen [19]. Figures 6 and 7 show two simulations of the pulsed-flow operation of the converter. The simulated concentration vs. time profiles show the same features as the measured profiles for pulsed-flow operation with constant as well as cycled λ-value during exhaust pulse. The agreement between calculated and measured profiles with regard to the time dependence is remarkable in view of the simplifying assumptions made. The oxygen equivalent peak heights in A and B were 21% and 1.75%, respectively (not shown). Note, however that the time integrals of these curves were about the same. In any case, the model was found to be useful for optimising the λ-value during pulsed-flow operation of the converter. The problem was to compensate for the oxygen introduced by the air-plug using rich exhaust gas without breaking through of the reducing components. The simulation of the optimised operation could be confirmed by experimental measurements, as illustrated in fig. 7.

5. CONCLUSION

A simple Langmuir-Hinshelwood model, extended by introducing CO and O_2 equivalents to mimic the complex exhaust gas, provided a satisfactory description of the dynamic behaviour of a honeycomb type Pt-Rh/CeO_2-Al_2O_3 catalyst. The suitability of the model was confirmed for steady-state, λ-cycling and pulsed-flow operation in the temperature range between 200°C and 600°C.

When combined with the use of quantitative oxygen sensors with short response time, the model should permit efficient control of the λ-value which is essential for the application of three-way catalysts in hybrid drive systems. Keeping the oxygen surface coverage at a suitable level would allow buffering both, rich and lean excursions. In addition, disturbances of the catalyst performance by exhaust gases strongly deviating from $\lambda = 1$ can be compensated for very efficiently.

ACKNOWLEDGEMENTS

Financial support of this work by the *Schweizerisches Bundesamt für Umwelt, Wald und Landschaft* is greatfully acknowledged.

REFERENCES

1. M. K. Eberle, Symposium Proceedings *Internationale Konferenz über hybride Automobilantriebe*, Eidgenössische Technische Hochschule, Zürich 1993.
2. L. Küng, A. Vezzini and K. Reichert, Symposium Proceedings 11th International Electric Vehicle Symposium (1992).
3. R. P. Canale, C. R. Carlson, S. R. Winegarden and D. L. Miles, SAE Trans., 87 (1978) No. 780205.
4. H. S. Gandhi, A. G. Piken, M. Shelef and R. G. Delosh, SAE Trans., 85 (1976) No. 760201.
5. J. C. Schlatter and P. J. Mitchell, Ind. Eng. Chem. Prod. Res. Dev., 19(3) (1980) 288-293.
6. R. K. Herz, Ind. Eng. Chem. Prod. Res. Develop., 20 (1981) 451-457.
7. R. K. Herz, J. B. Kiela and J. A. Sell, Ind. Eng. Chem. Prod. res. Dev., 22(3) (1983) 387-396.
8. K. C. Taylor and R. M. Sinkevitch, Ind. Eng. Chem. Prod. Res. Dev., 22(1) (1983) 45-51.
9. H. Muraki, H. Shinjoh, H. Sobukawa, K. Yokota and Y. Fujitani, Ind. Eng. Chem. Prod. Res. Dev., 24(1) (1985) 43-49.
10. B. K. Cho and L. A. West, Ind. Eng. Chem. Fundam., 25(1) (1986) 158-164.
11. H. Shinjoh, H. Muraki and Y. Fujitani, A. Crucq and A. Frennet (Editors), Catalysis and Automotive Pollution Control, Elsevier Science Publishers B.V., Amsterdam (1987) 187-197.
12. H. Shinjoh, H. Muraki and Y. Fujitani, Appl. Catal., 49 (1989) 195-204.
13. M. G. Goodman, M. B. Cutlip, C. N. Kenney, W. Morton and D. Mukesh, Surf. Sci., 120 (1982) L453.
14. W. R. C. Graham and D. T. Lynch, AIChE Journal, 36(12) (1990) 1796-1806.
15. B. N. Racine and R. K. Herz, J. Catal., 137 (1992) 158-178.
16. L. Padeste and A. Baiker, Ind. Eng. Chem. Res., 5 (1994).
17. E. Koberstein and G. Wannemacher, Stud. Surf. Sci. Catal., 30 (1987) 155-172.
18. E. Kreyszig, Statistische Methoden und ihre Anwendungen, 7th. ed., Vandenhoeck und Ruprecht, Göttingen 1982.
19. T. Engel and G. Ertl, Adv. in Catal., 28 (1979) Ch. 3.

A. Frennet and J.-M. Bastin (Eds.)
Catalysis and Automotive Pollution Control III
Studies in Surface Science and Catalysis, Vol. 96
1995 Elsevier Science B.V.

THE PERFORMANCE OF A MONOLITHIC CATALYTIC CONVERTER OF AUTOMOBILE EXHAUST GAS WITH OSCILLATORY FEEDING OF CO, NO, AND O_2: A MODELLING STUDY

A.J.L. Nievergeld, J.H.B.J. Hoebink and G.B. Marin*

Laboratorium voor Chemische Technologie, Eindhoven University of Technology, P.O. Box 513, NL-5600 MB Eindhoven, The Netherlands

ABSTRACT

An isothermal monolithic catalytic converter of automobile exhaust gas was modelled in order to assess the effects of oscillatory feeding of CO, O_2 and NO on the performance of the reactor. The influence of the temperature, frequency, and amplitude on the time average conversions was investigated. An improvement relative to the steady state conversion of 8% for CO and 30% for NO can be obtained. An explanation is given in terms of strongly changing surface coverage during cycling of the feed concentrations.

1. INTRODUCTION

Oscillatory feeding of a catalytic converter has a well-known potential to improve the reactor performance [1-5]. For automotive exhaust gas treatment the phenomenon has drawn wide attention [6-11] as in this case the oscillations occur as a result of applying a feedback controller with lambda sensor in order to maintain the composition of the exhaust gas close to stoichiometric point. Experimental work of Cho [9] showed a better performance of the exhaust gas converter with oscillating feed when operation took place below the light-off temperature. Model simulations by Lie *et al.* [12] concerning the CO oxidation in a monolith converter support these results and explain them from the calculated surface coverages under dynamic operation. The present work extends the model of Lie *et al.* towards reactions between CO, O_2, and NO at typical exhaust gas conditions to calculate the influence of cycling the feed on the time-average conversions of CO and NO. A substantial improvement compared to steady state kinetics would be an incentive to adapt the currently used controllers, as a part of a strategy towards ultra low emission vehicles.

---

* to whom correspondence should be addressed

2. KINETIC MODEL

The kinetic model for reaction between CO, O_2, and NO used in this study is based on the models of Oh *et al.* [13] for the CO-O_2 and the CO-NO reaction on a Rh/Al_2O_3 catalyst. The elementary reactions and the reaction paths are shown in table 1.

Table 1
Elementary reactions and reaction paths

elementary reaction	N1	N2	N3
$CO + * \underset{k_{d,CO}}{\overset{k_{a,CO}}{\rightleftarrows}} CO*$	2	2	2
$O_2 + 2* \xrightarrow{k_{a,O_2}} 2O*$	1	0	0
$NO + * \underset{k_{d,NO}}{\overset{k_{a,NO}}{\rightleftarrows}} NO*$	0	2	2
$NO* + * \xrightarrow{k_{diss}} N* + O*$	0	1	2
$CO* + O* \xrightarrow{k_{CO_2}} CO_2 + 2*$	2	2	2
$NO* + N* \xrightarrow{k_{N_2,1}} N_2 + O* + *$	0	1	0
$2N* \xrightarrow{k_{N_2,2}} N_2 + 2*$	0	0	1

N1: $2CO + O_2 \rightarrow 2CO_2$

N2, N3: $2CO + 2NO \rightarrow CO_2 + N_2$

where * stands for a catalytic site. Following Oh [13], the O_2 adsorption rate is taken proportional to the fraction of vacant sites θ_*, rather then the expected θ_*^2 dependence. Contrary to Oh the activation energy for the CO desorption and the recombination of two adsorbed N atoms are taken coverage-independent. Two different pathways lead to N_2 formation (N2, N3): the recombination of two adsorbed N atoms and the reaction of adsorbed N with adsorbed NO. At low temperatures (< 550 K) the CO-NO reaction kinetics is dominated by the dissociation of NO, while at higher temperatures the recombination of two adsorbed N atoms becomes the dominant path. In this model, the surface of the catalyst can become completely covered with oxygen, in contrast to, e.g. the model of Herz and Marin for the CO oxidation over Pt/Al_2O_3 catalyst in which the oxygen surface coverage is limited to 0.5 [14].

Table 2 lists the kinetic parameter values used in this study. Except for the activation energy for CO desorption and the reaction between two adsorbed N atoms, they are identical to those reported by Oh *et al.* [13].

Table 2
Kinetic parameters values used in this study

parameter	value	parameter	value
S_{CO}^{0}	0.5	E_{diss}	79.4
$A_{d,1}$	$1.6 \cdot 10^{14}$	A_{r,CO_2}	$1 \cdot 10^{12}$
$E_{d,1}$	112.0	E_{CO_2}	60.0
$S_{O_2}^{0}$	0.01	$A_{N_2,1}$	$2 \cdot 10^{9}$
S_{NO}^{0}	0.5	$E_{N_2,1}$	87.8
$A_{d,3}$	$5 \cdot 10^{13}$	$E_{N_2,2}$	120.0
$E_{d,3}$	108.7	$A_{N_2,2}$	$3 \cdot 10^{10}$
A_{diss}	$3 \cdot 10^{10}$	L_t	$2.7 \cdot 10^{-5}$

Table 3 shows the characteristic times of the elementary reactions in table 1 which indicate at which time scale a reaction proceeds. The calculation of the characteristic times was performed at 505 K and at concentrations corresponding to a stoichiometric inlet composition (table 5). Surface reactions are relatively slow processes when compared to ad- and desorption.

Table 3
Definitions and values of the characteristic times of the elementary reactions*

characteristic time	definition	value [s]	characteristic time	definition	value [s]
$\tau_{a,CO}$	$\left(k_{a,CO}\overline{C}_{CO}^{in}\right)^{-1}$	$2.5 \cdot 10^{-6}$	τ_{diss}	$(k_{diss}\theta_*)^{-1}$	7.2
$\tau_{d,CO}$	$k_{d,CO}^{-1}$	$2.4 \cdot 10^{-3}$	τ_{CO_2}	$(k_{CO_2}\theta_O)^{-1}$	0.15
τ_{a,O_2}	$\left(2k_{a,O_2}\overline{C}_{O_2}^{in}\right)^{-1}$	$1.6 \cdot 10^{-4}$	$\tau_{N_2,1}$	$(k_{N_2,1}\theta_N)^{-1}$	7.6
$\tau_{a,NO}$	$\left(k_{a,NO}\overline{C}_{NO}^{in}\right)^{-1}$	$1.5 \cdot 10^{-5}$	$\tau_{N_2,2}$	$(2k_{N_2,2})^{-1}$	43
$\tau_{d,NO}$	$k_{d,NO}^{-1}$	$3.5 \cdot 10^{-3}$			

* 505 K, $\theta_* = 7.6 \cdot 10^{-4}$, $\theta_O = 10^{-5}$, $\theta_N = 0.08$

3. REACTOR MODEL

A one-dimensional model [12] for the isothermal monolith is used which consists of a set of continuity equations for the reactants in the three phases considered: the gas phase, the pores of the washcoat and the catalyst surface. The dependent variables are expressed as C/ρ_f. The continuity equation for reactant i (i = CO, O_2 or NO) in the bulk gas phase is given by:

$$\varepsilon \rho_f \frac{\partial}{\partial t}\left[\frac{C_{f,i}}{\rho_f}\right] = -\phi_m^{sup} \frac{\partial}{\partial x}\left[\frac{C_{f,i}}{\rho_f}\right] - k_{f,i} a_v \left(C_{f,i} - C_{s,i}\right) \tag{1}$$

The continuity equation for reactant i in the pores of the washcoat is given by:

$$\varepsilon_w \rho_f \frac{4 \varepsilon d_w}{d_b} \frac{\partial}{\partial t}\left[\frac{C_{s,i}}{\rho_f}\right] = k_{f,i} a_v \left(C_{f,i} - C_{s,i}\right) - a_{cat} r_{a,i} + a_{cat} r_{d,i} \tag{2}$$

The continuity equations for the species CO*, O*, NO*, and N* on the surface are given by:

$$L_t \frac{\partial \theta_{CO}}{\partial t} = r_{a,CO} - r_{d,CO} - r_{CO_2} \tag{3}$$

$$L_t \frac{\partial \theta_{O}}{\partial t} = 2 r_{a,O_2} + r_{diss} + r_{N_2,1} - r_{CO_2} \tag{4}$$

$$L_t \frac{\partial \theta_{NO}}{\partial t} = r_{a,NO} - r_{d,NO} - r_{diss} - r_{N_2,1} \tag{5}$$

$$L_t \frac{\partial \theta_{N}}{\partial t} = r_{diss} - r_{N_2,1} - 2 r_{N_2,2} \tag{6}$$

The reactor is assumed to operate under steady state conditions before starting any oscillation of the feed. Hence, the initial conditions for the concentration of reactant i in the gas phase and in the pores of the washcoat are:

$$C_{f,i}(x,0) = C_{f,i}^{ss}(x) , \quad 0 \leq x \leq L \tag{7}$$

$$C_{s,i}(x,0) = C_{s,i}^{ss}(x) , \quad 0 \leq x \leq L \tag{8}$$

and for the degree of coverage of the species (j = CO*, O*, NO*, or N*):

$$\theta_j(x,0) = \theta_j^{ss}(x) , \quad 0 \leq x \leq L \tag{9}$$

The boundary conditions at x=0 result from the oscillating feed at the inlet of the reactor, for CO given by:

$$C_{f,CO}(0,t) = \overline{C}^{in}_{f,CO}\left[1 + B\sin(2\pi ft)\right] \tag{10}$$

CO is oscillating in counter phase with O_2 and NO, so the time dependent feed concentration of O_2 and NO is given by:

$$C_{f,i}(0,t) = \overline{C}^{in}_{f,i}\left[1 + B\sin(2\pi ft + \pi)\right] \tag{11}$$

Values for the parameters of the reactor model are shown in table 4.

Table 4.
Parameters of the reactor model used in this study (T = 505 K).

parameter	value	parameter	value
ρ_f	0.673	a_v	$2.4\cdot10^{3}$
ϕ_m^{sup}	5.833	d_w	$2.5\cdot10^{-5}$
$k_{f,CO}, k_{f,O_2}$	0.186	L	0.15
$k_{f,NO}$	0.216	ε	0.6
$\tau_{f,CO}^{*}, \tau_{f,O_2}^{*}$	$1.3\cdot10^{-4}$	ε_w	0.4
$\tau_{f,NO}^{*}$	$1.2\cdot10^{-4}$	a_{cat}	$1.25\cdot10^{4}$

* the characteristic time for mass transfer $\tau_{f,i}$ is defined as $\varepsilon/(k_{f,i}a_v)$.

The model is solved using the same mathematical methods as Lie *et al.* [12] except for the integration in the x direction, in which collocation with bicubic splines was applied.

4. RESULTS AND DISCUSSION

Simulations have been performed with time-invariant and cyclic feeding at typical exhaust gas conditions. The simulations with oscillatory feeding allow a comparison of the time average conversions with the conversions during time-invariant feeding as a function of temperature, frequency, and amplitude. The stoichiometric equivalence ratio of the reactants is defined relative to the stoichiometry of the global reaction:

$$\phi = \frac{2\overline{C}_{f,O_2} + \overline{C}_{f,NO}}{\overline{C}_{f,CO}} \tag{12}$$

Time-average inlet concentrations and corresponding values of ϕ are given in table 5.

Figure 1 shows the steady state gas phase concentrations of the reactants versus the axial reactor coordinate with CO as limiting reactant. Mass transfer limitation is negligible at this temperature, because the production rate of CO_2 and the production

Table 5
Inlet concentrations and corresponding values of ϕ.

ϕ	$C_{f,CO}$ (vol%)	$C_{f,O2}$ (vol%)	$C_{f,NO}$ (vol%)
0.5	0.6	0.1	0.1
1.0	0.6	0.25	0.1
1.5	0.6	0.4	0.1

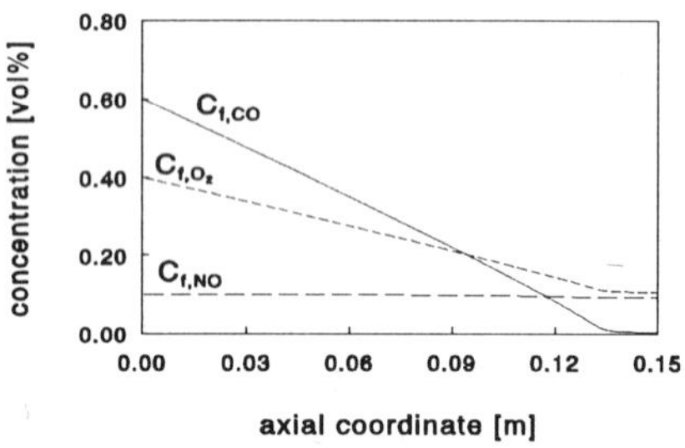

Figure 1. Gas phase concentrations vs. axial coordinate. T = 490 K, G = $5 \cdot 10^{-3}$ kg/s, ϕ = 1.5.

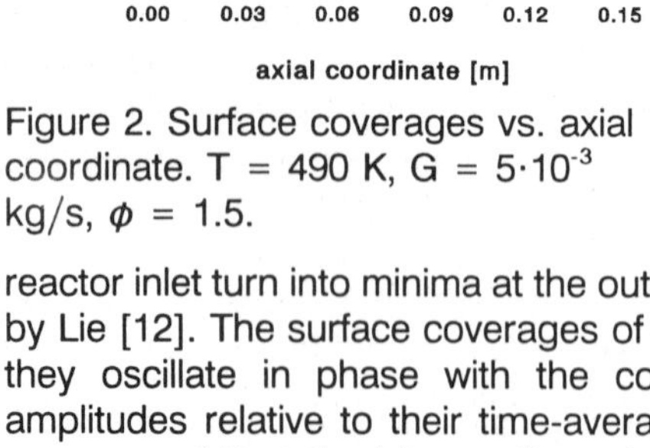

Figure 2. Surface coverages vs. axial coordinate. T = 490 K, G = $5 \cdot 10^{-3}$ kg/s, ϕ = 1.5.

rate of N_2 are low. The concentrations of the reactants in the pores of the washcoat 'follow' the gas phase concentrations at around 95% of their value, and are therefore not shown. The CO conversion near the outlet of the reactor is 100%, while the conversion of NO is only 9%. The surface coverages as a function of the axial reactor coordinate are shown in figure 2. The oxygen coverage does not exceed 10^{-5} as long as the CO conversion is not complete. As shown in figure 3, both the production rates of CO_2 and N_2 increase towards the outlet of the reactor until the surface concentration of CO* becomes zero. The relatively high surface coverage of CO at the inlet of the reactor is inhibiting the adsorption of the other reactants and the dissociation of NO, resulting in low production rates for CO_2 and N_2. Towards the outlet, the adsorption rate of CO decreases due to the decrease of the gas phase concentration, and as a result more free sites become available for the adsorption of O_2 and adsorption and dissociation of NO. The resulting higher oxygen coverage, causes the production rate of CO_2 to increase, while the increasing coverage of NO and N leads to a higher production rate of N_2.

Simulation results with cyclic feeding at a frequency of 0.1 Hz are shown in figures 4-9. Figure 4 shows the gas phase concentrations at the outlet of the reactor as a function of time. The amplitude of the CO oscillation relative to the time average value increases from 15% to 76% towards the outlet of the reactor, while the amplitude of NO and O_2 decrease from 15% to 10%. The maxima in the oxygen oscillation at the reactor inlet turn into minima at the outlet as discussed for CO oxidation over Pt/Al_2O_3 by Lie [12]. The surface coverages of CO, O, and NO at the inlet are not shown, but they oscillate in phase with the corresponding gas phase concentrations with amplitudes relative to their time-average values of 7%, 30% and 24%. The surface coverage of N at the inlet oscillates in phase with NO with an amplitude of only 4%. Figure 5 shows the surface coverages at the outlet. The oxygen coverage, which is not shown, oscillates in phase with NO* around a time average value of $1.6 \cdot 10^{-5}$ and an

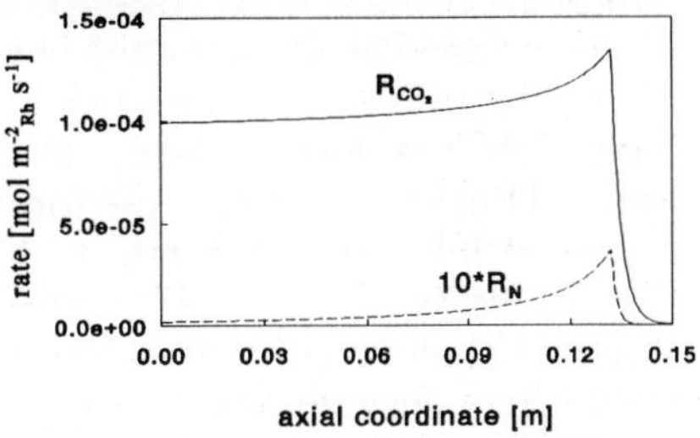

Figure 3. Production rates of CO_2 and N_2 vs. axial coordinate. T = 490 K, G = $5 \cdot 10^{-3}$ kg/s, ϕ = 1.5.

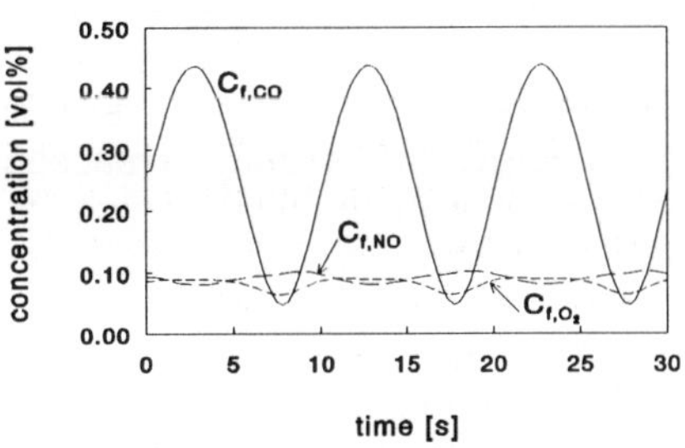

Figure 4. Outlet gas phase concentrations vs. time. T = 505 K, G = 10^{-2} kg/s, f = 0.1 Hz, B = 15%, ϕ = 1.0.

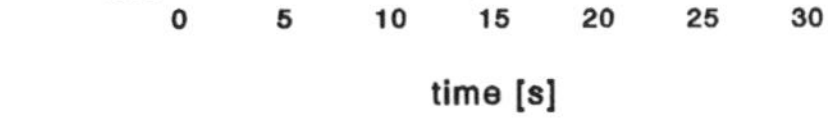

Figure 5. Outlet surface coverages vs. time. T = 505 K, G = 10^{-2} kg/s, f = 0.1 Hz, B = 15%, ϕ = 1.0.

amplitude of 151%. The amplitudes of all coverages increase towards the outlet, and the minima of CO* and the maxima of the other species become sharper, leading to relaxation type of oscillations [15]. Figure 6 shows the production rate of both CO_2 and N_2 at the outlet of the reactor. The steady state conversions of CO and NO, are 56% and 7.4% respectively, while the time-average conversions during cycling of the feed are 57.5% and 8.9%.

These differences are explained as follows. At steady state the high surface coverage of CO causes a relatively low production rate of CO_2 and N_2. When the gas phase concentration of CO increases owing to the oscillation, the surface becomes even more occupied with CO*, while the corresponding decrease of the concentrations of O_2 and NO results in a decrease of the other surface species. The result is a further decrease of the production rates of CO_2 and N_2. In contrast, when the CO gas phase concentration is decreased, free sites become available for O_2 and NO adsorption and for dissociation of NO, with higher production rates as a result. The fact that the CO decrease in the bulk gas is accompanied by an increase of O_2 and NO gives an extra enhancement of the production rates of CO_2 and N_2 and hence of the conversions of CO and NO. The positive feedback from the CO coverage to the CO gas phase concentration results in a reinforcement of the CO oscillation whereas the negative feedback from the O and NO coverages to the corresponding gas phase concentrations causes attenuation of the corresponding oscillations. The periodic enhancement of the reaction rates overcompensates the periodic decrease, leading to a positive effect on the time average conversions of both CO and NO when compared with the steady state conversions. This phenomenon is attributed to the non-linear character of the kinetics [16].

The influence of the modulation frequency is discussed in terms of the regimes of operation [16]. According to the characteristic times, the dissociation of NO and the formation of N_2 are the first reactions to enter the sliding regime. Simulations showed

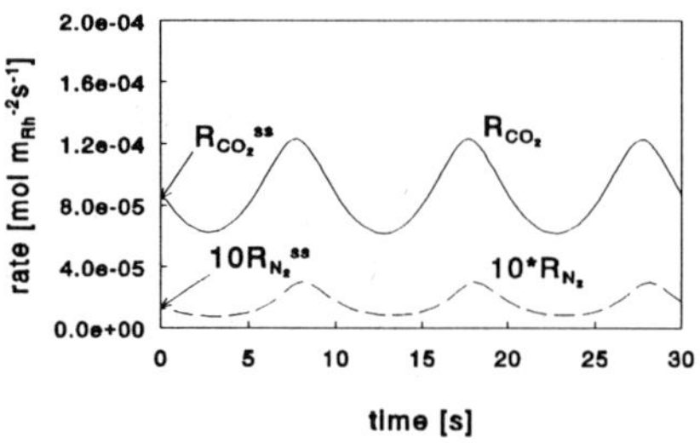

Figure 6. Outlet production rates vs. time. T = 505 K, G = 10^{-2} kg/s, f = 0.1 Hz, B = 15%, ϕ = 1.0.

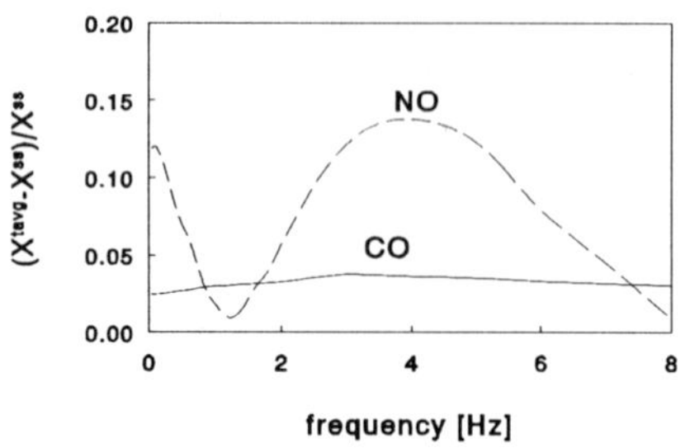

Figure 7. Time average conversions relative to steady state conversions vs. frequency. T = 500 K, G = 10^{-2} kg/s, B = 15%, ϕ = 1.0.

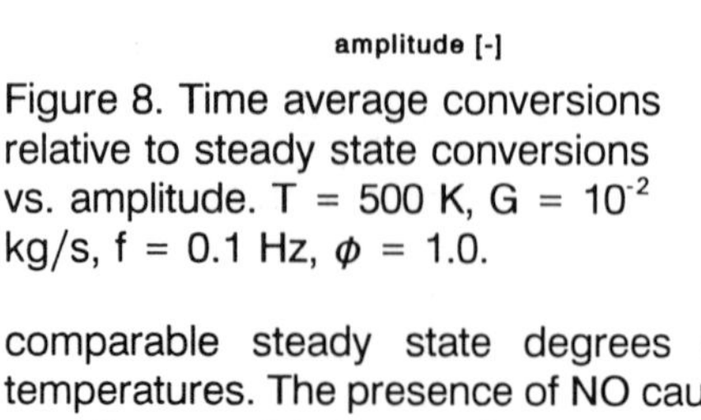

Figure 8. Time average conversions relative to steady state conversions vs. amplitude. T = 500 K, G = 10^{-2} kg/s, f = 0.1 Hz, ϕ = 1.0.

that the amplitude of the surface coverage of nitrogen atoms decreased at frequencies above 0.1 Hz, meaning that the time-average values approach the lower steady state coverage. Therefore the periodic enhancement of the production rate of N_2 shown in figure 6 disappears, which explains the decreasing beneficial effect of oscillating the feed in the frequency range 0.1 - 1 Hz for NO (figure 7) when comparing the increase of the time average conversion relative to the steady state value. The beneficial effect of oscillating for CO remains constant, because the coverages of CO and oxygen are still in the dynamic regime. The calculated rise between 1 and 4 Hz is not yet fully understood. Above 1 Hz all surface reactions approach operation in the sliding regime, but adsorption/desorption probably is still in the dynamic range causing on-going oscillations of the surface coverage of CO and NO. It was noticed, that during the periodically low NO gas phase concentrations NO desorbs from the surface, especially in the first part of the reactor, which leads to a gradual phase shift in the NO gas phase oscillation. Above 4 Hz all reactions enter the sliding regime, ultimately resulting in steady state conversions. Larger amplitudes enhance the advantage of the cyclic feeding at 0.1 Hz (figure 8) up to 30% for NO and 8% for CO. Higher amplitudes cause periodic stoichiometric limitations by CO resulting in decreasing time average conversions. Higher temperatures are beneficial when oscillating the feed (figure 9) as long as the local CO conversion is incomplete during a cycle. This is in contrast to CO oxidation in the absence of NO [12], where beneficial effects were only observed at temperatures below the light-off. This was attributed to the comparable steady state degrees of coverage by CO and oxygen at high temperatures. The presence of NO causes much lower degree of coverage by oxygen over the complete temperature range of interest. Further increase of temperature again leads to stoichiometric limitation by CO during a larger fraction of the cycle time, ultimately leading to time-average conversions below the steady state conversions.

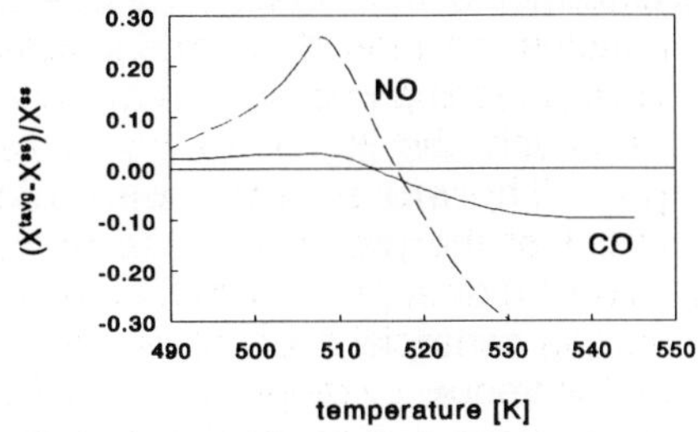

Figure 9. Time average conversions relative to steady state conversions vs. the temperature. $G = 10^{-2}$ kg/s, f = 0.1 Hz, B = 15%, ϕ = 1.0.

CONCLUSIONS

Oscillatory feeding increases the time-average conversions of both CO and NO relative to the steady state values until periodic complete conversion of CO occurs. The enhancement can be explained by strongly changed surface coverages during cyclic feeding compared to steady state values. A maximum enhancement of the NO conversion is obtained at a modulation frequency of 0.1 and 4 Hz. The CO conversion is rather frequency independent. Larger amplitudes and higher temperatures enhance the positive effects of cyclic feeding, as long as no stoichiometric limitations occur.

SYMBOLS

Roman letters

a_{cat}	catalytic surface area per unit reactor volume	$m_{Rh}^2\ m_R^{-3}$
a_v	geometric surface area per unit reactor volume	$m_i^2\ m_R^{-3}$
A	pre-exponential factor	-
B	amplitude in percentage of C_g^{in}	%
C	concentration	mol m_f^{-3}
d_b	internal diameter of channel	m_R
d_w	thickness of washcoat	m_R
E_A	activation energy	kJ mol^{-1}
f	frequency	Hz
G	mass flow	kg s^{-1}
$k_{a,i}$	adsorption coefficient for species i	$m_f^3\ mol^{-1}\ s^{-1}$
$k_{d,i}$	desorption coefficient for species i	s^{-1}
$k_{f,i}$	mass transfer coefficient for species i	$m_f^{-3}\ m_i^{-2}\ s^{-1}$
k_r	reaction rate coefficient	s^{-1}
L	reactor length	m_R
L_t	mole Rh per unit catalytic surface area	mol m_{Rh}^{-2}
r	reaction rate	mol $m^{-2}_{Rh}\ s^{-1}$
R	gas constant	kJ $mol^{-1}\ K^{-1}$
S_i	sticking coefficient for component i	-
t	time	s
T	temperature	K
x	axial coordinate	m_R
X	conversion	-

Greek letters

ε	void fraction of monolith	$m_f^3\ m_R^{-3}$
ε_w	washcoat porosity	$m_f^3\ m_w^{-3}$
θ	surface coverage	mol mol_{Rh}^{-1}
ρ_f	gas density	kg m_f^{-3}
ϕ	stoichiometric equivalence ratio	-
ϕ_m^{sup}	superficial mass flow	kg $m_R^{-2}\ s^{-1}$
τ	characteristic time	s

Subscripts

a	adsorption
cat	catalytic surface area
d	desorption
f	bulk gas phase
i	referring to reactant i or interface
s	pores in the washcoat
w	washcoat

Superscripts

in	inlet
ss	steady state
tavg	time average

REFERENCES

1. M.B. Cutlip, AIChE J., 25(3) (1979) p.502-508
2. G. Vaporciyan, A. Annapragada, E. Gulari, Chem. Eng. Sci., 43(11) (1988) p. 2957-2966
3. X. Zhou, E. Gulari, Chem. Eng. Sci., 41(4) (1986) p.883-890
4. X. Zhou, Y. Barshad, E. Gulari, Chem. Eng. Sci., 41(5) (1986) p. 1277-1284
5. R.C. Graham, D.T. Lynch, AIChE J., 36(12) (1990) p.1796-1806
6. J.C. Schlatter, R.M. Sinkevitch, P.J. Mitchell, Ind. Eng. Chem. Prod. Res. Dev., 22 (1983) p. 51-56
7. R.K. Herz, J.B. Kiela, J.A. Sell, Ind. Eng. Chem. Prod. Res. Dev., 22 (1983) p. 387-396
8. K.C. Taylor, R.M. Sinkevitch, Ind. Eng. Chem. Prod. Res. Dev., 22 (1983) p.45-51
9. B.K. Cho, Ind. Eng. Chem. Res., 27 (1988) p. 30-36
10. B.K. Cho, L.A. West, Ind. Eng. Chem. Fundam., 25 (1986) p. 158-164
11. J.A. Moulijn, F. Kapteijn, Chemisch Magazine (1989) p. 499-501
12. A.B.K. Lie, J. Hoebink, G.B. Marin, Chem. Eng. J., 53 (1993) p. 47-54
13. S.H. Oh, G.B. Fisher, J.E. Carpenter, D.W. Goodman, J. Catal., 100 (1986) p.360-376
14. R.K. Herz, S.P. Marin, J. Catal., 65 (1980) p. 281-296
15. P. Gray, S.K. Scott, Chemical Oscillations and Instabilities, Oxford Science Publications, 1990
16. Yu.Sh. Matros, Catalytic Processes Under Unsteady-state Conditions, Elsevier, Amsterdam, 1989

A. Frennet and J.-M. Bastin (Eds.)
Catalysis and Automotive Pollution Control III
Studies in Surface Science and Catalysis, Vol. 96

COLD START HYDROCARBON EMISSIONS CONTROL VIA ADMIXING THREE WAY CONVERSION CATALYSTS WITH HEAT EXCHANGE AND HYDROCARBON ADSORPTION PHENOMENA

P.L. Burk,[a] J.K. Hochmuth,[a] D.R. Anderson,[a] S. Sung,[a] A. Punke,[b] U. Dahle,[b] S.J. Tauster,[a] C.O.Tolentino,[a] J. Rogalo,[a] G. Miles,[a] M. Mignano[a] and M. Niejako[a]

[a]Engelhard Corporation, 101 Wood Avenue, Iselin, New Jersey 08830,U.S.A.

[b]Engelhard Technologies GmbH & Co. OHG, Hannover, Germany

ABSTRACT

The revisions in the United States Clean Air Act of 1990 and recent regulatory actions taken by the California Air Resources Board and European Economic Community require the development of automobiles with much lower tail pipe emissions. A significant portion of the total pollutants emitted to the atmosphere by motor vehicles occurs immediately following the startup of the engine when the engine block and exhaust manifold are cold, and the catalytic converter has not yet reached high conversion efficiencies. To meet these stringent, mandated emission levels, new technologies must be created that attack these "cold start" pollutants.
One intriguing strategy for handling cold start emissions employs hydrocarbon adsorbers. The sorber scavenges and withholds hydrocarbons from the exhaust gas until the catalyst becomes active, then the hydrocarbons are released and burned on the catalyst. Recent advances in materials technology have uncovered non-carbon-based sorbent alternatives to the classic carbon-based beds. These solids also trap via a physical interaction between the hydrocarbons and the scavenging constituent, but display better thermal stability, especially in the presence of oxygen. To take advantage of these new hydrocarbon scavenging constituents, an auto exhaust system known as the Low Hydrocarbon Emissions System (LHES) composed of a catalyzed heat exchanger and hydrocarbon trap has been designed and tested. The system was custom fitted onto several 1993 vehicles with a variety of engine management strategies. In each case U.S. FTP 75 total hydrocarbon emissions were reduced between 45 - 75% versus the vehicle's stock exhaust system. The LHES system also proved resistant to an accelerated thermal aging that was equivalent to 50,000 miles. A deactivation factor for all pollutants of 1.0 was calculated for the LHES following a 75 hr engine test bed aging with the heat exchanger inlet gas temperature set at 760°C.

1. Introduction

Over the next decade and in several regions of the world, vehicle emission regulations will become so stringent that Three Way Conversion (TWC) catalyst-only exhaust systems will not be sufficient to meet these standards. The reason is simple. During the time it takes for the exhaust gas to bring a TWC catalyst from room temperature to an active state, i.e., somewhere in excess of 300°C, pollutants escape to the atmosphere untouched. Such "cold start" emissions found in the U.S. FTP 75 and European tests account for >60% of all the hydrocarbons (HC's) emitted by the vehicle [1], and overwhelm a vehicle's chance for meeting California Air Resources Board's Low Emission Vehicle (LEV) or Ultra Low Emission Vehicle (ULEV) targets (Table 1). Diverse technologies for lowering these cold start hydrocarbon emissions can be found in the literature including exhaust gas burners [2 - 4], exhaust gas igniters [5 - 7] , and electrically heated catalysts [8 - 13] These technologies reduce the time it takes for a catalyst to reach light-off temperature by the injection of either chemical or electrical energy into the gas flow upstream of the TWC catalyst.

Table 1 California's Low Emission Vehicle Program Certification Standards

	FTP (grams/mile)[a]		
Category[c]	NMOG[b]	CO	NO_x
TLEV	0.125	3.4	0.4
LEV	0.075	3.4	0.2
ULEV	0.040	1.7	0.2
ZEV	0.000	0.0	0.0

[a] Emissions after 50,000 miles
[b] NMOG = Nonmethane organic gases including oxygenated species
[c] LEV = transition low emission vehicle; LEV = low emission vehicle; ULEV= ultra low emission vehicle; and ZEV = zero emission vehicle

An alternate strategy for dealing with cold start HC's makes use of HC traps. A HC trap first scavenges HC's from the cold exhaust via a physical adsorption process until catalyst light-off occurs, then returns the HC's to the activated catalyst wherein they are burned to CO_2 and H_2O. Since the HC adsorption and desorption for reliable traps are physical processes [14, 15], the rate of HC release from a trap depends both on inlet gas HC concentration and gas temperature. Several HC trap-based system patents and papers have recently surfaced, and make use of bypass valves for controlling when the exhaust flow is directed through or around the HC trap [16 - 20]. In each case the effectiveness of the design depends on reliably delivering at just the right time the trapped HC's to a hot, active TWC catalyst. Figure 1 shows a typical sequence for a valve-based system. During the cold start period, the exhaust gas flows untouched through the TWC catalyst and into the trap where the HC's are scavenged and held Once the catalyst heats up and begins converting the pollutants, the valves switch and the exhaust bypasses the HC trap. In the next step, the trap is purged so that the HC's reenter the gas stream in front of the TWC catalyst.

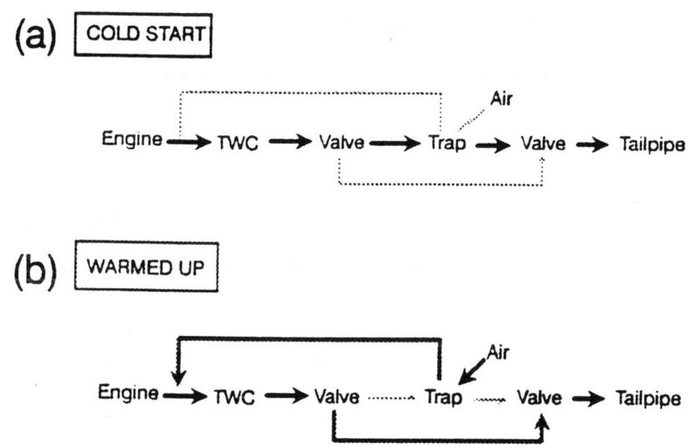

Figure 1. An exhaust system fitted with valves and a HC adsorber located downstream of the TWC catalyst bed. (a) During cold start operation, the exhaust gas flow passes through the TWC and into the HC trap wherein HC's are removed and held. (b) After TWC light-off, the valves switch and the exhaust gas bypasses the HC trap, and the trap is purged into the TWC bed.

In this paper we report on how heat exchange technology can be married to HC trap technology to give an auto exhaust system free of valves. Figure 2 shows such a system with a heat exchanger composed of two TWC catalyst beds in heat exchange relation and a HC trap. When the exhaust system is cold, the exhaust gas stream passes through the first catalyst bed then into the HC trap. The trap scavenges the HC's as in a valved system. The scrubbed gas returns to the heat exchanger, passes through the cold second catalyst bed and then out the tail pipe As the exhaust system heats up, the first TWC catalyst bed undergoes normal light-of. The cleaned gas exits the heat exchanger and then sweeps the HC's out of the HC Trap. At the same time the HC trap releases the HC's, the first pass TWC catalyst bed has been transferring heat via conduction to the second catalyst bed. The HC's purged from the trap reenter the heat exchanger and encounter a hot, activated TWC catalyst bed wherein they are converted to carbon dioxide and water. Our successful merging of hydrocarbon trap and heat exchanger technologies has resulted in a simple, passive Low Hydrocarbon Emissions System (LHES) able to transform diverse 1993 vehicles into either Low Emission or Ultra Low Emission Vehicles. Previously, we described a rudimentary heat exchanger prototype/hydrocarbon trap system on an eight year old 1985 2.3 L Volvo 740 GLE that gave an FTP-75 0.065 g/mile NMOG, a fresh LEV performance [21].

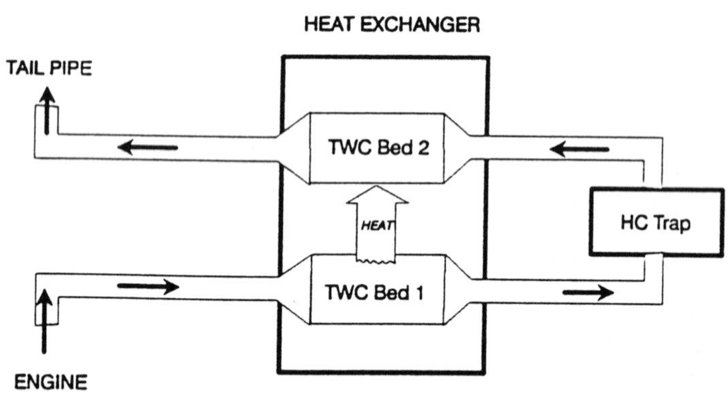

Figure 2. Gas flow through a system composed of a TWC catalyzed heat exchanger and an HC trap. The heat exchanger and HC trap remain on-line throughout the driving cycles.

2. EXPERIMENTAL

2.1. Vehicle evaluations

Evaluations of vehicle exhaust systems were performed on a dual roll Clayton dynamometer. The power absorption unit was 50 hp (37.3 kW). Water was used as the load circuit fluid. Vehicle emissions were collected over the course of US FTP 75 test (Federal Test Procedure) using a Horiba Instruments, Inc. Vehicle Emissions testing system. Hydrocarbon emissions were measured by flame ionization using a Horiba Model FIA-23A total hydrocarbon analyzer with electronics module OPE-135. Carbon monoxide emissions were measured by using nondispersive infrared absorption with a Horiba Model AIA-23 NDIR analyzer. Oxides of nitrogen were measured by chemiluminescence using a Horiba Model CLA-22A analyzer package Oxygen was measured by paramagnetism using a Horiba Model MPA-21 analyzer.

Vehicle FTP evaluations were conducted using both the stock commercial exhaust system and a Phase I Low Hydrocarbon Exhaust System (LHES). The stock systems had been on-road aged for 5,000 - 7,000 miles and were deemed essentially fresh. Installation of the Phase 1 system onto a vehicle consisted of

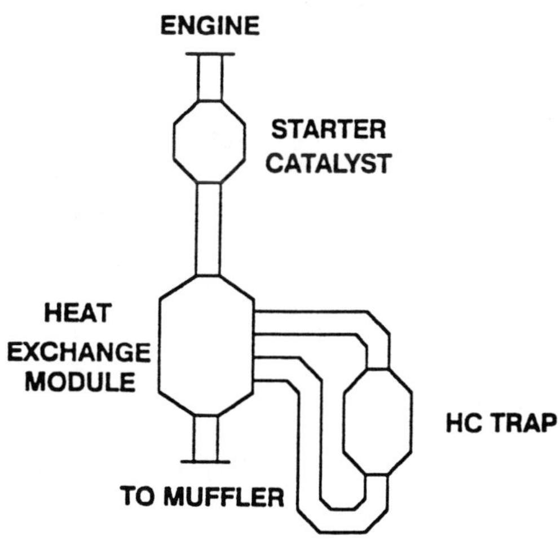

Figure 3. Phase 1 Low Hydrocarbon Emissions System. Start catalyst was 8.6 cm diameter by 9.3 cm long. Heat exchange module was 7.6 cm by 12.7 cm by 20.3 cm long. Hydrocarbon trap was 8.0 cm by 17 cm racetrack by 7.6 cm long.

removing the stock exhaust system then custom mounting an exhaust composed of a start catalyst (0.5 L, 5/1 Pt/Rh at 1.2 g/L)) followed by a catalyzed heat exchanger (2.0 L, 5/1 Pt/Rh at 1.4 g/L) and a hydrocarbon trap (1.8 L, washcoated 400 cpsi cordierite monolith) as shown in Figure 3. The Phase 1 system was fitted with thermocouples and exhaust gas sampling ports to fully characterize the performance across each catalyst or HC trap bed.
The HC trapping efficiency across the HC trap was calculated from equation (1):

$$\text{HC \% Trapping Efficiency} = [1-(HC_{outlet}/HC_{inlet}] \times 100 \qquad (1)$$

The vehicle testing sequence comprised of the following: The test vehicles as purchased from the dealers were evaluated by conducting FTP runs to assess both the engine out pollutant levels, and to measure the performance of the stock exhaust system. Next, the stock exhaust system was removed from the vehicle and the Phase 1 LHES was custom fitted to the vehicle underbody. FTP runs were then conducted. Comparisons of the FTP runs between the stock exhaust and the fresh Phase 1 LHES have been summarized in Table 2.

Table 2 FTP runs of stock converter systems versus the Phase 1 LHES

		FTP Emissions (grams/mile)		
1993 Vehicle Class	Exhaust System	THC[a]	CO	NOx
3.8 L Full Size[b] (U.S.)	Stock	0.246	1.08	0.07
	Phase 1 LHES	0.041	0.28	0.01
3.0 L Full Size (Europe)	Stock	0.227	1.10	0.12
	Phase 1 LHES	0.088	0.69	0.16
3.0 L Mid-Size (U.S.)	Stock	0.158	1.68	0.46
	Phase 1 LHES	0.087	1.29	0.14
2.4 L Mid-Size (Japan)	Stock	0.174	1.70	0.24
	Phase 1 LHES	0.054	0.75	0.10
2.2 L Mid-Size (Europe)	Stock	0.143	0.27	0.14
	Phase 1 LHES	0.040	0.05	0.19
2.0 L Mid-Size (Japan)	Stock	0.156	1.86	0.24
	Phase 1 LHES	0.049	0.37	0.06

[a] THC = Total Hydrocarbons including methane
[b] 1990 Model Year Vehicle

2.2. Accelerated aging of Phase 1 LHES

The Phase 1 LHES fitted to the 1993 3.0 L test vehicle was separated from the start catalyst. The remaining system, i.e., the heat exchanger/HC trap combination was mounted on an engine test bed for an accelerated aging to ascertain the thermal durability of the heat exchanger unit. During this aging, the exhaust gas inlet temperature to the first pass of heat exchanger was maintained at 760°C for 75 hr. The hydrocarbon trap was exposed to gas temperatures of 500 - 550°C. When the aging was complete, the Phase 1 system was remounted onto the test vehicle and FTP evaluations were conducted. The results of the fresh and aged systems have been summarized in Table 3.

Table 3 FTP evaluations of fresh and aged Phase 1 LHES on a 1993 3.0 L Mid-Size Vehicle

		FTP (grams/mile)		
	Runs	THC[a]	CO	NOx
Fresh	10	0.087	1.29	0.14
Aged[b]	5	0.085	1.24	0.14
Standard Deviation		0.006	0.09	0.03

[a] Total Hydrocarbons including methane
[b] Engine aging conditions: Inlet to first pass heat exchanger set at 760°C for 75 hrs

3. RESULTS AND DISCUSSION

This investigation began with the aim of establishing whether heat exchanger technology could be married to TWC catalyst and HC trap materials in a manner that would substantially reduce cold start HC emissions and would have the durability to survive on-road driving conditions. With this in mind several different 1993 test vehicles with mileage ranging from 5,000 - 7000 miles were purchased.FTP evaluations were completed on the stock systems, and the unremarkable results have been summarized in Table 2. Note should be made that the FTP grams/mile HC emissions are reported as total hydrocarbon (THC) including methane which varied from 10 - 20% depending on the vehicle. The stock exhaust systems were removed and each vehicle was custom fitted with a Phase 1 Low Hydrocarbon Emissions System (LHES) as described in the Experimental Section. The heat exchanger and HC trap were sized to fit comfortably on the preexisting underbodies of the test fleet, as shown in Figure 3The FTP evaluations with the Phase 1 system have been summarized in Table 2. Remarkably, the Phase 1 system dramatically improved FTP total hydrocarbon (THC) emissions on all vehicles tested, some by >65%. This unique combination of heat exchange and sorbent technologies resulted in a system that displayed a low sensitivity to a vehicle's particular engine management strategy and elevated all the test vehicles into either LEV or ULEV performance ranges. Our Phase 1 technical target of Fresh LEV/ULEV FTP performances had been realized without having to modify any other vehicle subsystems, such as, electrical, and fuel management.

The influence of each component on the overall system performance emerged on examination of both the FTP second-by-second and modal analysis data. For purposes of illustration data from the 2.0 L and 2.4 L 1993 Japanese vehicles has been summarized in the figures below. The data from the other test vehicles was consistent with these two automobiles. First, cumulative emissions (Figures 4 and 5) for the 2.4 L vehicle showed the improvements in NOx and CO

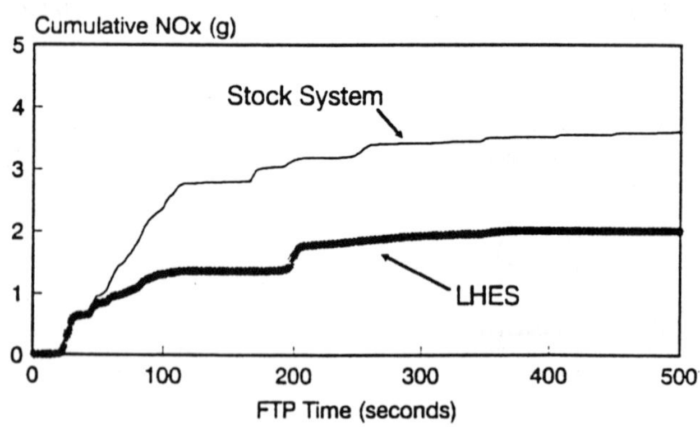

Figure 4. Comparison of NO_x tailpipe emissions for the Low Hydrocarbon Emissions System versus the stock exhaust system for the 2.4 L Mid-size vehicle.

emissions were directly attributable to the increased catalyst volume of the Phase 1 system relative to the vehicle's original exhaust. The rates of accumulation at the tailpipe for CO and NOx for the LHES system were simply displaced to a lower value for the LHES: the LHES TWC beds underwent light-off at substantially the same time as the stock system.

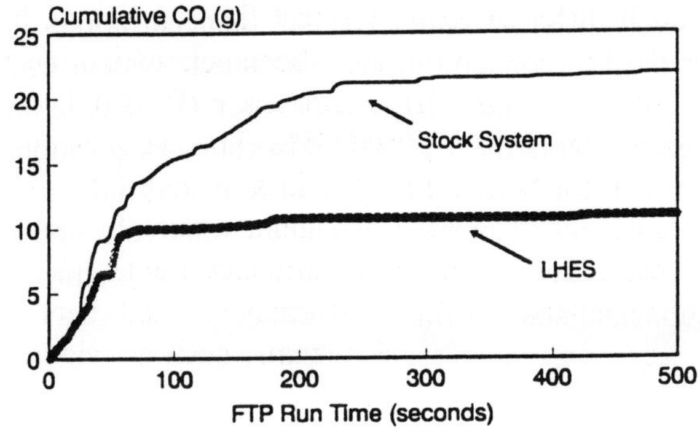

Figure 5. Comparison of CO tailpipe emissions for the Low Hydrocarbon Emissions System versus the stock exhaust system for the 2.4 L Mid-size vehicle.

In contrast to the CO and NOx emissions improvements, the relative drop in tail pipe THC's were linked directly to the LHES's HC trap. Figure 6 shows the stock exhaust and the LHES tail pipe HC emission traces of the 2.0 L and 2.4 L

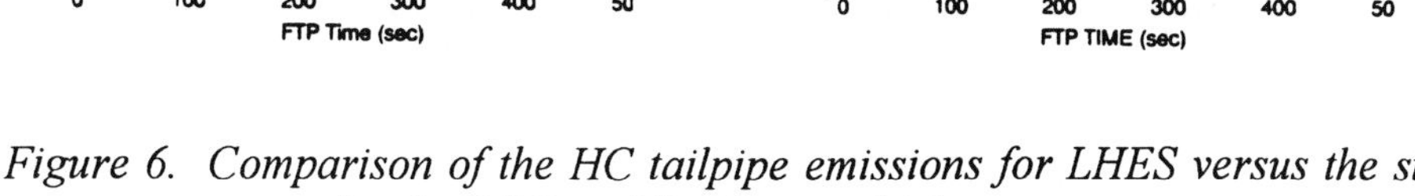

Figure 6. Comparison of the HC tailpipe emissions for LHES versus the stock system for the 2.0 L and 2.4 L test vehicles.

Japanese test vehicles. Engine out (EO) HC emissions have also been shown for reference.

After 60 - 70 seconds, the stock system traces stop tracking with and then drop below the EO line. At this point, the TWC catalyst has achieved light-off. The 2.4 L stock system trace briefly merges again with the EO line at 100 seconds. This is a relatively common occurrence for current underfloor converter systems, and reflects a drop in the exhaust system temperature. The LHES tailpipe trace follows a much different path: The HC concentrations begin and remain at all times below the EO concentrations. The stock system trace finally overlays the LHES line after 170 and 270 seconds for the 2.0 L and 2.4 L vehicles, respectively. Figure 7 plots the FTP HC adsorption/desorption function of the trap with respect to time for both vehicles, as calculated in equation 1. In the cold start period, the percent adsorption remained relatively constant until ca. 175 - 200 seconds even though the trap inlet gas velocities and HC concentrations were in a constant state of flux.

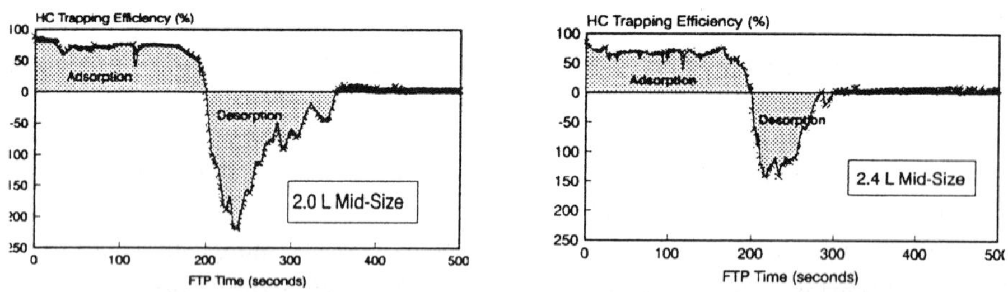

Figure 7. Performance of the HC trap portion of the LHES on the 2.0 L and 2.4 L vehicles.

Upon being heated by the clean gas exiting the first heat exchanger catalyst bed, the trap then smoothly shifted to a desorption mode at about 190 seconds as shown by the negative percent conversion in Figure 7 The desorption was complete by the end of Hill 2. On leaving the trap, the HC's encounter the second TWC catalyst bed in the heat exchanger. Unlike the first pass heat exchanger TWC bed which achieved light-off via heat transfer from the exhaust gas augmented by HC and CO combustion, the second pass catalyst bed must reach light-off by conduction of heat from the first catalyst bed through the heat exchanger's walls to the second catalyst bed. The efficiency of the Phase 1 LHES, thus depends upon the released HC's encountering a hot second pass. The Phase 1 LHES heat exchanger module design based on the above considerations was highly successful. The HC's released per Figure 7 were not detected in the LHES tail pipe trace in Figure 6. Figure 8 summarizes the HC conversion performance across the second pass TWC bed, and clearly shows that during the time the released HC's flowed into the catalyst bed they were converted in excellent yields. The heat exchanger thus functioned properly, delivering heat to the second catalyst in sufficient quantity and in time to deal with the trapped HC's.

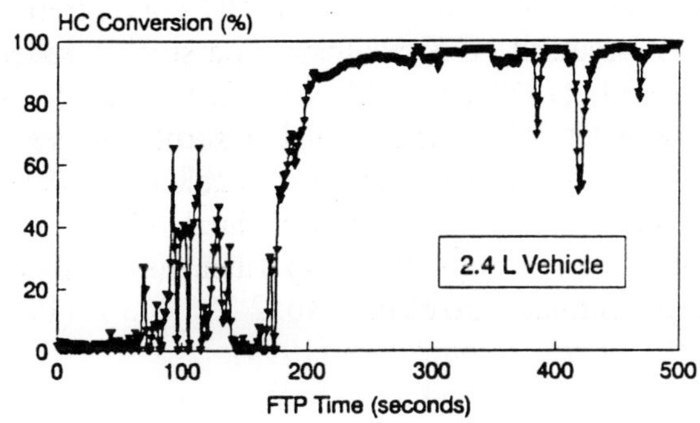

Figure 8. Hydrocarbon conversion across the second heat exchanger catalyst bed.

To ascertain the thermal durability of the heat exchanger, the Phase 1 LHES was removed after the fresh FTP's had been completed on the 1993 3.0 L U.S. vehicle. The start catalyst was left on the car. The heat exchanger module/ HC trap converter system was aged on an engine test bed at 760°C (first pass catalyst bed inlet gas temperature) for 75 hrs, the equivalent to 50,000 miles on-road aging. The system was remounted on the vehicle and FTP's were re-run. The data has been summarized in Table 3. Within experimental error the fresh and aged FTP performances were identical. The integrity of the Phase 1 LHES had been maintained. This data also points out another benefit of having two catalyst beds in heat exchange relation: removal of heat from the first bed mediates the thermal degradation of the TWC catalyst function.

4. Conclusion

A wide variety of vehicle types were fitted with an LHES system composed of a start catalyst, catalyzed heat exchanger, and an HC trap. This system substantially lowered the THC emissions (45 - 70%) via the trapping and subsequent conversion of cold start HC emissions such that all the test vehicles could qualify as either LEV's or ULEV's. The technologies that drove the performance improvements were an effective HC trap and a heat exchanger able to transfer sufficient heat in a timely manner to a TWC catalyst downstream of the HC trap.

REFERENCES

1 K.H. Hellman, R.I. Breutsch, G.K. Piotrowski, and W.D.Tallent, SAE Paper No. 890799 (1989).
2 R. Reiter and J. Wandler, DE Patent No. 4 141 116-C1 (1993).
3 L. Greiner and D.M. Moard, US Patent No. 5 207 185 (1993).
4 T.T-H. Ma, GB Patent No. 2 262 331-A (1993).
5 T. Ma, N. Collings, and T. Hands, SAE Paper No. 920400 (1992).
6 T.T-H. Ma, and N. Collings, GB Patent No. 2 256 603-A (1992).
7 T.T-H. Ma, GB Patent No. 2 260 279-A (1993).
8 I. Gottberg, J.E. Rydquist, O. Backlund, S. Wallman, R. Maus,R. Bruck, and H. Swars, SAE Paper 910840 (1991).
9 G.K. Piotrowski, EPA/AA/CCTAB/88-12 (1988).
10 R.G. Hurley, SAE Paper No. 912384 (1991).
11 L.S. Socha, and D.F.Thompson, SAE Paper No. 920093 (1992).
12 W.A. Whittenberger and J.E. Kubsch, SAE Paper No. 900503 (1990).
13 K. Yuuki, US Patent No. 5 191 763-A (1993).
14 K. Otto, C.N. Montreuil, O.Todor, R.W. McCabe, and H.S. Gandhi, Ind. Eng. Chem. Res., 30 (1991) 2333 - 2340.
15 H. Tsuchida, K. Ishihara, Y. Iwakiri, M. Matsumoto, SAE Paper No. 932718 (1993).
16 T. Minami, US Patent No. 4 985 210 (1991).
17 T. Minami, and T. Nagase, US Patent No. 5 140 811 (1992).
18 M.J. Heimrich, L.R. Smith, and J. Kitowski, SAE Paper No. 920847 (1992).
19 M.D. Patil, L.S. Socha, and I.M. Lachman, US Patent No. 5 125 231 (1992).
20 S. Dunne, US Patent No. 5 142 864 (1992).
21 J.K. Hochmuth, P.L. Burk, C.O. Tolentino, and M.J. Mignano, SAE Paper No. 930739 (1993).

AUTHOR INDEX

STUDIES IN SURFACE SCIENCE AND CATALYSIS